REVERSAL OF AGING
RESETTING THE PINEAL CLOCK

VLADIMIR DILMAN (1925–1994)

ANNALS OF THE NEW YORK ACADEMY OF SCIENCES

Volume 1057

REVERSAL OF AGING

RESETTING THE PINEAL CLOCK

Edited by Walter Pierpaoli

The New York Academy of Sciences
New York, New York
2005

Library of Congress Cataloging-in-Publication Data

Stromboli Conference on Aging and Cancer (4th : 2005)
    Reversal of aging : resetting the pineal clock / edited by Walter Pierpaoli.
        p. ; cm. — (Annals of the New York Academy of Sciences, ISSN 0077-8923 ; v. 1057)
    Fourth Stromboli Conference on Aging and Cancer held on June 6–11, 2005 in Stromboli, Italy.
    Includes bibliographical references and index.
    ISBN 1-57331-602-4 (cloth : alk. paper) — ISBN 1-57331-603-2 (pbk. : alk. paper)
    1. Longevity—Congresses. 2. Aging—Physiological aspects—Congresses. 3. Aging—Genetic aspects—Congresses. 4. Pineal gland —Physiology—Congresses. I. Pierpaoli, Walter. II. Title. III. Series.
    [DNLM: 1. Aging—physiology—Congresses. 2. Aging—genetics —Congresses. 3. Biological Clocks—physiology—Congresses. 4. Biomedical Enhancement—Congresses. 5. Pineal Gland—physiology—Congresses. W1 AN626YL v.1057 2005 / WT 104 S921 2005]
    Q11.N5 vol. 1057
    [QP85]
    500 s—dc22
    [612.6'8]

                                                                    2005026834

GYAT/PCP
Printed in the United States of America
ISBN 1-57331-602-4 (cloth)
ISBN 1-57331-603-2 (paper)
ISSN 0077-8923

ANNALS OF THE NEW YORK ACADEMY OF SCIENCES
Volume 1057
December 2005

REVERSAL OF AGING

RESETTING THE PINEAL CLOCK

Editor and Conference Organizer
WALTER PIERPAOLI

This volume is the result of the **Fourth Stromboli Conference on Aging and Cancer,** which was held on June 6–11, 2005, at Stromboli, Aeolian Islands, Sicily, Italy.

CONTENTS

Financial assistance was received from:

• THE WALTER PIERPAOLI FOUNDATION OF LIFE SCIENCES

Participants in the **Fourth Stromboli Conference on Aging and Cancer,** held in Stromboli, Aeolian Islands, Sicily, Italy on June 6–11, 2005.

Fourth Stromboli Conference on Aging and Cancer

Welcoming Remarks

Precisely 18 years ago I was standing here at the first Stromboli conference. We have a filmed record of that time, and perhaps Novera Spector can describe that event in his final "cocktail." However, we were so serious in our endeavor that even NATO Bruxelles was impressed, and I could fill enough paper to be compensated with the money needed for an NATO Advanced Research Workshop. They did not yet realize it was so important. If you look at the visual records, you can distinguish, among those who were here in 1987 and are here now, those who, thanks to or in spite of their lucky or unfortunate genetic heredity, were better able to preserve—or even improve—their juvenility without recourse to aesthetic surgery *et similia*. This is a pitiless examination, what I call "the ultimate defiance," is it not? Let us look at Novera, Marguerite, George, myself. Daily life, all kinds of environmental and social insults, bad nutrition, wrong diets, lack of money, anxiety, excessive night light, and thousands of other causes combined to make us, in spite of long-lived parents, what we are now. But a question: can we now not only arrest, but even reverse our progressive senescence and the aging program? My answer is "Definitely yes," and I do believe that this fact will be visible when (and if) we will be sitting in this same room at Stromboli 2007, 2009, 2011, and so on.

If you do not accept this defiance, I would invite you to get up and leave this room, and certainly I will not see you anymore in Stromboli. In fact, we have no time to waste discussing probabilities, but we must confirm our certainty! It is amazing how little we believe what we hope and say. Simply, the truth of non-aging seems blasphemous and, to many of us, utterly ridiculous. We are even ashamed to express shyly the proposition that we may be able sooner or later to laugh at those who persist in their "necrophilia." Is not aging an anthropological event, amenable to be modified by nature? As far as I am concerned, it is curiosity that brings me to Stromboli, curiosity about the significance of life and death, not different from the curiosity of Parmenides and the Greek philosophers. Stromboli 2005 is oceans far from Stromboli 1987. At that ancestral time, immunology was the dominant discipline of aging, with Roy Walford, thymologists, and joggers imposing their views and "therapies." Neuroimmunomodulation had different names, such as PNEI, PNI, and NI, as if putting the body together were a persecuted, underground activity. At that time cells and hormones were separate entities, and hardly anybody had heard of the obvious body–mind links. Several Nobel prizes were distributed to such great immunologists as Jerne, Benacerraf, and Adelman, investigators of the body's molecular mysteries.

The only question now is if we will be here for Stromboli 2007, 2009, 2011, and later. One thing is sure: you have the obligation to be and feel free to joyfully expose all the scientific madness you are able and courageous enough to tell us without blushing.

Ann. N.Y. Acad. Sci. 1057: xi–xii (2005). © 2005 New York Academy of Sciences.
doi: 10.1196/annals.1356.100

I deeply miss Vladimir Dilman to whom this conference is dedicated. That great man is for me the emblem of our struggle to survive as individuals and scientists in the most hostile conditions. He, too, was convinced that aging, together with the metabolic derangements undelying its course, can be reversed. Great souls and minds such as Vladimir Dilman, Ashley Montagu, Jean Choay, and Cook Kimball are here spiritually with us to help us elude the deceptive biological cage imprisoning us. On another side we must be ready to accept in peace whatever our destiny has prepared for us, even death, as liberation from self-deception. Let me conclude with a note of fun for all those who fight for intellectual freedom:

RIGHT AND WRONG

(With Special Reference to the Discovery of the "Aging Clock")

- Experiments supposed to be wrong are right.
- Experiments not replicated by others and not shown to be wrong are right.
- Experiments repeatedly replicated by others and shown to be wrong are wrong.
- Wrong experiments remain right until replicated and shown to be wrong.
- Chasing and challenging right experiments in the hope that they are wrong is devilish.
- The discovery of a simple natural truth is the most challenging endeavor of man because it is hidden in himself and he cannot see it. When it is discovered, it is violently opposed by the churches of the "nonbelievers."
- The most fashionable sport of scientists is to publish wrong experiments with the help of statistics.

In his beautiful article published in 1976 (*Preventive Medicine* **5**: 496–507) and entitled "The Illusion of Immortality and Health," Ashley Montagu writes: "Some of the more sophisticated among these doubters of immortality think of it as an interminable prospect of boring Sundays. But that is not the opinion of the major part of humanity." To conclude, without commenting of my personal doubts about our personal indestructibility, I simply try to stay on the side of Oliver Wendell Holmes, who wrote in a poem prepared for the National Sanitary Association:

> And lo! The starry folds reveal
> The blazoned truth we hold so dear:
> To guard is better than to heal,
> The shield is nobler than the spear!

This is what I strongly believe: it is curiosity about life, not fear of death, that should motivate us.

Last, but certainly not least, I wish here to express my deep gratitude to all those members of the Foundation—none excluded—for their tireless efforts to make this conference possible. In particular, we owe this renaissance of the Stromboli conferences to the unique vision of Giancarlo Parretti—his personal and convinced anticipation of a truly innovative medicine based on the rediscovery of those rhythmic biological mechanisms that strictly regulate our life program.

WALTER PIERPAOLI
Stromboli, June 6, 2005

Is Aging Truly Inevitable?

An Overview of the Fourth Stromboli Conference

LINDA HOTCHKISS MEHTA

New York Academy of Sciences, 2 East 63rd Street, New York, New York 10021, USA

Ever since humans have been aware of the inevitability of death and aging, we have attempted to impede the slow decline in abilities that is the usual course of aging and postpone the hour of death. Life itself guards against the forces that would end all life forms through its seemingly infinite adaptations and diversifications of its "portfolio" of organisms. Thus, the desire for eternal life began not with human awareness, but with the advent of life itself. And the striving toward the perfectibility of the aging of our organic selves, whether or not there are limits to actual longevity, is not merely the quaint, historical pursuit of explorers like Ponce de Leon, who searched the New World for the Fountain of Youth, but is as current and modern as the biomedical research techniques that form our current voyage of discovery.

The Fourth Stromboli Conference on Aging and Cancer, held on the volcanic island of Stromboli in the Aeolian archipelago, included an international and multidisciplinary group of scientists who are studying the mechanisms of the aging process and seeking the means for extending good health well into the oldest reaches of the human life span. The emphasis of the meeting was less on longevity than on extending the range of healthy, active living and on the potential for reversing the aging process. Through the extension of good health and the slowing or reversing of aging, an extension of longevity is hoped for. "A chain is no stronger than its weakest link, and life is after all a chain" is how William James laid out the problem, leaving the identity of that weakest link (or those weak links) to the scientists to discover.

Walter Pierpaoli, the organizer of this conference, likens the human condition to a cage, inasmuch as our vision is constrained in many ways by the limits of our own perceptions. It is his vision that our bodies are profoundly affected by the rotational forces that shape our life on this planet and that characterize the atoms that form our physical selves. The cyclical rhythmicity of our neuroendocrine–immune system can become desynchronized and thus cause the processes of disease and aging, which we observe after these damaging processes begin; Pierpaoli presents interventions that can reestablish its natural balance.

Address for correspondence: Linda H. Mehta, New York Academy of Sciences, 2 E. 63rd St., New York, NY 10021. Voice: 212-880-2913; fax: 212-880-2089.
lmehta@nyas.org

Ann. N.Y. Acad. Sci. 1057: xiii–xxii (2005). © 2005 New York Academy of Sciences.
doi: 10.1196/annals.1356.101

THE EARTHLY CAGE OF AGING: CAN WE ESCAPE FROM IT?

In the first section of this volume, the speakers address the basic concepts of aging and describe the current theories about its mechanisms. Robert Arking begins by raising the question, Is human longevity already approaching its limits? He observes that in mammals, three longevity phenotypes exist, only two of which are known in humans. The third, the delayed onset of senescence phenotype, can be induced in various ways. It is important to realize that the presence of multiple phenotypes means that determining potential life expectancy for a species requires that we consider all of the longevity phenotypes provided for in the genome. The weak link in William James's chain of life may turn out to be a stochastic growth factor/signal transduction failure, which brings about decreased efficiency of the cell's defenses with the result that senescence begins in earnest. Arking challenges the research community by pointing out that it can be argued that up until now public health improvements have contributed more than medical interventions to gains in life expectancy.

Bill Bushell reports on the behavioral interventions that humans have used in the effort to prolong good health and extend longevity, including the practice of meditation, nutritional manipulations, and exercise regimens. He expresses the concern that scientific standards have been relaxed too far in some investigations of what might be the mechanisms through which these sorts of interventions act. Nevertheless, because recent studies have shown that alternative medicine is gaining in popularity, it is important that these practices and their effects on the human organism continue to be studied in a scientifically rigorous way. Our increased knowledge of the molecular effects of stress, which can apparently be effectively mitigated by these interventions, provides an example of an area where good information is emerging through scientific research.

Across the known species, a wide range of longevity is manifest—from moments to centuries. In general, larger organisms (species, not individuals) have longer life spans. Within the human species, evidence from twin and centenarian studies show that the aging process is influenced by genes. José Lao identifies three processes that appear to be factors in determining life span: the insulin signaling pathway, the capacity for stress resistance, and the production of melatonin. The interplay between damaging effects and protective responses determines the rate and process of aging in an individual.

Samuel McCann revisits the nitric oxide theory of aging. Nitric oxide (NO) plays a ubiquitous role in the body in controlling organ functions. It is generated by endothelial, neuronal, and mitochondrial NO synthase (NOS). NO protects by defending against invading viruses and bacteria, but it can also be toxic to the host cells through the inactivation of enzymes, which can lead in turn to cell death. Aging of the anterior pituitary and pineal glands may be caused by NO, resulting in a decrease in production of pituitary hormones and melatonin. Induction of NOS (iNOS) by the temperature-regulating centers as a result of infection may cause the decreased febrile response observed in the aged through the loss of thermosensitive neurons. Fat is the major source of NO in the body; and as fat stores increase, leptin and NO increase in parallel in a circadian rhythm that peaks at night. Antioxidants, particularly melatonin, vitamin C, and vitamin E, may be capable of playing a role in reducing or eliminating the oxidant damage produced by NO.

As proteins age, the changes they undergo can cause them to change their function and become immunogenic. Marguerite Kay reports on studies showing that animals and humans have physiologic antibodies that bind to one of these "neoantigens," senescent cell antigen (SCA). SCA appears on senescent and damaged cells and initiates their removal by macrophages. This observation has led to the discovery that oxidation can generate neoantigens and, among other things, accelerate the aging and removal of red cells. Molecular recognition of senescent cells involves IgG autoantibody binding to a neoantigen, and this binding action allows us to identify its antigen. Using fast atom bombardment ionization mass spectrometry (FAB MS), scientists have found that the likely aging antigen is part of a family of proteins called *anion exchange proteins*. These "band 3" proteins, which are major structural proteins, are responsible for respiration and are present on all membranes, cellular, nuclear, mitochondrial, and Golgi. Kay's lab is using MS-MS spectroscopy to identify the posttranslational modifications in this molecule so they can design interventions to halt the senescence process. Another approach they are pursuing is to manipulate the immune response by enhancing the protection of natural antibodies, perhaps through immunization with select peptide-defined epitopes.

Alexy Olovnikov believes that we should reject the teleomere theory of aging. Although the shortening of teleomeres correlates with cell senescence, it is his contention that this correlation is circumstantial. A hypothetical nuclear organelle may turn out to be the key instead. This is the redumere, which shortens simultaneously with the telomeres. Redumeres are perichromosomal organelles in the interphase nucleus that exist in two forms: printomeres, which are found in dividing cells and interpret positional information of cells in morphogenesis; and chronomeres, which control development over time. Olovnikov posits that chronomeres are the life span "clock" of an individual.

Walter Pierpaoli was the first to demonstrate the role of the pineal gland in tracking the programs of growth, fertility, aging, and death. He reports here on experiments investigating the workings of the pineal "clock" through pinealectomy performed at the point that it begins to send signals that initiate the aging and death processes. The "pineal complex" of the brain clearly initiates the aging process, but the mechanisms by which it does so remain to be fully elucidated.

Vladimir Skulachev describes the mitochondria as being in charge of an atavistic program resulting in the death of the organism. A point mutation in a single gene for a protein involved in the death cascade may affect this process. Such point mutations have been demonstrated to accelerate or decelerate aging, even to the point of immortality, as in the case of the fungus *Podospora anserina*. Thus, it may be possible to switch off aging-related destructive systems.

An anti-aging approach that takes advantage of the body's own protective responses is presented by Joan Smith Sonneborn in her paper on the reversal of aging by hormesis. The basic principle is to harness the positive effects of stress by providing just enough stress to produce the organism's own stress-protective response. A wide range of these protective effects is available to an organism through hormetic agents, for which a low dose is protective and a high dose is harmful. (Exercise, antibiotics, diet, and alcohol can all behave as hormetic agents.) The teleomeres activate some of these protective responses, and in aging some of these responses diminish. For example, cell sensitivity to UV radiation increases as the cells age. Smith Sonneborn presents evidence that hormesis is a technique that can extend life span through altered *timing* of onset of senescence and duration of senescence.

MOLECULAR, CELLULAR, AND ORGANISMIC
MECHANISMS OF AGING

The cellular functions known to be affected by protein kinase C (PKC) signaling are ion channel modulation, receptor regulation, neurotransmitter release, and synaptic plasticity and survival, all of which are essential for healthy brain function. Furthermore, PKC-dependent B50/GAP43 phosphorylation is involved in memory formation. Fiorenzo Battaini presents evidence that the PKC adaptor protein RACK1 and the calcium-dependent and -independent kinases are deficient in physiological and pathological (Alzheimer's disease) brain aging. Studies in rodents and humans have provided data implicating changes in protein-to-protein interactions in the interference with mechanisms of PKC activation. Greater insight into the factors controlling PKC activation and its regulation may not only help elucidate the molecular basis of signal transmission, but also help identify new strategies for correcting or even preventing age-dependent alterations in cell-to-cell communication.

Alexander Boldyrev provides further evidence that protein damage may be more important than cellular lipid damage produced by reactive oxygen species (ROS) within the cells and tissues. For this reason, hydrophobic antioxidants are less effective in protecting tissues from oxidative damage because, although they protect the membrane lipids, they are ineffective in preventing protein oxidation. Anti-glycating agents, such as the neuropeptide carnosine, may prevent protein oxidation and modification. In tests with mice, carnosine was shown to decrease the blood levels of protein carbonyls and lipid peroxides, which demonstrates normalization of oxidative metabolism. Improvements in age-related behavioral changes were also observed in the animals.

Heat-shock proteins provide cellular defense, not only from elevated temperatures, as their name implies, but also from the effects of alcohol, toxic heavy metals, some forms of oxidative and postischemic stress, and other forms of environmental insult. One characteristic of senescent cells is their reduced capacity to respond to stress. These molecular chaperones protect proteins from the conformational changes and partial denaturation that are so damaging to their function by "chaperoning" the unfolded proteins and guiding them into the correct folding conformation, translocation, or oligomeric organization. Ezra Pierpaoli reports on the potential for HSPs in the prevention and reversal of many diseases associated with aging as well as their capacity to fight cancers. These proteins are found in organisms from bacteria to humans; they seem to have an important role in many of the age-related diseases such as Alzheimer's disease, Parkinson's disease, and prion-related diseases. But heat-shock protein synthesis, particularly Hsp70, is greatly reduced in old cells, and its activity is reduced by approximately half, so the cell will accumulate unfolded and misfolded proteins. Hsp70 is also central to the immune response, and Hsp70-peptide complexes have been under trial as cancer vaccines.

BIOMOLECULAR INTERVENTIONS FOR INTERPRETING, PREVENTING, OR "CURING" AGING

Vladimir Anisimov turns the focus on biomolecular interventions first to the putative link between aging and insulin/insulin growth factor (IGF)-1 signaling, the evidence for which is found in the age-related increase in the incidence of insulin resistance and type 2 diabetes and in the effectiveness of caloric restriction in extending life span in rodents. Hyperglycemia and hyperinsulinemia turn out to be important factors not only in aging, but also in the development of cancer. Treatment with antidiabetic drugs can increase survival and inhibit spontaneous carcinogenesis in mice and colon carcinogenesis in rats.

Lev Berstein adds changes in estrogen production to glucose tolerance/insulin sensitivity in the list of endocrine shifts associated with aging and age-related pathologic conditions. Estrogens exhibit both hormonal and, through their conversion into catecholestrogen free-radical mutagenic derivatives, genotoxic properties. Likewise, under the stress of postprandial hyperglycemia, reactive oxygen species are formed in addition to the hormonal effect glucose exerts in stimulating insulin secretion, setting off a chain of events similar to that in which estrogens convert into free-radical derivatives.

Immune function is essential to our survival, and the adaptive immune system is found only in jawed vertebrates. The shark is the most primitive animal with a complete immune response; take one phylogenetic step down the ladder to the hagfish and lamphrey, and no immune system is found; rather an innate system is present that relies on phagocytic-expressing receptors and circulating molecules distinct from antibodies. John Marchalonis and his collaborators have conducted investigations directed toward determining the molecular properties of antigen recognition molecules, particularly antibodies and T cell receptors, recombination activator genes, and cells of the innate and adaptive immune systems. Defining the functional natural antibody repertoire and elucidating the mechanisms of recognition and elaboration of the immune response may in turn increase understanding of these processes in autoimmunity and aging. The function of the immune system is a complex orchestration of specific self-recognition and non-self-recognition capacities mediated by cells of the innate system acting in coordination with T and B lymphocytes in a series of processes modulated by cytokines. Here, Marchalonis provides evidence for a natural immunomodulatory system that might be harnessed through administration of peptides (e.g., CDR1 peptides) to reverse cytokine dysfunction in immunosuppressed individuals and upregulate inflammatory activities mediated by TH1-type helper cells.

AGING AND DEGENERATION OF THE CENTRAL AND PERIPHERAL NERVOUS SYSTEM: PREVENTATIVE AND REGENERATIVE MEDICINE AND BIOENGINEERING INTERVENTIONS

Katrin Beyer's chapter addresses the importance of the age at onset of Alzheimer's disease (AD), which she demonstrates is essential for the definition of genetic risk factors for sporadic AD. AD features an accumulation of misfolded and otherwise-abnormal proteins. Over 30 potential risk factors have been tested for their impact on developing AD, but only the apolipoprotein E (APOE) gene has been demonstrated to indicate a major risk. Beyer's group genotyped AD patients and controls for polymorphisms of the APOE gene promoter and the cathepsin D (catD), butyrilcholinesterase (BChE), cystatin C (CST3), methionine sythase (ms), and cystathionine beta-synthase (CBS) genes. They then analyzed the association between age at onset of AD and the presence of these polymorphisms. Their results indicated that a clear association exists between age at onset of AD and specific gene polymorphisms—that is, different mechanisms lead to AD at different ages, such that including the age at onset of AD will help us define global and population-specific genetic risk factors for each age at onset–dependent AD subgroup. This in turn will be helpful in establishing a genotype pattern coincident with risk at each age range and permit early detection of those at risk.

The relationship between the memory capacity of the immune system and the acquisition of neuronal memory has been studied by Katica Jovanova-Nesic. Brain lesions can produce both cognitive and immune deficits, and several agents producing brain lesions exhibit both effects. In Javanova-Nesic's experiments, ethanol, kainic acid, and 6-OHDA all produced temporary lesions in the brain that rendered rats unable to learn and remember and also produced inhibition of both humoral immune reactions (antibody titer and Arthus reaction) and cell-mediated immune reactions (such as delayed hypersensitivity of skin). Ethanol is known to interfere with the transfer of information from short-term to long-term storage, and it may interfere with immune memory through similar molecular mechanisms.

Glia have a major role to play in the development in idiopathic Parkinson's disease (PD), one of the most common aging-associated chronic neurodegenerative disorders; the mechanisms of that role were presented by Bianca Marchetti. Both astrocytes and macrophages/microglial cells express receptors for 17 beta-estradiol (E2) and glucocorticoids (GC) and are both a source and a target of cytokines, growth, and neurotrophic factor activities in the brain. Both also exert neuroprotective functions. Nevertheless, in brain injury, astroglia can be both neurotoxic and neuroprotective. Neuroinflammation and oxidative stress play a role in the selective loss of nigrostriatal dopaminergic (DA) neurons in PD. At it turns out, glial–neuron cross-talk appears to play a pivotal role in dictating resistance versus vulnerability against various cytotoxic mediators. The expression of inducible nitric oxide synthase (iNOS) in astrocytes and microglia during brain inflammation coupled with a high nitric oxide output are predominant effectors of MTPT-induced neuronal cell death. The data presented here suggest that iNOS/NO is a key target of steroid hormone signaling in astrocytes and microglial cells and define glia and steroid receptor-iNOS/NO cross-talk as a common final pathway, orchestrating neuroprotection and neuronal repair in experimental PD.

Walter Pierpaoli concludes this section with a presentation postulating a common origin for all neurodegenerative diseases: the decline of T cell–dependent or delayed immunological resistance. Because immunity is closely controlled by cyclic hormonal output, a disruption in this cycle invariably precedes the onset of neurodegenerative disease.

THE PINEAL LIFE CLOCK: EXPERIMENTAL AND
CLINICAL ACTIVITIES OF MELATONIN

How does the body measure and track the passage of time? Daniel Cardinali addresses this question in his clinical report on the use of melatonin as a chronobiotic–

cytoprotective drug. He begins by noting that the cycle of sleep and wakefulness is the most obvious of the circadian rhythms. Treatment of circadian rhythm disorders with melatonin is a well-established intervention. It has been discovered to promote slow-wave sleep in the elderly and could be beneficial in Alzheimer's disease by augmenting the restorative phases of sleep, including secretion of growth hormone. Other neuroprotective actions attributed to melatonin include protection against oxyradical-mediated damage, apoptosis, and glutamate excitotoxicity. It also appears to be cytoprotective against the actions of osteoporosis, ischemia–reperfusion, and diabetic microangiopathy.

Synchronization of the neuroendocrine–reproductive axis (hypothalamic–pituitary–gonadal axis) is vital for normal reproductive function. Beatriz Díaz López points out that gonadotrophin-releasing hormone (GnRH) is obviously a key hormone for the regulation of the secretion of gonadotrophins LH and FSH. Melatonin can be demonstrated to be necessary for normal sexual maturation, and therefore it may also be implicated in the age-related decline of the reproductive system. Díaz López reports on studies in which melatonin treatment during aging of the female reproductive system enhanced the amount of hormones released during the surge in the proestrus and decreased gonadotrophin concentrations to levels similar to those observed in young rats. Melatonin administration has different effects on the reproductive systems of animals in different age groups, however.

George Roth, in collaboration with Mark Lane and Donald Ingram, has written about the efficacy of caloric restriction (CR) mimetics in reaching the goal of extending both the median and maximum life span across the phylogenetic spectrum. CR mimetics are agents or strategies that can mimic the beneficial effects of CR. He contends that biological disorder is ultimately lethal and caloric restriction boosts our defenses against disordering mechanisms (free redicals, glycolysis, and so forth). Body temperature is reduced by about 0.5 to 1.0°C during caloric restriction, DHEA is elevated, and insulin level is lowered. Because of the difficulty in maintaining long-term CR, mimetics may offer a more reliable means of reaping its benefits. Candidate mimetics include glycolytic inhibitors, lipid-regulating agents, antioxidants, and specific gene modulators.

Interestingly, melatonin levels are elevated at night as compared to daytime values in all species investigated, including those that are active at night. Christof Schomerus reports on the mechanisms regulating melatonin synthesis in the pineal gland. The regulatory cues that control melatonin biosynthesis vary among vertebrate species; one significant difference is that the pineals of mammals have lost direct photosensitivity and the ability to generate a circadian rhythm, unlike those of birds and fish, which have retained direct photosensitivity. Rather, mammals have an internal clock located in the suprachiasmatic nucleus of the hypothalamus, whose activity is modulated by retinal afferents; and it is this "clock" that drives melatonin production. Thus this common neurendocrine principle, the nocturnal rise in melatonin, is controlled by strikingly diverse regulatory mechanisms. This diversity reflects the high adaptive plasticity of the melatonin-generating pineal gland.

Changxian Yi conducted a case study involving melatonin treatment of patients with age-related macular degeneration (AMD). Fifty-five patients were followed for six months and showed a remarkable reduction in pathologic macular changes. Six months is not long, but it is sufficient to observe a slowing of the progress of this disorder. Melatonin at 3 mg daily appeared to delay macular degeneration. Yi postulates that this may be through melatonin's ability to scavenge hydroxyradicals and protect retinal pigment epithelium cells from oxidative damage. Dysfunction of retinal pigment epithelium is known to initiate AMD.

Evidence that melatonin affects nonspecific immunity is given by Monica Hriscu, who writes that phagocytosis, which is a crucial component of nonspecific immunity, is influenced by both endogenous and exogenous melatonin. The increase in phagocytotic response induced by bacterial lipopolysaccharides (LPS) in nonimmunized mice is extremely variable over 24 hours. Functional pinealectomy, achieved by two-week exposure to constant light, was followed by a marked reduction in the circadian average of basal phagocytosis compared to control. In another study, melatonin was able to counteract the negative effects of ethanol on neutrophil phagocytosis.

Muscle mass decreases with age, and a dramatic morphological change can been seen in the immobilized muscle of older animals. In a study conducted by Marina Bar Shai, the hindlimbs of young and old Wistar rats were immobilized, and the activation of various proteolytic systems in the immobilized muscles was measured. In young animals, these systems showed a marked increase in highly synchronized activation of proteolytic systems (MMPs, acid phosphatase, and ubiquitin proteasome) during the fourth week of immobilization; whereas in old animals this synchronization was disrupted. After the immobilization period, the young animals were observed to recover fully in terms of muscle mass and biochemical parameters, but the old animals did not return to preimmobilizatin levels of these measures.

REPLACEMENT THERAPIES OF A FAILING BODY: CLINICAL EVIDENCE AND RESTORATIVE MEDICINE

Restorative medicine is another term for replacement therapies, and several promising therapies are presented in this chapter. Thierry Hertoghe asks the question, Is aging the result of multiple hormone deficiencies? The list of hormone changes that come with aging is long: some of the affected hormones are the thyroid hormones, anabolic hormones, insulin, testosterone, growth hormone, and estrogen, all of which decrease. T3 receptors decrease, and metabolic clearance is reduced. Because reductions in so many hormones accompany aging, one is likely to presume that hormone replacement therapies may counter senescence. Treatments should consider the circadian cycles of the affected hormones.

Hormonal insufficiencies may also be labeled as psychiatric disorders. Suzie Schuder notes that patients suffering from posttraumatic stress disorder (PTSD), for example, are often found to have low cortisol levels. Cortisol insufficiency symptoms also closely parallel the symptoms of other psychiatric disorders, particularly those of addiction. Treatment of patients with physiologic doses of bio-equivalent hydrocortisone and other hormones found to be deficient has produced improvement in psychiatric symptoms, and Schuder recommends hormonal evaluations and treatment of any insufficiencies with physiologic doses of the hormone in question before using psychotropic medication.

Jerry Shay's chapter returns to the role of telomeres in inducing senescence. He characterizes cellular senescence as a general stress-response program that restrains cellular proliferation. Under optimal growth conditions, the onset of senescence depends on telomere status. If normal cell cycle checkpoints are altered, cells can bypass the normal senescence-signaling pathway and continue to grow until they reach a second growth-arrest state known as *crisis*. In both senescence and crisis, expression of telomerase (hTERT) leads directly to an immortal state, demonstrating that telomeres are important in both replicative senescence and crisis. In addition, although telomere shortening can be an initial blockade to cancer, in the presence of other genetic and epigenetic alterations, telomerase can be permissive of cancer. Shay believes that specific tissues and locations in which replicative aging occurs contribute to the decline in physiological function resulting in aging. Telomere shortening regulates replicative aging, and we can experimentally block it by expressing telomeres and maintaining telomere length. The production of hTERT-engineered tissues offers the possibility of producing tissues to treat a variety of diseases and age-related medical conditions that result from telomere-based replicative senescence.

Luis Vitetta looks at the complex interactions between the individual and his or her environment. He discusses actions that can have an impact on maintainance of good health and can also significantly affect longevity. These include the problem of undernutrition (even in a world of overconsumption). Advances in our knowledge about the interactions of the mind and the body will also enhance quality of life and may extend longevity.

A case study in which serum estriol, estradiol, and estrone were measured and tracked in nonpregnant, premenopausal women is presented in Jonathan Wright's contribution to the volume. Elevated urinary or serum estrone and estradiol concentrations in postmenopausal women are associated with a moderately elevated risk of breast cancer. Wright's conclusion is that this association may reflect overdose conditions. Furthermore, iodine and iodide can promote more complete metabolism and detoxification of estradiol and estrone to estriol, which has an anticarcinogenic effect.

STRIVING FOR HEALTH: ANTI-AGING MEDICINE AND ITS PRESENCE IN THE WORLD

Limits probably exist on our ability to prolong life, and our genetic inheritance provides the guidelines for an individual's potential longevity and susceptibility to the various morbidities associated with aging. Lifestyle choices, however, can greatly influence both a person's actual longevity and the quality of life enjoyed as one ages. Bill Anton addresses the ways in which clinicians need to consider the effects of aging in an increasingly elderly cohort of patients and how the aging process can alter the effectiveness of medical treatments. His chapter reviews the means for "compressing" the morbidity of aging and even reversing aging-related decline.

The United Nations predicts a steady increase in life expectancies for all countries, regardless of their level of economic development. Ronald Klatz calls this nothing less than a longevity revolution, and he cites as evidence the advances brought about in medical knowledge by stem cell research, therapeutic cloning potential, biomedical use of nanotechnology, and the possibility of artificial organs, which will fuel the longevity revolution.

In the closing chapter, Phil Micans urges the emulation of the six-month dental checkup in medical practice: doctors should create more welcoming environments, where access to preventative medicine can help patients avoid future degenerative diseases. This is a philosophical shift in medical practice, but one that must be made if the benefits of anti-aging medicine are to be achieved. He contends that the health of individuals would benefit, and healthcare costs for the larger society would also be reduced.

This volume is the fourth in a series of Stromboli meetings, held over a span of 17 years, in which advances in research on aging and the related diseases that are collectively called cancer have been shared among a group of dedicated scientists. Novera H. Spector provides historical perspective and future challenges in his closing summation of the meeting, along with a recipe for the most recently formulated "Stromboli cocktail," which contains the ingredients for extended good health and longevity proposed in the research reports. Jorge Luis Borges wrote that long life meant "the terror of being trapped in a human body whose abilities diminish." The hope of those engaged in this research is that this terror will be vanquished and recognized as an unnecessary prospect as we age.

The support of the Walter Pierpaoli Foundation for the meeting resulting in these proceedings is gratefully acknowledged. The New York Academy of Sciences, in publishing these reports and disseminating the information and advances presented at the meetings, has also promoted and furthered this research.

The Rotational Origin and State of the Whole

Its Relation to Growth, Fertility, Aging, Death, and Diseases

WALTER PIERPAOLI

Walter Pierpaoli Foundation of Life Sciences, Orvieto, Italy

Why do I worry about my own aging?
It is only curiosity of life! It is simply
more time to escape the traps, not of
nature, but of man-generated confusion.

Becoming is eternal!

ABSTRACT: The purpose of my report is to synthetically summarize the concept of the rotatory essence of the Whole and to bring evidence that while aging responds to a precise inner "program" of the mammalian and any other species' "brain," acceleration of aging and all diseases are simply the direct outcome of a desynchronization of our inner "clock" with respect to the precise periodicity and hormone-integrated rhythmicity of the solar system. Those neuroendocrine, hormonal derangements of our inner clock are easily detectable and inevitably anticipate even by decades the onset of all diseases (autoimmune, cardiovascular, neurodegenerative, neoplastic). I will introduce those interventions capable of detecting early alterations and of restoring hormonal rhythmicity, which will consequently restore immunological surveillance in a positive cascade sequence.

KEYWORDS: cyclicity; rotation; aging; hormonal rhythms; melatonin; restricted-calorie diet; pineal gland; pineal transplantation; immunity; cancer

PREAMBLE

In May 1977 some thoughts suddenly crossed my mind—probably elaborated in a remote section of my hippocampus as a kind of premonition—and the idea of the basic nature of aging slowly took shape. Long before the experimental facts demon-

Address for correspondence: Dr. Walter Pierpaoli, M.D., Via San Gottardo 77, CH-6596 Gordola, Switzerland. Voice: 41 91 7451940; fax: 41 91 7451946.
pierpaoli.fnd@bluewin.ch

Ann. N.Y. Acad. Sci. 1057: 1–15 (2005). © 2005 New York Academy of Sciences.
doi: 10.1196/annals.1356.007

strated that maintenance of circadian cyclicity of the neuroendocrine, hormonal network was essential for a disease-free and "natural" aging, I wrote the text of the following section.

Cyclicity as the Basis of Life in Our "Unus Mundus"

As a general formulation, it seems to me that the perception, if not the knowledge, of the cosmic conditions, and in particular those of the solar system, that have allowed and shaped the genesis of life such as we now feel and know it, is the necessary condition to approach that "unus mundus" where psyche and matter merge and mingle. In fact, there exist intrinsic conditions in the solar system that have determined the basic characteristics of all aspects of psychic and physical life, which coexist at atomic, molecular, and biologic–macroscopic levels. These conditions can be defined and condensed into one word: *cyclicity.*

This dynamic concept of cyclic, rhythmic rotation of our planet around the sun and around its own axis often escapes our imagination and perception because it is integrative. Consequently the comprehension of cyclicity—with ourselves existing inside the system itself—requires an immense endeavor of intuitive abstraction. Natural or spontaneous intuition sometimes permits a flight into the unconscious, but this is one of the many amateurish methods fondly described by philosophers, clergymen, and psychologists. It seems to me that one of the very few to combine integrated intuition at the level of consciousness was C. G. Jung.[1] His concept of synchronicity is certainly an overcoming of space and time barriers, but as far as I can appreciate—which is certainly a restricted personal view—Jung lacked a phylogenic and evolutionary basis that makes comprehensible the "model man" as a perceptive, sensorial mechanism for the phenomena of so-called reality and also of the subconscious. In other words, the breaking up and fragmentariness of the cognitive processes through the limiting filters of the senses are precisely the consequence of the inability, inside the prison of the "reality system," to return to how and why our receptivity model was emerging in evolution.

I think that the intuitive processes that enable us to occasionally and temporarily overcome the space-time-reality barrier are not at all pre-existent in the unus mundus; rather they have developed during evolution as cognitive biological processes. In fact, they are imprinted under different forms in the biological species. It is perhaps intuitively possible to connect the concept of intuition to that of cyclic, rotational periodicity, under the form of an oscillatory phenomenon linked to the same basic character of atomic energy. It is feasible that life seen as a temporary organization of matter, energy, and psyche—such as we perceive it—has been closely shaped and determined by the cyclical, repetitive influence of the solar system in all its expressions, originally physical and later biological (cyclic, circadian, day–night "wavelength" of hormonal rhythms).

There is nothing in our physical, biological, and psychic world that escapes this planetary determinant in the physical and biological evolution of life and in the emergence of consciousness in man. Because of this absoluteness, the concept of cyclic periodicity or rotation biology (the rotation–cyclicity principle) is hard to apprehend intuitively as the cause of all human affairs. Yet the concept of rotation–cyclicity should be the dynamic of all cognitive proceedings and be fully assimilated into our methodology. The concept eludes us because cyclicity itself is our reality and we are, in our wholeness, subject to its phylogenic and planetary laws. To abandon its evolutionary rules, even at a physiological level, leads inevitably to disease and biological death. We may thus be able to apprehend the rotation–cyclicity principle by complying with its laws and thus fluctuating synchronistically with it. In this way we can follow the sole natural command imposed on us in the course of life. In fact, if these laws of cosmic, planetary rotation periodicity are not followed, mental and physical confusion results simply because our mind, body, and psyche have been constructed and shaped in conformity with the "planetary–cyclic" model.

In my view, "synchronicity" as perceived by Jung is the temporary, casual tuning with the cyclic world that life itself originated from. It seems to me that the intuition of Jung was correct, but he did not place these phenomena into an evolutionary and cosmic setting. Accordingly, his method remains scientifically empirical and does not contemplate the mass of the subconscious. It seems to me that the formation of the receptive structures of man as a consequence of the whole array of cosmic, evolutionary influences (physical, chemical, biological, etc.) must first be evaluated and interpreted, then psychic and biological behaviors can be deciphered, which could lead to understanding the etiology of mental and biological pathology. Pathology often serves as a model for aberrations and helps us to trace the pathway of natural evolution resulting in an environmentally adapted, balanced, "healthy" psychic or biological state. Recent findings in medicine and biology point to a distinctive role of the rotation–rhythmicity and periodicity for the development of the body and the maintenance of health.[2–4]

Finally, I suggest that the concept of "constantly becoming" and the concept of the dominant role of the planetary physical laws of oscillatory forces, periodicity, and cyclicity, if not identical, are certainly very close to the definition of God in Exodus: "I am the One Who Becomes."[5–7] Therefore, the attributes of God and life itself share undefined boundaries. Life is certainly a "becoming" and not a "being," because "being" is a permanently static condition—namely, dissolution and "death." The idea of cosmic and planetary periodicity for the genesis and maintenance of life originates from the observation that a central, endogenous "aging clock" may exist in some structures of the brain. The progressive alteration and dysfunction of this endogenous central adapter and regulator of all our neuroendocrine and reproductive functions result in senescence of the body and then death.[2–4,8,9] We can now combine the concept of becoming with that of cyclically oscillating, remembering that all atomic and planetary motions are not casual, but rigidly oscillatory under physical laws. One could suggest that the notion of God is an integral component of our endogenous oscillatory system (the pineal gland, the ancestral light receptor now connected to the retina) because this is the main evolutionary device that constantly links and adapts us to the many variables (e.g., light and temperature) of the natural environment. The origin of life and life in general on our planet can thus be defined as a cyclical, oscillatory phenomenon imprinted in any atom, molecule, and biological expression.

In the Bible, we read in Genesis that sun and light were created by God in the context of a planetary solar system in order to generate life, but the maintenance of that kind of life in mammals requires a juvenile pineal gland! The circadian and seasonal life-generating light must be properly detected so that its signals are properly utilized by the body organs and systems.

INITIAL COMMENT

The more I read about aging, the less I understand its attributes, which are predominantly sociologic rather than biologic. First of all, the conviction that "physiologic senescence" is a kind of inevitable and inescapable fact of life (as is growth and puberty), precludes *a priori* a different view of aging. We are psychologically incapable of seeing what is in front of us, namely a genetically inherited and evolutionary rooted program in the brain and in the neuroendocrine system. We must understand it before we can interpret and possibly modify it.[2] This inability to develop a new view of aging is an innate mental inertia that escapes any "logic." The worst enemy of life and health is the myth of aging! If we possess a "genetic program" for growth, puberty, procreation, and aging, and finally death, the expression of this program is certainly interpretable and amenable to an external modulation.[8] This is as feasible as the interference with growth and puberty, of which there are many examples in animals and man. If puberty can be delayed or accelerated, so can aging, which in its typical expression is a progressive decay of central, adaptive neuroendocrine functions resulting in metabolic diseases, immunodepression, and cancer.

THE FACTS

The highly deceptive "simplicity" of nature eludes our capacity to appreciate what we are and where we are. The discovery of a life, aging, and death "clock" in the pineal gland[10–12] has revealed an immense and totally unexplored ground for exploration of dimensions and forces that regulate the course of our earthly life. These forces are inevitably interwoven with our earthly and temporary organic structures. It is in fact beyond any reasonable doubt that we are totally dependent and conditioned, at any moment of our daily life, by the cosmic forces that have shaped our planetary system and established a kind of temporary "solution" that has brought about the existence of *Homo sapiens*.

Without repeating what I have expressed in previous articles,[2–4,8–12] it is now time to initiate the analysis of the mechanisms and molecules that mediate the progression of the "aging and death program" and are closely linked to the evolution of the program itself. In fact, we are obviously curious to know if we can decipher and unravel the mysteries of the "program" in our brain, to the extent that we may shape and modify it, aiming at prevention of diseases and achievement of a long and healthy life.

THE MELATONIN MIRACLE

It is in fact a kind of wonder or "miracle" that I was confronted with the life-prolonging effects of nocturnal melatonin in rodents and the even more dramatic aging-postponing effects of young pineal grafting.[10–12] The questions arising from those striking observations forced me to consider that the facts observed cannot be considered in the context of any simplistic explanation of the present scientific knowledge. It seems to me that two intriguing questions must be understood and properly answered: (1) Is melatonin a "normal" chemical agent with classical reactivity, receptors, and so on? (2) How is melatonin prolonging life if it is not contained in the life-prolonging, transplanted young pineal gland?

Obviously the shock produced by the findings[13] and the scandalous outcry[14,15] deserves no answer unless we reply the above questions. However, we are not able to answer the two questions simply because we are in front of a totally new dimension, which requires a huge jump in the interpretation of the essence of life itself. Is life in our planetary system based on something that cannot be defined by chemistry or biochemistry? Does the ubiquitous molecule melatonin represent an element devised by evolution to maintain and/or protect life itself in all its expressions? Melatonin is in fact everywhere and nowhere! Its versatility and elusive behavior escapes a proper classification. My consideration of the many facts and claims surrounding melatonin brought me to the conclusion that we are in front of a "mysterious" natural molecule that has been developed at the same time as life itself and that carries or transfers an energy of unknown character and nature. Melatonin is a vehicle for energies or forces that are unknown to us and yet they exist! Their presence during life is essential to maintain a proper brain program and to protect the pineal gland itself from daily insults of all kinds (bacterial, viral, psychic, poisons, and environmental, for example). In the presence of melatonin, an aging pineal can reacquire a large degree of its pristine capacity to control hormonal cyclicity and thus immunity itself. Absence of melatonin causes diseases and death. It is likely that life would not exist at all on our planet without melatonin. Can it be replaced by a similarly "smart" molecule able to simultaneously detect energy of unknown nature and to transfer it to cells and tissues? Are acetylation and methylation and thus the molecular conformation of melatonin basic to this capacity? Obviously melatonin is now basic to decipher the secrets of life and of the forces that surround us, such as the mysteries of bird orientation, magnetism, gravity, and cosmic rays. In fact, melatonin is basic to the interpretation of what goes on beyond the speed of light: brain energy and telepathy, or premonition perhaps? A clue to the laws of the universe?

Certainly melatonin prolongs life in aging animals only because it protects the pineal gland from aging, thus allowing the pineal network to perform its program! As we have shown,[16] an aging pineal will accelerate aging in a normal young animal. We have thus answered our second main question. What to do now?

THE ALL-PERVADING CYCLIC-ROTATIONAL ENERGIES OF THE UNIVERSE

For the first time we have a model to study the basic essence of life as it has been generated and shaped on our planet. The intimate nature of energy, and thus life, is primarily rotational, from the atom to the molecules, to the double helix of DNA, to planetary cyclicity and gravity, even as far as the rotational conformation of galaxies! Not accidentally planets and their satellites are round or spherical. It would be unthinkable that this basic principle of life is not be reproduced in our own body: construction and maintenance of basic rotational cyclicity in all the atoms, molecules, cells, and tissues of the body, and its progressive disintegration or de-synchronization in the course of aging. Rotational energies, whatever their classification and dimensions (from the atom to the galaxies) are basic for life. It would also be inconceivable if a device did not exist in our brain suitable to maintain planetary, lunar, and cosmic cyclicity and periodicity, which is inscribed in the genes of all species, from the simplest forms of life such as unicellular organisms to the complexity of vertebrates. Yet, we ignore the kind of forces that shape our body and maintain our "brain program for growing, procreating and finally age and die!"[17] Chemistry and biochemistry represent a tiny and not even relevant piece of the complex mechanisms that created and maintain life on our planet. More pervasive forces, which have been determinant for the origin of life, escape our limited capacity inside the sensorial cage in which we live.

THE CENTRIPETAL AND THE CENTRIFUGAL FORCES OF LIFE: FROM ATOMS TO PLANETS TO GALAXIES

My son—who studies modern physics and all kind of theories—tells me that no one knows precisely what gravity is! Is this possible? My naive and innocent question is: How could gravity not exist when a planet turns on itself and centripetal forces are counter-balanced by the centrifugal forces in relation to the inner condensed mass of each planet? Gravity is in fact the boomerang of life! It generates life and brings back its end, namely death! There would be no life in the absence of gravity on the planet Earth. The planetary, centrifugal, and centripetal forces are in fact the basic generating life-elements, in so far as they reproduce the atomic, centripetal, and centrifugal forces of all atomic and subatomic particles in all inorganic and organic matter. It is thus unconceivable that life would be generated on our planet without the shaping energy of all rotating forces in the Whole. Consequently, the maintenance of such shaping forces is fundamental for our body. Such forces are precisely reproduced by the inborn periodicity forces, such as circadian and ultradian cyclicity shaped in close relation to the centripetal and centrifugal energies of our planet and the "magnetic" lunar attraction, which has molded female's reproductive cyclicity. Therefore, by considering the Whole: we have now the answer! The cosmic forces are identical in their nature, from the atom's electrons and subatomic entities to the galaxies: they strictly reproduce the essential physics of nature and consequently the basic shapes of life and its regulation according to the cosmic-planetary laws. If this wasn't the case, the instability of the planetary system (e.g., the solar system) would lead to its sudden collapse, in a fraction of a second. The delicate balance of all inorganic and organic matter is obeying the powerful and dictatorial order of the Whole: rotate and keep rotating!

WHAT IS AGING?

Life from Chaos: The Haphazard Patchwork of Nature for Survival of the Species

The cause of aging is simple: it is a program in the neuroendocrine (simply called "hormonal") system and it is independent of diseases. We age because we are genetically programmed to central pineal-hypothalamic (brain) desynchronization of planetary rhythmicity (circadian, lunar, seasonal).[17,18] We age exactly the way we grow and stop growing. The program is genetic for man, dog, pig, mouse, but we have the capacity to learn how to change, not the genetic program, but the manner and time of its expression. We can slow down aging, *we can stop aging*. We can also reverse aging and restore youth (provided no permanent damage or destruction of brain or other tissues has occurred). We will learn to see it in the course of the next few years, when the results of our capacity for resetting the periodicity clock will be visible. My way of "resetting the central clock" now is very primitive and empirical, but we shall learn progressively to be more knowledgeable and clever, more cautious and sophisticated, we shall use the ways of nature and the clever natural way: no jogging, no aerobic exercise, no stretching, no fanatic dieting, no wild and whimsical antioxidants, but simply by re-adjusting the body rhythms to those of the planetary system, namely the sun, the moon, the seasons, the day-night cyclicity. We age simply because we inadvertently lose the adaptation capacity to be and to remain "periodical" and "cyclical," we deregulate our neuroendocrine system and become refractory or insensitive to the laws of natural cyclicity.

Typical is the progressive or abrupt hormonal desynchronization resulting into menopause and andropause[19,20]: they are indicators of loss of periodicity and lack of periodicity is death. When we stop perceiving the rotational forces of the planetary system that have shaped our "life program," we simply die. Our body becomes insensitive to the basic impulses and messages from the Whole, we desynchronize and become cosmic dust, the way we were for eons. Immortality is not a myth: it is simply a permanent re-adaptation to sun and lunar cyclicity, and much more (see below). Now we know the way to go and we could achieve this aim within a few years without the prejudice and arrogance of the "saviors."

THE EXPERIMENTAL MODELS FOR AGING POSTPONEMENT AND REVERSAL

The neuro-hormonal and the immune systems are interdependent. The interdisciplinary medical research now named neuroimmunomodulation emerged in the last 40 years as an inevitable consequence of the demonstration that no separation exists, during ontogeny, embryogenesis, and adult life, in the function of the brain and the neuroendocrine and immune systems.[21] As many like to describe it, they "talk to each other." Earlier studies carried out in many experimental and natural models [such as neonatally thymectomized mice,[21] hypopituitary dwarf mice,[23,24] and in particular genetically thymusless (athymic) nude mice[25]] all indicated that many hormones control immunity and that immune cells affect the maturation and function of fundamental neuroendocrine organs, such as the hypothalamus and hypophysis, and basic endocrine glands, such as the thyroid, the adrenals, and the gonads.[26,27]

In addition, it was observed that many natural or induced diseases or "syndromes" resembling a "precocious aging" in mice were closely linked to a derangement of the bidirectional regulation of the neuroendocrine and of the immune systems in the course of postnatal growth. It was thus progressively understood that some basic neuroendocrine alterations in the course of early life or, conversely, some missing cellular or humoral elements of the thymo-lymphatic immune system during early ontogeny, were responsible for the emergence of physical and functional deficits that closely mimic "senescence."[28] This was also confirmed by studies of mice kept at a low-calorie diet.[29] The "story" of the work is now also summarized in a popular book.[13]

Data from our laboratory led us to the idea that the most crucial aspect of aging could be an aging-related, progressive blunting (and finally abrogation) of hormonal cyclicity, in particular of day-night circadian periodicity. This circadian periodicity is a basic rhythmicity that determines the daily fluctuation of all hormones and all physiological functions, including immunity, reproduction, and sleep.[31,32] Our experiments considered the use of melatonin simply because some early observations in athymic nude mice[25] and later in mice kept under permanent illumination for several generations had shown a reconstitution of immunity and an impairment of growth, "runting" or "wasting," respectively. These closely resembled aging and were a consequence of the abrogation of night melatonin production by the pineal gland.[30]

We started in 1985 with experiments that led to the present evidence indicating that the pineal gland and one of its products, melatonin, play a fundamental role in the initiation and in the progression of aging, namely that melatonin itself is a signal for pineal aging.[10]

The Pineal Gland and Neuroimmunomodulation of Aging: An Inborn Program

We constructed three models (described in detail in the next section) to evaluate the role of the progressive alteration of pineal function in the course of aging. These models were based on the consideration that if the pineal gland, thanks to its known bidirectional linkage with the entire neuroendocrine system, is the "master gland" of the body. Its own progressive "aging" in the course of life may result in the desynchronization of all functions and in particular those metabolic pathways that maintain the integrity of the neuro-hormonal (endocrine) and of the immune systems. Although we still do not know why the pineal gland itself ages, we have demonstrated that the pineal gland and at least one of its products, melatonin, are key elements to start understanding why and when we start aging.[8–10]

It is beyond any doubt that a very significant aging-postponing effect is achieved in aging mice (and rats) with nocturnal administration of melatonin or young-to-old pineal grafting.[10–13] However, an even more remarkable observation derived from the young-to-old and old-to-young pineal cross-transplantation model revealed that an acceleration of aging is achieved when a younger animal is grafted with an older pineal gland after removal of its own pineal.[12]

This striking finding led us to an unexpected field of investigation. It seems that after a certain age the pineal gland actively promotes aging, as if the inborn program for pubertal maturation and reproductive function would inevitably lead to another step of the maturational events, namely aging. This is even more evident if we consider that the implantation of a young pineal gland into the thymus of older mice does not result in any life prolongation if the recipient is too old (unpublished observations). It means that once the endogenous pineal has aged, no intervention can prolong life. There exists a critical age in pre-senescent or senescent mice when pineal grafting or melatonin administration positively affect the life span. But beyond a certain age, apparently the old mouse's pineal gland determines the termination of life. This is clearly shown by the acceleration of aging in pinealectomized young mice implanted with an old pineal gland.[12] This extraordinary observation forced us to recognize that there might exist a "death clock" in the pineal network of the brain whose "program" cannot be modified unless the aging pineal is removed in due time and replaced with a younger pineal.[31,32]

We must now consider a completely new element for the evaluation of the causes of aging and the significance of biological death. We can infer that the pineal gland, at least in our experimental models with rodents, is at the same time an "aging clock", a "life clock," and a "death clock", depending on its chronological stage from birth to death.[16,31] If this is the case, a large number of experiments are needed to assess these crucial temporary steps in mammals, man included. This is more relevant than the work aimed at the investigation of the mechanisms and they should proceed in parallel. We are prone to study the details before understanding how the pineal gland programs aging.

Proof for the Existence of a Life, Aging and Death "Clocks"

Recent experiments have proved beyond any reasonable doubt that a precise program exists in our mammalian brain that determines the time for growing, procreating, aging, and dying. It was highly doubtful that the experiments performed and published would be credible to the scientific audience.[31,32] They have shown that even a tiny, very old pineal gland conveys aging and death signals to a young body, possessing its own young pineal![32] This amazing observation demonstrated that even an isolated and dissected old pineal gland, obviously denervated and deprived of its neural connection and its normal links to the suprachiasmatic nuclei, optic nerves, hypophysis, ganglia—and consequently to the main hormone-producing glands—is able to deliver powerful and merciless aging and death signals and thus imposing its program to the young body![32] The solution to this riddle will allow us to understand at a cellular and molecular level how our program for aging and dying has been constructed in the course of phylogeny of vertebrates. We must learn how the message "age!" and "die!" is delivered from the very old pineal without need for neural impulses.[32,33] It is truly a metaphysical question and a mystery that leads us to a totally ignored level of interpretation of life processes.

THE EXPERIMENTAL EVIDENCE

Life Prolongation via a Restricted Calorie Diet

Since the first dramatic experimental evidence produced by McCay with rodents,[33] an immense literature is now become available that documents the different methods used to prove that reduced calorie intake will significantly delay aging and affect the many aging-related diseases and metabolic dysfunctions.[35] However, it has taken more than 60 years for the National Institute on Aging to the evaluate whether a restricted-calorie diet applied to non-human primates retards aging. There is now evidence that this is the case and many results from this long-term trial are now available.[35]

When, thanks to its basic developmental functions in ontogeny, we considered the thymus as a kind of "clock" for aging of the immune system, we demonstrated (in a rather neglected publication) that the thymus does not deserve such a primary role for initiation and progression of aging.[36] Removal of the thymus at different times after birth in mice did not significantly affect their life span.[36] However, it became clear that the thymus was deeply involved in the ontogenetic programming and maturation of the entire neuroendocrine system and that athymic nude mice suffered a kind of precocious senescence that could be completely prevented by thymic implantation. Thymus grafting resulted in a complete normalization of neuroendocrine functions.[25] On the basis of those findings and the subsequent observations on the ability of mature lymphocytes to restore growth and immunity and to prolong the life of dwarf mice,[24] the idea evolved that a different hormonal regulation must be responsible for the aging-postponing effects of a restricted-calorie diet. In other words, we suggested that a decreased calorie intake would produce permanent changes in the central, hypothalamic-pituitary hormonal functions, thus maintaining the body at a more juvenile level of endocrine and metabolic regulation.[29] This was particularly clear with regard to sexual functions of rodents maintained at a restricted-calorie diet.[29]

We conducted some experiments in which mice were kept at a restricted-calorie diet for a few weeks after weaning, and then fed again *ad libitum* and then examined to see if they maintained, in spite of this normal feeding, a permanently different pattern of hormonal regulation.[29] These data confirmed that feeding behavior at a time when the neuro-endocrine system is still immature permanently affects maturation and function of the entire neuroendocrine system.[29] This observation is relevant with regard to the onset of obesity in overfed children and the consequent irreversible derangement of their mature neuroendocrine and metabolic system. This environmentally induced obesity, now so dramatically evident in the affluent western society, is different from mild or severe fattening of "normal" metabolic aging in humans. Also, this environmentally acquired dietary obesity is different from genetically inherited obesity, which afflicts a relatively minor number of families and individuals.

These studies indicate that a restricted-calorie diet produces juvenility-oriented and permanent changes of neuroendocrine regulation, which are exactly the opposite outcome of environmentally induced and aging-accelerating dietary obesity. If endocrine and metabolic dysfunction are the expression of the program of aging, and if the pineal gland is a "life and aging clock," we must consider that a restricted-calorie diet affects mainly the pineal gland and its functional state. This seems to be the case. It has been reported that a restricted-calorie diet maintains juvenile levels of melatonin both in rodents and in primates.[37] In a collaborative project with Dr. George Roth and Mark Lane at the National Institute on Aging, Baltimore, USA, large groups of primates were fed a restricted-calorie diet for several years. Data indicate that a restricted-calorie diet very significantly maintains high levels of nocturnal melatonin in both male and female aging monkeys, comparable to the levels in young primates.[37] My personal conclusion is that a restricted-calorie diet, by setting the "neuroendocrine clock" at a more juvenile level, including obviously juvenile nocturnal production and levels of melatonin, protects the pineal gland from aging and thus protects from aging the whole pineal-controlled hormonal, circadian, and seasonal periodicity and rhythmicity, whose progressive decay leads to aging.[18,38] However, melatonin is only a signal from the pineal gland of an overall desynchronization of the whole neuroendocrine network leading to a progressive alteration of hormonal cyclicity and consequently surveillance of immune functions.[30]

The Anti-Aging Molecule Melatonin

In spite of the vulgarity of the defamatory campaign against the anti-aging properties of melatonin,[14,15] it is beyond any doubt that exogenous, nocturnal administration of melatonin to aging rodents postpones their aging and/or prolongs their life.[4,11] Unfortunately, for mysterious or tactical reasons, those experiments have not been properly replicated while the deceptive behavior against melatonin anti-aging properties continues. Research on melatonin indicates that the pineal gland is deeply involved with the aging process. Its aging-postponing activity suggested the pineal grafting experiments (see section on grafting below), and these disclosed a dramatic, new approach to aging-postponing strategies. These fundamental experiments also have not been replicated.[4,11,12] The pineal grafting experiments also served to indicate that the pineal gland, via its links to the entire neuroendocrine system, controls the "program of aging" and that an aging pineal can accelerate aging even in a normal young animal carrying his own young pineal.[16]

These striking observations helped us understand that other key mechanisms and/or molecules must be operative for the anti-aging and the aging-accelerating properties of pineal grafting. Whether or not the anti-aging and the pro-aging capacities of the young and old pineal gland depend on a unique mechanism, it is reasonably clear that other pineal components must play a prominent role.

That melatonin could significantly postpone aging thanks to its antioxidative and hydroxyl radical–scavenging properties, such as those of vitamin E or glutathione, is not supported either by logic or by any serious *in vivo* confirmation.[39] It does not seem that the many receptor-mediated effects of melatonin and the myriad of affinity-binding mechanisms can explain its anti-aging properties. The anti-stress, immuno-protecting effects of melatonin show a rather slow "buffer" mechanism.[39]

This reinforces my hypothesis that melatonin does not by itself exert the activities observed, but rather protects the pineal gland from aging. Nocturnal melatonin supply will not protect from aging when the age of the animals is too advanced. This has now been proven in another kind of placebo-controlled clinical trial, in which perimenopausal women aged 40 to 60 years, have been treated with melatonin. Already after six months the evidence emerges that younger women are more susceptible than older women to the anti-aging properties of melatonin.[20]

This fact strongly supports the view that the beneficial and pineal-protecting effects of melatonin are more pronounced at a time when the pineal is still relatively younger. This unexpected finding indicates that melatonin can exert a more pronounced anti-aging effect if the administration starts rather early in life, insofar as it protects the pineal from aging. This observation is fundamental for the prophylactic use of melatonin in anti-aging interventions and strengthens the suggestion that the mechanism of action of melatonin cannot be attributed to a "hormonal" effect on specific receptors but rather to a relatively simple nocturnal saturation of melatonin in the pineal gland, and consequent abrogation of night endogenous melatonin production.[39]

If this suggestion is true, it must be possible to drastically reduce or abrogate aging-dependent endocrine and metabolic dysfunctions by the administration of exogenous melatonin in the early, postpubertal life of mammals, man included, as hinted at by the emerging results in perimenopausal women.[20]

Young-to-Old Pineal Grafting Delays Aging

This third model, although not immediately suitable for a practical application in man as is a restricted-calorie diet and administration of melatonin, has been crucial for understanding that a precise "clock" exists in the brain that is genetically programmed and located in the "pineal network."[10–13,17–19] In addition, pineal grafting into old aging and also in young normal, non-pinealectomized rodents, has disclosed the amazing evidence that not only the pineal network programs all the chronological steps of our life (growth, puberty, fertility, aging), but also of our death at a time when the genetic program for our species is concluded.[16,32] It is truly hard to visualize the kind of signals from the aging pineal gland, even if dissected from all its neural connection, that will suddenly force on us, at age 120 and in good health, that we must die! Is there a "death hormone" or perhaps we are simply "going to sleep" similar to a deep hibernation or a sudden drop of all metabolic, energy-producing processes? Our experiments indicate that the pineal gland actively promotes aging and death according to a precise genetic program of mammals.[16,32]

CONCLUSION

Hormonal-Immunological Desynchronization Accelerates Aging and Anticipates Autoimmune Diseases, Cardiovascular Diseases, and Cancer

My experience with several hundred patients in the last five years have convinced me of the following.

One, that basically, all diseases have the same origin. Whatever the main co-factors responsible for the disease (genetical proneness, environmental, social, psychic, viral, etc.), I have seen that all diseases, including cardiovascular diseases, autoimmune diseases, and cancer, are anticipated and announced by rather repetitive and typical alterations of hormonal cyclicity and levels, whether or not they concern pituitary hormones (FSH, LH, TSH, ACTH, GH, prolactin), or peripheral gland hormones (T3, T4, cortisol, adrenalin, progesterone, estradiol, etc.). Those alterations are easily detected in the peripheral blood. Early symptoms do not escape an expert eye!

Two, that inevitably, all immune functions are deeply affected by the alteration of hormonal cyclicity, which totally controls and maintains the expression of immunity, be it antibacterial, antiviral or anticancer, immunological surveillance.

Three, that bacteria and viruses can attack all cells and organs of the body and more or less silently produce a deeper immunosuppression and a progression of the process. This process is particularly evident in all kind of neurodegenerative diseases, where the initiating neurohormonal derangement results into the autoimmune process initiated by a viral attack. The condition of immunosuppression is perpetuated and maintained and the viral-induced autoimmune damage results into the typical progression of the disease (e.g., multiple sclerosis, Alzheimer, Parkinson, etc.).

Four, that cancer has the same origin. Any virus can induce cancer at a time when the immune system is deranged by a chronic or acute alteration of hormonal cyclicity.[40]

And five, that aging is simply accelerated by diseases, whatever the cause and origin, and can be brought to the physiological expiration of the "biological clock program" only if we are able to prevent and avoid diseases. The clock in humans corresponds to the age of 120 to 140 years.

Are Diseases Preventable? Can We Cure All Diseases?
The Ultimate Defiance of Man

It is my firm conviction that we can prevent all kind of diseases by early detection of those hormonal, neuroendocrine alterations of their cyclicity that anticipate the onset of all diseases.

If a disease is already progressed, I try to correct the hormonal alteration and to rebuild a correct hormonal cyclicity, which conversely produces a violent and healthy restoration of immunity, both of the immediate (antibody production, etc.) and of the cell-mediated delayed type. The same is true for cancer, where autonomic auto-control of immunity is fundamental for re-acquisition of anti-cancer capacity.[40] I do believe that all diseases can be healed if the process is not too advanced.

Therefore the present-day "ultimate defiance" of man is prevention and cure of all diseases by acting on their cause. A disease-free man can live as long as the "pineal program" dictates. In addition, the further identification of the molecular system in the pineal gland that provides efficiency and function of the biological clock, will allow us to interfere with those genetically inherited and dominant elements of the aging and death program that promote aging and death, as illustrated in our pineal grafting experiments.

REFERENCES

1. JAFFE, A. 1972. From the Life and Work of C.G. Jung. English translation by R.F.C. Hull. Hodder & Stoughton. London.
2. PIERPAOLI, W. 1991. The pineal gland and melatonin: a circadian or seasonal aging clock? Aging **3:** 99–101.
3. PIERPAOLI, W. 1994. The pineal aging clock. Evidence, models, and an approach to aging-delaying strategies. In Aging, Immunity and Infection. D.D. Powers, J.E. Morley & R.M. Coe, Eds.: 166–175. Springer. New York.
4. PIERPAOLI, W. & W. REGELSON. 1994. The pineal control of aging. The effect of melatonin and pineal grafting on aging mice. Proc. Natl. Acad. Sci. USA **91:** 787–791.
5. BUDDHA. 1987. Thus have I heard, page 46. Translated from the Pali by M. Walshe. A new translation of the Digha Nikaya. Wisdom Publications. London.
6. LACOCQUE, A. 1967. Le Devenir de Dieu, page 100. Editions Universitaires. Paris.
7. THE NEW ENGLISH BIBLE. 1970. Exodus 3:14. Page 63 and corresponding note. Oxford University Press. Oxford.
8. PIERPAOLI, W. & C.X. YI. 1990. The pineal gland and melatonin: the aging clock? A concept and experimental evidence. In Stress and the Aging Brain. G. Nappi, E. Martignoni, A.R. Genazzani & F. Petraglia, Eds.: 171–175. Raven Press. New York.
9. PIERPAOLI, W., C.X. YI, & A. DALL'ARA. 1990. Aging-postponing effects of circadian melatonin: experimental evidence, significance, and possible mechanisms. Int. J. Neurosci. **51:** 339–340.
10. PIERPAOLI, W. & G. MAESTRONI. 1987. Melatonin: a principal neuroimmunoregulatory and anti-stress hormone: its anti-aging effects. Immunol. Lett. **16:** 355–362.
11. PIERPAOLI, W., A. DALL'ARA, E. PEDRINIS & W. REGELSON. 1991. The pineal control of aging. The effects of melatonin and pineal grafting on the survival of older mice. Ann. N.Y. Sci. **621:** 291–313.
12. LESNIKOV, V.A. & W. PIERPAOLI. 1994. Pineal cross-transplantation (old-to-young and vice versa) as evidence for an endogenous "aging clock." Ann. N.Y. Acad. Sci. **719:** 456–460.

13. PIERPAOLI, W., W. REGELSON & C. COLMAN. 1995. The Melatonin Miracle. Simon & Schuster. New York.
14. REPPERT, S.M. & D.R. WEAVER. 1995. Melatonin madness. Cell **83:** 1059–1062.
15. TUREK, F.W. Melatonin hype hard to swallow. 1996. Nature **379:** 295–296.
16. PIERPAOLI, W. & D. BULIAN. 2001. The pineal aging and death program. I. Grafting of old pineals in young mice accelerates their aging. J. Anti-Aging Med. **4:** 31–37.
17. PIERPAOLI, W. & V.A. LESNIKOV. 1997. Theoretical considerations on the nature of the pineal "Aging Clock." Gerontology **43:** 20–25.
18. PIERPAOLI, W. 1994. The pineal gland as ontogenetic scanner of reproduction, immunity and aging: the aging clock. Ann. N.Y. Acad. Sci. **741:** 46–49.
19. PIERPAOLI, W., D. BULIAN, A. DALL'ARA, *et al.* Circadian melatonin and young-to-old pineal grafting postpone aging and maintain juvenile conditions of reproductive functions in mice and rats. 1997. Exp. Gerontol. **32:** 587–602.
20. BELLIPANNI, G., P. BIANCHI, W. PIERPAOLI, *et al.* 2001. Effects of melatonin in perimenopausal and menopausal women: a randomized and placebo controlled study. Exp. Gerontol. **36:** 297–310.
21. FABRIS, N., M.M. MARKOVIC, N.H. SPECTOR & B.D. JANKOVIC, Eds. 1994. Neuroimmunomodulation: The State of the Art. Ann. N.Y. Acad. Sci. Vol. 741.
22. PIERPAOLI, W. & E. SORKIN. 1967. Relationship between thymus and hypophysis. Nature **215:** 834–837.
23. PIERPAOLI, W., C. BARONI, N. FABRIS & E. SORKIN. 1969. Hormones and immunological capacity. II. Reconstitution of antibody production in hormonally deficient mice by somatotropic hormone, thyrotropic hormone, and thyroxin. Immunology **16:** 217–230.
24. FABRIS, N., W. PIERPAOLI & E. SORKIN. 1972. Lymphocytes, hormones, and aging. Nature **240:** 557–559.
25. PIERPAOLI, W. & E. SORKIN. 1972. Alterations of adrenal cortex and thyroid in mice with congenital absence of the thymus. Nat. New Biol. **238:** 282–285.
26. PIERPAOLI, W. & H.O. BESEDOVSKY. 1975. Role of the thymus in programming of neuroendocrine functions. 1975. Clin. Exp. Immunol. **20:** 323–338.
27. PIERPAOLI, W., H.G. KOPP & E. BIANCHI. 1976. Interdependence of thymic and neuroendocrine functions in ontogeny. Clin. Exp. Immunol. **24:** 501–506.
28. PIERPAOLI, W., H.G. KOPP, J. MUELLER & M.KELLER. 1977. Interdependence between neuroendocrine programming and the generation of immune recognition in ontogeny. Cell. Immunol. **29:** 16–27.
29. PIERPAOLI, W. 1977. Changes of the hormonal status in young mice by restricted caloric diet. 1977. Experientia **33:** 1612–1613.
30. MAESTRONI, G. & W. PIERPAOLI. 1981. Pharmacological control of the hormonally mediated immune response. *In* Psychoneuroimmunology. R.A. Ader, Ed.: 404–428. Academic Press. New York.
31. PIERPAOLI, W. Integrated phylogenetic and ontogenetic evolution of neuroendocrine and identity-defense immune functions. *In* Psychoneuroimmunology. R.A. Ader, Ed.: 575–606. Academic Press. New York.
32. PIERPAOLI, W. & D. BULIAN. 2005. The pineal aging and death program. II. Life prolongation in pre-aging pinealectomized mice. Ann. N.Y. Acad. Sci. In press.
33. MCCAY, C.M. & M.F. CROWELL. 1934. Prolonging the life span. Sci. Monthly **39:** 405–414.
34. WEINDRUCH, R. & R.L. WALFORD. 1988. The Retardation of Aging and Disease by Dietary Restriction. Charles C. Thomas, Ed. Thomas. Springfield, IL.
35. INGRAM, D.K. & G.S. ROTH. 1997. Beyond the rodent model: caloric restriction in rhesus monkey. Age **20:** 45–56.
36. PIERPAOLI, W., M. HAEMMERLI, E. SORKIN & H. HURNI. 1977. Role of thymus and hypothalamus in aging. *In* Fifth European Symposium on Basic Research in Gerontology. U.J. Schmidt *et al.,* Eds.: 141–150. Verlag Dr. Med. D. Straube. Erlangen, Germany.
37. ROTH, S.G., V.A. LESNIKOV, M. LESNIKOV, *et al.* 2001. Dietary caloric restriction prevents the age-related decline in plasma melatonin levels of Rhesus monkeys. J. Clin. Endocrin. Metab. **86:** 3292–3295.

38. PIERPAOLI, W & V.A. LESNIKOV. 1994. The pineal aging clock. Evidence, models, mechanisms, interventions. Ann. N.Y. Acad. Sci. **719:** 456–460.
39. PIERPAOLI, W. & V.A. LESNIKOV. 1997. Pineal control of stress, distress, and aging: the melatonin evidence. Dev. Brain Dysfunct. **10:** 528–537.
40. PIERPAOLI, W. 2005. Our endogenous "pineal clock" and cancer. Cancer and non-cancer strategies for early detection, prevention, and cure of neoplastic processes. *In* Hormones, Age, and Cancer. L.M Berstein, Ed.: 207–225. Nauka. St. Petersburg.

Multiple Longevity Phenotypes and the Transition from Health to Senescence

ROBERT ARKING

Department of Biological Sciences, Wayne State University, Detroit, Michigan 48202, USA

ABSTRACT: Three different longevity phenotypes exist in *Drosophila* and other model systems, but only two are known in humans. The "missing" phenotype is the delayed onset of senescence phenotype, which can be induced by various interventions, including pharmaceuticals. The lability of the onset of senescence indicates that the mechanisms involved are plastic and can be altered. Only interventions that involve the upregulation of stress resistance genes, probably via the JNK pathway and/or dFOXO3a transcription factor, seem capable of generating a delayed onset of senescence phenotype. The data suggest that the cellular mechanisms responsible for maintaining the cell in a healthy state are under constant attack by ROS and/or abnormal protein accumulation. A stochastic growth factor/signal transduction failure may be the proximal event responsible for the decreased efficiency of the cell's defenses, resulting in the onset of senescence, degradation of the gene interaction network, and continuing loss of function.

KEYWORDS: longevity phenotypes; health span; senescent span; delayed onset of senescence; cell senescence; stress resistance

INTRODUCTION

There is an ongoing debate as to whether or not human longevity is approaching its limits. We add our empirical observation that there exist multiple longevity phenotypes, each of which arises from the alteration of fundamental processes. Only two of these three phenotypes are now known in humans. The "missing" phenotype is the "delayed onset of senescence" phenotype, which can be induced by various interventions, including pharmaceuticals. The onset of senescence can be experimentally accelerated or decelerated. The observed lability of the age of onset of senescence indicates that we need to know more about the mechanisms responsible for allowing a healthy cell to transit into a state of senescence and eventual organismal death. A review of the various extended longevity strains and mutants generated in *Drosophila* leads to the view that only interventions that involve the upregulation of stress resistance genes, probably via the dFOXO3a transcription factor, are capable of generating a delayed onset of senescence phenotype. The proximal cause of

Address for correspondence: Robert Arking, 3103 BSB, Department of Biological Sciences, 5470 Gullen Mall, Wayne State University, Detroit, MI 48202. Voice: 313-577-2891/2850; cell: 248-376-4849; fax: 313-577-6891.
aa2210@wayne.edu

Ann. N.Y. Acad. Sci. 1057: 16–27 (2005). © 2005 New York Academy of Sciences.
doi: 10.1196/annals.1333.001

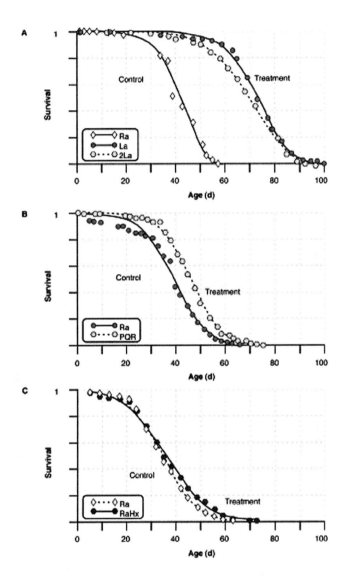

FIGURE 1. (**A**) Survival curves of the normal-lived Ra strain and of two long-lived strains (La and 2La) that are significantly different (log-rank test = 281.46, df = 1, $P < 0.00005$). See Arking et al.[8] for experimental details. (**B**) Survival curves of the normal-lived Ra strain and the PQR strain selected from it by direct selection for paraquat resistance. The two curves are significantly different (log-rank test = 24.76, df = 1, $P < 0.00005$). See Vettraino et al.[32] for experimental details. (**C**) Survival curves of the normal-lived Ra control strain and the longer-lived Ra heat-treated strain. The two curves are significantly different (log rank test = 17.84, df = 1, $P < 0.00005$). See Keuther and Arking[14] for experimental details. (Reproduced from Arking[23] with permission.)

transition to the senescent phase may stem from a (stochastic?) degradation of regulatory signals.

THE THREE LONGEVITY PHENOTYPES OF *DROSOPHILA*

The existence of conserved public mechanisms makes aging comprehensible without, however, removing its complexity. The work that my colleagues and I have done on *Drosophila* longevity bears this out. We reported that aging in our *Ra* strain of wild-type flies is characterized by at least three different extended longevity phenotypes, each of which is induced by specific stimuli and has different demographic mortality and survival profiles.[1] Thus the aging profile of a given wild-type genome is not fixed but can take one of three different paths. As shown in FIGURE 1A, the first longevity phenotype (type 1) is a delayed onset of senescence that leads to a significant increase in both mean and maximum life span of the experimental strain. The second longevity phenotype (type 2), shown in FIGURE 1B, is an increased early survival that leads to a significant increase in mean, but not in maximum, life span. The third longevity phenotype (type 3), shown in FIGURE 1C, is an increased later survival that leads to a change in maximum (LT_{90}), but not in mean, life span.

Analysis of the mortality data supports these statements. The type 1 phenotype yields a Gompertz curve that is significantly different from that of the control strain (FIG. 2A). Analysis of the data suggests that the type 1 phenotype involves a ~50% reduction in the mortality rate doubling time (MRDT) of the long-lived strain (8.8 days) relative to the normal-lived strain (5.8 days). The MRDT is a commonly used proxy indicator of comparative aging rates. However, neither the type 2 long-lived populations (FIG. 2B) nor the type 3 long-lived populations (FIG. 2C) show any sustained alteration in aging rates relative to their controls, but rather show only a transient decrease in either early or late life, but not in both. These likely represent a change in morbidity but not an alteration of aging.

Examples of these three longevity phenotypes are found in all model organisms; but only the latter two phenotypes (types 2 and 3) are found in humans.[2] Since aging mechanisms are conserved across species, it seems reasonable to suggest that the current absence of the extended longevity phenotype in humans more likely indicates the absence to date of the appropriate stimulus than it does the impossibility of extending human longevity.

Moreover, it should be pointed out that the same phenotype may be induced by different stimuli. For example, the type 1 delayed onset of senescence phenotype may be induced in flies by (a) caloric restriction,[3] (b) downregulation of the insulin-like signaling pathway,[4–6] (c) upregulation of the antioxidant defense system (ADS) plus altered mitochondrial properties,[1] (d) drugs that inhibit histone deacetylases,[7] and by other mechanisms, as discussed below. Finally, the fact that forward selection can create delayed senescence, while reverse selection is capable of nullifying the type 1 phenotype by moving the transition point from an advanced age back to a young age,[8] indicates that the molecular basis of the transition point is labile and susceptible to manipulation. This genomic plasticity raises a set of three nested questions: (1) What mechanisms enable the organism to shift the age of its transition from the health span to the senescent span? (2) How is it that such a varied assortment of stimuli and mechanisms can bring about a seemingly identical type 1

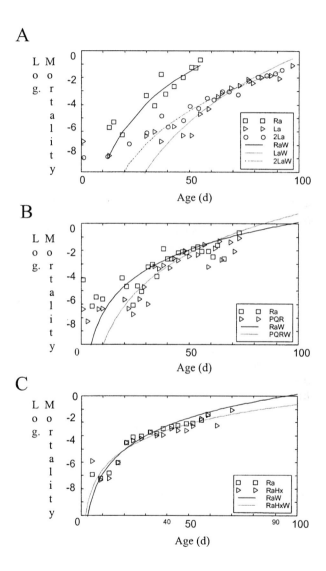

FIGURE 2. (**A**) Age-specific logarithmic mortality curves of the normal-lived Ra strain and of two long-lived strains (La and 2La) shown in FIG. 1A. (**B**) Age-specific mortality curves of the normal-lived Ra strain and the PQR strain selected from it by direct selection for paraquat resistance shown in FIG. 1B. (**C**) Age-specific mortality curves of the normal-lived Ra control strain and the longer-lived Ra heat-treated strain shown in FIG. 1C. (Reproduced from Arking et al.[33] with permission.)

phenotype? (3) What is it that distinguishes these stimuli from the stimuli that yield only the type 2 or 3 phenotypes? This paper is our initial attempt to answer these questions.

WHAT TRAITS DIFFERENTIATE LONG-LIVED
FROM NORMAL-LIVED ANIMALS?

Much of our efforts over the past two decades has been devoted to characterizing the differences between the normal- and long-lived strains we created by artificial selection.[9,10] Certain of these seem to play a critical role in delaying the transition from health to senescence, and their characterization allows us to identify some of the molecular species and pathways involved, as follows:

(1) Our biomarker data suggested that the earliest detectable event in the expression of the extended longevity phenotype took place early in adult life at about 5–7 days of age.[11] The La animals exhibit a coordinated and specific upregulation of many different antioxidant defense genes (and proteins) beginning at ~5–7 days after eclosion,[12] and this increased ROS scavenging ability plays a major role in restricting the extent of oxidative damage. The animal's metabolism is altered so as to support these antioxidant activities.[13]

(2) Heat shock experiments showed that the Ra but not the La animals expressed a hormetic response, suggesting that the heat shock proteins (hsps) of the latter were already constituitively expressed.[14] Kurapati et al.[15] showed that hsp22 was significantly elevated in a different line (M. Rose's O strain) of flies selected for extended longevity using the same protocol as we used.[16]

(3) We showed that the mean daily metabolic rates of the Ra and La animals were statistically identical.[17] This finding demonstrated that the classic rate of living theory was not correct. Ross[18] confirmed these findings, using inbred animals from sister strains to our own; and Khaezali et al.[19] used more sensitive techniques to extend these findings to other strains.

(4) Ross[18] also showed that the La mitochondria have significantly lower age-specific rates of mitochondrial H_2O_2 production. Young La mitochondria are ~20% less leaky than are comparably aged Ra mitochondria. This difference only grows with age, the old La mitochondria showing a ~40% diference in their age-specific rate of mitochondrial H_2O_2 production. In addition to this difference in absolute values, the La mitochondria show a much slower rate of increase in H_2O_2 production as a function of age. The simplest interpretation of these data is that the La mitochondria generate more useful ATP/calorie expended than do those of the Ra strain. The mitochondrial differences in the two strains are consistent with those traits observed to be significantly related to life-span differences.

(5) The combination of decreased ROS production and increased ROS scavenging ability should lead to a decreased level of oxidative damage in the animal, and this has been empirically determined to be the case.[8]

(6) The La mitochondria affect longevity in a nutrition-dependent manner. Animals containing a Ra-derived nucleus and La-derived mitochondria

(designated herein as RaLa) have a ~25% increase in longevity when compared to the RaRa control strain.[20] We confirmed and extended this experiment by creating and testing both RaLa and LaRa isogenic lines and find that longevity is altered in the direction predicted by the source of the mitochondria. The effect is less obvious as the number of generations between the creation of the cybrid stock and its assay increases, suggesting a possible systemic and progressive alteration of nuclear and/or mitochondrial gene expression.

(7) The ability to mount a caloric restriction (CR) response seems to be correlated with the absence of a La nucleus and/or mitochondria (Jung Won Soh and R. Arking, unpublished data). Control w^{1118} males show a maximum mean life span at a 25% reduction in the caloric content of their food (i.e., 0.75× food), with significantly decreased values at both lower and higher food concentrations. Ra males show a similar pattern except that their maximum plateaus over the 0.75×–1.0× levels before decreasing at higher levels. LaLa, LaRa, and RaLa animals express their maximum mean life span over the 0.75×–1.25× range with no sign so far of a decrease at high food concentrations. A detailed biochemical and genetic analysis of the mechanisms underlying these observations needs to be done so as to test if the dependency of longevity on mitochondrial type arises from the increase in the La mitochondrial efficiency.

(8) Our gene mapping studies are fully consistent with these biochemical findings. We mapped the loci responsible for the extended longevity of the La strain to region 65–75 of chromosome 3 Left (c3L).[20] A more precise quantitative trait locus (QTL) mapping experiment was done by Khaezili and Curtsinger[22] on recombinant inbred lines derived from sister strains to our La and Ra lines. Their experiment yielded novel data in that there was only one QTL for longevity located on c3L at regions 66–67 (in addition to four other scattered loci essential for female fecundity). The independently determined QTL for paraquat resistance colocated to region c3L 66–67, thus suggesting that the extended longevity was due in large part to the animal's enhanced resistance to reactive oxygen species (ROS), as suggested above. CuZnSOD is known to be located within region 66, while several small hsps (known to be localized in the mitochondria) are located within region 67. These mitochondrial chaperones are believed to play an important role in minimizing mitochondrial damage. Analyses of longevity in other strains generally shows a complex assemblage of multiple sex- and environment-dependent QTLs. The simple genetic architecture of the La extended longevity is unusual and may hasten our achieving an understanding of the mechanisms and pathways involved in delaying the onset of senescence.

(9) Finally, epistasis data suggest that the extended longevity of the La strain is not dependent on signaling pathways involving the insulin receptor (InR) (unpublished data), a finding consistent with our mapping data. Preliminary data (unpublished) suggests that it may be dependent on the dFOXO3 transcription factor.

Taken together, the above data suggest that the decreased ROS levels, increased ADS activities, and the hsp activities must be acting cooperatively so as to minimize the ROS damage to important macromolecules and allow their efficient repair and/ or replacement. The decreased structural damage, particularly in the mitochondria, allows for both a more efficient use of energy as well as a slower decline in energy production. The extended maintenance of various functions (e.g., fecundity) and the delayed transition from the health span to the senescent span likely flow out of this cooperative reduction in damage.

EXPRESSION OF THE TYPE 1 LONGEVITY PHENOTYPE APPEARS TO BE dFOXO DEPENDENT

The extended longevity phenotype can be induced by caloric restriction as well as other forms of nutritional manipulation involving the insulin-like signaling pathway (ISP). In addition, there is an antagonistic relationship between reproduction and longevity such that high levels of juvenile hormone (JH) or 20-hydroecdysone

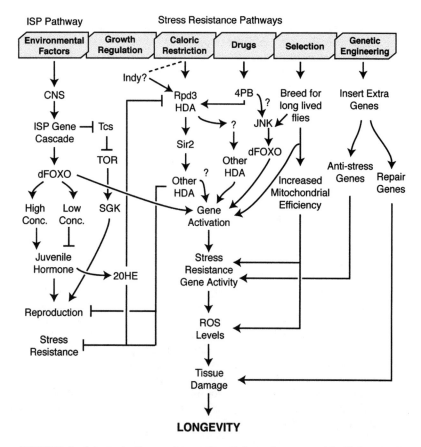

FIGURE 3. Schematic diagram integrating all the pathways empirically known to extend longevity in *Drosophila*. (Reproduced from Arking *et al.*[23] with permission).

(20HE) repress extended longevity. Many, but not all, animals that are known to be stress resistant are also long-lived; the difference seems to be that stress resistance involving the activation of signaling pathways affecting the dFOXO transcription factor can generate a type 1 longevity phenotype. Genetic engineering, in the form of either mutational inactivation of specific candidate genes or transgenic insertion of extra copies of candidate genes, has led to a similar conclusion. (A detailed review of the pertinent literature may be found in chapter 7 of reference 23.) FIGURE 3 summarizes the known activation and/or inhibition signals generated by each of the several longevity-extending mechanisms known to be operative in flies. Although this schematic is somewhat speculative in detail, its illustration of the convergence of a few specific input pathways (ISP, CR, JNK) onto a common dFOXO-dependent stress gene activation output pathway is likely correct in concept and is easily testable.

IS THERE EVIDENCE IN OTHER SYSTEMS THAT THE TRANSITION POINT IS AFFECTED BY THESE PROTEINS?

Data from *C. elegans* suggests that certain ADS enzymes (e.g., *sod-3*) as well as the small hsps are coordinately regulated by *daf-2* (i.e., *InR*) and/or thermal stress such that both types of molecules are induced by either signal.[24,25] This not only confirms the *Drosophila* data, but it illustrates the existence of cross-pathway coordinate signaling.

Huntington disease (HD) is due to the accumulation in the cell of a pathogenic protein (Htt) containing an abnormal polyglutamine (CAG). Genetic manipulation has allowed the construction of both *Drosophila* and *C. elegans* strains containing different numbers of CAG repeats (i.e., polyQ). The age of onset of the disease—and of the onset of senescence—is positively correlated with the repeat length, and there is a sharp threshold in the polyQ number separating the normal and the disease-state animals in both *Drosophila*[26] and *C. elegans*.[25] One implication of these data is that the various hsps confer protection against the deleterious effects of these abnormal aggregates characteristic of all polyQ diseases, presumably by refolding the abnormal proteins and/or targeting the aggregates for destruction. In some polyQ diseases, such as Alzheimer disease (AD), there is evidence that the abnormal amyloid proteins kill the neurons by increasing ROS damage.[27] The association of increased ROS with abnormal proteins suggests that the cross-pathway coordinate signaling of the ADS and hsps proteins has a functional purpose, protecting the cell against both types of damage. It is known that treating mouse models of HD with histone deaccytlase inhibitors (HDACs) slows neurodegeneration and preserves function. This seems counterintuitive, but it has been speculated that this finding means that what is important for normal functioning of the cell is a balance between histone acetylation (HAT) and HDAC activities. If the polyQ abnormal protein inhibits HAT activity and brings about downregulation of important genes, then the proportional inhibition of HDACs may restore the balance and once again permit gene expression to proceed.[26] The important point is the evidence suggesting that the cell's transition from health to senescence is largely determined by the cell's ability to minimize the deleterious effects of abnormal proteins and elevated ROS by activating a FOXO3-dependent gene pathway. This pathway is highly conserved, for the FOXO3 gene is active in humans, and its repression leads to the rapid senescence of human fibro-blast cells in culture.[28]

Both the worm and fly are postmitotic organisms. In actively dividing vertebrate cells, there is some evidence that the proteins responsible for monitoring the transition from health to senescence might also include certain early DNA repair genes, since genetic instability may be a major senescence factor in mitotically active tissues.

SO WHAT IS THE PROXIMAL EFFECT THAT CAUSES THE CELL TO TRANSIT FROM HEALTH TO SENESCENCE?

Our cells and tissues are thermodynamically unstable structures and are under continual attack by stochastic degradative events and processes. During the health span, the cell's various defenses manage to keep these adversaries at bay. The transition into senescence might arise from an increase in the level of the stressors affecting the cell and/or the result of a decrease in the level of the cell's defenses. The latter case might come about via a breakdown in inter- or intracellular signaling. Examples of the former case are known in flies. The aging of the *Drosophila* heart can be halted by downregulating the insulin-like signaling pathway (ISP) and/or dFOXO3 activity either systemically or within the organ, without affecting the life span.[29] This may be interpreted as indicating that the treated heart cells have a de-layed onset of senescence but that these old flies die as a result of a failure in non-cardiac critical tissue (neurons?). It is also known that downregulation of the FOXO3a transcription factor accelerates cellular senescence in human fibroblasts,[28] again supporting the evolutionary conservatism of a particular stress resistance pathway. What is needed is evidence tying together an age-related signaling defect with the decreased effectiveness of the cell's resistance to the stresses arising from ROS and/or abnormal proteins, as described above.

INTEGRATED THEORY OF AGING OVER THE LIFE SPAN

TABLE 1 presents the outlines of an integrated theory of aging based on the evidence presented above. Organisms attain a state of maximum functional ability early in life, characterized by the processes and states indicated in the left-hand column of the table. Whether the organism maintains that healthy state for a long or a short time depends on the nature and activity of its genetically based longevity determinant mechanisms and their interaction with its environment(s). Eventually these mechanisms lose their effectiveness, as the level of unrepaired damage increases to some critical level. The increased stochastic damage brings about the loss of regulatory functions both within and between cells. Bad things begin to happen, as detailed in the right-hand column of the table; and the only available interventions available are those symptomatic treatments that retard (but do not necessarily reverse) the loss of function characteristic of the senescent span. Eventually the functional level of the system falls below that minimum threshold compatible with life, and the organism dies. The foregoing adequately describes the aging process of most, but not all, in-

TABLE 1. Model of a cell's transition from the health span phase to the senescent phase, based on the oveerwhelming of its cellular defenses by accumulated unrepaired damage

Life span =	Health span	+	Transition	+	Senescent span
	Longevity determinant mechanisms		Altered balance of cell's defenses due to accumulated damage		Degradation of gene/ protein interaction networks
	Homeostatic ability sensitive and reliable		Abnormal proteins aggregate, exceeds chaperone capacity		Cell's regulatory ability decreases Tissue/systemic functions deteriorate Feedback cascades ruin homeostatic ability Critical thresholds passed
	Low ISP levels		Damaged cells survive		High OxDam level
	High antioxidant levels		Apoptosis decreases		High inflammation level
	High repair levels		Tumors increase		Low stress resistance

NOTE: The cell's major defenses include the various antioxidant stress resistance proteins as well as the various heat shock proteins (HSPs). These, along with the upstream genes regulating the expression of these stress resistance genes, constitute the highly conserved (or public) major genes that exhibit a Mendelian phenotype but collectively account for only a small portion of the variance normally observed in the phenotype. Their effectiveness is impaired both by the gradual accumulation of unrepaired oxidative and protein damage, and by the loss of connectivity of the numerous modifier (or minor) genes, which together account for most of the variance in the phenotype. When the accumulated unrepaired damage reaches a threshold at which it saturates the cell's stress resistance genes, then any further accumulation of such damage will push the cell from equilibrium into a positive feedback cascade in which each increment of damage causes a loss of function and the onset of senescence. See text for details. Based on the work of multiple authors. (From Arking.[23])

dividuals. The phenomenon of epigenetic stratification[30] suggests that some small proportion of the population will undergo connectivity changes that will actually increase or at least maintain their functional level. Investigation of these outliers may prove useful. Perhaps some of the non-genetically based human centenarians are an example of such positive changes.

Taken together, the data suggests that the cellular mechanisms responsible for maintaining the cell in a healthy state are under constant attack by ROS and/or abnormal protein accumulation. Our ability to manipulate longevity will be dependent on our ability to manipulate this transition phase of the life span. In any event, the model presented here has the advantage of focusing attention on the molecular species and signaling pathways responsible for causing the cell to transit from health to senescence. There is no reason to believe that the unique biocultural adaptations of humans[31] are in any way contradictory to the successful development of cell level interventions aimed at maintaining the signal strength and network integrity essential for the extension of the human health span.

[*Competing interests*: The author declares that he has no competing financial interests.]

REFERENCES

1. ARKING, R., S. BUCK, V.N. NOVOSELTSEV, *et al.* 2002. Genomic plasticity, energy allocations, and the extended longevity phenotypes of Drosophila. Aging Res. Rev. **1:** 209–228.
2. ARKING, R., V. NOVOSELTSEV & J. NOVOSELTSEVA. 2004. The human life span is not that limited: effects of multiple longevity phenotypes. J. Gerontol. A Biol. Sci. Med. Sci. **59:** 697–704.
3. PLETCHER, S.D., S.J. MACDONALD, R. MARGUERIE, *et al.* 2002. Genome-wide transcript profiles in aging and calorically restricted Drosophila melanogaster. Curr. Biol. **12:** 712–723.
4. HWANGBO, D.-S., B. GERSHMAN, M.-P. TU, *et al.* 2004. Drosophila dFOXO controls lifespan and regulates insulin signalling in brain and fat body. Nature **434:** 118.
5. TATAR, M., A. KOPELMAN, D. EPSTEIN, *et al.* 2001. A mutant Drosophila insulin receptor homolog that extends life-span and impairs neuroendocrine function. Science **292:** 107–110.
6. CLANCY, D.J., D. GEMS, L.G. HARSHMAN, *et al.* 2001. Extension of life-span by loss of CHICO, a Drosophila insulin receptor substrate protein. Science **292:** 104–107.
7. KANG, H.-L., S. BENZER & K.-T. MIN. 2002. Life extension in Drosophila by feeding a drug. Proc. Natl. Acad. Sci. USA **99:** 838–843.
8. ARKING, R., V. BURDE, K. GRAVES, *et al.* 2000. Selection for longevity specifically alters antioxidant gene expression and oxidative damage patterns in Drosophila. Exp. Gerontol. **35:** 167–185.
9. LUCKINBILL, L.S., ARKING, R. CLARE, *et al.* 1984. Selection for delayed senescence in *Drosophila melanogaster.* Evolution **38:** 996–1003.
10. ARKING, R. 1987. Successful selection for increased longevity in *Drosophila*: analysis of the survival data and presentation of a hypothesis on the genetic regulation of longevity. Exp. Gerontol. **22:** 199–220.
11. ARKING, R. & R.A. WELLS. 1990. Genetic alteration of normal aging processes is responsible for extended longevity in *Drosophila.* Dev. Genet. **11:** 141–148.
12. DUDAS, S.P. & R. ARKING. 1995. A coordinate up-regulation of antioxidant gene activities is required prior to the delayed onset of senescence characteristic of a long-lived strain of *Drosophila.* J. Gerontol. A Biol. Sci. Med. Sci. **50:** B117–B127.
13. BUCK, S.A. & R. ARKING. 2002. Metabolic alterations in genetically selected Drosophila strains with different longevities. J. Am. Aging Assoc. **24:** 151–162.
14. KUETHER, K. & R. ARKING. 1999. Drosophila slected for extended longevity are more sensitive to heat shock. J. Am. Aging Soc. (AGE) **22:** 175–180.
15. KURAPATI, R., H.B. PASSANANTI, M.R. ROSE & J. TOWER. 2000. Increased hsp22 RNA levels in *Drosophila* lines genetically selected for increased longevity. J. Gerontol. A Biol. Sci. Med. Sci. **55:** B552-B559.
16. ROSE, M. 1984. Laboratory evolution of postponed senescence in Drosophila melanogaster. Evolution **38:** 1004–1010.
17. ARKING, R., S. BUCK, R.A. WELLS & R. PRETZLAFF. 1988. Metabolic rates in genetically based long-lived strains of *Drosophila.* Exp. Gerontol. **23:** 59–76.
18. ROSS, RE. 2000. Age-specific decreases in aerobic efficiency associated with increase in oxygen free radical production in *Drosophila melanogaster.* J. Insect Physiol. **46:** 1477–1480.
19. KHAZAELI, A.A., W. VAN VOORHIES & J.W. CURTSINGER. 2005. Longevity and metabolism in *Drosophila melanogaster*: genetic correlations between life span and age-specific metabolic rate in populations artificially selected for long life. Genetics **169:** 231–242.
20. DRIVER, C. & N. TAWADROS. 2000. Cytoplasmic genomes that confer additional longevity in *Drosophila melanogaster.* Biogerontology **1:** 255–260.

21. BUCK, S., S. DUDAS, R. WELLS, *et al.* 1993. Chromosomal localization and regulation of the longevity determinant genes in a selected strain of *D. melanogaster.* Heredity **71:** 11–22.
22. CURTSINGER, J.W. & A.A. KHAZAELI. 2002. Lifespan, QTLs, age-specificity, and pleiotropy in *Drosophila.* Mech. Ageing Dev. **123:** 81–93.
23. ARKING, R. 2005. Biology of Aging: Observations and Principles, 3rd edit. Oxford University Press. Oxford. In press.
24. WALKER, G.A. & G.J. LITHGOW. 2003. Lifespan extension in C. elegans by a molecular chaperone dependent upon insulin-like signals. Aging Cell. **2:** 131–139.
25. MORLEY, J.E. & R.I. MORIMOTO. 2004. Regulation of longevity in *Caenorhabditis elegans* by heat shock factor and molecular chaperones. Mol. Biol. Cell **15:** 657–664.
26. MARSH, J.L. & L.M. THOMPSON. 2004. Can flies help humans treat neurodegenerative disease? BioEssays **26:** 485–496.
27. SMITH, W.W., D.D. NORTON, M. GOROSPE, *et al.* 2005. Phosphorylation of p66Shc and forkhead proteins mediates A{beta} toxicity. J. Cell Biol. **169:** 331–339
28. KIM, H.K., Y.K. KIM, I.-H. SONG, *et al.* 2005. Downregulation of a forkhead transcription factor, FOXO3a, accelerates cellular senescence in human dermal fibroblasts. J. Gerontol. A Biol. Sci. Med. Sci. **60:** 4–9.
29. WESSELLS, R.J., E. FITZGERALD, J.R. CYPSER, *et al.* 2004. Insulin regulation of heart function in aging fruit flies. Nat. Genet. **36:** 1275–1281.
30. JAZWINKSI, S.M., S. KIM, C.-Y. LAI & A. BENGURIA. 1998. Epigenetic stratification: the role of individual change in the biological aging process. Exp. Gerontol. **33:** 571–580.
31. CREWS, D.E. 2003. Human Senescence: Evolutionary and Biocultural Perspectives. Cambridge University Press. Cambridge.
32. VETTRAINO, J., S. BUCK & R. ARKING. 2001. Direct selection for paraquat resistance in Drosophila results in a different extended longevity phenotype. J. Gerontol. Biol. Sci. **56A:** B415–B425.
33. ARKING, R., J. NOVOSELTSEV, D.S. HWANGBO, *et al.* 2002. Different age-specific demographic profiles are generated in the same normal-lived Drosophila strain by different longevity stimuli. J. Gerontol. Biol. Sci. **57A:** B390–B399.

From Molecular Biology to Anti-aging Cognitive–Behavioral Practices

The Pioneering Research of Walter Pierpaoli on the Pineal and Bone Marrow Foreshadows the Contemporary Revolution in Stem Cell and Regenerative Biology

W. C. BUSHELL

Anthropology Program, Massachusetts Institute of Technology, Cambridge, Massachusetts 02139, USA

ABSTRACT: Evidence is accruing that a cognitive–behavioral regimen integrating cognitive techniques (meditation-based anti-stress, anti-inflammatory techniques, others), dietary modification ("dietary restriction" or modified dietary restriction), and certain forms of aerobic exercise, may prolong the healthy life span in humans. Recent research has identified some of the likely molecular mediators of these potentially broad-ranging, health-enhancing and anti-aging effects; these include DHEA, interleukins -10 and -4 (IL-10, 1L-4), and especially melatonin. Relatedly, what some are calling a revolution in biology and medicine has been emerging from research on stem cells and regeneration processes more generally. Dogma regarding limitations on the regenerative capacities of adult vertebrates is being cautiously yet enthusiastically revised in the wake of rapidly accumulating discoveries of more types of adult stem cells in mammals, including humans. For example, a recent review by D. Krause of Yale concluded that "in the [adult] bone marrow, in addition to hematopoietic stem cells and supportive stromal cells, there are cells with the potential to differentiate into mature cells of the heart, liver, kidney, lungs, GI tract, skin, bone, muscle, cartilage, fat, endothelium and brain." In addition, very recent studies have shown that DHEA, ILs-10 and -4, and melatonin all possess potential regenerative, including stem cell–activating, properties. More than a quarter of a century ago, Walter Pierpaoli initiated a series of extraordinary studies that demonstrated in experimental animals the potential for *dramatic* regeneration associated with changes in the pineal gland and bone marrow. This appeared to be not only *retardation* of aging, but also its *reversal*. Furthermore, as Pierpaoli was attempting to understand both anti-aging regeneration *and* oncogenesis, he was focusing on *both pro- and anti*-mitotic mechanisms: recent research now suggests that there is a nonpathologic, "healthy" form of regeneration that is actually antagonistic to oncogenesis, and that melatonin may be important in this form of regeneration. This paper ex-

Address for correspondence: W.C. Bushell, Ph.D., Anthropology Program, Massachusetts Institute of Technology, 16-241, 77 Massachusetts Avenue, Cambridge, MA 02139. Voice: 617-253-3065.

wcbphd@att.net

Ann. N.Y. Acad. Sci. 1057: 28–49 (2004). © 2004 New York Academy of Sciences.
doi: 10.1196/annals.1322.002

plores Pierpaoli's pioneering studies in light of recent developments in stem cell and regenerative biology, particularly as related to the regenerative potential associated with certain cognitive–behavioral practices, and includes evidence on this subject presented for the first time.

KEYWORDS: melatonin; pineal; bone marrow; regeneration; stem cells; aging reversal; longevity; transferrin; interleukin-10; negligible senescence; meditation; calorie restriction; exercise; yoga; cognitive–behavioral; anti-inflammatory; antioxidant; anti-stress; anticancer; peripheral nerve regeneration; Parkinson's disease

INTRODUCTION

In this paper I present evidence for the following: (1) that cognitive–behavioral (C-B) practices (in particular a "classical" traditional regimen of certain forms of meditation, diet, and physical exercises) appear to possess properties that *retard* aging and that are broadly, and probably comprehensively, health enhancing; (2) that this regimen may also possess properties that *reverse* aging, a possibility that becomes compelling when the regimen is evaluated in terms of recent developments in stem cell biology and more generally regenerative biology; and (3) that research on melatonin, the pineal gland, bone marrow, and related topics by Walter Pierpaoli and colleagues, spanning several decades, foreshadows these recent developments and with them provides a compelling basis for understanding what is emerging as a perhaps previously unimagined potential, in the modern scientific context, for tissue regeneration and longevity in humans.

COGNITIVE–BEHAVIORAL PRACTICES: MEDITATION, CALORIE RESTRICTION, AND EXERCISE ENHANCE HEALTH SPAN AND LIFE SPAN

Several studies on meditation have demonstrated effects that are consistent with a lengthening of both the (mean) life span and the "health span." Most recently, a large-scale epidemiologic study comparing hypertensive practitioners of Transcendental Meditation (TM) with matched hypertensive controls over two decades revealed a 23% decrease in all-cause mortality (primary outcome), while secondary analyses showed a 30% decrease in the rate of cardiovascular mortality and a 49% decrease in cancer-related mortality.[1] Other studies on TM that have attempted to evaluate practitioners against controls with respect to putative aging-related biomarkers, including blood pressure,[2] lipid peroxide levels,[3] and standardized sensory and cognitive measures[2] have found meditation to be a successful anti-aging intervention according to such criteria. Other studies have proposed a role for meditation in significantly reducing age-related increases in blood pressure[4] and in prolonging survival,[5] but these studies did not explicitly parse out the effects of meditative practice from other related factors, such as cloistered seclusion and social support.

The mechanisms by which meditation may retard aging and prolong survival and the health span are currently under investigation. As with a number of other putative anti-aging interventions, evidence has pointed to the counteracting of oxidative and other (e.g., nitrosative, mental) forms of stress, and inflammation, and to general promotion of innate reparative and regenerative processes.[6,7]

McEwen,[8] Sapolsky,[9] and others[10] have summarized evidence for the role of psychological or emotional stress in the acceleration or exacerbation of fundamental aging processes. Recently, a landmark study conducted by Elizabeth Blackburn and colleagues[11] demonstrated that the chronic and severe emotional stress induced in caregivers of severely ill family members actually led to *telomere shortening* in the caregivers; moreover, the data indicated that the degree of shortening actually correlated with the degree of *perceived* stress. Furthermore, another recent (pilot) study by Waelde *et al.* at Stanford University[12] demonstrated that meditative practice could significantly counteract the deleterious effects of caregiver stress in practitioners, as evaluated with standard scales of anxiety and depression.

Oxidative stress is considered a primary agent of aging in many leading theories of aging, and is conceptualized as a product of oxygen radical formation during normal metabolic functioning and/or resulting from deficiencies in protective endogenous antioxidant, free radical scavenging capacities.[13] Several studies on meditation have demonstrated reduced indices of oxidative stress during meditation as measured by reduced levels of lipid peroxides,[3] malondialdehyde concentration,[14] and urinary vinyl mandelic acid levels,[15] and such reduced levels of oxidative stress have been hypothesized to reflect a slower rate of aging.[13] In particular, the idea that normal metabolic functioning, which causes oxidative radical byproducts, in turn causes aging over time (i.e., the "rate of living" theory of aging), is currently a subject of broad and intense debate, with proponents[16,17] and critics[18] offering many explanations to resolve apparent inconsistencies and contradictions in the data. If, as some contend, reduced metabolic rate leads to the (relative) retardation of aging, then the practice of the classic meditative regimen may indeed be associated with significant retardation of aging and/or extension of life span and health span on the basis of much data on the metabolism-lowering effects of meditation. Of particular interest in this regard are studies by researchers from Stanford,[19] Harvard,[20] Rockefeller University,[21] and other research institutions on the special ability of some long-term, *virtuoso* yoga practitioners to induce and maintain profoundly lowered metabolic rates during meditation, with reductions ranging from 38–64% below resting levels! Such lowered metabolism is actually within the range of hibernating animals,[22] and hibernation has been associated with extended longevity in some studies because of lowered metabolism, according to some investigators[23] (although not others[24]).

Several studies of meditation and meditation-like practices (self-hypnosis; systematic stylized forms of relaxation, etc.) have also demonstrated a range of anti-inflammatory effects and dampening of inflammation-like immune processes, and much recent research in gerontology has focused on elucidating the aging-accelerating effects of chronic inflammation and related immune processes. According to these studies, meditation and such meditation-like practices, may lead to reductions in levels of tumor necrosis factor-α (TNF-α),[25] interleukin-2 (IL-2),[26] interferon-γ (IFN-γ),[26,27] as well as reduced inflammatory/immune responses to a range of antigens, allergens, and irritants, including lipopolysaccharide (LPS),[28] histamine,[29] capsaicin,[30] varicella zoster,[31] Mantoux antigen,[32,33] dinitrochlorobenzene (DNCB) and diphenylcyclopropenone (DCP),[34] ultraviolet B (UVB),[35] and standardized thermal stimulation.[36] In addition, anti-inflammatory substances such as IL-4 may be upregulated during meditation.[27]

In fact, recent studies have shown that meditative practices upregulate other important endogenous substances possessing not only anti-inflammatory properties, but antioxidant and antistress, as well as general longevity-enhancing, regenerative properties, namely melatonin[37,38,39,40] and DHEA.[41,42,43] The anti-inflammatory,[44,45] antioxidant,[46,47] and antistress[48,49] effects of these latter two substances are well established.

Furthermore, the other primary components of the regimen—physical exercise and special dietary programs—appear in general to possess anti-aging effects that are similar to those achieved by meditation. Aerobic exercise of moderate intensity studied in the Western clinical/laboratory setting is physiologically similar to that of the classic yoga-meditational regimen; that is, we can cite in particular the dozens to hundreds of rapidly performed repetitions of prostrations, in which the practitioner from an erect standing position bows, then "prostrates" completely on the ground, before rising back up to a fully erect standing position.[50,51] While I know of no studies to date that indicate that aerobic exercise can extend the *maximum* life span in animals or humans,[52] studies have shown that exercise can increase *mean* life span in laboratory animals, and certainly the health span in animals and humans.[53,54] Exercise intensity, duration, frequency, and other factors appear to play important roles in anti-aging outcomes, as does the role of training.[55,56] These salubrious effects appear to be most likely due to stress-reducing,[57] anti-inflammatory,[58] and antioxidant[59] effects. Such effects may in turn be mediated by increases in the activity of melatonin,[60] DHEA,[53] and ILs–10[61] and –4[62], among others; while the exercise-induced enhancement of the latter three substances appears to be robust as well as replicated, the data with respect to exercise and melatonin appear to be more equivocal.

It has been well known for decades in gerontological research that restriction of caloric intake (calorie restriction or CR) by between roughly one- and two-thirds, while maintaining nutritional balance ("undernutrition without malnutrition," maintaining necessary vitamins, minerals, protein, etc), can extend the mean and *maximum* life span of model organisms, and delay or prevent degenerative physiological changes and age-related disease.[63] It is less well-known that the "classic" monastic or ascetic diet is actually similar or identical to the standard CR diet, and is focused on nutrient-dense staples including beans and legumes (such as "lentils"), nuts and seeds, milk, yoghurt, and other nutritious dairy products, and fruits and vegetables, often taken in one or two meals per day (caloric restriction of roughly one- to two-thirds[50,51]). While most research has been conducted on model organisms (yeast, nematodes, *Drosophila*, rodents), nonhuman primate studies, and several recent human studies, have recently been undertaken. While these latter studies are not of sufficient duration yet to determine whether CR will extend the mean or maximum life span of nonhuman primates and humans, data already generated have demonstrated apparent aging-retarding changes consistent with the results in model organisms (including reductions in blood glucose, insulin resistance or its analogues, blood pressure in higher organisms, etc.) in which maximum life span *has* been extended.[64–66]

More recent studies have determined that modifications of the CR paradigm, including decreased restriction of caloric intake (i.e., more food permitted), increased time between meals rather than meal-skipping (independently of total calories), and "intermittent fasting" (alternating daily eating schedule), can also produce many of the salutary effects of the standard CR diet.[67,68] And such modifications can also be found in the range of ascetic dietary regimens.

Although the mechanisms responsible for the life span–extending and health-enhancing effects of CR remain a topic of intensive investigation, much research points to the importance of anti-inflammatory[69] and antioxidant (decreased radical oxygen species [ROS] generation and/or increased activation of endogenous free radical scavenging[70]) mechanisms. Furthermore, more recent research has shown that CR can enhance regenerative stem cell activation in several tissues,[71] although it also possesses significant and robust anticancer properties as well, and both regenerative and oncostatic properties have an obvious significance for health enhancement and life span extension. In this regard, CR has been found to retard the age-related decline in melatonin and DHEA,[72–74] both of which possess anti-inflammatory, antioxidant, and regenerative properties.

THE REVOLUTION IN STEM CELL AND REGENERATIVE BIOLOGY AND THE WORK OF WALTER PIERPAOLI

The late 1990s saw the beginnings of new discoveries concerning the regenerative potential of adult mammalian tissues through activation of endogenous stem cells in a way that was previously unexpected.[75] Around the same time these new understandings of adult stem cell biology, coupled with technological and methodological advances in molecular and genetic biology, began to transform another, related stream of research.This research was devoted to the study of the "virtuoso" regenerative capacities (e.g., limb regeneration, heart regeneration) of organisms such as urodele amphibians, zebrafish, and others, and this field that had been making slow and steady, if relatively unheralded, progress for decades. The upshot of this research has been similar, in that recent discoveries regarding such regeneration-competent species reveals previously unsuspected (potential) biological commonalities with adult mammals, possibly even including humans, that appear to exist in a genetically latent or suppressed state, according to many leading experts in the field.[76]

Regarding continual, "youth-maintaining," multiple-tissue, *global organismic* regenerative activity in such organisms (also known as "maintenance regeneration") —rather than the ability to regenerate a circumscribed tissue or organ in response to acute injury—there are far less empirical data. However, one apparent and extremely interesting exception to this generalization *may* be represented by the pioneering, if inadequately recognized, studies conducted by Walter Pierpaoli and his colleagues, on apparent aging reversal and tissue regeneration in chronic melatonin-treated and pineal-transplanted (i.e., young-to-old) mice. Although not originally framed in terms of either stem cell or "classic" regenerative biology, this series of studies,[77–86] commencing nearly a quarter-century ago, appears to have foreshadowed some of these recent (and undoubtedly still-to-come) developments, and may reflect highly significant implications for the future of regenerative biology and medicine, and for potential aging reversal and longevity.

Melatonin: Cytoprotection, Longevity, Aging Reversal, and Regeneration

Beginning in the early 1980s, Pierpaoli and colleagues began intensive investigation of the bone marrow as a key site for both pro- and antimitotic activity;[77–79] they were interested in both regenerative and cancer-protective capacities. Around the same time, he and his colleagues also began what would be an extensive and penetrating series of investigations into the key role of melatonin and the pineal gland in these capacities.[80–86]

In terms of anti-aging, regenerative, and possibly even aging-*reversing* capacities, the primary culmination of knowledge came from studies conducted in the mid-1980s through the mid-1990s. Through chronic melatonin administration, as well as (young-to-old and old-to-young) grafting and cross-transplantation of pineal glands in laboratory mice, Pierpaoli and colleagues discovered surprising, extraordinary results. In their words,[80–82] these manipulations appeared to induce startling changes in the mice (see Fig. 2 in Ref. 85), and "exerted…extraordinary positive actions on their performance and reversed or delayed the symptoms of age-related debility, disease, and cosmetic decline in a dramatic fashion." Differences in debility and cosmetic decline were further characterized as revealing "a striking difference in vigor, fur quality, motility, and weight" in "general body and pelage conditions." Moreover, there were "astonishing differences in the fur and in the general conditions of the two groups (vigor, activity, posture).":

> To our surprise, chronic, circadian, night administration of melatonin resulted in a progressive, striking improvement of the general state of the mice and …a remarkable prolongation of their life.
>
> In fact, starting at 5 months from the initiation of melatonin administration, the body weight of the untreated mice still surviving started to decrease rapidly.…Melatonin treatment preserved completely optimal pelage conditions and the body weight was maintained at the original values (about 36 grams). Furthermore, the mean survival time ± standard deviation was 931 ± 80 days in the melatonin-treated group versus 752 ± 81 days in the untreated controls. This difference is significant with $P < 0.01$ (analysis of variance).[82]

What could possibly account for such dramatic changes? The answer in part may lie in a recent and growing spate of research studies which, although not well known, have nevertheless been revealing that melatonin may actually possess *true regenerative* properties in a number of tissues.

To begin with, several years ago it was found that pinealectomy inhibits, whereas pineal transplantation then restores, liver regeneration in rats[87]; liver of course is one of several tissues in adult mammals that is already known to be regenerative.

More recently, several studies have shown that melatonin appears to tip the remodeling process in bone towards a regenerative, rather than degenerative balance, as revealed in an excellent review by Cardinali.[88] It may be that several mechanisms are involved in melatonin's bone-regenerating effects, including stimulation of proteins required for bone matrix, such as procollagen type 1 c-peptide, as well as antagonistic actions on osteoclast-based bone resorption, through melatonin's stimulation of osteoprotegerin, as well as its potent antioxidant properties.

Similarly, multiple properties of melatonin may be involved in both its protective and possibly regenerative effects on various age-related pathologic processes in muscle. Melatonin enhances microcirculatory structure and hence function in muscle,[89] and also protects it from potentially damaging ischemic, oxidative stress-related, and hormonal (eg, hyperthyroid) effects.[89,90] Moreover, melatonin's stimulatory effects on other key substances, including DHEA,[91] growth hormone (and consequently IGF-1),[92] and zinc[93] may play an important role in protective and regenerative effects on both bone and muscle.

Several recent studies have also shown that melatonin may have regenerative as well as protective effects on the nervous system. It is well known that peripheral nerves in most mammals cannot regenerate after most forms of severing lesions, despite experimental indications of considerable regenerative potential in this nervous tissue. However, only in the past year have several labs in the United States and abroad determined that melatonin treatment can lead to actual regeneration of severed sciatic nerves in animal models of sciatic nerve injury.[94,95] In 2004, moreover, it was discovered that melatonin receptors exist on stem cells (C17.2NSC cells) of the CNS, are co-expressed with the stem/progenitor cell marker nestin, and that melatonin (in low physiological concentrations in the experimental culture medium) leads to robust induction of glial cell line–derived neurotrophic factor (GDNF), brain-derived neurotrophic factor (BDNF), and nerve growth factor, NGF, in undifferentiated cells.[96] The latter discovery led the authors to speculate on the potential significance of melatonin in adult stem cell stimulation in the nervous system in general, and in the protection/regeneration of midbrain dopaminergic neurons, as in Parkinson's disease, in particular.

The potential significance of these exciting new findings for the tissue-protective and regenerative potential of melatonin for the studies of Pierpaoli and colleagues is obvious. Melatonin, which is now known to have multiple sites of synthesis and release, also "crosses all morphophysiological barriers and is distributed equally in all cells in the organism,"[97] has access to tissues and organs of the entire organism and, according to mounting evidence, does have protective and/or regenerative effects on multiple tissues. Moreover, as Pierpaoli et al. expressed, the "striking," "astonishing," comparatively *youthful appearance* of the melatonin-treated mice, also implies that *other* tissues would have been influenced by the treatment; in fact, several other relevant tissues have recently been found to possess regenerative capacities in response to melatonin. In particular, melatonin has very recently been found to be synthesized in hair follicles,[98] and to stimulate growth and regeneration (e.g., to stimulate the anagen phase of the hair cycle) in hair and fur, respectively, of humans[99] and various animal species, including dogs,[100] raccoon dogs,[101] ferrets,[102] mink,[103] and goats.[104]

Furthermore, along with regeneration of hair and fur (pelage), melatonin *may* reverse canities, that is, the age-related graying or whitening of hair/fur. As recently reviewed by Van Meste and Tobin,[105] canities (1) results at least in part from the effects of ROS on hair follicles, and (2) has recently been found to be reversible, indicating that the capacity to regain youthful hair/fur coloring is not irreversibly lost in older age. Therefore, the fact that melatonin (1) is synthesized in abundance in hair follicles; (2) possesses abundant receptors in hair follicles; (3) is one of the most potent antioxidants known; and (4) exhibits widespread tissue-regenerative properties, suggest that it may indeed be responsible for the changes in fur quality (pelage) found in the studies of Pierpaoli et al.

Moreover, the hormone appears to also possess protective and regenerative properties with respect to the skin as well: recent studies have shown that it can protect dermal fibroblasts and epidermal keratinocytes against UVB and other forms of oxidative stress.[106] In addition, melatonin may stimulate regeneration in epidermal keratinocytes and melanocytes. Although the literature is not entirely consistent, it appears that "the cutaneous serotoninergic/melatoninergic system"[106] occupies a role of major protective importance in the skin, the outer barrier of which acts as the organism's first line of defense against the threat of environmental insults. Along with an antioxidant-based tissue-protective role, melatonin appears to exhibit regenerative effects in some self-renewing epidermal tissue (keratinocytes, hair follicles), and also, importantly, substantial oncostatic effects. Melatonin also stimulates regeneration in another epithelial tissue involved in barrier protection against outer as well as inner challenges, the epithelium of the GI tract where it also appears to provide protection against potentially cancer-producing damage.[107]

This theme of protection/regeneration in healthy tissue, combined with antimitotic, apoptosis-stimulating properties with respect to malignant tissue (as noted above) will be developed further below. In this respect it is important here also to mention that in a very recent study,[97] in which pinealectomy was used as a model of aging, melatonin was found to prevent the typical skin morphometric changes of both pinealectomy and aging. Melatonin counteracted epidermal and dermal thinning, loss of dermal papillae, loss of hair follicles, loss of antioxidant enzymes catalase and glutathione peroxidase (normally upregulated by melatonin), and also counteracted the increase in lipid peroxidation common to both pinealectomy and aging.

Hence, the "dramatic," "striking," and "astonishing" changes found in lab mice subjected to long-term melatonin administration (or pineal transplantation or grafting) appear to represent *global organismic* changes; and recent research points compellingly to both powerful protective and regenerative properties of melatonin in multiple tissues and organs, including skin, hair/fur, muscle, bone, peripheral nervous system, central nervous system, and visceral organ (the liver; on other organs, including the heart, see the work of D. P. Cardinali, this volume; and below). Protection against the powerful, continuous, unrelenting oxidative stress–based (as well as other) agents of aging, as well as tissue- and organ-regenerative properties associated with melatonin, could then potentially account for the changes seen in the mice studied by Pierpaoli and colleagues. In fact, this set of global changes encompasses the "outer, external" (cosmetic) dimension (skin, fur/hair), a more "structural" dimension (bone, skeletal muscle), and a more "functional" dimension (PNS, CNS). While such a schema is obviously deliberately simplified for explicative purposes (in which, of course, such dimensions and factors actually overlap), it is further supported by other recent studies on the physiologically widespread salubrious effects of melatonin. In particular, these other recent studies to which I refer, on behavior, locomotor capacity, posture, and related phenomena, can also potentially help to account for the global organismic profile suggestive of aging reversal observed by Pierpaoli *et al.* in these animals (e.g., "extraordinary positive actions on…performance"; "reversed…debility"; "vigor"; "activity"; "posture").

I speak here of a spate of recent studies that have demonstrated enhancing effects of melatonin on balance, equilibrium, coordination, locomotor activity, and related functions. For example, in tests of age-related dysfunction in the vestibular system, Alpini *et a.l*[108] found that decreased melatonin plays a role in the age-related increase in dizziness, while Pompeiano *et al.*[109] determined through posturographic investigations that melatonin appears to be important for fundamental postural mechanisms generally, and with respect to scoliosis and several other disorders specifically. Paredes *et al.*[110] discovered that melatonin increases acetylcholine release in the nucleus accumbens, enhancing both horizontal and vertical motor capacity; acetylcholine in the nucleus accumbens is critical for all aspects of motor function, and its levels are of importance in the etiopathogenesis of disorders such as Parkinson's disease. In fact, studies published over the last two decades have determined that melatonin may counteract degenerative changes in a range of neurological and neuromuscular disorders;[111] along with Parkinson's disease these include multiple sclerosis,[112] periodic limb movement disorder,[113] and dyskinesia.[111]

Furthermore—and quite significantly—researchers from another lab have recently published studies that in many ways appear to replicate the results of the Pierpaoli studies on chronic melatonin administration. In their studies of chronic melatonin administration in mice, Bondy, Sharman, and colleagues present data along with striking photographic evidence of dramatic differences in appearance quite similar to those published by Pierpaoli and co-workers (see Fig. 7 in Bondy et al.[114]). The authors state:

> Using chronic administration of melatonin, good evidence has been found for the utility of melatonin in partially restoring the biochemical profile of older animals to that more closely resembling younger animals. ... It is noteworthy that the overall appearance of aged animals chronically treated with melatonin is clearly superior to that of untreated mice in that the fur is less patchy and more glossy ... [and] restoration of a biochemical profile characterizing younger animals was paralleled by changes in behavioral indices so as to resemble those of younger animals.[114]

In these studies as well the impressive photographic evidence of distinctly youthful versus distinctly aged appearance is also associated with functional or behavioral changes. For both "outer" appearance and "functional," behavioral changes, there would be, presumably, underlying parallel "structural" (bone, muscle, etc.) changes as well.

It is obvious that the changes in the studies of Pierpaoli and Bondy et al. are relevant to the concept of *frailty*, a concept of central importance in contemporary gerontological research.[115] Interestingly, there appears to be very little research on the relationship of melatonin activity to frailty per se—even though, as reviewed above, there are significant findings with respect to melatonin activity and what might be considered the "components" of frailty: that is, loss of bone and muscle tissue volume and viability, central nervous system dysfunction, locomotor dysfunction, problems in balance, coordination, posture, and others. In fact, a search on Medline brought up only one reference for the keywords, "melatonin and frailty," a study by Anisimov et al.[116] which apparently produced results indicating an exacerbating role for melatonin in the syndrome. (Although it should be noted that the authors interpreted the negative, conflicting results as being "unique to the transgenic model used.")

Bone Marrow, Regeneration, Melatonin, and Cancer

Pierpaoli's studies on the anti-aging effects of melatonin and related pineal manipulations, of course, appear to both predate as well as presage the recent revolutions in stem cell and regenerative biology and medicine, but that current conceptual framework and terminology were not available to him in their present state. However, not only does his work with melatonin reflect his focus on the same subject of regenerative, aging-reversing phenomena under investigation by contemporary stem cell/regenerative biologists, but also his intensive research on the key role of the bone marrow likewise reflects the same forward-looking instincts and concerns as those already described.

The bone marrow (BM) has emerged as a central locus in stem cell biology. A review by Krause of Yale Medical School summarizing recent studies finds that "in the [adult mammalian] bone marrow...in addition to hematopoietic stem cells and supportive stromal cells, there are cells with the potential to differentiate into mature cells of the heart, liver, kidney, lungs, GI tract, skin, bone, muscle, cartilage, fat, endothelium, and brain."[117] Moreover, not only is the BM central to the production and/or residence of stem cells of multiple lineages, but it is also central for the *delivery* of stem cells as well. As Goolsby *et al.* note, "stem cells from the bone marrow are the only known source of stem cells that circulate in the blood and [thereby] have access to all tissues of the body...."[118] While the normal physiological relevance of stem cells from the BM, as well as other sites, to tissue maintenance and repair, is currently a subject of intense investigation and controversy,[119] recent studies from Stanford, utilizing parabiotically joined pairs of genetically labeled and wild-type mice, have demonstrated that "bone marrow–derived cells contribute to myofibers in response to physiologic stresses encountered by the healthy organism throughout life."[120] Other studies have recently been demonstrating the physiological relevance of adult stem cells of the BM to other mature tissues as well.[121]

Some twenty years before the emergence of this evidence regarding the central importance of the BM in stem cell–based tissue regeneration, Pierpaoli and several collaborators possessed an intense interest in the role of the BM in both salubrious and pathologic forms of regeneration.[77–79] Originating in his work in the field of bone marrow transplantation in cancer treatment, Pierpaoli was intrigued by the fact that genetically incompatible BM transplantation could lead to various forms of derangement in tissues that possessed regeneration-competence under normal physiological conditions. In a 1987 paper Pierpaoli *et al.* reviewed evidence of irradiation and other forms of disruption to the BM as leading to pathologic outcomes not only in lymphohemopoietic tissues (including leukopenia and anemia), but also in self-renewing (regenerative) tissues such as epithelia and mucosae (exfoliative dermatitis, hemorrhagic and ulcerative enterocholitis), and also in the liver (veno-occlusive disease, focal necrosis). A series of studies in the lab found that in the BM were "DNA-promoting factors [important for] tissue repair processes":

> The bone marrow and its own internal microenvironment, including a large variety of endothelial, reticular fat cells and other types of cells, represents in vertebrates a central organ for monitoring and regulating of identity-defense mechanisms and for controlling proliferation of its own cells and *of all those tissues that preserve a regeneration potential like the liver, the mucosae, the skin and all epithelial tissues in which proliferation is an ongoing process* [my italics].[78]

Here the specific resonance with the review above by Krause is notable. And through Pierpaoli's series of studies investigating the regenerative response of the BM to stresses such as the insult of laparotomy, they concluded that the BM possessed a fundamental, physiologically relevant "latent proliferative response," a highly significant capacity for "tissue regeneration in general."[78,79]

"Regeneration vs. Cancer": Role of Bone Marrow and Cancer

Pierpaoli *et al.* also discovered an *anti*-proliferative, regeneration-inhibiting capacity in the BM, specifically in BM supernatant; this was also particularly relevant to their concerns with BM transplantation for cancer treatment. Because of its regenerative response to tissue *loss* (and derangement), and inhibitory response to malignant tissue *growth*, along with its immunologic response to *foreign* (as well as compatible) antigens, Pierpaoli *et al.* were led to define the BM as a kind of "morphostatic brain."[79]

From the framework of current developments in stem cell and regenerative biology and medicine, again it is not hard to see the prescience of Pierpaoli's instincts. This is even more apparent in light of several other, related developments, in particular (1) the recent discovery that the BM is a significant site of melatonin synthesis and activity (both receptor-dependent and receptor-independent); and (2) the recent[122,123] (but less well-known) set of discoveries concerning the putative *antagonism* between "true" regeneration and the malignant regeneration of cancer.

The late 1990s saw the initiation of a series of investigations demonstrating the synthesis and high concentration of melatonin in the BM. In fact, it was determined that melatonin levels in the BM were more than two orders of magnitude higher than in the circulation,[122] and, in fact, concentrated more highly than in the pineal gland itself.[123] While these ongoing investigations are still in the process of unraveling the details and significance of BM melatonin, several studies have already revealed that melatonin in this site (1) protects the especially vulnerable hematopoietic cells from oxidative stress,[122,124] and (2) appears to augment the regenerative functions of both hematopoietic and bone tissues.[80]

Furthermore, very recent studies have shown that melatonin actually selectively inhibits the development of malignant cells in the BM[125] and elsewhere in the body. A recent review by Karasek[126] reveals that the oncostatic effects of melatonin are strong and widespread throughout the tissues of the body, apparently including in the breast, lungs, stomach, endometrium, ovaries, cervix, prostate, GI tract, and other tissues. On the other hand, melatonin is highly protective and anti-apoptotic in many tissues. Incredibly, as recently summarized by Reiter and colleagues, a large literature reveals that melatonin is "bimodal" with respect to healthy and malignant cells, inhibiting and augmenting apoptosis in such cells, respectively. When this literature is considered in the context of the growing literature on melatonin's apparent regenerative effects in multiple tissues, a highly desirable profile emerges, of a molecule at once regenerative and protective of healthy tissues, and selectively destructive to malignant and perhaps other kinds of unhealthy cells.

In fact, melatonin may be a (key?) part of a larger and even more fundamental biological "blueprint," one in which true regeneration is actually *antagonistic* to oncogenesis, according to another large body of research that has been developing for several decades. This corpus of information has been developing within a lesser-known domain of biology—regeneration biology—which has focused on special regenerating organisms of other species, such as newts and salamanders and zebrafish, among others, and on special regenerative capacities in mammalian, including human, tissues, such as fetal scarless wound healing.

Research on regeneration and carcinogenesis in this context actually dates back centuries, but in particular over the last five decades it has discovered what appears to be a fundamental antagonistic relationship between regenerative and oncogenetic properties.[127–133] For example, comparative biologists have found that particularly competent regenerating species such as urodele amphibians appear to possess among the lowest incidence of spontaneous carcinogenesis throughout the animal kingdom.[130–133] Moreover, attempts to induce cancer in these species through (1) application of potent carcinogens or (2) implantation of tumors tend to *neutralize* malignancy. In some experiments, implanted tumor cells will actually revert to a normal cellular phenotype, or in the case of some axolotl amphibians, the formation of extra limbs or other body parts such as the retina may be induced by the implanted tumors or application of potent carcinogens. In the mammalian fetus, likewise, application of carcinogens or implantation of tumors may result in the stimulation of various nonmalignant regenerative phenomena.[132] Hence research on spontaneous and (attempts at) induced cancer in fetal/embryonic mammalian tissues, as well as in the regeneration-competent tissues of virtuoso regenerators such as urodele amphibians, leads to the conclusion that "the presence in a tissue of the capacity for true epimorphic regeneration seems to be associated with a resistance to tumor induction,"[132] that "amphibian tissues seem to resist oncogenesis in proportion to their regenerative capacity"[132] and that, in the words of one leading investigator in the field,

> Even application of carcinogens onto [amphibian] tissues capable of regeneration like eyes or limbs only results in normal morphogenesis and tissue differentiation in these animals, not in the formation of tumors. To our surprise, regenerative capacity and tumorigenicity are inversely correlated....[133]

STATE OF THE FIELD: MAMMALS AS "LATENT AMPHIBIANS"— REGENERATION COMPETENT?

Within this domain of regenerative biology new developments in molecular biology, genetics, and other fields have been leading many investigators to new conclusions challenging the fundamental position that dominated the field for many years—that regeneration-competent species, especially *virtuoso* regenerators such as urodele amphibians (newts and salamanders), anuran larvae ("tadpoles"), and others (e.g., teleost fish such as the zebrafish), are fundamentally and profoundly different from mammals in terms of regenerative capacity. Whatever discoveries were to be made with these unique and fascinating animals, the hope was that some sort of artificial means could be developed to aid mammals (humans) in limited and highly specialized ways.[76]

Within the last few years exciting discoveries have led to what amounts to a sea change in this position. In fact, many leaders in the field now believe that such regenerative capacities exist in mammals, including humans, in *latent* form, and that the fundamental biological "abyss" supposedly separating mammals from these "special animals" is not nearly as wide as hitherto believed; moreover, this gap is potentially bridgeable.[76] Recent exciting studies provide striking evidence potentially supportive of such a position. For example, Mark T. Keating, a Howard Hughes Investigator at Harvard, and recent inductee into the National Academy of Sciences, recently conducted a study that illustrates this new perspective in dramatic fashion. He and his colleagues isolated a factor from the plasma of amputated and regenerating newt limbs, and upon culturing mammalian (murine) myotube cells with the so-called "newt regeneration extract," actually induced the initial stages of regeneration in those mammalian cells.[134]

Melatonin itself has been found to play a significant role in the regenerative capacity of some of the "classic" virtuoso regenerators (although the literature is admittedly mixed). Melatonin appears to be crucial for tail regeneration in the gekkonid lizard *Hemidactylus flavivindis*,[135] as well as for limb regeneration in the fiddler crab *Uca pugilator*.[136] Recently, the crucial role of melatonin in developmental cell proliferation in another "classic" regenerator, the zebrafish, was demonstrated.[137] And finally, melatonin is involved in the regeneration of another organism much used experimentally in the field, the phylogenetically primitive planaria.[138] While I am aware of no studies addressing the possible role of melatonin in newt regeneration, according to a very recent study on *Triturus carnifex*, it is critical for the ability of the newt to survive severe hypoxic conditions in an estivation-like state.[139]

The list of key molecules—both those that are upregulated as well as downregulated during epimorphic and related forms of regeneration—has recently been reviewed by Stocum and others.[140] Included among key molecules upregulated during regeneration are members of the fibroblast growth factor (FGF) family, retinoids, and two other substances that Pierpaoli has focused on, transferrin and IL-10, while among key downregulated substances are proinflammatory cytokines (IL-1β, IL-6, TNF-α) and transforming growth factor (TGF-β).

Of these key molecules cited, all but one, the newt regeneration factor, have been identified in both the mammal and the amphibian. It should be recalled here that regeneration is not completely "foreign" to mammalian tissues and organs. As discussed elsewhere in this paper, it is, of course, already well-established that several adult mammalian tissues undergo several types of regeneration, including "compensatory hyperplasia" in liver, pancreas, and blood vessels, as well as stem cell–based regeneration in blood cells (hematopoiesis), epithelia (including epidermis), liver, skeletal muscle, and some parts of the brain (olfactory bulb and dentate gyrus). True epimorphic regeneration, also as briefly discussed above, also takes place in the mammalian fetus, up until a critical juncture in gestation.[141] Furthermore, some new discoveries have also been made in this field, including the demonstration that regeneration of complex, multi-tissue regions such as the fingertip in children, and more recently in adults, is possible.[133] And several recent exciting discoveries have shown that ear and even heart muscle regeneration can occur in a mammal, a special strain of mouse (the MRL mouse).[142]

As mentioned above, investigators from many different labs, working on many different regeneration models, varying from mammalian fetal scarless wound models, adult self-renewing tissues including brain, to virtuoso regenerators such as larval anurans and adult urodeles, have been zeroing in on the key antagonistic role of inflammation and scarring in regeneration.[143] For example, in the well-established fetal scarless wound healing model, up until a critical point in gestation, wounds in the fetus exhibit true regeneration, complete with total restoration of complex tissue, and an absence of inflammation and scarring. However, when proinflammatory, scar formation–enhancing molecules (such as TGF-β) are experimentally added to the wound microenvironment in wounded fetal tissues, the regeneration process will be disrupted, and normal "repair" processes consisting of inflammation and scar tissue will result.[143,144] Conversely, when such molecules are "neutralized" (by the use of antagonists, antisense, or other techniques) in adults, the healing process will more closely resemble a true regenerative phenotype.[143,145] Likewise, in pre-metamorphic anurans, the addition of proinflammatory and scar-enhancing mediators will disrupt true regeneration, while the antagonism of such substances in *post*-metamorphic, normally regeneration-incompetent anurans, will *restore* regeneration competence.[143,146] Moreover, not only do such direct, receptor-mediated agonistic and antagonistic actions determine these outcomes, but also, quite importantly, pro- and anti-inflammatory mediators appear to decisively influence the expression of genes related to regenerative tissue patterning and differentiation in both fetal mammalian and anuran systems.[143,146,147]

RECENT WORK BY PIERPAOLI AND COLLEAGUES AND THE FUTURE OF REGENERATIVE BIOLOGY AND MEDICINE

As briefly mentioned above, Pierpaoli and his colleagues have more recently focused on several other molecules which appear to be key in regeneration biology, along with melatonin. Transferrin (Tf) has been found to play a critical role in mammalian regeneration, growth, and development in rat liver[148] and peripheral nerve[149] as well as in chicken skeletal muscle[150] and chick embryonic CNS and retina.[151] It is also crucial—that is, "necessary and sufficient"—for amphibian limb regeneration, especially in the initial nerve-dependent stages.[152]

This latter finding concerning the nervous system-dependence of transferrin action, along with other findings, has begun to "connect" the nervous system to regenerative phenomena in general in new ways. It has been known for some time that early regeneration in urodele amphibians is nervous system–dependent, presumably owing to the fact that in initial stages of regeneration, the vascular system is not developed sufficiently to allow for plasma-based perfusion of such critical regenerative factors. More recently, investigators from various related fields have been contributing to an emerging picture in which the nervous system appears to possess greater potential for influence on regeneration than once suspected. For example, Bhatt et al.[153] have recently demonstrated that epileptic activity, including kindled seizures in the limbic system of laboratory animals, produces "hyperproliferation of bone marrow progenitor cells." Moreover, Steidl et al.[154] showed that primary human CD34+ hematopoietic progenitor and stem cells express a previously unexpected range of active neuromediator receptors, including orexin/hypocretin 1 and 2, adenosine A2B, opioid kappa1 and mu1, as well as CRH 1,2, 5-HT 1F, and GABA-B.

Regarding the latter in particular, GABA-B, recent work has shown that human Tf receptors are associated with GABA (A) receptors.[155] Some 15 years ago Watanabe et al. demonstrated that vagally mediated stimulation of the parasympathetic nervous system increased the synthesis and release of Tf.[156] More recently, Viola et al. found that the Tf derivative apotransferrin induced prolonged "anxiolytic-like behavior in rats," which was abolished by GABA antagonism,[157] while Pierpaoli et al. found that Tf has glucocorticoid-antagonizing (and thereby anti-stress) effects in old mice which appear to be associated with anti-aging, restorative effects on the immune and endocrine systems.[158] Multiple, reciprocal, and also probably overlapping pathways appear to be involved, and future research will, it is hoped, more precisely elucidate them.

At this point, it seems likely, on the basis of these studies, that Tf possesses true regenerative, anti-stress, and also longevity-enhancing effects; in fact, a recent study on Sicilian centenarians provides some support for this assertion.[159] The anti-stress effects of Tf may be initiated through vagal stimulation, such as that produced by the meditation practices of the C-B regimen discussed above, which not only exhibit melatonin-enhancing, but GABA-enhancing properties,[160] as well. Both melatonin and GABA[161] have been associated with enhanced longevity, and the glucocorticoid/stress-antagonizing effects of both melatonin and Tf may also have significant anti-aging effects: As will be recalled, McEwen, Sapolsky, and others have shown that age-related increases of glucocorticoids appear to produce age-related degenerative effects in the brain and, probably, subsequently in the soma ("the glucocorticoid theory of aging"); and according to the important study of Blackburn *et al.*, stress appears to "dose-dependently" lead to the depletion of telomerase and the shortening of telomeres!

Furthermore, Tf, possibly by itself[162] and most likely through IL-10-enhancing effects,[163] possesses significant anti-inflammatory properties, and IL-10 has been linked to both regeneration and longevity.[164] Recently, Pierpaoli and colleagues found that Tf-induced inhibition of the alloantigen response, which is critical for graft-versus-host disease (GVHD), is mediated by IL-10.[163]

SYNTHESIS, CONLUSIONS, AND THOUGHTS FOR THE FUTURE

A most compelling study that resonates with the aging reversal studies of Pierpaoli and colleagues was quite recently published by one of the "deans" of contemporary stem cell research, Irving Weissman of Stanford University.[165] Utilizing a brilliant experimental design—reminiscent of but different from Pierpaoli's cross transplantation (young-to-old and vice versa) studies—Weissman and colleagues created parabiotic pairings of young and old mice in which a shared circulatory system is surgically established through vascular anastomoses. In this design a paired young and old mouse (heterochronic parabiosis) are each exposed to the circulating factors of the other.

Testing the effects of heterochronic parabiosis on both skeletal muscle and liver, the investigators found that resident stem/progenitor cells in the old mice were influenced by the circulating factors to become rejuvenated, and to regenerate muscle and liver tissue (regeneration of old-mice tissues by young-mice circulating stem/progenitor cells was ruled out through the use of special cell markers, including green fluorescent protein, GFP). Weissman and colleagues concluded:

> These results indicate that the impaired regenerative potential of aged satellite cells can be improved by a modification of the systemic environment, by means of an increase in positive factors in young mouse serum, a decrease or dilution of inhibitory factors present in old mouse serum, or both....

> Our studies also demonstrate that the decline of tissue regenerative potential with age can be reversed through the modulation of systemic factors, suggesting that tissue-specific stem and progenitor cells retain much of their intrinsic proliferative potential even when old, but that age-related changes in the systemic environment and niche in which progenitor cells reside preclude full activation of these cells for productive tissue regeneration.

While Weissman and colleagues are no doubt now in the process of attempting to determine the identity of these factors, it seems quite possible that melatonin and perhaps Tf and IL-10, along with others reviewed in this paper, may be among them. Furthermore, it seems likely that the C-B regimen described earlier in this paper may be capable of restoring an older systemic environment to a more youthful one, through associated neuroendocrine and other biochemical changes. It also seems likely that a synthesizing of the conceptual and perhaps methodologic frameworks of Weissman and Pierpaoli could lead to significant advances in the understanding, and possible achievement, of important forms of tissue regeneration in humans.

ACKNOWLEDGMENTS

This paper is dedicated to the memory of another pioneer in longevity research, Roy L. Walford, M.D.—friend, mentor, collaborator. I would like to very gratefully acknowledge the various forms of assistance provided by Linda H. Mehta, Novera Herbert Spector, Jean Jackson, and Ganden, Robert, and Nena Thurman.

[*Competing interests*: The author declares that he has no competing financial interests.]

REFERENCES

1. SCHNEIDER, R.H. *et al.* 2005. Long-term effects of stress reduction on mortality in persons > or = 55 years of age with systemic hypertension. Am. J. Cardiol. **95:** 1060–1064.
2. WALLACE, R.K. *et al.* 1982. The effects of the transcendental meditation and TM-Sidhi program on the aging process. Int. J. Neurosci. **16:** 53–58.
3. SCHNEIDER, R.H. *et al.* 1998. Lower lipid peroxide levels in practitioners of the transcendental meditation program. Psychosom. Med. **60:** 38–41.
4. TIMIO, M. *et al.* 1988. Age and blood pressure changes: a 20-year follow-up study in nuns in a secluded order. Hypertension **12:** 457–461.
5. HELM, H.M. *et al.* 2000. Does private religious activity prolong survival? A six-year follow-up of 3,851 older adults. J. Gerontol. A Biol. Sci. Med. Sci. **55:** M400–405.
6. BUSHELL, W.C. 2001. Evidence that a specific meditational regimen may induce adult neurogenesis [abstract]. Devel. Brain Res. **132:** A26.
7. BUSHELL, W.C. 2005. Model: Potential cognitive-behavioral stem cell activation in multiple niches [poster]. Presented at the Stem Cell Biology & Human Disease conference, UCSD/Salk Institute/Nature Medicine. La Jolla, CA. March 17, 2005.
8. MCEWEN, B.S. 1999. Stress and the aging hippocampus. Front. Neuroendocrinol. **20:** 49–70.
9. SAPOLSKY, R.M. *et al.* 1986. The neuroendocrinology of stress and aging: the glucocorticoid cascade hypothesis. Endocr. Rev. **7:** 284–301.
10. FROLKIS, V.V. 1993. Stress-age syndrome. Mech. Ageing Dev. **69:** 93–107.
11. EPEL, E.S. *et al.* 2004. Accelerated telomere shortening in response to life stress. Proc. Natl. Acad. Sci. USA **101:** 17312–17315.
12. WAELDE, L.C. *et al.* 2004. A pilot study of a yoga and meditation intervention for dementia caregiver stress. J. Clin. Psychol. **60:** 677–687.
13. GREDILLA, R. & G. BARJA. 2005. Minireview: the role of oxidative stress in relation to caloric restriction and longevity. Endocrinology **146:** 3713–3717.
14. KIM, D.H. *et al.* 2005. Effect of Zen meditation on serum nitric oxide activity and lipid peroxidation. Prog. Neuropsychopharmacol. Biol. Psychiatry **29:** 327–331.
15. PANJWANI, U. *et al.* 1995. Effect of Sahaja yoga practice on stress management in patients with epilepsy. Indian J. Physiol Pharmacol. **39:**111–116.
16. WEYER, C. *et al.* 2000. Energy metabolism after two years of energy restriction: the Biosphere 2 experiment. Am. J. Clin. Nutr. **72:** 946–953.
17. BLANC, S. *et al.* 2003. Energy expenditure of rhesus monkeys subjected to 11 years of dietary restriction. J. Clin. Endocrinol. Metab. **88:** 16–23.
18. BORDONE, L. & L. GUARENTE. 2005. Calorie restriction, SIRT1 and metabolism: unDerstanding longevity. Nat. Rev. Mol. Cell Biol. **6:** 298–305.

19. HELLER, H.C. *et al.* 1987. Voluntary hypometabolism in an Indian yogi. J. Therm. Biol. **12:** 171–173.
20. BENSON, H. *et al.* 1990. Three case reports of the metabolic and electroencephalographic changes during advanced Buddhist meditation techniques. Behav. Med. **16:** 90–95.
21. YOUNG, J.D. & E. TAYLOR. 1998. Meditation as a voluntary hypometabolic state of biological estivation. News Physiol. Sci. **13:** 149–153.
22. DAVENPORT, J. 1992. Animal Life at Low Temperature. Chapman & Hall. London.
23. LYMAN, C.P. *et al.* 1981. Hibernation and longevity in the Turkish hamster *Mesocricetus brandti*. Science **212:** 668–670.
24. AUSTAD, S.N. & K.E. FISCHER. 1991. Mammalian aging, metabolism, and ecology: evidence from the bats and marsupials. J. Gerontol. **46:** B47–53.
25. WEBER, C. *et al.* 2002. Impact of a relaxation training on psychometric and immunologic parameters in tinnitus sufferers. J. Psychosom. Res. **52:** 29–33.
26. WOOD, G.J. *et al.* 2003. Hypnosis, differential expression of cytokines by T cell subsets, and the hypothalamo–pituitary–adrenal axis. Am. J. Clin. Hypn. **45:** 179–196.
27. CARLSON, L.E. *et al.* 2003. Mindfulness-based stress reduction in relation to quality of life, mood, symptoms of stress, and immune parameters in breast and prostate cancer outpatients. Psychosom. Med. **65:** 571–581.
28. JONES, B.M. 2001. Changes in cytokine production in healthy subjects practicing Guolin Qigong: a pilot study. BMC Complement. Altern. Med. **1:** 8–12.
29. LAIDLAW, T.M. *et al.* 1996. Reduction in skin reactions to histamine after a hypnotic procedure. Psychosom. Med. **58:** 242–248.
30. LUTGENDORF, S. *et al.* 2000. Effects of relaxation and stress on the capsaicin-induced local inflammatory response. Psychosom. Med. **62:** 524–534.
31. SMITH, G.R. *et al.* 1985. Psychologic modulation of the human immune response to varicella zoster. Arch. Intern. Med. **145:** 2110–2112.
32. BLACK, S. *et al.* 1963. Inhibition of Mantoux reaction by direct suggestion under hypnosis. Br. Med. J. **1:** 1649–1652.
33. ZACHARIAE, R. *et al.* 1989. Modulation of type I immediate and type IV delayed immunoreactivity using direct suggestion and guided imagery during hypnosis. Allergy **44:** 537–542.
34. ZACHARIAE, R. *et al.* 1993. Increase and decrease of delayed cutaneous reactions obtained by hypnotic suggestions during sensitization. Studies on dinitrochlorobenzene and diphenylcyclopropenone. Allergy **48:** 6–11.
35. ZACHARIAE, R. *et al.* 1994. Effects of hypnotic suggestions on ultraviolet B radiation-induced erythema and skin blood flow. Photodermatol. Photoimmunol. Photomed. **10:** 154–160.
36. CHAPMAN, L.F. 1959. Changes in tissue vulnerability induced during hypnotic suggestion. J. Psychosom. Res. 499–505.
37. TOOLEY, G.A. *et al.* 2000. Acute increases in nighttime plasma melatonin levels following a period of meditation. Biol. Psychol. **53:** 69–78.
38. MASSION, A.O. *et al.* 1995. Meditation, melatonin, and breast/prostate cancer: hypothesis and preliminary data. Med. Hypoth. **44:** 39–46.
39. COKER, K.H. 1999. Meditation and prostate cancer: integrating a mind/body intervention with traditional therapies. Semin. Urol. Oncol. **17:** 111–118.
40. HARINATH, K. *et al.* 2004. Effects of hatha yoga and Omkar meditation on cardiorespiratory performance, psychologic profile, and melatonin secretion. J. Altern. Complement Med. **10:** 261–268.
41. GLASER, J.L. *et al.* 1992. Elevated serum dehydroepiandrosterone sulfate levels in practitioners of the transcendental meditation (TM) and TM-Sidhi programs. J. Behav. Med. **15:** 327–341.
42. WALTON, K.G. *et al.* 1995. Stress reduction and preventing hypertension: preliminary support for a psychoneuroendocrine mechanism. J. Altern. Complement Med. **1:** 263–283.
43. CARLSON, L.E. *et al.* 2004. Mindfulness-based stress reduction in relation to quality of life, mood, symptoms of stress and levels of cortisol, dehydroepiandrosterone sulfate (DHEAS) and melatonin in breast and prostate cancer outpatients. Psychoneuroendocrinology **29:** 448–474.

44. CARRILLO-VICO, A. *et al.* 2005. A review of the multiple actions of melatonin on the immune system. Endocrine **27:** 189–200.
45. SCHWARTZ, A.G. & L.L. PASHKO. 2004. Dehydroepiandrosterone, glucose-6-phosphate dehydrogenase, and longevity. Ageing Res. Rev. **3:** 171–187.
46. HARDELAND, R. 2005. Antioxidative protection by melatonin: multiplicity of mechanisms from radical detoxification to radical avoidance. Endocrine **27:** 119–130.
47. TUNEZ, I. *et al.* 2005. Treatment with dehydroepiandrosterone prevents oxidative stress induced by 3-nitropropionic acid in synaptosomes. Pharmacology **74:** 113–118.
48. MAESTRONI, G.J. *et al.* 1988. Role of the pineal gland in immunity. III. Melatonin antagonizes the immunosuppressive effect of acute stress via an opiatergic mechanism. Immunology **63:** 465–469.
49. MORGAN, C.A., 3RD. *et al.* 2004. Relationships among plasma dehydroepiandrosterone sulfate and cortisol levels, symptoms of dissociation, and objective performance in humans exposed to acute stress. Arch. Gen. Psychiatry **61:** 819–825.
50. BUSHELL, W.C. 1995. Psychophysiological and comparative analysis of ascetico-meditational discipline: toward a new theory of asceticism. *In* Asceticism. V. L. Wimbush & R. Valantasis, Eds.: 553–575. Oxford University Press [Oxford Reference Series]. New York.
51. THURMAN, R.A.F. 1995. Tibetan Buddhist perspectives on asceticism. *In* Asceticism. V. L. Wimbush & R. Valantasis, Eds. Oxford University Press [Oxford Reference Series]. New York.
52. HOLLOSZY, J.O. 1993. Exercise increases average longevity of female rats despite increased food intake and no growth retardation. J. Gerontol. **48:** B97–100.
53. LEE, I.M. *et al.* 1995. Exercise intensity and longevity in men. The Harvard Alumni Health Study. J. Am. Med. Assoc. **273:** 1179–1184.
54. HOLLOSZY, J.O. & K.B. SCHECHTMAN. 1991. Interaction between exercise and food restriction: effects on longevity of male rats. J. Appl. Physiol. **70:** 1529–1535.
55. TREMBLAY, M.S. *et al.* 2004. Effect of training status and exercise mode on endogenous steroid hormones in men. J. Appl. Physiol. **96:** 531–539.
56. SALMON, P. 2001. Effects of physical exercie on anxiety, depression, and sensitivity to stress: a unifying theory. Clin. Psychol. Rev. **21:** 33–61.
57. GOTO, S. *et al.* 2004. Regular exercise: an effective means to reduce oxidative stress in old rats. Ann. N.Y. Acad. Sci. **1019:** 471–474.
58. PEDERSEN, B.K. & H. BRUUNSGAARD. 2003. Possible beneficial role of exercise in modulating low-grade inflammation in the elderly. Scand. J. Med. Sci. Sports. **13:** 56–62.
59. JI, L.L. 2002. Exercise-induced modulation of antioxidant defense. Ann. N.Y. Acad. Sci. **959:** 82–92.
60. BUXTON, O.M. *et al.* 1997. Acute and delayed effects of exercise on human melatonin secretion. J. Biol. Rhthyms **12:** 568–574.
61. CADET, P. *et al.* 2003. Cyclic exeercise induces anti-inflammatory signal molecule increases in the plasma of Parkinson's patients. Int. J. Mol. Med. **12:** 485–492.
62. SMITH, J.K. *et al.* 1999. Long-term exercise and atherogenic activity of blood mononuclear cells in persons at risk of developing ischemic heart disease. J. Am. Med. Assoc. **281:** 1722–1727.
63. WEINDRUCH, R. & R.L. WALFORD. 1988. The Retardation of Aging and Disease by Dietary Restriction. Charles C Thomas. Springfield, Ill.
64. ROTII, G.S. *et al.* 2004. Aging in rhesus monkeys: relevance to human health interventions. Science **305:** 1423–1426.
65. WALFORD, R.L. 2002. Calorie restriction in biosphere 2: alterations in physiologic, hematologic, hormonal, and biochemical parameters in humans restricted for a two-year period. J. Gerontol. A Biol. Sci. Med. Sci. **57:** B211–224.
66. FONTANA, L. *et al.* 2004. Long-term calorie restriction is highly effective in reducing the risk for atToscerosis in humans. Proc. Natl. Acad. Sci. USA **101:** 6659–6663.
67. MATTSON, M.P. 2005. Energy intake, meal frequency, and health: a neurobiological perspective. Annu. Rev. Nutr. **25:** 237–260.
68. WILLCOX, B.J. *et al.* 2004. How much should we eat? The association between energy intake and mortlaity in a 36-year follow–up study of Japanese-American men. J. Gerontol. A Biol. Sci. Med. Sci. **59:** 789–795.

69. CHUNG, H.Y. *et al.* 2002. Molecular inflammation hypothesis of aging based on the antiaging mechanism of calorie restriction. Microsc. Res. Tech. **59:** 264–272.
70. KIM, H.J. *et al.* 2002. Modulation of redox-sensitive transcription factors by calorie restriction during aging. Mech. Ageing Dev. **123:** 1589–1595.
71. LEE, J. *et al.* 2000. Dietary restriction increases the number of newly generated neural cells, and induces BDNF expression, in the dentate gyrus of rats. J. Mol. Neurosci. **15:** 99–108.
72. MATTISON, J.A. *et al.* 2003. Calorie restriction in rhesus monkeys. Exp. Gerontol. **38:** 35–46.
73. ROTH, G.S. *et al.* 2001. Dietary caloric restriction prevents the age-related decline in plasma melatonin levels of rhesus monkeys. J. Clin. Endocrinol. Metab. **86:** 3292–3295.
74. LANE, M.A. *et al.* 1997. Dehydroepiandrosterone sulfate: a biomarker of primate aging slowed by calorie restriction. J. Clin. Endocrinol. Metab. **82:** 2093–2096.
75. FORBES, S.J. *et al.* 2002. Adult stem cell plasticity: new pathways of tissue regeneration become visible. Clin. Sci. (Lond.) **103:** 355–369.
76. STOCUM, D.L. 1999. Regenerative biology: a millenial revolution. Sem. Cell Dev. Biol. **10:** 433–440.
77. PIERPAOLI, W. 1984. The bone marrow, our autonomous morphostatic "brain." *In* Breakdown in Human Adaptation to Stress, Vol. 2. :713–721. Martinus Nijhoff. The Hague.
78. PIERPAOLI, W. 1987. Neuroendocrine and bone marrow factors for control of marrow transplantation and tissue regeneration. Ann. N.Y. Acad. Sci. **496:** 27–38.
79. PIERPAOLI, W. *et al.* 1988. Bone marrow: a "morphostatic brain" for control of normal and neoplastic growth. Experimental evidence. Ann. N.Y. Acad. Sci. **521:** 300–11.
80. PIERPAOLI, W. & G.J.M. MAESTRONI. 1987. Melatonin: a principal neuroimmunoregulatory and anti-stress hormone: its antiaging effects. Immunol. Lett. **16:** 355–362.
81. PIERPAOLI, W. & G.J.M. MAESTRONI. 1988. Animals rejuvenated: melatonin extends rat lives. Brain/Mind Bulletin **13:** 1, 8.
82. MAESTRONI, G.J.M. *et al.* 1988. Pineal melatonin, its fundamental immunoregulatory role in aging and cancer. Ann. N.Y. Acad. Sci. **521:** 140–148.
83. PIERPAOLI, W. *et al.* 1991. The pineal control of aging: the effects of melatonin and pineal grafting on the survival of older mice. Ann. N.Y. Acad. Sci. **621:** 291–313.
84. PIERPAOLI, W. & W. REGELSON. 1994. Pineal control of aging: effect of melatonin and pineal grafting on aging mice. Proc. Natl. Acad. Sci. USA **91:** 787–791.
85. LESNIKOV, V.A. & W. PIERPAOLI. 1994. Pineal cross-transplantation (old-to-young and vice versa) as evidence for an endogenous "aging clock." Ann. N.Y. Acad. Sci. **719:** 456–460.
86. PIERPAOLI, W. 1998. Neuroimmunomodulation of aging: a program in the pineal gland. Ann. N.Y. Acad. Sci. **840:** 491–497.
87. ABBASOGLU, O. *et al.* 1995. The effect of the pineal gland on liver regeneration in rats, J. Hepatol. **23:** 578–581.
88. CARDINALI, D.P. *et al.* 2003. Melatonin effects on bone: experimental facts and clinical perspectives. J. Pineal Res. **34:** 81–87.
89. WANG, W.Z. *et al.* 2005. Microcirculatory effects of melatonin in rat skeletal muscle after prolonged ischemia. J. Pineal Res. **39:** 57–65.
90. ONER, J. & E. OZAN. 2003. Effects of melatonin on skeletal muscle of rats with experimental hyperthyroidism. Endocr. Res. **29:** 445–455.
91. HAUS, E. *et al.* 1996. Stimulation of the secretion of dehydroepiandrosterone by melatonin in mouse adrenals *in vitro*. Life Sci. **58:** 263–267.
92. VALCAVI, R. *et al.* 1993. Melatonin stimulates growth hormone secretion through pathways other than the growth hormone-releasing hormone. Clin. Endocrinol. **39:** 193–199.
93. MOCCHEGIANI, E. *et al.* 1994. The immuno-reconstituting effect of melatonin or pineal grafting and its relation to zinc pool in aging mice. J. Neuroimmunol. **53:** 189–201.
94. STAVISKY, R.C. *et al.* 2005. Melatonin enhances the *in vitro* and *in vivo* repair of severed rat sciatic axons. Neurosci. Lett. **376:** 98–101.

95. TURGUT, M. *et al.* 2005. Pinealectomy exaggerates and melatonin treatment suppresses neuroma formation of transected sciatic nerve in rats: gross morphological, histological, and stereological analysis. J. Pineal Res. **38:** 284–291.

96. NILES, L.P. *et al.* 2004. Neural stem cells express melatonin receptors and neurotrophic factors: colocalization of the MT1 receptor with neuronal and glial markers. BMC Neurosci. **5:** 41–49.

97. ESREFOGLU, M. *et al.* 2005. Potent therapeutic effect of melatonin on aging skin in pinealectomized rats. J. Pineal Res. **39:** 231–237.

98. KOBAYASHI, H. *et al.* 2005. A role of melatonin in neuroectodermal-mesodermal interactions: the hair follicle synthesizes melatonin and expresses functional melatonin receptors. FASEB J. **19:** 1710–1712.

99. FISCHER, T.W. *et al.* 2004. Melatonin increases anagen hair rate in women with androgenetic alopecia or diffuse alopecia: results of a pilot randomized controlled trial. Br. J. Dermatol. **150:** 341–345.

100. FRANK, L.A. *et al.* 2004. Adrenal steroid hormone concentrations in dogs with hair cycle arrest (AlopeciaX) before and during treatment with melatonin and miotane. Vet. Dermatol. **15:** 278–284.

101. XIAO, Y. *et al.* 1995. Effects of melatonin implants in spring on testicular regression and moulting in adult male raccoon dogs (*Nyctereutes procynoides*). J. Reprod. Fertil. **105:** 9–15.

102. NIXON, A.J. *et al.* 1995. Seasonal fiber growth cycles of ferrets (*Mustela putorius furo*) and long-term effects of melatonin treatment. J. Exp. Zool. **272:** 435–445.

103. ROSE, J. *et al.* 1987. Apparent role of melatonin and prolactin in initiating winter fur growth in mink. Gen. Comp. Endocrinol. **65:** 212–215.

104. NIXON, A.J. *et al.* 1993. Fiber growth initiation in hair follicles of goats treated with melatonin. J. Exp. Zool. **267:** 47–56.

105. VAN NESTE, D. & D.J. TOBIN. 2004. Hair cycle and hair pigmentation: dynamic interactions and changes associated with aging. Micron **35:** 193–200.

106. SLOMINSKI, A. *et al.* 2005. The cutaneous sertoninergic/melatoninergic system: securing a place under the sun. FASEB J. **19:** 176–194.

107. BUBENIK, G.A. 2002. Gastrointestinal melatonin: localization, function, and clinical relevance. Dig. Dis. Sci. **47:** 2336–2348.

108. ALPINI, D. *et al.* 2004. Aging and vestibular system: specific tests and role of melatonin in cognitive involvement. Arch. Gerontol. Geriatr. Suppl. **9:** 13–25.

109. POMPEIANO, O. *et al.* 2002. Pineal gland hormone and idiopathic scoliosis: possible effect of melatonin on sleep-related postural mechanisms. Arch. Ital. Biol. **140:** 129–158.

110. PAREDES, D. *et al.* Melatonin acts on the nucleus accumbens to increase acetylcholine release and modify the motor activity pattern of rats. Brain Res. **850:** 14–20.

111. SANDYK, R. 1990. MIF-induced augmentation of melatonin functions: possible relevance to mechanisms of action of MIF-1 in movement disorders. Int. J. Neurosci **52:** 59–65.

112. SANDYK, R. 1997. Resolution of sleep paralysis by weak electromagnetic fields in a patient with multiple sclerosis. Int. J. Neurosci. **90:** 145–157.

113. KUNZ, D. & F. BES. 2001. Exogenous melatonin in periodic limb movement disorder: an open clinical trial and a hypothesis. Sleep. **24:** 183–187.

114. BONDY, S.C. *et al.* 2004. Retardation of brain aging by chronic treatment with melatonin. Ann. N.Y. Acad. Sci. **1035:** 197–215.

115. LENG, S. *et al.* 2002. Serum interleukin-6 and hemoglobin as physiological correlates in the geriatric syndrome of frailty: a pilot study. J. Am. Geriatr. Soc. **50:** 1268–1271.

116. SEMENCHENKO, G.V. *et al.* 2004. Stressors and antistressors: how do they influence life span in *HER-2/neu* transgenic mice? Exp. Gerontol. **39:** 1499–1511.

117. KRAUSE, D.S. 2002. Plasticity of marrow-derived stem cells. Gene Therapy. **9:** 754–758.

118. GOOLSBY, J. *et al.* 2003. Hematopoietic progenitors express neural genes. Proc. Natl. Acad. Sci. **100:** 14926–14931.

119. KUCIA, M. *et al.* 2004. Tissue-specific muscle, neural and liver stem/progenitor cells reside in the bone marrow, respond to an SDF-1 gradient and are mobilized into peripheral blood during stress and tissue injury. Blood Cells Mol. Dis. **32:** 52–57.
120. PALERMO, A.T. *et al.* 2005. Bone marrow contribution to skeletal muscle: a physiological response to stress. Dev. Biol. **279:** 336–344.
121. KOMORI, M. *et al.* 2005. Involvement of bone marrow-derived cells in healing of experimental colitis in rats. Wound Rep. Regen. **13:** 109–118.
122. TAN, D.X. *et al.* 1999. Identification of highly elevated levels of melatonin in bone marow: its origin and significance. Biochim. Biophys. Acta **1472:** 206–214.
123. CONTI, A. *et al.* 2000. Evidence for melatonin synthesis in mouse and human bone marrow cells. J. Pineal Res. **28:** 193–202.
124. YU, Q. *et al.* 2000. Melatonin inhibits apoptosis during early B cell development in mouse bone marrow. J. Pineal Res. **29:** 86–93.
125. WOLFLER, A. *et al.* 2001. Prooxidant activity of melatonin promotes fas-induced cell death in human leukemic Jurkat cells. FEBS Lett. **502:** 127–131.
126. SAINZ, R.M. *et al.* 2003. Melatonin and cell death: differential actions on apoptosis in normal and cancer cells. Cell. Mol. Life Sci. **60:** 1407–1426.
127. PREHN, R.T. 1971. Immunosurveillance, regeneration, and oncogenesis. Prog. Exp. Tumor Res. **14:** 1–24.
128. EGUCHI, G. & K WATANABE. 1973. Elicitation of lens formation from the "ventral iris" epithelium of the newt by a carcinogen, N-methyl-N'-nitro-N-nitrosoguanidine. J. Embryol. Exp. Morphol. **30:** 63–71.
129. TSONIS, P.A. 1983. Effects of carcinogens on regenerating and nonregenerating limbs in amphibia. Anticancer Res. **3:** 195–202.
130. TSONIS, P.A. & K. DEL RIO-TSONIS. 1988. Spontaneous neoplasmas in amphibia. Tumour Biol. **9:** 221–224.
131. DEL RIO-TSONIS, K. & P.A. TSONIS.1992. Amphibian tissue regeneration: a model for cancer regulation (review). Int. J. Oncol. **1:** 161–164.
132. PREHN, R.T. 1997. Regeneration versus neoplastic growth. Carcinogen **18:** 1439–1444.
133. LOWENHEIM, H. 2003. Regenerative medicine for diseases of the head and neck: principles of *in vivo* regeneration. DNA Cell Biol. **22:** 571–592.
134. McGANN, C.J. *et al.* 2001. Mammalian myotube dedifferentiation induced by newt regeneration extract. Proc. Natl. Acad. Sci. **98:** 13699–13704.
135. RAMACHANDRAN, A.V. & P.I. NDUKUBA. 1989. Parachlorophenylalanine retards tail regeneration in the gekkonid lizard *Hemidactylus flaviviridis* exposed to continuous light. J. Exp. Biol. **143:** 235–243.
136. TILDEN, A.R. *et al.* 1997. Melatonin cycle in the fiddler crab *Uca pugilator* and influence of melatonin on limb regeneration. J. Pineal Res. **23:** 142–147.
137. DANILOVA, N. *et al.* 2004. Melatonin stimulates cell proliferation in zebrafish embryo and accelerates its development. FASEB J. **18:** 751–753.
138. CSABA, G. 1993. Presence in and effects of pineal indoleamines at very low level of phylogeny. Experientia **49:** 627–634.
139. FRANGIONI, G. *et al.* 2003. Melatonin, melanogenesis, and hypoxic stress in the newt, *Triturus carnifex*. J. Exp. Zool. **296:** 125–136.
140. STOCUM, D.L. 2004. Amphibian regeneration and stem cells. Curr. Top. Microbiol. Immunol. **280:** 1–70.
141. STOCUM, D.L. 2004. Regenerative biology and medicine: an overview. Cellsci. Rev. **1:** 1.
142. LEFEROVICH, J.M. *et al.* 2001. Heart regeneration in adult MRL mice. Proc. Natl. Acad. Sci. USA **97:** 2830–2835.
143. HARTY, M. *et al.* 2003. Regeneration or scarring: an immunological perspective. Dev. Dyn. **226:** 268–279.
144. McCALLON, R.L. & M.W.J. FERGUSON. 1996. Fetal wound healing and the development of antiscarring therapies for adult wound healing. *In* The Molecular Biology of Wound Repair. R. A. F. Clark, Ed. Plenum. New York.
145. HUANG, J.S. *et al.* 2002. Synthetic TGF-beta antagonist accelerates wound healing and reduces scarring. FASEB J. **16:** 1269–1270.

146. GHOSH, A.K. 2002. Factors involved in the regulation of type I collagen gene expression: implication in fibrosis. Exp. Biol. Med. **227:** 301–314.
147. VERRECCHIA, F. & A. MAUVEL. 2002. Transforming growth factor-beta signaling through the Smad pathway: role in extracellular matrix gene expression and regulation. J. Invest. Dermatol. **118:** 211–215.
148. FERNANDEZ, M.A. *et al.* 2004. Intracellular trafficking during liver regeneration. Alterations in late endocytic and transcytotic pathways. J. Hepat. **40:** 132–139.
149. RAIVICH, G. *et al.* 1991. Transferrin receptor expression and iron uptake in the injured and regenerating rat sciatic nerve. Eur. J. Neurosci. **3:** 919–927.
150. BEACH, R.L. *et al.* 1983. The identification of neurotrophic factor as a transferrin. FEBS Lett. **156:** 151–156.
151. BRUININK, A. *et al.* 1996. Neurotrophic effects of transferrin on embryonic chick brain and neural retinal cell cultures. Int. J. Dev. Neurosci. **14:** 785–795.
152. MESCHER, A.L. *et al.* 1997. Transferrin is necessary and sufficient for the neural effect on growth in amphibian limb regeneration blastemas. Dev. Growth Differ. **39:** 677–684.
153. BHATT, R. *et al.* 2003. Effects of kindled seizures upon hematopoiesis in rats. Epilepsy Res. **54:** 209–219.
154. STEIDL, U. *et al.* 2004. Primary human CD34+ hematopoietic stem and progenitor cells express functionally active receptors of neuromediators. Blood. **104:** 81–88.
155. GREEN, F. *et al.* 2002. Association of human transferrin receptor with GABARAP. FEBS Lett. **518:** 101–106.
156. WATANABE, Y. *et al.* Neural control of biosynthesis and secretion of serum transferrin in perfused rat liver. Biochem. J. **267:** 545–548.
157. VIOLA, H. *et al.* 2001. Anxiolytic-like behavior in rats is induced by the neonatal intracranial injection of apotransferrin. **63:** 196–199.
158. PIERPAOLI, W. *et al.* 2000. Transferrin treatment corrects aging-related immunologic and hormonal decay in old mice. Exp Gerontol. **35:** 401–408.
159. LIO, D. *et al.* 2002. Association between the MHC class I gene HFE polymorphism and longevity: a study in a Sicilian population. Genes Immun. **3:** 20–24.
160. ELIAS, A.N. *et al.* 2000. Ketosis with enhanced GABAergic tone promotes physiological changes in transcendental meditation. Med. Hypotheses **54:** 660–662.
161. MATTSON, M.P. *et al.* 2002. Modification of brain aging and neurodegenerative disorders by genes, diet, and behavior. Physiol. Rev. **82:** 637–672.
162. MCGAHAN, M.C. *et al.* 1994. Transferrin inhibits the ocular inflammatory response. Exp. Eye Res. **58:** 509–511.
163. LESNIKOVA, M. *et al.* Upregulation of interleukin-10 and inhibition of alloantigen responses by transferrin and transferrin-derived glycans. J. Hematother. Stem Cell Res. **9:** 381–392.
164. LIO, D. *et al.* 2003. Inflammation, genetics, and longevity: further studies on the protective effects in men of IL-10-1082 promoter SNP and its interaction with TNF-gamma-308 promoter SNP. J. Med. Genet. **40:** 296–299.
165. CONBOY, I.M. *et al.* 2005. Rejuvenation of aged progenitor cells by exposure to a young systemic environment. Nature **433:** 780–784.

Genetic Contribution to Aging

Deleterious and Helpful Genes Define Life Expectancy

J. I. LAO,[a] C. MONTORIOL,[a] I. MORER,[a] AND K. BEYER[b]

[a]Clinical Analysis Laboratory Dr. Echevarne, Unit of Molecular Genetics,
Barcelona, Spain

[b]Hospital German Trias I Pujol, Pathology Department, Badalona, Spain

ABSTRACT: For the best understanding of aging, we must consider a genetic
pool in which genes with negative effects (deleterious genes that shorten the life
span) interact with genes with positive effects (helpful genes that promote longevity) in a constant epistatic relationship that results in a modulation of the
final expression under particular environmental influences. Examples of deleterious genes affecting aging (predisposition to early-life pathology and disease) are those that confer risk for developing vascular disease in the heart,
brain, or peripheral vessels (APOE, ACE, MTFHR, and mutation at factor II
and factor V genes), a gene associated with sporadic late-onset Alzheimer's disease (APOE E4), a polymorphism (COLIA1 Sp1) associated with an increased
fracture risk, and several genetic polymorphisms involved in hormonal metabolism that affect adverse reactions to estrogen replacement in postmenopausal
women. In summary, the process of aging can be regarded as a multifactorial
trait that results from an interaction between stochastic events and sets of epistatic alleles that have pleiotropic age-dependent effects. Lacking those alleles
that predispose to disease and having the longevity-enabling genes (those beneficial genetic variants that confer disease resistance) are probably both important to such a remarkable survival advantage.

KEYWORDS: genetics; DNA; gerontogenes; longevity assurance genes; longevity; aging; life span; centenarians; epistasia; thrombophilia; polymorphism;
age-related disorders; life expectancy

BACKGROUND

Aging is not simply a matter of destiny. Several researchers have provided convincing evidence that longevity, at least in part, is genetically determined. We currently believe that genetic factors account for about 30% of the variance in life
expectancy. Genetic mechanisms that possibly trigger and determine the rate at
which we age have been well documented in lower organisms. Although less is
known in humans, commonality in molecular and biological processes, evolutionary

Address for correspondence: J.I. Lao, Clinical Analysis Laboratory Dr. Echevarne, Unit of
Molecular Genetics, Barcelona, Spain.
jilao@echevarne.com
jilaov@tele2.es

Ann. N.Y. Acad. Sci. 1057: 50–63 (2005). © 2005 New York Academy of Sciences.
doi: 10.1196/annals.1356.003

arguments, and epidemiological data would strongly suggest that similar mechanisms also apply.[1]

So far it seems that eternity is not possible because our cells have a predetermined limit. This is the Hayflick's limit, defined as the number of cell divisions that will take place in a cell culture before it dies out.

It has been demonstrated that the number of cell divisions can be related to each species' life span. For example, in mice (with a life span of 3 years) we could expect about 15 cell divisions but in the Galapagos tortoise (a long-lived species) we could expect about 110 cell divisions. In humans, 50 cell divisions has been estimated.[2] According to the Hayflick's limit, the life span of human beings could reach ±120 years.

Several human longevity studies have looked at genes involved in the cell-aging process (how cells lose their ability to reproduce), as well as in apoptosis (the process by which cells are programmed to die).[3,4]

Studying cell culture samples from Alzheimer's disease (AD) patients in comparison with cell cultures from healthy control samples, we observed some disturbances mainly affecting telomeric regions in the AD group. It seems that AD cells were not able to keep both the length and the terminal structure of telomeres, which is crucial for genomic integrity and, consequently, for long-term cell viability.[5]

It is clear that with aging the cell becomes less resistant to stress and more vulnerable to damage. In some pathological conditions, this phenomenon of aging may reflect gene dysregulation

Maybe we are designed not to achieve old age. Yet some people—centenarians— seem to escape this design limitation. Centenarians represent an extremely rare group (the proportion of people who reach age 100 is only about 1 per 5,000– 10,000). Centenarians demonstrate not only extreme longevity but also escape the effects of those diseases commonly associated with the aging process (cardiovascular and cerebrovascular disease, cognitive decline, and cancer).

SOME EVIDENCE OF GENETIC CONTRIBUTIONS TO AGING

Evidence of genes that promote longevity are found in this exceptional group. We are convinced that they have a genetic potential for a healthy long life. Some researchers have been looking for common characteristics among centenarians and they have described the following:

- Obesity is rare among centenarians because they usually control their weight well.[6,7]

- At least 50% of centenarians have first-degree relatives that also achieve very old age and many have exceptionally old siblings.

- According to these studies we could say that there is a very strong relative risk (fourfold to 17-fold) among siblings of centenarians for becoming a centenarian. Such relative risks, especially those greater than eightfold, are consistent with an important genetic component to exceptional longevity.[8]

- Descendants of centenarians (age range of 65 to 82 years) are similar to their parents and show significantly lower rates of age-related diseases such as hypertension, coronary disease, and diabetes as well as a lower susceptibility to stress.[6–8]

- Many centenarian women had a delayed menopause and extensive fertility (bearing children after the age of 35 years and even 40 years). For that reason, some researchers say that a woman who naturally has a child after the age of 40 will have a four times greater chance of living to 100 compared to women who do not. This prolonged fecundity could be a good indicator that the woman's reproductive system is aging slowly and maybe the rest of her is as well.[9]

- Neuropsychological evaluations of centenarians have shown that adaptability is a common personality trait among them. They are optimistic people that helps them manage stress.[6–8]

- It has been noted that 90% of the centenarians are able to function independently until the average age of 92 years and 75% were still independent at an average age of 95 years.[6–8]

- Some alleles have been described as more prevalent among centenarians and are therefore catalogued as protective alleles.[10,11]

In addition to these data emerging from centenarian studies, we have further evidence that gives support to the statement of the existence of genetic influences on aging:

- Results from major twin studies indicate that approximately 25–50% of the variations in life span are genetically determined.[12]

- Findings on genetic-related disorders reveal the role of particular loci predisposing or protecting against the disorders.[10,11,13–15]

- Longevity-assurance genes have been described in several lower organisms.[16–18]

EXAMPLES OF HUMAN GENES WITH INFLUENCES ON LIFE SPAN

In humans, a number of genetic and environmental factors have been identified that decrease length of life. Unfortunately, much less progress has been made in identifying helpful genes, which protect organisms from common diseases or slow the aging process). We do have several documented examples of genes that alter longevity by modulating the risk of early-life pathology and disease.[19]

A number of age-related diseases have been traced to particular versions of certain genes (called alleles). Indeed, many older adults free of these diseases have been found to possess the desirable, apparently disease-protective, variants of those same genes. Thus, the same gene that in one form is a longevity-assurance gene might in another form be one that decreases longevity. In this sense we have to consider two main groups: genes that promote longevity (helpful alleles) and genes that decrease length of life (deleterious alleles).

DELETERIOUS GENES

In human populations, we have more documented examples of genetic markers associated with aging that decrease longevity than those that promote longevity. We present four categories of markers strongly associated with age-related diseases: markers for cardiovascular disease, for cancer, for cognitive impairment, and for bone mineralization disturbances.

Although monogenic conditions can result in most of these abnormalities, the vast majority of incidences of these diseases in patients is due to the contribution of many genes and gene–environment interactions (as a multifactorial trait). This is our main focus in predictive and preventive medicine.

Global Risk for Vascular Disease

Association studies have suggested the involvement of candidate genes in suscep-tibility to vascular disease.[20] Most of these genes are associated with lipid metabo-lism, such as apolipoproteins (APOE and APOB), cholesterol ester transfer protein (CETP), sterol element binding transcription factor (SREBF2); endothelial damage and vascular remodeling, such as 5,10-methylenetetrahydrofolate reductase (*MTH-FR*), endothelial nitric oxide synthase (*NOS3*), GAP junction protein alpha4 (GJA4), matrix metalloproteinase 3 (MMP3); thrombogenesis, such as, factor II (F2); coag-ulation disorders, such as factor II (*F2*), factor V (*F5*), factor XIII (*F13*), plasmino-gen activator inhibitor-1 (PAI*1*), platelet glycoprotein IIIA (GPIIIa); blood pressure regulation, such as angiotensin converting enzyme (ACE), angiotensinogen (AGT), alpha2B-adrenergic receptor (ADRA2B), beta1 adrenergic receptor (ADRB1); and oxidative stress, such as paraoxogenase 1 (PON1).

To a correct evaluation of a global vascular risk we have to consider all these com-mon genetic variants of genes coding proteins related to the molecular etiopathogen-esis of the vascular disease so that it is possible to differentiate our patients according to their specific vulnerability.

In the complexity of a polygenic disorder such as vascular disease, we have to consider the constant gene–gene and gene–environment interactions.Coexistence of some of these previously mentioned genes could act in a synergistic way, thus en-hancing a pathogenic effect (e.g., combination of alleles 4G at PAI-1 and PI2 at locus GPIIIa confers more thrombogenic risk than the coexistence of allele 4G at PAI-1 locus with PI1 at GPIIIa). On the other hand, an antagonistic effect due to combina-tion of these prothrombotic alleles (4G at PAI-1 and PI2 at locus GPIIIa) with the allele T at position 34 of exon 2 in the factor XIII gene (which enhances fibrinolysis) confers a natural compensatory effect minimizing thromboembolic complications.

Two genes with effects on heart disease have been linked to human longevity. The best studied of these genes produces a protein called apolipoprotein E that circulates in our blood. Researchers have found that those people who carry at least one copy of the E4 variant of the APOE gene have a higher risk of heart disease[21] than those who carry an E2 variant. *APOE E4* is also related to stroke. Researchers working in animal models have described an increased risk of stroke in mice that inherit the un-desirable E4 version of the APOE gene.[22] Other animal studies have shown that genes play a role in recovery from stroke. Older rats with deliberately induced strokes have less ability to turn on the genes that could promote recovery than do younger rats with induced strokes.[23] In a previous analysis carried out in a Spanish population, we have found that *APOE E4* is also associated with vascular demen-tia.[24] Additional genes that are likely to predispose individuals to stroke include ACE and nitric oxide synthase.[25]

The *ACE* gene encodes the angiotensin-converting enzyme (ACE), which affects blood pressure regulation. In studying *ACE* polymorphisms, it has have been established that allele D (deletion instead of insertion or allele I) is strongly associated with hypertension and higher risk for myocardial infarction. Nevertheless, analysis of this gene polymorphism in older human populations has shown paradoxical results: people who carry the variant of this gene associated with an increased risk of heart disease (allele D) also have a longer life expectancy. Clearly, more work is needed to understand the relationship between ACE and aging.[26]

This research on both the *APOE* and *ACE* suggests some of the difficulty of untangling the effects of individual genes on complex disease and aging processes. Namely, genes can have both positive and negative effects on specific diseases and aging in general.

The paraoxonase 1 (PON1) is hypothesized to protect serum lipoproteins from oxidative stress. In this case, it has been established that the influence of the deleterious genotype may be taken into account only in older male patients.[27]

Cancer

In cancer predisposition, as well as all other conditions, we have to consider two main groups of inherited cancers, those of monogenic origin and those of multifactorial origin.

Cancers that are monogenic in origin have a clear pattern of familial transmission, as seen in *BRCA1* and *BRCA2* genes for familial breast cancer; *APC* gene for polyposic colorectal cancer; and *MSH1* and *MSH2* genes for non-polyposic colorectal cancer.

Breast cancer often occurs in many women within the same family, leading researchers to conclude that these women carry one or more "bad" gene (alleles) that promote cancer progression. Two genes, *BRCA1* and *BRCA2*, are strongly linked to hereditary breast and ovarian cancer.[28] It is estimated that women with one nonfunctional allele of *BRCA1* have a 44% likelihood of getting breast or ovarian cancer by age 70. That likelihood is 27% for those women with *BRCA2*. There are clearly other alleles that predispose women to breast cancer that remain to be identified.

Colorectal cancer is one of the most common cancers worldwide. About 5% of the population will develop colorectal cancer late in life. It is estimated that approximately 20% of the cases of colorectal cancer can be attributed to hereditary influences and some of the responsible genes have been identified.[29] For instance, most people diagnosed with familial adenomatous polyposis inherit one bad copy of the *APC* gene. Similarly, people diagnosed with hereditary non-polyposis colorectal cancers often inherit a bad version of one of a group of mismatch repair genes whose protein products repair DNA damage (*MLH1* and *MSH2*).

While much cancer is caused by external factors (smoking, toxins, repeated sunburns, for example), scientists are uncovering an important role for genetics in both susceptibility and resistance to various cancers. It has become clear that genes coding for some phase I and II enzymes (COMT, CYP1A1, CYP1B1, CYP17, GTS) and genes coding for enzymes associated to steroid metabolism must be considered, among others.

Proteins involved in the metabolism of drugs, toxins, and cancers are lumped into the category of drug-metabolizing enzymes. Some of the variants of these enzymes are associated with a higher risk of cancers. Japanese studies noted that certain variants of the genes responsible for the presence of these enzymes are associated with an increased risk of prostate cancer.[30]

In some cases, a single susceptibility genotype may increase disease risk after exposure to certain environmental factors, while at the same time protecting against the effects of exposure to others. In this sense, we have the example of the impaired function of glutathione-*S*-transferase, which has been associated with a higher risk of lung and bladder cancer among male smokers, but at the same time it may reduce the risk of stomach, colon, and liver cancer among individuals exposed to haloalkanes and haloalkenes. For that reason, we say that genetic susceptibility testing does not necessarily define "at-risk" groups in general, but instead identifies groups that are at higher risk of suffering certain adverse effects from particular environmental exposures.

Detoxification Enzymes That Help Us Get Rid of Carcinogens

Worldwide, tobacco-related cancers cause the death of up to 1,000,000 people annually. A large number of genes have been linked to lung cancer susceptibility.[31] According to the example of glutathione-*S*-transferase P1 (GSTP1), which plays a crucial role by catalyzing the conjugation of many hydrophobic and electrophilic compounds containing glutathione. There is a polymorphism within the GSTP1 gene that correlates with reduced enzyme activity that confers increased risk for lung cancer in individuals exposed to environmental tobacco smoke. Another polymorphism, present in the glutathione-*S*-transferase M1 (GSTM1), could act in a synergistic manner with GSTP1, thus increasing the risk for developing lung cancer.[32] Another example of genetic risk factors for lung cancer is the polymorphisms T3801C in the cytochrome P450 1A1 (CYP1A1) gene.

For breast and prostate cancer, we also consider genes coding detoxification enzymes as the cytochrome P450 1B1 (CYP1B1) gene. CYP1B1 plays a role in the process of activation of many carcinogens, like polycyclic aromatic hydrocarbons (PAHs). PAHs are deposited in adipose tissues, such as those in the breast. PAH is altered in a two-phase reaction to detoxify and excrete it from the body. Phase I of the reaction involves enzymes, particularly cytochrome P450 1B1 (CYP1B1) in breast tissue, which converts PAH to a water-soluble, carcinogenic intermediate. Phase II involves enzymes that detoxify this intermediate for excretion.[33]

When studying oxidative stress as an important carcinogenic agent, we must consider superoxide dismutases (Mn, Cu, ZnSOD) that catalyze the dismutation of two superoxide radicals, producing hydrogen peroxide and oxygen. MnSOD is a major enzyme involved in the scavenging of free radicals. MnSOD with alanine at the 9 position could be related to breast cancer risk by having an altered capacity to reduce oxidative stress.[34]

Enzymes Involved in the Metabolism of Steroid Hormones

Many of the enzymes involved in estrogen metabolism are polymorphically distributed within the human population (i.e., CYP17, CYP19, CYP1A1, CYP1B1, MnSOD, COMT, and GST). Inherited alterations in the activity of any of these enzymes have the potential to define differences in breast cancer risk associated with estrogen carcinogenesis. However, it is evident that no single genotype can be linked to all breast cancers because we are studying a complex, multifactorial disease.[35]

Several Phase II enzymes either inactivate catecholestrogens or protect against estrogen carcinogenesis by detoxifying products of oxidative damage that may arise on redox cycling of catecholestrogens. As with the CYP1B1 gene, the COMT gene codes for catechol-O-methyltransferase, which is one of the key enzymes of catecholestrogen biosynthesis and metabolism. Their polymorphisms determine the variation of enzyme activities.[36]

Cognitive Impairment

Early onset familial Alzheimer's disease (EOAD) arises before age 60. Mutations in genes for proteins called presenilins are responsible for 50% of familial AD. AD that develops after age 60 (late-onset AD) is associated with the same E4 variant of apolipoprotein E gene that increases the risk of heart disease.[24,37,38] Some cases of familial LOAD have been associated with mutations on TM2 domain of the PS2 gene.[39]

Studies carried out among elderly people (without AD diagnosis) have shown that APOE gene polymorphism appears to affect cognitive function among those who smoke or drink. In smokers and light drinkers, the APOE E4 allele seems to decrease the risk of cognitive impairment,[40] while increasing cognitive decline among heavier drinkers.[41]

Bone Mineralization Disturbances

Osteoporosis is another multifactorial polygenic disease in which genetic influences are modulated by several non-genetic factors (hormonal, nutritional factors, and life-style factors in general). For a better understanding of genetic contribution to bone mineralization we have to consider alleles at a combination of genes that are interacting with environmental factors. Today a few genes have been identified including the vitamin D receptor (VDR), the gene for type 1 collagen (COLIA1), the gene for the estrogen receptor (ER), and others.[42]

Polymorphism in VDR gene has been reported to play a major role in variations for genetic regulation of bone mass. An example is the BsmI VDR polymorphisms where the B allele has been associated to the annual rates of bone loss during early and late postmenopausal periods in several populations.[43,44] At the same time, it has been demonstrated that response to both calcium and calcitriol therapy would be dependent on genetic variation at the VDR locus.[45]

The first intron of the gene for type 1 collagen (COLIA1) has an Sp1 binding site that was identified as a polymorphism occurring in about 22% of Caucasians and causes reduced binding affinity for the Sp1 protein. This allele has been shown to be associated with postmenopausal bone mineralization disturbances in several populations worldwide. Since the effects of the Sp1 allele on bone mineralization disturbances were found to be dependent on age post menopause, it is thought that this allele may play a role in determining the rate of postmenopausal bone loss.[46]

Estrogens play an important role in regulating bone homeostasis and preventing postmenopausal bone loss. They act through binding to two different ERs, ERα and ERβ. Different polymorphisms have been described in both the ERα and ERβ genes.

Although a large number of association studies have been performed, the individual contribution of these polymorphisms to the pathogenesis of osteoporosis remains to be universally confirmed.[47]

HELPFUL GENES

There are some alleles that confer the best adaptative response. The best known and well-established examples are the "disease-protective alleles," as such as APOE and APOE2, APO CIII, and some HLA variants, which have been described in higher frequency in centenarians. They might be catalogued as longevity-assurance genes in humans.

Apolipoprotein E

Statistically, people who survive to age 100 have been found to be about half as likely to carry the E4 gene and somewhat more likely to carry the E2 gene.[26] Interestingly, investigators have found that people possessing two copies of the apparently protective E2 (i.e., one from each parent) have an increased likelihood of high blood triglyceride levels (hyperlipoproteinemia type III), a predictor risk factor for heart disease.[19] In a study we carried out in 2002, we detected a Th1/E47cs-T allele accumulation in healthy individuals over 75 years of age, which suggests it might play a protective role against AD. The Th1/E47cs-T allele may provide greater protection against AD than APOE2, although this awaits proof of Th1/E47cs-T allele overrepresentation in healthy individuals of other populations.[48]

Apolipoprotein C-III

Russian researchers reported another gene in the apolipoprotein family that has been found to be associated with increased longevity. At the level of the APO CIII T −455C polymorphism they found an increased incidence of the C allele in people over the age of 80. Their included 137 people between the ages of 70 and 106.[49] The increased incidence of the C allele with advanced age indicates that this variant promoter could be associated with longevity.

HLA Variants

Certain inherited variants of the HLA system known as ancestral haplotypes have been associated with increased longevity, particularly in males.[50] Bolognese scientists reporting on a study of more 1,000 Italian centenarians found that the frequency of specific haplotypes varied between old females and males. They also reported that the ratio of females to males who survived to over age 100 ranged from about 2:1 in Sardinia to 7:1 in Mantova, a province in northern Italy. They found that male centenarians were generally healthier than their female counterparts. They suggested that their results indicated that female longevity was less dependent on genes than male longevity was and that women probably had healthier lifestyles and fewer toxic environmental exposures during their lifetimes. Men who reached age 100 were more likely to be able to thank their genes.[51]

In a study performed in Milan, Italy, scientists compared the frequency of the loss of the drug-metabolizing enzyme gene designated GSTT1 in 94 nonagenarians and 418 younger controls. The frequency of the absence of the GSTT1 gene was 28% in the older subjects and only 19% in the younger people studied. The researchers commented that these kinds of studies could help sort out the interactions between genes and environmental influences on aging and cancer formation.[52]

The Insulin/IGF-1 Pathways

In humans, insulin sensitivity normally declines during aging, and insulin resistance is an important risk factor associated with a variety of intermediate phenotypes (hypertension, arteriosclerosis, obesity) strongly affecting morbidity, disability, and mortality among the elderly.[53]

In 2001, a study including 466 healthy subjects with a wide age range (range 28–110 years) demonstrated a significant reduction of insulin resistance in subjects 90–100 years old.[54] These results indicate that an efficient insulin response has an important impact on human longevity.

One of the investigated polymorphisms is the IGF-IR locus, where the IGF-IR A allele carriers show lower plasma IGF-I levels than the rest of the population. Also, IGF-IR A carrier subjects are found in increased proportion in long-lived individuals.

In a study that compared aging parameters of young (up to 39 years) and old (over 70 years) individuals having similar IGF-1 blood levels, the result was that old males with IGF-1 levels similar to young ones do not show the age-dependent decrease in serum testosterone and lean body mass, nor the increase in fat body mass.[55]

MITOCHONDRIAL GENES AND LONGEVITY

Most of the genes associated with longevity are in the nuclei of our cells. Nevertheless, mitochondria have their own genetic material that we also inherit and there are some findings that have indicated some mitochondrial genes might play an important role in longevity.As the power source of the cell and the locus of oxidative metabolism, mitochondria are quite vulnerable to oxidative damage.

An ongoing study involving multiple research centers in Italy has looked at many facets of longevity. A report issued by some of the scientists involved in that study points out that certain patterns of mitochondrial gene inheritance could easily lead to increased longevity, if these mitochondria conferred more resistance to the damage of free radicals. Conversely, a less favorable group of mitochondrial genes might lead to decreased longevity.[56] They have preliminary findings from a study of more than 800 Italian centenarians of both sexes showing that some mitochondrial gene variants were more common among them. One such variant, named the J variant, was notably overrepresented in centenarians.[57]

These same scientists looked at a gene pattern in the regular chromosomes that occur in the nucleus of the cells that predisposes to either longevity or shortened life span. This gene is called THO, and is a stress-responder gene. Some patterns of inheritance of THO variants favor longevity while others do not. The Italian researchers noted that some centenarians were found to possess the unfavorable version of THO, but that many of them also carried a favorable mitochondrial gene pattern, called the U haplotype. Therefore, variants of the THO gene may promote long life span in the context of some mitochondrial haplotypes but not in others. They concluded that there is more interplay between the nuclear and mitochondrial genes than had previously been realized in determining the genetic contribution to longevity.[58]

Some mitochondrial gene variants are more common in centenarians. One such variant, called J variant, is notably higher in centenarians. On the contrary, certain mitochondrial genes have been linked to a shortening of life span conferring risk to develop Parkinson's disease and AD.[59]

Today, we are sure that any single limited theory of aging is certain to be insufficient. Aging is a very complex, multifactorial trait that must be analyzed by taking into account genetic as well as environmental factors. We have to consider these complex interactions (gene–gene, genes–environment) in order to differentiate our patients according to their particular risks because we have genetic markers that permit us to predict vulnerability at different levels.

Several studies in worms, fruit flies, mice, and other animals have shown that those organisms that ate less than they would have eaten ad libitum, lived up to 50% longer. This is an example of gene expression modulated by some particular environmental influences (food stress). Thus, caloric restriction seems to induce in our bodies an over expression of all molecular defenses to survive better and increase our life span.[60]

In summary, as practitioners of predictive medicine we are faced with several challenges to identify and characterize genetically sensitive subpopulations, especially those that are easily identified at the first exam. This is the basic principle of predictive medicine: to forecast the appearance of certain diseases before their symptoms are expressed.

MOLECULAR GENETICS AND PREDICTIVE MEDICINE

Molecular genetics have supplied new tools that have gradually become the basis for predictive medicine and that, it is hoped, will open the way to preventive medicine.

When we study aging in human populations we should observe a normal distribution with two extreme points: one extreme negative represented by premature aging diseases and one extreme positive represented by centenarians. Both extremes are very rare. The vast majority of people are in the middle. Here aging is the result of a net balance between the effects exerted by positive and negative gene variants. According to this balance we are more or less far away from each extreme.

Determining this balance is one of the main objectives in antiaging medicine, especially identifying people at higher risk to develop early onset pathologies commonly associated with aging, which increase their probabilities for a premature death.

In this sense, the two pillars of antiaging medical practice are the fields of predictive and preventive medicine. Predictive medicine seeks to identify the individuals at increased risk and preventive medicine maps out a therapy to prevent or slow down the pathological process in these patients identified to be at high risk.

In order to have a more appropriate interpretation of DNA tests applied to predictive medicine, we must take into account several considerations. First, the tests we use to determine a genetic predisposition are not absolute but probabilistic in nature. In this sense, we are able to identify individuals at higher risk (not at risk). While some people may have polymorphisms that confer only moderate risk of disease, others may have genetic variants that make them substantially more sensitive than others in front of certain environmental factors. Second, because genetic susceptibility is often influenced by many genes, we must take into account gene–gene interactions that may be synergistic or antagonistic. Synergistic actions are the effects of two or more alleles that are susceptibility factors for some disease and that increase the risk for those individuals who carry all these respective genes beyond the risk of those who carry only one of them (allele 4G at PAI1+ allele H2 at FGNB). Antagonistic actions are dictated by the presence of one factor that is able to antagonize the effect of another (pro-thrombogenic effect of the allele H2 at FGNB and anti-thrombogenic effect of the allele 34T at FXIII gene). Finally, because we are testing multifactorial diseases, we must consider a direct influence of environmental factors on the molecular events that govern gene expression. This influence could be negative (enhancing a pathogenic effect) or positive (modulating lifestyle factors in order to make prevention).

We are in the age of molecular medicine, and we have to be prepared to use all these tools to diagnose and treat a disorder. With DNA tests it is possible to predict—even in a symptom-free state—whether a person is at higher risk for developing a particular disease. If we are able to use these tests as early as possible we will have sufficient time to design successful prevention strategies. Pharmacogenetics offers us the promise of more direct, effective, and safe treatment of our patients.

[*Competing interests*: The authors declare that they have no competing financial interests.]

REFERENCES

1. BARZILAI, N. & R. SHULDINER. 2001. Searching for Human Longevity Genes. The Future History of Gerontology in the Post-genomic Era. J. Gerontol. A. Biol. Sci. Med. Sci. **56:** 83–87.
2. MAGALHAES, J.P. 2004. From cells to aging: a review of models and mechanisms of cellular senescence and their impact on human aging. Exp. Cell Res. **300:** 1–10.
3. WONG, T.P. 2001. An Old Question Revisited: Current Understanding of Aging Theories. McGill J. Med. **6:** 41–47.
4. WANG, E. 1995. Senescent human fibroblasts resist programmed cell death, and failure to suppress bcl2 is involved. Cancer Res. **55:** 2284–2292.
5. LAO, J.I., K. BEYER & R. CACABELOS. 1998. Centromeric disturbances affecting all chromosomes of cultured lymphocytes obtained from Alzheimer's disease patients. *In* Progress in Alzheimer's and Parkinson's Diseases. A. Fisher, I. Hanin & M. Yoshida, Eds.: 793–798. Plenum Press. New York.
6. PERLS, T. 2001. Guest editorial: genetic and phenotypic markers among centenarians. J. Gerontol. A. Biol. Sci. Med. Sci. **56:** 67–70.
7. YASHIN, A.I. *et al.* 2000. Genes and longevity: lessons from studies of centenarians. J. Gerontol. A Biol. Sci. Med. Sci. **55:** 319–328.
8. PERLS, T.T. *et al.* 2000. Exceptional familial clustering for extreme longevity in humans. J. Am. Geriatr. Soc. **48:** 1483–1485.
9. PERLS, T.T., L. ALPERT & R.C. FRETTS. 1997. Middle-aged mothers live longer. Nature **389:** 133.

10. TAKATA, H. *et al.* 1987. Influence of major histocompatibility complex region genes on human longevity among Okinawan-Japanese centenarians and nonagenarians. Lancet **2:** 824–826.
11. FRISONI, G.B. *et al.* 2001. Longevity and the E2 allele of apolipoprotein E: the Finnish Centenarians Study. J. Gerontol. A. Biol. Sci. Med. Sci. **56:** 75–78.
12. MCGUE, M. *et al.* 1993. Longevity is moderately heritable in a sample of Danish twins born 1870–1880. J. Gerontol. **48:** B237–B244.
13. DUFOUIL, C. *et al.* 2000. Influence of apolipoprotein E genotype on the risk of cognitive deterioration in moderate drinkers and smokers. Epidemiology **11:** 280–284.
14. SALAMONE, L.M. *et al.* 2000. Apolipoprotein E gene polymorphism and bone loss: estrogen status modifies the influence of apolipoprotein E on bone loss. J. Bone Miner. Res. **15:** 308–314.
15. RALSTON, S.H. 2002. Genetic control of susceptibility to osteoporosis. J. Clin. Endocrinol. Metab. **87:** 2460–2466.
16. SHMOOKLER, R.J. & R.H. EBERT. 1996. Genetics of aging: current animal models. Exp. Gerontol. **31:** 69–81.
17. JAZWINSKI, S.M. 1998. Genetics of longevity. Exp. Gerontol. **33:** 773–783.
18. MURAKAMI, S. & T. JOHNSON. 2001. The OLD-1 positive regulator of longevity and stress resistance is under DAF regulation in *Caenorhabditis elegans*. Curr. Biol. **11:** 1517–1523.
19. ROTHSCHILD, H. & S.M. JAZWINSKI. 1998. Human longevity determinant genes. J. La State Med. Soc. **150:** 272–4.
20. LAO, J.I. 2004. Polimorfismos Genéticos a Considerar en la evaluación del riesgo vascular. Medicina Antienvejecimiento **4:** 55–63.
21. DAVIGNON, J., R.E. GREGG & C.F. SING. 1998. Apolipoprotein E polymorphism and atherosclerosis. Atherosclerosis **8:** 1–21.
22. CHIANG, A.N. *et al.* 1999. Differential distribution of apolipoprotein E in young rats and aged spontaneously hypertensive and stroke-prone rats. J. Hypertens. **17:** 793–800.
23. POPA-WAGNER, A. *et al.* 1999. Upregulation of MAP1B and MAP2 in the rat brain after middle cerebral artery occlusion: effect of age. J. Cerebr. Blood Flow Metab. **19:** 425–434.
24. BEYER, K. *et al.* 1997. APOE epsilon 4 allele frequency in Alzheimer's disease and vascular dementia in the Spanish population. Ann. NY Acad. Sci. **826:** 452–455.
25. CARR, F.J. *et al.* 2002. Genetic aspects of stroke: human and experimental studies. J. Cereb. Blood Flow Metab. **22:** 767–773.
26. SCHACHTER, F. *et al.* 1994. Genetic associations with human longevity at the APOE and ACE loci. Nat. Genet. **6:** 29–32.
27. LEVIEV, I. *et al.* 2002. High expressor paraoxonase PON1 gene promoter polymorphisms are associated with reduced risk of vascular disease in younger coronary patients. Atherosclerosis **161:** 463–467.
28. PHILLIPS, K.A. 2001. Current perspectives on BRCA1- and BRCA2-associated breast cancers. Intern. Med. J. **31:** 349–356.
29. FEARNHEAD, N.S., J.L. WILDING & W.F. BODMER. 2002. Genetics of colorectal cancer: hereditary aspects and overview of colorectal tumorigenesis. Br. Med. Bull. **64:** 27–43.
30. MURATA, M. *et al.* 2001. Genetic polymorphisms in cytochrome P450 (CYP) 1A1, CYP1A2, CYP2E1, glutathione S-transferase (GST) M1, and GSTT1 and susceptibility to prostate cancer in the Japanese population. Cancer Lett. **165:** 171–177.
31. KIYOHARA, C. *et al.* 2002. Genetic polymorphisms and lung cancer susceptibility: A review. Lung Cancer **37:** 241–256.
32. STÜCKER, I. *et al.* 2002. Genetic polymorphisms of glutathione S-transferases as modulators of lung cancer susceptibility. Carcinogenesis **23:** 1475–1481.
33. GOTH-GOLDSTEIN, R., C. ERDMANN & M. RUSSELL. 2000. CYP1B1 expression in normal human breast tissue specimen. Proc. Am. Ass. Cancer Res. **41:** 807.
34. AMBROSONE, C.B. *et al.* 1999. Manganese superoxide dismutase (MnSOD) genetic polymorphisms, dietary antioxidants, and risk of breast cancer. Cancer Res. **59:** 602–606.

35. THOMPSON, P.A & C. AMBROSONE. 2000. Molecular epidemiology of genetic polymorphisms in estrogen metabolizing enzymes in human breast cancer. J. Natl. Cancer Inst. Monogr. **27:** 125–134.
36. ZIMARINA, T.S. *et al.* 2004. Polymorphisms of *CYP1B1* and *COMT* in breast and endometrial cancer. Mol. Biol. **38:** 322–328.
37. STRITTMATTER, D. *et al.* 1998. Apolipoprotein E: high avidity binding to β-amyloid and increased frequency of type 4 allele in late-onset Alzheimer's disease. Proc. Natl. Acad. Sci. **90:** 1971–1981.
38. MEYER, M.R. *et al.* 1998. APOE genotype predicts when—not whether—one is predisposed to develop Alzheimer disease. Nat. Genet. **19:** 321–322.
39. LAO, J.I. *et al.* 1998. A novel mutation in the predicted TM2 domain of the presenilin 2 gene in a Spanish patient with late-onset Alzheimer's disease. Neurogenetics **1:** 293–296.
40. CARMELLI, D. *et al.* 1999. The effect of apolipoprotein E epsilon4 in the relationships of smoking and drinking to cognitive function. Neuroepidemiology **18:** 125–133.
41. DUFOUIL, C. *et al.* 2000. Influence of apolipoprotein E genotype on the risk of cognitive deterioration in moderate drinkers and smokers. Epidemiology **11:** 280–284.
42. RALSTON, S.H. 2002. Genetic control of susceptibility to osteoporosis. J. Clin. Endocrinol. Metab. **87:** 2460–2466.
43. UITTERLINDEN, A.G. *et al.* 1996. A large-scale association study of the association of vitamin D receptor gene polymorphisms with bone mineral density. J. Bone Min. Res. **11:** 1241–1248.
44. KIKUCHI, R. *et al.* 1999. Early and late postmenopausal bone loss is associated with BsmI vitamin D receptor gene polymorphism in Japanese women. Calcif. Tissue Int. **64:** 102–106.
45. MORRISON, N.A. *et al.* 2005. Vitamin D receptor genotypes influence the success of calcitriol therapy for recurrent vertebral fracture in osteoporosis. Pharmacogenet. Genomics **15:** 127–135.
46. LANGDAHL, B.L. *et al.* 1998. An Sp1 binding site polymorphism in the COLIA1 gene predicts osteoporotic fractures in both men and women. J. Bone Miner. Res. **13:** 1384–1389.
47. GENNARI, L. *et al.* 2005. Estrogen receptor gene polymorphisms and the genetics of osteoporosis: a HuGE review. Am. J. Epidemiol. **161:** 307–320.
48. BEYER, K. *et al.* 2002. Identification of a protective allele against Alzheimer disease in the APOE gene promoter. Neuroreport **13:** 1403–1405.
49. ANISIMOV, S.V. *et al.* 2001. Age-associated accumulation of the apolipoprotein C-III gene T-455 polymorphism C allele in a Russian population. J. Gerontol. A. Biol. Sci. Med. Sci. **56:** 27–32.
50. TAKATA, H. *et al.* 1987. Influence of major histocompatibility complex region genes on human longevity among Okinawan-Japanese centenarians and nonagenarians. Lancet **2:** 824–826.
51. RICCI, G. *et al.* 1998. Association between longevity and allelic forms of human leukocyte antigens (HLA): population study of aged Italian human subjects. Arch. Immunol. Ther. Exp. **46:** 31–34.
52. TAIOLI, E. *et al.* 2001. Polymorphisms of drug-metabolizing enzymes in healthy nonagenarians and centenarians: difference at GSTT1 locus. Biochem. Biophys. Res. Commun. **280:** 1389–1392.
53. FACCHINI, F.S. *et al.* 2001. Insulin resistance as a predictor of age-related diseases. J. Clin. Endocrinol. Metab. **86:** 3574–3578.
54. PAOLISSO, G. *et al.* 2001. Low insulin resistance and preserved beta-cell function contribute to human longevity but are not associated with TH-INS genes. Exp. Gerontol. **37:** 149–156.
55. RUIZ-TORRES, A & M. SOARES DE MELO KIRZNER. 2002. Aging and longevity are related to growth hormone/insulin-like growth factor-1 secretion. Gerontology **48:** 401–407.
56. DE BENEDICTIS, G. *et al.* 2000. Inherited variability of the mitochondrial genome and successful aging in humans. Ann. NY Acad. Sci. **908:** 208–218.

57. DE BENEDICTIS, G. *et al.* 1999. Mitochondrial DNA inherited variants are associated with successful aging and longevity in humans. FASEB J. **13:** 1532–1536.
58. DE BENEDICTIS, G. *et al.* 2000. Does a retrograde response in human aging and longevity exist? Exp. Gerontol. **35:** 795–780.
59. CALABRESE, V. *et al.* 2000. Mitochondrial involvement in brain function and dysfunction: relevance to aging, neurodegenerative disorders, and longevity. Neurochem. Res. **26:** 739–764.
60. WILLCOX, B.J. *et al.* 2004. How much should we eat? The association between energy intake and mortality in a 36-year follow-up study of Japanese-American men. J. Gerontol. A. Biol. Sci. Med. Sci. **59:** 789–795.

The Nitric Oxide Theory of Aging Revisited

S. M. McCann,[a] C. MASTRONARDI,[b] A. DE LAURENTIIS,[a] AND V. RETTORI[a]

[a]Centro de Estudios Farmacológicos,
Consejo Nacional de Investigaciones Científicas y Técnicas (CEFYBO-CONICET),
School of Medicine, UBA, Paraguay 2155 piso 16, 1121, Buenos Aires, Argentina

[b]Division of Neuroscience, Oregon National Primate Research Center,
Beaverton, Oregon 97006, USA

ABSTRACT: Bacterial and viral products, such as bacterial lipopolysaccharide (LPS), cause inducible (i) NO synthase (NOS) synthesis, which in turn produces massive amounts of nitric oxide (NO). NO, by inactivating enzymes and leading to cell death, is toxic not only to invading viruses and bacteria, but also to host cells. Injection of LPS induces interleukin (IL)-1β, IL-1α, and iNOS synthesis in the anterior pituitary and pineal glands, meninges, and choroid plexus, regions outside the blood–brain barrier. Thereafter, this induction occurs in the hypothalamic regions (such as the temperature-regulating centers), paraventricular nucleus (releasing and inhibiting hormone neurons), and the arcuate nucleus (a region containing these neurons and axons bound for the median eminence). Aging of the anterior pituitary and pineal with resultant decreased secretion of pituitary hormones and the pineal hormone melatonin, respectively, may be caused by NO. The induction of iNOS in the temperature-regulating centers by infections may cause the decreased febrile response in the aged by loss of thermosensitive neurons. NO may play a role in the progression of Alzheimer's disease and parkinsonism. LPS similarly activates cytokine and iNOS production in the cardiovascular system leading to coronary heart disease. Fat is a major source of NO stimulated by leptin. As fat stores increase, leptin and NO release increases in parallel in a circadian rhythm with maxima at night. NO could be responsible for increased coronary heart disease as obesity supervenes. Antioxidants, such as melatonin, vitamin C, and vitamin E, probably play important roles in reducing or eliminating the oxidant damage produced by NO.

KEYWORDS: nitric oxide synthase; cyclic GMP; cyclooxygenase; bacterial lipopolysaccharide; cytokines; hypothalamus; brain; pituitary gland; pineal gland; degenerative diseases; inflammation; infection; stress; coronary heart disease

INTRODUCTION

The causes of aging are undoubtedly multifactorial. One of the most prominent current theories of aging is the free radical theory. According to this theory, free rad-

Address for correspondence: Samuel M. McCann M.D., Av. Cordoba 2465, piso 13 "A," 1120 Buenos Aires, Argentina. Voice/fax: 54-11-49634473.
smmccann2003@yahoo.com

Ann. N.Y. Acad. Sci. 1057: 64–84 (2005). © 2005 New York Academy of Sciences.
doi: 10.1196/annals.1356.004

icals generated through mitochondrial metabolism can act to cause abnormal function and cell death. Various toxins in the environment can injure mitochondrial enzymes, leading to increased generation of free radicals that, over the life span, eventually play a major part in aging.[1]

At the Third International Symposium on the Neurobiology and Neuroendocrinology of Aging, we[2] presented evidence to suggest that excessive production of the free radical nitric oxide (NO) in the central nervous system (CNS) and its related glands, such as the pineal and anterior pituitary, may be the most important factor in the aging of these structures. Evidence for this hypothesis has been accruing rapidly. Because of the fact that the synthesis of inducible NO synthase (iNOS) following injection of bacterial lipopolysaccharide (LPS) in the rat was much greater outside the blood–brain barrier,[3] for example, in the anterior pituitary and pineal gland, than inside this barrier, it occurred to us that NO might play a role in aging of every organ system of the body. The evidence for this concept is particularly well developed to explain the pathogenesis of coronary arteriosclerosis.

NO has been found to play an ubiquitous role in the control of physiological functions throughout the body. The toxic effects of the soluble gas occur following infection or inflammations that cause the production of iNOS, which floods the tissue with toxic concentrations of NO. Here, we briefly review the methods of formation of NO in the body and its physiological role in the organs where we believe it also plays an important role in aging, and then present the evidence that NO is largely responsible for the aging process.

PRODUCTION OF NO IN THE BODY

Three isoforms of NO synthase (NOS) occur in the body.[4,5] The first of these is iNOS. LPS or cytokines [interleukins (IL)-1, IL-2, IL-6 and tumor necrosis factor alpha (TNF-α)] act via receptors on the cell surface of immune cells, particularly macrophages and other cells, such as endothelial cells, to activate DNA-directed mRNA synthesis which induces synthesis of iNOS. These LPS receptors are toll-like receptors (TLRs).[6] They activate nuclear factor κ B (NFκB), which activates DNA-directed synthesis of not only pro-inflammatory cytokines (such as IL-1, IL-6, and TNF-α), but also iNOS. A single injection of LPS leads to the release of massive amounts of NO, beginning within a few hours, peaking at 18 h, and then declining by 24 h. This enzyme is active as soon as it is induced, because it contains within itself calcium and calmodulin, which are required for activation of all isoforms of the enzyme. Like all forms of NOS, iNOS converts arginine and equimolar molecules of oxygen in the presence of various cofactors, such as NADPH and tetrahydrobiopterin, into NO and equimolar amounts of citrulline. The NO decays in solution with a half time of 5 to 10 sec, whereas citrulline remains in the cell and can even be recycled into arginine to provide further substrate for conversion by NOS into NO and citrulline. There is also a transport mechanism that carries arginine into the cell to provide substrate.

NO is a soluble gas and diffuses to neighboring cells, bacteria, and viruses. The high concentration formed after the induction of iNOS interferes with metabolism, leading to death of viral and bacterial invaders and also host cells. NO blocks cellular enzymes required in metabolism and also activates soluble guanylate cyclase (sGC), a soluble enzyme present in the cytoplasm of cells. The activation occurs via interaction of NO with the Fe^{2+} in the heme portion of the molecule, thereby altering its conformation and activating it. This causes conversion of guanosine triphosphate (GTP) to cyclic guanosine monophosphate (cGMP), which mediates many of the physiological actions of NO in mammalian cells. NO also activates cyclooxygenase, generating prostaglandins and lipoxygenase, and forming leukotrienes, which are toxic in high concentrations.

Endothelial NOS (eNOS) is a constitutive enzyme present in vascular endothelium.[7] Cholinergic stimulation by parasympathetic innervation of the vessels activates the enzyme by increasing the intracellular free calcium concentration $[Ca^{2+}]$, which combines with calmodulin. This interacts with eNOS, activating it. The NO produced diffuses to overlying smooth muscle and activates sGC. The cGMP released reduces intracellular $[Ca^{2+}]$, thereby relaxing the vascular smooth muscle. NO exercises tonic vasodilator tone in the vascular system since blockade of NO synthase leads to rapid development of hypertension.

A third isoform of the enzyme, neural NOS (nNOS) is a constitutive enzyme, like eNOS, present in many neurons in the central and autonomic nervous systems.[7] It is present in high concentrations in the cerebellum, cerebral cortex, and hippocampus, and in very high concentrations in the hypothalamus, particularly in the paraventricular and supraoptic neurons. The axons of these neurons project to the median eminence and neural lobe of the pituitary gland, where very large amounts of enzyme are present. This form of the enzyme, like eNOS, requires activation by synaptic input, causing elevation of intracellular Ca^{2+} concentrations, which combines with calmodulin, which activates the enzyme leading to the production of NO.

Recently, a mitochondrial form of the enzyme has been discovered localized in the inner mitochondrial membrane. The NO produced by this enzyme has the potential to produce cellular death by oxidizing intramitochondrial structures.[8,9]

ROLE OF NO IN THE CNS

The first evidence that NO played a role in the CNS was derived from experiments using cerebellar or hippocampal explants.[7,10] These showed that NO plays an important role in cerebellar function via activation of sGC and induction of cGMP formation. However, because of the complexity of function in the cerebellum, it is difficult to say what particular physiologic functions of the cerebellum are altered by NO. Garthwaite and colleagues[11] showed that NO may be important in induction of long-term potentiation in the hippocampus, which is thought to play a role in memory.[4–7] Palmer and colleagues[10] showed by mass spectroscopy that the substance was indeed NO. Earlier work had previously pointed clearly to NO; however, it is possible that the major product is not NO itself, but nitroso compounds formed by combination of NO with various other compounds. Indeed, slow release of NO from nitroso compounds may account for a longer duration of action of NO than can be explained on the basis of the free radical itself, which, as indicated above, decays within seconds in solution.

ROLE OF NO IN HYPOTHALAMIC FUNCTION

It has been clearly established that NO plays a key role in controlling the physiological release of a number of hypothalamic peptides and classical neurotransmitters. It has been shown to control the release of corticotropin-releasing hormone (CRH) by the CRH neurons in the paraventricular nucleus (PVN).[12] The stimulatory effect is mediated by cholinergic neurons that act on muscarinic receptors to stimulate the release of NO from nNOS-containing neurons, termed NOergic neurons, in the PVN. This NO diffuses to the CRH-containing neurons, and activates the release of CRH. The mechanism involves stimulation of release of prostaglandin E_2 (PGE_2) and leukotrienes in the CRH neurons via activation of cyclooxygenase and lipoxygenase. In the case of the cyclooxygenase (COX), it is clear that NO activates constitutive COX (COX-1) by interaction of the free radical with the heme group of the enzyme, altering its conformation. Lipoxygenase also contains Fe^{2+}, but a heme group has yet to be demonstrated. Nonetheless, this enzyme is definitely activated because sodium nitroprusside (NP), which releases NO, increases conversion of ^{14}C arachidonic acid to leukotrienes.[13]

NO is thought to act physiologically, mainly by activation of sGC.[5] Indeed, NP increases the release of cGMP from hypothalamic explants.[14] We hypothesize that NO releases CRH and other releasing hormones via the activation of GC, which releases cGMP, which in turn increases intracellular $[Ca^{2+}]$, leading to activation of phospholipase-A2. This converts membrane phospholipids into arachidonate, which is then converted to PGE_2 and leukotrienes by the activated cyclooxygenase and lipoxygenase, respectively.[7] The result is activation of adenylate cyclase, conversion of ATP to cAMP, which activates protein kinase A, causing extrusion of CRH granules. The CRH then enters the hypophyseal portal vessels, reaches the anterior lobe of the pituitary, and causes release of adrenocorticotropic hormone (ACTH).[12]

ROLE OF NO IN LUTEINIZING HORMONE-RELEASING HORMONE RELEASE

Release of LH is pulsatile and controlled by pulsatile release of luteinizing hormone-releasing hormone (LHRH) into hypophyseal portal vessels that deliver it to the sinusoids of the anterior pituitary, where it acts on the gonadotropes to promote the release of LH. LHRH also increases follicle-stimulating hormone (FSH) release to a lesser extent.

Release of FSH can occur without LH release and vice versa, but both hormones are released concomitantly in about 50% of the pulses. Immunocytochemistry revealed a different localization in the hypothalamus of the two peptides. Most neurons that contain LHRH or FSH-releasing hormone (FSHRH) contain only one of the peptides, but a few contain both peptides.[15] Specific receptors for each peptide activate NOS, which activates release of each peptide by cGMP.[16]

On the basis of *in vitro* studies, we postulate that the pathway of activation of LHRH release is as follows: either direct activation of noradrenergic terminals in the region of the arcuate-median eminence region or their activation by glutamergic neurons, via N-methyl-D-aspartic (NMDA) receptors, causes the release of norepinephrine, which acts by α_1 adrenergic receptors on the NOergic neurons to cause the release of NO.[17] This diffuses to the LHRH terminals intermingled with NOergic neurons nearby in the median eminence–arcuate region, stimulating the release of LHRH by the identical mechanism described above for CRH—namely, by activation of sGC, COX-1 and lipoxygenase—leading to the release of LHRH secretory granules into the hypophyseal portal vessels.[18] The intermingling of NOergic and LHRH terminals in this region has been demonstrated.[18]

NO controls LHRH release, which mediates mating behavior in both male and female rats by activating brain stem neurons controlling lordosis in the female and penile erection in the male rat. The role of NO in mediating LHRH release controlling both of these activities has been demonstrated *in vivo*.[19] The penile erection is controlled by pelvic nerve cholinergic activation of nNOS. This NO activates guanylyl cyclase in the corpora cavernosa penis, causing relaxation of the corpora and erection. When stimulation stops, the cGMP is degraded, causing loss of erection. Sildenafil citrate (Viagra) inhibits the phosphodiesterase responsible for degrading cGMP, thereby prolonging erection.[20]

We have demonstrated that oxytocin stimulates the release of LHRH via NO. The low concentrations required (10^{-8}–10^{-10} M) probably fall within the physiological range in view of the high concentrations of oxytocin in the median eminence in juxtaposition to LHRH terminals.[21] There is an ultrashort-loop negative feedback by which NO released from NOergic neurons feeds back to inhibit the release of oxytocin.

Not only does NO stimulate the release of many hypothalamic peptides, but it also stimulates the release of the inhibitory neurotransmitter, gamma amino-butyric acid (GABA).[22] This blocks the response of the LHRH terminal to NO. Therefore, GABA mediates a feed-forward negative feedback to inhibit pulsatile LHRH release. Furthermore, NO also inhibits the release of both norepinephrine and dopamine from the medial basal hypothalamus, constituting another negative feedback of pulsatile LHRH release by feeding back on the terminals of the noradrenergic and dopaminergic neurons to inhibit the release of both of these transmitters, one of which, and probably both of which, stimulate the release of NO that drives LHRH release.[23]

Interestingly, 30 min after addition of norepinephrine or NMDA, both stimulators of LHRH release, there is an increase in content of NOS in the hypothalamus measured by the citrulline method.[14] This method is an index of the quantity of NOS in the tissue because the labeled arginine is added to the homogenate and the labeled citrulline formed is measured. That this represents *de novo* synthesis of NOS is indicated by the fact that this effect is blocked by the inhibitor of DNA-directed RNA synthesis, actinomycin D (Rettori and colleagues, unpublished observations).

Thus, pulsatile LHRH release is mediated by noradrenergic neuronal terminals that activate NOergic neurons. The NO synthesized diffuses to the LHRH neurons and activates LHRH release. The pulses of LHRH are terminated by NO-induced release of GABA that blocks the response to the LHRH neuron to NO and NO-induced inhibition of further norepinephrine and dopamine release.

ROLE OF NO IN CONTROL OF OTHER HYPOTHALAMIC PEPTIDES AND NEUROTRANSMITTERS

Growth hormone (GH) release is also pulsatile, but the greatest release occurs during early sleep at night in humans. *In vivo* studies have shown that the inhibitor of NO synthase, Nγ-monomethyl-L-arginine (NMMA), can block pulsatile release of GH in the rat.[24] Furthermore, NO can also stimulate somatostatin release and its messenger RNA levels in the PVN by activation of sGC as determined in *in vitro* studies utilizing explants of the paraventricular region.[25] We hypothesize that the pulsatile release of GH that occurs under normal conditions is brought about princi- pally by NO stimulation of GH-releasing hormone (GHRH) release. At the same time, somatostatin release is probably inhibited. In the interpulse interval when GHRH release is absent, somatostatin release mediated by NO increases. IL-1 not only inhibits GHRH release, but also stimulates somatostatin release, thereby inhib- iting GH release during infections. The IL-induced prolactin release is also mediated by NO[26] probably by NO stimulation of prolactin-releasing peptides, such as oxyto- cin,[21] and by inhibition of the release of dopamine, a potent prolactin release-inhib- iting hormone, into the hypophyseal portal vessels.[23,27]

All of these results indicate that under physiological conditions, NO plays a fun- damental part in the control of neurotransmitter and neuropeptide release in the hy- pothalamus. Therefore, these areas where NO is released physiologically will be subjected to low levels of pulsatile NO throughout the life of the individual. It is not clear whether such levels can be toxic, but they may be. No studies have been done to determine whether long-term exposure to low levels of NO is damaging to neu- rons and/or glia. Interestingly, Schultz and colleagues[28] reported neurofibrillary de- generation in nerve fibers in the arcuate nucleus of aged men but not women. Possibly, pulsatile release of NO driving LHRH release over the life span may have caused this degeneration. Because aged women failed to show the degeneration, there appears to be a sex difference, probably mediated by sex hormones such as es- trogen, which has been shown to alter NOS concentrations.

THE ROLE OF NO IN CONTROL OF POSTERIOR PITUITARY FUNCTION

The neural lobe of the pituitary is one of the regions richest in NOS in the rat, with a predominance of the nNOS isoform, suggesting that NO may play a role in controlling the release of neuropeptides and neurotransmitters from the posterior pi- tuitary.[7,29] NADPH-diaphorase, used as a marker of NOS, was co-localized with va- sopressin (VP) and oxytocin (OT) in the hypothalamic-neurohypophyseal system.[30,31] The synthesis of NO by oxytocinergic or vasopressinergic neurons sug- gests that it may participate in auto- and/or cross-regulation of OT and VP secretion. Our studies indicate that NO donors reduce OT secretion from the neural pituitary lobe and we postulated that released NO may suppress OT secretion through an ul- trashort-loop negative feedback mechanism.[21] It has been reported that intracere- broventricular administration of L-NAME enhanced plasma levels of both OT and VP peptides.[32] Since NOS activity increases following salt loading and dehydration, it has been suggested that this increase may provide a negative feedback to prevent overstimulation of OT and VP release.[33]

Tachykinins belong to a family of peptides that includes substance P and neuro- kinin A (NKA). They are contained in hypothalamic neurons and nerve fibers and secretory cells of the posterior and anterior pituitary lobes, suggesting that these pep- tides may have a physiological role in the control of pituitary function.[34] Some actions of tachykinins are known to be exerted through NO release,[35,36] and NOS immunoreactivity was detected in some nerve terminals with NK-2 receptors,[37] the receptor subtype to which NKA binds with preferential affinity. Although we observed an inhibitory effect of NKA on OT release by activation of NOergic neurons, when NOS activity was blocked, NKA stimulated OT release, thus suggesting that NKA may have a dual effect on OT release, decreasing it through NO and increasing OT release by an NO-independent mechanism.[38] Such opposite responses could result from different signaling events linked to NK-2 receptors. Alternatively, these NKA effects could occur following paracrine activation of local modulators that could play a crucial role in the biological actions of NKA.[39] Nevertheless, the net effect of NKA on OT release from posterior pituitary seems to be inhibitory and may intervene in the control of OT response to osmotic stimuli.

Our study also shows that NO can inhibit the release of GABA from the posterior pituitary. Furthermore, the inhibition of NOS activity by L-NMMA and L-NAME increased GABA release, indicating that endogenous NO has an inhibitory effect on GABAergic activity in posterior pituitary. However, we used 8-Br-cGMP, which did not affect GABA release, suggesting that the inhibitory effect of NO on GABA release from posterior pituitary may be exerted through a cGMP-independent mechanism.[38] Support for this is in the report that NO activates K^+ channels in posterior pituitary nerve terminals by a cGMP-independent signaling pathway.[40] GABA release could result from vesicular exocytosis by terminals of tuberohypophyseal GABA system. Also, a carrier-mediated release from nerve terminals or pituicytes could be another source of GABA in the posterior pituitary.[41,42]

GABAergic terminals in the neural and intermediate lobes participate in the control of OT, VP, and α-MSH release[43–45] raising the possibility that NO modulation of GABAergic activity in the posterior pituitary may be involved in the regulation of the secretory function of this pituitary lobe. However, since NO decreased both OT and GABA release from the posterior pituitary and GABA was reported to inhibit OT release from nerve terminals of the neural lobe, it is unlikely that GABA mediates the inhibitory effect of NO on OT release from posterior pituitary. It is possible that NO might influence the release of OT at the level of the cell bodies and on nerve terminals of the neural lobe by different mechanisms.

NKA inhibits GABA release from posterior pituitary without affecting hypothalamic GABA release. We showed that the inhibition of NOS activity by L-NAME completely blocked the inhibitory effect of NKA on GABA release. The stimulatory effect of NKA on NO synthesis, together with the blockade of the inhibitory effect of NKA on GABA release by L-NAME, indicates that NO is involved, at least partially, in the reduction of the GABAergic activity induced by NKA. GABA arising from the posterior pituitary may arrive at the anterior pituitary and directly decrease prolactin secretion. The reduction in GABA release may contribute to the stimulatory effect of NKA on lactotroph function.[46]

THE ROLE OF NO IN CONTROL OF ANTERIOR
PITUITARY FUNCTION

Neural NOS has been localized in anterior pituitary cells. At least two cell types contain the enzyme; one of these, the folliculostellate cells, are modified macrophages and known to secrete IL-6 and other cytokines. The other type is the gonadotropes that secrete LH. The *in vitro* secretion rates of most pituitary hormones are low because of withdrawal of hypothalamic stimulation, but the secretion of prolactin is greatly enhanced because of withdrawal of hypothalamic inhibition by dopamine. In the case of prolactin, its secretion can be increased by inhibiting NOS with NMMA, a competitive inhibitor of NOS or NAME, another inhibitor of the enzyme. On the other hand, NP spontaneously releases NO and lowers prolactin release. The prolactin-inhibiting action of dopamine, the principal prolactin-inhibiting hormone, appears to be mediated via NO, because the action of dopamine to lower prolactin release was blocked by inhibition of NOS. We hypothesize that DA released into portal vessels reaches the anterior lobe, where it acts on D_2 receptors in folliculostellate cells and or gonadotropes to activate NOS, which activates NO release in turn activating sGC, increasing cGMP, which then suppresses secretion of prolactin from the lactotrophs. Indeed, cGMP decreased prolactin secretion.[27]

Adenosine is secreted by the folliculostellate cells and is the most powerful stimulant of prolactin secretion from anterior pituitaries *in vitro* yet identified, increasing release at concentrations of 10^{-10}–10^{-5} M with maximal release of three times basal levels at 10^{-8} M. The action appears to be mediated by an autocrine activation of adenosine 1 receptors on the surface of the folliculostellate cells, which activates inhibitory G proteins (G_i) that lower intracellular $[Ca^{2+}]$, thereby inhibiting nNOS within the folliculostellate cells that decreases NO production. The reduced paracrine NO inhibition of the lactotropes increases prolactin release.[47]

In contrast to prolactin, the release of which is inhibited by the hypothalamus, the release of the gonadotropins LH and FSH is stimulated by the hypothalamic peptide LHRH, which stimulates LH and to a lesser extent FSH, and by FSHR factor (FSHRF) (lamprey III LHRH or a closely related peptide), which preferentially stimulates FSH release. The mechanism is by stimulation of LHRH and FSHRF receptors, respectively, leading to increased $[Ca^{2+}]$, and activation of nNOS in the gonadotropes with resultant generation of cGMP, which in turn, activates PKG, leading to extrusion of gonadotropin secretory granules.[48]

Gonadotropin secretion is pulsatile. Pulses can consist of the simultaneous release of FSH and LH, brought about by prior simultaneous release of FSHRF and LHRH, or selective pulses of FSH or LH, brought about by prior release of the respective releasing hormone. The relative abundance of the pulses of each type is governed by sex hormones. Thus, on the basis of the research done so far, it appears that the pituitary gland is exposed to NO throughout normal life. Again, whether or not these concentrations could be toxic is not clear.

THE EFFECT OF INFECTION ON CYTOKINE AND NO FORMATION IN BRAIN, PITUITARY, AND PINEAL GLANDS

CNS infection is a powerful inducer of cytokine production in glia and neurons of the brain. It causes induction of iNOS and production of potentially toxic quantities of NO. Injection of bacterial LPS also induces the pattern of pituitary hormone secretion that characterizes infection. Intravenous LPS induces a dose-related release of ACTH and prolactin, a transient release of GH followed by profound inhibition, decreased secretion of thyrotropin-releasing hormone (TSH), and inhibition of LH, and to a lesser extent FSH release, in rats. It is believed that this pattern is caused by effects of LPS directly on the brain because, after intravenous injection of an intermediate dose of LPS, this pattern of pituitary hormone response occurred. Also there was an induction of IL-1α immunoreactive neurons in the preoptic-hypothalamic region. These cells were shown to be neurons by the fact that double staining revealed the presence of neuron-specific enolase. The neurons are present in saline-injected control animals, but increased in number by a factor of two within two hours after LPS injection. They were located in a region that also contains the thermosensitive neurons. They may be the cells that are stimulated to induce fever after LPS injection. They have short axons that did not clearly project to the areas containing the various hypothalamic releasing and inhibiting hormones, but they could also be involved in the stimulation or inhibition of their release, which occurs following infection.[49]

This study led to further research in which we determined the effect of intraperitoneal injection of a moderate dose of LPS on the induction of Il-1β and iNOS mRNA in the brain, pituitary, and pineal gland. The results were very exciting, because an induction of IL-1β and iNOS mRNA occurred with the same time course as found in the periphery following injection of LPS, namely, clear induction of iNOS mRNA within two hours, reaching a peak in two or six hours, followed by a decline to near basal levels at the next measurement by 24 hours after the single injection of LPS. The induction of both mRNAs occurred in the meninges, the choroid plexus, the circumventricular organs (such as the subfornical organ and median eminence) in the ependymal cells lining the ventricular system, and, very surprisingly, in parvocellular neurons of the PVN and arcuate nucleus (AN), areas of particular interest because they contain, not only the hypothalamic releasing and inhibiting hormones, but also other neurotransmitters controlled by NO.[3]

The greatest induction occurred in the anterior lobe of the pituitary, where the iNOS mRNA was increased at two hours by a factor of 45, and the pineal, where the activity was increased by a factor of 7 at six hours, whereas the increase in the PVN was five-fold. At six hours, the medial basal hypothalamus was found to have an increased content of NOS measured *in vitro*, and the collected cerebrospinal fluid (CSF) had increased concentrations of the NO metabolite, nitrate. These results indicate that the increase in iNOS mRNA was followed by *de novo* synthesis of iNOS that liberated NO into the tissue and also into the CSF. Presumably, LPS was bound to its receptors in the circumventricular organs and in the choroid plexus. These receptors, as in macrophages, activated DNA-directed IL-1β mRNA synthesis, which in turn caused the synthesis of IL-1β. IL-1β then activated iNOS mRNA and synthesis.

How can neurons in the AN and PVN—neurons inside the blood–brain barrier—be activated? In the case of the AN, the neurons may have axons that project to the median eminence. These neurons may have LPS receptors on their cell surface, which then induce IL-β mRNA and IL-β synthesis. This may then induce iNOS mRNA. Alternatively, LPS acting on its receptors may simultaneously induce IL-β mRNA and iNOS mRNA.

Active transport mechanisms for IL-1 and other cytokines,[50] and perhaps LPS, are present in the choroid plexus. The cells of the choroid plexus on the basis of our results must have LPS receptors on them. LPS must stimulate IL-1β and iNOS mRNA followed by synthesis of IL-1β and iNOS in the choroid plexus. LPS and IL-1β are then transported into the CSF. LPS is carried by CSF flow to the third ventricle, where it either crosses the ependyma or acts on terminals of PVN neurons in the ependyma to induce IL-1β and iNOS mRNA.

These results raise the possibility that even moderate infection, without direct CNS involvement, can increase iNOS levels and lead to production of toxic levels of NO. Therefore, it is possible that repeated infections over the life span could lead to brain damage in areas where there is large induction of iNOS in neurons, such as the PVN—the site of the cell bodies of most of the releasing and inhibiting hormone neurons—and the AN–median eminence region, which is also the site of production of GHRH, many neurotransmitters, and the site of passage of axons of many of the releasing hormone neurons, such as LHRH neurons, which project to the median eminence. There may also be damage to glial elements, meninges, and to the choroid plexus over the life span. The induction of IL-1α neurons in the temperature-regulating regions of the preoptic area should also be followed by induction of iNOS. Exposure to high levels of NO in this region may kill thermosensitive neurons and thus be responsible for the decreased febrile response to infection in the elderly. Measurement of iNOS activity in the hypothalamus of aged male rats (greater than two years of age) revealed a significant increase in NOS activity in comparison with that in young adults, which provided the first experimental support for this concept (Rettori, unpublished data).

The greatest increase in iNOS mRNA after LPS injection occurred in the anterior pituitary gland. Therefore, the likelihood of damage to the cells in this gland during infection is great. This, coupled with the damage to the releasing hormone neurons, could account for aging changes in secretion of pituitary hormones. For example, GH and prolactin are released largely at night, and nocturnal GH release is known to be impaired with age.

The massive induction of iNOS in the pineal could very well contribute to the gradual reduction in function of this gland associated with decreased nocturnal melatonin levels and finally even calcification of the gland, which occurs with aging.[51] Melatonin is an antioxidant, and has been shown to reduce oxidative damage produced by brain ischemia and reperfusion. There is evidence that exogenous melatonin increases the life span of mice.[51] Therefore, NO-induced pineal "aging" may play a role in aging. The pineal hormone melatonin releases LHRH from MBH of male rats that was prevented by blockade of NOS or GC activation.[52] Not only are proinflammatory cytokines found in tissue following injection of LPS but they are also present in blood partly by activation of NFκB.[6]

EFFECT OF LPS ON TNF-α LEVELS IN PLASMA OF ADULT MALE RATS

On the morning following insertion of a jugular catheter the night before, there was no detectable plasma TNF-α. TNF-α concentrations rose within 30 min after withdrawal of the first blood sample (0.6 mL). By 2 h, the concentration had risen 40-fold to 1,000 pg/mL. At 3 h, it had dropped precipitously to 125 pg/mL. It was zero the following morning but rose to 125 pg/mL at 1 h after the initial blood sample only to fall rapidly back to zero at 4 h. These results indicate that the rat, under resting conditions, has little or no circulating TNF-α, but can respond with a rapid synthesis and release of this cytokine into the circulation under bleeding stress. Previously, the rat had only been thought to respond to bacterial products.[53]

The release of TNF-α is mediated by the brain by inhibition of dopamine release leading to release of prolactin, which acts on its receptors on the immune cells to stimulate synthesis and release of TNF-α. There is inhibitory β-adrenergic control and stimulatory α-adrenergic control of TNF-α release, as indicated by studies with α- and β-adrenergic drugs.[54] TNF-α may well induce iNOS. If so we would have potentially damaging effects of stress. Indeed, Kishimoto and colleagues[55] reported that immobilization stress increased iNOS, another indication that stress may produce damaging levels of NO.

ROLE OF NO IN OTHER NEURODEGENERATIVE DISEASES

There is already considerable evidence that NO plays a role in neuronal cell death, which brings on Parkinson's disease by loss of the neurons of the nigrostriatal dopaminergic system. Indeed, the toxin 1-methyl-4-phenyl-1,2,3,4-tetrahydropyridine (NPTP) induces a parkinsonism-like syndrome. It is transported into the substantia nigra dopaminergic neurons by the dopamine transporter. It then interferes with mitochondrial metabolism, leading to the production of oxygen free radicals. Apparently, the basal production of NO can then be sufficient to cause toxicity via its diffusion into the dopaminergic neurons and combination with superoxide to generate peroxynitrite, a much more potent free radical than either superoxide or NO itself.[56] Another probable mechanism for toxicity of NO in all sites is in combination with the heme groups, with various enzymes, thereby inactivating them and blocking cellular respiration leading to cell death.[5,56]

These findings provide an explanation for the high incidence of early-onset Parkinson's disease in many people who served in World War I and developed influenza. At that time there was a major epidemic of influenza with encephalitis, which presumably led to generation of large amounts of NO in the region of the substantia nigra that then caused loss of dopaminergic neurons and eventual development of Parkinson's symptoms many years before it would have appeared as a result of normal aging. The appearance of parkinsonism with age is probably related to the quite rapid decline, beginning at age 45, in dopaminergic neurons in this region, even in normal individuals,[57] which may also be caused by enhanced NO generation during infection.

In Alzheimer's disease, an important neurodegenerative disease, plaques form consisting of amyloid. Surrounding these plaques are many abnormal astrocytes, which are seen on immunocytochemical study to contain IL-1.[58] The IL-1 should cause the induction of iNOS and production of NO, which may be a large factor in neuronal cell death in the vicinity of the plaques in this condition. Indeed, prostanoids, presumably formed by action of NO, accumulate adjacent to these plaques.

Even in normally aging brain, there is an increased incidence of these abnormal astrocytes.[58] Their production of IL-1 and NO could be partly responsible for the general neuronal cell loss that occurs with aging. NO is probably also involved in producing cell death around any area of inflammation in the CNS, for example, in multiple sclerosis, or after brain trauma.[58]

In Huntington's chorea, there is a mutation of the huntingtin protein that is associated with the selective loss of basal ganglion neurons that characterize this disease. Recently, a brain-specific protein that is associated with huntingtin has been identified, and has been termed huntingtin-associated protein (HAP-1). The location of this protein with neurons containing nNOS mRNA, with dramatic enrichment in both the pseudopedunculopontine nuclei, the accessory olfactory bulb, and the supraoptic nucleus of the hypothalamus, with co-localization of HAP-1 and nNOS in some of these neurons, suggests that, here again, NOS could generate sufficient NO to produce the neuronal cell loss responsible for Huntington's disease.[59]

THE ROLE OF NO IN CORONARY ATHEROSCLEROSIS

Because we had found such profound induction of iNOS in the anterior pituitary and pineal glands, areas outside the blood–brain barrier, it occurred to us that LPS would probably induce similar changes in all organs outside the blood–brain barrier, a prime example being the coronary arteries. Indeed, we have determined that there is induction of IL-1β and iNOS mRNA in the endothelium of the vascular system in the same rats given LPS. Induction of IL-1β and iNOS was also dramatic in renal vessels.[60]

A great deal of evidence has accrued, suggesting the possibility that chronic infections may have a relationship with coronary heart disease (CHD).[61] In the 1970s, experimental infection of germ-free chickens with avian herpes virus induced pathologic changes resembling those in human CHD.[62] There have been many studies showing the presence of high titers of antibodies against various organisms in patients with CHD. Although there is always some question about such studies, the incidence is such as to make it appear very likely that antibodies against *Helicobacter pylori, Chlamydia pneumonia,* cytomegalovirus, or other herpes viruses are very common in these patients. There is even an association with severe dental caries.[61]

Stimulated by these reports, there have now been two reports of treatment of patients with CHD with tetracycline derivatives.[63,64] In both studies, further complications of CHD were significantly reduced in the treated groups. In one study, treatment reduced the complications ten-fold.[63]

Tetracyclines have now been studied in chondral cell cultures from patients with osteoarthritis and in cell cultures from animals with experimentally produced arthritis. They have been shown to have chondro-protective effects.[65] NO is spontaneously released from human cartilage affected by osteo- or rheumatoid arthritis in quantities sufficient to cause cartilage damage. In a recent report, tetracyclines have been shown to reduce the expression and function of human osteoarthritis-effected NOS (iNOS).[65] It appears that in addition to the antibacterial action of these drugs, tetracyclines inhibit the expression of NOS, leading to reduction in the toxic consequences of production of NO. It is likely that these compounds will be beneficial in the treatment of osteoarthritis, as well as CHD. They will also probably be of therapeutic value in rheumatoid arthritis and cardiomyopathy, both thought to be autoimmune diseases caused largely by excess NO.

The current theory of CHD is that it is induced by an elevation of plasma cholesterol above the normal limit of 200 mg%. However, if one looks at the incidence of CHD versus the concentration of plasma cholesterol, one finds that as cholesterol passes the 200 mg% concentration, there is only a very slight increase in the incidence of the disease. The slope of incidence begins to rise between 250 and 300 and then rises quite rapidly as it approaches 400 mg%. There are many cases of CHD in patients with perfectly normal cholesterol. Indeed, increased LDL cholesterol has been considered particularly ominous, whereas HDL cholesterol has been thought to be protective. However, in many cases, CHD develops and has its downward progression in the presence of normal cholesterol and other lipids.

In a reported case study, a 72-year-old male had gradually rising cholesterol values from 200 mg% at the age of 40 to 220 at the age of 60, but had no symptoms of CHD. In late November, 1996, 72 h after return from a trip to England, he contracted influenza, followed by pneumonia, with a fever of 103.5°F and a pulse of 120. The pneumonia was probably bacterial, because it was responsive to ampicillin. This severe infection was followed by the development of angina pectoris within five weeks, which progressed to the point that he finally sought medical attention five months later. At this time, he had advanced CHD. His weight was normal. His cholesterol was 240 mg%, with a slightly elevated LDL and low HDL, but normal triglycerides. Angiography revealed extensive disease that was judged unsuitable for either balloon angioplasty or coronary bypass surgery, and he was placed on a cholesterol synthesis inhibitor that normalized his total cholesterol, LDL, and HDL within one month. The angina gradually improved over the next five months.

At that point in mid-November 1997, the subject took a trip to Europe and the Middle East. Beginning in Frankfurt, Germany, then to Cairo, Egypt, Luxor, and by Nile steamer to Aswan, back to Cairo, then to Israel and back to Germany. He developed a bad cold, necessitating antibiotic treatment just before departure. He reactivated his osteoarthritis while tramping through the ruins of Egypt. The cold recurred one week after it had ceased, and there was additional activation of the arthritis by walking the streets of Jerusalem. Additional stress was occasioned by the marked time changes, alternating cold and hot temperatures, dry weather, and the stress of the political tensions in the Middle East at that time. However, his angina was still ameliorating. In Germany, it was cold and damp, and his cold and arthritis were active. On return to this country, there was an extraordinary exacerbation of the osteoarthritis followed by a rapid downhill progression of angina pectoris. Because of the success with zithromycin in reducing the complications of CHD, he was treated with this drug. The treatment was followed by marked amelioration of arthritis, but the angina continued to worsen, and he developed angina during sleep, unrelieved by nitroglycerin.

Finally, he was admitted to the hospital. Angiography demonstrated that his coronary arteries had deteriorated greatly since first examined, even in the presence of perfectly normal lipids (cholesterol of 170 mg%, normal HDL and LDL, and low triglycerides). Fortunately, there was no evidence of myocardial infarction, and he survived and recovered from quadruple bypass surgery.[2]

ROLE OF LEPTIN IN THE RESPONSE TO LPS
AND ITS ROLE IN NO PRODUCTION

Leptin release from the adipocytes is also under neural control. Its release is stimulated by prolactin and inhibited by bromocriptine, a stimulator of dopamine release that inhibits prolactin release. It is decreased by ketamine anesthesia and inhibited by activation of β- and α-adrenergic agonist drugs with the antagonists having the opposite effects. Leptin release is also stimulated within 10–30 min by LPS. It is stored in microcytic vesicles lying just beneath the plasma membrane and is rapidly released from them. The relative increase of leptin release in response to LPS is less than that of TNF-α, but the absolute release is greater. Not only is the release of leptin stimulated by LPS, but also its synthesis is stimulated, as indicated by a dramatic increase in leptin mRNA in epididymal fat pads at 6 h.[66]

In both male rats and humans, plasma leptin levels are directly related to body weight and fat mass. There is a diurnal release of leptin, most apparent in humans, but also present in the rat. Amazingly, there is a highly significant direct correlation of NO, as evidenced by its metabolites NO_2-NO_3, with plasma leptin throughout the 24-h period in male rats. The quantity of NO metabolites (NO_2-NO_3) in plasma is roughly 1,000 times greater than that of leptin and reached high millimolar concentrations. Since micromolar concentrations of NO have been considered to be cytotoxic, one wonders if these concentrations of NO in fact could be detrimental and of course they would be increased further after LPS or during infection.

To determine whether leptin could stimulate NO production in hemisected epididymal fat pads, we incubated them *in vitro* for 1 h and found a high production of NO_2-NO_3 (380 μM NO_2-NO_3 in the medium). Incubation with 10^{-7} or 10^{-6} M leptin did not alter this production, but 10^{-5} M leptin highly significantly increased it to 750 μM. Surprisingly, leptin evoked a significantly dose-related *in vitro* stimulation of TNF-α. Stimulation was small but significant at the lowest dose of leptin (10^{-7} M) and was increased 25-fold at the 10^{-5} M dose.

In vivo leptin (6 or 60 nM/kg) was tested and it evoked a significant, nearly doubling of the area under the plasma curve NO_3-NO_2 (mM $\times$ min) with the 60 nM/kg but not the ten-fold lower dose. There was a dose-related significant increase in plasma TNF-α peaking at 90 min but still significant at 120 min. The ability of leptin to increase NO release may be a means by which the adipocyte increases blood flow when it is active metabolically.[67]

The large amounts of NO_3-NO_2 *in vitro* could be toxic since it is known that NO concentrations greater than 1 μM are toxic, whereas nanomolar concentrations are associated with the physiological functions of NO. The stimulation of TNF-α by leptin both *in vitro* and *in vivo* was unexpected and raises the question of a possible contribution of TNF-α to NO secretion in stress.[67]

NO ACTIVATES SECRETION OF CORTICOSTERONE

All sorts of stresses activate ACTH secretion, which then acts to evoke release of pinocytotic vesicles containing corticosterone by activation of NO. NO activates cyclooxygenase-1, which produces PGE_2, evoking exocytosis of corticosterone-containing vesicles.[68]

Toll-like receptors (TLRs) are located in many organs including the adrenal cortex, and TLR-2–deficient mice have a deficient corticosterone response to inflammatory stress, probably mediated by a decreased response to proinflammatory cytokines, which act directly on cortical cells to induce corticosterone release.[6]

Overwhelming infection or stress may cause massive production of NO in the adrenal, leading to excessive dilation of adrenal vessels and excessive permeability, leading to hemorrhage into the adrenal, which can lead to permanent adrenal insufficiency (Waterhouse Freidrichson's syndrome). In overwhelming infection massive production of NO and similar hemorrhage into the anterior pituitary gland may cause anterior pituitary gland insufficiency, known as Sheehan's syndrome.

Ascorbic Acid and Vitamin E

Ascorbic acid is synthesized by most mammals, the exceptions being man, the great ape, and the guinea pig. It is present in highest amount in the adrenal cortex followed by the corpus luteum and the brain. It and vitamin E are the principal antioxidants in the body. Ascorbic acid is present in far greater quantities than vitamin E and must be the major antioxidant in the body.

The pineal hormone melatonin releases hypothalamic ascorbic acid and LHRH presumably stored in synaptic vesicles by activation of NOS. The released NO activates GC, which in turn activates LHRH and ascorbic acid release by cGMP. Inhibitors of NOS and GC can block the release of both.[52]

Similarly, corticosterone and ascorbic acid are released together from the adrenal cortex following stimulation by ACTH released in response to stress. The concentration of ascorbic acid in the adrenal cortex is by far the highest in the body. Further study of its role in corticosterone release is urgently needed. Since NO and GC bring about ascorbic acid release from the hypothalamus, it is likely that they are similarly involved in ascorbic acid release from the adrenals.

Mitochondrial NOS

Recently a NOS has been shown to be associated with the inner mitochondrial membranes. It appears to be identical to nNOS.[8] If activated by infection or cell damage, it could produce mitochondrial damage and damage to other cell components, leading to cell death.[9] More work is necessary to determine the significance of mitochondrial NOS.

A significant 20% prolongation of life occurred in transgenic mice with overexpression of human catalase to the peroxisome, nucleus or mitochondria, supporting a pathophysiological role of mitochondrial NO inactivated by catalase.[69] Cardiac pathology and cataract formations were delayed and oxidative damage reduced and mitochondrial deletions reduced. This study provides direct support of the free radical theory of aging.

DISCUSSION

The data presented above indicate that there are many areas in the brain where there is regular periodic physiological release of NO throughout the life span. This probably occurs in the hippocampus, cerebellum, and, in particular, in the hypothalamus, in which NO controls most of the hypothalamic peptidergic neurons (such as CRH, LHRH, GHRH, somatostatin, oxytocin, and vasopressin) and also activates the release of GABA and inhibits that of norepinephrine and dopamine. It is not definite that this physiological release could ever reach levels that would produce neuronal cell damage; however, Schultz and colleagues[28] have reported that in men, but not women, over 75 years of age, there were neurodegenerative changes in the arcuate-median eminence region associated with an increase in neurofibrillary protein. This is the site of the interaction between NO and LHRH neurons.

The fact that injection of moderate amounts of LPS to mimic the effect of bacterial infection induces increased numbers of IL-1α immunoreactive neurons in the region of the thermosensitive neurons in the preoptic hypothalamic region, plus increased IL-β mRNA and iNOS mRNA in the PVN, AN, median eminence, choroid plexus, meninges, and in massive amounts in the anterior pituitary and pineal with consequent release of NO, suggests that toxic amounts of NO could exist in these regions during moderate infections, even though there is no direct involvement of the brain.

Destruction of neurons in the temperature-regulating centers and in the paraventricular and arcuate median-eminence region after multiple infections during the life span may be responsible for the reduced febrile response and pituitary hormone secretion in response to infection, respectively. Presumably, LPS acts on receptors in the choroid plexus and after its transport into the CSF on neuronal terminals in the wall of the third ventricle, which induce IL-1 and iNOS mRNA with resultant production of NO at their cell bodies in the PVN. Repeated bouts of infection over the life span of the individual, even without direct CNS involvement, could lead to neuronal cell loss in the hippocampus, because NO is important in memory formation. Cerebellar and cortical dysfunction could also ensue.

The evidence that NO is involved in a number of neurodegenerative diseases, among them Huntington's and Parkinson's disease, is already impressive. In chronic CNS infections such as AIDS, there would certainly be even greater responses in the aforementioned areas, plus increased iNOS in glia producing large amounts of NO. Indeed, CNS AIDS has led to Alzheimer-like changes in the brain.[58] Therefore, NO may cause much of the neuropathologic changes in CNS AIDS.

Even more impressive is the induction of iNOS in the pituitary after LPS, which, with age, could alter the responses of the pituitary to infection. The dramatic induction of iNOS mRNA in the pineal should lead to high concentrations of NO that could result in the death of pineal cells and reduction of melatonin secretion, leading to impaired resistance to free radicals normally scavenged by melatonin and acceleration of aging.

In humans, it is already well known, as indicated before, that infections with release of bacterial or viral products, such as LPS, causes the induction of cytokines, which are released and travel through the bloodstream. LPS and the released cytokines combine with their receptors on the coronary artery endothelial cells. They induce iNOS in the endothelial cells and in macrophages that might be adherent to or resident within the vessel. The result would be production of 1,000 times more NO than would be released by eNOS. NO would oxidize LDL and cause the production of prostaglandins, leukotrienes that are damaging to the vessel. NO itself would have toxic effects to bring about cell injury and death. There would be generation and enlargement of the atherosclerotic plaques, producing a rapid, downhill course of the coronary disease. It is also known that inflammation, as for example in severe osteoarthritis[65] (as in the patient described earlier), causes the induction of cytokines that would circulate to the coronary vessels and also induce iNOS.

Finally, in rats, it has been shown that stress itself, even without tissue damage, can cause the induction of nNOS in the same areas that have been studied in the case of iNOS induction by LPS,[55] and therefore, presumably also in the vascular system, although this has not yet been studied. Also, removal of blood from conscious rats produced a dramatic elevation of TNF-α that might induce NOS.[53] Relatively large concentrations of NO_2-NO_3 with a diurnal rhythm that parallels that of leptin have been described that increases with fat mass, suggesting that these could be toxic at least in obese subjects. Whether or not psychological or physical stress can cause induction of NOS in the coronary endothelium of humans has not been determined, but it is a well-known fact that stress predisposes to CHD and myocardial infarction. In fact, executives who fired their employees were twice as prone to have a heart attack around that time as on ordinary days. Even if stress-induced heart attacks are not directly caused by NO, they may be caused by increased vasoconstriction associated with the stress-induced withdrawal of NOergic vasodilator tone or augmented adrenergic vasoconstrictor tone. Further studies are needed to determine which of these possibilities is correct in this case; however, the evidence is rapidly mounting that the final mediator of the effects of inflammation and infection on the coronaries is the massive amounts of NO released by iNOS that cause a rapid progression of CHD, leading to the development of angina pectoris and finally myocardial infarction. It appears that the triad of stress, infection, and inflammation are the main factors that precipitate rapid deterioration of the coronary vessels mediated in large part by NO. NO can cause CHD even in the presence of a normal lipid profile.

OTHER ORGAN SYSTEMS AND THERAPEUTIC IMPLICATIONS

Space does not allow development of this hypothesis in other organ systems, but quite clearly it is probable that the aging effects of sunlight on the skin are also mediated by the inflammatory response and production of massive amounts of NO. Similarly, infections could mediate aging changes in the gonads, digestive system, and every other organ of the body. The reduction in incidence of infectious disease via public health and sanitation measures, from immunization and from their successful treatment with chemo- and antibiotic therapy, may account for the increased longevity in developed countries by reducing exposure to toxic concentrations of NO.

In conclusion, although much work needs to be done, it is already known that treatment of patients with antioxidants, vitamins C and E, which would reduce the toxic effects of NO, is of value in patients with CHD. This is probably the mechanism of their protective effects against CHD. Melatonin, as indicated above, is a naturally occurring antioxidant that has been shown to increase the life span of mice. Finally, compounds that inhibit the production of NO directly, such as inhibitors of NOS or agents that inhibit the production of NOS, such as corticoids, the tetracyclines, and α-MSH, may prove useful in slowing the aging process. Aspirin blocks cyclooxygenase 1, thereby reducing production and toxicity of prostanoids produced by NO, accounting for its protective effect in CHD. It may even be beneficial to decrease the production of NO in infections by the use of inhibitors of NOS, such as NAME, or if they can be developed, specifically of iNOS.

REFERENCES

1. KIRKWOOD, T.B.L. & A. KOWALD. 1997. Network theory of aging. Exp. Gerontol. **32:** 395–399.
2. McCANN, S.M., J. LICINIO, M.L. WONG, *et al.* 1997. The nitric oxide hypothesis of brain aging. Exp. Gerontol. **32:** 431–440.
3. WONG, M.L., V. RETTORI, A. AL-SHEKHELEE, *et al.* 1996. Inducible nitric oxide synthase gene expression in the brain during systemic inflammation. Nat. Med. **2:** 581–584.
4. McCANN, S.M. & V. RETTORI. 1996. The role of nitric oxide in reproduction. Proc. Soc. Exp. Biol. Med. **211:** 7–15.
5. McDONALD, L.J. & F. MURAD. 1996. Nitric oxide and cyclic GMP signaling. Proc. Soc. Exp. Biol. Med. **211:** 1–6.
6. BORNSTEIN, S.R., R.R. SCHUMANN, V. RETTORI, *et al.* 2004. Toll-like Receptor 2 and toll-like receptor 4 expression in human adrenals. Horm. Metab. Res. **36:** 470–473.
7. BREDT, D.S., P.M. HWANG, S.H. SNYDER. 1990. Localization of nitric oxide synthase I indicating a neural role for nitric oxide. Nature **347:** 768–770.
8. GIUVILI, C., J.J. PODEROSO & A. BOVERIS. 1998. Production of nitric oxide by mitochondria. J. Biol. Chem. **273:** 11038–11043.
9. NAVARRO, A., M.J. SÁNCHEZ DEL PINO, C. GÓMEZ, *et al.* 2002. Behavioral dysfunction, brain oxidative stress, and impaired mitochondrial electron transfer in aging mice. Am. J. Physiol. Regul. Integr. Comp. Physiol. **282:** 985–992.
10. PALMER, R.M.J., A.G. FERRIGE & S. MONCADA. 1987. Nitric oxide release accounts for the biological activity of endothelium-derived relaxing factor. Nature **327:** 524–526.
11. GARTHWAITE, J., S.L. CHARLES & R. CHESS-WILLIAMS. 1988. Endothelium-derived relaxing factor release on activation of NMDA receptors suggests role as intercellular messenger in the brain. Nature **336:** 385–388.
12. KARANTH, S., K. LYSON, & S.M. McCANN. 1993. Role of nitric oxide in interleukin 2-induced corticotropin releasing factor release from incubated hypothalami. Proc. Natl. Acad. Sci. USA **90:** 3383–3387.
13. RETTORI, V., M. GIMENO, K. LYSON & S.M. McCANN. 1992. Nitric oxide mediates norepinephrine-induced prostaglandin E2 release from the hypothalamus. Proc. Natl. Acad. Sci. USA **89:** 11543–11546.
14. CANTEROS, G., V. RETTORI, A. FRANCHI, *et al.* 1995. Ethanol inhibits luteinizing hormone releasing hormone (LHRH) secretion by blocking the response of LHRH neuronal terminals to nitric oxide. Proc. Natl. Acad. Sci. USA **92:** 3416–3420.
15. HINEY, J.K., W.H. SOWER, W.H. YU, *et al.* 2002. Gonadotropin releasing hormone neurons in the preoptic-hypothalamic region of the rat contain lamprey gonadotropin releasing hormone III, mammalian luteinizing hormone-releasing hormone, or both peptides. Proc. Natl. Acad. Sci. USA **99:** 2386–2391.
16. McCANN, S.M., S. KARANTH, C.A. MASTRONARDI, *et al.* 2002. Hypothalamic control of gonadotrophin secretion. Prog. Brain Res. **141:** 153–166.
17. KAMAT, A., W.H. YU, V. RETTORI & S.M. McCANN. 1995. Glutamic acid stimulated luteinizing-hormone releasing hormone release is mediated by alpha adrenergic stimulation of nitric oxide release. Brain Res. Bull. **37:** 233–235.
18. RETTORI, V., N. BELOVA, W.L. DEES, *et al.* 1993. Role of nitric oxide in the control of luteinizing hormone-releasing hormone release *in vivo* and *in vitro*. Proc. Natl. Acad. Sci. USA **90:** 10130–10134.
19. MANI, S.K., J.M.C. ALLEN, V. RETTORI, *et al.* 1994. Nitric oxide mediates sexual behavior in female rats by stimulating LHRH release. Proc. Natl. Acad. Sci. USA **91:** 6468–6472.
20. McCANN, S.M., C. MASTRONARDI, A. WALCZEWSKA, *et al.* 2003. The role of NO in the control of LHRH release that mediates gonadotrophin release and sexual behavior. Curr. Pharmacol. Des. **9:** 381–390.
21. RETTORI, V., G. CANTEROS, R. REYNOSO, *et al.* 1997. Oxytocin stimulates the release of luteinizing hormone–releasing hormone from medial basal hypothalamic explants by releasing nitric oxide. Proc. Natl. Acad. Sci. USA **94:** 2741–2744.
22. SEILICOVICH, A., B.H. DUVILANSKI, D. PISERA, *et al.* 1995. Nitric oxide inhibits hypothalamic luteinizing hormone–releasing hormone release by releasing γ-aminobutyric acid. Proc. Natl. Acad. Sci. USA **92:** 3421–3424.

23. SEILICOVICH, A., M. LASAGA, M. BEFUMO, *et al.* 1995. Nitric oxide inhibits the release of norepinephrine and dopamine from the medial basal hypothalamus of the rat. Proc. Natl. Acad. Sci. USA **92:** 11299–11302.
24. RETTORI, V., N. BELOVA, W.H. YU, *et al.* 1994. Role of nitric oxide in control of growth hormone release in the rat. Neuroimmunomodulation **1:** 195–200.
25. AGUILA, M.C. 1994. Growth hormone-releasing factor increases somatostatin release and mRNA levels in the rat periventricular nucleus via nitric oxide by activation of guanylate cyclase. Proc. Natl. Acad. Sci. USA **91:** 782–786.
26. RETTORI, V., N. BELOVA, M. GIMENO & S.M. MCCANN. 1994. Inhibition of nitric oxide synthase in the hypothalamus blocks the increase in plasma prolactin induced by intraventricular injection of interleukin-1a in the rat. Neuroimmunomodulation **1:** 116–120.
27. DUVILANSKI, B.H., C. ZAMBRUNO, A. SEILICOVICH, *et al.* 1995. Role of nitric oxide in control of prolactin release by the adenohypophysis. Proc. Natl. Acad. Sci. USA **92:** 170–174.
28. SCHULTZ, C., H. BRAAK & E. BRAAK. 1997. A sex difference in neurodegeneration of the human hypothalamus. Abstract presented at the 3rd International Symposium on Neurobiology and Neuroendocrinology of Aging. Bregenz, Austria.
29. SAGAR, S.M. & D.M. FERREIRO. 1987. NADPH diaphorase activity in the posterior pituitary: relation to neuronal function. Brain Res. **400:** 348–352.
30. CALKA, J. & C.H. BLOCK. 1993. Relationship of vasopressin with NADPH-diaphorase in the hypothalamo-neurohypophyseal system. Brain Res. Bull. **32:** 207–210.
31. MIYAGAWA, A., H. OKAMURA & Y. IBATA. 1994. Coexistence of oxytocin and NADPH-diaphorase in magnocellular neurons of the paraventricular and supraoptic nuclei of the rat hypothalamus. Neurosci. Lett. **171:** 13–16.
32. KADEKARO, M., M.L. TERRELL, H. LIU, *et al.* 1998. Effects of L-NAME on cerebral, vasopressin, oxytocin, and blood pressure responses in hemorrhaged rats. Am. J. Physiol. **274:** 1070–1077.
33. LIU, Q.S., Y. JIA & G. JU. 1997. Nitric oxide inhibits neuronal activity in the supraoptic nucleus of the rat hypothalamic slices. Brain Res. Bull. **43:** 121–125.
34. NUSSDORFER, G.G. & L.K. MALENDOWICZ. 1998. Role of tachykinins in the regulation of the hypothalamo-pituitary-adrenal axis. Peptides **19:** 949–968.
35. EUTAMENE, H., V. THEODOROU, J. FIORAMONTI & L. BUENO. 1995. Implication of NK1 and NK2 receptors in rat colonic hypersecretion induced by interleukin 1 beta: role of nitric oxide. Gastroenterology **109:** 483–489.
36. FIGGINI, M., C. EMANUELI, C. BERTRAND, *et al.* 1996. Evidence that tachykinins relax the guinea pig trachea via nitric oxide release and by stimulation of a peptide-insensitive NK1 receptor. Br. J. Pharmacol. **117:** 1270–1276.
37. PORTBURY, A.L., J.B. FURNESS, B.R. SOUTHWELL, *et al.* 1996. Distribution of neurokinin-2 receptors in the guinea pig gastrointestinal tract. Cell Tissue Res. **286:** 281–292.
38. DE LAURENTIIS, A., D. PISERA, B. DUVILANSKI, *et al.* 2000. Neurokinin A inhibits oxytocin and GABA release from the posterior pituitary by stimulating nitric oxide synthase. Brain Res. Bull. **53:** 325–330.
39. EISTETTER, H.R., A. MILLS & S.J. ARKINSTALL. 1993. Signal transduction mechanisms of recombinant bovine neurokinin-2 receptor stably expressed in baby hamster kidney cells. J. Cell. Biochem. **52:** 84–91.
40. AHERN, G.P., S.F. HSU & M.B. JACKSON. 1999. Direct actions of nitric oxide on rat neurohypophysial K^+ channels. J. Physiol. Lond. **520:** 165–176.
41. THEODOSIS, D.T. & B. MACVICAR. 1996. Neurone-glia interactions in the hypothalamus and pituitary. Trends Neurosci. **19:** 363–367.
42. VIZI, E.S. & B. SPERLAGH. 1999. Separation of carrier mediated and vesicular release of GABA from rat brain slices. Neurochem. Int. **34:** 407–413.
43. CROWLEY W. & W. ARMSTRONG. 1992. Neurochemical regulation of oxytocin secretion in lactation. Endocr. Rev. **13:** 33–65.
44. SCHNEGGENBURGER, R. & A. KONNERTH. 1992. GABA-mediated synaptic transmission in neuroendocrine cells: a patch-clamp study in a pituitary slice preparation. Pflugers Arch. **421:** 346–373.

45. SLADEK, C. & W. ARMSTRONG. 1987. Gamma-aminobutyric acid antagonists stimulate vasopressin release from organ cultured hypothalamo-neurohypophyseal explants. Endocrinology **120:** 1576–1580.
46. PISERA, D., A. DE LAURENTIIS, B. DUVILANSKI, *et al.* 1996. Neurokinin A affects the tubero-hypophyseal GABAergic system. Neuroreport **7:** 2236–2240.
47. YU, W.H., M. KIMURA, A. WALCZEWSKA, *et al.* 1998. Adenosine acts by A1 receptors to stimulate release of prolactin from anterior-pituitaries *in vitro*. Proc. Natl. Acad. Sci. USA **95:** 7795–7798.
48. YU, W.H., A. WALCZEWSKA, S. KARANTH & S.M. MCCANN. 1997. Nitric oxide mediates leptin-induced luteinizing hormone releasing hormone (LHRH) and LHRH and leptin-induced LH release from the pituitary gland. Endocrinology **138:** 5055–5058.
49. RETTORI, V., W.L. DEES, J.K. HINEY, *et al.* 1994. An interleukin-1a-like neuronal system in the preoptic-hypothalamic region and its induction by bacterial lipopolysaccharide in concentrations which alter pituitary hormone release. Neuroimmunomodulation **1:** 251–258.
50. BANKS, W.A., A.J. KASTIN & R.D. BROADWELL. 1995. Passage of cytokines across the blood–brain barrier. Neuroimmunomodulation **2:** 241–248.
51. PIERPAOLI, W., D. BULIAN, A. DALL'ARA, *et al.* 1997. Circadian melatonin and young-to-old pineal grafting postpone aging and maintain juvenile conditions of reproductive functions in mice and rats. Exp. Gerontol. **32:** 587–602.
52. KARANTH, S., W.H. YU, C.A. MASTRONARDI & S.M. MCCANN. 2004. Inhibition of stimulated ascorbic acid and luteinizing hormone-releasing hormone by nitric oxide synthase or guanylyl cyclase inhibitors. Exp. Biol. Med. **229:** 72–79.
53. MASTRONARDI, C.A., W.H. YU & S.M. MCCANN. 2001. Comparisons of the effects of anesthesia and stress on release of tumor necrosis factor–alpha, leptin, and nitric oxide in adult male rats. Exp. Biol. Med. **226:** 296–300.
54. MASTRONARDI, C.A., W.H. YU, S.M. MCCANN. 2001. Lipopolysaccharide-induced tumor necrosis factor-alpha release is controlled by the central nervous system. Neuroimmunomodulation **9:** 148–156.
55. KISHIMOTO, J., T. TSUCHIYA, P.C. EMSON & Y. NAKAYAMA. 1996. Immobilization-induced stress activates neuronal nitric oxide synthase (nNOS) mRNA and protein in hypothalamic–pituitary–adrenal axis in rats. Brain Res. **720:** 159–171.
56. PRZEDBORSKI S., V. JACKSON-LEWIS, R. YOKOYAMA, *et al.* 1996. Role of neuronal NO in 1-methyl-4-phenyl-1,2,3,4-tetrahydropyridine (MPTP)-induced dopaminergic toxicity. Proc. Natl. Acad. Sci. USA **93:** 4565–4571.
57. KNOLL, J. 1997. Sexual performance and longevity. Exp. Gerontol. **32:** 539–552.
58. GRIFFIN, S. 1998. Neuroimmunomodulation Symposium. Am. Assoc. Immunol. Meet. San Francisco. April, 1998.
59. XIAO-JIANG, L., A.H. SHARP, L. SHI-HUA, *et al.* 1996. Huntington-associated protein (HAPI): discrete neuronal localizations in the brain resemble those of neuronal nitric oxide synthase. Proc. Natl. Acad. Sci. USA **93:** 4839–4844.
60. WONG, M.L., F. O'KIRWAN, N. KHAN, *et al.* 2003. Identification, characterization, and gene expression profiling of endotoxin-induced myocarditis. Proc. Natl. Acad. Sci. USA **100:** 14241–14246.
61. DANESH, J., R. COLLINS & R. PETO. 1997. Chronic infections and coronary heart disease: is there a link? Lancet **350:** 430–436.
62. FABRICANT, C.G., J. FABRICANT, M.M. LITRENTA & C.R. MINICK. 1978. Virus-induced atherosclerosis. J. Exp. Med. **148:** 335–340.
63. GURFINKEL, E., G. BOZOVICH, A. DAROCA, *et al.* 1997. For the ROXIS Study Group. Randomised trial of roxithromycin in non-Q-wave coronary syndromes: ROXIS pilot study. Lancet **350:** 404–407.
64. GUPTA, S., E.W. LEATHAM, D. CARRINGTON, *et al.* 1997. Elevated *Chlamydia pneumoniae* antibodies, cardiovascular events, and azithromycin in the male survivors of myocardial infarction. Circulation **96:** 404–407.
65. AMIN, A.R., M.G. ATTUR, G.D. THAKKER, *et al.* 1996. A novel mechanism of action of tetracyclines: effects on nitric oxide synthases. Proc. Natl. Acad. Sci. USA **93:** 14014–14019.

66. MASTRONARDI, C.A., W.H. YU, V.K. SRIVASTAVA, *et al.* 2001. Lipopolysaccharide-induced leptin release is neurally controlled. Proc. Natl. Acad. Sci. USA **98:** 14720–14725.
67. MASTRONARDI, C.A., W.H. YU & S.M. MCCANN. 2002. Resting and circadian release of nitric oxide is controlled by leptin in male rats. Proc. Natl. Acad. Sci. USA **99:** 5721–5726.
68. MOHN, C., J. FERNANDEZ SOLARI, A. DE LAURENTIIS, *et al.* 2005. The rapid release of corticosterone from the adrenal induced by ACTH is mediated by NO acting by PGE2. Proc. Natl. Acad. Sci. USA **102:** 6213–6218.
69. SCHRINER, S.E., N.J. LINFORD, G.M. MARTIN, *et al.* 2005. Extension of murine lifespan by overexpression of catalase targeted to mitochondria. Science **308:** 1909–1911.

Immunoregulation of Cellular Life Span

MARGUERITE KAY

Institute of Biomedical Sciences and Technology, University of Texas at Dallas, Richardson, Texas 75083-0688, USA, and

The International Foundation for Biomedical Aging Research, Temple, Texas 76503 USA

ABSTRACT: Our current studies focus on the molecular changes induced by aging. During aging, changes in proteins occur that alter their function and render them immunogenic. These "neoantigens" are recognized by physiologic autoantibodies. Physiologic autoantibodies and their corresponding antigens offer therapeutic strategies for disease intervention through the innate immune response. Early studies done in the 1970s showed humans and animals to have physiologic antibodies that bind to a neoantigen called *senescent cell antigen* (SCA), which appears on senescent and damaged cells and initiates their removal by macrophages. These studies led to the discovery that oxidation can generate a new antigen *in situ*. Oxidation accelerated aging of red cells, generated SCA and IgG binding, and triggered removal of red cells by macrophages. Since then, a number of laboratories have found that oxidation can generate other neoantigens. For example, oxidized LDL (OxLDL) induces antibodies that can modify the natural progression of atherosclerosis. Apoptotic cells express oxidatively modified moieties on their surfaces that are involved in macrophage recognition and phagocytosis. Physiologic autoantibodies were used to isolate SCA from brain tissue. HPLC and fast atom bombardment ionization mass spectrometry (FAB-MS) of the isolated antigen suggested that the aging antigen is a subset of band 3, a family of proteins also called *anion exchange proteins* (AE1-3). FAB-MS results indicate that residues matching all three band 3 isoforms (AE1, AE2, and AE3) are detected in aging antigen fractions. Among the fragments identified with FAB-MS was a sequence corresponding to an aging epitope, human band 3 sequence LFKPPKYHPDVPYVKR, residue 812–830 in AE1; HHPDVTYVK, residue 1144–1152 in AE2; or HH-PEQPYVTK, residue 1135–1144 in AE3. A residue that is close to that region was identified in mouse AE1 ASGPGAAAQIQEVK, residue 762–775. The potential for altering the natural progression of diseases using select peptide-defined epitopes within or overlapping the aging antigenic site (547–553 and 824–829) is discussed using the innate immune response to band 3 in malaria as an example.

KEYWORDS: physiologic autoantibodies; cellular aging; anion exchange proteins; band 3; oxidation; neoantigen; malaria; senescent cell antigen

Address for correspondence: Marguerite Kay, Institute of Biomedical Sciences and Technology, University of Texas at Dallas, P.O. Box 830688, Richardson, Texas 75083-0688. Voice: 972-883-4872; fax: 972-883-2409.

margueritekay@mail.com

Ann. N.Y. Acad. Sci. 1057: 85–111 (2005). © 2005 New York Academy of Sciences.

doi: 10.1196/annals.1356.005

PRELUDE AND DEDICATION

You see things; and you say, "Why?" But I dream
things that never were; and I say, "Why not?"
GEORGE BERNARD SHAW, *Back to Methuselah*

"When I am an old woman, I shall wear purple ..."[a] This is the first line of a poem by Jenny Joseph that exhorts the old to do as they please. A very dear and good friend of mine, an eccentric member of the Nuveen family, wears purple every day. I think that I am now a "grand dame" of aging. Next conference, I shall wear purple ....

I value this Fourth Stromboli Conference more for the discussions and personal interactions than for the papers presented at the sessions. The presentations were, as usual, spectacular, interesting, and deserve all of the positive accolades you can imagine. Every one did a great job. The most important interactions, however, occurred during lunch, dinner, coffee break .... I appreciated catching up on the lives of some, getting to know people better, and meeting new friends, and...Walter is to be congratulated again for yet another conference, for bringing such an interesting and diverse group together, for being a gracious host, and the too-numerous-to-mention things at which Walter excels—the excellence stemming from his unbridled enthusiasm for the topic and his innate intellectual curiosity.

One of the major contributions of aging research to medicine and science is the demonstration of a neuroendocrine-hypothalamic axis (reviewed in refs. 1–3). The concept of a "master" gland or glands functioning as an aging clock was another major contribution to the biology and immunobiology of aging. Today, the existence of a neuroendocrine axis under hypothalamic control is accepted as a fact. Walter, Bill Regelson, and Takashi Makinodan were the pioneers who discerned this knowledge. Another major contribution of aging research to medicine and science is the demonstration of physiologic autoantibodies. This also "pushed the envelope" and was considered "heresy."[4] Physiologic autoantibodies are part of our normal immune repertoire. They function to maintain homeostasis by performing physiologic functions. The role of physiologic autoantibodies in removing senescent and damaged cells is probably the best example of a physiologic autoantibody, complete with well-established function. IgG autoantibodies bind to altered band 3 anion exchanger protein on senescent cells and trigger their removal by macrophages. Band 3 isoforms are found in all cells, tissues, and membranes, and in all species examined. Some of this will be discussed later in this paper.

Ron Klatz asked me an interesting and insightful question when we were biding time before departure. He said that Bill Regelson told him that the original intent of the creators of the National Institute on Aging (NIA) was to do what the American Academy of Anti-Aging Medicine (A4M) is now trying to do. He asked me if I thought that was true. I pondered for a moment, ... thought back all those years to the beginning of this dream, when a group of us broke off from National Institute of Child Health and Human Development (NICHD) to form the new institute of aging, the NIA, many years ago. We believed that we could change things ... the suffering, the accumulation of losses, many of the unnecessary negative medical and physical

---

[a]See <www.wheniamanoldwoman.com/>.

changes that accompany aging. Our goal was to extend productive life span. We believed. We believed at a time when aging research was not popular, when no one believed that we could do anything that made a difference. We proselytized. We evangelized. I, who do not like to travel, traveled to spread the word. There was no internal medicine subspecialty in geriatrics. People told me that there would never be. They patiently explained to me that geriatrics was not like pediatrics. There were no basic differences between the middle-aged and the elderly medically speaking (yes, they really said that). Soooo, I spent my time trying to convince them that there were differences by presenting the data, reviewing the work, and doing the research that showed that there *were* differences and that many of the changes associated with aging were reversible or at least could be delayed. I taught house staff, I taught medical students. As a result of my research, I knew what to look for and could suspect and diagnose infections based on subtle changes in the activities of daily life. Makinodan's and my work clearly showed that the immune system could be reconstituted with stem cells and hormones. There was no possibility of applying this knowledge to humans at that time. Therefore, I moved on to studying molecular and cellular aging and senescent cell antigens. So, yes, as I remember, the people who formed the NIA intended to do what A4M is now trying to do including developing and applying interventions that change the progression of aging and prevent or delay diseases associated with aging.

Regelson—creative, visionary, and persistent—strived toward that dream of extending health. He believed.

So what happened? Did the NIA lose the vision? Were they sidetracked? Was it usurped by more conventional thoughts and activities?

Fortunately, there are others who are carrying the dream forward—the next lap of the relay. A4M is now trying to accomplish the original goal. As this conference has demonstated interventions that can be put into daily practice can truly make a difference in people's lives.

This paper remembers and honors some of the early dreamers and major contributors to aging and anti-aging research without whom the dream would not have moved forward: Tuck Finch, Joseph Goodman, Takashi Makinodan, Walter Pierpaoli, Bill Regelson, Bernie Strehler, Anddrus Viidik, Roy Walford, and all the others that I have not mentioned. I have known these people, talked and laughed with them at conferences in out-of-the-way places, and, eventually, at the Gordon Conferences on Aging, when this work became more acceptable and "mainstream."

So, next time we meet, I shall wear purple ....

As a salute to Bill (Regelson) and Joe (Goodman), this paper is also dedicated to the man who made the statement "All that is necessary for the triumph of evil is that good men do nothing."[b] And it is dedicated to Dietrich Bonhoeffer, author of *The Cost of Discipleship*, who was hanged on April 9, 1945 in the Flossenbürg concentration camp; his last known words were, "This is the end—for me, the beginning of life."[c] And it is dedicated to Martin Niemoller, to whom these words are attributed:

First, they came for the socialists, and I did not speak out because I was not a socialist.
Then they came for the trade unionists, and I did not speak out because I was not a trade

---

[b]The origin and exact wording of this sentence are unknown; it is attributed to Edmund Burke.
[c]See, e.g., <http://www.forerunner.com/forerunner/X0528_Bios-_Martin_Niemlle.html>.

unionist. Then they came for the Jews, and I did not speak out because I was not a Jew. Then they came for me, and there was no one left to speak for me.[d]

These statements, the latter of which stayed with me since I first read them as a 12-year-old, have had a tremendous influence on my life and actions, as those who are acquainted with my travails can attest. Bill and Joe understood these influences. Both were defined by a strong sense of right and wrong and opposition to evil.

I was acutely aware of Bill's absence at this conference because of his large presence at the others. Thus, I dedicate this paper to Bill Regelson and Joseph R. Goodman,[5] two men of great personal convictions who made a difference in many lives, and of whom these lines seem appropriate:

> These hearts were woven of human joys and cares,
> Washed marvellously with sorrow, swift to mirth.
> The years had given them kindness. Dawn was theirs,
> And sunset, and the colours of the earth.
> These had seen movement, and heard music; known
> Slumber and waking; loved; gone proudly friended;
> Felt the quick stir of wonder; sat alone;
> Touched flowers and furs and cheeks. All this is ended.
>
> There are waters blown by changing winds to laughter
> And lit by the rich skies, all day. And after,
> Frost, with a gesture, stays the waves that dance
> And wandering loveliness. He leaves a white
> Unbroken glory, a gathered radiance,
> A width, a shining peace, under the night.

RUPERT BROOKE, *IV. The Dead*

INTRODUCTION

All individuals have physiologic antibodies to band 3/SCA that bind to senescent and damaged cells and initiate their removal by macrophages[5–29] (for a recent review of the innate immune response to band 3, see Kay and Goodman[29]). These physiologic autoantibodies are part of the primitive innate immune network that operates in the background to maintain homeostasis. Antibody response to the antigenic residues of band 3 are found at least as far back in evolution as the lamprey.[30]

This article is a brief survey of work on senescent cell antigen and band 3. Details and more extensive reference lists are available in other publications.[29]

MATERIALS AND METHODS

Materials and methods are described in previous papers.[30]

---

[d]The origin and exact wording of this paragraph are unknown. See <http://internet.ggu.edu/university_library/if/Niemollertml#note>.

RESULTS AND DISCUSSION

Binding of physiologic immunoglobulin (Ig) G autoantibodies *in vivo* to senescent cells is a trigger mechanism for their removal by macrophages through the process of phagocytosis (FIGS. 1–3).[5–29] For a review of physiologic autoantibodies to red cells and band 3, see Kay and Goodman.[29]

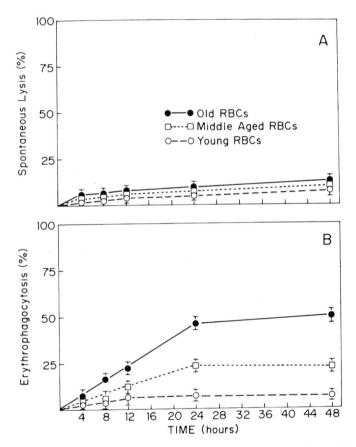

FIGURE 1. Erythrophagocytosis *in vitro*: splenic macrophages phagocytize significantly more senescent red cells than young or middle-aged red cells *in vitro* (**B**, $P \leq 0.001$). Spontaneous lysis (%) of young, middle-aged, or old red cells during the same culture period are shown in **A** for comparison. Freshly obtained red cells were separated on calibrated Percoll gradients and incubated with syngeneic macrophages at 37°C for 48 hours in humidified air containing 5% CO_2. Aliquots of 10 μL were counted at various times after initiation of culture as indicated. The number of red cells phagocytized reached a plateau at about 24 hours of culture, suggesting metabolic limitations to the culture technique. Since syngeneic cells were used for all of these studies, the only difference was the age of the cell. (Reproducted from Bennett and Kay[11] with permission.)

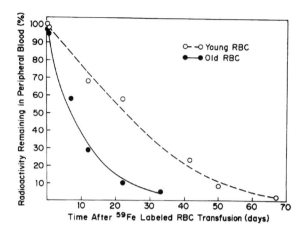

FIGURE 2. Erythrophagocytosis *in vivo* I: Senescent red cells ($P \leq 0.001$) are removed more rapidly from the circulation than young red cells as determined by the decrease of circulating ^{59}Fe-labeled cells in mice. ^{59}Fe-labeled young and old red cells were isolated on Percoll gradients and transfused into syngeneic mice as described elsewhere.[11] Standard deviations for each point are less than one. (Reproducted from Bennett and Kay[11] with permission.)

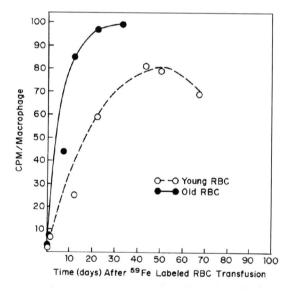

FIGURE 3. Erythrophagocytosis *in vivo* II: ^{59}Fe-labeled senescent red cells appear more rapidly within macrophages following transfusion than do young red cells ($P \leq 0.01$). ^{59}Fe-labeled young or old red cells isolated on Percoll gradients were injected intravenously into BALB/C mice. At time intervals thereafter, spleens were removed and spleen cell suspensions separated on calibrated Percoll gradients as described previously.[11] Macrophages are isolated from the least dense fraction and red cells from the densest.[11] The radioactivity was inside the macrophages because it could be recovered only by lysis of the cells. (Reproducted from Bennett and Kay[11] with permission.)

Neoantigen Appears on Old Cells

Binding of a physiologic autoantibody to senescent cells indicates that a new antigen appears.[5–11,17] This autoantibody binds via the Fab region and induces phagocytosis of cells by autologous macrophages *in vitro* and *in situ*[8] (FIGS. 4 and 5, ref. 8). Autoantibodies isolated from senescent cells were used to identify its antigen. Investigations into that "neoantigen" revealed that it was a ~62-kDa protein, and it was named *senescent cell antigen* (SCA) (FIG. 6).[10,16–18]

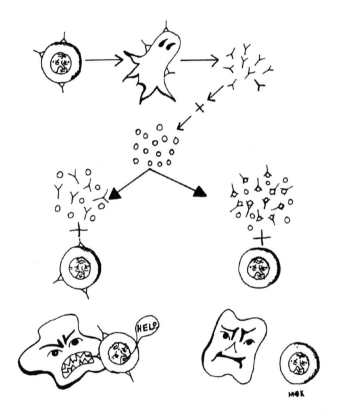

FIGURE 4. Cartoon depicting the phagocytosis inhibition assay as used in these studies. Starting in the **upper left-hand corner**, freshly isolated old cells with IgG on their surface (shown as "Ys" bound to the surface) are lysed, generating "ghosts." The senescent cell IgG (SCIgG) is eluted using glycine-HCl, pH 2.3, a method that was originally used to isolate Rh antibodies. SCIgG is incubated with a molecule or cell to determine if it carries the antigenic determinants recognized by SCIgG. If the molecule or cell carries SCA epitopes, IgG binds via its Fab region to the antigen (**right side**). Since the antigen binding region of the IgG is occupied, it is not free to bind to the target cell when mixed with it. Therefore, the target red cell is not phagocytized when mixed with macrophages (**lower right**). On the other hand, if the molecule or cell that is mixed with the SCIgG does *not* carry SCA antigenic determinants, the antigen binding region of the IgG is *not* occupied and is therefore free to bind to the target red cell (**lower left**). The IgG-bound red cells are phagocytized by macrophages. (Reproduced from Kay[32] with permission.)

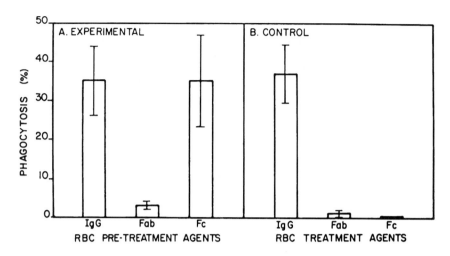

FIGURE 5. The autoantibody to senescent cell (SC) IgG binds to old cells via its antigen binding fragment (Fab) region and induces phagocytosis of cells by autologous macrophages *in vitro* and *in situ*.[8] A modified version of the phagocytosis inhibition assay described in FIGURE 4 was used for these studies. Pretreatment of target cells with SCIgG Fab prior to their incubation with SCIgG blocks the intact IgG binding; and as a result no phagocytosis occurs when the cells are mixed with macrophages (**A**). This is because the macrophages bind to the Fc (fragment crystalline) domain of IgG that was removed. Preincubation with the Fc fragment that does not bind to target RBCs does not interfere with phagocytosis. Preincubation with the whole IgG molecule containing both the Fab and Fc fragments hooked together results in phagocytosis. (**B**) Controls show that IgG by itself induces phagocytosis, whereas the Fab and Fc fragments alone do not. Autologous cells and antibody were used for these and the earlier experiments to avoid antigenic differences. (Reproduced from Kay[8] with permission.)

SCA Is Present on Non-Erythroid Cells

Since other cells are removed from the body at the end of their life span, we examined a number of cell types for the presence of SCA following forced aging *in vitro* (FIG. 6, TABLE 1). SCA was found on all cells and tissues examined including platelets, lymphocytes, neutrophils, liver, and kidney.[10]

SCA Is Derived from Band 3

Band 3, also called *anion exchange* (AE) *protein*, is a ubiquitous family of proteins that is widely distributed in the membranes of diverse cell types[34–39] as well as in subcellular organelles, such as nuclei,[34] mitochondria,[40] and Golgi.[41] Identification of SCA as band 3 was based upon immunological reactions, inhibition assays, and two-dimensional peptide mapping.[42,43] In a functional assay, the presence of free senescent cell antigen purified from old red cells or synthetic peptides of band 3 inhibited binding of physiologic autoantibodies to old cells and their phagocytosis by autologous macrophages.[44] Later studies described below and those done in many other laboratories confirm that SCA is from band 3.[23,29,45–50]

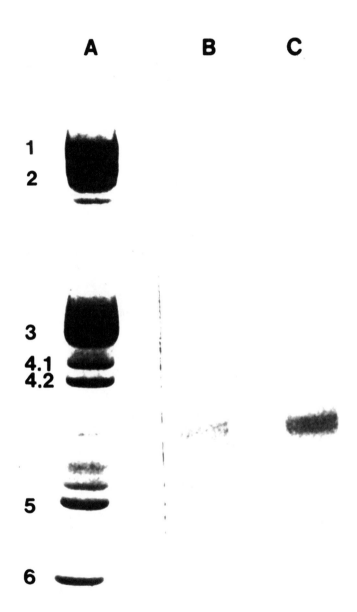

FIGURE 6. Senescent cell antigen isolated from red cells and lymphocytes. **(A)** RBC membranes; **(B)** ~62-kDa polypeptide isolated with SCIgG affinity column from red cell membranes; **(C)** ~62-kDa polypeptide isolated with SCIgG affinity column from lymphocytes. Lymphocytes (8×10^8) were isolated from 500 mL of blood by centrifugation on Percoll gradients. Cells were washed three times and lysed in hypotonic phosphate buffer. Membranes were obtained by differential centrifugation, washed, and solubilized; and senescent cell antigen was isolated with a column to which senescent cell IgG was covalently bound. (Reproduced from Kay[10] with permission.)

TABLE 1. SCA is present on nucleated cells

Cell set	Cell type used for absorption	Phagocytosis (%) (mean ± SD)
A	none	37 ± 4
	platelets	2 ± 2
	lymphocytes	0
	neutrophils	0
B	none	38 ± 6
	liver	0
	kidney	7 ± 7

NOTE: SCAs appeared on stored nucleated cells and platelets because they reduced or abolished the phagocytosis-inducing ability of IgG directed against them. SCIgG from senescent red cells was absorbed with each of the cell types or platelets listed and then tested with target red cells. Cells in set A were freshly obtained from human blood and then stored to induce SCA formation. Cells in set B were cultured cells. The percentage of phagocytosis was determined as described previously. (Modified from Kay,[10] and Kay and Goodman.[29])

TABLE 2. Functions of band 3 anion exchange proteins[a]

Ankyrin binding
Generation of SCA
Anion transport
Acid base balance
Cell volume regulation
Structural stability and integrity

[a]From Kay and Goodman.[29]

Band 3 performs crucial physiological functions (TABLE 2). Human erythrocyte band 3, a M_r ~92-kDa protein, exists in about 1.2×10^6 copies per red cell. This ubiquitous family of proteins participates in or is implicated in a number of physiological activities such as cell volume and osmotic homeostasis, HCO_3^-/Cl^- exchange, IgG binding and cellular removal, and the maintenance of the structural integrity of cells. Erythrocyte band 3 is the prototype for band 3 isoforms. More is known about its structure and function than about those of other isoforms. Band 3 malfunction has been associated with neurological disease, neoplastic transformation, renal disease, and cellular and organismal aging.

Band 3 alterations associated with aging and possible mechanisms of generation of SCA have been reviewed.[29] Changes in band 3 observed with aging include a decrease in efficiency of anion transport (decreased V_{max}), in spite of an increase in the number of anion binding sites (increased K_m), decrease in glucose transport, binding of antibodies to aged band 3, increase in band 3 degradation to smaller fragments as detected by quantitative binding of antibodies to band 3 breakdown products and amino acid residues 812–830, and *in situ* binding of physiological IgG autoantibodies resulting in cellular removal (TABLE 3).[7–19,29,51–56]

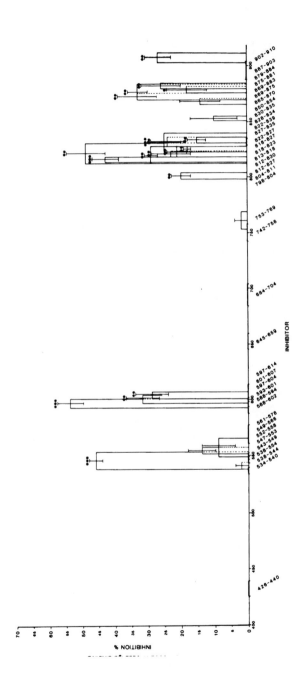

FIGURE 7. Mapping the aging antigenic site on the band 3 molecule. Synthetic peptides of erythroid band 3 were tested in the phagocytosis inhibition assay to determine which ones carried SCA epitopes. Immunoblotting studies showed binding to peptides $*P \leq 0.05$, $**P \leq 0.01$, $***P \leq 0.001$ compared to control without peptide. See previous publications[44,59,60,61] for amino acid sequences corresponding to the residue number. (Modified from Kay[61] with permission.)

TABLE 3. Summary of cellular aging changes[a]

Decreased anion transport
Decreased glucose transport
Generation of "aged," immunologically altered, band 3
Band 3 degradation
Appearance of senescent cell antigen
IgG binding
Phagocytosis of cell
Enzyme alterations

[a]"Biological profile" of an old cell, also called "red cell aging panel."[13] From Kay and Goodman.[29]

Band 3 Is Present on All Cells

Since SCA is present on non-erythroid cells, we examined other cells and tissues for band 3. Band 3 is present on all cells examined by a number of techniques including immunoflorescence, immunoelectronmicroscopy, and immunoblotting.[10,34] This has been confirmed by a number of laboratories.[38–41,57,58]

Identification of Antigenic Sites (Epitopes) Recognized by SCIgG

Senescent cell antigen contains band 3 residues 538–554 and 788–827 based on studies using peptide "walking" and binding assays (FIG. 7). [44,59-63] Two peptides combined, peptides 538–554 and 812–830, inhibited binding of the specific IgG autoantibody and phagocytosis of cells by macrophages (FIG. 8). [44] The two antigenic peptides 538–554 and 812–830 reside on regions of band 3 that are conserved across tissues and species (see, for example, refs. 30, 64). Antibodies to erythroid band 3 or its peptides react with band 3 isoforms in other cell types and other species including lamprey.[30,38–41,57,58,64]

Several band 3 models show these residues on the outside of the membrane or capable of accessing the outside.[44,59,65]

Senescent Cell Antigen and Band 3 Homology

We used high-pressure liquid chromatography (HPLC) and fast atom bombardment ionization mass spectrometry (FAB-MS) to compare SCA to band 3 obtained from brain to determine whether SCA was homologous to band 3 by a method independent of those previously employed, and to identify segments of band 3 included in SCA.[66]

For these studies, band 3 and SCA were isolated from brain since earlier studies showed that both band 3 and SCAs are present in all tissues. SCIgG from human red cells was used for isolating SCA.

HPLC and (FAB-MS) suggest that the aging antigen is a subset of band 3, and includes all band 3 isoforms (AE1-3). HPLC peptide maps of band 3 and SCA showed substantial homology, suggesting that SCA includes an estimated ≥45% of the band 3 molecule (FIG. 9). FAB-MS results indicate that residues matching all three band 3 isoforms (AE1, AE2, and AE3) are detected in SCA fractions (TABLE 4).[66] Among the fragments identified with FAB-MS was a sequence corresponding

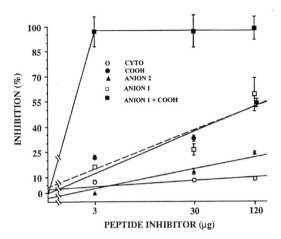

FIGURE 8. Senescent cell IgG binding to defined peptides of band 3 as determined by a peptide binding inhibition assay. A combination of peptide COOH (residues 812–827) and anion 1 (residues 538–554) showed synergy resulting in close to 100% inhibition of binding of SCIgG to target red cells. The other peptides are: CYTO (residues 129–144), anion 2 (residues 588–602); COOH (residues 812–827); anion 1 (residues 538–554). For an explanation of the assay, see FIGURE 4. In this study, binding was determined by radioactive protein A binding to the Fc region of SCIgG instead of by phagocytosis. (Modified from refs. 44, 59, and 61)

to an aging epitope, human band 3 sequence LFKPPKYHPDVPYVKR, residue 812–830 in AE1, HHPDVTYVK, residue 1144–1152 in AE2, or HHPEQPYVTK, residue 1135–1144 in AE3. A residue that is close to that region was identified in mouse AE1 ASGPGAAAQIQEVK, residue 762–775.

These findings suggest that other isoforms of band 3 may undergo the same aging changes that AE1 on red blood cells undergoes to generate SCA, and provides confirmation that SCA is on non-erythroid cell types and/or that SCIgG can bind other band 3 isoforms present on non-erythroid cells.

Oxidation Generates SCA

Generation of SCA results from oxidation.[18] This was determined in studies of erythrocytes from vitamin E–deficient rats.[18] The importance of vitamin E as an antioxidant, providing protection against free radical–induced membrane damage, has been well documented (for example, see ref. 18). Vitamin E is localized primarily in cellular membranes, and a major role of vitamin E is the termination of free radical chain reactions propagated by the polyunsaturated fatty acids of membrane phospholipids. Vitamin E–deficient erythrocytes are defective in their ability to scavenge free radicals. There is a correlation between life span and natural antioxidant levels in a variety of species. The level of such antioxidants appears to correlate with metabolic activity of individual species. Specific biochemical alterations in the membrane of erythrocytes from vitamin E–deficient rhesus monkeys have been described (see ref. 18). Band 3 crosses the lipid bilayer multiple times and therefore would be expected to be vulnerable to free radical damage.

TABLE 4. Murine band 3 anion exchange protein (AE) isoforms identified in HPLC fractions of SCA isolated from mouse brain using SCIgG isolated from old human red cells[a]

			Mass	
Isoform	Peptide sequence	Residue no.	Average	Database
AE1 (B3AT_mouse)	VFSK	127–130	481	480.5842
	LR	619–620	288	288.3696
	ASGPGAAAQIQEVK	762–775	1327	1327.4796
AE2 (B3A2_mouse)	SYGEEDFEYHR	64–74	1432	1432.4451
	SYNLQER	228–234	910	909.9738
	IGSMTVEQALLPR	237–250	1473	1472.7424
	EGR	298–300	361	361.3776
	LR	533–534	288	288.3696
	FFPGR	922–926	624	623.7320
	IR	927–928	288	288.3696
	AVAPGDKPK	1070–1078	883	883.0350
	HHPDVTYVK	1144–1152	1096	1096.2306
	TMR	1156–1158 (MSO: 1157	424	423.5073
	MVVLTR	1197–1202 (MSO: 1197	735	734.9319
	EMK	1208–1210 (MSO: 1209)	424	423.5043
AE3 (B3A3_mouse)	SYSER	65–69	642	641.6581
	RPPPTSAR	97–104	882	882.0099
	EKPLHMPGGDGHR	486–498	1432	1431.6117
	EQTK	639–642	506	505.5481
	DVK	687–689	361	361.4180
	AQDLEYLTGR	783–792	1166	1166.2756
	LYK	842–844	424	423.5322
	HHPEQPYVTK	1135–1144	1236	1236.3723

[a]From Kay and Goodman.[66]

Since vitamin E deficiency represents a "physiological" method for rendering cells susceptible to free radical damage and might simulate conditions encountered *in situ*, we used vitamin E deficiency as a model for studying what role, if any, oxidation played in generating SCA.[18] Vitamin E deficiency simulated conditions encountered *in situ* more closely than did chemical treatment of cells *in vitro*. Middle-aged cells from vitamin E–deficient rats behaved like old erythrocytes in the phagocytosis assay and in anion transport and glyceraldehyde-3-phosphate dehydrogenase activity (TABLE 5). In addition, increased breakdown products of band 3 were observed in erythrocyte membranes from vitamin E–deficient rats as was observed in old cells from normal animals and humans. Vitamin E–deficient rats developed a

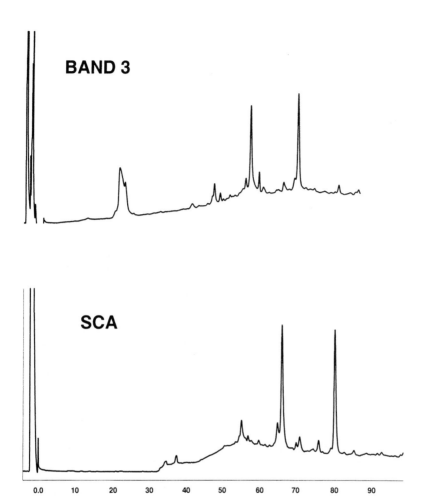

FIGURE 9. Peptide maps of band 3 and SCA show homology. Limited tryptic digests of band 3 and SCA were separated by reverse-phase HPLC on a C18 column. The eluting peptides, detected at 215 nm, indicate that peaks observed for SCA are a subset of those derived from band 3. (Reproduced from Kay[66] with permission.)

compensated hemolytic anemia, as was observed in vitamin E–deficient humans. Thus, vitamin E deficiency led to accelerated red cell aging, presumably through oxidation. This suggested that oxidation caused band 3 aging and was the first suggestion that oxidation generated neoantigens recognized by the immune system. As will be discussed below, there is now additional evidence that oxidation generates neoantigens to which immunoglobulins bind *in vivo*.

Since vitamin E deficiency caused accelerated red cell aging, the effect of high levels of vitamin E on cellular aging in mice was examined. Two different and in-

TABLE 5. Vitamin E deficiency accelerates red cell aging[a]

Vitamin E diet	RBC age	Phagocytosis (%)	Anion transport K_m	V_{max}
None	middle aged		0.5 ± 0.2	41.8 ± 1.9
	old		1.6 ± 0.4	17.3 ± 3.8
Normal (50 mg/kg)			0.7 ± 0.1	41.2 ± 3.7
	young	2 ± 3		
	middle aged	1 ± 1		
	old	72 ± 3		
Deficient (0mg/kg)			2.0 ± 0.3	16.2 ± 2.0
	young	89 ± 1		
	middle aged	88 ± 1		
	old	99 ± 0		

[a]In these studies, plasma levels of vitamin E were measured by HPLC. Results are presented as mean ± 1 standard deviation. The phagocytosis assay is performed with U927 cells. Red blood cells (RBCs) are incubated with macrophages overnight at 37°C in a humidified atmosphere containing 5% CO_2. For the phagocytosis assay, $n = 3$. There is no statistical difference between deficient and old cells. K_m, concentration at half-maximal exchange corresponding to an apparent Michaelis-Menten constant; V_{max}, maximal flux, determined at 37°C and pH 7.2. (Modified from Kay.[18])

terconnected systems were examined—namely, the immune system and the central nervous system. High vitamin E treatment (200 mg/kg) increased anion transport by splenic lymphocytes from old mice compared to those that received standard vitamin E treatment (50 mg/kg) and decreased binding of antibodies that recognize aged band 3 to brain tissue.[67]

Data from the studies of band 3 alterations in malaria are consistent with oxidative alterations generating SCA. First, infection with *Plasmodium falciparum* resulted in degradation of band 3.[68–70] Degraded forms of band 3 block cytoadherence of red cells involved in the sequestration that blocks capillaries.[71] Second, malaria infection generates free radicals and results in the generation of two SCA antigenic sites.[71] Interestingly, monkeys injected with these peptides suffered no ill effects, and micromolar amounts of the peptides were effective as adhesion blockers.

Recently, work in other systems has provided additional evidence that oxidation can generate neoantigens to which immunoglobulins bind *in vivo*. For example, oxidation of LDL (OxLDL) rendered them immunogenic.[72–74] A high levels of ox-LDL-specific antibody was protective in that it decreased the extent of atherosclerosis in hypercholesterolemic mice.[74] Anti-phospholipid antibodies found in anti-phospholipid antibody syndrome bound to oxidized but not native phospholipids.[73] Human atherosclerotic lesions contained IgG that recognized epitopes of oxidized LDL, suggesting that the neoantigens generated by oxidation played a role in pathophysiology *in situ*.[72]

EFFECT OF ORGANISMAL AGING AND DISEASE ON RED CELL LIFE SPAN AND IGG BINDING

Red cells age more rapidly in old individuals, suggesting an influence of environmental and/or humoral factors. The mean life span of RBCs in old, specific pathogen–free rats was reduced by 44% when compared to the life span in young rats, as determined by[61] Fe pulse-labeling *in vivo*.[75] The proportion of young cells circulating in the blood of old animals was increased. In old rats, young as well as old RBCs were heavily labeled with IgG; whereas predominantly old cells carried IgG in young rats. In humans, there was significant increase in reticulocytes in healthy elderly humans even though their hematocrits were the same as those of younger individuals.[76] This indicated that homeostasis was maintained by increased red cell production to keep pace with their destruction. A significant increase in autologous IgG on lower middle-aged cells was consistently demonstrated in these elderly individuals. These findings suggested that young cells aged prematurely in old individuals. Bartosz *et al.*[77] found increased amounts of cell membrane–bound IgG on RBCs from patients with Down's syndrome and suggested that accelerated red cell aging occurred in these individuals. Accelerated aging of other cells and systems, including the immune system, was reported in patients with Down's syndrome. Bosman *et al.*[78] have found that erythrocytes from patient's with Alzheimer's disease but not multi-infarct dementia showed characteristics of accelerated aging.

It was shown that red cells were increasingly susceptible to oxidation as they aged, probably secondary to decreased enzymatic activity.[75,79] Red cell band 3 from old human donors was degraded more by calpain, a Ca^{++} protease, than was band 3 from young individuals.[80] Calpain in older individuals translocated to the red cell membrane and was activated by autolysis, resulting in degradation of membrane proteins. This process seemed to be Ca^{++} dependent. It is interesting the Heidrick found that macrophage protease activity increased in older animals as compared to young ones,[81] which suggested increased cell destruction in older individuals. Perhaps the environment had cell life–shortening effects. In the 1970s, Makinodan showed that cellular aging could be influenced by the microcellular environment. Processing of red cells by dendritic cells could result in their antigens being presented to the immune system, resulting in increased autoantibody production.

In summary, changes in band 3 with aging included decreased efficiency of anion transport (decreased V_{max}) in spite of an increased number of anion binding sites (increased K_m), decreased glucose transport, increased band 3 degradation to smaller fragments as detected by quantitative binding of antibodies to band 3 breakdown products and amino acid residues 812–830, and *in situ* binding of physiological IgG autoantibodies. The latter findings indicated that a post-translational modification of band 3 has occurred. Evidence suggested that oxidation was a mechanism for generating SCA, perhaps through degradation of band 3 and involvement of proteases. Accelerated aging of erythrocytes was observed in old individuals.[75,76,82]

SENESCENT CELL ANTIGEN AND ANTI-BAND 3 ANTIBODIES IN DISEASES

Besides their role in the removal of senescent and damaged cells, antibodies to band 3 also appear to be involved in the removal of erythrocytes in clinical hemolytic anemias[51,83,84] and sickle cell anemia,[85–88] and the removal of malaria-infected erythrocytes[68–70,89] and stored cells.[8,10,13,90] Since oxidation generates senescent cell antigen *in situ*,[18] it may have contributed to red cell damage in these situations.

In a study of a severe familial autoimmune hemolytic anemia (AIHA), antibodies to band 3 were demonstrated in plasma by immunoblotting and on red cells by quantification with radioactive protein A.[51] The red cells also had a band 3 alteration that preceded formation of SCA as determined by binding of R980 antibodies. It was suspected that red cells in this patient were particularly susceptible to oxidative or proteolytic damage. Antibodies to band 3 were involved in other cases of AIHA in humans.[91,92]

The AIHA of New Zealand black (NZB) mice involved an antibody response against band 3.[93–95] This response was a T helper–dependent autoantibody response to band 3 peptide 861–874.[95–98] The core epitope spanned residues 862–870 based on its ability to elicit NZB splenic T cell proliferation.[96] This response was present in 6-week-old NZB. T cells responded to an increasing number of band 3 peptides with age, and the magnitude of the response increased.[97] Peptides that comprised the active antigenic sites of SCA as described above were not tested in those studies. It is not yet known whether the antibodies to band 3 in NZB AIHA were responding to conformational changes in the band 3 molecule in a pathway leading to cell death or whether the antibodies themselves initiated the terminal events.

AIHA can be elicited in mice by repeated injections of rat red cells. Splenic T cells from these mice proliferated in response to many murine red cell membrane components, including band 3, spectrin, and a low molecular weight glycophorin. In contrast, T cells from healthy control mice did not respond to antigens except band 3, and, in one experiment, spectrin.[99]

IgG autoantibodies to band 3 bind to banked human RBCs and render them susceptible to removal.[13] However, there is some evidence that banked red cells can recover at least partially, although some were removed shortly after transfusion. This suggested that cells may be able to recover from some of the changes induced by storage.

Senescent cell antigen–band 3 or its fragments are in fibrillary structures and processes in old but not young brains.[63] Brain sections from 10-year-old and 96-year-old individuals were examined with senescent cell antigen–band 3 antibodies in that study.[35] In normal brains from elderly individuals, band 3 antibodies reacted with cortex neurons in layers III and IV, with Purkinje cells and their dendrites extending into the molecular layer, and with cerebellar dentate nucleus neurons. Aged band 3 (presenescent cell antigen) antibodies reacted with astrocytes in the white matter, with a "mossy fiber" distribution in the cerebellum, and with select Purkinje cells. Dentate neurons were strongly reactive, especially those containing lipofuscin; but the staining did not resemble that of lipofuscin. There was a moderately strong reaction with many, but not all, large neurons in the cerebrum. Aged band 3 antbodies recognized old band 3 before senescent cell antigen was formed. They bound to band 3 in old but not young or middle-aged red cells.[51] In brains from patients with Alzheimer's disease (AD), aged band 3 antibodies labeled the amyloid core of classical plaques and the microglial cells located in the middle of the plaque. Adjacent neurons displayed a stronger and more widespread reaction than normals. In contrast, band 3 antibodies labeled the neuritic components of plaques, with some reaction noted in microglial cells, adjacent astrocytes, and neurons.

Some antibodies to band 3 and synthetic peptides of band 3 and some antibodies to aged band 3 differentiate band 3 proteins in brain membranes of normal individuals from those of patients with AD (FIG. 9).[14,64,78,102,103] One of these, an antibody to band 3 synthetic peptide pep-COOH, detected a triplet in the 60–70 kDa range in brains from patients with AD but not from age-matched controls (FIGS. 10A and 10B).[64] A number of studies indicate that AD has an immunological component. Human B cells have been shown capable of producing antibodies that bind to the AD brain. It would be interesting to know what effect aging has on these antibodies and whether they exert a protective effect on the brain and other tissues and cells.

Santos-Silva[104] found that band 3 changes, including fragmentation and IgG binding to erythrocyte membranes, were a risk marker for cardiovascular disease.

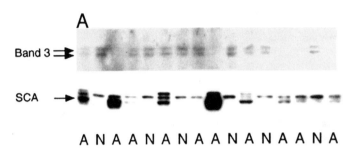

A N A A N A N A A N A N A A N A

FIGURE 10A. Antibody to band 3 synthetic peptide pep-COOH, part of an SCA antigenic site, detects a triplet in the 60–70 kDa range in brains from patients with AD but not from age-matched controls. Immunoblots of intact band 3 (**top**) and degraded band 3 (**bottom**) detected by anti pep-COOH in membranes from Alzheimer's disease (A) or normal (N) brains. Pep-COOH antibodies were used because COOH is included in senescent cell antigen.

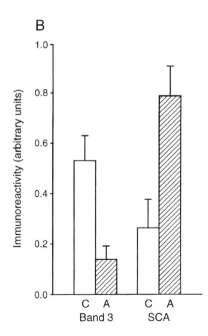

FIGURE 10B. Quantitation of band 3 and its degradation products in normal (C) and Alzheimer's disease (A). In other studies, antibodies to pep-COOH were used for immunohistochemical analysis of Alzheimer's disease (AD) and normal frontal cortex. Pep-COOH antibodies only weakly stained neurons in normal cortex. Only a few large neurons were strongly reactive. In AD, the number of neurons labeling with antibodies to pep-COOH increased 260% ($P \leq 0.01$). Because of a 50% neuronal loss in AD compared to controls ($P \leq 0.002$), the proportion of neurons remaining that were pep-COOH–positive increased 540%. In AD, antibodies to pep-COOH labeled glial cells found around plaques. This suggested that band 3 existed in normal neurons in, predominately, an unaltered or uncleaved state; and that degradation of band 3 occurred in AD, thus allowing binding of antibodies to pep-COOH. In other studies, antibodies to band 3 and its peptides bound to structures in AD but not normal brain, as determined by immunoelectron microscopy.[100,101] (Reproduced from Saitoh et al.[64] with permission.)

TABLE 6. Seropositivity in nonimmune and immune donors and children[a]

		Seropositivity (%)	
	Peptide:	DHPLQKTYNY	DVPYVKRVKTWRMH
Donor group	n	546–555	821–834
Nonimmune (Danish)	137	2.9	5.1
Adult immune (Liberian) +	217	20.3	43.8
Adult immune (Liberian) −	127	10.2	45.7
Children (Liberian)	202	15.3	36.1

[a]Modified from Hogh et al.[107]
+, Easy access to antimalarial drugs.
−, Antimalarial drugs not easily accessible.

Marchalonis et al.[105] found an increase in autoantibodies to band 3 segments in patients with systemic lupus erythematosis (SLE) and rheumatoid arthritis. Two of these peptides are SCA antigenic determinants.

THERAPEUTIC STRATEGIES: BAND 3 AND BAND 3 ANTIBODIES IN MALARIA

The immune response can be manipulated to change the natural progression of a disease by enhancing the protection afforded by natural antibodies or by using peptide antigens to block adhesion. Immunization with select peptide-defined epitopes would be one approach. For example, some laboratories have shown that peptides within or overlapping the aging antigen residues (547–553 and 824–829) can provide protection from malaria infection.[71,106,107] In areas where malaria is endemic, the immune response to band 3 appears beneficial. Malaria-immune children with higher levels of antibody reactive with conserved band 3 peptides, relative to nonimmune children, had a lower mean parasite density (TABLE 6). [107] This led some to propose the development of a vaccine for malaria based on the augmentation of the innate autoimmune response to band 3. Such a vaccine might reduce the morbidity and mortality of the initial infection.[108,109]

Plasmodium falciparum has infected between 300 and 500 million people and has resulted in more than 3 million deaths. Antibodies to band 3 are involved in clearance of malaria-infected red cells.[110] Band 3 residues 547–553 and 824–829 seem to be key residues. Monoclonal antibodies to band 3 external loops 3 and 7 block cytoadherence.[111] Synthetic band 3 peptides residues 547–553 and 824–829 from these loops, which are included in those that contain the active antigenic sites of SCA,[44] inhibit cytoadherence/sequestration of the malaria parasite *Plasmodium falciparum*[71,112] and are involved in antibody-mediated clearance of infected red cells.

Malaria infection of red cells results in degraded forms of band 3 protein.[68–70] Studies suggest that SCA is generated by degradation of band 3.[17,54]

It has been postulated that part of the malarial protection imparted by hemoglobinopathies such as sickle cell and beta thalassemia results from enhanced antigenic stimulation by altered band 3 molecules.[108,109] The reasoning is that antibodies to band 3 block or decrease the adhesiveness of erythrocytes in sickle cell and thus prevent some of the vaso-occlusive pathology such as severe and disabling pain and organ infarction. Band 3 is involved in adhesion of malaria-infected and sickle erythrocytes. Band 3 synthetic peptide residues 476–485 and 547–553 block the adherence of sickle erythrocytes to human umbilical vein endothelium, and antibodies to peptide 542–555 decrease the adhesiveness of malaria-infected erythrocytes.[113] The natural adherence between *Plasmodium falciparum*–infected erythrocytes and CD36 on amelanotic melanoma cells is blocked in a dose-depen-

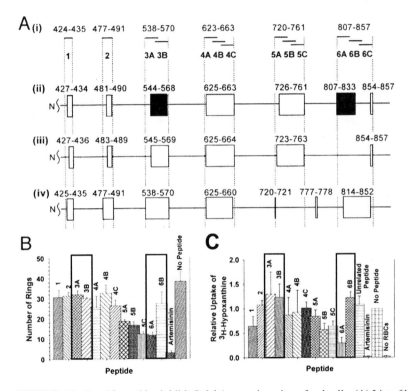

FIGURE 11. Band 3 peptides inhibit *P. falciparum* invasion of red cells. (A) List of human band 3 peptides used for the malaria experiments. (i) Overlapping 12- to 20-aa peptides shown in *solid bars* were prepared based on the inclusive ectodomain boundaries in three topology models: the Casey model (ii), the Reithmeier model (iii), and the Sherman model (iv). *Boxes* denote ectodomains. The *darkened boxes* indicate peptide residues 544–568 and 807–833. These are indicated in **B** and **C** with a *dark outline* to focus the reader's attention. (B) Invasion inhibition assay by the visual counting method. The number of ring-stage parasites counted from Giemsa-stained thin smears is plotted. No Peptide, negative control. (C) Effect of peptides on intraerythrocyte parasite maturation. The total number of schizonts and trophozoites was scored from the same smears in **B**. These peptides did not significantly retard the growth of trophozoites and schizonts. The authors of the study used artemisinin (25 M) as a negative control for parasite development in the culture. Parasitemia was calculated as the mean of three experiments (SE), and Student's *t* test was used to compare the mean. (Modified from Goel *et al.*[114])

dent manner by band 3 peptides.[71] Decreased adhesiveness of red cells imparted by band 3–specific antibodies could allow the infected cells to reach the spleen, where they are eliminated, and decrease sequestering of cells in postcapillary venules that can adversely affect the function of vital organs such as the brain, heart, lung, and kidneys. Other studies suggest that protection in hemoglobin AS are related to increased IgG response to *P. falciparum*–infected red cell membrane antigens that lead to immune phagocytosis of parasitized red cells. This reduces parasite density *in vivo*.

As expected, band 3 peptides block binding of the malaria parasite. Recent studies have shown that band 3 is a host receptor for the malaria merozoite stage (via merozoite-binding surface protein 1).[114] Select band 3 peptides inhibit *P. falciparum* invasion of human red cells in culture as determined by visual counting and [^{3}H] hypoxanthine incorporation methods (FIG. 11). Band 3 peptides residues 720–739, 731–750, 742–761, and 807–826 showed significant levels of inhibition of invasion at 500 mM concentration compared to other band 3 peptides and the control without peptide. Inhibition by these peptides was dose dependent. These peptides did not significantly retard the growth of trophozoites and schizonts. Peptide 807–826 overlaps SCA. In that study, the peptides were designed by randomly dividing the putative ectodomains of band 3, not by optimal composition for physiological activity.

CONCLUDING REMARKS

Molecular recognition of senescent cells involves oxidation of a crucial membrane protein leading to generation of a neoantigen and binding of physiologic autoantibodies. These autoantibodies trigger macrophage removal of the cell prior to its lysis at a time when anion transport has decreased but the membrane is still grossly intact. Autoantibodies such as this one contribute to the maintenance of homeostasis either by performing regulatory functions like cellular removal and/or protective ones such as removal of antigen or cells before they cause a problem either by disintegration or the evoking of a potentially damaging immune response. The physiologic autoantibody response to neoantigens generated by oxidation is a protective one.

SCA offers the potential for the development of specific interventions in aging by targeting the regions of band 3 that are affected. It also offers the possibility of treating other proteins that are damaged during aging since they are probably similarly affected. SCA epitopes and SCIgG also have the potential for altering the natural progression of diseases. Manipulation of the natural immune response to SCA would be expected to prevent or alter the course of certain diseases, of which malaria is an example.

[*Competing interests*: The author declares that she has no competing financial interests.]

REFERENCES

1. KAY, M.M.B. 1996. Aging and immunity. *In* The Thymus. J. Marsh & M. Kendall, Eds.: 395-411. CRC Press. Baton Rouge, LA.
2. KAY, M.M.B. & T. MAKINODAN. 1981. Relationship between aging and the immune system. Prog. Allergy **29:** 134–181.

3. KAY, M.M.B. 1979. The thymus: a clock for immunologic aging? J. Invest. Dermatol. **73:** 29–38.
4. SCHECHTER, B. 1981.The heresy of Dr. Kay. Discover 37–39.
5. KAY, M.M.B. 2005. Molecular and cellular biology of health and disease: colleagues celebrating the life and work of Joe Goodman. Cell. Mol. Biol. **51:** 113–175.
6. KAY, M.M.B. 1974. Mechanisms of macrophage recognition of senescent red cells. Gerontologist **14:** 1.
7. KAY, M.M.B. 1975. Mechanism of removal of senescent cells by human macrophages *in situ*. Proc. Natl. Acad. Sci.**72:** 3521–3525.
8. KAY, M.M.B. 1978. Role of physiologic autoantibody in the removal of senescent human red cells. J. Supramol. Struct. **9:** 555–567.
9. KAY, M.M.B. 1980. Cells, signals, and receptors: the role of physiological autoantibodies in maintaining homeostasis. Adv. Exp. Med. Biol. **129:** 171–200.
10. KAY, M.M.B. 1981. Isolation of the phagocytosis inducing IgG-binding antigen on senescent somatic cells. Nature Lond. **289:** 491–494.
11. BENNETT, G. & M.M.B. KAY. 1981. Homeostatic removal of senescent murine erythrocytes by splenic macrophages. Exp. Hematol. **9:** 297–307.
12. KAY, M.M.B., K. SORENSEN, P. WONG & P. BOLTON. 1982. Antigenicity, storage & aging: Physiologic autoantibodies to cell membrane and serum proteins and the senescent cell antigen. Mol. Cell. Biochem. **49:** 65–85.
13. BOSMAN, G.J.C.G.M. & M.M.B. KAY. 1988. Erythrocyte aging: a comparison of model systems for simulating cellular aging *in vitro*. Blood Cells **14:**19–35.
14. BOSMAN, G.J., K. RENKAWEK, F.P. VAN WORKUM, *et al.* 1997. Involvement of neuronal anion exchange proteins in cell death in Alzheimer's disease. Gerontology **43:**67–78.
15. BOSMAN, G.J., J.G. JANZING, I.G. BARTHOLOMEUS, *et al.* 1997. Erythrocyte aging characteristics in elderly individuals with beginning dementia. Neurobiol. Aging **18:** 291–295.
16. KAY, M.M.B. 1983. Appearance of a terminal differentiation antigen on senescent and damaged cells and its implications for physiologic autoantibodies. Biomembranes **11:**119–150.
17. KAY, M.M.B. 1985. Aging of cell membrane molecules leads to the appearance of an aging antigen and the removal of senescent cells. Gerontology **31:** 215–235.
18. KAY, M.M.B., G.J.C.G.M. BOSMAN, S.S. SHAPIRO, *et al.* 1986. Oxidation as a possible mechanism of cellular aging: vitamin E deficiency causes premature aging and IgG binding to erythrocytes. Proc. Natl. Acad. Sci. USA **83:** 2463–2467.
19. BARTOSZ, G., M. SOSYNSKI & A. WASILEWSKI. 1982. Aging of the erythrocyte XVII. Binding of autologous immunoglobulin. Mech. Ageing Dev. **20:** 223–232.
20. WALKER, W.S., J.A. SINGER, M. MORRISON & C.W. JACKSON. 1984. Preferential phagocytosis of *in vivo* aged murine red blood cells by a macrophage-like cell line. Br. J. Haemat. **58:** 259–266.
21. SINGER, J.A., L.K. JENNINGS, C. JACKSON, *et al.* 1986. Erythrocyte homeostasis: antibody-mediated recognition of the senescent state by macrophages. Proc. Natl. Acad. Sci. USA **83:** 5498–5501.
22. ZIMMERMANN, N., K.J. HALBHUBER, W. LINSS & H. FEUERSTEIN. 1986. Immunocytochemical investigations of the membrane of experimentally altered and physiologically aged erythrocytes. Acta Histochem. **33**(suppl): 61–67.
23. ZIMMERMANN, N., K.J. HALBHUBER, H. FEUERSTEIN & W. LINSS. 1987. Detection of IgG bound at the erythrocyte membrane by means of an immunohistochemical gold–silver technique. Folia Haematol. Int. Mag. Klin. Morphol. Blutforsch. **114:** 471–472.
24. CHRISTIAN, J.A., A.H. REBAR, G.D. BOON & P.S. LOW. 1993. Senescence of canine biotinylated erythrocytes: increased autologous immunoglobulin binding occurs on erythrocytes aged *in vivo* for 104 to 110 days. Blood **82:** 3469–3473.
25. GIGER, U., B. STICHER, R. NAEF, *et al.* 1995. Naturally occurring human anti-band 3 autoantibodies accelerate clearance of erythrocytes in guinea pigs. Blood **85:** 1920–1928.
26. LUTZ, H., R. FLEPP & G. STRINGARO-WIPF. 1984. Naturally occurring autoantibodies to exoplasmic and cryptic regions of band 3 protein, the major integral membrane protein of human red blood cells. J. Immunol. **133:** 2160–2168.

27. KHANSARI, N. & H.H. FUDENBERG. 1983. Immune elimination of autologous senescent erythrocytes by Kupffer cells *in vivo*. Cell. Immunol. **80:** 426–430.
28. KHANSARI, N., G.F. SPRINGER, E. MERLER & H.H. FUDENBERG. 1983. Mechanisms for the removal of senescent human erythrocytes from circulation: specificity of the membrane-bound immunoglobulin G. J. Mech. Ageing Dev. **21:** 49–58.
29. KAY, M.M.B. & J. GOODMAN. 2003. Immunoregulation of cellular lifespan: physiologic autoantibodies and their peptide antigens. Cell. Mol. Biol. **49:** 217–243.
30. KAY, M.M.B., C. COVER, S. SCHLUTER, *et al.* 1995. Band 3, the anion transporter, is conserved during evolution: implications for aging and vertebrate evolution. Cell. Mol. Biol. **41:** 883–892.
31. KAY, M.M.B. 1988. Immunologic techniques for analyzing red cell membrane proteins. *In* Methods in Hematology: Red Cell Membranes. S. Shohet & N. Mohandas, Eds.: 135–170. Churchill Livingston, Inc. New York.
32. KAY, M.M.B. 1983. Appearance of a terminal differentiation antigen on senescent and damaged cells and its implications for physiologic autoantibodies. Biomembranes **11:** 119–150.
33. KAY, M.M.B. 2004. Band 3 and its alterations in health and disease. Cell. Mol. Biol. **50:** 117–138.
34. KAY, M.M.B., C.M. TRACEY, J.R. GOODMAN, *et al.* 1983. Polypeptides immunologically related to erythrocyte band 3 are present in nucleated somatic cells. Proc. Natl. Acad. Sci. USA **80:** 6882–6886.
35. KAY, M.M.B., G.J.C.G.M. BOSMAN, M. NOTTER & P. COLEMAN. 1988. Life and death of neurons: the role of senescent cell antigen. Ann. N.Y. Acad. Sci. **521:** 155–169.
36. KAY, M.M.B., J. HUGHES, I. ZAGON & F. LIN. 1991. Brain membrane protein band 3 performs the same functions as erythrocyte band 3. Proc. Natl. Acad. Sci. **88:** 2778–2782.
37. DEMUTH, D.R., L.C. SHOWE, M. BALLANTINE, *et al.* 1986. Cloning and structural characterization of a human non-erythroid band 3-like protein. EMBO J. **5:** 1205–1214.
38. DRENCKHAHN, D., K. ZINKE, U. SCHAUER, *et al.* 1984. Identification of immunoreactive forms of human erythrocyte band 3 in nonerythroid cells. Eur. J. Cell Biol. **34:** 144–150.
39. KARNISKI, L.P. & M.L. JENNINGS. 1989. Identification and partial purification of a band 3-like protein from rabbit renal brush border membranes. J. Biol.Chem. **264:** 4564–4570.
40. OSTEDGAARD, L.S., M.L. JENNINGS, L.P. KARNISKI & V.L.SCHUSTER. 1991. A 45-kDa protein antigenically related to band 3 is selectively expressed in kidney mitochondria. Proc. Natl. Acad. Sci. USA **88:** 981–985.
41. KELLOKUMPU, S, L. NEFF, S. JAMSA-KELLOKUMPU, *et al.* 1988. A 115-kD polypeptide immunologically related to erythrocyte band 3 is present in Golgi membranes. Science **242:** 1308–1311.
42. KAY, M.M.B., S. GOODMAN, K. SORENSEN, *et al.* 1983.The senescent cell antigen is immunologically related to band 3. Proc. Natl. Acad. Sci. USA **80:** 1631–1635.
43. KAY, M.M.B. 1984. Localization of senescent cell antigen on band 3. Proc. Natl. Acad. Sci. **81:** 5753–5757.
44. KAY, M.M.B., J.J. MARCHALONIS, J. HUGHES, *et al.* 1990. Definition of a physiologic aging auto-antigen using synthetic peptides of membrane protein band **3**: Localization of the active antigenic sites. Proc. Natl. Acad. Sci. USA **87:** 5734–5738.
45. BEPPU, M., A. MIZUKAMI, M. NAGOYA & K. KIKUGAWA. 1990. Binding of anti-band 3 autoantibody to oxidatively damaged erythrocytes. Formation of senescent antigen on erythrocyte surface by an oxidative mechanism. J. Biol. Chem. **265:** 3226–3233.
46. FERRONI, L., A. GIULIANI, S. MARINI, *et al.* 1991. A new monoclonal antibody to an age-sensitive band 3 transmembrane segment. Adv. Exp. Med. Biol. **307:** 351–356.
47. LUTZ, H.U., F. BUSSOLINO, R. FLEPP, *et al.* 1987. Naturally occurring anti-band-3 antibodies and complement activation together mediate phagocytosis of oxidatively stressed human erythrocytes. Proc. Natl. Acad. Sci. **84:** 7368–7372.
48. LUTZ, H.U., P. STAMMLER & S. FASLER. 1993. Preferential formation of C3b-IgG complexes *in vitro* and *in vivo* from nascent C3b and naturally occurring anti-band 3 antibodies. J. Biol. Chem. **268:** 17418–17426.

49. Lutz, H.U., M. Pfister & R. Hornig. 1996. Tissue homeostatic role of naturally occurring anti-band 3 antibodies. Cell. Mol. Biol. **42:** 995–1005.
50. Low, P.S., S.M. Waugh, K. Zinke & D. Drenckhahn. 1985. The role of hemoglobin denaturation and band 3 clustering in red blood cell aging. Science **227:** 531–553.
51. Kay, M.M.B., N. Flowers, J. Goodman & G.J.C.G.M. Bosman. 1989. Alteration in membrane protein band 3 associated with accelerated erythrocyte aging. Proc. Natl. Acad. Sci. USA **86:** 5834–5838.
52. Bosman, G.J.C.G.M. & M.M.B. Kay. 1990. Alterations of band 3 transport protein by cellular aging and disease: erythrocyte band 3 and glucose transporter share a functional relationship. Biochem.Cell Biol. **68:** 1419–1427.
53. Bartosz, G., M. Gaczynska, E. Grzelinska, et al. 1987. Aged erythrocytes exhibit decreased anion exchange. Mech. Ageing Dev. **39:** 245–250.
54. Kay, M.M.B. 1984. Band 3, the predominant transmembrane polypeptide, undergoes proteolytic degradation as cells age. Monogr. Dev. Biol. **17:** 245–253.
55. Christian, J.A., A.H. Rebar, G.D.Boon & P.S. Low. 1996. Methodologic considerations for the use of canine *in vivo* aged biotinylated erythrocytes to study RBC senescence. Exp. Hematol. **24:** 82–88.
56. Christian, J.A., J. Wang, N. Kiyatkina, et al. 1998. How old are dense red blood cells? The dog's tale. Blood **92:** 2590–2591.
57. Allen, D.P., P.S. Low, A. Dola & H. Maisel. 1987. Band 3 and ankyrin homologues are present in eye lens: evidence for all major erythrocyte membrane components in same non-erythroid cell. Biochem. Biophys. Res. Commun. **149:** 266–275.
58. Alper, S.L., J. Natale, S. Gluck, et al. 1989. Subtypes of intercalated cells in rat kidney collecting duct defined by antibodies against erythroid band 3 and renal vacuolar H+-ATPase. Proc.Natl.Acad.Sci. **86:** 5429–5433.
59. Kay, M.M.B., F. Lin, G.J.C.G.M. Bosman, et al. 1991. Human erythrocyte aging: cellular and molecular biology. Trans. Med. Rev. **5:** 173–195.
60. Kay, M.M.B. 1991. Molecular mapping of human band 3 anion transport regions using synthetic peptides. Fed. Proc. **5:** 109–115.
61. Kay, M.M.B. 1992. Molecular mapping of human band 3 aging antigenic sites and active amino acids using synthetic peptides J. Protein Chem. **11:** 595–602.
62. Kay, M.M.B., T. Wyant & J. Goodman. 1994. Autoantibodies to band 3 during aging and disease and aging interventions. Ann. N.Y. Acad. Sci. **719:** 419–447.
63. Kay, M.M.B., D. Lake & C. Cover. 1995. Band 3 and its peptides during aging, radiation exposure, and Alzheimer's disease: alterations and self-recognition. Adv. Exp. Med. Biol. **383:** 167–193.
64. Saitoh, T., E. Masliah, L. Baum, et al. 1992. Degradation of proteins in the membrane-cytoskeleton complex in Alzheimer's disease. Ann. N.Y. Acad. Sci. **674:** 180–192.
65. Zhu, Q., D. Lee & J.R. Casey. 2003. Novel topology in C-terminal region of the human plasma membrane anion exchanger, AE1. J. Biol. Chem. **278:** 3112–3120.
66. Kay, M.M. B. & J. Goodman. 2004. Mapping of senescent cell antigen on brain anion exchanger (AE) protein isoforms using fast atom bombardment ionization mass spectrometry. J. Mol. Recogn. **17:** 33–40.
67. Poulin, J., M. Guftafson & M.M.B. Kay. 1996. Vitamin E prevents oxidative modification of Band 3 related proteins and subsequent generation of senescent cell antigen during aging. Proc. Natl. Acad. Sci.USA **93:** 5600–5603.
68. Winograd, E., J.R. Greenan & I.W. Sherman. 1987. Expression of senescent antigen on erythrocytes infected with a knobby variant of the human malaria parasite *Plasmodium falciparum*. Proc. Natl. Acad. Sci. USA **84:** 1931–1935.
69. Winograd, E. & I.W. Sherman. 1989. Naturally occurring anti-band 3 autoantibodies recognize a high molecular weight protein on the surface of *Plasmodium falciparum*–infected erythrocytes. Biochem. Biophys. Res. Commun. **160:** 1357–1363.
70. Roggwiller, E., M.E. Betoulle, T. Blisnick & B.C. Breton. 1996. A role for erythrocyte band 3 degradation by the parasite gp76 serine protease in the formation of the parasitophorous vacuole during invasion of erythrocytes by *Plasmodium falciparum*. Mol. Biochem. Parasitol. **82:** 13–24.
71. Crandall, I., W.E. Collins, J. Gysin & I.W. Sherman. 1993. Synthetic peptides based on motifs present in human band 3 protein inhibit cytoadherence/sequestration

of the malaria parasite *Plasmodium falciparum*. Proc. Natl. Acad. Sci. USA **90:** 4703–4707.

72. YLA-HERTTUALA, S., W. PALINSKI, S.W. BUTLER, *et al.* 1994. Rabbit and human atherosclerotic lesions contain IgG that recognizes epitopes of oxidized LDL. Arterioscler. Thromb. **14:** 32–40.

73. SHAW, P., S. HARKKA, S. TSIMIKAS, *et al.* 2001. Human-derived anti-oxidized LDL autoantibody blocks uptake of oxidized LDL by macrophages and localizes to atherosclerotic lesions *in vivo*. Arter. Thromb. Vasc. Biol. **21:** 1333–1339.

74. BINDER, C.J., S. HORKKO, A. DEWAN, *et al.* 2003. Pneumococcal vaccination decreases atherosclerotic lesion formation: molecular mimicry between *Streptococcus pneumoniae* and oxidized LDL. Nat. Med. **9:** 736–743.

75. GLASS, G.A., D. GERSHON & H. GERSHON. 1983. The effect of donor and cell age on several characteristics of rat erythrocytes. Exp. Hematol. **11:** 987–995.

76. GLASS, G.A., D. GERSHON & H. GERSHON. 1985. Some characteristics of the human erythrocyte as a function of donor and cell age. Exp. Hematol.**13:** 1122–1126.

77. BARTOSZ, G., M. SOSYNSKI & J. KREDZIONA. 1982. Aging of the erythrocyte. VI. Accelerated red cell membrane aging in Down's syndrome. Cell. Biol.Int. Rep. **6:** 73–77.

78. BOSMAN, G.J., I.G. BARTHOLOMEUS, A.J. DE MAN, *et al.* 1991. Erythrocyte membrane characteristics indicate abnormal cellular aging in patients with Alzheimer's disease. Neurobiol. Aging **12:** 13–18.

79. GLASS, G.A. & D. GERSHON. 1984. Decreased enzymic protection and increased sensitivity to oxidative damage in erythrocytes as a function of cell and donor aging. Biochem. J. **218:** 531–537.

80. GLASER, T., N. SCHWARZ-BENMEIR, S. BARNOY, *et al.* 1994. Calpain (Ca(2+)-dependent thiol protease) in erythrocytes of young and old individuals. Proc. Natl. Acad. Sci. USA **91:** 7879–7883.

81. HEIDRICK, M.L. 1972. Age-related changes in hydrolase activity of peritoneal macrophages. Gerontologist **12:** 28.

82. KOSOWER, N.S. 1993. Altered properties of erythrocytes in the aged. Am. J. Hematol. **42:** 241–247.

83. KAY, M.M.B., J. GOODMAN, S. GOODMAN & C. LAWRENCE. 1990. Membrane protein band 3 alteration associated with neurologic disease and tissue reactive antibodies. Clin. Exp. Immunogenet. **7:** 181–199.

84. KAY, M.M.B., J. GOODMAN, C. LAWRENCE & G. BOSMAN. 1990. Membrane channel protein abnormalities and autoantibodies in neurological disease. Brain Res. Bull. **24:** 105–111.

85. PETZ, L.D., P. YAM, L. WILKINSON, *et al.* 1984. Increased IgG molecules bound to the surface of red blood cells of patients with sickle cell anemia. Blood **64:** 301–304.

86. HEBBEL, R.P. & W.J. MILLER. 1984. Phagocytosis of sickle erythrocytes. Immunologic and oxidative determinants of hemolytic anemia. Blood **64:** 733–741.

87. GREEN, G.A. 1993. Autologous IgM, IgA, and complement binding to sickle erythrocytes *in vivo*. Evidence for the existence of dense sickle cell subsets. Blood **82:** 985–992.

88. BOSMAN, G. 2004. Erythrocyte aging in sickle cell disease. Cell. Mol. Biol. **50:** 81–86.

89. WINOGRAD, E. & I.W. SHERMAN. 1989. Characterization of a modified red cell membrane protein expressed on erythrocytes infected with the human malaria parasite *Plasmodium falciparum*: possible role as a cytoadherent mediating protein. J. Cell. Biol. **108:** 23–30.

90. ANDO, K., M. BEPPU, K. KIKUGAWA & N. HAMASAKI. 1993. Increased susceptibility of stored erythrocytes to anti-band 3 IgG autoantibody binding. Biochim. Biophys. Acta. **1178:** 127–134.

91. LEDDY, J.P., J.L. FALANY, G.E. KISSEL, *et al.* 1993. Erythrocyte membrane proteins reactive with human (warm-reacting) anti-red cell autoantibodies. J. Clin. Invest. **91:** 1672–1680.

92. LEDDY, J.P., S.L.WILKINSON, G.E. KISSEL, *et al.* 1994. Erythrocyte membrane proteins reactive with IgG (warm-reacting) anti-red blood cell autoantibodies. II. Antibodies coprecipitating band 3 and glycophorin A. Blood **84:** 650–656.

93. BARKER, R.N., K.M. CASSWELL & C.J. ELSON. 1993. Identification of murine erythrocyte autoantigens and cross-reactive rat antigens. Immunology **78:** 578–573.

94. BARKER, R.N., G.G. DE SA OLIVEIRA, C J. ELSON & P.M. LYDYARD. 1993. Pathogenic autoantibodies in the NZB mouse are specific for erythrocyte band 3 protein. Eur. J. Immunol. **23:** 1723–1726.
95. PERRY, F.E., R.N. BARKER, G. MAZZA, et al. 1996. Autoreactive T cell specificity in autoimmune hemolytic anemia of the NZB mouse. Eur. J. Immunol. **26:** 136–141.
96. SHEN, C.R., F.J. WARD, A. DEVINE, et al. 2002. Characterization of the dominant autoreactive T-cell epitope in spontaneous autoimmune haemolytic anaemia of the NZB mouse. J. Autoimmun. **18:** 149–157.
97. SHEN, C.R., D.C. WRAITH & C.J. ELSON. 1999. Splenic but not thymic autoreactive T cells from New Zealand black mice respond to a dominant erythrocyte Band 3 peptide. Immunology **96:** 595–599.
98. SHEN, C.R., G. MAZZA, E.F. PERRY, et al. 1996. T-helper 1 dominated responses to erythrocyte Band 3 in NZB mice. Immunology **89:** 195–199.
99. BARKER, R.N., L.P. ERWIG, K.S. HILL, et al. 2002. Antigen presentation by macrophages is enhanced by the uptake of necrotic, but not apoptotic, cells. Clin. Exp. Immunol. **127:** 220–225.
100. KAY, M.M.B. & J. GOODMAN. 1997. Brain and erythrocyte anion transporter protein, band 3, as a marker for Alzheimer's disease: posttranslational changes detected by electronmicroscopy, phosphorylation, and antibodies. Gerontology **43:** 44–66.
101. KAY, M., S. RAPCSAK, G. BOSMAN & J. GOODMAN. 1996. Posttranslational modifications of brain and erythrocyte band 3 during aging and disease. Cell. Mol. Biol. **42:** 919–944.
102. BOSMAN G.J., F.P. VAN WORKUM, K. RENKAWEK, et al. 1993. Proteins immunologically related to erythrocyte anion transporter band 3 are altered in brain areas affected by Alzheimer's disease. Acta Neuropathol. (Berl) **86:** 353–359.
103. BOSMAN, G.J., A. ENGBERSEN, C.H. VOLLAARD, et al. 1996. Implications of aging- and degeneration-related changes in anion exchange proteins for the maintenance of neuronal homeostasis. Cell. Mol. Biol. **42:** 905–918.
104. SANTOS-SILVA, A., E. CASTRO, N. TEIXEIRA, et al. 1995. Altered erythrocye band 3 profile as a marker in patients at risk for cardiovascular disease. Atherosclerosis **116:** 199–209.
105. MARCHALONIS J., S. SCHLUTER, L.WILSON, et al. 1993. Natural human autoantibodies to synthetic peptide autoantigens: correlations with age and autoimmune disease. Gerontology **39:** 65–79.
106. THEVENIN, B.J., I. CRANDALL, S.K. BALLAS, et al. 1997. Band 3 peptides block the adherence of sickle cells to endothelial cells *in vitro*. Blood **90:** 4172–4179.
107. HOGH, B., E. PETERSEN, I. CRANDALL, et al. 1994. Immune responses to band 3 neoantigens on *Plasmodium falciparum*–infected erythrocytes in subjects living in an area of intense malaria transmission are associated with low parasite density and high hematocrit value. Infect. Immun. **62:** 4362–4366.
108. KENNEDY, J. 2002. Modulation of sickle cell crisis by naturally occurring band 3 specific antibodies: a malaria link. Med. Sci. Monit. **8:** 10–13.
109. KENNEDY, J. 2001. Malaria: a vaccine concept based on sickle haemoglobin's augmentation of an innate autoimmune process to band 3. Int. J. Parasitol. **31:** 1275–1277.
110. CRANDALL, I., N. GUTHRIE & I. SHERMAN. 1995. *Plasmodium falciparum*: sera of individuals living in a malaria-endemic region recognize peptide motifs of the human erythrocyte anion transport protein. Am. J. Trop. Med. Hyg. **52:** 450–455.
111. CRANDALL, I. & I. SHERMAN. 1994. Cytoadherence-related neoantigens on *Plasmodium falciparum* (human malaria)–infected human erythrocytes result from the exposure of normally cryptic regions of the band 3 protein. Parasitology **108:** 257–267.
112. CRANDALL, I., N. GUTHRIE & I. SHERMAN. 1996. *Plasmodium falciparum*: naturally occurring rabbit immunoglobulins recognize human band 3 peptide motifs and malaria-infected red cells. Exp. Parasitol. **82:** 45–53.
113. GUTHRIE, N., I. CRANDALL, S. MARINI, et al. 1995. Monoclonal antibodies that react with human band 3 residues 542–555 recognize different conformations of this protein in uninfected and Plasmodium falciparum infected erythrocytes. Mol. Cell Biochem. **144:** 117–123.
114. GOEL, V.K., X. LI, H. CHEN, et al. 2003. Band 3 is a host receptor binding merozoite surface protein 1 during the *Plasmodium falciparum* invasion of erythrocytes. Proc. Natl. Acad. Sci. USA **100:** 5164–5169.

Lunasensor, Infradian Rhythms, Telomeres, and the Chronomere Program of Aging

ALEXEY OLOVNIKOV

Institute of Biochemical Physics of RAS, Chernyakhovskogo, 5-94, Moscow 125319, Russia

ABSTRACT: According to the redusome hypothesis, the aging of an organism is determined by the shortening of chronomeres (small perichromosomal linear DNA molecules). In this paper, a presumptive role for infradian hormonal rhythms is considered. Endogenous infradian rhythms are supposed to actively interact with those hormonal shifts which are governed by an exogenous infradian gravitational lunar rhythm. As a result of this interaction, the so-called T-rhythm is formed. Peaks of T-rhythms are used as the pacemaker signals to keep the life-long "clockwork" of the brain running. The "ticking" of this clock is realized by the periodically repeated shortening of chronomeres in postmitotic neuroendocrine cells, which occurs just at the maxima of T-rhythms. Shortening of telomeres in mitotic cells *in vivo* is a witness of the aging of the organism, but not the cause of aging. The primary cause of aging is shortening of chronomeres, the material carriers of a temporal program of development and aging. To recognize exogenous gravitational infradian rhythms, a special physiological system—the "lunasensor" system—evolved. It is assumed that it is a necessity to have a lunasensor as a particular variant of sensors of gravitation.

KEYWORDS: telomere; chronomere; infradian rhythms; gravitation; brain sand; pineal gland; aging; chronobiology

INTRODUCTION

Aging is a universal chronic molecular disease of the soma determined by a regular shortening of the length of chronomeres with time. In the redusome hypothesis of aging,[1,2] which replaces my earlier proposed telomere hypothesis of aging,[3-5] an important role is given to the long-period infradian rhythms (periods are several days or weeks long). A key role of the infradian rhythms and their interplay with gravitation influence is considered after a brief account of the essence of a chronomere program of development and aging as a basis for the redusome hypothesis.

Address for correspondence: Alexey Olovnikov, Institute of Biochemical Physics of RAS, Chernyakhovskogo, 5-94, Moscow 125319, Russia. Voice: 7(095)1511784; fax: 7(095)2123521. olovnikov@dol.ru

Ann. N.Y. Acad. Sci. 1057: 112–132 (2005). © 2005 New York Academy of Sciences. doi: 10.1196/annals.1356.006

CHRONOMERE PROGRAM OF AGING

Redusomes

It is assumed that there exist special perichromosomal organelles called redusomes (also named *reducemeres* or *redumeres*), which are represented by small linear DNA double-helix molecules. They are called redusomes because their size is assumed to decrease owing to DNA-end undercopying. Reduction of their size in dividing cells occurs in parallel to (but independent of) the shortening of telomeres. Redusomes arise in somatic cells during morphogenesis and differentiation. They are perichromosomal copies of segments of a chromosomal DNA. These chromosomal segments are almost always in a tightly compacted state. They are activated for a short time in order to create redusomes. Once formed, a redusome is kept on the chromosome beside that segment, or a chromosomal nest of redusomes, which codes for this redusome. On their ends, redusomes have buffer sequences or acromeres, but they do not have their own centromeres.

Being kept in their nests, redusomes share with chromosomes their mitotic destiny during distribution among daughter cells. Redusomes contain the genes coding mainly for regulatory untranslatable RNAs, though some redusome genes can code for regulatory proteins, too. The regulatory factors of redusome origin participate in control of the specific reconfigurations of chromatin, which are necessary for differentiation. Besides, the redusome-related products are able to modulate tissue-specific syntheses by controlling the activity of chromosomal genes. Redusomes of different cellular differentiations are distinct in sets of their specific redusomal genes. Some of these genes can be repeated, while other genes of redusomes can be unique. The number of redusomes located at their nests on the surface of individual chromosomes is dissimilar in cells of different specialization.

In the process of shortening, redusomes lose bit by bit the parts of their terminal DNA and, because of this, they stepwise lose their activity. Loss of genetic activity of redusome due to its linear DNA truncation can be termed as "genoptosis," or "acroptosis." This can alter, both qualitatively and quantitatively, the expression of chromosomal genes and, therefore, the activity of differentiated cells. Owing to genoptosis, which proceeds in cells of a physiologically mature organism, the body should gradually and inevitably lose the activity of many its systems, and this persistent enfeebling causes aging.

There are two types of redusomes: printomeres, which are necessary for interpretation of positional information of the cells in morphogenesis, and chronomeres, which are involved in control of the individual development in time (including puberty and menopause). Chronomeres function in postmitotic neuroendocrinal cells of a brain and their shortening is caused by a hypothetical process of "scrupting."[1] Scrupting is a super-high-speed transcription-dependent rupture of the chronomere terminus immobilized in chromosomal DNA. Rupture of this terminus is performed by transcription machinery that runs along the chronomere at a super-high speed under the influence of a hormonal peak of T-rhythm. Scrupting leads to the abovementioned genoptosis. A growing shortage of the products encoded in redusomes leads to a chronic abnormal production of free radicals, to increasing dysfunction in various cell systems, and, hence to the aging of the whole organism. Shortening of telomeres in dividing cells is merely witness to redusome shortening. Only printomere shortening (printomere is a redusome of a mitotic cell) is responsible for cellular senescence and for the Hayflick limit.

Hypotheses of Aging

Although a series of predictions made by the telomere hypothesis of aging,[3,4] including the prediction of telomerase existence and the telomere shortening in dividing cells, was confirmed internationally I suppose now that the telomere model of aging should be rejected. The free radical hypothesis of aging should also be abandoned. Both of these competing hypotheses could be replaced by the redusome theory of aging, which puts forth a novel paradigm of gerontology, and suggests a research program that could yield future practical results that hopefully could enable the protracted maintenance of the young and healthy state of a mature organism.

Thus, the replicative senescence of dividing cells is caused by the shortening of redusomes, rather than telomeres; hence the Hayflick limit is a result of shortening of redusomes rather than telomeres. A stepwise loss of DNA from redusomes evokes a stepwise decline in cellular activity and leads to cellular senescence. The generally accepted statement that the elongation of telomeres leads to cell immortalization is erroneous! In reality, artificial immortalization of dividing cells by telomerase is owed to a telomerase-dependent extension of redusomes, rather than telomeres. Just as telomere attrition is only an innocent witness of redusome shortening, telomere elongation by telomerase is an innocent bystander of redusome elongation.

Cells become vulnerable to free radicals only when a deficit of the products coded by redusomal genes occurs. Detrimental effects of free radicals are themselves a consequence of the redusome-dependent aging, rather than the primary cause of aging itself. Free radicals exist even in cells of immortal clones. Telomerase immortalizes cells, preventing their aging, though this enzyme is not an antioxidant agent, contradicting the foundations of the free radical hypothesis. That is why both the free radical hypothesis of aging and the telomere-shortening hypothesis of aging should yield to a new theory.

The carrier of the redusome program of development and aging is a set of redusomes (chronomeres and printomeres). Aging of mammals is mainly based on the aging of a brain that is caused by a regular shortening of chronomeres and a consequent diminution of their genetic activity. It should be emphasized that redusome genes exist only for healthy life, not for aging. There is a program of aging but there are no genes for it.

Chronomeres

Chronomeres of various specificities are genetic "top managers" located at the top of the pyramid of hierarchically subordinated management systems. Perhaps the most effective variant of the regulation of development (and aging) of the organism is the following: chronomeres control the activity of major hormones and receptors to major hormones, and these receptors are located in the neuroendocrinal and neurotrophic centers of the CNS. Because of this, chronomeres would be able to control thousands of various processes in the organism. The number of chronomeres required to govern the functioning of the mammalian organism doesn't have to be very large; it is quite possible that, even for the human organism, just 10–20 types of chronomeres of different specificity is enough. Nonaging mammalian species do not exist—not because it is impossible genetically, but because selective pressure on the ecological niche of any species limits numbers. From the genetic point of view, there are no principal vetoes, not only on the immortalization of cells (on the basis of work already done in many labs using telomerase), but also on the possibility of the long-term maintenance of a "young state" of the organism. To proceed from the proposed chronomere conception, biological age and calendar age of an individual could be uncoupled. It will be possible to prevent and partly reverse the age-related changes of the organism as soon as one is able to prevent the shortening of chronomeres, recreate chronomeres, or compensate by the vectors carrying chronomere gene activity for the lost activity of a patient's chronomere genes.

T-Rhythm

Longevity may also be governed in another way, based on the extension of the period of hormonal infradian rhythm—T-rhythm, where T stands for time because it is responsible for the performance of the temporal program of individual development—the peaks of which are presumably coupled with chronomeres' truncation. This way is aimed at economizing chronomeres by increasing the length of this period, that is, at decreasing the frequency of the acts of chronomere shortening. This aspect of the problem demands a more detailed elucidation of the nature of factors that could influence the organismal infradian rhythms and, through them, the destiny of chronomeres and the rate of changes in the biological age of the organism as well. Consideration of the problems related to infradian rhythms enables one to reach conclusions beyond just gerontological interests.

PINEAL SAND

Infradian Biological Rhythms

Infradian biological rhythms with periods of different length are infrequently studied, but they undoubtedly exist. It is assumed here that the pattern of infradian rhythms can be influenced by fluctuations of the gravitational field and that mineral inclusions in the brain are created just for the purpose of regulating these processes.

The well-known example is a female menstrual cycle. Ambulatory, wheel-running, and drinking activities oscillate, showing 4-day rhythmicity in female rats, reflecting the behavioral changes in sexual cycles.[6] Studies of pineal gland size performed over a period of 10 successive days, at 6-h intervals under a lighting regimen of LD 12:12 have revealed that changes in its volume could not be correlated with the light/dark phases. Instead, the infradian rhythm was revealed with a 2-day period of changes in pineal gland volume.[7] Periodic changes in food intake and body weight (at 3- to 4-day intervals) were observed in adult control female rats.[8,9] An infradian rhythm with the average cycle of 9.6 days was recognized in all of the subjects in terms of the weight increment velocity in patients with Crohn's disease; the cycle of the rhythm was unaffected by energy intake, contents of nutritional therapy, and medical examination during the period.[10] Ant lion (Myrmeleontidae) might catch more prey on moonlit nights, so it has worked out a corresponding circalunar rhythm that includes pit building in the sand. In the laboratory environment, it was found out that ant lions dig bigger pits at a full moon than at a new moon, with a 29.53-day synodic lunar cyclicity. Observations were made in the conditions of laboratory isolation under all available constant outside influences, except a possible gravitational influence of the moon.[11]

The moon exerts a gravitational force on every object on and in the earth. In this respect, a contribution, though less than that of the moon, is also made by the sun, to which the earth comes somewhat closer twice a year. There is also the correlation of the semi-annual periodicity of average monthly data on the acceleration of tide changes in gravity and semi-annual changes in levels of activity of liver monooxygenase system in rats.[12] From a possible influence of periodical changes of the gravitational moon's pull, it does not at all follow that the period of all infradian biological rhythms should strictly correspond, for example, to new and full moon's phases of the lunar cycle, since endogenous infradian cycles should be orientated to a whole series of ecological and intrinsic requirements. It seems reasonable to assume that the length of period of endogenous infradian rhythm would be most stably maintained provided this length were equal to the length of gravitational period or were less or more by multiple times so that the peak of exogenous rhythm might coincide with one of the peaks of endogenous rhythm, serving as a guideline in maintaining rhythmicity.

Pineal Sand

In contrast to circadian rhythms,[13] the problem of infradian rhythms and their mechanism is insufficiently worked out in the biorhythmological literature. Offered below are some novel hypothetical mechanisms participating in their regulation. The pineal gland is considered as one of the participants in forming the infradian rhythms. Hypophysis generally plays a significant role in development and aging.[14-22] An important histological peculiarity of this gland is the presence of mineral inclusions. Larger inclusions are sometimes called "brain sand" and otherwise the inclusions are considered pineal sand. In this paper, I postulate a key role for pineal sand and its amorphous predecessor—very small scattered granules or microsands—in forming the infradian rhythms of mammals necessary for their proper development in time and in the process of aging.

Brain sands (also known as corpora arenacea, or acervuli) are calcified intercellular concretions in the pineal gland. Their presence has been known since the discovery of the pineal gland itself more than 400 years ago.[23,24] Interspersed in the tissue of the pineal gland are multiple biomineral granules, which can be somewhat different in their content and markedly different in form and size. They are much denser than cells. Mineral deposits are also found in pineal gland as intracytoplasmic scattered accumulations, or microacervuli.[25] Electron microscopic studies of calcification revealed both amorphous and crystalline deposits. The latter type had needle-shaped crystals.[26] Enlarged crystalline profiles of hydroxyapatite were observed within functionally stimulated pinealocytes and in the interstitial concretions.[27] The intensity of the growth of large inclusions is influenced by various physiological factors. For example, in pineal glands of South Indians, corpora arenacea appear at around 12 years of age,[23] though sand in a pineal gland has been noted in other human subjects at the age of 7 and even 2 years of age. The number of concretions significantly increases after puberty, and they are more frequently found in premenopausal women than in men.[28,29] The largest frequency of pineal acervuli was detected in Uganda's population, where there is the largest circannual light intensity.[30]

More-intensive presence of acervuli in older pineal glands was stated to be a result of a normal process of progressive degeneration of pinealocytes, which leads to the decrease in the amount of normal uncalcified tissue and consequently the decline of the pineal gland function.[31] The decrease in the level of melatonin secreted daily that occurs during aging was correlated with the age-related increase in the amount of calcified pineal tissue and with the corresponding decrease in the amount of normal uncalcified pineal tissue.[32] Small acervuli with regular round or oval shapes are localized at the gland's periphery. Their size increases during aging and their shape becomes more irregular.[31] However, smaller biomineral inclusions of irregular shape with uneven surface are dispersed across the whole pineal parenchyma.[26,31]

Pineal acervuli have had a practical importance for radiologists as a landmark during neuroimaging analysis. In spite of a long history of studies, the influence of pineal mineral inclusions on pineal function remains a genuine mystery and this is why they attract much research attention.[31] "Stones" in the brain not only may be beneficial when participating in the assumed infradian rhythms and in their coordination with the phases of the moon, but also may be harmful when these structures push the pineal parenchyma out to the gland's periphery[31] and cause calcification of this organ. With age, large inclusions, evolving from ultramicroscopic grains (which are supposed to be of vital importance for the organism) increase in number and, due to aggregation, become so large that they cannot be moved to the periphery.

Pineal tissue is mainly represented by the "light" pinealocytes, but it also contains "dark" pinealocytes, glial cells, etc. Light pinealocytes have a focal accumulation of secretory vesicles with synaptophysin (calcium-binding glycoprotein), a common constituent of these clear vesicles, and some other factors.[33] Vesicles locate at the cell periphery and fill the bulbous elongated cell protrusions denoted as bright cell processes. In some of these light pinealocytes, the sporadic presence of the needle-like crystalline profiles identified as hydroxyapatite is usually observed.[34] In some pinealocytes there are solitary concretions consisting of a crystalline core and a homogeneous or lamellar amorphous body. The latter aggregation, apparently, originates as modification of the former, "younger" needle-like inclusions. Concretions consisting of a crystalline core and amorphous bodies are also present between pinealocytes. If animals are stressed by immobilization, certain stimulated light cells are characterized by the increased presence of needle-like crystalline profiles,[27] although the numbers of bright pinealocytes with concretions does not differ among the stressed and not stressed animal groups. The pineal organs of lower vertebrates exhibit high calcium content as seen on ultrastructural calcium histochemistry, although concrements were not found.[35] It is not improbable that mineral ultramicrograins are present in the pineal glands of many species since it is possible that these micrograins are capable of dissolving irreversibly before becoming too large or that they do not grow to large sizes.

On the whole, in higher animals, the two mineralization forms are present in the pineal gland: polycrystalline concretions (which are up to several millimeters long, and which drew the attention of anatomists as a proper "brain sand") and microcrystals, the size of which varies from 1 to 20 micrometers.[36,37] Small grains and concretions, although probably biologically important, have for a long time slipped from observation. A possibility of reversibility of some concretions is clear from the example of the later disappearance of the "sand" present in the human embryonic anterior pituitary.[38,39] In some studies, cytoplasmic microacervuli within human pinealocytes were found from the postnatal age of 2 days to 86 years.[23] Various mineral pineal compounds that increase with age were predominantly composed of calcium and magnesium salts, and it was emphasized that though acervuli were especially numerous in old pineal glands, they did not reflect a pathologic state.[35] It is these deposits that put the gland into the shape of a pine cone, from which the pineal gland gets its name. The form of concretion relates to its location. Where the concretion was found adjacent to pinealocytes, it was crystalline, needle-shaped, and globular, but when adjacent to glial fibers or other connective tissue it was amorphous, concentric, and lamellar. In advanced years, a portion of crystalline inclusions decreases (it is they that are necessary in the young and mature organisms for the pineal lunasensor mechanism considered below), whereas amorphous concentric lamellar pattern concretions increase.[23]

Proposed "Lunasensor" Hypothesis

The proposed "lunasensor" hypothesis assumes that: (1) the pineal gland is an organ specializing in the sensing of periodic changes of gravitational attraction of the moon (and the sun), using pineal biomineral grains; (2) the pineal gland is involved in the formation of endogenous hormonal infradian rhythms, whose pattern is maintained by interaction with those hormonal changes, which are periodically evoked by the moon-related gravitational impacts; and (3) the resultant infradian hormonal rhythms are used as an element of the work of the "lifelong clock" located in the brain.

Thus, I postulate that ultramicroscopic biomineral grains are necessary for the pineal gland to perform its main function as a hypothetical lunasensor device. According to this suggestion, the lunasensor helps to use moon's gravitational pull to regulate intrinsic infradian rhythms of the organism. In some groups of animals, the pineal gland combines two vitally important functions (sensing of light and sensing of gravitation), but in mammals it reserves to itself only the infradian biorhythmological function because of its paramount role in regulating the lifelong clock of the organism. The need for such a "clock," its relation to the hypothalamus, and the importance of age-related changes of properties of the hypothalamus were first clearly pointed out by Dilman,[40] though the mechanism of the work of the clock remains largely unknown.

The current explanation of the role of the crystallization of Ca^{2+} in a brain is as follows. Inclusions are the means of maintenance of a noradrenalin-stimulated Ca^{2+} influx at an optimal level during attenuated pinealocyte turnover.[27] To assign a key role to this mechanism seems, however, erroneous. This function can at best be subsidiary rather than crucial. It is assumed that the biomineral grains, which are denser than cells, during definite phases of the moon are able to induce mechanical damage to adjacent pinealocytes. The main role of calcium-rich inclusions is, according to the present hypothesis, that they act as a needle or dagger that is able (under circumstances that are considered below) to plunge into the walls of pineal vesicles containing highly active substances. This should induce the cascade of further reactions, including activation of other cells of the pineal gland itself, as well as of the other neural and endocrine centers. Thus, an organism can sense the influence of the gravitational field of the moon (and the sun). The postulated pineal lunasensor enables one to differentiate the effect of weak gravitational extraterrestrial influence acting at the background of a stronger earth's gravitation.

The aforementioned function can be realized owing to the combination of several factors. First, each micrograin, as a miniature biomineral inclusion evolved by means of crystallization, often has pricking or cutting projections on its surface. Second, such micrograins may have numerous and immediate contacts with the surface of vesicles inside the pinealocyte; besides, the interstitially located grain also may have contacts with the outer surface of the vesicle-bearing long processes of pinealocytes. Third, the force of attraction of a grain towards the moon is higher as compared to that of a cell since this force is proportional to the mass of the two bodies in each pair, sand particle and the moon or cell and the moon, respectively. When changing its orientation in space relative to the adjacent pinealocyte (and its vesicles) under the influence of the gravitational field of the moon, the grain, under the particular conditions considered below, is able to disintegrate the pineal vesicles, thereby participating in the formation or modification of the pattern of the organism's infradian rhythms.

The force of the earth's gravitational field acting on the cell and on the nearby micrograin should not lead on its own to the damage of the vesicles because it is not useful for the organism, as this would create constant irritation of cells and no rhythmicity. Intracellular vesicles and the cell as a whole should have a certain degree of resistance to mechanical damage by microsands under the conditions of the background action of the earth's gravity. For this purpose, gaskets of some sort inside and outside pinealocytes could resist the damaging action of the earth's gravitation and hamper the untimely injury of a vesicle and cell by a grain. In this connection it should be noted that in photomicrographs of pineal slices one can see that the large mineral inclusions are usually surrounded with some material. Krstic detected fibrous baskets around the pineal sand particles, which connected acervuli with adjacent pineal parenchyma.[41] One can assume, however, that when the force of the earth's gravity is added to the gravitational force of extraterrestrial origin, then the resistance of vesicles and their cells will be insufficient, that is, the integrity of their gaskets or lining will be impaired. As a result, the devastating interaction of the rugged surface of a grain with the intrapinealocytic vesicle or the whole cell inevitably occurs, depending on the size and position of the mineral granule.

Depending on the degree of resistance of these gaskets or lining gadgets formed in the course of evolution of a given species, animals will react by pouring out the bioactive content of their pinealocytic vesicles at different phases of lunar cycle. For example, they could respond only to the strongest extraterrestrial gravitational influence during the phase of the new moon or their vesicles could be activated also by the full moon, etc. It is significant that, as seen on histologic study of slices of the pineal gland, grains lay near some groups of cells and far away from others. That is why, even under the strongest gravitational influence of the moon, grains are not able to deactivate all pinealocytes simultaneously. Moreover, it is even possible that the slow displacement of micrograins and cells relative to each other takes place in order to make the next clusters of pinealocytes accessible to gravitational influence and is strictly controllable by pineal gland itself. By the way, if so, it is necessary to accept that intracellular micrograins are forming not simultaneously around all pinealo-cytes, but successively in definite cell clusters. This would help to avoid the total si-multaneous activation and subsequent destruction of all pinealocytes of the gland. Humbert and Pevet established that the two aforementioned populations of pinealo-cytes present in pineal gland differ in the following way. Light pinealocytes are nor-mal, functional pinealocytes, whereas dark pinealocytes are cells in different stages of degeneration. Using lanthanum nitrate, which is solely an extracellular cation, they demonstrated that it was present inside the cytoplasm of dark pinealocytes. That indicated the breakdown of dark pinealocytes plasma membrane.[26]

By virtue of the addition of the lunar attraction to the terrestrial one, the biomin-eral micrograins whose density could be about 1.5–3 times higher than that of the cytoplasm must move by gravitation to the adjacent net of pinealocytes immobilized by the connections between cells of pineal organ. This situation partly resembles how the ocean water is shifted relative to the rigid mainland under the influence of gravitational forces of the moon and sun during tides. The abovementioned long pi-nealocytes' processes, which are filled with vesicles containing bioactive substanc-es, are mostly clustered.[27] When clustered (i.e., mechanical tie-ups of cells and their long processes), components of a cluster under the influence of gravitational chang-es do not alter their mutual orientation. As opposed to them, biomineral inclusions represent individual bodies capable of changing their position in space. A very small shift in the cells' position relative to the nearby mineral granule should exert a great influence on the whole organism through a gravity-dependent mechanically induced excretion of pinealocytic factors, which act on uninjured cells of the gland. The re-sulting signal could be transmitted across the gland and outside both in humoral way and even through nerves, taking into account the abundance of the nerve fibers there.[42]

Immobilization of pinealocytes plays the role of a precondition in that the grain becomes capable of inflicting pricks on their targets when the environmental gravi-tational situation changes. The shift of the grain relative to the cell serves as a trigger, or detonator, of a cascade of subsequent endocrine events. As a result, the organism perceives and senses the gravitational influence of an extraterrestrial origin even at the background of the permanent gravitational field of the earth. It is pertinent to make some interpretations regarding the physical and biological aspects of the pineal organ as a hypothetical device—the lunasensor.

As mentioned above, it uses mineral granules that are denser than cells, and hence granules must be able to behave as sediment that falls under the influence of the gravitational field of the moon (and sun), even at the background of the invariable earth's gravitation, and be engaged in the process of sedimentation. The velocity of a spherical particle settling in a viscous fluid is known to be described according to the Stokes law by the equation:

$$Vg = D^2 g(\rho_1 - \rho_2)/18\eta,$$

where Vg is the sedimentation velocity, D is the particle diameter, g is gravitational attraction, ρ_1 and ρ_2 are particle and liquid densities, respectively, and η is fluid shear viscosity. This equation clearly demonstrates a potential ability of the particles denser than cells to move over the tissue under the influence of the gravitational field, and it follows from the equation that the greater the difference between densities of the moving particle and the surrounding medium, the more effective the process of the particles' displacement. It should be emphasized that the dense biomineral particle does not need to travel a long distance to initiate a biological effect via breaking the integrity of biological membranes, provided that one of the cutting edges of the micrograin may already be in direct contact with the gasket that separates it from the target vesicle. A minimal shift is enough for this "ampule" to break. Since the settling rate is proportional to the square of the particle size (D^2), any increase in the size of the particle will have a considerable enhancing effect on the rate of the shift of the particle's position relative to the target. Although pineal mineral particles have various forms, the process of their gravitational "intratissue sedimentation" can obey the Stokes law. If so, a grain should sink in tissue like a stone (density is about 2.7 g/cm^3) that sinks in water (density of 1 g/cm^3) owing to the gravitational attraction of the earth. Average density of mammalian cells, cytoplasm included, usually varies around 1.04–1.10 g/cm^3; therefore the biomineral grain (its assumed density is about 1.5–3 g/cm^3) must also "sink" in the gland.

Why does this not occur? The reason is, apparently, that mineral inclusions in the pineal gland remain in their positions with the help of the abovementioned gaskets and fibrous baskets, which function as suspension clips that mechanically hold up the "stone" particles. Because of these gadgets, the particles are suspended and they do not settle down and behave as particles in the medium of high viscosity. Properties of these gadgets were selected in the course of evolution so that sinking wouldn't occur without the influence of the extraterrestrial gravitational field, so the pineal organ does get an opportunity to recognize the moon's gravitational pull and to respond to it in a proper way. In order to fulfill their triggering in a still-young pineal gland, the mineral inclusions need not grow constantly. A working size of high-density granules, sufficient for their triggering mission to be realized, does not demand to have even the size of their cellular targets. If so, why then does the pineal gland have mineral concretions that are many times larger than every pinealocyte and that are scattered across the tissue of an old pineal gland as boulders across the field? According to accepted opinion, the appearance of large inclusions and their gradual accumulation over time, especially in advanced age, occurs merely because the pineal gland is not able to stop the process of calcification, which it itself gives birth to.

Since the pineal gland is filled up by pineal sand with age, this process is sometimes considered as one of the factors of age-related involution of pineal gland itself. I propose, however, an entirely different explanation for this interesting phenomenon: In addition to the other functions, glial cells of the brain are known to fulfill a function of connective tissue cells, and glial activation is one of the manifestations of brain aging.[43] Factors from the brain of aging mice are able to stimulate the proliferation of cells of glial origin in vitro and can potentially be responsible for gliosis of an old brain.[44] In this context, one may suppose that enlargement of biomineral granules with age is not at all a result of violation of calcium-related metabolic processes, but the result of homeostatic mechanisms compensating for the age-related increase in rigidity of pineal tissue occurring in parallel with the common age-related augmentation of the connective tissue in aged brain, pineal gland included. Increased stiffness of pineal tissue should directly hinder the proper function of the lunasensor, because this increasing stiffness can be accompanied by an increasing stiffness and/or excessive thickness of the gaskets that protect cells from being pricked by spiny grains. I propose a compensatory response of the pineal organ as a lunasensor device under these conditions. As one can deduce from Stokes law, an excessive rigidity of the tissue can hinder gravitation-dependent movements of the grain relative to the adjacent cell, and can be overcome (necessary for the full functioning of the lunasensor) in three ways: The first way would be to decrease the viscosity of the medium (namely, decrease of the stiffness of tissue) and the second would be to increase the density of the mineral particles. In all probability, both of these possibilities are not available for the pineal gland. However, the third way, namely increasing the size of the mineral inclusions, is possible. Since the sedimentation velocity of a particle is directly proportional to the square of the particle's diameter, a two-fold increase of the particle diameter, for example, will produce a four-fold increase in the velocity. That is, the larger the particle, the more viscous medium can be and yet not affect the time required to pass the same distance, other conditions being equal. Thus, the pineal organ, owing to increase in the size of its mineral concretions, continues to function as a lunasensor despite the fact that the viscosity of its working medium is changing with age in an unfavorable direction.

As to the presumptive destruction of cellular vesicles by a grain, it is pertinent to take into account the equation $F = P * A$, where pressure (P) is the application of force (F) to an area (A). The smaller the area of the target to which the pressure is applied, the more devastating the result, since the concentration of force in a given area increases. For example, a finger can be pressed against a wall without consequence; however, the same finger pushing a thumbtack can damage the wall, because the point concentrates that force into a smaller area, though the force applied was the same. An even more impressive effect can be seen in the action of a feeble mosquito when it easily pierces the skin of a blood donor using its fine proboscis. In the case of pineal grains, the higher the density of the particle, the more the gravitational force of its attraction to the moon. Correspondingly, the greater this gravitational force, the stronger the pressure of a mineral grain's spikes on the target. That is why, even extremely low lunar gravitation can exert a huge effect on an organism via its lunasensor device. The release of the vesicles' bioactive substances after perforation of only one or several pinealocytes is probably quite enough to launch a cascade of the reactions activating a lot of uninjured cells of the pineal gland and other regions of a brain. The pineal gland as a lunasensor organ has two very strong amplifying cascades, one is physical (based on mineral granules) and the other is biochemical (liberation from the pinealocyte of its hormones, Ca^{2+}-regulated factors, etc.). It is possible that nerves are involved in this cascade as well, because specific contact areas are frequently observed between the nonmyelinated nerve fibers and the terminal clubs of interstitial cell processes.[45] Nerve signaling could be used for the most selective activation of those hypothalamic neurons that supervise production of certain factors, such as a growth hormone, which could play an important role in infradian rhythms and chronomeres' shortening (see below for a possible growth hormone role). Because of the lunasensor of the pineal organ, our eternal celestial satellite can be involved in the organization of the pattern of infradian rhythms necessary for "ticking" of the lifelong clock of the organism.

It is important to emphasize that for the functioning of the proposed lunasensor, a mutual orientation of the pineal gland and the moon is of no importance. Wherever the head of the animal may turn at its movements, the lunar gravitation field is always capable of inducing an intratissue displacement of some grain respective to its cellular neighbors as cells and grains alternate with each other like various sweets in a pudding. The work of the lunasensor is similar to the workings of tides. The pull of the planet's tides is known to be almost the same whether the sun and moon are lined up on opposite sides of the earth (full lunar phase) or on the same side (new lunar phase). When the moon is at the first quarter or last quarter phase (i.e., the moon is located at the right angle to the earth-sun line), the sun and moon partly compensate each other, and tides turn out much lower. It is also known that the moon returns to the same phase every 29.53 days (the synodic lunar month) and it returns to the same background stars every 27.3 days (the tropical lunar month). As applied to the work of the lunasensor apparatus of a pineal gland, all this should correspond to the statement that the maximum release of the content of pinealocytes' vesicles under exogenous gravitational influence should happen approximately every two weeks, rather than once a synodic or tropical lunar month.

In order that the chronomeres of the lifelong clock are not exhausted too quickly, the peaks of the infradian rhythm participating by means of their hormones in the process of chronomere shortening must be as rare as possible. This does not mean that the length of a period in short-living and in long-living mammals must always be two weeks. The function of lunasensor is not only to help a species to adapt to a rhythmically varying gravitational environment, but also, first and foremost, to serve as a sensor of the periodic signals used in long-term maintenance of its own endogenous infradian rhythmicity. Exogenous gravitational periodicity perceived by means of the lunasensor is used by the organism as a guideline, much as a metronome guides a musician's playing. Periodic coincidences of maxima of the exogenous and endogenous rhythms are used by organisms for collation (i.e., as checkpoints). These checkpoints are necessary to support and control the accuracy of organismal endogenous rhythmicity. Depending on a ratio of the lengths of periods of exogenous and endogenous rhythms, the frequency of these checking events can be distinct in different animal species. Correspondingly, the length of the endogenous period of infradian T-rhythm can be species specific. The period length of the T-rhythm can change with age in some limits, but it is nevertheless a strictly species-specific trait. For example, an average period length of T-rhythm of long-living humans should be considerably longer than in short-living mice. This enables long-living species to much more sparingly spend, in the course of time, the length of their chronomeres. Peaks of T-rhythms are used as the pacemaker signals of the lifelong clockwork of the brain. Ticking of this clock is realized by the periodically repeated shortening of chronomeres in postmitotic neuroendocrine cells which occurs just at the maxima of T-rhythms. The presence of sand particles is a marker that indicates a precise region of the brain involved in the generation of infradian pacemaker signals. The question about how endogenous infradian rhythms are organized with the periods of different length in various species is beyond the framework of the present communication and will be considered elsewhere.

The lunasensor principally differs from the devices using mechanosensitive ion channels ("gravisensors"[46]), from the animal organs that use otoliths to sense position relative to the earth's gravitation axis,[47] and from the mechanisms used in plants for their gravitropism.[48–50] The mechanism used by *Paramecium* and *Euglena* for perceiving terrestrial gravitation which uses the difference in densities of the whole cytoplasm and the extracellular medium is similar to but is a distinct invention.[51] As opposed to such kind of gravisensors, the lunasensor is a device that allows the perception of the extraterrestrial gravitation of the moon (and sun) at the permanent background of the terrestrial gravitation in order to reliably maintain the intrinsic infradian rhythmicity, which is vitally necessary to control the flow of biological time in a developing, mature, and aging organism.

TWO LUNASENSORS

The "Adult" Lunasensor of the Pineal Gland and the Embryonal Lunasensor of Anterior Pituitary

The lunasensor-like mechanism could work in other hormonally active tissue(s) besides the pineal gland. It should be noted that calcified deposits are seen in tissues other than pineal tissue. They are observed in the choroid plexus of lateral ventricles, meninges, and in the habenular commissure.[23,35] It would be of interest to elucidate whether the grains play the same role as the lunasensor or are merely a sequel of drifts of the pineal factors that are responsible for calcification. Answering this question is especially worthwhile because the choroid plexus plays a wide range of roles in brain development, maturation, the aging process, and endocrine regulation.[52]

One can suppose that relatively primitive animals using chronomeres and their own variant of lunasensor could not have their own endogenous infradian rhythms. Instead they could use only exogenous gravitational rhythm as a source of T-rhythm, necessary to regulate the rate of chronomere shortening and aimed at arranging the individual development in time. As a basis for their lunasensor they could use, in place of pineal calcified granules, any other structures drastically differing in density from their adjacent cells.

This is apparently true even for the embryonal stages of human development: at the early development of the human brain, when there is no mineralization in the pineal gland (and thus the pineal lunasensor cannot function), mineralization is found in the adenohypophysis (anterior lobe of the pituitary gland). In studying anterior pituitary glands of hundreds of human fetuses and infants ranging in age from 15 weeks of gestation to 1 year of life, calcified concretions were found in all cases up to 1 month of life.[38,39] These mineral concretions decreased in incidence postnatally and were rarely seen after 6 months of age. Most were round to ovoid and measured between 5 and 30 micrometers in diameter. It is significant that the calcifications followed no particular pattern of distribution among the most prevalent pituitary cell types. Ultrastructural examination revealed small single or multiple intracellular calcified deposits, and larger extracellular calcifications. The lamellar nature of the calcified bodies suggested that waves of calcium deposition occurred during morphogenesis.[53] Calcification in early anterior pituitary gland represents a normal finding in fetuses and young infants.[38] The anterior pituitary gland in humans makes up 70–80% of the pituitary.

Besides growth hormone, which is required for normal growth in the developing child, the anterior pituitary gland produces and secretes major hormones, such as adrenocorticotropic hormone, follicle-stimulating hormone, luteinizing hormone, prolactin, and thyroid-stimulating hormone. Do the dense mineral inclusions play a role similar to that of the components of the lunasensor? In the pituitary, the embryonic lunasensor, with its ovoid biomineral granules, could not prick neighboring cells, but could squeeze them and thus modify the intensity of synthesis and secretion of hormones. It is extremely interesting that sand appears in the pineal gland approximately just when it disappears in the anterior pituitary gland!

The shortening of chronomeres is indispensable for temporal control, but endogenous infradian T-rhythm, without an exogenous guideline, is likely not to keep its periodicity stable. Therefore the organism responds to exogenous infradian rhythm (orientating to the lunar cycle). The major hormones of anterior pituitary are likely candidates for the role of pivotal participant of T-rhythm. It seems likely that, because of its very important general significance for the organism, growth hormone is the most likely candidate.

Pituitary "stones" can stimulate the emission of other hormones and calcifications follow no pattern of distribution among the pituitary cell types. Such distribution is the manifestation of the imperfection of the embryonic lunasensor. The pineal lunasensor operates much more precisely, activating, possibly with participation of the nerves, a particular group of cells in the hypothalamus.

Thus, it is predicted here that during evolution a lunasensor physiological system developed. It uses pineal sand in adult higher animals and anterior pituitary gland sand in embryos. The pineal lunasensor acts more selectively than an embryonal lunasensor of an anterior pituitary, which is why it replaces the more primitive pituitary lunasensor as the organism develops.

THE ROLE OF GROWTH HORMONE

Growth hormone can serve as a basis of infradian rhythms in a developing organism for both the embryonic lunasensor of anterior pituitary gland and the pineal lunasensor in postnatal and adult life of the mammalian organism. A chain of events relating the involvement of growth hormone to the work of the lifelong clock is surely longer in the adult lunasensor than the embryonal lunasensor. Liberation of the content of pinealocyte secretory granules is realized with the aid of exocytosis.[54] By means of the lunasensor's identifying gravitational changes, changes in the position of the moon are able to considerably increase the release of the factors of pinealocyte vesicles in a corresponding time interval. Exceeding a certain threshold in the level of excreted pinealocyte vesicular factor(s) should trigger a surge of endocrine activity in the corresponding cells of the hypothalamus and other brain centers. The pineal gland can in particular influence the ventromedial hypothalamic nucleus, which in turn regulates various autonomic, endocrine, and behavioral activities.[55] In the long run, the corresponding range of events leads to the anterior pituitary and evokes the peak release of growth hormone, especially into the cells of the lifelong clock of a brain. Although other participants of the event are admissible, it is growth hormone that seems to be the basis of the hormonal cocktail, the oscillations of which correspond to the infradian T-rhythm. In this scenario, the pineal gland is thus given the role of a pacemaker of the brain's lifelong clock. However, it does not follow from this that chronomeres should localize in the same place as a pacemaker, namely, in the pineal gland.

The lifelong clockwork of the brain is not less complicated than, say, the heart with its pump and pacemaker functions located in distinct structures. The lifelong clock itself with its chronomeres localizes outside the pituitary gland as well as outside the pineal gland, although it is unable to function without these two glands. The pacemaker's infradian signals, emanating from the pineal gland, achieve their targets in two pathways. One of them uses cerebrospinal fluid, what is favored by the absence of a hematoencephalic barrier in the pineal gland. This pathway permits the direct humoral contact of the pacemaker's signal with those brain targets, which are parts of the circumventricular system and, as such, have a direct contact with the cerebrospinal fluid. The second pathway is neural. As to intrapineal nerve cells giving rise to pinealofugal neuronal projections, they are reduced in the course of phylogeny.[56] Nevertheless, together with other factors, the nervous signaling from the pineal gland can potentially also be used for the aiming activation of those neurons of the hypothalamus that control production of the growth hormone–releasing hormone (GHRH), whereupon the burst of GHRH can evoke the intensive liberation of growth hormone by anterior pituitary, whose level is then abruptly increased and corresponds to a peak of T-rhythm.

Thus, within the framework of the developed notion, the main executive hormone of T-rhythms is nothing but growth hormone. It acts both in an embryo and in an adult, impelling during the peak of T-rhythm the super-high-speed transcription of chronomeres of the lifelong clock, thus forcing the clock to tick. In the framework of the proposed conception, the pineal gland and the hypophysis participate, at different periods of ontogeny and in different ways, in the generation of T-rhythms, without which the lifelong clock of the organism can not operate. The pineal gland is necessary for the organism as a pacemaker of the lifelong clock. However, it produces other vitally important factors (melatonin, peptides, etc.), as shown by the work of Pierpaoli and his coauthors. They found that old animals implanted with a young pineal gland had a considerably increased life span and young animals implanted with an old pineal gland had a considerably reduced life span.[14–22] If the pineal pacemaker forces the lifelong clock to tick, why doesn't pinealectomy create an immortal organism? After all with pinealectomy, the flow of time is stopped when the ticking is stopped. It is likely that the removal of the pineal gland also removes from the organism factors (including melatonin) required for normal functioning.

BRAIN ANATOMY AND THE LIFELONG CLOCK

In considering the proposed roles for the anterior pituitary and pineal glands as components of the lifelong clock, certain aspects of brain anatomy and blood circulation in the hypothalamus–anterior pituitary gland complex should be noted.

It is commonly accepted that cerebrospinal fluid serves several functions in the brain: as a shock absorber to avoid concussion, as a conduit of nourishment, and as a remover of toxic factors. Yet, the transport functions are well handled by the blood system and so this fluid is likely only used for shock absorption. The central nervous system starts forming as a system of cisterns and cavities (ventricular system). What is the reason for such strange anatomy? Is it possible that one of the reasons is to create the opportunity of functioning of the mechanism of the lifelong clock that needs peaks of infradian rhythms? To achieve this purpose, it is necessary for the infradian humoral signals generated by the pineal gland to achieve their targets at the peak concentration of pineal factors. Providing that these pineal factors merely come into blood circulation, their concentration would decrease because of dilution in the great volume of blood. Moreover, before attaining their targets in the hypothalamus, signal substances of the pineal organ would be subjected to metabolic activity and destruction. To exclude such variables and to be able to immediately deliver the proper high level of pineal infradian signal factors, the targets of these factors should be easily attainable and bypass the blood circulation. The ventricular system provides this bypass. The cerebrospinal fluid is both the transport and the integrator system, delivering infradian signals at once to all exposed targets.

For the successful work of the lifelong clock it is necessary to prevent the dilution of factors entering the hypothalamus from the pineal gland and the anterior pituitary which can produce the peak level of growth hormone as a basis of T-rhythm. The peak of growth hormone secretion would fail to accomplish its mission of abrupt activation of cells of the hypothalamic lifelong clock if hormones from the anterior pituitary were carried only in the general circulation of blood. The evolutionary invention made here was a creation of a special local blood circulation system. Growth hormone–rich blood should go from the pituitary not only into the general blood flow, but also directly into the hypothalamus. Only then can cells of the hypothalamus get the maximum concentration of growth hormone. Thus growth hormone is necessary to initiate in hypothalamic cells of the lifelong clock the hypothetical process of the "scrupting" of chronomeres. *Scrupting* is defined as the super-high-speed transcription-dependent shortening of the chronomeric linear DNA molecule occurring even in a nondividing cell due to the rupture of the chronomere terminus immobilized in a chromosomal DNA. Details of scrupting are presented elsewhere.[1,2] The main requirement for scrupting is a powerful speeding up of the transcription process under the influence of a short-term high concentration of the factor that could initiate super-high-speed transcription. The role of such a factor is attributed here to growth hormone.

The anterior pituitary, where growth hormone is produced, has two capillary plexuses, primary and secondary. It is worth noting that such a trait is absent from the posterior pituitary, since it does not need to send a high dose of its factors to the hypothalamus. The primary capillary plexus collects the neurosecretions, GHRH included, of hypothalamus and transmits them to secretory cells of the anterior pituitary, which responds by excretion of their own hormones including growth hormone. Adrenohypophyseal hormones now go into the secondary capillary plexus. Then growth hormone–rich blood goes at last, via efferent vessels, into the general circulation, including the hypothalamus itself. It is important that this growth hormone–rich blood enters the hypothalamus immediately. By virtue of its blood supply, the hypothalamus gets the highest available dose of growth hormone. This occurs at the peak of infradian rhythm. Just at this moment, in the cells of hypothalamus, the act of chronomeres' shortening occurs, and in that way the act of ticking of the lifelong clockwork is put into effect.

METABOLIC SYNDROME AND GRAVITATION-DEPENDENT EPIGENETIC PROGRAMMING OF PHENOTYPE

Interestingly, there is another side to the embryonic anterior pituitary lunasensor. The activation of the pituitary lunasensor (as opposed to pineal lunasensor) affects all types of cells in the anterior pituitary, thus provoking an intense release of all its hormones (including ACTH), rather than just GH. A special sensitivity of the brain to the "metabolic programming," committed by the shifts in hormonal levels[57] (especially by glucocorticoids), creates a certain dependence for the developing brain of the embryo on the changes of extraterrestrial gravitation during different phases of the moon. Too great a hormonal shift in the blood of a fetus induced by the pituitary lunasensor during a critical period of embryonic development could evoke changes in the metabolism of the adult organism—in effect "gravitational metabolic programming."

Barker's conception (i.e., that environmental factors acting during fetal development might be partly responsible for some state of health in adult life[58,59]) and the idea of endocrine disruptors as its endocrine foundation now constitute the cutting edge of a modern endocrinology.[57,60] Epigenetic processes (changes in DNA methylation and genetic imprinting, histone modifications, etc.) have been shown to be highly sensitive to the influence of abrupt deviation in levels of some major hormones. This is why even a brief period of hormonal manipulation in early life may have implications for health outcomes, which become evident in adulthood as phenotypic deviations that range from barely perceptible to very pronounced.[57,61-65] For example, an artificially elevated insulin concentration in neonatal rats results in increased weight gain in juvenile life and adulthood accompanied by hyperinsulinemia, impaired glucose tolerance, and increased systolic blood pressure in adulthood.[66] Fetal glucocorticoid programming results in low birth weight with subsequent adult hyperinsulinemia and hyperglycemia. Moreover the "programming phenotype" also includes intergenerational inheritance, observed even in a second generation.[67]

Hormonal shifts caused by any source, including those induced in the embryonic hypothalamus by anomalously high concentration of hormones from the anterior pituitary (for example, during an unusual gravitational environment), can be epigenetically "memorized" by the organism. In altering the hypothalamic-pituitary-gonadal axis in early life, some phenotypes could be altered between norm and disease, and sometimes even lead to fortunate deviations from the most common phenotypes. As a "double-edged weapon," there can be overt conflict between the two activities of the pituitary lunasensor: it can generate infradian rhythm and it can generate gravity-dependent variants of the metabolic programmed polymorphisms. If we take into account the possibility of variable gravitational influences of the moon, sun, and planets perceived by the organism through the embryonic lunasensor, then we might consider that some traditional astrological statements were discarded not so much due to lack of observation or incompatibility with science, but rather to the total absence of any scientifically feasible interpretation.

MOON'S GRAVITATIONAL INFLUENCE
ON LENGTH OF GESTATION

With its homeostatic mechanisms, an embryo might be able to adjust the time-table of its development with an evenly changing gravitational environment. In doing so, its homeostatic mechanisms are trying to maximally protect the organism when passing through critical period(s)—those developmental stages when the fetus is especially sensitive to the hormonal shifts induced by the moon's pull and perceived by the lunasensor. The homeostatic systems responding to exogenously induced hormonal shifts could accelerate or retard the rate of almost any developmental stage. This is of special value with regard to the gravitational situation during development. Changing this rate, the embryo optimizes the course of its morphogenesis. But it pays for this by deviation of the gestational span from an average value typical for a given species. At the same time, the fetus gets a more optimal hormonal environment for its development. This statement is in line with the data on the wide variability of the length of pregnancy within a species. For example, gestation in mice lasts 17–21 days, in cats it continues from 63–71 days, in the African buffalo *Syncerus caffer* it can last 10–11 months, and in the giraffe 395–425 days. In humans, the length of pregnancy varies in different women and even from pregnancy to pregnancy in one woman. The birth of a mature human fetus after 280 days occurs only in 5% of all births. More than 10% of all pregnancies last more than 42 weeks, 14% last more than 43 weeks. The variation is quite wide—from 221 to 328 days.[68] Summarizing, one may conclude that other conditions (absence of disease, normal nutrition, etc.) being equal, the cause of this variability springs from the capacity of the supersystem of mother-fetus to "choose" the optimal duration of every stage of embryonal development, depending on gravitational (and, hence, hormonal) situations.

CONCLUSION

Further elaboration of the redusome theory of aging is presented. Redusomes, or redumeres, are hypothetical perichromosomal organelles. These transient structures are created in the course of morphogenesis. Redusomes are represented in the organism by two types—printomeres (in mitotic cells) and chronomeres (in postmitotic neuroendocrinal cells).[1] Chronomeres are pivotal components of the lifelong clock of a brain. Printomere shortening occurs in the course of cell doublings, and it is solely responsible for the existence of the Hayflick limit.[69–72] In contrast, chronomere shortening is responsible for the temporal control of individual development and aging. That is, the primary cause of aging is the shortening (that occurs under influence of infradian hormonal rhythm) of the material carrier of the temporal program, which is vitally necessary for development. Continuation of this shortening after maturation of the organism leads to molecular, chronic, and universal disease— aging. Telomere shortening in mitotic cells *in vivo* is merely a harmless and naïve witness of organismal aging, rather than its cause. Success with immortalization of dividing cells[73,74] by telomerase was caused by extension of their redusomes (printomeres), rather than telomeres. Telomere elongation is also an innocent bystander of printomere elongation performing by telomerase.

It is worth emphasizing that nothing but the analysis of the Hayflick limit led in 1971 to predictions of telomere shortening and the existence of telomerase,[3–5] and recently the consideration of the weakness and the advantages of the telomere hypotheses led to the redusome theory of aging.[1] This work was inspired by the pioneering studies of Pierpaoli on the pineal gland.

For a long time, a sand glass has been used as a symbol of the passing of time and as a means of its measurement. It is symbolic that the sand of a brain is used by Nature itself as a "sand clock" to control the flow of biological time.

ACKNOWLEDGMENT

This study was performed with partial support of the Russian Foundation for Basic Research (Grant No. 04-04-49600.).

[*Competing interests*: The author declares that he has no competing financial interests.]

REFERENCES

1. OLOVNIKOV, A.M. 2003. Redusome hypothesis of aging and biological time control in individual development. Biochemistry (Moscow) **68:** 2–33. http://protein.bio.msu.ru/biokhimiya/contents/v68/full/68010007.html
2. OLOVNIKOV, A.M. 2004. Role of hypothetical nuclear organelles—redusomes—in morphogenesis, aging, and cancer. *In* Frontiers of Neurodegeneration and Disorders of Aging: Fundamental Aspects, Clinical Perspectives, and New Insights. NATO Science Series. Series I: Life and Behavioural Sciences. T. Ozben & M. Chevion, Eds.: **358:** 89–98. IOS Press. Amsterdam.
3. OLOVNIKOV, A.M. 1971. Principle of marginotomy in template synthesis of polynucleotides. Dokl. Akad. Nauk. SSSR **201:** 1496–1469.
4. OLOVNIKOV, A.M. 1973. A theory of marginotomy. The incomplete copying of template margin in enzymic synthesis of polynucleotides and biological significance of the phenomenon. J. Theor. Biol. **41:** 181–190.
5. OLOVNIKOV, A.M. 1996. Telomeres, telomerase, and aging: origin of the theory. Exp. Gerontol. **31:** 443–448.
6. SHINODA, M. *et al.* 1988. Differences of behavioral rhythms observed by flat cage and running-wheel cage in female rat. Jikken Dobutsu **37:** 463–468.
7. DIEHL, B.J. *et al.* 1984. Day/night changes of pineal gland volume and pinealocyte nuclear size assessed over 10 consecutive days. J. Neural Transm. **60:** 19–29.
8. FERNANDEZ-GALAZ, C. *et al.* 1990. Body weight rhythmicity in the unweaned female rat following neonatal estrogen treatment. Physiol. Behav. **48:** 273–276.
9. MERCER, L.P. *et al.* 1993. Weanling rats display bioperiodicity of growth and food intake rates. J. Nutr. **123:** 1356–1362.
10. AYABE, T. *et al.* 1994. An infradian rhythm of weight increment velocity in patients with Crohn's disease during nutritional therapy. Nippon Shokakibyo Gakkai Zasshi **91:** 250–256.
11. GOODENOUGH, J. 1993. Perspectives on Animal Behavior. John Wiley & Sons. New York.
12. BORTNIKOVA, G.I. & I.R. MAVLIANOV. 1998. Correlation of fluctuations in the activity of the rat monooxygenase system with dynamics of tidal gravitational changes. Biofizika **43:** 600–602.
13. REDDY, A.B. *et al.* 2005. Circadian clocks: neural and peripheral pacemakers that impact upon the cell division cycle. Mutat. Res. **574:** 76–91.
14. BULIAN, D. & W. PIERPAOLI. 2000. The pineal gland and cancer. I. Pinealectomy corrects congenital hormonal dysfunctions and prolongs life of cancer-prone C3H/He mice. J. Neuroimmunol. **108:** 131–135.
15. PIERPAOLI, W. 1998. Neuroimmunomodulation of aging. A program in the pineal gland. Ann. N.Y. Acad. Sci. **840:** 491–497.
16. PIERPAOLI, W. *et al.* 1997. Circadian melatonin and young-to-old pineal grafting postpone aging and maintain juvenile conditions of reproductive functions in mice and rats. Exp. Gerontol. **32:** 587–602.
17. PIERPAOLI, W. & V. LESNIKOV. 1997. Theoretical considerations on the nature of the pineal "ageing clock." Gerontology **43:** 20–25.
18. PIERPAOLI, W. & V.A. LESNIKOV. 1994. The pineal aging clock: evidence, models, mechanisms, interventions. Ann. N.Y. Acad. Sci. **719:** 461–473.
19. LESNIKOV, V.A. & W. PIERPAOLI. 1994. Pineal cross-transplantation (old-to-young and vice versa) as evidence for an endogenous "aging clock." Ann. N.Y. Acad. Sci. **719:** 456–460.
20. PIERPAOLI, W. & W. REGELSON. 1994. Pineal control of aging: effect of melatonin and pineal grafting on aging mice. Proc. Natl. Acad. Sci. USA **91:** 787–791.
21. PIERPAOLI, W. 1994. The pineal gland as ontogenetic scanner of reproduction, immunity, and aging. The aging clock. Ann. N.Y. Acad. Sci. **741:** 46–49.

22. PIERPAOLI, W. *et al.* 1991. The pineal control of aging. The effects of melatonin and pineal grafting on the survival of older mice. Ann. N.Y. Acad. Sci. **621:** 291–313.
23. KOSHY, S. & S.K. VETTIVEL. 2001. Varying appearances of calcification in human pineal gland: a light microscopic study. J. Anat. Soc. India **50:** 17–18.
24. HALBERG, F. *et al.* 1999. The story behind: pineal mythology and chronorisk. The swan song of Brunetto Tarquini. Neuroendocrinol. Lett. **20:** 91–100.
25. GALLIANI, I. *et al.* 1989. Histochemical and ultrastructural study of the human pineal gland in the course of aging. J. Submicrosc. Cytol. Pathol. **21:** 571–578.
26. HUMBERT, W. & P. PEVET. 1995. Calcium concretions in the pineal gland of aged rats: an ultrastructural and microanalytical study of their biogenesis. Cell Tissue Res. **279:** 565–573.
27. MILIN, J. 1998. Stress-reactive response of the gerbil pineal gland: concretion genesis. Gen. Comp. Endocrinol. **110:** 237–251.
28. ZIMMERMANN, R.A. & L.T. BILANIUK. 1982. Age-related incidence of pineal calcification detected by computed tomography. Radiology **142:** 659–661.
29. TAPP, E. & M. HUXLEY. 1972. The histological appearance of the human pineal gland from puberty to old age. J. Pathol. **108:** 137–144.
30. MUGONDI, S.D. & A.A. POLTERA. 1976. Pineal gland calcification in Ugandans. A radiological study of 200 isolated pineal glands. Br. J. Radiol. **49:** 594–599.
31. ANTIC, S. *et al.* 2004. Morphology and histochemical characteristics of human pineal gland acervuli during the aging. Facta Universitatis Series. Med. Biol. **11:** 63–68.
32. KUNZ, D. *et al.* 1999. A new concept for melatonin deficit: on pineal calcification and melatonin excretion. Neuropsychopharmacology **21:** 756–772.
33. REDECKER, P. & G. BARGSTEN. 1993. Synaptophysin—a common constituent of premptive secretory microvesicles in the mammalian pinealocyte: a study of rat and gerbil pineal glands. J. Neurosci. Res. **34:** 79–96.
34. HUMBERT, W. & P. PEVET. 1991. Calcium content and concretions of the pineal glands of young and old rats. Cell Tissue Res. **262:** 593–596.
35. VIGH, B. *et al.* 1998. Comparative histology of pineal calcification. Histol. Histopathol. **13:** 851–870.
36. KODAKA, T. *et al.* 1994. Scanning electron microscopy and electron probe microanalysis studies of human pineal concretions. J. Electron Microsc. (Tokyo) **43:** 307–317.
37. BACONNIER, S. *et al.* 2002. Calcite microcrystals in the pineal gland of the human brain: first physical and chemical studies. Bioelectromagnetics **23:** 488–495.
38. GROISMAN, G.M. *et al.* 1996. Calcified concretions in the anterior pituitary gland of the fetus and the newborn: a light and electron microscopic study. Hum. Pathol. **27:** 1139–1143.
39. BARSON, A.J. & J. SYMONDS. 1977. Calcified pituitary concretions in the newborn. Arch. Dis. Child. **52:** 642–645.
40. DILMAN, V.M. 1981. The Law of Deviation of Homeostasis and Diseases of Aging. H. T. Blumenthal, Ed. John Wright PSG, Inc. Boston.
41. KRSTIC, R. 1976. A combined scanning and transmission electron microscopic study and electron probe microanalysis of human pineal acervuli. Cell. Tissue Res. **174:** 129–137.
42. MOLLER, M. *et al.* 1996. The chemical neuroanatomy of the mammalian pineal gland: neuropeptides. Neurochem. Int. **28:** 23–33.
43. FINCH, C.E. 2002. Neurons, glia, and plasticity in normal brain aging. Adv. Gerontol. **10:** 35–39.
44. ZUEV, V.A. 2001. From prion diseases to the problem of aging and death. Vestn. Ross. Akad. Med. Nauk. **11:** 46–49.
45. REGODON, S. *et al.* 2001. Postnatal maturation of parenchymal cell types in sheep pineal gland: an ultrastructural and immuno-electron-microscopic study. Rev. Méd. Vét. **152:** 325–333.
46. TORDAY, J.S. 2003. Parathyroid hormone-related protein is a gravisensor in lung and bone cell biology. Adv. Space Res. **32:** 1569–1576.
47. KONDRACHUK, A.V. 2001. Simulation and interpretation of experiments with otoliths. J. Gravit. Physiol. **8:** 101–104.

48. TASAKA, M. *et al.* 2001. Genetic regulation of gravitropism in higher plants. Int. Rev. Cytol. **206:** 135–154.
49. CHEN, R. *et al.* 1999. Gravitropism in higher plants. Plant Physiol. **120:** 343–350.
50. SONG, I. *et al.* 1988. Do starch statoliths act as the gravisensors in cereal grass pulvini? Plant Physiol. **86:**1155–1162.
51. HEMMERSBACH, R. & D.P. HADER. 1999. Graviresponses of certain ciliates and flagellates. FASEB J. **13:** 69–75.
52. ZHENG, W. & Q. ZHAO. 2002. Establishment and characterization of an immortalized Z310 choroidal epithelial cell line from murine choroid plexus. Brain Res. **958:** 371–380.
53. TERADA, T. & L. STEFANEANU. 1996. Vanishing psammoma bodies in the anterior pituitary of the human newborn: an immunohistochemical and histometric study. Endocr. Pathol. **7:** 151–157.
54. NOTEBORN, H.P. *et al.* 1986. Ultrastructural demonstration of secretion by exocytosis in rat pinealocytes with the use of the tannic acid method. Cell. Tissue Res. **245:** 223–225.
55. PESCHKE, D. *et al.* 2000. Pineal influence on annual nuclear volume changes in ventromedial hypothalamic nucleus (VMH) neurons of the male Wistar rat. Chronobiol. Int. **17:** 15–28.
56. KORF, H.W. 1994. The pineal organ as a component of the biological clock. Phylogenetic and ontogenetic considerations. Ann. N.Y. Acad. Sci. **719:** 13–42.
57. MCMILLEN, I.C. & J.S. ROBINSON. 2005. Developmental origins of the metabolic syndrome: prediction, plasticity, and programming. Physiol. Rev. **85:** 571–633.
58. BARKER, D.J.P. 1995. Fetal origins of adult coronary heart disease. Br. Med. J. **311:** 171–174.
59. OSMOND, C. & D.J.P. BARKER. 2000. Fetal, infant, and childhood growth are predictors of heart disease, diabetes, and hypertension in adult men and women. Environ. Health Perspect. **108:** 545–553.
60. VICKARYOUS, N. & E. WHITELAW. 2005. The role of early embryonic environment on epigenotype and phenotype. Reprod. Fertil. Dev. **17:** 335–340.
61. ANDREWS, M.H. & S.G. MATTHEWS. 2004. Programming of the hypothalamo-pituitary-adrenal axis: serotonergic involvement. Stress **7:** 15–27.
62. DRAKE, A.J. & B.R. WALKER. 2004. The intergenerational effects of fetal programming: nongenomic mechanisms for the inheritance of low birth weight and cardiovascular risk. J. Endocrinol. **180:** 1–16.
63. SECKL, J.R. & M.J. MEANEY. 2004. Glucocorticoid programming. Ann. N.Y. Acad. Sci. **1032:** 63–84.
64. SECKL, J.R. 2004. Prenatal glucocorticoids and longterm programming. Eur. J. Endocrinol. **151:** 49–62.
65. PATEL, M.S. & M. SRINIVASAN. 2002. Metabolic programming: causes and consequences. J. Biol. Chem. **277:** 1629–1632.
66. HARDER, T. *et al.* 1998. Syndrome X-like alterations in adult female rats due to neonatal insulin treatment. Metabolism **47:** 855–862.
67. DRAKE, A.J. *et al.* 2005. Intergenerational consequences of fetal programming by in utero exposure to glucocorticoids in rats. Am. J. Physiol. Regul. Integr. Comp. Physiol. **288:** 34–38.
68. JORDANIA, I.F. 1964. Handbook of obstetrics. Moscow (in Russ.).
69. HAYFLICK, L. & P.S. MOORHEAD. 1961. The serial cultivation of human diploid cell strains. Exp. Cell. Res. **25:** 585–621.
70. HAYFLICK, L. 1965. The limited *in vitro* lifetime of human diploid cell strains. Exp. Cell. Res. **37:** 614–636.
71. HAYFLICK, L. 1997. Mortality and immortality at the cellular level: a review. Biochemistry (Moscow) **62:** 1180–1190.
72. HAYFLICK, L. 2002. DNA replication and traintracks. Science **296:** 1611–1612.
73. BODNAR, A.G. *et al.* 1998. Extension of life span by introduction of telomerase into normal human cells. Science **279:** 349–352.
74. VAZIRI, H. & S. BENCHIMOL. 1998. Reconstitution of telomerase activity in normal human cells leads to elongation of telomeres and extended replicative life span. Curr. Biol. **8:** 279–282.

The Pineal Aging and Death Program

Life Prolongation in Pre-aging Pinealectomized Mice

WALTER PIERPAOLI [a] AND DANIELE BULIAN [b]

[a]Walter Pierpaoli Foundation of Life Sciences, Orvieto, Italy

[b]Jean Choay Institute for Neuroimmunomodulation, Riva San Vitale, Switzerland

ABSTRACT: A precise temporal program for growth, fertility, aging, and death exists in the "pineal complex" of the brain. It tracks, like a "clock," the ontogenetic phases of our life program. Transplantation of a very old pineal gland into the thymus or under the kidney capsule of a young mouse produces acceleration of aging and early death. We investigated the existence of such an inner biological clock on the assumption that a time exists in the pineal program when the pineal gland actively starts to deliver aging and death "signals" to the body, thus accomplishing its genetically inscribed sequence. Groups of BALB/c male or female mice were surgically pinealectomized (PX) at the age of 3, 5, 7, 9, 14, and 18 months, and their life span was evaluated. Periodical measurements of blood and hormonal and metabolic parameters were taken. Results showed that while PX at the age of 3 and 5 months promotes acceleration of aging, no relevant effect of PX is observed in mice PX at 7 or 9 months of age. On the contrary, a remarkable life prolongation was observed when mice were PX at the age of 14 months. No effects were seen when the mice were PX at 18 months of age. The same aging-promoting or -delaying effects were confirmed in the hematological and hormonal-metabolic values measured. The findings demonstrate the existence of an evolutionary–developmental role for the pineal complex during growth, fertility, and aging. The dominant role of the pineal in the initiation and progression of aging as a death signal is clear, but its nature and mechanism are totally unknown. In fact new experiments showed that an additional pineal gland from a young donor, when grafted into a young mouse, induces acceleration of aging. The significance of these intriguing findings is discussed.

KEYWORDS: pinealectomy; aging; aging clock; aging program; death program; hormone cyclicity; pineal grafting; zinc; thyroxin

INTRODUCTION

We have repeatedly demonstrated the existence of a precise life, aging, and death "clock" in the pineal network of the brain, in those neural structures controlling and modulating our genetically inherited life "program."[1–4] In a previous work we observed the remarkable life-shortening effect of grafting very old pineals into young

Address for correspondence: Walter Pierpaoli, M.D., Via San Gottardo 77, CH-6596 Gordola, Switzerland. Voice: +41 91 7451940; fax: +41 91 7451946.

pierpaoli.fnd@bluewin.ch

Ann. N.Y. Acad. Sci. 1057: 133–144 (2005). © 2005 New York Academy of Sciences.
doi: 10.1196/annals.1356.008

hosts.[5] Later we conducted experiments in which we tried to determine, again in rodents, the time when the pineal gland starts to deliver the so-called aging and death signals to the body. As hinted in the previous work,[5] if the aging pineal gland actively promotes aging at the end of the adult and fertile life, it must be possible to establish the time when removal of the aging pineal would interfere with the aging process and somehow prolong life.

Our experiments were divided into two groups. One group of experiments were carried out over the course of three years and used groups of BALB/c female or male inbred mice, pinealectomized (PX) or sham-operated (SO) surgically by using a stereotaxis equipment. The mice were 3, 5, 7, 9, 14, and 18 months of age at the time of the surgery. They were kept under observation as long as they lived. Peripheral blood was taken periodically to evaluate hormonal, hematologic, and metabolic changes. A second set of experiments involved similar groups of 4-month-old young mice of the same strain. Pineal gland from young donors of the same age were implanted under the kidney capsule. This unusual experiment was suggested by the idea that an additional exogenous young pineal may influence the function (and aging) of the endogenous young pineal.

MATERIALS AND METHODS

Animals

Balb/c, inbred, genetically H-2–compatible male and female mice, bred and maintained in air-conditioned rooms under conventional conditions in our laboratory, were used as donors and recipients. Donors and recipients of a pineal gland were of the age indicated in the experiments. The mice were housed in plastic cages (4–6 mice per cage) and fed with standard maintenance pellets (NAFAG, Gossau, Switzerland) and tap water *ad libitum*. Room temperature was 20°C. Illumination was 7 P.M. lights off and 7 A.M. lights on.

Pineal Grafting

The pineal glands were removed from 4-month-old donors sacrificed by rapid cervical dislocation, and implanted into the thymus of the 4-month-old recipients. The mice were implanted (PG) or sham-operated (SO) with the procedure described in detail in previous reports.[3] Briefly, the mice were anesthetized and surgery was performed between 9 and 11 A.M. The animals were housed in groups of five and controlled daily in the first week, and then weekly for mortality rates and body weight.

Pinealectomy

Pinealectomy was performed in a manner similar to the method described in detail previously.[4] Briefly, anesthetized (Hexenal, 200 mg/kg) mice were fastened in a stereotaxic apparatus.[4] After shaving, the head skin was cut sagittally along the midline, and the aponeurosis and other soft tissues were removed to make visible the fissures of the calvarium. A triangle-shaped skull fragment located at the intersection of the sagittal and occipital fissures was cut with the help of a dental drilling machine. The next steps of the operation were carried out with of a dissection stereomicroscope. After turning the animal's head around its midline horizontal axis, we lifted the skull fragment, including its adherent pineal. The pineal ligaments connected with dura mater were cut with fine scissors, and the pineal gland was removed. Bleeding from the brain vessels was stopped by repeated rinsing with saline and drying with tissue pads. The skull fragment without the pineal gland was then cemented with polymer. After drying, the scalp skin was sutured with silk stitches, and the sealed operation field was sprayed with antibiotic powder. Sham-operated mice underwent the same procedure except that the pineal was exposed but not removed.

Plasma Zinc Determination

Zinc was determined by atomic absorption spectrophotometry (AAS) from individual plasma, previously frozen at $-70°C$, according to the method of Fernandez and Kahn[6] and the reference standard procedure suggested by Evenson and Warren.[7]

Blood Cell Counts

All mice were bled from the retroorbital venous plexus under rapid ether anesthesia. White blood cell (WBC) numbers were evaluated by direct counting in a Bürker chamber. In addition, blood smears were stained with May-Grünwald-Giemsa for differential counting of white cells. Absolute leukocyte and lymphocyte number and relative lymphocyte number were measured.

Lipid Measurement

The assays were performed with 10-µL plasma samples obtained from peripheral blood of individual mice by using enzymatic methods and a Johnson & Johnson Vitros 250 Dry-Chemistry Analyzer.

Thyroid Hormone Measurement

Thyroid hormones (total T3, total T4) were measured in plasma samples from individual mice. T3 and T4 values were determined by the MEIA-method (microparticle-enzyme-immunoassay).

Statistical Analysis

Results are expressed as mean ± SD. Statistical significance was determined by using analysis of variance (ANOVA). When significant differences were found, statistical analysis was made by paired Student's t-test. Log rank test was used to evaluate the difference of survival rates between pineal grafted or pinealectomized mice and sham-operated mice. Differences were considered statistically significant when $P < 0.05$.

TABLE 1. Effect of surgical pinealectomy in 3- to 5-month-old Balb/cJ male mice on plasma zinc level, peripheral blood leukocytes, lymphocytes, thyroid hormones (T3 and T4), and plasma triglycerides

Age of mice (mo): Mo after pinealectomy:	11 7		15 11		19 15	
Parameter measured	Pinealectomized ($N=14$)	Sham operated ($N=14$)	Pinealectomized ($N=10$)	Sham operated ($N=13$)	Pinealectomized ($N=6$)	Sham operated ($N=6$)
Blood leukocytes (No./mm$^3\times10^3$)	9.25 ± 15.4	99.4 ± 18.6	76.2 ± 20.2	96.1 ± 28.7	98.8 ± 22.8	81.4 ± 9.5
% Lymphocytes	$56.8 \pm 8.5^{***}$	69.9 ± 8.4	$55.3 \pm 3.6^{***}$	68.5 ± 8.8	$44.5 \pm 3.4^{*}$	55.2 ± 8.6
Blood lymphocytes (No./mm$^3\times10^3$)	$51.3 \pm 10.7^{**}$	84.0 ± 15.5	$42.9 \pm 13.4^{*}$	67.5 ± 25.5	42.4 ± 16.3	42.2 ± 11.5
Zn plasma level (µg/dL)	$67.2 \pm 4.7^{*}$	72.1 ± 7.5	68.2 ± 14.9	69.3 ± 17.6	ND	ND
T3 (mmol/L)	ND	ND	0.51 ± 0.12	0.45 ± 0.09	0.95 ± 0.55	0.92 ± 0.46
T4 (nmol/L)	ND	ND	50.4 ± 13.6	54.9 ± 15.3	$27.7 \pm 14.2^{*}$	58.5 ± 15.7
Triglycerides (mmol/L)	ND	ND	$2.94 \pm 0.57^{*}$	2.38 ± 0.53	ND	ND

NOTE: Mean ± SD. *$P < 0.05$ when compared to sham operated; **$P < 0.01$ when compared to sham operated; ***$P < 0.001$ when compared to sham operated (Student's t-test). Balb/cJ male mice were pinealectomized at 3–5 months of age.

RESULTS

Pinealectomy in 3- and 5-Month-Old Mice

The results illustrated in FIGURE 1 and on TABLE 1 show that PX performed in 3- and 5-month-old mice significantly accelerated aging and consequently shortened their life span. The aging-accelerating effects of PX were also clearly expressed by the drop of lymphocyte counts in the peripheral blood, by the consistent decrease of zinc levels, by the negative effects on thyroxin (T4) production, and by the increased level of triglycerides (TABLE 1).

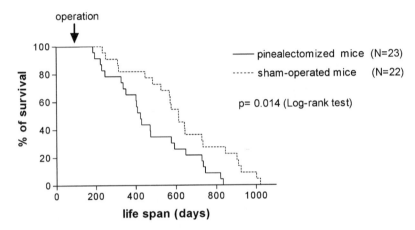

FIGURE 1. Pinealectomy in 3–5-month-old Balb/cJ male mice induces an earlier onset of aging. Survival curves (Kaplan-Meier) are different; $P < 0.05$ by log-rank test.

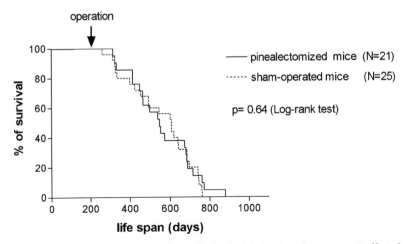

FIGURE 2. Pinealectomy in 7–9-month-old Balb/cJ male mice does not affect their life span. Survival curves (Kaplan-Meier) are not different; $P > 0.05$ by log-rank test.

TABLE 2. Effect of surgical pinealectomy in 7-month-old Balb/cJ male mice on plasma zinc level, peripheral blood leukocytes, lymphocytes, thyroid hormones (T3 and T4), and plasma triglycerides

| Age of mice (mo): | 14 | | 18 | | 22 | |
| Mo after pinealectomy: | 7 | | 11 | | 15 | |
Parameter measured	Pinealectomized ($N=8$)	Sham operated ($N=7$)	Pinealectomized ($N=6$)	Sham operated ($N=6$)	Pinealectomized ($N=4$)	Sham operated ($N=4$)
Blood leukocytes (No./mm³×10³)	76.0 ± 14.9	91.3 ± 10.8	84.5 ± 29.8	99.3 ± 27.7	98.5 ± 25.0	79.5 ± 27.2
% Lymphocytes	52.5 ± 5.2*	59.3 ± 6.5	62.7 ± 9.6	56.9 ± 5.0	40.5 ± 10.6	38.7 ± 8.7
Blood lymphocytes (No./mm³×10³)	39.7 ± 7.2**	54.3 ± 9.7	46.0 ± 10.3	54.0 ± 13.0	37.8 ± 5.9	30.8 ± 12.9
Zn plasma level (µg/dL)	59.0 ± 9.6*	73.6 ± 7.9	56.8 ± 13.4	49.7 ± 6.4	ND	ND
T3 (mmol/L)	ND	ND	0.59 ± 0.34	0.67 ± 0.17	ND	ND
T4 (nmol/L)	ND	ND	43.7 ± 13.7	35.3 ± 6.3	ND	ND
Triglycerides (mmol/L)	ND	ND	3.07 ± 0.58	3.72 ± 0.69	ND	ND

NOTE: Mean ± SD. *$P < 0.05$ when compared to sham operated; **$P < 0.01$ when compared to sham operated (Student's t-test). Balb/cJ male mice were pinealectomized at 7 months of age.

TABLE 3. Effect of surgical pinealectomy in 9-month-old Balb/cJ male mice on plasma zinc level, peripheral blood leukocytes, lymphocytes, thyroid hormones (T3 and T4), and plasma triglycerides

Age of mice (mo):	13		16	
Mo after pinealectomy:	4		7	
Parameter measured	Pinealectomized ($N = 7$)	Sham operated ($N = 7$)	Pinealectomized ($N = 6$)	Sham operated ($N = 6$)
Blood leukocytes (No./mm^3×10^3)	100.0 ± 20.5	95.1 ± 28.6	105.5 ± 20.4	97.2 ± 20.7
% Lymphocytes	59.0 ± 7.8	64.9 ± 9.7	54.8 ± 8.0	63.5 ± 13.4
Blood lymphocytes (No./mm^3×10^3)	59.0 ± 14.0	60.6 ± 19.5	58.2 ± 15.3	62.8 ± 20.0
Zn plasma level (μg/dL)	62.3 ± 4.6*	71.8 ± 4.5	ND	ND
T3 (mmol/L)	ND	ND	0.97 ± 0.45	0.97 ± 0.21
T4 (nmol/L)	ND	ND	50.2 ± 14.2	53.6 ± 18.6
Triglycerides (mmol/L)	ND	ND	2.74 ± 0.33	2.59 ± 0.44

NOTE: Mean ± SD. *$P < 0.01$ when compared to sham operated; **$P < 0.01$ when compared to sham operated (Student's t-test). Balb/cJ male mice were pinealectomized at 9 months of age.

Pinealectomy in 7- and 9-Month-Old Mice

PX in 7- and 9-month-old mice did not significantly affect their longevity and life span (FIG. 2). As shown on TABLES 2 and 3, no relevant and durable effects were also observed in their peripheral blood.

Pinealectomy in 14-Month-Old Mice

PX in 14-month-old mice produced a considerable delay of aging and/or a life prolongation (FIG. 3). Also PX at that age resulted in highly significant and positive effects on peripheral blood parameters, such as lymphocyte counts, plasma zinc level, and thyroxin (T4) production (TABLE 4). In contrast and unexpectedly, four months after PX an increased level of triglycerides was observed, but they consistently decreased at 8 months after PX (TABLE 4). The positive effects of PX were particularly evident at 4 months after PX (TABLE 4).

Pinealectomy in 18-Month-Old Aging Mice

In spite of some transitory early and late positive effects on the peripheral blood values measured (lymphocytes and zinc in TABLE 5), and on survival rate (FIG. 4), no significant prolongation of their life span was observed in PX 18-month-old mice.

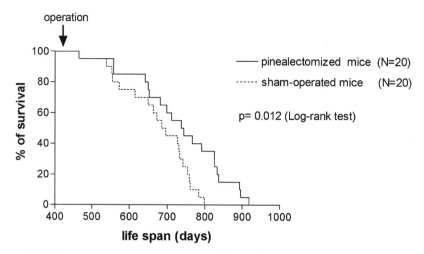

FIGURE 3. Pinealectomy in 14-month-old Balb/cJ male mice induces a delay of their aging and a consequent prolongation of their life. Survival curves (Kaplan-Meier) are different; $P < 0.05$ by log-rank test.

TABLE 4. Effect of surgical pinealectomy in 14-month-old Balb/cJ female mice on plasma zinc level, peripheral blood leukocytes, lymphocytes, thyroid hormones (T3 and T4), and plasma triglycerides

| Age of mice (mo): | 18 | | 22 | |
| Mo after pinealectomy: | 4 | | 8 | |
Parameter measured	Pinealectomized ($N = 9$)	Sham operated ($N = 10$)	Pinealectomized ($N = 6$)	Sham operated ($N = 6$)
Blood leukocytes (No./mm^3×10^3)	48.5 ± 7.0	47.1 ± 7.7	61.5 ± 14.4	44.8 ± 15.2
% Lymphocytes	67.9 ± 4.7***	55.3 ± 6.8	55.6 ± 4.2	53.0 ± 6.7
Blood lymphocytes (No./mm^3×10^3)	33.1 ± 5.9*	26.0 ± 4.7	34.1 ± 8.4	25.2 ± 9.8
Zn plasma level (µg/dL)	78.7 ± 17.6*	57.8 ± 11.0	ND	ND
T3 (mmol/L)	0.79 ± 0.26	0.60 ± 0.14	1.20 ± 0.41	0.94 ± 0.57
T4 (nmol/L)	71.3 ± 19.0**	48.1 ± 10.6	66.0 ± 18.2	56.6 ± 9.9
Triglycerides (mmol/L)	3.47 ± 0.50**	2.42 ± 0.41	1.95 ± 0.30	2.26 ± 0.30

NOTE: Mean ± SD. *$P < 0.05$ when compared to sham operated; **$P < 0.01$ when compared to sham operated; ***$P < 0.001$ when compared to sham operated (Student's t-test). Balb/cJ female mice were pinealectomized at 14 months of age.

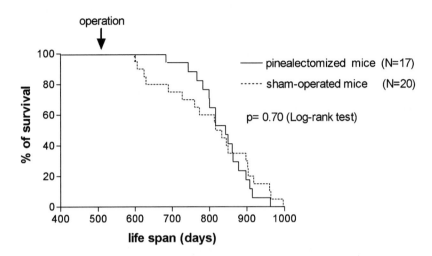

FIGURE 4. Pinealectomy in 18-month-old Balb/cJ male mice does not affect their life span. Survival curves (Kaplan-Meier) are not different; $P > 0.05$ by log-rank test.

TABLE 5. Effect of surgical pinealectomy in old mice on plasma zinc level, peripheral blood leukocytes, and lymphocytes

Age of mice (mo):	22		25	
Mo after pinealectomy:	4		7	
Parameter measured	Pinealectomized ($N = 8$)	Sham operated ($N = 9$)	Pinealectomized ($N = 8$)	Sham operated ($N = 9$)
Blood leukocytes (No./mm^3×10^3)	103.1 ± 11.5	94.1 ± 25.5	95.0 ± 15.0	92.4 ± 16.2
% Lymphocytes	62.5 ± 5.1**	41.9 ± 1.7	57.1 ± 2.9**	46.9 ± 4.8
Blood lymphocytes (No./mm^3×10^3)	62.2 ± 6.8**	39.4 ± 10.7	54.4 ± 9.8*	43.4 ± 9.3
Zn plasma level (μg/dL)	86.1 ± 14.9**	64.6 ± 9.1	76.0 ± 10.0	65.5 ± 10.0

NOTE: Mean ± SD. *$P < 0.05$ when compared to sham operated; **$P < 0.01$ when compared to sham operated (Student's t-test). Balb/cJ male mice were pinealectomized at 18 months of age.

TABLE 6. Effect of young pineal grafting into young recipients on plasma zinc level, peripheral blood leukocytes, lymphocytes, thyroid hormones (T3 and T4), and plasma triglycerides

| Age of mice (mo): | 8 | | 12 | |
| Mo after pineal grafting: | 4 | | 8 | |
Parameter measured	Young+young pineal ($N = 6$)	Young sham operated ($N = 6$)	Young+young pineal ($N = 7$)	Young sham operated ($N = 7$)
Blood leukocytes (No./mm^3×10^3)	56.1 ± 5.8	59.0 ± 10.1	57.3 ± 8.0	57.6 ± 13.8
% Lymphocytes	66.0 ± 13.2	80.5 ± 5.2*	66.4 ± 8.4	70.2 ± 5.9
Blood lymphocytes (No./mm^3×10^3)	36.5 ± 4.5	47.2 ± 6.3*	37.3 ± 8.2	40.6 ± 6.7
Zn plasma level (µg/dL)	54.7 ± 3.3*	67.7 ± 9.4	ND	ND
T3 (mmol/L)	0.71 ± 0.13	0.66 ± 0.10	0.73 ± 0.25	0.60 ± 0.14
T4 (nmol/L)	82.8 ± 12.6**	60.7 ± 5.2	72.0 ± 8.3*	60.8 ± 10.2
Triglycerides (mmol/L)	3.28 ± 0.51*	2.59 ± 0.15	2.65 ± 0.50	2.16 ± 0.43

NOTE: Mean ± SD. *$P < 0.05$ when compared to sham operated; **$P < 0.01$ when compared to sham operated (Student's t-test). A pineal gland from 4-month-old donors was implanted into the thymus of 4-month-old Balb/cJ recipients. Donor and recipient mice were female.

Pineal Grafting from Young Donors into Young Recipients

As shown in FIGURE 5, transplantation of a "young" pineal gland into the thymus or under the kidney capsule of "young" recipients of the same age resulted into a significant early onset of aging in the grafted mice and into a shortening of their life span. The aging-accelerating effects of young-pineal grafting into young recipients was confirmed by the measurement of several parameters in their peripheral blood. At 4 months after pineal grafting, lymphocyte counts and plasma zinc were remarkably reduced, triglycerides were significantly increased, while T4 and T3 were increased, hinting to a durable hyperthyroidism condition (TABLE 6). The effects were still visible but less evident at 8 months after pineal grafting (TABLE 6).

DISCUSSION

The findings shown above have demonstrated once more the existence of an "aging program" in the pineal gland and that a very precise time exists when the aging pineal actively promotes aging. In the mouse strain used, we identified this age to be 14 months. Pinealectomy at this age significantly prolonged the life span of the mice (FIG. 3). Evidence from the blood measurements showed that removal of the pineal in mice at the age of 14 months resulted in maintenance of more juvenile hormonal and metabolic values at 4 and 8 months after PX (TABLE 4). On the contrary, a dele-

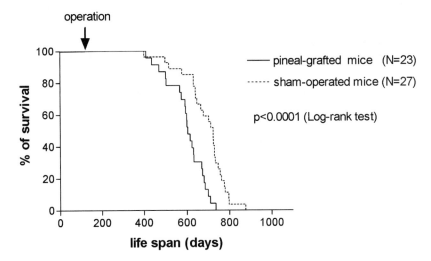

FIGURE 5. Transplantation of a young pineal gland (pineal gland from 4-month-old Balb/cJ female mice) into normal young, 4-month-old Balb/cJ female mice induces an earlier onset of aging. Survival curves (Kaplan-Meier) are different; $P < 0.0001$ by log-rank test.

terious effect of pinealectomy was seen in mice PX at the age of 3 and 5 months, with acceleration of aging parameters and an earlier death (FIG. 1). This observation confirms the developmental role of the pineal gland in early life and that its "program" includes several steps ranging from growth, to fertility, and on to aging and death.[1-5] Our experiments also demonstrate that PX at an age greater than 14 months does not affect the life span of the mice. Apparently this is the time when the pineal gland has accomplished its "aging program" and prevention of and/or recovery from aging becomes impossible (FIG. 4). The experiments with young mice grafted with a pineal from young mice of the same age were most intriguing. Young pineal–grafted young mice showed a shortened life span, as if the grafted young pineal gland promoted aging in the young mouse (FIG. 5). We attribute this amazing observation to accelerated aging of the grafted pineal gland, obviously deprived of its neural connections. However different interpretations are possible for this phenomenon.

Altogether these data clearly confirm once more the undeniable existence of a "clock" for aging and also of a signal for dying in the pineal gland; and they confirm that the "program" can be studied, interpreted, and possibly modified. The cellular and molecular mechanisms underlying this most fundamental aspect of our life program deserves intensive investigations in different species of mammals, man included.

[*Competing interests*: The author declares that he has no competing financial interests.]

REFERENCES

1. PIERPAOLI, W. 1991. The pineal gland: a circadian or seasonal aging clock? Editorial Aging **3:** 99–101.

2. PIERPAOLI, W. 1994. The pineal gland as ontogenetic scanner of reproduction, immunity, and aging. Ann. N.Y. Acad. Sci. **741:** 46–49.
3. PIERPAOLI, W. & W. REGELSON. Pineal control of aging: effect of melatonin and pineal grafting on aging mice. Proc. Natl. Acad. Sci. USA **91:** 787–791.
4. LESNIKOV, V.A. & W. PIERPAOLI. 1994. Pineal cross transplantation (old-to-young and vice versa) as evidence for an endogenous "Aging Clock." Ann. N.Y. Acad. Sci. **719:** 456–460.
5. PIERPAOLI, W. & D. BULIAN. 2001. The pineal aging and death program. I. Grafting of old pineals in young mice accelerates their aging. J. Anti-Aging. Med. **4:** 31–37.
6. FERNANDEZ, F.J. & H.L. KHAN. 1971. Clinical methods for atomic absorption spectroscopy. Clin. Chem. Newslett. **3:** 24–29.
7. EVENSON, M.A. & B.L. WARREN. 1975. Determination of serum copper by atomic absorption with use of graphite cuvette. Clin. Chem. **21:** 619–624.

Aging as a Mitochondria-Mediated Atavistic Program

Can Aging Be Switched Off?

VLADIMIR P. SKULACHEV[a] AND VALTER D. LONGO[b]

[a]Belozersky Institute of Physico-Chemical Biology, Moscow State University, Moscow 119992, Russia

[b]Andrus Gerontology Center and Department of Biological Sciences, University of Southern California, Los Angeles, California 90095-1555, USA

ABSTRACT: Programmed death phenomena have been demonstrated on subcellular (mitoptosis), cellular (apoptosis), and supracellular (collective apoptosis) levels. There are numerous examples of suicide mechanisms at the organismal level (phenoptosis). In yeast, it was recently shown that the death of aging cells is programmed. Many of the steps of programmed cell death are shown to be common for yeast and animals, including mammals. In particular, generation of the mitochondrial reactive oxygen species (ROS) is involved in the suicide programs. Aging of higher animals is accompanied by an increase in damage induced by mitochondrial ROS. Perhaps prevention of such damage by scavenging of mitochondrial ROS might slow down or even switch off the aging programs.

KEYWORDS: programmed death; mitochondria; reactive oxygen species

Since the 19th century, two alternative concepts of aging—optimistic and pessimistic—have been discussed. The former suggests that aging is the final step of an ontogenetic program and, hence, can be prevented by switching the step off. The latter assumes that this process is an inevitable result of life, being a consequence of the accumulation of errors and injuries in biomolecules, the exhaustion of vital forces, and the malfunctioning of some genes that are useful in the beginning of life but become harmful at its end. If the second view is correct, there is no chance to greatly increase the span of life, since such a complicated system as an organism must of necessity break at some point, like an old car. Comfort once mentioned that it is difficult to believe that a horse ages in the same way as a cart.[1] However, the optimistic concept of aging never dominated the discussion. Only quite recently, significant observations have been made that strongly favor the programmed aging hypothesis.

Address for correspondence: Vladimir P. Skulachev, Belozersky Institute of Physico-Chemical Biology, Moscow State University, Moscow 119992, Russia.

skulach@belozersky.msu.ru

Ann. N.Y. Acad. Sci. 1057: 145–164 (2005). © 2005 New York Academy of Sciences.
doi: 10.1196/annals.1356.009

BRIEF HISTORY

It was Alfred Russel Wallace, the codiscoverer of the natural selection law, who first mentioned that death can be programmed. He wrote in one of his letters (around 1865–1870): "... when one or more individuals have provided a sufficient number of successors, they themselves, as consumers of nourishment in a constantly increasing degree, are an injury to those successors. Natural selection therefore weeds them out, and in many cases favors such races as die almost immediately after they have successors."[2] This principle was later developed by August Weismann: "Work-out individuals are not only valueless to the species but they are even harmful, for they take the place of those, which are sound ... I consider that death is not a primary necessity but it has been secondarily acquired as an adaptation." Weismann was immediately attacked as an anti-Darwinist even though Darwin wrote: "There can be no doubt that a tribe including many members who ... were always ready to aid one another, and to scarify themselves for the common good would be victorious over most other tribes; and this would be natural selection"[3] (for discussion see Ref. 112). Recently this aspect was comprehensively analyzed by Goldsmith.[4]

Today Weismann critics are usually cite Medawar,[5] who assumed that aging could not be invented by biological evolution. He stressed that, under natural conditions, the majority of organisms die before they become old. Such an assumption is hardly correct since the aging starts long before it appears to be an immediate cause of death.[4,6–8] Indirectly, age-dependent weakening of an organism can be the reason of death due, for example, to attack by predators, pathogens, etc. [4,6–9] Recently Loison and colleagues[10] and Bonduriansky and Brassil[11] clearly showed that both long-lived ungulates (roe deer, bighorn sheep, izards) and the short-lived antler fly suffer senescence under natural conditions. This is not surprising because a decrease in, for example, skeletal muscle strength starts, as a rule, rather early, that is, at the age when organism stops growing.

THE PRECEDENT OF THE PROGRAMMED DEATH MECHANISM: APOPTOSIS

The phenomenon of the programmed death was directly proved in the cell. We mean discovery of apoptosis. In 1972, Kerr, Wyllie, and Currie published their famous paper "Apoptosis: a basic biological phenomenon with wide-ranging implication in tissue kinetics."[12] Later, numerous observations clearly demonstrated that apoptosis is involved in ontogenic development, anticancer defense, immune response, etc. In 1996–1997, the role of mitochondrial proteins, such as apoptosis-inducing factor (AIF)[13] and cytochrome c,[14–16] as well as of mitochondrial ROS,[17] in amplification of apoptotic stimuli was elucidated. In 2002, a Nobel Prize in Physiology and Medicine was given to Brenner, Horvitz, and Sulston for studies on *C. elegans*, including identification of genes of an apoptotic program responsible for elimination of about 6% of cells during ontogenesis of the worm.[18,19]

MITOPTOSIS, COLLECTIVE APOPTOSIS, AND ORGANOPTOSIS

During the last decade, programmed death mechanisms were shown to operate at both sub- and supracellular levels. In 1992, Zorov and colleagues[20] suggested that animal mitochondria possess a mechanism of self-elimination. This function was attributed to so-called permeability transition pore (PTP), a large, unspecific channel in the inner mitochondrial membrane. Normally PTP is closed but under certain (as a rule, pathological) conditions it opens causing dissipation of mitochondrial transmembrane electric potential difference ($\Delta\Psi_m$). It is known that $\Delta\Psi_m$ is required for the import of cytoplasmic precursors of mitochondrial proteins and for proper arrangement of mitochondria-synthesized proteins in the inner membrane. Thus, repair of the PTP-bearing mitochondrion ceases, and such an organelle perishes.[21,22] Reactive oxygen species (ROS) are one of the most powerful PTP-opening stimuli. If ROS are overproduced by a mitochondrion, it will be killed by these ROS due to the PTP opening. This mechanism was suggested to rid the intracellular mitochondrial population of those that have become dangerous to the cell because their ROS generation exceeded their ROS scavenging capacity. Such a process of self-elimination of malfunctioning mitochondria was termed *mitoptosis*.[23] Recently Lemasters and coworkers[24] have shown that glucagon added to hepatocyte cell culture causes mitoptosis followed by autophagia of dead mitochondria. The effect was arrested by cyclosporin A, the PTP inhibitor. It was found that a similar process accompanies apoptosis induced by the tumor necrosis factor.[25] Tolkovsky and coworkers[26,27] have demonstrated disappearance of all the mitochondria in the cell in response to apoptotic stimuli provided that apoptosis was blocked downstream of mitochondria. A type of progeria (accelerated aging) accompanied by mitoptosis was described by Jazwinski and coworkers in yeast.[28,29] The authors showed that a point mutation in the *ATP2* gene coding for β-subunit of mitochondrial H$^+$-ATP-synthase resulted in a situation where the daughter yeast cell is no longer born young and instead possesses the age of its mother. Such a defect was manifested only when glycolysis was the sole energy source, which suggests that, under conditions used, H$^+$-ATP-synthase operated as H$^+$-ATPase being required to form $\Delta\Psi_m$ at the expense of hydrolysis of glycolytic ATP while respiratory phosphorylation was impossible. In line with the concept of mitoptosis, Jarwinski and colleagues observed progressing decline in $\Delta\Psi_m$ and reduction of number of mitochondria in the mutant. The net result was the generation of cells totally lacking mitochondria, which become dominant cell type as yeast clones became extinct.[28,29]

Mitoptosis exemplifies a self-elimination program operating at a subcellular level. As to supracellular level, several cases were described when apoptotic cells formed clusters in tissues *in vivo* or in cell cultures *in vitro* (so-called death of bystander cells[8,30,31]). It was suggested that H$_2$O$_2$ produced in an apoptotic cell serves as an apoptogenic signal to the bystander cells surrounding the apoptotic cell. In this way, a tissue antiviral defense may be organized, assuming that propagation of H$_2$O$_2$ is faster than that of a virus. As a result, a "dead area" may be organized around the infected cell, as observed when a leaf of the so-called hypersensitive strain of tobacco plant is infected by the tobacco mosaic virus.[32]

The hypothesis of an H$_2$O$_2$-mediated bystander killing was confirmed by Bakalkin and coworkers.[30] The authors reported that serum deprivation resulted in formation of clusters of cultured apoptotic osteosarcoma cells. The clustering was abolished by adding catalase, the H_2O_2 scavenger. The long-distance transmission of an apoptotic signal was quite recently demonstrated by Pletjushkina and colleagues.[31] The human cervical carcinoma cells of a HeLa line was grown on two glass coverslips. One of them was treated with an apoptogen (TNF, staurosporine, or H_2O_2). Then this coverslip was removed from the apoptogen-containing medium, washed by a medium without apoptogen, and put side by side with the second coverslip (which was not treated by the apoptogen). It was found that numerous apoptotic cells appear on the second (nontreated) coverslip, the effect being sensitive to catalase. The cells on the treated coverslip were shown to produce some H_2O_2. Apoptosis induced by added H_2O_2 or by HeLa cell–produced H_2O_2 proved to be abolished by the mitochondria-addressed antioxidant MitoQ, whereas apoptoses caused by TNF or staurosporine were not.

It is obvious that massive apoptosis of cells composing an organ that should be eliminated during ontogenesis can be a mechanism of such an elimination, being defined as *organoptosis*.[33] Consider the disappearance of the tail of a tadpole when it converts to a frog. Addition of thyroxine (the hormone known to cause regression of the tadpole tail) to severed tails surviving in a special medium was shown to cause shortening of the tails, occurring on the time scale of hours. The following chain of events was elucidated:[34]

thyroxine $\rightarrow$ NO synthase induction $\rightarrow [NO\cdot]\uparrow \rightarrow [H_2O_2]\uparrow \rightarrow$

apoptosis $\rightarrow$ organoptosis

The mechanism of the $NO\cdot$-induced H_2O_2 increase may consist of inhibition by $NO\cdot$ of the main H_2O_2 scavengers, catalase and glutathione peroxidase,[35] as well as in the NO-induced arrest of the respiratory chain in the cyanide and antimycin A–sensitive sites.[8]

PROGRAMMED DEATH OF ORGANISM: PHENOPTOSIS

Bacteria and Unicellular Eukaryotes

Mechanisms of self-elimination are clearly operative at subcellular, cellular, and supracellular levels. Life seems to follow a Samurai principle: "It is better to die than to be wrong."[35] This principle can be stated as: "Any complex biological system is equipped with programs of self-elimination that are actuated when the system in question appears to be dangerous or unnecessary for the system(s) of higher position in biological hierarchy."[36]

If such a principle operates at an organismal level, we may speak about the programmed death of organism. Such a phenomenon can be defined it as *phenoptosis*, by analogy with mitoptosis and organoptosis.[33] Existence of phenoptosis, if it were experimentally proved, would mean that an organism, at least in some cases, would control its own death. This might be a consequence of operation of one more biological principle: "Living creatures try to avoid spontaneous process controlling all events occurring in their bodies."[8] Within the framework of such a concept, it is not surprising that death, the most dramatic event in the life history of any individual, is also under its control.

In bacteria, a suicide mechanism may represent the last line of defense of the population gene pool, assuming that a cell that cannot guarantee maintenance of intactness of its genome—hence, if survives, can generate antisocial monsters among its progeny—commits suicide.[8,33] Such a process of self-elimination should be referred to phenoptosis since we deal here with a unicellular organism (as compared to apoptosis, which is the self-elimination of a cell in multicellular organism). A scheme of phenoptosis induced by damage to bacterial DNA was considered by Lewis,[37] who followed the path of signal transduction from damaged DNA to autolysin-mediated death of a bacterium.

Moreover, these events may be referred to bacterial phenoptoses: (1) active lysis of the mother cell of *B. subtilis* and *Streptomyces* during sporulation, which is required to release spores; (2) development of bacteroids in *Rhizobium*; (3) lysis of some cells of *S. pneumoniae* to release DNA, which is picked up by other cells that did not lyse; (4) toxin/antitoxin systems killing bacterium when protein synthesis is inhibited; (5) lysis of the colicin forming *E. coli* to release a colicin killing bacteria of other strains; (6) suicide of a phage-infected *E. coli* cell to prevent propagation of the phage infection (in the latter case, three different suicide mechanisms, all resulting in arrest of the protein synthesis, were described).[8,37] In these events, the suicide mechanisms proved to be quite different from those in the animal cells.

Mechanisms of self-elimination in unicellular eukaryotes are studied mainly in the yeast *Saccharomyces cerevisiae*. Here it was found that pro- or anti-apoptotic mammalian proteins, such as Bax or Bcl-2 expressed in the yeast, induce the cell death or rescue the cells, respectively. Certain mutations in the *S. cerevisiae* genome entail death showing features of the mammalian cell apoptosis.[8, 38–40] Harsh treatments (H_2O_2, acetic acid, etc.) proved to be inducers of the yeast death also resembling apoptosis.[41,42] Indications were published that in *S. cerevisiae* there are proteases involved in the apoptotic cascade, namely, a caspase[43] and Omi.[44] As was shown by Allis and coworkers,[45] the H_2O_2-induced programmed death of yeast is accompanied by phosphorylation of histone H2B at serine 10, catalyzed by Ste20 protein kinase. This process proved to be responsible for chromatin condensation, a typical trait of the programmed cell death both in animals and yeast. In humans, the same histone is phosphorylated at serine 14 by Mst1 kinase, a mammalian homolog of yeast Ste20. This fact strongly points to mechanistic similarity of animal apoptosis and yeast phenoptosis.

In 2001, Narasimhan and colleagues[46] showed that plant antibiotic osmotin caused an apoptosis-like death of yeast. Surprisingly, components of the pheromone cascade were reported to be involved in the osmotin killing.[46,47] The reason for this became clear when Severin and Hyman[48] revealed that killing *S. cerevisiae* by high concentration of pheromone (α-factor) is programmed. In fact, this observation created the precedent of a signal compound produced by a unicellular eukaryote, which induces death of cells of the same species. Analysis of the mechanism of this effect showed[48,49] that it includes the following consecutive components: (1) pheromone receptor, (2) Ste20 kinase, (3) synthesis of some protein(s), (4) an increase in cytosolic [Ca^{2+}], (5) stimulation of mitochondrial respiration and an increase in energy coupling, (6) a $\Delta\Psi_m$ elevation, (7) a burst in the ROS production in the middle span of the mitochondrial respiratory chain, (8) decomposition of mitochondrial filaments to small roundish mitochondria, and (9) $\Delta\Psi_m$ collapse and cytochrome *c* release from mitochondria to cytosol (FIG. 1).

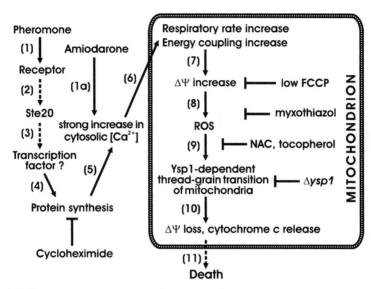

FIGURE 1. Mechanism of programmed death of yeast (phenoptosis), initiated by pheromone or amiodarone. Pheromone (α-factor) combines with a receptor on the outer surface of the plasma membrane (step 1). The binding initiates a cascade of events (MAPK cascade) including activation of the Ste20 kinase (step 2). Ste20 activates (apparently via a transcription factor, step 3), synthesis of protein(s) (step 4), inducing a strong increase in cytosolic Ca^{2+} due to the Ca^{2+} release from the intracellular depot (step 5). Amiodarone can substitute for these protein(s) (step 1a). Cytosolic Ca^{2+} activates respiration and increases its coupling to generation of mitochondrial membrane potential ($\Delta\Psi$) (step 6). This in turn increases $\Delta\Psi$ (step 7), an effect elevating ROS production by the respiratory chain (step 8). ROS initiate fission of mitochondrial filaments to small, roundish mitochondria (thread-grain transition), which requires $ysp1$ protein (step 9). Then a $\Delta\Psi$ collapse and release to cytosol of proteins hidden in the mitochondrial intermembrane space (e.g., cytochrome c) take place (step 10), events involved with the cell death (step 11). (Reproduced from Pozniakovsky et al.[49] with permission.)

Multicellular Organisms That Reproduce Only Once

As was noticed by Wallace and Weismann, some of organisms of this kind are constructed in a way predetermining death shortly after reproduction. For instance, imagos of mayflies die within a few days since they cannot eat due to lack of functional mouth and their intestines are filled with air.[2] In the mite *Adactilidium*, the young hatch inside the mother's body and eat their way out.[50] The male of some squids dies just after transferring his spermatophore to a female.[51] The female octopus stops eating when her children are hatched. This does not occur if her optical glands are removed. Such an operation results in a four-fold increase in life span of the animal.[52] Bamboo can live for 15–20 years reproducing vegetatively but then, in the year of flowering, dies at the height of the summer time immediately after the ripening of the seeds (see Skulachev[8] for discussion).

Striking observations were made in studies of salmon. The Pacific salmon was shown to die immediately after spawning as a result of accelerated aging (progeria), which develops when the fish leaves the ocean and swims along a river to its upper reaches. The traditional explanation of this kind of death was that the animal spends too much energy when swimming in the river for a long distance against current. However, this point of view proved to be wrong since (1) aging and death did not occur if gonads or adrenal glands were removed[53] and (2) progeria was observed even when the river was very short and current was weak. In the Far East of Russia, two populations of salmon were compared, one spawning in the upper reaches of the Amur river (thousands of kilometers long) and another spawning in a very small river on the Sakhalin island (only 0.2 km long). In both cases, the spawning fish showed typical traits of aging that resulted in death. A signal for progeria proved to be change from the sea to fresh water. In this example, a biological function of suicide seems to be that the remains of the old fish become food for river invertebrates who, in turn, are food for the young fish.[54]

The Atlantic salmon, in contrast to its Pacific relative, after spawning in a river returns from river to ocean. If it is the summer generation of the fish, it often dies in the fall. A Russian ichthyologist V.V. Ziuganov has recently studied larvae of a mollusk (pearl mussel *Margaritifera margaritifera*) that develops in gills of the Atlantic salmon. He found that larvae can somehow switch off the fish's "death program" so the larvae-infected fish live at least one season more than the majority of non-infected salmon (some of infected salmon live up to 13 years). An increase in the host's life span is needed for the larvae to complete their own development. It was shown that the infected fish had fewer tumors and were more resistant to wounds and burns.[54] Leng and colleagues[55] reported that a peptide from another mollusk related to the mussel, a *Mercenaria meretrix*, activates superoxide dismutase but inhibits tyrosinase and proliferation of carcinoma cells. Earlier it was shown that a *Mercenaria* extract possesses anticancer activity and decreases the blood sugar and fat.[55]

AGING AS AN EVOLUTION-ACCELERATING MECHANISM

The Case of the Hare versus the Fox

It is a time to revisit the old Wallace-Weismann idea concerning aging as a program.[2] Two questions should be addressed: What is the physiological function(s) of aging if it represents a particular case of slow phenoptosis? (2) How is the mechanism of the aging program constructed to avoid substitutions of non-aging phenotype for an aging one by means of natural selection?

The answer to the first question is that it seems reasonable to assume that slow aging serves as a specialized mechanism to accelerate evolution. Let us consider an example. Two young hares differing "intellectually" have almost equal chances to escape from a fox since both of them run much faster than a fox. However, with age the clever hare acquires some advantage, which is of crucial importance as the older hare slows down and runs as fast as the fox. Now the clever hare has a better chance to escape and, hence, to produce leverets, than the stupid hare. This is turn will be favorable for selection of clever hares.[8]

This answer presumes that muscles weaken with age when reproduction is still possible. This is apparently the case at least for humans. Here the age-dependent atrophy of muscles seems to begin around 25 years.[56] Initially, this process is slow but it is activated in an age-dependent fashion. Nowadays, measurable loss of muscles strength in Swedish men was shown to start at age 50[57] and in Saudi Arabian men in the fourth decade of life.[58] On the other hand, at the beginning of nineteenth century the decline of this parameter started just after 25 years (in a study of Belgian men[57]). Such a dynamics may be explained by improvement of living conditions during studied period of time. In any case, as is noticed by Goldsmith, "because even a relative minute deterioration will cause a statistically significant increase in death rate, one suspects that the evolutionary effects of aging in wild animals begin at relatively young ages."[4] As Loison and colleagues noticed, "observed death rates in wild animals increase beginning at puberty."[10]

In line with the above reasoning, the better adapted creatures live, as a rule, longer. Species were described that do not age at all. The pearl mussel lives up to 200 years without any traits of aging. Its weight and reproductive ability continue to increase for its entire life. The mollusc suddenly dies when the shell weight appears to be too great for the muscles of the mollusc to maintain a vertical position at the bottom of a river.[59] Pike has no natural enemies and lives for a century without losing the ability to reproduce. This is also true for some big ocean fish.[59] Some big ocean birds live about 50 years and then suddenly die.[6] Among birds, there are examples of very long-lived species with constant (or even increasing with age) ability to reproduce, showing no age-dependent increase in the death rate.[59] Bat lives 17 times longer than shrew of the same weight with the same diet of insects.[60] It seems that species occupying a new area (like the air or the ocean) can stop paying a price for a fast evolution—aging and a short life span that results in a high rate of the change of generations.[8]

CHRONOLOGICAL AGING OF YEAST

In yeast, two kinds of aging are described, a replicative and a chronological. The first one means that each yeast cell can form a limited number of daughter cells (for *S. cerevisiae* the number is about 30). The second type means that the yeast can survive for a limited amount of time. In both types, death shows many typical apoptotic markers (see Breitenbach and colleagues[40] for review). For replicative aging, the physiological role of cell suicide is not obvious, since old mother cells compose an insignificant part of population. In this case, phenoptosis may be the result of the aforementioned "Samurai" law.

As for chronological aging, it proved to be precedent for the adaptive (evolution-accelerating) role of aging recently directly demonstrated. It was found that yeast strains isolated from grapes[61] and from laboratory strains[61,62] undergo an age-dependent death that has features of animal cell apoptosis. Moreover, it was shown that such a death was prevented by maintaining of the medium pH at the initial (6.5–7) level (normally, pH lowered to 3.5–4 in the stationary phase at day 12 of cultivation) (FIG. 2A) or by exclusion of nutrients from the medium (FIG. 2B).[61] In rich medium, acidification resulted in the death of 90–99% of the population despite the fact that nutrients are still available. Then a small subpopulation showed a regrowth (FIG. 2C).[61,62] The regrowth is accompanied by a strong increase in the mutation rate, being stimulated by deletions in the superoxide dismutase (SOD1) and catalase genes and inhibited by overexpression these ROS-scavenging enzymes.[61] Disruption of the death program through deletion of yeast caspase YCA1 initially resulted in better survival of aged cultures. However, surviving cells have lost regrowth capacity.[62]

A four-step explanation of these data is: (1) Long-term cultivation of yeast results in lowering of pH due to accumulation of acidic endproducts of metabolism. (2) The pH decrease entails protonation of the primary ROS, the superoxide anion ($O_2^{-\cdot}$), so that much more aggressive $HO_2^\cdot$ is formed[63] that easily penetrating through membranes (the pK value for the $O_2^{-\cdot}$ protonation is 4.8). (3) $HO_2^\cdot$ attacks (i) DNA, resulting in an increase of mutagenesis, and (ii) mitochondria that react to ROS by

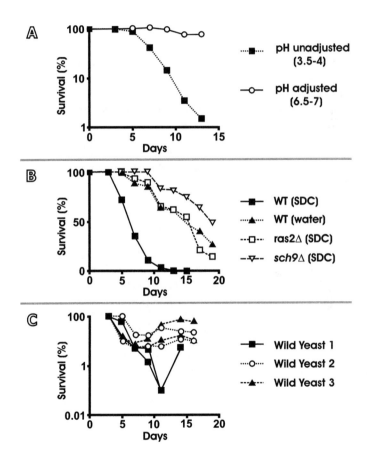

FIGURE 2. Chronological aging of yeast *Saccharomyces cerevisiae*. (**A**) Death is delayed when the pH value in the medium, normally lowering to 3.5–4 on day 12, is adjusted at 6.5–7. The growth medium contained 2% synthetic dextrose (SDC). (**B**) Death is delayed when water substitutes for a rich (SDC) medium or genes encoding regulatory proteins (*ras2* or *sch9*) are mutated. (**C**) The regrowth phenomenon: a small fraction of old yeast starts regrowth on days 10–15 in the SDC medium. Three samples of wild yeast isolated from grapes were studied. (Reprinted from Fabrizio *et al.*[61] with permission.)

committing suicide (mitoptosis). (4) Massive mitoptosis initiates cell suicide. However, a very small fraction of cells proved to be resistant to massive mitoptosis due to the increased mutagenesis that gave rise to the disruption of a gene involved in the cell suicide mechanism. If the ambient conditions improve, the wild type genetic pattern is recovered in certain mutant cells by formation of a diploid from two haploids differing in mutated proteins involved in the phenoptotic cascade.

This four-step explanation may describe how suicide mechanisms appear as a result of biological evolution. It is obvious that such systems as SOD–catalase or mitoptotic cascade are favorable for cell survival under more or less normal conditions. In the above experiments, viability of *SOD1* deletant in the young culture was about ten time lower than that of the wild type—that is, 7 versus 70 cells $\times\ 10^6$/mL. However, at day 12 of cultivation, the ratio of these values was reversed—that is, 7×10^{-2} versus 1×10^{-5} cells $\times\ 10^6$/mL, respectively.[61]

It should be stressed that such a scheme can operate only if the death is programmed to occur *before* the conditions are worsened so strongly that *all* the cells in populations die in a non-controlled fashion. As a result, a chance appears that, due to increased mutagenesis, the cell suicide system is damaged, allowing a mutant to survive. In the same study,[61] some mutations preventing the programmed death were identified. One of them is localized in the gene encoding the Sch9 protein (analog of animal Akt, or protein kinase B). Knock-out of this gene was shown to result in an increase in SOD1 level.[64] Another mutation inactivated the gene encoding Ras2 protein (analog of the oncogen Ras of mammals). Both mutations markedly lengthen the life span of yeast (FIG. 1B).[61,64,65]

THE p53 PARADOX

It was recently demonstrated in mammals that the same mechanism can help a young organism survive but also shortens the life of an old one. In mice, Donehower and coworkers found that a mutation giving rise to an increase in activity of a protein, p53, resulted in disappearance of cancer from causes of death. Normally, almost 50% of mice die from cancer so one could expect the mutation in question to prolong life. Surprisingly, the life span of the mutants was shortened by 20%.[66] Later Garcia-Cao and colleagues, who apparently obtained lower level of the p53 activation than Donehower's group (the death from cancer decreased not to zero but to 17%) did not observed any statistically significant effect on the life span.[67] Such a result, however, is in line with Donehower's observation because a big decrease in cancer death rate would result in an increase in the life span if p53 did not have an effect on aging. Quite recently, Scrable and colleagues[68] have activated p53 in one more fashion that seems to result in the highest level of its activity so the life span decreased by factor 3. In both Donehower's and Scrable's experiments, the mutant mice died from progeria.[66,68]

The p53 protein is known as a "guard of genome." It is involved in stimulation of DNA repair, cell cycle arrest, and mitochondria-mediated apoptosis, occurring in response to the increasing DNA damage. Moreover, p53 able to stimulate aging (as is clear from the above-cited works). This means that for p53 deletants, where presumably aging should be absent, it is impossible to win competition with the wild-type animals since all of them die from DNA damage–linked pathologies, most of all to cancer. In fact, p53 deletants were found to have much shorter life spans and died exclusively from cancer.[66] Thus, today we can answer the question posed more than a century by Weismann: "There cannot be the least doubt that the higher organisms, as they are now constructed, contain within themselves the germs of death ... The question arises as to how this has come to pass."[2]

MUTATIONS PROLONGING LIFE

Examples of the aging program discussed illustrate possibilities of how the mechanism of aging could have evolved. Mitoptosis in yeast, which is employed to remove mitochondria that overproduce ROS, ultimately kills the cell when ROS become much more aggressive due to acidification of cytoplasm so that too many mitochondria commit suicide and release proapoptotic proteins to cytosol.[49,69] The p53 protein, which initiates apoptosis in malignant cells with damaged DNA, initiates suicide of cells damaged only slightly when is activated too much. This creates difficulties in selecting mutant organisms who are immortal. However, in certain mutants an interplay of these factors results in a prolongation of life span, sometimes very significantly.

In an ascomycete, filamentous fungus *Podospora anserina*, mutation in a cytochrome oxidase subunit arrests activity of this enzyme, which entails induction of the alternative cyanide-resistant oxidase, an event strongly lowering the rate of ROS formation by the respiratory chain (FIG. 3). This results in a strong decrease in the ROS level and, by some unknown mechanism, in a switch from sexual to vegetative reproduction. Such a switch is accompanied by disappearance of typical morphological age-dependent changes in the fungus. Death due to aging, normally occurring on day 25 since germination of spore, disappears so the mutant survives for years. The same result can be achieved by any other mutation (or by adding inhibitors) making operation of the canonical respiratory chain impossible.[70,71]

Hekimi and his colleagues succeeded in extending by a factor of 5.5 the life span of *Caenorhabditis elegans*, if two genes were inactivated—those coding for an insulin-receptor-like protein DAF-2 and for an enzyme catalyzing a final step of CoQ synthesis.[72–74]

In mice, Pelicci and coworkers[75–77] have reported that animals lacking a particular 66 kDa protein (p66Shc) live 30% longer and are less sensitive to paraquat-induced oxidative stress, and that fibroblasts derived from these mice did not respond more to an H_2O_2 addition by initiating apoptosis. The following chain of events seems to be responsible for these effects:

$$ROS \rightarrow DNA\ damage \rightarrow p53 \rightarrow stabilization\ of\ p66Shc \rightarrow [p66Shc]\uparrow \rightarrow [ROS]\uparrow \rightarrow$$
$$PTP\ opening\ in\ mitochondria \rightarrow massive\ mitoptosis \rightarrow apoptosis.$$

In vivo knock-out of the p66Shc gene resulted in a decrease in oxidative damage to both mitochondrial and nuclear DNA in lung, liver, spleen, skin, skeletal muscles, and kidney but not in brain and heart. This correlated with the p66Shc content in various organs, which is lowest in brain and heart.[75–77]

Quite recently, Rabinovich and coworkers[78] have demonstrated an increase in the medium (by 5 months) and maximal (by 5.5 months) life span of mice overexpress-

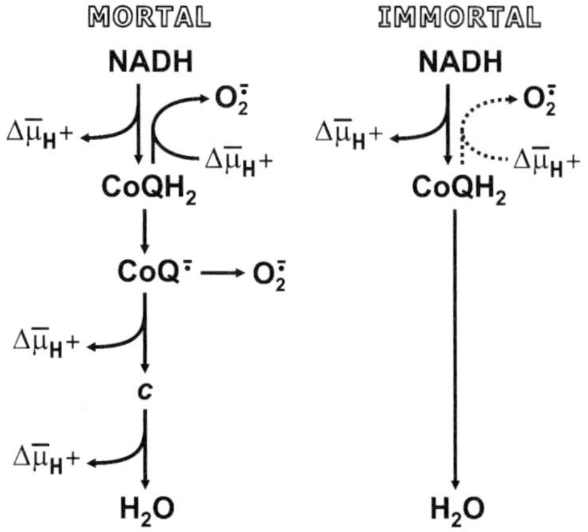

FIGURE 3. The respiratory chain inherent in the mortal and immortal modes of life of fungus *Podospora anserina*. (**Left**) The mortal mode. "Canonical" respiratory chain forms the transmembrane electrochemical H^+ potential difference ($\Delta\bar{\mu}_H^+$) in three energy coupling sites. Superoxide is generated by reverse electron transfer in the first coupling site and by oxidation of semiubiquinone (CoQ^-·) in the second site. (**Right**) The immortal mode. Cytochrome c oxidase operating in the third coupling site is replaced by noncoupled alternative $CoQH_2$ oxidase without the second and third coupling sites being involved. The second place of the O_2^-· generation in the middle span of the chain is omitted, whereas that in the initial span is partially inhibited due to the lower rate of $\Delta\bar{\mu}_H^+$ production. As a result, the rate of mitochondrial O_2^-· generation is strongly decreased.

ing catalase targeted to the mitochondrial matrix, which normally contains no catalase. Cardiac pathologies and cataract development were delayed, oxidative damage of DNA was reduced, H_2O_2 production and H_2O_2-induced aconitase inactivation were attenuated, and age-related accumulation of mitochondrial DNA deletions was decelerated. Targeting of catalase to nucleus or peroxisome was of smaller or negligible effects.

HORMONAL CONTROL OF LIFE SPAN

Programmed aging needs a time-measuring device. In higher animals, such a biological clock might be localized in the same place as other biotimers responsible for initiation and maintenance of periodic processes in their bodies, such as circadian rhythm, menstrual cycle, and season changes. In all the cases listed, the pineal gland hormone melatonin is involved. The pineal gland is primarily responsible for generation of the circadian rhythm signals in birds and is the second step in transmission of these signals in mammals, with suprachiasmatic nucleus of hypothalamus being the signal generator.

In 1971, Dilman postulated that aging is under hormonal control and involves hormones of hypothalamus, pineal gland, hypophysis and pancreas.[79–81] Later, this idea was confirmed by studies carried out on eukaryotes of various taxonomic positions. In 1974, Denckla showed that pituitary ablation prevents the age-dependent decline of O_2 consumption by mice and of sensitivity of O_2 consumption to thyroid hormones.[82] Later, Pierpaoli succeeded in demonstrating that (1) substitution of pineal gland in old mice by that from the young mice prolongs life whereas the old-to-young substitution has the opposite effect, (2) transplantation of the old pineal gland to thymus of a young mouse shortens the life span, (3) removal of pineal gland on the fourteenth month of the life has an effect similar to substitution of old pineal gland in young mice, (4) the nocturnal treatment of old mice with melatonin is favorable for longevity (see Pierpaoli, this volume).[83–88]

Blüher and colleagues[89] reported that knockout of the insulin receptor in adipocytes resulted in an 18% increase in the mouse life span. Independently, Holzenberger and colleagues[90] showed that heterozygote $Igf1r^{+/-}$ mice (Igf1r is for insulin-like growth factor receptor 1) lived 26% longer than the wild type. This was accompanied by 50% decrease in p66Shc (for reviews on other participants of the hormonal life-shortening cascade, see Bowles,[6,7] Longo and Finch,[91] and Anisimov, this volume).

In flies, indications of involvement of an insulin-type hormone in shortening the life span were reported.[92] In *C. elegans*, a single mutation in the *daf-2* gene coding for an insulin receptor prolonged life by two to three times without arresting development of the worm at the dauer stage.[93,94] In the same animal, NAD^+-dependent histone deacetylase (Sir2.1) was shown to downregulate the insulin pathway and to increase the life span.[8] High doses of gene *SIR2* in yeast were shown to cause some rise in duration of life.[8] Also in yeast, Fabrizio and colleagues[65] showed that deletion in genes encoding RAS2 protein or a Akt/PKB-like protein kinase entailed a two- to threefold increase in the life span (see above).

WHY ARE ROS AN OBLIGATORY COMPONENT OF THE AGING PROGRAM?

There are numerous pieces of evidence that intramitochondrial ROS are specifically involved in the life-shortening cascade responsible for aging of both yeast and animals.[8,61,64] We already mentioned that targeting of catalase to the mitochondrial matrix lowers the rate of accumulation of the mitochondrial DNA deletions and is favorable for longevity. Opposite effect on the life span was shown to be inherent in when the probability of accumulation of mistakes in the mitochondrial DNA increases. Recently, Larsson and colleagues[95] and Zassenhaus and coworkers[96] independently reported that expression of a proof reading-deficient mitochondrial DNA polymerase results in (1) a strong increase in mutations in mitochondrial DNA, especially in the region of cytochrome *b* of the respiratory chain complex III (inhibition of this cytochrome by antimycin A is known to strongly increase ROS production), (2) a strong decrease in the mouse life span, and (3) the appearance of many features typical for aging. Zasselhaus and coworkers,[96] who modified the polymerase only in the heart, succeeded in preventing of such changes by adding cyclosporin A, an inhibitor of the mitochondrial PTP and related ROS formation.

It is obvious that destructive role of ROS is still the great problem for any aerobe. This is apparently why aging as a specialized mechanism of evolution is arranged in such a way that it is favorable for improvement of the multifaceted antioxidant system of organism. To some degree, ROS operate like the fox in our hare-versus-fox case, so evolution is always directed toward more robust antioxidant defense. This appears to be a consequence of the fact that execution of the aging signal results in lowering of the antioxidant defense in organelles, cells, tissues, and organs.[8] Such a lowering could be a consequence of an increase in ROS generation and/or a decrease in the amount of ROS scavengers. The above relationships explain why longevity correlates with low level of ROS and high resistance to the oxidative stress, This correlation was revealed at levels of the cell cultures[97] and isolated mitochondria. In the latter case, it was shown that mitochondria from long-lived birds produce less ROS than from short-lived mammals of the same body weight.[98–99] Similar effect was found to be inherent in mammals i.e. flying (bat) and non-flying (shrew).[60] It is noteworthy that both longevity and low mitochondrial ROS production correlated with slower rate of accumulation of oxidized proteins and damaged DNA in tissues of aging animals.[98]

PRESENT STATE OF THE ART AND SOME PERSPECTIVES

In 2005, Thomas Kirkwood, one of the most prominent modern gerontologists, treated the programmed aging concept as a mistaken view.[100] In favor of such reasoning, Kirkwood noticed three major items. (1) *Aging cannot contribute significantly to the mortality in natural populations since most animals die comparatively young.* However, as was discussed above, aging seems to start at puberty and aging-induced weakness of the middle-aged organisms can well contribute to death caused by predators, infections, etc. (2) *Any mutant in whom the aging process was inactivated would enjoy an advantage and the mutant genotype would spread.* This item ignores at least three essential facts. First, the death of many organisms sexually reproducing only once occurs immediately after the reproduction process is over (e.g., bamboo or octopus). Undoubtedly, this takes place due to execution of a suicide (phenoptotic) program that could be lost by a mutation but evolution somehow preserved such a program. Second, aging of unicellular eukaryotic organisms such as yeast is shown to be programmed. This program cannot be eliminated by a mutation since such a mutation should be lethal (e.g., mutation in the mitoptotic program). Third, the latter is also valid for mammals. Here aging program includes, e.g., p53 which is also involved in the anticancer defense so mutation in p53, instead of resulting in immortality due to arrest of aging, gives rise to a strong decline of the life span because of stimulation of carcinogenesis. (3) *If there is a program of aging, why are immortal organisms lacking this program not described?* All the vegetatively reproducing organisms are, in fact, immortal. *Podospora anserina* is an example when immortality accompanying the sexual-to-vegetative reproduction switch is experimentally shown (see above). Moreover, switching off the death program does not necessarily mean immortality. It seems quite possible that organisms lacking the programmed death systems will live longer but after some period of time die because of non-programmed aging.

Thus, at present there are no unequivocal arguments excluding the possibility that aging is programmed. On the other hand, there is a precedent for programmed aging in yeast[61,62] and numerous pieces of indirect evidence that this may also be the case for organisms other than unicellular fungi.

More and more scientists now recognize that the generally accepted ideas concerning aging are insufficient to explain this phenomenon. Nemoto and Finkel,[101] when analyzing the present state of the Harman's hypothesis on aging as a simple result of accumulation of the ROS-induced injuries,[102] wrote: "If ageing represents the legacy of the combustible mixture of food and oxygen in our mitochondria, then oxidant-related damage should be, to some degree, cumulative. But when fruitflies were caloric-restricted for just two days, their mortality rate became equivalent to that of flies that spent their life hungry."[103] Within the framework of the phenoptotic concept of aging, caloric restriction can prolong the life of animals by such a change in hormonal balance, which inhibits execution of the aging program to some degree. Perhaps that practice of religious fasts was empirically invented as a tool to prolong the lives of the believers, provided that mechanisms of caloric restriction effect are similar in fly and man.

Few years ago, modern proponents of the Wallace-Weismann concept were very rare.[4,6-8,82,88,104-108] However, today specialists from various fields of research and different countries are likely to accept such a point of view. Curiously, some of them apologize for attacking the paradigm of "non-adaptive aging" still accepted by the great majority of gerontologists, by saying that their observations deal with some specific conditions only. For example, Hekimi (the world champion in prolonging the life of *C. elegans*) and Guarente wrote in 2003: "life span, therefore, appears to be regulated in these situations in spite of the fact that it is not the feature shaped adaptively by natural selection."[73] Certainly, at present we cannot guarantee that all the age-related events in all the living creatures should be regarded as programmed death. On the other hand, as soon as such a possibility seems rather probable at least in some cases, it must be taken into account in consideration of strategy of the twenty-first century medicine. The phenoptotic concept of aging opens the possibility of a dramatic increase in the human life span whereas the opposite point of view, still traditional for gerontologists, excludes such a perspective (if aging is an inevitable breakage in a complicated system, its improvement will be followed by a next breakage).

As a perspective on the longevity studies within the framework of our concept, we can point to the necessity for further analysis of major constituents of the aging programs. When identified, these constituents or the products of their activity should be arrested by specific small molecule ligands to interrupt execution of the phenoptotic program. Among them, mitochondria-targeted antioxidants look promising. Being constructed from a penetrating cation[109] and an antioxidant (e.g., ubiquinol[110]), such compounds can accumulate in mitochondria at concentrations up to 1000 times higher than outside mitochondria, strongly increasing the antioxidant capacity of these organelles responsible for apoptosis-inducing ROS formation. They were found to completely arrest the H_2O_2-induced apoptosis[84,110,111] like deletion of the gene encoding p66Shc, which entails prolongation the life span of a mammal.[75]

As was mentioned above, the larvae of the pearl mussel succeed in switching off the aging program in Atlantic salmon. Perhaps now is the right time to consider the possibility of doing the same in man.

ACKNOWLEDGMENTS

The authors are grateful to the Kerr Program (A. Simpson, Program Director), Ludwig Institute for Cancer Research; Paritet Foundation (O.V. Deripaska, sponsor); Vital Spark Foundation, and grant "Leading Scientific Schools" 1710.2003.04, Russian Ministry of Education and Science for financial support.

[*Competing interests*: The authors declare that they have no competing financial interests.]

REFERENCES

1. COMFORT, A. 1979. The Biology of Senescence, 3rd edit. Churchill Livingstone. Edinburgh.
2. WEISMANN, A. 1889. Essays Upon Heredity and Kindred Biological Problems. Claderon Press. Oxford.
3. DARWIN, C. 1871. The Descent of Man. Murray. London.
4. GOLDSMITH, T.C. 2003. The Evolution of Aging. iUniverse. New York. Lincoln. Shanghai.
5. MEDAWAR, P.B. 1952. An unsolved problem of biology. H.K. Lewis. London.
6. BOWLES, J.T. 1998. The evolution of aging: a new approach to an old problem of biology. Med. Hypotheses **51**: 179–221.
7. BOWLES, J.T. 2000. Sex, kings, serial killers and other group-selected human traits. Med. Hypotheses **54**: 864–894.
8. SKULACHEV, V.P. 2003. Aging and the programmed death phenomena. *In* Topics in Current Genetics. T. Nystrom & H.D. Osiewacz, Eds. 3: 191–238. Model systems in ageing. Springer-Verlag, Berlin.
9. SKULACHEV, V.P. 2001. The programmed death phenomena, aging, and the Samurai law of biology. Exp. Gerontol. **36**: 995–1024.
10. LOISON, A., M. FESTA-BLANCHET, J.M. GAILLARD, *et al.* 1999. Age-specific survival in five populations of ungulates: evidence of senescence. Ecology **80**: 2539–2554.
11. BONDURIANSKY, R. & C.E. BRASSIL. 2002. Senescence: rapid and costly ageing in wild male flies. Nature **420**: 377.
12. KERR, J.F., A.H. WYLLIE & A.R. CURRIE. 1972. Apoptosis: a basic biological phenomenon with wide-ranging implications in tissue kinetics. Br. J. Cancer **26**: 239–257.
13. SUSIN, S.A, N. ZAMZAMI, M. CASTEDO, *et al.* 1996. Bcl-2 inhibits the mitochondrial release of an apoptogenic protease. J. Exp. Med. **184**: 1331–1341.
14. LIU, X., C.N. KIM, J. YANG, *et al.* 1996. Induction of apoptotic program in cell-free extracts: requirement for dATP and cytochrome *c*. Cell **86**: 147–157.
15. YANG, J., X. LIU, K. BHALLA, *et al.* 1997. Prevention of apoptosis by Bcl-2: release of cytochrome *c* from mitochondria blocked. Science **275**: 1129–1132.
16. KLUCK, R.M., E. BOSSY-WETZEL, D.D. GREEN & D.D. NEWMEYER. 1997. The release of cytochrome *c* from mitochondria: a primary site for Bcl-2 regulation of apoptosis. Science **275**: 1132–1136.
17. SKULACHEV, V.P. 1996. Why are mitochondria involved in apoptosis? Permeability transition pores and apoptosis as selective mechanisms to eliminate superoxide-producing mitochondria and cell. FEBS Lett. **397**: 7–10.
18. HORVITZ, H.R. 2003. Nobel lecture. Worms, life, and death. Biosci. Rep. **6**: 239–303.
19. SULSTON, J.E. 2003. *C. elegans*: the cell lineage and beyond. Biosci. Rep. **3**: 49–66.
20. ZOROV, D.B., K.W. KINNALLY & H. TEDESCI. 1992. Voltage activation of heart inner mitochondrial membrane channels. J. Bioenerg. Biomembr. **24**: 119–124.
21. SKULACHEV, V.P. 1994. Decrease in the intracellular concentration of O_2 as a special function of the cellular respiratory system. Biochem. Moscow **59**: 1433–1434.
22. SKULACHEV, V.P. 1996. Role of uncoupled and noncoupled oxidations in maintenance of safely low levels of oxygen and its one-electron reductants. Q. Rev. Biophys. **29**: 169–202.

23. SKULACHEV, V.P. 1998. Uncoupling: new approaches to an old problem of bioenergetics. Biochim. Biophys. Acta **1363:** 100–124.
24. ELMORE, S.P., T. QIAN, S.F. GRISSOM & J.J. LEMASTERS. 2001. The mitochondrial permeability transition initiates autophagy in rat hepatocytes. FASEB J. **15:** 2286–2287.
25. SHCHEPINA, L.A., O.Y. PLETJUSHKINA, A.V. AVETISYAN, *et al.* 2002. Oligomycin, inhibitor of F_o part of H^+-ATP-synthase, suppresses the TNF-induced apoptosis. Oncogene **21:** 8149–8157.
26. FLETCHER, G.C., L. XUE, S.K. PASSINGHAM & A.M. TOLKOVSKY. 2000. Death commitment point is advanced by axotomy in sympathetic neurons. J. Cell Biol. **150:** 741–754.
27. XUE, L., G.C. FLETCHER & A.M. TOLKOVSKY. 2001. Mitochondria are selectively eliminated from eukaryotic cells after blockade of caspases during apoptosis. Curr. Biol. **11:** 361–365.
28. LAI, C.-Y., E. JARUGA, C. BORGHOUTS & S.M. JAZWINSKI. 2002. A mutation in the ATP2 gene abrogates the age asymmetry between mother and daughter cells of the yeast *Saccharomyces cerevisiae*. Genetics **162:** 73–83.
29. JAZWINSKI, S.M. 2003. Mitochondria, metabolism, and aging in yeast. *In* Topics in Current Genetics: Model Systems in Aging. T. Nystrom & H.D. Osiewacz, Eds.: 39–59. Springer-Verlag. Berlin Heidelberg.
30. REZNIKOV, K., A.L. KOLESNIKOVA, A. PRAMANIK, *et al.* 2000. Clustering of apoptotic cells via bystander killing by peroxides. FASEB J. **14:** 1754–1764.
31. PLETJUSHKINA, O.Y., E.K. FETISOVA, K.D. LYAMZAEV, *et al.* 2005. Long-distance apoptotic killing of cells is mediated by hydrogen peroxide in a mitochondrial ROS-dependent fashion. Cell Death Diff. **12:** 1442–1444.
32. SKULACHEV, V.P. 1998. Possible role of reactive oxygen species in antiviral defence. Biochem. Moscow **63:** 1438–1440.
33. SKULACHEV, V.P. 1999. Mitochondrial physiology and pathology: concepts of programmed death of organelles, cells, and organisms. Mol. Asp. Med. **20:** 139–184.
34. KASHIWAGI, A., H. HANADA, M. YABUKI, *et al.* 1999. Thyroxine enhancement and the role of reactive oxygen species in tadpole tail apoptosis. Free Radic. Biol. Med. **26:** 1001–1009.
35. IZYUMOV, D.S., A.V. AVETISYAN, O.Y. PLETJUSHKINA, *et al.* 2004. "Wages of fear": transient threefold decrease in intracellular ATP level imposes apoptosis. Biochim. Biophys. Acta **1658:** 141–147.
36. SKULACHEV, V.P. 2000. Mitochondria in the programmed death phenomena: a principle of biology: "It is better to die than to be wrong." IUBMB Life 49: 365–373.
37. LEWIS, K. 2000. Programmed death in bacteria. Microbiol. Mol. Biol. Rev. **64:** 503–514.
38. SKULACHEV, V.P. 2002. Programmed death in yeast as adaptation? FEBS Lett. **528:** 23–26.
39. JIN, C. & J.C. REED. 2002. Yeast and apoptosis. Nat. Rev. Mol. Cell Biol. **3:** 453–459.
40. BREITENBACH, M., F. MADEO, P. LAUN, *et al.* 2003. Yeast as a model for aging and apoptosis research. *In* Topics in Current Genetics: Model Systems in Aging. T. Nystrom & H.D. Osiewacz, Eds.: 61–96. Springer-Verlag. Berlin.
41. MADEO, F., E. FRÖHLICH, M. LIGR, *et al.* 1999. Oxygen stress: a regulator of apoptosis in yeast. J. Cell Biol. **145:** 757–767.
42. LUDOVICO, P., M.J. SOUSA, M.T. SILVA, *et al.* 2001. *Saccharomyces cerevisiae* commits to a programmed cell death process in response to acetic acid. Microbiology **147:** 2409–2415.
43. MADEO, F., E. HERKER, C. MALDENER, *et al.* 2002. A caspase-related protease regulates apoptosis in yeast. Mol. Cell **9:** 911–917.
44. FAHRENKROG, B., SAUDER, U. & AEBI U. 2004. The *S. cerevisiae* HtrA-like protein Nma111p is a nuclear serine protease that mediates yeast apoptosis. J. Cell Sci. **117:** 115–126.
45. AHN, S.-H., W.L. CHEUNG, J.-Y. HSU, *et al.* 2005. Sterile 20 kinase phosphorylates histone H2B at serine 10 during hydrogen peroxide-induced apoptosis in *S. cerevisiae*. Cell **120:** 25–36.
46. NARASIMHAN, M.L., B. DAMSZ, M.A. COCA, *et al.* 2001. A plant defense response effector induces microbial apoptosis. Mol. Cell **8:** 921–930.

47. NARASIMHAN, M.L., M.A. COCA, J. JIN, et al. 2005. Osmotin is a homolog of mammalian adiponectin and controls apoptosis in yeast through a homolog of mammalian adiponectin receptor. Mol. Cell **17:** 171–180.
48. SEVERIN, F.F. & A.A. HYMAN. 2002. Pheromone induces programmed cell death in *S. cerevisiae*. Curr. Biol. **12:** 233–235.
49. POZNIAKOVSKY, A.I., D.A. KNORRE, O.V. MARKOVA, et al. 2005. Role of mitochondria in the pheromone- and amiodarone-induced programmed death of yeast. J. Cell Biol. **168:** 257–269.
50. KIRKWOOD, T.B.L. & T. CREMER. 1982. Cytogerontology since 1881: a reappraisal of August Weismann and a review of modern progress. Hum. Genet. **60:** 101–121.
51. NESIS, K.N. 1997. Cruel love among the squids. *In* Russian Science: Withstand and Revive. A.V. Byalko, Ed.: 358–372. Nauka-Physmatlit. Moscow.
52. WODINSKY, J. 1977. Hormonal inhibition of feeding and death in the Octopus: control by optic gland secretion. Science **198:** 948–951.
53. ROBERTSON, O.H. & B.C. WEXLER. 1962. Histological changes in the organs and tissues of senile castrated kokanee salmon (*Oncorhynchus nerka kennerlyi*). Gen. Comp. Endocrinol. **2:** 458–472.
54. ZIUGANOV, V.V. 2005. Long-lived parasite extending the life span of his host. The pearl mussel *Margaritifera margaritifera* turns out the program of rapid senescence in Atlantic salmon *Salmo salar*. Doklady RAS (in Russian) **403:** 701–705.
55. LENG, B., X.D. LIU & Q.X. CHEN. 2005. Inhibitory effects of anticancer peptide from *Mercenaria* on the BGC-823 cells and several enzymes. FEBS Lett. **579:** 1187–1190.
56. LEXELL, J., C.C. TAYLOR & M. SJÖSTRÖM. 1988. What is the cause of the aging atrophy? Total number, size, and proportion of different fiber types studied in whole *vastus lateralis* muscle from 15- to 83-year-old men. J. Neurol. Sci. **84:** 275–294.
57. LARSSON, L., G. GRIMBY & J. KARLSSON. 1979. Muscle strength and speed of movement in relation to age and muscle morphology. J. Appl. Physiol. **46:** 451–456.
58. AL-ABDULWAHAB, S.S. 1999. Effects of aging on muscle strength and functional ability of healthy Saudi Arabian males. Ann. Saudi Med. **19:** 211–215.
59. ZIUGANOV, V.V. 2004. Arctic and southern freshwater pearl mussel *Margaritifera margaritifera* with long and short life span as a model system for testing longevity mechanisms. Adv. Gerontol. Russ. **14:** 21–30.
60. BRUNET-ROSSINNI, A.K. 2004. Reduced free-radical production and extreme longevity in the little brown bat (*Myotis lucifugus*) vs. two nonflying mammals. Mech. Ageing Dev. **125:** 11–20.
61. FABRIZIO, P., L. BATTISTELLA, R. VARDAVAS, et al. 2004. Superoxide is a mediator of an altruistic aging program in *Saccharomyces cerevisiae*. J. Cell Biol. **166:** 1055–1067.
62. HERKER, E., H. JUNGWIRTH, K.A. LEHMANN, et al. 2004. Chronological aging leads to apoptosis in yeast. J. Cell Biol. **164:** 501–507.
63. SARAN, M. 2003. To what end does nature produce superoxide? NADPH oxidase as an autocrine modifier of membrane phospholipids generating paracrine lipid messengers. Free Radic. Res. **37:** 1045–1059.
64. FABRIZIO, P., F. POZZA, S.D. PLETCHER, et al. 2001. Regulation of longevity and stress resistance by Sch9 in yeast. Science **292:** 288–290.
65. FABRIZIO, P., L.L. LIOU, V.N. MOY, et al. 2003. SOD2 functions downstream of Sch9 to extend longevity in yeast. Genetics **163:** 35–46.
66. TYNER, S.D., S. VENKATACHALAM, J. CHOI, et al. 2002. p53 mutant mice that display early aging-associated phenotypes. Nature **415:** 45–53.
67. GARCIA-CAO, I., M. GARCIA-CAO, J. MARTIN-CABALLERO, et al. "Super p53" mice exhibit enhanced DNA damage response, are tumor resistant, and age normally. EMBO J. **21:** 6225–6235.
68. MAIER, B., W. GLUBA, B. BERNIER, et al. 2004. Modulation of mammalian life span by the short isoform of p53. Genes Dev. **18:** 306–319.
69. LUDOVICO, P., F. RODRIGUES, A. ALMEIDA, et al. 2002. Cytochrome *c* release and mitochondria involvement in programmed cell death induced by acetic acid in *Saccharomyces cerevisiae*. Mol. Biol. Cell **13:** 2598–2606.

70. DUFOUR, E, J. BOULAY, V. RINCHEVAL & A. SAINSARD-CHANET. 2000. A causal link between respiration and senescence in *Podospora anserina*. Proc. Natl. Acad. Sci. USA **97:** 4138–4143.
71. OSIEWACZ, H.D. 2003. Aging and mitochondrial dysfunction in the filamentous fungus *Podospora anserine*. *In* Topics in Current Genetics: Model Systems in Aging, Vol. 3. T. Nystrom & H.D. Osiewacz, Eds.: 17–38. Springer-Verlag. Berlin.
72. LAKOWSKI, B. & S. HEKIMI. 1996. Determination of life-span in *Caenorhabditis elegans* by four clock genes. Science **272:** 1010–1013.
73. HEKIMI, S. & L. GUARENTE. 2003. Genetics and the specificity of the aging process. Science **299:** 1351–1354.
74. ARANTES-OLIVEIRA, N., J.R. BERMAN & C. KENYON. 2003. Healthy animals with extreme longevity. Science **302:** 611.
75. MIGLIACCIO, E., M. GIORGIO, S. MELE, *et al.* 1999. The p66shc adaptor protein controls oxidative stress response and life span in mammals. Nature **402:** 309–313.
76. TRINEI, M., M. GIORGIO, A. CICALESE, *et al.* 2002. A p53-p66Shc signaling pathway controls intracellular redox status, levels of oxidation-damaged DNA, and oxidative stress-induced apoptosis. Oncogene **21:** 3872–3878.
77. NAPOLI, C., I. MARTIN-PADURA, F. DE NIGRIS, *et al.* 2003. Deletion of the p66Shc longevity gene reduces systemic and tissue oxidative stress, vascular cell apoptosis, and early atherogenesis in mice fed a high-fat diet. Proc. Natl. Acad. Sci. USA **100:** 2112–2116.
78. SCHRINER, S.E., N.J. LINFORD, G.M. MARTIN, *et al.* 2005. Extension of murine life span by overexpression of catalase targeted to mitochondria. Science **308:** 1909–1911.
79. DILMAN, V.M. 1971. Age-associated elevation of hypothalamic threshold to feedback control, and its role in development, aging, and disease. Lancet **1:** 1211–1219.
80. DILMAN, V.M. 1978. Aging, metabolic immunodepression, and carcinogenesis. Mech. Ageing Dev. **8:** 153–173.
81. DILMAN, V.M. & V.N. ANISIMOV. 1979. Hypothalamic mechanisms of aging and of specific age pathology. I. Sensitivity threshold of hypothalamo-pituitary complex to homeostatic stimuli in the reproductive system. Exp. Gerontol. **14:** 161–174.
82. DENCKLA, W.D. 1974. Role of the pituitary and thyroid glands in the decline of minimal O_2 consumption with age. J. Clin. Invest. **53:** 572–581.
83. PIERPAOLI, W. & C.X. YI. 1990. The pineal gland and melatonin: the aging clock? A concept and experimental evidence. *In* Stress and the Aging Brain. G. Nappi, E. Martignoni, A.R. Genazzani & F. Petraglia, Eds.: 172–175. Raven Press. New York.
84. PIERPAOLI, W., A. DALL'ARA, E. PEDRINIS & W. REGELSON. 1991. The pineal control of aging. The effects of melatonin and pineal grafting on the survival of older mice. Ann. N.Y. Acad. Sci. **621:** 291–313.
85. PIERPAOLI, W. & W. REGELSON. 1994. Pineal control of aging: effect of melatonin and pineal grafting on aging mice. Proc. Natl. Acad. Sci. USA **91:** 787–791.
86. LESNIKOV, V.A. & W. PIERPAOLI. 1994. Pineal cross-transplantation (old-to-young and vice versa) as evidence for an endogenous "aging clock." Ann. N.Y. Acad. Sci. **719:** 456–460.
87. PIERPAOLI, W., D. BULIAN, A. DALL'ARA, *et al.* 1997. Circadian melatonin and young-to-old pineal grafting postpone aging and maintain juvenile conditions of reproductive functions in mice and rats. Exp. Gerontol. **32:** 587–602.
88. PIERPAOLI, W. & D. BULIAN. 2001. The pineal aging and death program. I. Grafting of old pineals in young mice accelerates their aging. J. Anti-Aging Med. **4:** 31–37.
89. BLÜHER, M., B.B. KAHN & C.R. KAHN. 2003. Extended longevity in mice lacking the insulin receptor in adipose tissue. Science **299:** 572–574.
90. HOLZENBERGER, M., J. DUPONT, B. DUCOS, *et al.* 2003. IGF-1 receptor regulates lifespan and resistance to oxidative stress in mice. Nature **421:** 182–187.
91. LONGO, V.D. & C.E. FINCH. 2003. Evolutionary medicine: from dwarf model systems to healthy centenarians. Science **299:** 1342–1346.
92. GIANNAKOU, M.E., M. GOSS, M.A. JUNGER, *et al.* 2004. Long-lived Drosophila with overexpressed dFOXO in adult fat body. Science **305:** 361.
93. KENYON, C., J. CHANG, E. GENSCH, *et al.* 1993. A *C. elegans* mutant that lives twice as long as wild type. Nature **366:** 461–464.

94. MURPHY, C.T., S.A. McCARROLL, C.I. BARGMANN, *et al.* 2003. Genes that act downstream of DAF-16 to influence the lifespan of *Caenorhabditis elegans*. Nature **434:** 277–283.
95. TRIFUNOVIC, A., A. WREEENBERG, M. FALKENBERG, *et al.* 2004. Premature aging in mice expressing defective mitochondrial DNA polymerase. Nature **429:** 417–423.
96. MOTT, J.L., D. ZHANG, J.C. FREEMAN, *et al.* 2004. Cardiac disease due to random mitochondrial DNA mutations is prevented by cyclosporin A. Biochim. Biophys. Res. Commun. **319:** 1210–1215.
97. KAPAHI, P., M.E. BOULTON & T.B.L. KIRKWOOD. 1999. Positive correlation between mammalian life span and cellular resistance to stress. Free Radic. Biol. Med. **26:** 495–500.
98. BARJA, G. 1998. Mitochondrial free radical production and aging in mammals and birds. Ann. N.Y. Acad. Sci. **854:** 224–238.
99. SKULACHEV, V.P. 2004. Mitochondria, reactive oxygen species, and longevity: some lessons from the Barja group. Aging Cell **3:** 17–19.
100. KIRKWOOD, T. 2005. Understanding the odd science of aging. Cell **120:** 437–447.
101. NEMOTO, S. & T. FINKEL. 2004. Aging and the mystery at Arles. Nature **429:** 149–152.
102. HARMAN, D. 1956. Aging: a theory based on free radical and radiation chemistry. J. Gerontol. **11:** 298–300.
103. MAIR, W., P. GOYMER, S.D. PLETCHER & L. PARTRIDGE. 2003. Demography of dietary restriction and death in Drosophila. Science **301:** 1731–1733.
104. SKULACHEV, V.P. 1997. Aging is a specific biological function rather than the result of a disorder in complex living systems: biochemical evidence in support of Weismann's hypothesis. Biochem. Moscow **62:** 1191–1195.
105. MITTELDORF, J. & D.S. WILSON. 2000. Population viscosity and the evolution of altruism. J. Theor. Biol. **204:** 481–496.
106. CLARK, W.R. 2004. Reflections on an unsolved problem of biology: the evolution of senescence and death. Adv. Gerontol. Russ. **14:** 7–20.
107. TRAVIS, J.M.J. 2004. The evolution of programmed death in a spatially structured population. J. Gerontol. **59:** 301–305.
108. BREDESEN, D.E. 2004. The nonexistent aging program: how does it work? Aging Cell **3:** 255–259.
109. LIBERMAN, E.A., V.P. TOPALI, L.M. TSOFINA, *et al.* 1969. Mechanism of coupling of oxidative phosphorylation and the membrane potential of mitochondria. Nature **222:** 1076–1078.
110. KELSO, G.F., C.M. PORTEOUS, G. HUGHES, *et al.* Prevention of mitochondrial oxidative damage using targeted antioxidants. Ann. N.Y. Acad. Sci. **959:** 263–274.
111. SKULACHEV, V.P. 2005. How to clean the dirtiest place in the cell: cationic antioxidants as intramitochondrial ROS scavengers. IUBMB Life **57:** 305–310.
112. LONGO, V.D., J. MITTELDORF & V.P. SKULACHEV. 2005. Programmed and altruistic ageing. Nat. Rev. Genet. **6:** 7–13.

The Myth and Reality of Reversal of Aging by Hormesis

JOAN SMITH SONNEBORN

Department of Zoology and Physiology, University of Wyoming, Laramie, Wyoming 82071, USA

ABSTRACT: Hormesis is an adaptive response to low doses of otherwise harmful agents by triggering a cascade of stress-specific resistance pathways. Evidence from protozoa, nematodes, flies, rodents, and primates indicate that stress-induced tolerance modulates survival and longevity. "Reality" is that hormesis can prolong the healthy life span. Genetic background provides the potential for longevity duration induced by stress. Senescence, or aging, is generally thought to be due to a different impact of selection for alleles positive for reproduction during early life but harmful in later life, a process called *antagonistic pleiotropy* (multiple phenotypic changes by a single gene). After reproduction, life span is "invisible" to selection. I propose the revision that mutations selected for survival until reproduction in early life may also extend later life (*protagonistic pleiotropy*). The *protagonist candidate genes* for extended life span are *hormetic response genes*, which activate the protective effect in both early and later life. My revision of the earlier evolutionary theory implies that *natural selection* of genes critical for early survival (life span until reproduction) can also be beneficial for extended longevity in old age, tipping the evolutionary balance in favor of a latent inducible life span extension unless excess stressor challenge exceeds the protection capacity. Mimetic triggers of the stress response promise the option of tricking the induction of metabolic pathways that confer resistance to environmental challenges, increased healthy life span, rejuvenation, and disease intervention without the danger of overwhelmiong damage by the stressor. Public policy should anticipate an increase in healthy life span.

KEYWORDS: hormesis; evolution; longevity; rejuvenation

EVOLUTION OF AGING

Evidence from the use of diverse model systems to identify genetic and environmental determinants of survival and life span provides insight into fundamental mechanisms of aging.[1–4] The hormesis phenomenon has been championed by Calabrese in biomedical sciences as a dose–response model and in toxicology as a pivotal element in the determination of environmental risk assessment.[5] Hormesis is now emerging as a major factor in health and longevity.[6–12] The critical role of stress response in health and disease was recognized in medical endocrinology pioneered by

Address for correspondence: Joan Smith Sonneborn, Ph.D., Zoology & Physiology Department, University of Wyoming, Box 3663, 16th & Gibbon Street, Laramie, WY 82071. Voice: 307-766-2341; fax: 307-766-5625.
cancun@uwyo.edu

Ann. N.Y. Acad. Sci. 1057: 165–176 (2005). © 2005 New York Academy of Sciences.
doi: 10.1196/annals.1356.010

Seyle[13] in 1946. Here, the role of stress in longevity and the implications for the future are outlined.

CELLULAR IMMORTALITY

Aging is not an obligate characteristic of all organisms and cells. On a cellular level, immortality, or the ability to maintain function and unlimited cell division, is found in lower organisms such as bacteria, algae, and some protozoa. In multicellular organisms, the succession of germ-line cells through generations implies their potential immortality. The existence of immortal cells establishes the fact that there is no intrinsic physical or biological law that requires the onset of senescence. Therefore it is useful to investigate options for intervention in the aging process. Indeed, the ability to clone a sheep from adult cells demonstrates a latent potential in adult cells to respond to environment epigenetic cues.[14]

CELLULAR MORTALITY AND LONGEVITY DETERMINATION

Sonneborn[15] proposed that aging arose with the origin of diploids. Organisms with only one chromosome of each kind, haploid organisms lack natural aging because bearers of a detrimental mutation die. A redundant chromosome allows diploids the luxury of accumulation of recessive mutations in the same individual. Diploids, sex, fertilization, recombination, and assortment allow accumulated mutations to be in all possible combinations including some combinations in which both genes of a pair are mutant and can be expressed, adaptive for environments and thus preserved by natural selection. It took two to three billion years to evolve unicellular diploids, but only a sixth as long for the diversity of species seen today.[16,17]

Williams[18] proposed that genes beneficial for early life are strongly selected but can be harmful in later life when selection wanes and introduced the term *antagonistic pleiotropy* (multiple effects of a single gene). Indeed, harmful expression of early genes in later life may contribute to loss of function. The theory does not include the option that alleles strongly selected for early life and reproductive success could also be beneficial in later life, an effect I call *protagonist pleiotropy.*

Environmental conditions that tip the balance toward benefit in later life (hormetic agents) would activate latent repair and extend life span unless the challenge overwhelmed the system. The candidate protagonist genes are the hormetic or stress genes induced at threshold levels of physical and chemical environmental toxins that provide tolerance to these otherwise harmful agents. The theory predicts the environmentally induced plasticity of lifespan observed from protozoa to man.[4] Longevity is clearly determined by genetic and environmental factors.

The genetic potential for stress-induced resistance to environmental challenges is evident throughout the phylogenetic spectrum of species ranging from protozoa to man. In protozoa, outbreeding ciliates have a long life cycle and greater potential of accumulation of longevity assurance loci, while inbreeders have short life spans and less opportunity to accumulate mutations beneficial in combinations of genes.[19] In yeast, mutations that favored resistance to ultraviolet light (UV)[20] also extended longevity and a telomere-associated site mutation increased resistance to starvation, heat, and ethanol.[21] In the mold *Neurospora* genetic manipulation of increased anti-oxidant expression increased longevity.[22] Mutations that increase resistance to stress increase longevity in the roundworm *Caenorhabditis elegans*.[23–25] The single-gene mutations that confer resistance to stress and increase longevity represent an elegant demonstration that single-gene mutations can have pleiotropic beneficial effects. Other determinants of nematode longevity have been found associated with germ line stem cells, thought to interact with steroids as well as altered biological clock genes.[26,27] The role of stress-induced steroid interactions on longevity is implicated by the well known imminent death after spawning in salmon and after mating in marsupial mice.

Genetic manipulations that led to underexpression of alleles coding for antioxidants in flies shortened lifespan,[28,29] while overexpression of antioxidants superoxide dismutase and catalase extended life span.[30] Selective breeding, which increased life span, required outbreeders to provide the genetic reservoir for longevity selection. [4,31–34] Epigenetic effects on gene expression in eggs from *Drosophila* selected from older mothers may have contributed to the selection process.

Variations in heat shock proteins are found among different species of fish. Hybrids formed from low heat resistance strains are more resistant than either of the parental strains,[35] implying additive resistance loci.

In mammals, one of the best examples of a genetic basis for stress-increased resistance is the correlation between the hormetic agent UV-induced repair and mammalian life span. The greater the latent inducible repair, the longer the life span of the mammal.[36] Alteration of the IGF-1 receptor in mice has extended longevity in mean and maximal life span as well as resistance to oxidative stress.[37] Thus the genetic background provides the potential for optimal resistance to environmental challenge. Those organisms that survive environmental challenge until reproduction are selected. A potent stress resistance response is of value not only in young but in old animals as well.

ENVIRONMENTAL MODULATION OF LONGEVITY

Stress Response in Aging and Disease

Beneficial effects induced by exposure to low doses of otherwise toxic agents have been reviewed using protozoa, molds, roundworms, rodents, and mammalian tissue.[6,7,38–40] Many if not all hormetic agents induce the classic stress response, which refers to an agent-specific induced cascade of protective processes including a subset of family members from the heat shock proteins and those that respond to glucose, oxygen deprivation, DNA damage, or calcium disruption. There is considerable homology between the two families.[41] Specialized responses to environmental stresses like heat, cold, nutrient limitation, salinity, and osmolarity have been characterized and reviewed.[42,43]

A major role for the induced heat shock Hsp70 family members is in chaperoning the refolding of denatured protein, which can rejuvenate protein function. Abnormal proteins can be refolded by this ATP-driven process and irreparable proteins can be degraded by the ATP ubiquitin pathway. Higher doses of the stressing agent, on the other hand, can overload the system and result in uncontrolled degradation of necessary regulatory molecules.[44]

Different stressors are related in the sense that they share member genes or protein products that interact. In *Escherichia coli,* ethanol and heat initiate only a heat shock response; hydrogen peroxide elicits only the oxidative response; and cadmium chloride stimulates all three responses to heat, oxidative, and UV damage.[42] Likewise, stressors that trigger increased longevity may activate multiple stress cascades with additive beneficial effects on the healthy life span interval.

Resistance to UV-induced stress is observed universally. In prokaryotes, the stress of UV damage results in induction of unlinked genes that can be coordinately controlled by common regulator genes called regulons in an SOS response.[45,46] The bacterial SOS response is activated by a derepressor protein interacting with damaged DNA with subsequent transcriptional upregulation of DNA repair enzymes.

Evidence for UV-induced repair by low dose irradiation in mammalian cells has been shown to enhance the repair rate for subsequent DNA damage in repair-proficient and -deficient cells.[47–49] UV irradiation in human skin results in upregulation of DNA repair genes elicited in part by the guardian of the human genome, the p53 transcription factor and tumor suppressor. The human damage response includes melanogenesis (tanning), transient immunosuppression, resistance to subsequent radiation, and antioxidant enzymes.[47] In aging human skin, there is increased mutation rate, reduced DNA repair, and increased cancer, which appear to be due to slow basal levels of repair. These levels can be restored to youthful levels by induction of a latent repair system by thymidine dinucleotides, an obligate substrate for UV damage and for induced repair.[48] An oligonucleotide mimic of a damaged telomere also triggers a subset of genes for repair or apoptosis. It is known that a yeast mutation associated with a telomeric site showed increased resistance to starvation, heat, and ethanol,[21] linking loss of telomere integrity with induction of multiple stress response cascades. Topical DNA oligonucleotide therapy was found to reduce UV and photocarcinogenesis in mice.[49] The evidence that oliognucleotide signals can trigger multigene response provides direct evidence of the potential power of hormetic agents to induce triggers for UV damage response that restores latent repair in aged cells. Single-gene mutations, like transcription factors, are candidate genes for longevity modulation.

Caloric restriction may elicit multiple stress-specific cascades that are additive in maintenance of the healthy life span duration using both the heat shock family and glucoregulatory pathway, which may be triggered by mimetic agents (see below).

Hormetic Response and Longevity

Application of hormesis for life span extension has been extensively reviewed using UV, ionizing radiation, temperature, hypergravity effects, and dietary restriction in diverse species.[12] In the ancient ciliated protozoan, *Paramecium tetraurelia,* repeated treatment with UV and photoreactivation induced resistance to UV and extended clonal life span.[50] *P. tetraurelia* has a finite clonal life span.[51] In this ciliate, fertilization is an obligate stage to initiate a new life cycle when the somatic macronucleus is discarded and a new one is formed from the zygote germ line micronucleus. The number of vegetative divisions (the micronucleus divides mitotically and the macronucleus amitotically) after fertilization represent the age of the cells in fis-

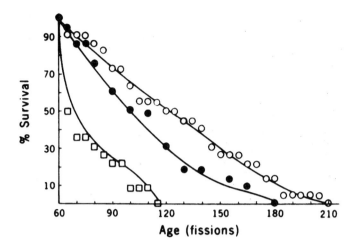

FIGURE 1. Effect of UV only and UV plus photoreactivation on clonal life span. At 60 fissions, one member of a dividing cell was given UV only at 5400 erg/mm^2 (*squares*), then the other cell member was given UV plus photoreactivation (*open circles*). Control cells (*closed circles*) were taken from the same population. The UV treatment was given 1.5 h after cell division since only those cells that had attained that age were used. (Reprinted from Smith Sonneborn[50] with permission.)

sions, and the number of fissions or days from birth at fertilization to death of the last cell in the clone is the fission or calendar age of the clone, respectively. Clonal senescence is characterized by predictable, species-specific longevity. Maximal life span is the highest number of cell divisions before the death of all members of the clone. The mean life span is the average cell divisions or days from fertilization origin to death of all members of the clone.

As *P. tetraurelia* ages, there is a concomitant increase in age-related losses, including DNA strand breaks[52] and sensitivity to UV irradiation, with no change in the basal level of DNA polymerase.[53,54] When stressed with UV radiation, life span was shortened, but when followed by photoreactivation repair treatment, increased resistance to UV and life span extension were found (FIG. 1).[50] The cells, pretreated with radiation and photoreactivation, were more resistant to radiation than naïve cells at the same dose and age (FIG. 2). Repeated treatments of UV irradiation and photoreactivation of cells in their senescent phase were additive (FIGS. 3 and 4) and resulted in a very significant 296% increase in remaining life span. The increase in life span both in cell divisions and days induced by UV and photoreactivation is most likely due to induced repair capacity for both induced UV as well as age damage since (1) cells previously treated with UV and photoreactivation were more resistant to UV and photoreactivation than naïve cells, and (2) extension was only found in older cells, which already had accumulated damage. In this 1979 study, I noted that "The results provide a new approach for regulation and reduction of mutation frequency; the activation of a reserve repair or protection process." The treated cells had a more youthful cell division rate, resistance to UV, and delayed increase in mortality rate; that is, the healthy life span was extended by rejuvenation of the senescent cells. The

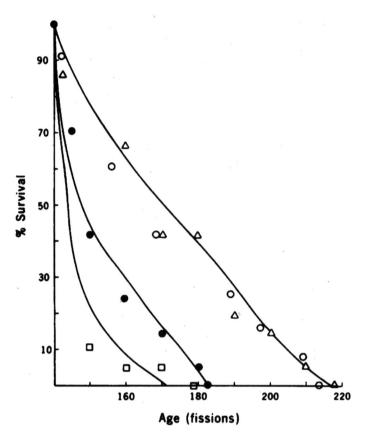

Age (fissions)

FIGURE 2. Induced resistance to ultraviolet. The effect of the same dose of UV (2700 erg/mm²) on clonal life span varied when cells 140 fissions old were previously untreated (*open circles*) or had received UV plus photoreactivation when 80 fissions old (*triangles*). The respective untreated controls (*closed circles*) and control cells that received UV plus photoreactivation when 80 fissions old (*open circles*) are included. (Reprinted from Smith Sonneborn[50] with permission.)

use of repeated treatments with the hormetic agent is an effective strategy to increase longevity.

Modification of the cell membrane of paramecia with low level inductively coupled pulsating currents with specific signals used to promote bone healing, increased mean and maximal life span when cells were shifted to age-specific optimal signals. Our studies detected significant increases in cell division rates of transport-deficient Ca^{2+} mutant pawn, not found in another mutant, paranoia, with adequate transport of this ion.[55]

The ability to alter the life span in the senescent stage is not unique to paramecia. When older flies were placed on a calorie-restriction diet their mortality rates shifted to that of younger animals[56] and older human fibroblasts treated with UV mimetic oligonucleotides showed a more youthful repair (see above). Paramecia, flies, and

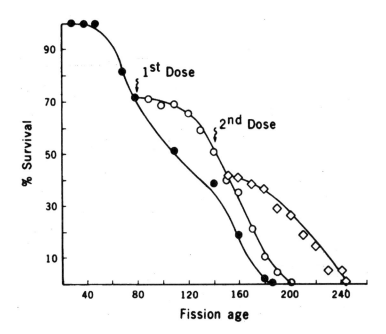

FIGURE 3. Life span extension. The normal decline in survival of untreated controls (*closed circles*) and their sensitivity to 3600 erg/mm^2 UV only (*open squares*) is compared with the extended life span of cells receiving their second dose of UV plus photoreactivation at 140 cell divisions old (*open circles*). The first UV plus photoreactivation was given at 80 fissions. (Reprinted from Smith Sonneborn[50] with permission.)

humans exhibit a latent inducible cascade effective in old animals that can rejuvenate or delay senescent decline using multiple hormetic agents.

In humans, exercise, a hormetic stressor, induces heat shock and ubiquitin transcripts[57] and increased mean life span profile in humans.[4] Since different exercise regimens are known to induce different physiological effects, exercise as a pharmaceutical agent to maximally increase longevity has not been sufficiently investigated.

The evolution of the antiaging action and retardation of disease by diet restriction is well documented.[58,59] Diet restriction may trigger different stressors (ischemia and nutrient limitation) whose effects are additive. The importance of the stress response in diet restriction as a unifying theory has been recently updated and reviewed[60] using data from yeast, worms, flies, and mammals to support the idea that caloric restriction is not simply a passive effect, but an active, highly conserved response to stress to increase survival in adverse conditions—as is proposed in this study. Since caloric-restriction stress response is an ancient adaptive mechanism found from simple organisms to man, applications of mimetics to trigger the hormetic response in humans should be effective.

The conservation of the ancient ability to adapt to nutrient limitation, the discovery of mutations in glucose/insulin pathway that lengthen life span and alter resis-

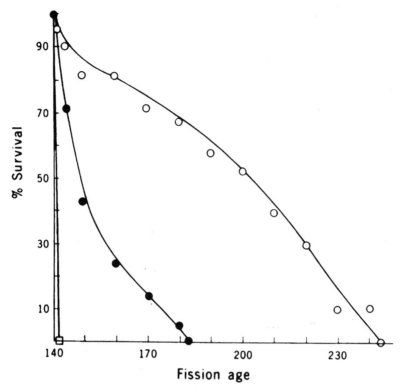

FIGURE 4. Effect of UV plus photoreactivation on clonal life span. The clonal life span of the untreated controls (*closed circles*) is compared with the effect of the first UV and photoreactivation treatment at 80 fissions (*open circles*) and the second UV plus photoreactivation at 140 fissions. The percentage survival was corrected for the percentage of sublines still viable. For example, at 80 cell fissions, only 72% of the sublines were viable. Thus all survival values for cells at that fission age were multiplied by 0.72 to permit direct comparison of treatment effects throughout the clonal life span. (Reprinted from Smith Sonneborn[50] with permission.)

tance to oxidative stress, places caloric restriction in the category of a hormetic agent capable of triggering latent protective stress responses within their genetic background potential.[60]

The protagonistic pleiotropy theory of evolution of aging is consistent with the evidence that different hormetic agents show stressor-specific longevity phenotypic patterns of survival expressed within their genetic potential.

Anti-Aging Supplements and Mimetics

Paramecia were used as a longevity bioassay. The bioassay reveals that agents that damage DNA, UV[61] or the repair inhibitor novobiocin,[64] reduced life span, while induced photoreactivation repair[50] and anti-aging candidates vitamin E,[62] melatonin,[63] DHEA,[64] and kinetin[65] show longevity extension in a hormetic pattern, i.e., low doses are beneficial, high doses are harmful. Melatonin slowed cell division rate and showed a significant increase in calendar days, not cells divisions.[63] DHEA, on the other hand squared the survival curve, significantly increasing mean healthy life span without an increase in maximal life span.[64] Kinetin showed a modest but significant increased mean and maximal life in paramecia.[61] Other mimetics are presently being considered for heat shock, damage response, and glucoregulatory stress for promotion of increased healthy life span and intervention in disease.[4,12,49]

Hibernation

Recently, another strategy has been implemented to mimic the ability of animals to tolerate the stress of hibernation. The hibernating animal tolerates the stress of depleted energy stores, intracellular acidosis, hypoxia, hypothermia, and cell volume shifts and resists muscle wasting.[65] The nature of the trigger for hibernation is an 88 kD peptide with opioid characteristics.[66] Endogenous opioids have been implicated in a pathway to stress and are considered potential mimics of the cardioprotective effects in ischemic preconditioning.[67] The delta opioid, deltorphin D, with other peptides, found on frog skin, has been used topically on humans by Peruvian Indians to increase physical strength, heighten senses, resist hunger, and to induce fearless emotions before a hunt.[68] Hibernation and deltorphin D conferred cardioprotection when challenged by surgically induced ischemia.[68–70] Both hibernation and deltorphin D elicited ubiquitin response, a known stress responder, when rats were subjected to surgically induced ischemia.[71]

Deltorphin D stimulated facilitation of recovery from blood pressure crash induced by hemorrhagic shock in conscience rats when administered during blood pressure crash[72] and modulates the baroreceptor complex. Hibernation may trigger multiple ancient stress pathways known to promote stress survival. Identification of mimics of the triggers could be useful in prevention and recovery from shock.

SUMMARY

In summary, the evolutionary theory proposed here predicts that extension of the healthy interval, even after the onset of senescence, is possible using hormetic agents to stimulate the latent life span by activating the genetic potential of that species. The ability to respond to hormetic agents is selected for early life and is beneficial for survival in later life. The multiple longevity profiles implying plasticity in longevity[4] are consistent with the assertion that different hormetic agents induce common as well as unique responses to stressors used from bacteria to man to preserve longevity. The existence of longevity profiles with extended mean and maximal life span extrapolated to humans implies that interventions may prolong the healthy life span until 90 years old.[4] Since the threshold level of stress may exceed the tolerance to stress, the present strategy is to identify triggers used to elicit the stress response to provide mimetic signals. These advances fulfill Seyle's proposition of the neuroimmunological pathway in aging and disease—that is, that we must learn to imitate, and if necessary complement, the body's own autopharmacologic efforts to combat disease. This prediction seems equally applicable to the role of stress in determination of life span. Remarkable progress has been made in the identification of mimetics of diet restriction, UV irradiation, and heat stress, which can be effective at all intervals of the life cycle in animals, including humans. Hormesis reversal of aging is not at all a myth but an evolutionarily selected mechanism to promote survival of the species.

[*Competing interests*: The author declares that she has no competing financial interests.]

REFERENCES

1. SMITH SONNEBORN, J. 1981. Genetics and aging in the protozoa. Int. Rev. Cytol. **73:** 319–354.
2. SMITH SONNEBORN, J. 1986. Longevity if protozoa. *In* Evolution of Longevity in Animals. A.D. Woodhead & K.H. Thompson, Eds.: 101–110. Plenum. New York.
3. MARTIN, G., S.N. AUSTAD & T.E. JOHNSON. 1996. Genetic analysis of aging: role of oxidative damage and environmental stresses. Nat. Genet. **13:** 25–34.
4. ARKING, R., V. NOVOSELTSEV & J. NOVOSELTSEV. 2004. The human lifespan is not that limited: The effect of multiple longevity phenotypes. J. Gerontol. **59A:** 697–704.
5. CALABRESE, E.J. 2005. Toxicological awakening: the rebirth of hormesis as a central pillar of toxicology. Toxicol. Appl. Pharmacol. **204:** 1–8.
6. SMITH SONNEBORN, J. 1992. The role of the "stress protein response." *In* Biological Effects of Low Level Exposures to Chemicals and Radiation. E.J. Calabrese, Ed.: 41–52. Lewis. Chelsea, MI.
7. SMITH SONNEBORN, J. 1992. Overview lecture. The role of the "stress protein response" in hormesis. *In* Low Dose Irradiation and Biological Defense Mechanisms. T. Sugahara, L.A. Sagan & T. Aoyama, Eds.: 399–404. Excerpta Medica. Elsevier. Amsterdam.
8. BOURG, E. 2003. Delaying aging: could the study of hormesis be more helpful than that of a genetic pathway used to survive starvation? Biogerontology **4:** 319–324.
9. POLLYCOVE, M. & L.E. FEINENDEGEN. 2001. Biologic responses to low doses of ionizing radiation: detriment versus hormesis. Part II. Dose responses of organisms. J. Nucl. Med. **42:** 26–37.
10. NEAFSEY, P.J. 1990. Longevity hormesis: a review. Mech. Ageing Dev. **51:** 1–31.
11. JOHNSON T.E., G.E. LITHGOW & S. MURAKAMI. 1996. Hypothesis: interventions that increase the response to stress offer the potential for effective life prolongation and increased health. J. Gerontol. **51:** 392–395.
12. RATTAN, S.I.S. 2004. Aging intervention, prevention, and therapy through hormesis. J. Gerontol. **59:** 705–709.
13. SEYLE, H. 1946. The general adaptation syndrome and the disease of adaptation. J. Clin. Endocrinol. **6:** 117–230.
14. WILMUT, I., A.E. SCHNIEKE, J. MCWHIR, *et al.* 1997. Viable offspring derived from fetal and adult mammalian cells. Nature **385:** 810–813.
15. SONNEBORN, T.M. 1978. The origin, evolution, nature, and causes of aging. *In* The Biology of Aging. J.S. Behnke, C.E. Finch & G.B. Moment, Eds.: 361–374. Plenum. New York.
16. MARGULIS, L. 1970. Origin of Eukaryote Cells. Yale University. New Haven.
17. OHNE, S. 1970. Evolution and Gene Duplication Springer Verlag, Berlin, Germany.
18. WILLIAMS, G.C. 1957. Pleiotropy, natural selection and the evolution of senescence. Evolution **11:** 398–411.
19. SONNEBORN, T.M. 1957. Breeding systems, reproductive methods, and species problems. *In* The Species Problem. Mayr, Ed.: 155–324. AAAS. Washington, D.C.
20. KALE, SP. & S.M. JAZWINSKI. 1996. Differential response to UV stress and DNA damage during the yeast replicative life span. Dev. Genet. **18:** 54–60.
21. KENNEDY, B.K., N.R AUSTRIACO, JR., J. ZANG & L GUARENTE. 1995. Mutation in the silencing gene SIR can delay aging in *S. cerevisiae*. Cell **80:** 485–496.
22. MUNKRES, K.D. 1990. Genetic coregulation of longevity and antioxienzymes in *Neurospora crassa*. Free Rad. Biol. Med. **8:** 355–361.

23. KLASS, M.R. 1983. A method for the isolation of longevity mutants in the nematode *Caenorhabditis elegans* and initial results. Mech. Ageing Dev. **22**: 279–286.

24. JOHNSON, T.E., S. HENDERSON, S. MURAKAMI, *et al.* 2002. Longevity genes in the nematode *Caenorhabditis* also mediate increased resistance to stress and prevent disease. J. Inherit. Metab. Dis. **25**: 197–206.

25. LITHGOW, G.J. & G.A. WALKER. 2002. Stress resistance as a determinant of *C. elegans* life span. Mech. Ageing Dev. **123**: 563–571.

26. ARANTES-OLIVEIRA, N., J. APFELD, A. DILLON & C. KENYON. 2003. Regulation of life span by germ-line stem cells in *Caenorhabditis elegans*. Science **295**: 502–505.

27. LAKOWSKI, B. & S. HEKIMI. 1996. Determination of life span in *Caenorhabditis elegans* by four clock genes. Science **272**: 1010–1013.

28. HILLIKER, A.J., B. DUFY, D. EVANS & J.P. PHILIPS. 1992. Urate-null rosy mutants of *Drosophila melanogaster* are hypersensitive to stress. Proc. Natl. Acad. Sci. USA **89**: 4343–4347.

29. GRISWOLD, C.M., A.L. MATTHEWS, K.E. BEWLEY & J.W. MAHAFFEY. 1993. Molecular characterization and rescue of acatalasemic mutants of *Drosophila melanogaster*. Science **134**: 781–788.

30. ORR, W.C. & R.S. SOHAL. 1994. Extension of life-span by overexpression of superoxide dismutase and catalase in *Drosophila melanogaster*. Science **263**: 1128–1130.

31. LUCKINBILL, L.S., R. ARKING, M.J. CLARE, *et al.* 1984. Selection for delayed senescence in *Drosophila melanogaster*. Evolution **38**: 996–1003.

32. ROSE, M.R. 1984. Laboratory evolution postpones senescence in *Drosophila melanogaster*. Evolution **38**: 1004–1010.

33. PARTRIDGE L. & K. FOWLER. 1992. Direct and correlated responses to selection at reproduction in *Drosophila melanogaster*. Evolution **46**: 76–91.

34. ZWAAN B., R. BULSMA & R.F. HOEKSTRA. 1995. Direct selection on life span in *Drosophila melanogaster*. Evolution **49**: 649–659.

35. WHITE, C.N., L.E. HIGHTOWER & R.J. SCHULTZ. 1994. Variations in heat shock proteins among species of desert fishes Poeciliidae, *Poeciliopsis*. Mol. Biol. Evol. **11**: 106–119.

36. HART R.W. & R.B. SETLOW. 1975. DNA repair and life span in mammals. Basic Life Sci. **5**: 80–84.

37. HOLZENBERGER, M, J. DUPONT, B. DUCOS, *et al.* 2003. IGF-1 receptor regulates lifespan and resistance to oxidative stress in mice. Nature **421**: 182–187.

38. SAGAN, L. 1987. Radiation hormesis. Health Physics **52**: 517–680.

39. FRITA-NIGGLI, H. 1995. One hundred years of radiobiology: implications for medicine and future perspectives. Experimentia **51**: 652–654.

40. VAN WYNGAARDEN, K.E. & E.K. PAUWELS. 1995. Hormesis: are low level doses of ionizing radiation harmful or beneficial? Eur. J. Nucl. Med. **22**: 481–486.

41. WELCH, W.J., L.A. MIZZEN & A.P. ARRIGO. 1989. Structure and function of mammalian stress proteins. *In* Stress-Induced Proteins. M.L. Pardue, J.R. Feramisco & S. Lindquist, Eds.: 187–202. Alan Liss. New York.

42. BHAGWAT, A.A. & S.K. APTE. 1989. Comparative analysis of protein induced by heat shock, salinity, and osmotic stress in nitrogen fixing *Cyanobacterium anabaena* species strain L31. J. Bacteriol. **171**: 5187–5189.

43. ELESPURU, R.K. 1987. Inducible responses to DNA damage in bacteria and mammalian cells. Environ. Mol. Mutagen **10**: 97–116.

44. ROTHMAN, J.E. 1989. Polypeptide chain binding proteins: catalyst of protein folding and related processes in cells. Cell **59**: 591–601.

45. WITKIN, E.M. 1977. Ultraviolet mutagenesis and inducible DNA repair in *Escherichia coli*. Bacteriol. Rev. **40**: 869–907.

46. WALKER, G.C. 1984. Mutagenesis and inducible responses to DNA damage in *Escherichia coli*. Microbiol. Rev. **48**: 60–93.

47. GILCHREST, B.A. & M.S. ELLER. 2001. Evidence in man for an evolutionary conserved protective adaptation in DNA damage. Comments Theor. Biol. **6**: 483–504.

48. GOUKASSIAN, S., L. BAGHERI, L. EL-KEEB, *et al.* 2002. DNA oligonucleotide treatment corrects the age-associated decline in DNA repair capacity. FASEB J. **16**: 754–756.

49. GOUKASSIAN, D.A., E. HELMS, H. VAN STEEG, *et al.* 2004. Topical DNA oliognucleotide therapy reduces UV induced mutations and photocarcinogenesis in hairless mice. Proc. Natl. Acad. Sci. USA **101:** 3933–3938.

50. SMITH SONNEBORN, J. 1979. DNA repair and longevity assurance in *Paramecium tetraurelia*. Science **203:** 1115–1117.

51. SONNEBORN, T.M. 1954. The relation of autogamy to senescence and rejuvenescence in *Paramecium aurelia*. J. Protozool. **1:** 38–53.

52. HOLMES G.E. & N.R. HOLMES. 1986. Accumulation of DNA fragments in aging *Paramecium tetraurelia*. Gerontology **32:** 252–260

53. SMITH SONNEBORN, J. 1971. Age-correlated sensitivity to ultraviolet radiation in Paramecium. Radiat. Res. **46:** 64–69.

54. WILLIAMS, T.J. & J. SMITH SONNEBORN. 1980. DNA polymerase activity in aged clones of *Paramecium tetraurelia*. Exp. Gerontol. **15:** 353–357.

55. DIHEL, L. & J. SMITH-SONNEBORN. 1985. Effects of low frequency electromagnetic field on cell division and the plasma membrane. Bioelectromagnetics **6:** 61–71.

56. MAIR, W., P. GOYMER, S.D. FLETCHER & L. PARTRIDGE. 2003. Demography of dietary restriction and death in Drosophila. Science **301:** 1731–1733.

57. SMITH SONNEBORN, J. & S.A. BARBEE. 1998. Exercise induced stress response as an adaptive tolerance strategy. Environ. Health Persp. **106**(Suppl.): 325–330.

58. MASORO, E.J. & S. AUSTED. 1996. The evolution of the anti-aging action of dietary restriction: a hypothesis. J. Gerontol. **51:** 387–391

59. WEINDRUCH, R. & R. WALFORD. 1988. The Retardation of Aging and Disease by Dietary Restriction. Charles C. Thomas. Springfield, IL.

60. SINCLAIR, D.A. 2005. Toward a unified theory of caloric restriction and longevity regulation. Mech. Ageing Dev. In press.

61. SMITH SONNEBORN, J., P.D. LIPETZ & R.E. STEPHENS. 1983. Paramecium bioassay of longevity modulating agents. *In* Intervention in the Aging Process. W. Regelson & F. M. Sinex, Eds. Alan R. Liss. New York.

62. THOMAS, J. & D. NYBERG. 1988. Vitamin E supplementation and intense selection increase clonal life span in *Paramecium tetraurelia*. Exp. Gerontol. **23:** 501–512.

63. THOMAS, J. & J. SMITH SONNEBORN. 1997. Supplemental melatonin increases clonal lifespan in the protozoan *Paramecium tetraurelia*. J. Pineal Res. **23:** 123–130.

64. SMITH SONNEBORN, J. 1983. Unpublished results.

65. HARLOW, H.J., T. LOHUIS, T.D. BECK & P.A. IAIZZO. 2001. Muscle strength in overwintering bears. Nature **409:** 997.

66. HORTON, N.D., D.J. KAFTANI, D.S. BRUCE, *et al.* 1998. Isolation and partial characterization of an opioid-like 88 kDa hibernation-related protein. Comp. Biochem. Physiol. **119:** 787–805.

67. FRYER, R.M., J.A. AUCHAMPACH & G.J. GROSS. 2002. Therapeutic receptor targets of ischemic preconditioning. Cardiovasc. Res. **55:** 520–525.

68. ERSPAMER, V., G.F. ERSPAMER, C. SEVERINI, *et al.* 1993. Pharmacological studies of "sapa" from *Phyllomedusa bicolor* skin: a drug used by the Peruvian Matses Indians in shamanic hunting practices. Toxicon. **9:** 99–111.

69. BOLLING, S.F., N.L. TRAMONTINI, K.S. KILGORE, *et al.* 1997. Use of "Natural" hibernation induction triggers for myocardial protection. Ann. Thorac. Surg. **64:** 623–627.

70. SIGG, D.C., J.A. COLES, JR., P.R. OELTGEN & P.A. IAIZZO. Role of β opioid receptor agonists on infarct size reduction in swine. Am. J. Physiol. Heart Circ. Physiol. **282:** 19553–19560.

71. SMITH SONNEBORN J., H. GOTTSCH, E. CUBIN, *et al.* 2004. Alternative strategy for stress tolerance: opioids. J Gerontol. **59:** 433–40.

72. MCBRIDE, S.M., J. SMITH SONNEBORN, P. OELTGEN & F.W. FLYNN. 2005. Delta2 opioid receptor agonist facilitates mean arterial pressure recovery after hemorrhage in conscious rats. Shock **23:** 264–268.

Protein Kinase C Signal Transduction Regulation in Physiological and Pathological Aging

FIORENZO BATTAINI[a] AND ALESSIA PASCALE[b]

[a]Department of Neuroscience, University of Roma "Tor Vergata," Rome, Italy

[b]Department of Experimental and Applied Pharmacology, University of Pavia, Pavia, Italy

ABSTRACT: Calcium/phospholipid-regulated protein kinase C (PKC) signalling is known to be involved in cellular functions relevant to brain health and disease, including ion channel modulation, receptor regulation, neurotransmitter release, synaptic plasticity, and survival. Brain aging is characterized by altered neuronal molecular cascades and interneuronal communication in response to various stimuli. In the last few years we have provided evidence that in rodents, despite no changes in PKC isoform levels (both calcium dependent and calcium independent), the activation/translocation process of the calcium-dependent and -independent kinases and the content of the adaptor protein RACK1 (receptor for activated C kinase-1) are deficient in physiological brain aging. Moreover, human studies have shown that PKC and its adaptor protein RACK1 are also interdependent in pathological brain aging (e.g., Alzheimer's disease); in fact, calcium-dependent PKC translocation and RACK1 levels are both deficient in an area-selective manner. These data point to the notion that, in addition to a well-described lipid environment alteration, changes in protein–protein interactions may impair the mechanisms of PKC activation in aging. It is interesting to note that interventions to counteract the age-related functional loss also restore PKC activation and the adaptor protein machinery expression. A better insight into the factors controlling PKC activation may be important not only to elucidate the molecular basis of signal transmission, but also to identify new strategies to correct or even to prevent age-dependent alterations in cell-to-cell communication.

KEYWORDS: protein kinase C; PKC; isoforms; isozymes; aging; brain; RACK1; anchoring; scaffolding; translocation; Alzheimer's disease; pineal transplant

SIGNAL TRANSDUCTION IN BRAIN AGING

Intracellular communication represents a key element of the molecular alterations associated with senescence, resulting in modifications of cell function and, at

Address for correspondence: Fiorenzo Battaini, Ph.D., University of Roma, Tor Vergata School of Medicine, Department of Neurosciences, Via Montpellier 1, 00133 Rome, Italy. Voice: +39 06 7259 6304/6310; fax: +39 06 7259 6302.

Battaini@med.uniroma2.it

Ann. N.Y. Acad. Sci. 1057: 177–192 (2005). © 2005 New York Academy of Sciences.
doi: 10.1196/annals.1356.011

a more integrated level, in cognitive and neurological impairments. In this context, the aging process can affect signal transduction at different levels from receptor availability and coupling with effectors systems to the production of second messengers and the activation of protein kinases and phosphatases. Current research is focusing on the molecular mechanisms of neuronal communication as possible targets for intervention to ameliorate age-related functional decay and to prevent cell degeneration.[1]

An important concept that has emerged in recent studies utilizing stereological techniques of neuronal count during physiological brain aging in humans, is that contrary to the original morphological studies showing age-related neuronal loss in cortical structures, neuronal loss is not evident between ages 60 and 90.[2] Moreover, no neuronal decline is present in the hippocampus of aged rats showing age-related deficits in the water-maze task, providing evidence that the behavioral impairments observed during senescence are not associated with neuronal degeneration in the hippocampus.[3] It therefore appears that although cortical and hippocampal neurons may not degenerate during aging, other alterations—such as those implying changes in cell-to-cell communication—may affect cerebral functions. The *primum movens* toward age-related changes may thus be related to alterations in signal transduction cascades and not to morphological deficits.

Protein modification by phosphorylation is a relevant regulatory tool utilized by all cells within various aspects of signal transduction under both physiological and pathological conditions. It represents the "final common pathway" in the actions of various transmitters and it can also constitute an important regulatory feedback mechanism for some of the steps involved in the production of intracellular second messengers. Several protein kinases have been reported to play critical roles in brain functions and protein kinase C is widely recognized as one of the relevant players in the regulation of learning and memory phenomena.[4,5]

FOCUS ON PROTEIN KINASE C ISOZYMES

The serine/threonine kinases, protein kinases C (PKCs) are a family of enzymes (at least 10 isoforms) highly expressed and functionally operating in brain tissues, transducing signals coupled to receptor-mediated hydrolysis of membrane phospholipids.[6] These enzymes transduce signals that regulate short-term events (ion fluxes, neurotransmitter release), mid-term events (receptor modulation), as well as long-term events (synaptic remodeling and gene expression). Brain tissues express all PKCs[7] and, in spite of sequence and structural homologies among the different PKCs, each isoform may have unique roles resulting from specific activation-dependent subcellular localization and substrate phosphorylation.

While the C-terminal catalytic domain of PKC shows homology among the different isoforms, the regulatory N-terminal sequence confers enzyme diversity and is the site where interaction with activators occurs. The regulatory domain binds and it is activated by calcium, DAG, phosphatidylserine (PtdSer), and other lipids (see below). Activation of PKCs is associated with the movement or translocation of each single isoform from a compartment (where PKC is catalytically inactive) to another (where it becomes catalytically active) where PKC finds the specific activators. PKC is kept in an inactive "folded" conformation by binding of the pseudosubstrate sequence (in the regulatory region) to the substrate binding site (in the catalytic region).[8,9]

PKCs are classified as calcium-dependent, calcium-independent, and atypical isozymes according to their sensitivity to calcium and diacylglycerol. The membrane lipid PtdSer sustains the activity of all PKCs. The conventional or calcium-dependent PKCs (cPKCs: α, βI, βII, and γ) require additional free calcium and diacylglycerol (DAG) for complete activation. The calcium-independent novel PKCs (nPKCs: δ, ε, η, and θ) are fully activated by DAG only, whereas the atypical PKCs (aPKCs: ζ and λ/ι) are calcium- and DAG-independent.[8] The tumor promoters phorbol esters can mimic DAG in PKCs activation. Two DAG-dependent kinases formerly described as additional members of the PKC family (μ and ν) are now classified as protein kinases D, sharing better homology with the catalytic domain of calcium/calmodulin kinases.[10]

REGULATION OF PKC ACTIVATION

Phosphorylation

The structure, subcellular localization, and function of PKCs are regulated by upstream and downstream events involving the inactive different isozymes. Upstream events are characterized by a cascade of phosphorylating reactions taking place on the catalytic domain of the nascent PKCs.[8] These "priming" phosphorylations are important to make PKCs responsive to activators. The newly synthesized "immature" kinases are initially phosphorylated as membrane-bound enzymes by a phosphoinositide-dependent kinase-1(PDK-1), which is modulated by 3′ phosphoinositides produced by phosphoinositide 3-kinase (PI 3-kinase).[9] The PDK-1 phosphorylation permits the subsequent autophosphorylation of two additional residues in the catalytic domain: at this point PKC is "mature" and released as the soluble form (inactive) able to respond to activators. The importance of PDK-1 for native PKC is underscored by the observation that deletion of PDK-1 dramatically decreases the levels of PKC isozymes.[11]

Downstream events are characterized by the binding of PKCs with their activators.

Protein–Lipid Interactions

A variety of lipid second messengers produced by phospholipases (PL) activate PKCs. The first described and well known pathway involves the receptor-mediated PLC activation that induces the hydrolysis of the membrane phosphatidylinositol 4,5-bisphosphate (PIP2), thus generating inositol 1,4,5-triphosphate (IP3) and DAG: IP3 releases free calcium ions from intracellular stores and DAG is also generated by receptor-mediated activation of PLD. In addition, PLA2 activation produces arachidonic acid, lysophosphatidylcholine, and cis-unsaturated fatty acids reported as activators and/or enhancing activators of several PKCs.[12]

PKC activation is not only regulated by protein–lipid interactions described so far, but also by protein–protein interactions.

Protein–Protein Interactions

The receptor-regulated production of DAG and other lipids induce in PKC a change in localization and the phosphorylation of specific substrates. How can individual activated PKCs find their way close to the relevant substrates? In this case protein–protein interactions play an important role. Several proteins interact with PKCs with isoform selectivity;[13] some proteins are direct substrates while others facilitate the interaction with the relevant substrates. For aPKCs, activation can be driven by intracellular localization through regulatory proteins and nuclear localization/export signals.[8,9] Among the PKC-interacting proteins, scaffolding/adaptor proteins termed RACKs (receptors for activated C kinases) bind with isozyme selectivity PKC only after activation and direct the enzyme close to the relevant substrates.[14] RACKs are proteins bearing seven repeats of the WD40 motif identified originally in the β subunit of G proteins: WD40 motifs regulate protein–protein interactions. Up to now, two RACK proteins have been identified and characterized namely RACK1 (selective for PKCβII)[15] and RACK2 (selective for PKCε).[16] RACKs are not substrates of the respective PKCs, upon PKC activation they colocalize and increase substrate phosphorylation by several fold. Because of the presence of seven WD repeats, RACKs bind and regulate a variety of other signaling enzymes.[17] One of the most interesting breakthroughs within this field is the recognition that it is possible to inhibit the translocation and the function of individual PKC isozymes by peptides mimicking the sites of interaction between PKC and RACK. Moreover pseudo-RACK peptides can activate PKC without production of second messengers.[18] By utilizing these peptides, specific functions have been recognized for individual PKC isoforms in both cardiac and neuronal homeostasis.[19] As for the regulation of "inactive" PKCs, other scaffolding proteins termed RICKs (receptors for inactive C kinases) may influence their localization before activation.[20]

PKC DOWNREGULATION

In the membrane-bound form, PKC is two orders of magnitude more sensitive to dephosphorylation when compared to the soluble enzyme. Long-term activation of PKC results in dephosphorylation by phosphatases and thus an interaction with the chaperone heat shock protein 70 (Hsp70) occurs.[21] At this point two pathways can be taken: one way in which Hsp70-bound PKC is rephosphorylated and recycled back as soluble newly reactivable enzyme; the other in which Hsp70 separates from PKC resulting in its association with cytoskeletal elements and eventual downregulation by proteolysis.[8]

PKC ISOFORM–SPECIFIC BRAIN FUNCTIONS

The roles of PKC isozymes in the brain are largely unknown, but experiments using knockout mice and isoform-specific inhibitors/activators seem to indicate that some functions are indeed regulated by specific PKC isoforms (TABLE 1). Studies in mouse lacking the neuron-specific PKCγ were the first to be described. In this animal model, spatial memory was mildly affected and long-term synaptic potentiation

TABLE 1. Selected functions of cPKC isozymes in neuronal tissues: knockout and peptide studies

PKC isozyme	Knockout (reference)	Peptide inhibitors (reference)	Peptide activators (reference)
PKCα	Cerebellar LTD impairment (24)	ND	ND
PKCβ	Loss in fear conditioning (25)	Astrocyte degeneration (29)	ND
PKCγ	Hippocampal LTP impairment (22) Loss of estrogen protection from ischemia (23) Ethanol/sedative hyposensitivity (26)	Ethanol withdrawal hyper-responsiveness (33)	ND
PKCδ	ND	Reduction of cerebral ischemic reperfusion injury (28)	Attenuation of kainate degeneration (31)
PKCε	Ethanol/sedative supersensitivity (26)	Hyperalgesia (30)	Protection from oxygen/glucose deprivation (29)
PKCθ	Synapse elimination impairment (27)	ND	ND
PKCζ	ND	Decreased excitotoxicity (32)	ND

NOTE: ND = not determined; LTD = long-term depression; LTP = long-term potentiation.

in hippocampus was impaired;[22] these animals also present a reduced neuroprotection by estrogens after ischemia.[23] In the PKCα null mice long-term synaptic depression is not inducible in the cerebellum[24] while mice lacking PKCβ suffer from a deficit in both cued and contextual fear conditioning.[25] The PKCε null mice indicate an involvement of this isoform in pain states and in the action of sedative drugs and in alcohol consumption.[26] In the PKCθ knockout, a decreased synapse elimination is observed.[27]

Investigations with isoform-specific inhibitors/activators have shown that PKCδ inhibition reduces reperfusion-induced injury after stroke[28] while inhibition of PKCβI enhances astrocyte degeneration.[29] PKCε and γ inhibition are also involved in hyperalgesia,[30] while activation of PKCδ attenuates the kainate-induced cell death of cortical neurons.[31] Additionally, PKCζ inhibition may protect from excitotoxic neuronal death[32] while ethanol withdrawal hyper-responsiveness may be sensitive to selective inhibition of PKCγ.[33] Moreover, mutational approaches point out that the PKCγ gene is associated with animal models of Parkinson's disease[34] and with cerebellar ataxia in humans.[35]

All these data suggest the possibility that interfering with PKC in an isoform-selective manner could control several aspects of neuronal homeostasis under both physiologic and pathologic conditions.

cPKC IN PHYSIOLOGICAL BRAIN AGING

A variety of functions related to PKC have been reported to be modified as a consequence of brain aging. In addition to changes at the molecular level, such as calcium ion homeostasis, neurotransmitter release, and receptor regulation, important modifications occur in long term synaptic changes and thus in some forms of memory. In addition behavioral, genetic, elecrophysiological, and pharmacological evidence underscore the importance of PKC in learning and memory processes.[4] Within this context, in the last few years we have investigated brain PKC homeostasis and its regulation during aging.

Kinase Activity Shows Strain and Substrate-Dependent Changes

We started by asking whether PKC function, evaluated with the conventional assay of histone phosphorylation in the presence of optimal concentrations of activators (calcium, PtdSer, DAG), was modified in different brain areas. We found that Sprague-Dawley rat cortex shows age-related decrease in spite of an increase of PKC activity detected in the hippocampus and no change in cerebellum.[36] This effect proved to be strain-dependent because it was also observed in Fisher 344,[37] but not in Wistar rats, where kinase activity was unchanged in all brain areas (cortex, cerebellum, and hippocampus) and fractions (soluble and membrane) investigated[38] (FIG. 1A, for the activity in cortex). Utilizing a selective brain substrate for PKC (i.e., purified protein B-50/GAP43) we were able to detect a significant impairment in

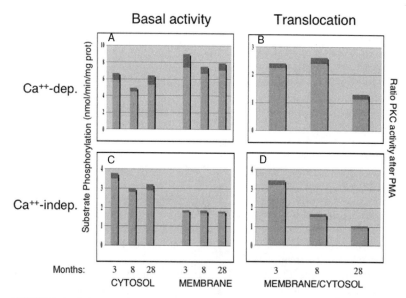

FIGURE 1. Calcium-dependent and -independent PKC activity and translocation in cortex from Wistar rats during aging. Calcium dependent PKC activity (**A**) and translocation (**B**) was assayed utilizing histone as substrate as described.[38] Calcium-independent PKC activity (**C**) and translocation (**D**) was assayed utilizing (Ser^{25}) PKC$_{19\text{-}36}$ as substrate as described.[41] Translocation is reported as ratio of activity in membranes and soluble fractions after treatment with PMA 160 nM for 15 min. Data are the mean values ± SE (darker) from 7–9 individual measurements.

PKC-dependent phosphorylation in cortical membranes from middle-aged (8 month old) and aged (28 month old) Wistar rats. B-50 phosphorylation was instead increased in the hippocampal membranes from the same aged Wistar rats.[38]

cPKCs Translocation Is Impaired in Aged Rats

In the previously described studies, PKC activity measurements were assessed *in vitro* under optimal conditions of activators and substrates, but this may not be the physiological milieu during aging. It is known that calcium ions, phospholipids, and substrates display changes during senescence that in some cases may be also area-dependent.

To circumvent these variables, we analyzed PKC function in terms of activation/translocation in response to phorbol esters challenge utilizing tissue slices. Upon short-term (15 min) low concentration (160 nM) phorbol ester (PMA: phorbol 12-myristate-13-acetate) treatment, the activation of PKC bypasses the availability of activators and the enzyme translocates from the cytosolic to the membrane compartment. This process is observed as a decrease in the cytosolic and an increase in the membrane-bound PKC activity. cPKC translocation appeared to be preserved up to middle age but was impaired in aged animals (Wistar) in both cortex (FIG. 1B) and hippocampus.[38] This effect is independent of the strain utilized because it is observed also in aged Fisher 344 rats. [39]

Transcription and Translation of cPKCs in Aging

Given that the activity and translocation of PKC might be influenced by a differential expression of the PKC isoforms during aging, we analyzed the existence of changes at transcriptional level. The presence and regional distribution of cPKCs (α, β, and γ) mRNA isoforms were assessed by Northern hybridization. No modifications in the amount pattern of the different mRNAs were observed in aged Wistar rats, only a decrease in the α and β signal that is already present at 8 months of age in cortex.[38] Northern blot analysis allows quantification of mRNA in tissue homogenate but localized changes in specific brain regions may remain undetected due to the limitation of this approach. To further investigate any potential changes in PKC mRNA levels in discrete hippocampal and cortical regions, we performed *in situ* hybridization: aging did not modify in a significant manner the relative mRNA of the three calcium-dependent PKC isoforms in any area and subfield analyzed.[38]

The possibility that translational changes in PKC isoforms might occur during aging was then addressed by looking at the content of each specific isoform by Western analysis, in both soluble and particulate fractions. PKCα, βI, βII, and γ immunoreactivities were not modified in cytosolic and membrane fractions of cortex and cerebellum; only hippocampal PKCγ was significantly increased at membrane level in aged animals. As we have speculated,[38] this observation may be directly related to learning capabilities in aged animals. In fact, when analyzing learning performance on a water-maze task as a function of PKCγ (as immunohistochemistry signal) in aged rats, Colombo and Gallagher recently found, in accord with our observations, that the worst performance was coupled to higher levels of PKCγ in hippocampal CA1 and dentate gyrus regions. [40] From our data we can add that this effect may be due to posttranslational changes because PKCγ mRNA in the CA1 and dentate gyrus regions does not change in aged animals.[38]

TABLE 2. RACK1 levels in cortical membranes fractions in physiological and pathological brain aging

Rat age (months)	RACK1 immuno-reactivity (ODUs/µg protein)	Percent of young	Percent of controls	Ratio immuno-reactivity RACK1 (human cortex/ internal standard)	Patients
Young (3)	208 ± 22	100	100	0.65 ± 0.06	controls
Middle-aged (8)	180 ± 33	86	53	0.35 ± 0.08*	Alzheimer's disease
Aged (28)	94 ± 8*	45			

NOTE: ODUs = optical density units; values are expressed as means ± SEM; *$P < 0.01$ Dunnett's (rats) or Student's (patients) t test; internal standard = rat cerebellum membranes.

Anchoring/Scaffolding of cPKCs

The most consistent change in cPKC function during brain senescence (irrespective of strain and area investigated) appeared to be the lack of enzyme activation/ translocation. Changes in membrane lipid composition and impairments in transduction could explain such functional deficit. Owing to the reputed role of RACK proteins in directing activated PKC to relevant targets, we studied whether the age-related impaired PKC translocation could be associated with changes in protein–protein interactions.[41] We observed that cortical levels of RACK1 were decreased in membranes prepared from aged rats (TABLE 2). Because of the reported specificity of RACK1 for PKCβII, we measured the translocation of this isoform after PMA treatment. PKCβII translocation to the membrane compartment was also deficient in tissue slices from aged animals. Recent findings have confirmed this age-related decrease in cortical RACK1 in aged Sprague-Dawley rats and also demonstrated a reduction in mRNA levels.[42] This effect is sex specific, not being observed in female animals. Taken together, these data indicate that a deficit in RACK1 may contribute to the functional impairment in cPKC translocation detected in aged rats.

nPKC AND aPKC IN PHYSIOLOGICAL BRAIN AGING

In order to complete the study of PKC during brain aging we also investigated the calcium-independent novel (nPKC) and atypical (aPKC) isoforms in the cortex. In agreement with the findings on cPKCs, PKCδ, ε, η, and ζ remain constant as immunoreactive proteins in both soluble and membrane fractions.[43] Accordingly, basal calcium-independent kinase activity did not change as function of age (FIG. 1C). On the other hand, calcium-independent PMA-dependent PKC translocation resulted in deficits in aged animals (FIG. 1D) as also confirmed by specific Western blotting analysis for PKCδ and ε. [41] We have also investigated the levels of another anchoring protein reported to interact preferentially with the calcium-independent PKCε.[16] However the levels of RACK2 (utilizing antibodies to βCOP kindly provided by Dr. K.J. Harrison Lavoie)[44] remain constant in membrane fractions from aging rats (unpublished results). Considering that RACK1 may interact, although with less affinity when compared to βII, with PKCδ and ε,[15] it is thus possible that the decrease in RACK1 may affect calcium-independent PKC translocation as well.

Collectively all the reported data on physiological aging point to the general observation that at least in rat cortex during aging:

- calcium-dependent and -independent kinase activity under optimal conditions of activators remains constant throughout life;

- all PKC isoforms are preserved as basal immunoreactive protein expression, in spite of a decreased mRNA for PKCα and β already present in adulthood;

- the translocation process upon PMA challenge is deficient for both calcium-dependent and -independent PKC isoforms; and

- the anchoring protein RACK1 deficit in the membrane may contribute to the impairment in calcium-dependent and -independent PKC translocation.

PKC IN PATHOLOGICAL BRAIN AGING

PKC has been implicated in the neurodegeneration that occurs in Alzheimer's disease (AD). Besides deficits in neurotransmitters and growth factors activating PKC, the enzyme itself and its responsiveness is modified in AD in central and peripheral tissues.[43,45–48] The kinase C family is involved in the regulation of the metabolism of the amyloid precursor protein (APP) towards the non-amyloidogenic pathway.[46] Of relevance to our previous studies, it was reported that there is a defect in PKC activation in AD brain in terms of marked loss of redistribution of cortical cytosolic PKC to the membrane fraction in response to activation with phorbol esters and K[+] depolarization.[48] This was an observation with analogies with our previous investigations in aging rats that could be utilized to prove the relevance of the animal studies for human pathologies. We reasoned that, accordingly to our findings in rat brain during senescence, anchoring proteins could also be involved in the regulation of PKC activation process in Alzheimer's disease. The analysis of RACK1 in human frontal cortex autopsies indicate that when compared to age-matched controls AD patients have reduced expression of RACK1 in both soluble and membrane compartments (TABLE 2).[49] The basal levels of PKC βII were also assessed and again, in agreement with our previous results, no pathology-related changes were observed in AD tissues,[49] suggesting that it is not the expression of PKC that is deficient AD, but rather the mechanism of activation-anchoring that undergoes pathology-dependent changes.

AGE-RELATED PKC ACTIVATION/ANCHORING
IN NON-NEURONAL TISSUES

We further explored whether the age-associated changes in PKC activation/compartmentalization could be extended to non neuronal tissues. For this purpose we utilized the model of rat alveolar macrophages. These cells secrete tumor necrosis factor α (TNFα) in response to lipopolysaccharide (LPS) through a PKC-mediated transduction machinery. Macrophages from old rats secrete less TNFα in the presence of LPS; the translocation of PKCβII upon LPS is impaired in aged animals, where immunoreactive RACK1 appears to be decreased by roughly 50%.[50]

Other studies have reported in senescent myocardium a disrupted PKCα and PKCε translocations to membrane in response to α1 receptor activation, associated with a reduction in the respective anchoring proteins RACK1 and RACK2.[51]

Moreover, recent reports on peripheral blood human leukocytes from aged donors have demonstrated a decreased expression of RACK1 that impairs the PKCβ-dependent LPS-induced TNFα production and the mitogen-dependent proliferation.[52]

All these data strengthen the notion that an age-dependent RACK1 deficit may impair PKC translocation and related functions in peripheral tissues as well.

INTERVENTIONS ON PKC SIGNAL TRANSDUCTION IN AGING

A number of conditions have been reported to counteract the blunted PKC transmission and/or anchoring mechanisms during aging. One of the first reports dealt with exogenous PtdSer that appeared to counteract the blunted PKC-dependent phosphorylation of hippocampal protein B50/GAP43.[53] In the same brain area, caloric restriction and antioxidant (N-tert-butyl-α-phenylnitrone) treatment can restore phorbol ester–dependent synapsin I phosphorylation.[54] In peripheral tissues a smaller reduction in lymphocyte PKC activation can be shown in aged men practicing aer-

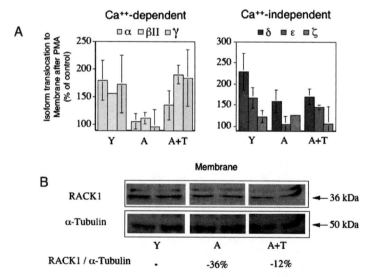

FIGURE 2. PKC isoforms translocation (**A**) and RACK1 levels (**B**) from cortex of aged rats after pineal transplant. Membrane PKC isozyme translocation and RACK1 levels were assayed as described.[41] PKC translocation was induced with treatment with PMA 160 nM for 15 min in cortical slices from 4–5 rats for each group. Values are means ±SD of 3 determinations. RACK1 IR was assayed in frozen aliquots under unstimulated conditions. Y = young, 3-month-old rats; A = aged, 25-month-old rats; A + T = aged 23-month-old rats subjected, at 18 months of age, to a transplant of the pineal gland from young animals. In **A** each *group of three bars* represents, from left to right, α, βII, and γ, respectively, for Ca++-dependent; and δ, ε, and δ, respectively, for Ca++-independent.

obic fitness[55] while chronic exercise training in rats partially reverses $\alpha1$-mediated activation of cardiac PKC.[56]

In other studies it has been reported that the age-related loss of RACK1, as both messenger RNA and protein, observed in the hippocampus of aged rats can be restored by a two weeks treatment with dehydroepiandrosterone.[57] Accordingly with this findings, this steroid restores RACK1 levels in aged rat alveolar macrophages and splenocytes.[58]

Recently, in collaboration with Drs. Walter Pierpaoli and Daniele Bulian, we have obtained preliminary results on PKC activation in the brain of aged rats (25 months old) in which implantation of a pineal gland from young rats (3 months old) into the thymus was performed at 18 months. Our findings seem to indicate that pineal implant partially restores the aging-associated impairment of PKC translocation in cortex, at least for βII and ε isoforms (FIG. 2A). Additionally, pineal transplant also induces a partial recovery (–12% with respect to young animals values) in RACK1 levels that were decreased by 36% in aged animals (FIG. 2B).

In rodents this model previously also showed a prolongation of survival and of T cell–mediated immune function.[59] Moreover a normalization of brain cortical β1 adrenergic receptor density[60] and of hippocampal LHRH receptors[61] has been reported in aged transplanted animals in comparison with controls with no transplants.

DISCUSSION

We[1,36,38,41] and several other groups[37,39,53,54] have provided evidence that PKC is a signal transduction system that is affected in aged animals in neuronal tissues and this observation can be extended to non-neuronal tissues as well,[50,51] underlining the importance of this finding for aging mechanisms in general terms.

Calcium-dependent PKC translocation is blunted as consequence of aging irrespective of animal strain and area investigated (at least in brain cortex, hippocampus, and cerebellum[38]) and in general no changes at isoform basal levels are evident. In cortical tissues, calcium-independent PKC translocation is also deficient despite the absence of changes at isoform basal levels.

The only exception is presented by PKCγ isoform, which increases in the hippocampal membranes of aged animals.[38] This may be related to the control of plasticity phenomena at brain level and thus to memory. This observation was at first sight unexpected mainly because in young animals hippocampal PKCγ levels are related to better performance in different memory tasks.[62] However, analyzing PKCγ levels in aged rats as function of learning performance on a water-maze paradigm, Colombo and Gallagher found that in the animals with the worst memory, performance parallels higher levels of hippocampal membrane-bound PKCγ in CA1 and dentate gyrus regions.[40] Taken together these data underline the possibility that memory impairment in aged rats may result, at least partially, from a dysregulation of PKCγ. Considering that PKCγ is specifically expressed in the nervous system, higher levels of membrane-bound PKCγ in unstimulated conditions may represent an attempt to compensate a functional deficit occurring in the translocation process of this specific isoform during senescence. In agreement with this concept, during aging one of the key factors controlling PKC activation/relocation—RACK1 protein—decreases thus contributing to the inability to sustain PKC translocation.[41] These observations can be extended to non neuronal tissues in physiological animal aging[50] and to pathological brain aging—Alzheimer's disease—where a deficient PKC translocation process occurs,[48] concomitantly with a decrease in RACK1 anchoring protein, indicating that the enzyme is preserved—but not the mechanism for its activation.[49] It is interesting to note that a defective RACK1 system is observed also in brain cortex of Down syndrome fetuses (early second trimester of gestation).[63]

The relevance of RACK1 changes with respect to PKC deserves further comment. Van der Zee and coworkers have investigated in great detail cPKCs in aged rabbit hippocampus through immunohistochemical and subcellular fractionation studies.[64] With the former, they observed an increase in immunoreactivity in selected subregions for β and γ, and with the latter they found a dramatic decrease for all the isoforms at membrane level associated with a robust loss of RACK1 protein (–96%). Their interpretation is that the increase in immunoreactivity is due to the loss of RACK1 (interacting with PKC) thus allowing more antigenic sites on PKC to be detected immunohistochemically.[64] Nonetheless they report unpublished findings on increased PKCγ in hippocampus of aged rats and emphasize the relevance of RACK1 loss in the dysregulation of PKC system in aged rabbits as well. These data suggest that, in addition to strain-specific changes, there are also species-specific changes in PKC isoforms during aging.

The relevance of RACK1 in regulating PKC function is further underscored by results of studies on brain PKC/RACK1 interdependence in morphine-dependent rats.[65] In addition, in a condition in which PKC is hyperfunctional (bipolar patients) PKC activation by PMA is upregulated, and this phenomenon goes in parallel with an increased association of various PKCs with RACK1.[66]

Various conditions affecting PKC-lipid interactions (treatment with exogenous phosphatidylserine[53] and antioxidants,[54] for example) can counteract the age-related changes in PKC detected during brain aging. Additionally, the PKC translocation deficit occurring during senescence may be prevented by hormonal (dehydroepiandrosterone) treatment[57] or by pineal transplant (FIG. 2) acting through a recovery in brain RACK1 levels. Concerning pineal transplant, we may only speculate that in transplanted aged animals a release of some peptides that influence positively brain functional homeostasis could be involved.[59] As far as the recovery in RACK1 levels is concerned, it is not known how this process can be regulated, but recent data have documented that NF-κB is a relevant transcription factor for RACK1 gene.[67] The same authors provided evidence that RACK1 promotes neuronal cell survival. It is interesting to note that inhibition of NF-κB increases beta amyloid neuronal apoptosis,[68] suggesting that RACK1 deficit in Alzheimer's disease may be dependent on a defect in NF-κB activation. Along this line, it is important to note that the activation of PKC with bryostatin 1 (a non-tumor promoter activator of the enzyme) decreases beta amyloid accumulation and premature death and improves behavioral outcomes in mice models of Alzheimer's disease.[69] Moreover rasagiline, a monoamine oxidase B inhibitor, facilitates *in vivo* non amyloidogenic processing of APP in rat hippocampus through increase in PKC/RACK1 levels.[70] It is thus tempting to postulate that it is possible to interfere directly with the PKC activation/anchoring machinery to counteract age-related physiological or pathological changes in brain functions. FIGURE 3 summarizes how PKC is activated in physiological conditions and how a decrease in RACK1 anchoring protein may affect such a process and induce a decreased response.

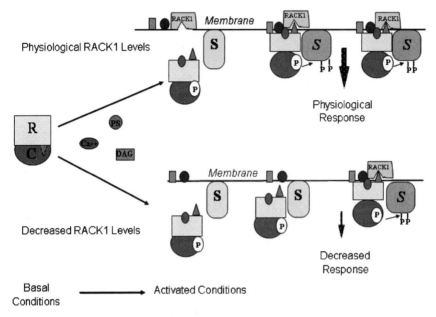

FIGURE 3. Activation of PKC and the role of anchoring/scaffolding protein RACK1. The figure depicts the involvement of RACK1 anchoring protein in the mechanism of cPKC translocation and phosphorylation of a relevant substrate. A decrease in RACK1 is associated with blunted translocation of the activated PKC with a consequent decrease in the functional response. Under basal conditions RACK1 binding site on PKC (*triangle*) is masked and is exposed only after activation. C = PKC catalytic subunit, R = PKC regulatory subunit, Ca^{2+} = calcium, PS = phosphatidylserine, DAG = diacylglycerol, S = substrate, S = substrate phosphorylated (P).

Moreover, RACK–PKC interactions have made it possible to introduce into preclinical studies PKC isoform–selective probes (inhibitors and activators) that have demonstrated isoform-specific effect in cardiac[40] and neuronal functions (TABLE 1). Such compounds are also increasing our understanding of pathological states and look promising for a therapeutic use. For instance, PKCε and PKCδ interacting peptides show encouraging results in the control of ischemic phenomena at both cardiac and neuronal levels.[19,28,29,31]

Modulating the PKC isoform activation/anchoring mechanism offers an interesting additional approach for developing innovative therapeutic tools in diverse physiologic and pathologic conditions.[71,72]

ACKNOWLEDGMENTS

This work was partially supported by a grant from Italian Ministero Sanità/Regione Lazio (Progetto Alzheimer) to F.B.

[*Competing interests*: The authors declare that they have no competing financial interests.]

REFERENCES

1. BATTAINI, F. *et al.* 1997. The role of anchoring protein RACK1 in PKC activation in the aging rat brain. Trends Neurosci. **20:** 410–415.
2. WICKELGREN, I. 1996. For the cortex, neuron loss may be less than thought. Science **273:** 48–50.
3. RAPP P.R. & M. GALLAGER. 1996. Preserved neuron number in the hippocampus of aged rats with spatial learning deficits. Proc. Natl. Acad. Sci. USA **93:** 9926–9930.
4. NOGUES, X. *et al.* 2003. Protein Kinase C. *In* Memories Are Made of These: From messengers to molecules. G. Riedel & B. Platt, Eds.: 383–410: Landes Bioscience. Georgetown, TX.
5. AMADIO, M. *et al.* 2004. Emerging targets for the pharmacology of learning and memory. Pharmacol. Res. **50:** 111–122.
6. NISHIZUKA, Y. 2003. Discovery and prospect of protein kinase C research: epilogue. J. Biochem. (Tokyo). **133:** 155–158.
7. TANAKA, C & Y. NISHIZUKA. 1994. The protein kinase C family for neuronal signalling. Ann. Rev. Neurosci. **17:** 551–567.
8. NEWTON, A.C. 2003. Regulation of the ABC kinases by phosphorylation: PKC as a paradigm. Biochem. J. **370:** 361–371.
9. PAREKH, D.B.,W. ZIEGLER & P.J. PARKER. 2000. Multiple pathways control PKC phosphorylation. EMBO J. **19:** 496–503.
10. RYKX, A. *et al.* 2003. Protein kinase D: a family affair. FEBS Lett. **546:** 81–86.
11. BALENDRAN, A. *et al.* 2000. Further evidence that 3-phosphoinositide-dependent protein kinase-1 (PDK1) is required for the stability and phosphorylation of protein kinase C (PKC) isoforms. FEBS Lett. **484:** 217–223.
12. NISHIZUKA, Y. 1995. Protein kinase C and lipid signaling for sustained cellular responses. FASEB J. **7:** 484–496.
13. POOLE, A.W. *et al.* 2004. PKC-interacting proteins: from function to pharmacology. Trends Pharmacol. Sci. **25:** 528–535.
14. MOCHLY-ROSEN, D. 1995. Localization of protein kinases by anchoring proteins: a theme in signal transduction. Science **268:** 247–251.
15. RON, D. *et al.* 1994. Cloning of an intracellular receptor for protein kinase C: a homolog of beta subunit of G proteins. Proc. Natl. Acad. Sci. USA **91:** 839–843.
16. CSUKAI, M. *et al.* 1997. The coatomer protein β'-COP, a selective binding protein (RACK) for protein kinase C epsilon. J. Biol. Chem. **272:** 29200–29206.
17. MCCAHILL, A. *et al.* 2002. The RACK1 scaffold protein: a dynamic cog in cell response mechanisms: Mol. Pharmacol. **62:** 1261–1273.
18. SCHECHTMAN, D. & D. MOCHLY-ROSEN. 2001. Adaptor proteins in protein kinase C-mediated signal transduction. Oncogene **20:** 6339–6347.
19. CSUKAI, M. & D. MOCHLY-ROSEN. 1999. Pharmacologic modulation of protein kinase C isozymes: the role of RACKs and subcellular localisation. Pharmacol. Res. **39:** 253–259.
20. MOCHLY-ROSEN, D. & A. GORDON. 1998. Anchoring proteins for protein kinase C: a means for isozyme selectivity. FASEB J. **12:** 35–42.
21. GAO, T. & A.C. NEWTON. 2002. The turn motif is a phosphorylation switch that regulates the binding of Hsp70 to protein kinase C. J. Biol. Chem. **277:** 31585–31592.
22. ABELIOVICH, A. *et al.* 1993. Modified hippocampal long term potentiation in PKC gamma–mutant mice. Cell, **75:** 1253–1262.
23. HAYASHI, S. *et al.* 2005. Involvement of gamma protein kinase C in estrogen-induced neuroprotection against focal brain ischemia through G protein-coupled estrogen receptor. J Neurochem. **93:** 883–891.
24. LEITGES, M. *et al.* 2004. A unique PDZ ligand in PKCα confers induction of cerebellar long-term synaptic depression. Neuron **44:** 585–594.

25. WEEBER, E.J. *et al.* 2000. A role for the b isoform of protein kinase C in fear conditioning. J. Neurosci. **20:** 5906–5914.
26. HODGE, C.W. *et al.* 1999. Supersensitivity to allosteric GABA(A) receptor modulators and alcohol in mice lacking PKC epsilon. Nat. Neurosci. **2:** 245–254.
27. LI, M.X. *et al.* 2004. The role of the theta isoform of protein kinase C (PKC) in activity-dependent synapse elimination: evidence from the PKC theta knock-out mouse *in vivo* and *in vitro*. J. Neurosci. **24:** 3762–3769.
28. BRIGHT, R. *et al.* 2004. Protein kinase C delta mediates cerebral reperfusion injury in vivo. J. Neurosci. **24:** 6880–6888.
29. WANG, J. *et al.* 2004. Cell-specific roles for epsilon and beta1-PKC isozymes in protecting cortical neurons and astrocytes from ischemia-like injury. Neuropharmacology **47:** 136–145.
30. SWEITZER, S.M. *et al.* 2004. Exaggerated nociceptive responses on morphine withdrawal: roles of protein kinase C epsilon and gamma. Pain **110:** 281–289.
31. JUNG, Y.S. *et al.* 2005. Activation of protein kinase C-delta attenuates kainate-induced cell death of cortical neurons. Neuroreport **16:** 741–744.
32. KOPONEN, S. *et al.* 2003. Prevention of NMDA-induced death of cortical neurons by inhibition of protein kinase C zeta. J. Neurochem. **86:** 442–450.
33. LI, H.F., D. MOCHLY-ROSEN & J.J. KENDIG. 2005. Protein kinase C gamma mediates ethanol withdrawal hyper-responsiveness of NMDA receptor currents in spinal cord motor neurons. Br. J. Pharmacol. **144:** 301–307.
34. CRAIG, N J. *et al.* 2001. A candidate gene for human neurodegenerative disorders: a rat PKC gamma mutation causes a Parkinsonian syndrome. Nat. Neurosci. **4:** 1061–1062.
35. YABE, I. *et al.* 2003. Spinocerebellar ataxia type 14 caused by a mutation in PKC gamma. Arch. Neurol. **60:** 1749–1751.
36. BATTAINI, F. *et al.* 1990. Regulation of phorbol ester binding and protein kinase C activity in aged rat brain. Neurobiol. Aging **11:** 563–566.
37. MEYER, M. *et al.* 1994. Effect of peroxidation and aging on rat neocortical Ach release and PKC. Neurobiol. Aging **15:** 63–67.
38. BATTAINI, F. *et al.* 1995. Protein kinase C activity, translocation and conventional isoforms in aging rat brain. Neurobiol. Aging **16:** 137–148.
39. FRIEDMAN, E. & H.-Y. WANG. 1989. Effect of age on brain cortical PKC and its mediation of 5-hydroxytryptamine release. J. Neurochem. **52:** 187–192.
40. COLOMBO, P.J. & M. GALLAGER. 2002. Individual differences in spatial memory among aged rats are related to hippocampal PKC gamma immunoreactivity. Hippocampus **12:** 285–289.
41. PASCALE, A. *et al.* 1996. Functional impairment in PKC by RACK1 (receptor for activated Kinase C1) deficiency in aged rat brain cortex. J. Neurochem **67:** 2471–2477.
42. SANGUINO, E. *et al.* 2004. Prevention of age-related changes in rat cortex transcription factor activator protein-1 by hypolipidemic drugs. Biochem. Pharmacol. **68:** 1411–1421.
43. PASCALE, A., S. GOVONI & F. BATTAINI. 1998. Age-related alteration of PKC, a key enzyme in memory processes. Mol. Neurobiol. **16:** 49–62.
44. HARRISON-LAVOIE, K.J. *et al.* 1993. A 102 kDa subunit of a Golgi associated particle has homology to beta subunits of trimeric G proteins. EMBO J. **12:** 2847–2853.
45. BATTAINI, F. 2001. Protein kinase C isoforms as therapeutic targets in nervous system disease states. Pharmacol. Res. **44:** 353–361.
46. RACCHI, M. & S. GOVONI. 1999. Rationalizing a pharmacological intervention on the amyloid precursor protein metabolism. Trends Pharmacol. Sci. **20:** 418–423.
47. ETCHEBERRIGARAY, R. & S. BHAGAVAN. 1999. Ionic and signal transduction alterations in Alzheimer's disease: relevance of studies on peripheral cells. Mol. Neurobiol. **20:** 93–109.
48. WANG, H.-Y., M.R. PISANO & E. FRIEDMAN. 1994. Attenuated PKC activity and translocation in Alzheimer's disease brain. Neurobiol. Aging **15:** 293–298.
49. BATTAINI, F. *et al.* 1999. Protein kinase C anchoring deficit in postmortem brains of Alzheimer's disease patients. Exp. Neurol. **159:** 559–564.

50. CORSINI, E. *et al.* 1999. A defective PKC anchoring system underlying age-associated impairment in TNF-alpha production in rat macrophages. J. Immunol. **163:** 3468–3473.
51. KORZICK, D.H. *et al.* 2001. Diminished alpha 1-adrenergic–mediated contraction and translocation of PKC in senescent rat heart. Am. J. Physiol. Heart Circ. Physiol. **281:** H581–H589.
52. CORSINI, E. *et al.* 2005. Age-related decline in RACK-1 expression in human leukocytes is correlated to plasma levels of dehydroepiandrosterone. J. Leukoc. Biol. **77:** 247–255.
53. GIANOTTI, C. *et al.* 1993. B-50/GAP43 phosphorylation in hippocampal slices from aged rats: effects of phosphatidylserine administration. Neurobiol. Aging **14:** 401–406.
54. ECKLES, K.E. *et al.* 1997. Amelioration of age-related deficits in the stimulation of synapsin phosphorylation. Neurobiol. Aging **18:** 213–217.
55. WANG, H.-Y. *et al.* 2000. Age-related decreases in lymphocyte protein kinase C activity and translocation are reduced by aerobic fitness. J. Gerontol. A Biol. Sci. Med. Sci. **55:** B545–555
56. KORZICK, D.H. *et al.* 2004. Chronic exercise improves myocardial inotropic reserve capacity through alpha1-adrenergic and protein kinase C–dependent effects in senescent rats. J. Gerontol. A Biol. Sci. Med. Sci. **59:** 1089–1099.
57. RACCHI, M. *et al.* 2001. Dehydroepiandrosterone and the relationship with aging and memory: a possible link with protein kinase C functional machinery. Brain Res. Rev. **37:** 287–293.
58. CORSINI, E. *et al.* 2002. *In vivo* dehydroepiandrosterone restores age-associated defects in PKC signal transduction pathway and related functional responses. J. Immunol. **168:** 1753–1758.
59. PIERPAOLI, W. & W. REGELSON. 1994. Pineal control of aging: effect of melatonin and pineal grafting on aging mice. Proc. Natl. Acad. Sci. USA **91:** 787–791.
60. VITICCHI, C. *et al.* 1994. Brain cortex alpha and beta adrenoceptors are differentially modulated by pineal graft in aging mice. Ann. N.Y. Acad. Sci. **719:** 448–453.
61. PIERPAOLI, W. *et al.* 1997. Circadian melatonin and young to old pineal grafting postpone aging and maintain juvenile conditions of reproductive functions in mice and rats. Exp. Gerontol. **32:** 587–602.
62. VAN DER ZEE, E.A., P.G. LUITEN & J.F. DISTERHOFT. 1997. Learning-induced alterations in hippocampal PKC-gamma immunoreactivity: a review and hypothesis of its functional significance. Prog. Neuropsychopharmacol. Biol. Psych. **3:** 531–572.
63. PEYRL, A. *et al.* 2002. Aberrant expression of signalling-related proteins 14-3-3 gamma and RACK1 in fetal Down syndrome brain (trisomy 21). Electrophoresis **23:** 152–157.
64. VAN DER ZEE, E.A. *et al.* 2004. Aging-related alterations in the distribution of Ca(2+)-dependent PKC isoforms in rabbit hippocampus. Hippocampus **14:** 849–860.
65. ESCRIBA, P.V. & J.A. GARCIA-SEVILLA. 1999. Parallel modulation of receptor for activated C kinase 1 and protein kinase C-alpha and beta isoforms in brains of morphine-treated rats. Br. J. Pharmacol. **127:**343–348.
66. WANG, H.-Y. & E. FRIEDMAN. 2001. Increased association of brain PKC with the receptor for activated C kinase 1 (RACK1) in bipolar affective disorder. Biol. Psych. **50:** 364–370.
67. CHOI, D.-S. *et al.* 2003. The mouse RACK1 gene is regulated by nuclear factor-kB and contributes to cell survival. Mol. Pharmacol. **64:** 1541–1548.
68. KALTSCHMIDT, B. *et al.* 1999. Inhibition of NF-kappa B potentiates amyloid beta-mediated neuronal apoptosis. Proc. Natl. Acad. Sci. USA. **96:** 9409–9414.
69. ETCHEBERRIGARAY, R. *et al.* 2004. Therapeutic effects of PKC activators in Alzheimer's disease transgenic mice. Proc. Natl. Acad. Sci. USA **101:** 11141–11146.
70. BAR-AM, O. *et al.* 2004. Regulation of protein kinase C by the anti-Parkinson drug, MAO-B inhibitor, rasagiline, and its derivatives *in vivo*. J. Neurochem. **89:** 1119–1125.
71. WAY, K.J., E. CHOU & G.L. KING. 2000. Identification of PKC-isoform-specific biological actions using pharmacological approaches. Trends Pharmacol. Sci. **21:** 181–187.
72. DEMPSEY, E.C. *et al.* 2000. Protein kinase C isozymes and the regulation of diverse cell responses. Am. J. Physiol. Lung Cell. Mol. Physiol. **279:** 429–438.

Protection of Proteins from Oxidative Stress

A New Illusion or a Novel Strategy?

ALEXANDER A. BOLDYREV

*M.V. Lomonosov Moscow State University and Institute of Neurology,
Russian Academy of Medical Sciences, Moscow, Russia*

ABSTRACT: Proteins damaged by oxidative stress have the most dangerous consequences. Oxidized protein derivatives inveigle lipids and carbohydrates into metabolic transformations that result in loss of protein functions and accumulation of glycated proteins and advanced glycated end products, which are difficult to remove from living tissues. Hydrophobic antioxidants are not very effective in protecting proteins from oxidative modification. At the same time, the natural hydrophilic antioxidant and anti-glycating agent carnosine efficiently prevents oxidative modification of proteins and increases the life span of experimental animals under unfavorable conditions. It can be considered a potent natural geroprotector.

KEYWORDS: oxidative stress; protein carbonyls; glycation; carnosine

INTRODUCTION

One of the most important problems in modern gerontology is the regulation of reactive oxygen species (ROS) in tissues, to equilibrate between prevention of their damaging effects to the cells and the support of their signaling functions.[1] The overproduction of ROS is known to result in the oxidative damage of a variety of cellular components, including nucleic acids, proteins, and lipids. However, lipophilic antioxidants are only partially protective. In this article, I draw attention to oxidative damage of proteins during oxidative stress and postulate that the use of natural hydrophilic protectors of proteins may provide a novel strategy to prevent tissue damage from oxidative stress.

INVOLVEMENT OF CELLULAR MACROMOLECULES IN OXIDATIVE STRESS

Among markers of oxidative stress, there are several products of oxidation of lipids (hydroperoxides, hydroxyl-nonenales, malonic dialdehyde, MDA), nucleic acids (8-hydroxy-2′-deoxyguanosine), and proteins (carbonyl derivatives).

Address for correspondence: Alexander Boldyrev, Department of Biochemistry, School of Biology, Lenin's Hills, 119992 Moscow, Russia. Voice/fax: 7-095-939-1398.
aaboldyrev@mail.ru
aaboldyrev@yahoo.com

Ann. N.Y. Acad. Sci. 1057: 193–205 (2005). © 2005 New York Academy of Sciences.
doi: 10.1196/annals.1356.013

Traditionally, lipid peroxidation is considered the most damaging to cellular integrity and viability because unsaturated fatty acid tails of membrane phospholipids are easily oxidized. For this reason, molecular products of lipid oxidative modification (like MDA) are routinely used to estimate the depth of oxidative injury. Cellular DNA is also susceptible to ROS attack. It is necessary to take in account that cellular defense includes several modes to protect or repair both lipids and nucleic acids damaged by oxidative attack. Also, proteins are not only easily oxidized by a variety of cellular ROS, but form a number of derivatives that can interact with lipids, carbohydrates, and other cellular components. The result is an accumulation of very stable compounds that are functionally inactive and resistant against degradation (ubiquitination, etc.).[1]

Several reactions are known that involve protein molecules in the oxidative pathway, including direct oxidation, glycosylation by carbohydrate moiety, and crosslinking with both carbohydrates and lipids (or products of their oxidation), carbonyl groups of oxidatively modified proteins being the most effective partners able to interact with native lipids and carbohydrates.

Schematically, the involvement of carbohydrates in the formation of non-metabolized products with oxidized proteins is presented in FIGURE 1. Oxidation of proteins by ROS results in accumulation of protein carbonyls, which can easily interact with amino groups of lipids and non-oxidized protein molecules, and thus results in the accumulation non-functional cross-linked products. The same ability is demonstrated by ketoaldehydes after oxidation of aldo-forms of carbohydrates.

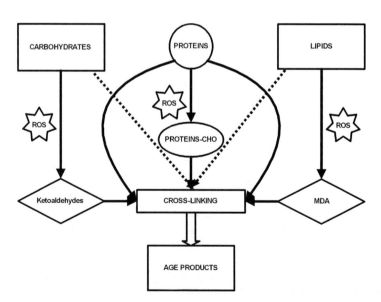

FIGURE 1. Involvement of proteins, lipids, and carbohydrates in oxidative modification. (*Solid lines*) Interaction of proteins with oxidized lipids and carbohydrates. (*Dotted lines*) Interaction of lipids and carbohydrates with oxidized protein molecules.

Moreover, native (non-oxidized) proteins can be attacked by either native carbohydrates or products of lipid peroxidation (MDA) and the result is multiple cross-links between each others. Thus proteins provide the route for involvement of both lipids and carbohydrates in oxidative modification. Accumulation of protein carbonyls provokes this process. As a result, a variety of stable products of cross-linking accumulate in tissues that can't be destroyed by intracellular degrading enzymes. While not all of them are identified, these are combined in the group of advanced glycated end products (AGEs), which consist of stable cross-linked agglomerates of glycated proteins, carbohydrates, and products of lipid oxidative modification, lipofuscins.

Taking these finding into consideration, we suggest that proteins are the most important ROS target during oxidative stress, more important than nucleic acids or even lipids. The illustration of this point of view is the well known fact the accumulation of oxidized proteins (protein carbonyls) accelerates with ageing of living species.[2]

Thus it is understandable why the hydrophobic antioxidants are relatively poor protectors of tissues from age-induced oxidative stress because they successfully protect lipids from oxidation but cannot stop accumulation of products of protein oxidation.[3] Yet anti-glycating agents could be useful in protecting proteins from oxidative modification and thus be a useful tool to protect living organisms against the accumulation of features of aging.

ANTI-GLYCATING AGENTS

In studies, aminoguanidine is the most commonly used of the compounds known to possess potent anti-glycating properties. It can interfere with Amadori products, thus preventing their glycation and formation of irreversible AGEs. The latter have been identified in age-related intracellular deposits of Alzheimer's disease (amyloid plaques and neurofibrillary tangles) and Parkinson's disease (Lewy bodies), suggesting that oxidized protein deposits have been exposed to AGEs' precursors, such as reactive dicarbonyl compounds, MDA, or methylglyoxal (MG).

MG detoxification is impaired when reduced glutathione levels are low, which corresponds to oxidative stress conditions. However, there is less known about the toxicity of MG itself. Recently it was shown that MG applied extracellularly has been toxic to human neuroblastoma cells in concentrations above 150 μM with a LD_{50} of approximately 1.25 mM. Preincubation of MG with carbonyl scavengers such as aminoguanidine or tenilsetam and the thiol antioxidant lipoic acid significantly reduced its toxicity.[4]

In Alzheimer's disease, age-related cellular changes are worsened by the presence of AGEs accumulation (most likely derived from MG). They cross-link with membrane-bound and cytoskeletal proteins and render them insoluble. These aggregates inhibit cellular functions (including transport processes) and contribute to neuronal dysfunction and death. Extracellular AGEs, which accumulate in aging tissue (but most prominently on long-lived protein deposits, such as senile plaques), exert chronic oxidative stress on excitable cells. Drugs that prevent the formation of AGEs by specific chemical mechanisms—including prevention of transformation of Amadori products into AGEs (aminoguanidine, tenilsetam, OPB-9195, and pyridoxamine)—attenuate the development of AGEs-mediated diabetic complications.[5]

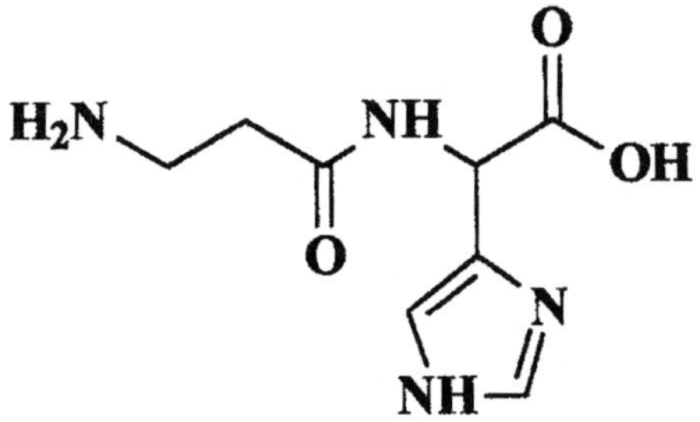

FIGURE 2. Structural formula of carnosine (β–alanyl-L-histidine).

Assuming that "carbonyl stress" contributes significantly to the aging processes, AGE inhibitors might also become potential geroprotectors.

Thus, carbonyl scavengers might offer a promising therapeutic strategy to reduce the neurotoxicity of reactive carbonyl compounds, providing a potential benefit for patients with age-related neurodegenerative diseases. Among those, natural neuropeptide carnosine is especially attractive because it has extremely low toxicity (LD$_{50}$ for this natural compound is about 20 g/kg body mass) and provides powerful protection of proteins from both oxidation and modification by a number of aldehydes under *in vitro* conditions.

BIOLOGICAL ACTIVITY OF CARNOSINE

Carnosine was found in the nitrogen-containing non-protein extracts from skeletal muscles by V. Gulevitsch as early as in 1900.[6] Carnosine is the dipeptide consisted of β-alanine and L-histidine (FIG. 2). It was shown to be a specific constituent of excitable tissues of all vertebrates accumulating in amounts exceeding that of ATP.[7] The antioxidant capacity of this compound is well documented as well as its pH buffering, osmoregulating, and metal-chelating abilities.[8,9] A potentially useful characteristic of carnosine is its ability to act as an anti-glycating agent.[10–13]

Protection of Nucleic Acids from ROS Action

Carnosine's ability to protect nucleic acids from oxidation by free radicals is very important. FIGURE 3 demonstrates that carnosine acts as a protector of DNA structure toward ROS attack. Such protective effect of carnosine on DNA is accompanied by an increase in the life span of T-cells.[14]

Effects on Lipid Peroxidation

Since 1984 when suppression of lipid peroxidation by carnosine was first described,[15] many studies have documented such activity.[7,16,17] As seen in FIGURE 4, carnosine suppresses accumulation of MDA induced by incubation of sarcoplasmic reticulum vesicles with ascorbate and ferrous ions, following with inhibition of both

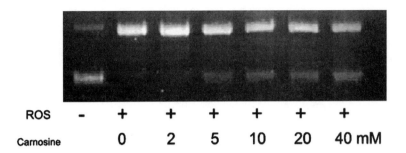

ROS   -   +   +   +   +   +   +

Carnosine   0   2   5   10   20   40 mM

FIGURE 3. Electrophoresis data illustrating distribution between supercoil (*upper line*) and open (*lower line*) forms of purified DNA after 30-min incubation under different conditions. DNA isolated from pBluescript II SK plasmid (2961 bp) was exposed to oxidative degradation (ROS) in the presence of 2 μM $FeSO_4$, 0.5 mM H_2O_2, and 0.5 mM ascorbate. Concentrations of carnosine added are noted below each electrophoresis spot. Samples were analyzed by 1.5% agarose gel electrophoresis, and gels were stained by ethidium bromide.

Ca-ATPase and Ca-pumping activity in the same vesicles. Such an effect indicates the ability of carnosine to quench superoxide anion and hydroxide radical[16,18,19] and to neutralize 4-hydroxy-nonenal (HNE) and other toxic aldehydes.[12,20,21] The most important result is that by trapping HNE in a stable adduct, carnosine prevents protein–protein cross-linking.[21]

Protection of Proteins

The important conclusion from FIGURE 4 is that carnosine not only suppresses lipid peroxidation and accumulation of MDA but also protects the Ca-ATPase protein from oxidative modification. It is clearly seen from comparison of Ca-pump activity in the presence or absence of carnosine at the same level of MDA accumulation (10 min in control sample and 30 min in carnosine-containing sample, FIG. 4). Despite of the same level of MDA in the presence of carnosine, the Ca-pump still operates suggesting carnosine protection is not restricted by its ability to suppress MDA accumulation.

Similar protective action of carnosine was demonstrated recently toward superoxide dismutase both *in vitro*[22] and *in vivo*,[23] and for telomerase[24]—both enzymes are important for the protection of cellular machinery from damaging effects of ROS.

Besides enzymes, some membrane-bound receptors are also protected by carnosine. Recently, glutamate receptors of NMDA class (*N*-methyl-D-aspartate activated receptors, NMDA receptors) in the brain of senescence accelerated mice (SAM) were shown to be suppressed with age and their normal level was restored when animals were treated with carnosine (100 mg/kg body mass, daily for 4 months).[25] Using male rats exposed to γ-irradiation, it was demonstrated that estradiol levels in serum and progesterone receptor density in uterus were decreased. Carnosine normalized both parameters whereas melatonin was not effective.[26]

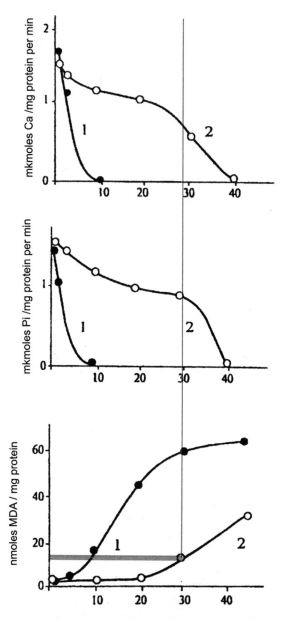

FIGURE 4. Effect of 20 mM carnosine on Ca-pumping activity (**top**) and Ca-ATPase (**middle**) of sarcoplasmic reticulum and accumulation of MDA (**bottom**) during progressive lipid peroxidation induced by ascorbic acid (100 μM) and $FeSO_4$ (1 nmole per 1 mg protein). *Horizontal line* shows time (about 10 min and 30 min) when the same amount of MDA is accumulated with the absence (*curve 1*) and presence (*curve 2*) of carnosine. *Vertical line* shows that in the presence of carnosine Ca-pumping and Ca-ATPase activities are still working despite the same level of MDA accumulation. (Modified from Dupin *et al.*[15])

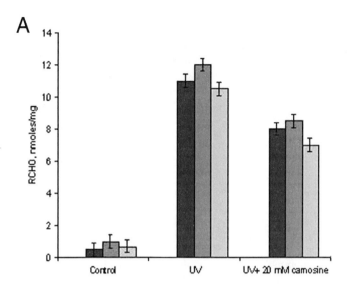

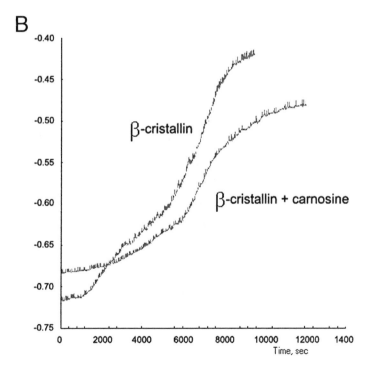

FIGURE 5. Accumulation of protein carbonyls after 30 min UV irradiation of bovine β-cristallin in the absence or presence of 20 mM carnosine (**A**) and prevention of UV-induced β-cristallin aggregation in the presence of the same concentration of carnosine (**B**). (Unpublished data obtained in collaboration with A. Tumofeeva and M. Ostrovsky).

CARNOSINE AS A NATURAL ANTI-GLYCATING AGENT

The first evidence demonstrating *in vitro* protein protection by carnosine against MDA and hypochlorite attack was published in 1997 and 1998.[11,13] These oxidants were considered oxygen metabolites and their attack on ovalbumin, bovine serum albumin, or crystallin was found to create protein carbonyls and high molecular weight non-soluble cross-linked proteins. Carnosine inhibited such protein modification, being more effective against hypochlorite than against MDA. The ability of carnosine to interact with hypochlorite to form non-oxidative complex thus preventing its damaging action was shown earlier.[27] The protective effect of MDA was explained by ability of carnosine to interact with aldehydes and ketones.[11] This suggestion was confirmed very recently by other investigators.[28]

We have also shown that carnosine can prevent accumulation of carbonyl groups of β-cristallin oxidized by UV illumination (FIG. 5A). Oxidized protein is aggregated fast enough and carnosine decreased not only β-cristallin oxidation but also the rate of its aggregation (FIG. 5B). The ability of carnosine to react with protein carbonyls was recently reviewed by Hipkiss and Brownson.[12] They showed that carnosine protects protein molecules from carbonylation much more effectively than does lysine and that this may related to both rejuvenation of human fibroblast cultured (propagation of Hayflick limit described by McFarland and Holiday[29]) and correlation between life span of different species and content of carnosine in their tissues.[12]

Moreover, carnosine was found to react with glycated proteins partially restoring protein structure and decreasing amount of carbonyl groups previously formed because of incubation of proteins with oxidizing agent (MG).[30] Carnosine was also able to prevent cross-linking between MG-treated ovalbumin and non-treated native β-cristallin. The imidazole ring of histidine was found to play important role in such protection.[31]

Two recent observations throw the light on the biological meaning of such properties of carnosine. In a publication by Yergans and Seidler,[32] it was shown that carnosine promotes heat denaturation of glycated protein. This is very useful effect can be directed to accelerated removal of protein molecules damaged by oxidative modification. The other observation demonstrated that carnosine is able to disaggregate glycated protein (β-cristallin) under conditions where guanidine chloride or urea had no effect.[33] In FIGURE 6, a very speculative scheme is presented based on data from Hipkiss and from Boldyrev and demonstrates how carnosine can act as protector of proteins under oxidative stress. All the data taken together suggest that carnosine can

---

FIGURE 6. Hypothetical scheme of carnosine interaction with ROS and native or oxidized proteins. Interaction of carnosine with ROS results in nonsaturated aldimine (*step 1*) and than aldehyde (*step 2*) formation. The aldehyde can interact with ROS transforming into acidic form (*step 3*). The latter can interact with protein molecules forming the product of carnosinylation (*step 4*) being directed to accelerated degradation (*step 5*) or restoration of native protein structure after interaction with another carnosine molecule (*step 6*) followed by formation of nontoxic carnosine derivative (*circled by dotted line*). Oxidized protein (*step 7*) can interact with free carnosine molecule forming carnosinylated derivative (*step 8*) undergoing the above-mentioned transformation. Thus, carnosine can prevent proteins from oxidation by ROS neutralization, by forming of carnosinylated derivatives able to accelerate degradation, and by restoration of native protein structure.

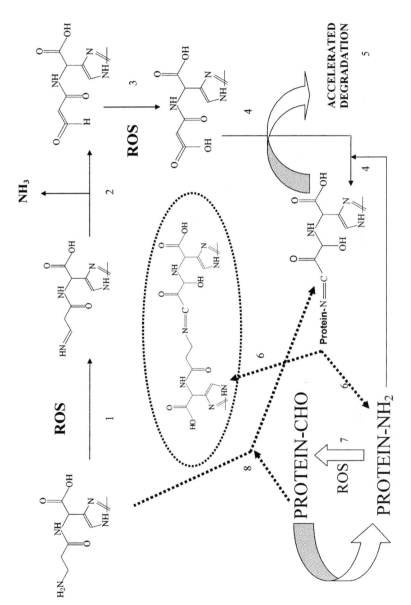

FIGURE 6. *See preceding page for legend.*

be a useful natural protector of protein moiety against oxidative stress and this feature may be expressed in prolongation of life span.

CARNOSINE AND MODULATION OF LIFE SPAN

In order to estimate the efficiency of carnosine as geroprotector, senescence accelerated mice (SAM), which have increased level of ROS and deficiency of antioxidant capacity, were used.[34,35] Using carnosine (100 mg/kg body weight) as a daily food additive, we demonstrated sufficient increase (up to 22%) of average life span of the mice with simultaneous improvement of a number of behavioral features of the animals studied (FIG. 7A). At the same time, the content of protein carbonyls and lipid peroxides in their blood was decreased, demonstrating normalization of oxidative metabolism in SAM tissues as a cause of increased life span.

It was possible to explain such a beneficial effect by action of the products of carnosine degradation in the bloodstream, which is provided by specific carnosinase resulting in accumulation of histidine and β-alanine. In order to prove this ability, another living organisms were used in the following experiments where *Drosophila melanogaster* was used, which possess no carnosine-metabolizing enzymes.[35] In these experiments a similar and even stronger life-sustaining effect of carnosine was demonstrated (FIG. 7B). This suggests that carnosine itself (not the products of its enzymatic degradation) demonstrates apparent life-sustaining action. In conclusion, use of a natural anti-oxidant and anti-glycating agent carnosine can open the new strategy in modern geroprotection.

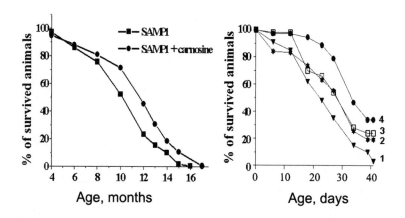

FIGURE 7. Effect of carnosine on life span of SAMP1 mice (50 mg/kg body weight daily, **left**) and *Drosophila melanogaster* (**right;** 1, 2, 3, and 4 correspond to 0, 0.2, 2, and 20 mg/L).

CONCLUDING REMARKS

Analyses of oxidative stress and exhaustion of antioxidant defense system permit us to advance the hypothesis that oxidative modification of proteins is most damaging for cell viability. Oxidized proteins not only inveigle carbohydrates and lipids into oxidative modification but form mixed AGEs, which are difficult to remove from the tissues and thus accumulate with age.

Various glycated protein derivatives are one of the intermediates of this process. For these reasons, natural anti-glycating agents can act as successful protectors against accumulation of AGE products and accelerated aging. Moreover, several diseases that accompany middle and elderly age of human beings should be successfully treated by anti-glycating compounds—a new strategy in modern gerontology.

Actually, carnosine is one example of natural compounds that can be used successfully to increase the life span of human beings and to treat some diseases of old age.[36,37] It was demonstrated recently that carnosine efficiently protects the brain of rats and Mongolian gerbils from experimental ischemia[38,39] and prevents acute renal failure in rats.[40] Carnosine provides significant decreases in protein carbonyl accumulation induced *in vitro* by treatment of brain homogenate with H_2O_2.[41] Carnosine eye drops have a beneficial effect on eye function damaged by senile cataract.[42] The facts make a convincing argument that this compound can help protect organisms from early aging.

[*Competing interests*: The author declares that he has no competing financial interests.]

REFERENCES

1. BOLDYREV, A.A. 2002. Significance of reactive oxygen species for neuronal function. *In* Free Radicals, Nitric Oxide, and Inflammation: Molecular, Biochemical, and Clinical Aspects. A. Tomasi, T. Ozben & V. Skulachev, Eds.: 153–169. IOS Press. Amsterdam.
2. LUBEC, B., J. GOLEJ, M. MARX, *et al.* 1995. L-Arginine reduces kidney lipid peroxidation, glycoxidation, and collagen accumulation in the aging NMRI mouse. Ren. Physiol. Biochem. **18:** 97–102.
3. GUTTERIDGE, J.M. 1999. Does red/ox regulation of cell function explain why antioxidants perform so poorly as therapeutic agents? Redox Rep. **4:** 129–131.
4. WEBSTER, J., C. URBAN, K. BERBAUM, *et al.* 2005. The carbonyl scavengers aminoguanidine and tenilsetam protect against the neurotoxic effects of methylglyoxal. Neurotox. Res. **7:** 95–101.
5. DUKIC-STEFANOVIC, S, R. SCHINZEL, P. RIEDERER & G. MUNCH. 2001. AGES in brain ageing: AGE-inhibitors as neuroprotective and anti-dementia drugs? Biogerontology **2:** 19–34.
6. GULEVITCH, V. & S. AMIRADGIBI. 1900. Uber das carnosine, eine neue organische base des fleischextraktes. Ber. Deutsch. Chem. Gesellschaft. **33:** 15504–15509.
7. BOLDYREV, A.A. & S.E. SEVERIN. 1990. The histidine-containing dipeptides, carnosine, and anserine: distribution, properties, and biological significance. Adv. Enz. Reg. **30:** 175–193.
8. BOLDYREV A.A. 1990. Retrospectives and perspectives on the biological activity of histidine-containing dipeptides. Int. J. Biochem. **22:** 129–132.
9. ABE, H. 2000. Role of histidine-related compounds as intracellular proton buffering constituents in vertebrate muscle. Biochemistry (Mosc.) **65:** 757–765.
10. BOLDYREV, A. 2002. Carnosine as natural antioxidant and neuroprotector: biological functions and possible clinical use. *In* Free Radicals, Nitric Oxide, and Inflamma-

tion: Molecular, Biochemical, and Clinical Aspects. A. Tomasi, T. Ozben & V. Skulachev, Eds.: 202–217. IOS Press. Amsterdam.

11. HIPKISS, A., V. WORTHINGTON, D. HIMSWORTH & W. HERWIG. 1998. Protective effect of carnosine against protein modification mediated by malondialdehyde and hypochlorite. Biochim. Biophys. Acta **1380:** 46–54.

12. HIPKISS, A. & C. BROWNSON. 2000. Carnosine reacts with protein carbonyl groups: another possible role for the anti-ageing peptide? Biogerontology **1:** 217–223.

13. KULEVA, N.V. & Z.S. KOVALENKO. 1997. Change in the functional properties of actin by its glycation *in vitro*. Biochem. Moscow **62:** 1119–1123.

14. HYLAND, P, O. DUGGAN, A. HIPKISS, *et al.* 2000. The effects of carnosine on oxidative DNA damage levels and *in vitro* lifespan in human peripheral blood derived CD4+T cell clones. Mech. Ageing Dev. **121:** 203–215.

15. DUPIN, A., A. BOLDYREV, Y. ARKHIPENLO & V. KAGAN. 1984. Carnosine protection of Ca transport from damage induced by lipid peroxidation. Biul. Exp. Biol. Med. **97:** 186–188.

16. ARUOMA, O., M. LAUGHTON & B. HALLIWELL. 1989. Carnosine, homocarnosine, and anserine: could they act as anti-oxidants *in vivo*? Biochem. J. **264:** 863–869.

17. HARTMAN, P.E, Z. HARTMAN & K.T. AULT. 1990. Scavenging of singlet molecular oxygen by imidazole compounds: high and sustained activities of carboxy terminal histidine dipeptides and exceptional activity of imidazole-4-acetic acid. Photochem. Photobiol. **51:** 59–66.

18. RUBTSOV, A.M, M. SCHARA, M. SENTJURC & A. BOLDYREV. 1991. Hydroxyl radical-scavenging activity of carnosine: a spin trapping study. Acta Pharm. Jugosl. **41:** 401–407.

19. PAVLOV, A., A. REVINA, A. DUPIN, *et al.* 1993. The mechanism of interaction of carnosine with superoxide radicals in water solutions. Biochim. Biophys. Acta **1157:** 304–312.

20. ALDINI, G., M. CARINI, G. BERETTA, *et al.* 2002. Carnosine as a quencher of 4-hydroxynonenal: through what mechanism of reaction? Biochem. Biophys. Res. Commun. **298:** 699–706.

21. LIU, Y., G. XU & M. SAYRE. 2003. Carnosine inhibits (E)-4-hydroxy-2-nonenal-induced protein cross-linking: structural characterization of carnosine-HNE adducts. Chem. Res. Toxicol. **16:** 1589–1597.

22. UKEDA, H, Y. HASEGAWA, Y. HARADA & M. SAWAMURA. 2002. Effect of carnosine and related compounds on the inactivation of human Cu,Zn-superoxide dismutase by modification of fructose and glycolaldehyde. Biosci. Biotechnol. Biochem. **66:** 36–43.

23. STVOLINSKY, S., T. FEDOROVA, M. YUNEVA & A. BOLDYREV. 2003. Protection of Cu/Zn-SOD by carnosine under disordering of brain metabolism in vivo. Biul. Exp. Biol. Med. **135:** 151–154.

24. SHAO, L., Q.H. LI & Z. TAN. 2004. L-Carnosine reduces telomere damage and shortening rate in cultured normal fibroblasts. Biochem. Biophys. Res. Commun. **324:** 931–936.

25. YUNEVA, M., E. BULYGINA, S. GALLANT, *et al.* 1999. Effect of carnosine on age-induced changes in senescence accelerated mice (SAM). J. Anti-Aging Med. **2:** 337–342.

26. BERSTEIN, L., E. TSIRLINA, T. POROSHINA, *et al.* 2003. Induction of the phenomenon of switching off estrogen effect and approaches to its removal. Russian Physiol. J. **89:** 964–971.

27. QUINN, P., A. BOLDYREV & V. FORMAZYUK. 1992. Carnosine: its properties, functions, and potential therapeutical applications. Mol. Aspects Med. **13:** 379–444.

28. YAN, H. & J.J. HARDING. 2005. Carnosine protects against the inactivation of esterase induced by glycation and a steroid. Biochim. Biophys. Acta **1741:** 120–126.

29. MCFARLAND, G.A. & R. HOLLIDAY. 1994. Retardation of senescence in cultured human diploid fibroblasts by carnosine. Exp. Cell Res. **212:** 167–175.

30. BROWNSON, C. & A. HIPKISS. 2000. Carnosine reacts with a glycated protein. Free Rad. Biol. Med. **28:** 1564–1570.

31. HOBART, L., I. SEIBEL, G. YEARGANS & N. SEIDLER. 2004. Anticross-linking properties of carnosine: significance of histidine. Life Sci. **75:** 1379–1389.

32. YEARGANS, G. & N. SEIDLER. 2003. Carnosine promotes the heat denaturation of glycated protein. Biochem. Biophys. Res. Commun. **300:** 75–80.
33. SEIDLER, N., G. YEARGANS & T. MORGAN. 2004. Carnosine disaggregates glycated α-cristallin: an in vitro study. Arch. Biochem. Biophys. **427:** 110–115.
34. BOLDYREV, A., M. YUNEVA, E. SOROKINA, *et al.* 2001. Antioxidant systems in tissues of senescence accelerated mice (SAM). Biochem. Moscow **66:** 1157–1163.
35. YUNEVA, A.O., G.G. KRAMARENKO, T.V. VETRESHCHAK, *et al.* 2002. Effect of carnosine on *Drosophila melanogaster* lifespan. Biul. Exp. Biol. Med. **133:** 559–561.
36. SZWERGOLD, B.S. 2005. Intrinsic toxicity of glucose, due to nonenzymatic glycation, is controlled *in vivo* by deglycation systems including FN3K-mediated deglycation of fructosamines and transglycation of aldosamines. Med. Hypotheses **65:** 337–348.
37. HIPKISS, A.R. 2005. Glycation, aging, and carnosine: are carnivorous diets beneficial? Mech. Ageing Dev. **126:** 1034–1039.
38. STVOLINSKY, S., M. KUKLEY, D. DOBROTA, *et al.* 2000. Carnosine protects rats under global ischemia. Brain Res. Bull. **53:** 445–448.
39. FEDOROVA, T.N., S.L. STVOLINSKY, D. DOBROTA & A.A. BOLDYREV. 2002. Therapeutic effect of carnosine under experimental brain ischemia. Prob. Biol Med. Pharmacol. Chem. **1:** 41–44.
40. FUJII, T., M. TAKAOKA, T. MURAOKA, *et al.* 2003. Preventive effect of L-carnosine on ischemia/reperfusion-induced acute renal failure in rats. Eur. J. Pharmacol. **474:** 261–267.
41. ZHOU, H., F. SHEN & A. WANG. 2003. The oxidative carbonyl protein and carnosine protection. Chin. J. Gerontol. **5:** 310–312.
42. WANG, A., C. MA, Z. XIE & F. SHEN. 2000. Use of carnosine as a natural antisenescence drug for human beings. Biochem. Moscow **65:** 869–871.

The Role of Hsp70
in Age-Related Diseases and the
Prevention of Cancer

EZRA V. PIERPAOLI

Chemin du Verger 16, 1752 Villars-sur-Glâne, Switzerland

ABSTRACT: This article focuses on the structure, function, mode of action of one of the major heat shock proteins, Hsp70, and its implication in age-related diseases and cancer.

KEYWORDS: molecular chaperone; Hsp70; neurodegenerative disease; cancer

INTRODUCTION

From bacteria to men, heat-shock proteins (Hsps), also known as molecular chaperones, constitute a major defense system that allows recovery from adverse modes of stress such as elevated temperatures, alcohol, toxic heavy metals (e.g., cadmium), some forms of oxidative and post-ischemic stress, and other forms of environmental insults. Much of the molecular damage exerted by these forms of stress involves conformational changes and partial denaturation of protein molecules in the affected cells leading to aggregation and inactivation of the molecules.[1] Hsps are ubiquitous, and different chaperone systems have been found from bacteria to mammalian cells.

The reduced capacity to respond to stress is one of the major characteristics of the senescent organism. It is known that aging mammalians show reduced expression and induction of heat-shock proteins in response to physiological stresses. The most frequent stress-induced form Hsp is Hsp70, named according to its molecular mass of 70 kDa. In particular, it has been shown that the induction of Hsp70 expression by heat shock as well as other stresses declines significantly with age in a variety of tissues from rats as well as in mononuclear cells from humans. Hepatocytes freshly isolated from old rats showed an approximately 50% reduced Hsp70 expression by heat shock as compared to those from young rats.[2] Moreover, molecular chaperones seem to have an important role in many age-related diseases such as Alzheimer's dis-

Address for correspondence: Ezra V. Pierpaoli, Ph.D., Chemin du Verger 16, 1752 Villars-sur-Glâne, Switzerland. Voice: +41 (26) 401 08 94.
epierpaoli@hotmail.com/ezra_pierpaoli@yahoo.de

Ann. N.Y. Acad. Sci. 1057: 206–219 (2005). © 2005 New York Academy of Sciences.
doi: 10.1196/annals.1356.014

ease, Parkinson's disease, and prion-related diseases.[3] This article focuses on the function, structure, mode of action of one of the major heat shock proteins, Hsp70, and its implication in age-related diseases and cancer.

FOLDING OF A PROTEIN IN THE CELLULAR ENVIRONMENT REQUIRES HEAT SHOCK PROTEINS

The folding of a protein in the cellular environment is a precarious event. The cellular cytosol is a highly viscous solution composed of proteins and other macromolecules at concentrations as high as ~340 mg/mL.[4] During the translation of a protein on a ribosome or the translocation of a polypeptide chain through the envelope of a mitochondrion, the endoplasmic reticulum or a chloroplast, the protein exposes hydrophobic sequences, which, in the native state, are buried in its interior. In the aqueous environment, such hydrophobic sequences tend to aggregate leading to unproductive folded products. A similar situation occurs under stress conditions, e.g. at elevated temperatures, when heat-sensitive proteins unfold and can form insoluble aggregates. To prevent such off-pathway reactions, nature has developed a whole set of proteins, termed molecular chaperones, which transiently bind to and stabilize unfolded or partially folded proteins. By controlled binding and release of the target protein, they promote its correct fate *in vivo*, be it folding, oligomeric assembly, transport to a particular subcellular compartment, or disposal by degradation.[4,5] It has to be emphasized that molecular chaperones increase the yield rather than the rate of protein folding. Molecular chaperones are subdivided into several families, the most prominent of them are heat-shock proteins of the Hsp104, Hsp90, Hsp70, and Hsp60 classes and the small heat-shock proteins with a molecular mass of 25 kDa, the Hsp25s.[6] A feature of many heat-shock proteins is their increased synthesis under cellular stress conditions. However, many Hsps are also constitutively expressed and essential under normal growth conditions.[7]

THE Hsp70 CLASS

The Hsp70s form an ubiquitous protein family, representatives of which are found in all organisms and in most cellular compartments of eukaryotic cells (TABLE 1).

Hsp70s have a vast spectrum of functions in the cell. Bacterial Hsp70s, e.g., DnaK of *Escherichia coli*, are involved in the prevention of aggregation during the unfolding of proteins, the repair of heat-damaged proteins,[8–10] and the degradation of intracellular proteins.[11] Cytosolic forms of Hsp70s, such as SSA1,2 in yeast and Hsp73 (Hsc70) in mammals, bind to nascent polypeptide chains and promote the folding of proteins.[4] Hsp70 in the lumen of mitochondria (mHsp70/SSC1) or of the endoplasmic reticulum (Kar2/BiP) are required for the translocation of polypeptide chains through the corresponding membranes and for their folding in the organellar space. One model for the import of polypeptides through these membranes proposes that Hsp70 works as an ATP-driven protein translocation motor, which captures the incoming chain and subsequently undergoes an ATP-driven conformational change that actively pulls a segment of the bound chain across the membrane.[12] BiP, an

TABLE 1. Members of the Hsp70 family in bacteria and eukaryotic cells

Organism	Cytosol	Mitochondria	Endoplasmic reticulum	Chloroplasts
Bacteria	DnaK[a]			
Yeast	SSA1-4[a]	SSC1	Kar2/BiP	
	SSA1, 2			
Mammals	Hsp72 (Hsp70)[b]		BiP	
	HSP73 (Hsc70)[c]			
Plants				ctHsp70

[a]Heat inducible and constitutive; [b]stress inducible; [c]constitutive.

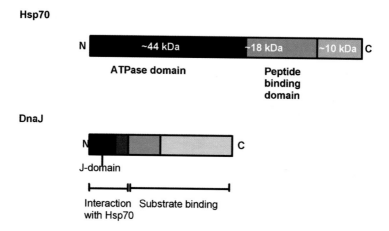

FIGURE 1. Domain structure of Hsp70 and DnaJ. The C-terminal 10 kDa subdomain in mammalian cytosolic forms of Hsp70 is involved in regulating interactions with the ATPase domain and with Hsp40.[24] The scheme is based on Hartl.[4]

abundant Hsp70 in the endoplasmic reticulum of higher eukaryotes is supposed to chaperone the folding and assembly of antibody molecules.[13]

THE STRUCTURE OF Hsp70 AND IT'S CO-CHAPERONES

Hsp70s contains a characteristic domain structure (FIG. 1). The N-terminal domain (~44 kDa) is relatively conserved and possesses ATPase activity[14] with a very slow turnover number (~1 ATP per 10 min at 37°C for DnaK). The crystal structure of the ADP-liganded ATPase domain of bovine heat-shock cognate protein 70 (Hsc70) has been resolved some years ago to a resolution of 2.2 Å (FIG. 2A).[15] Like all Hsp70s, DnaK does not work alone, but in conjunction with helper-proteins, the co-chaperones. Recently, the structure of the nucleotide-free ATPase domain complexed to a dimer of its co-chaperone GrpE has been determined to 2.8 Å resolution

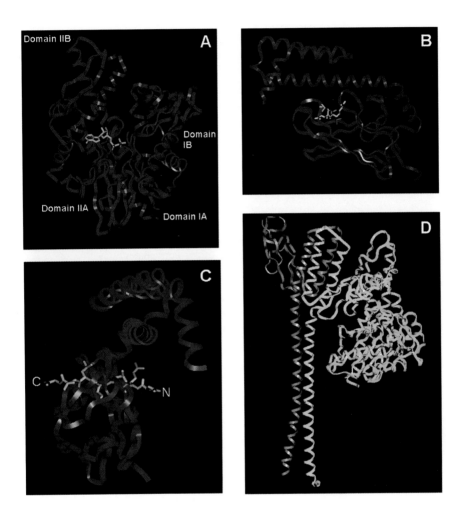

FIGURE 2. Structures of the 44-kDa ATPase domain of bovine Hsc70, the 18-kDa peptide binding domain of DnaK in complex with a target peptide, and the dimer of the co-chaperone GrpE, complexed to the ATPase domain of DnaK. **(A)** The ribbon diagram of Hsc70[15] shows the four subdomains of the ATPase domain with liganded ADP in the middle. **(B)** Standard view of the substrate binding unit of DnaK.[20] The peptide NRLLLTG is bound to the β-sandwich subdomain. The α-helical subdomain forms a lid encapsulating the peptide in the peptide binding channel. **(C)** View rotated 90° counterclockwise around the vertical axis of the standard view. **(D)** The ribbon drawing represents the two monomers of GrpE in the complex with the nucleotide-free ATPase domain of DnaK (on the right side).[16] All representations were produced with INSIGHT II from Biosym using the coordinates from the Brookhaven Protein Data Bank.

(FIG. 2D).[16] Both structures of the ATPase domain show two lobes, each of which is composed of two subdomains that make up the sides of the nucleotide binding site. Correct positioning of ATP requires one Mg^{2+} and two K^+ ions.[17] The C-terminal domain of Hsp70 comprises the ~27-kDa peptide binding site (FIG. 1).[18,19] The separated peptide binding domain of DnaK, complexed to the short peptide NRLLLTG has been determined at 2.0 Å resolution (FIG. 2B, C).[20] The structure consists of a β-sandwich subdomain followed by an α-helical subdomain. The peptide is bound to DnaK in an extended conformation through a channel defined by loops of the β-sandwich. The α-helical subdomain, which functions as a lid, stabilizes the complex but does not contact the peptide directly. Peptide binding to DnaK is mediated by several types of interactions. These are mainly seven main-chain hydrogen bonds between the peptide backbone and the cavity-forming loops of DnaK and numerous van der Waals interactions of peptide side chains.

Hsp70s generally bind short polypeptides of 5–7 residues,[21,22] containing hydrophobic amino acids[19,21,22] and possessing an extended conformation.[20,22,23] Such stretches of polypeptides are normally buried in the interior of a protein because they are not, or are only poorly, water-soluble. The capability of Hsp70 to bind relatively unspecifically to hydrophobic stretches of proteins is also very important for its role in immunological functions where they bind tumor-derived peptides.[24]

The N- and C-terminal domains of Hsp70 interact in an allosteric manner (FIG. 3). In the ATP-liganded state (T state), the α-helical lid, which locks the peptide into the binding site, is thought to be displaced, facilitating the release of the peptide. Hydrolysis of ATP by Hsp70 produces the R state, which might be brought about by a rotation of the α-helical lid, resulting in closure of the peptide binding domain[20] and high-affinity binding of the target peptide.[25–27]

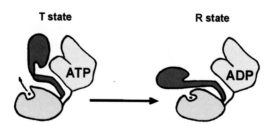

T state   **R state**

FIGURE 3. ATP- and ADP-induced conformations of Hsp70 (DnaK). The model represents the two functional states of Hsp70. In the T state, the α-helical lid (*dark gray*) of the peptide binding domain (*medium gray scale*) is displaced and the peptide (*dot*) interacts with the peptide-binding channel with high on and off rates. Conformational changes, due to hydrolysis of ATP in the ATPase domain (*light gray*), are transmitted to the peptide binding domain and result in closure of the lid and locking in of the target peptide (R state). In this state, peptides are bound and released with low on and off rates. The scheme is adapted from Zhu and colleagues.[20]

MODE OF ACTION OF THE HSP70 CHAPERONE SYSTEM OF *E. COLI*: THE DnaK/DnaJ/GrpE MACHINERY

The only representative of the Hsp70 family in *E. coli* is termed DnaK. DnaK was initially identified by its function in the replication of bacteriophage λ DNA in *E. coli*.[28] DnaK⁻ is constitutively expressed but is overexpressed under heat-shock and other stress conditions.[29] *E. coli* DnaK mutants are unable to grow at both high and low temperatures and grow at intermediate temperature very poorly, displaying an abnormally high expression of other Hsps.[30] DnaK shares 40–50% sequence identity with its eukaryotic homologs. The cycle of ATP binding and hydrolysis by DnaK is regulated by two co-chaperones, termed DnaJ and GrpE.[31] DnaJ (41 kDa) is a member of a large family of homologous proteins.[32] DnaJ-like proteins control Hsp70s by stimulating the rate of γ-phosphate cleavage of Hsp70s-bound ATP.[33,34]

The second co-chaperone, GrpE, interacts with an exposed loop of the ATPase domain of DnaK.[35] Recently, the crystal structure of GrpE in complex with the nucleotide-free ATPase domain of DnaK was determined to 2.8 Å resolution (FIG. 2D).[16] A dimer of GrpE binds asymmetrically to a single molecule of DnaK. The GrpE-dimer consists of two long, paired helices that lead into a small four-helix bundle to which each monomer contributes two helices. While DnaJ stimulates phosphate cleavage of DnaK up to 1,200-fold, GrpE accelerates the exchange of the bound nucleotide up to 5,000-fold.[36] DnaK, DnaJ, and GrpE form a machinery that in the presence of ATP is able to suppress the aggregation of proteins[8,9] or repair heat-denatured protein damage. In the course of the last couple of years, a cycle for the interaction of DnaK, DnaJ, GrpE, and its target sequences has been proposed by many groups. From work conducted in the group of Prof. Philipp Christen from the Biochemical Institute of the University of Zürich, Switzerland, the following hypothetical model of Hsp70 action for the bacterial DnaK/DnaJ/GrpE molecular chaperone system has been suggested (FIG. 4).[37–39] (1) The fast-binding and releasing ATP-liganded open state of DnaK (T state) binds an entangled hydrophobic segment of a misfolded protein or an aggregated polypeptide. (2) The co-chaperone DnaJ

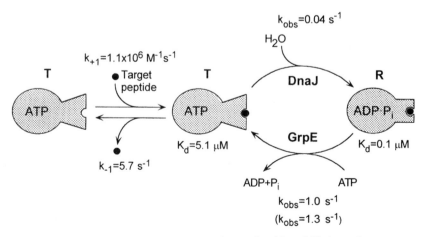

FIGURE 4. DnaK/DnaJ/GrpE reaction cycle with unfolded proteins.

triggers the hydrolysis of DnaK-bound ATP. This conformational change, mainly responsible for the chaperone effect of the system, is slow and represents the power stroke of the system. The entangled polypeptide segment is stretched and accommodated into the peptide binding cavity of DnaK. In the resulting ADP state of DnaK (R state), the peptide has adopted an extended conformation. (3) GrpE facilitates the exchange of ADP and P_i against ATP. Upon binding of ATP, DnaK releases the disentangled polypeptide segment. The fraction of DnaK that sequesters the polypeptide is controlled by the joint action of the co-chaperones. In the case of eukaryotic homologs of DnaK, this conformational change, driven by the hydrolysis of ATP, may result in the translocation of polypeptide chains through membranes.

Very importantly, this mechanism includes an energy-consuming step in which the aggregated protein is actively disassembled by Hsp70. A recent *in vitro* study with aggregated glucose-6-phosphate dehydrogenase has confirmed that in this case the DnaK chaperone machinery is in fact capable of solubilizing and refolding small protein aggregates by disaggregation.[40]

THE DNAK CHAPERONE CYCLE REACTS IMMEDIATELY TO HEAT SHOCK

Temperature directly controls the functional properties of the DnaK chaperone system. At temperatures higher than 40°C, the long helix pair of GrpE acts as an thermosensor that undergoes an extensive, fully reversible thermal transition. As described above, the GrpE-protein is a dimer that contains two long helices. At elevated temperatures, these helices unzip and the protein no longer catalyzes DnaK's ATP/ADP exchange. Therefore, the ADP state of the heat shock protein with the bound substrate protein is stabilized.[41,42] Thereby, the target protein remains sequestered, and its further aggregation is prevented until the temperature increase is over. This mechanism is very quick and reversible as soon as the temperature drops into the normal range.[43]

Hsp70 FUNCTION IS REDUCED DURING AGING

The reduced capacity to respond to stress is one of the major characteristics of senescent organisms. To investigate this notion on a molecular level, Shpund and Gershon from the Department of Biology at the Israel Institute of Technology[44] tested the major stress-inducible form, Hsp70, in young and old rats. The proteins were purified to homogeneity from livers of young (6 months) and old (27 months) rats, and their functional capacity was examined by the ability to protect the two enzymes creatine kinase (CPK) and aldolase A (Ald A) treated at high temperatures (56°C for CPK and 51°C for Ald A). When Hsp70 from young rats was present during the heat treatment, it protected CPK by 80–90% and Ald A by 50–60%. However, Hsp70 from old rats provided only 50% protection of CPK as compared to the protein of young animals. This experiment shows very well that the function of one of the major stress proteins is altered considerably during age. Since Hsp70s are involved in many fundamental tasks in cellular physiology, this may have a strong impact on the entire aged organism ability to cope with stress. At this moment, it is unclear where the age-related alterations occur in the Hsp70 molecule.

Hsp70 EXPRESSION IS ALSO REDUCED DURING AGING

It has been shown by many investigators that the induction of Hsp70 expression by heat shock as well as other stresses is reduced in many cell types. For example, the synthesis in rat hepatocytes from 26–28 month-old rats compared to 4–6 month-old rats is reduced by 45%.[45] In cultured human T-lymphocytes there is a 50% decrease of Hsp70 synthesis of the cells with 80% compared to those with 11% *in vitro* life span.[46]

The expression of the heat shock proteins is regulated primarily at the transcriptional level. The increased transcription of Hsp70 upon heat shock requires the binding of a protein, called heat shock transcription factor (HSF), to a conserved DNA sequence, the heat shock element, which is located in the promoter region of all heat-inducible genes. Normally, HSF is present in a monomeric form that does not bind to the DNA. Under heat shock, this HSF monomers form trimers and are transported into the nucleus to start the synthesis of Hsps. When hepatocytes from male rats are exposed to a temperature increase of only 2°C, from 39°C to 41°C, the binding of the HSF to the heat shock element is already maximal after 10 to 15 min, and at 20 to 30 min the transcription of Hsp70 is maximal. Obviously, a mere increase of a few degrees over the normal temperature of the individual cell type switches on a strong heat shock response. In order to investigate on which molecular level the production of Hsp70 is reduced during aging, several experiments were conducted. These showed that during aging there may be a problem in converting the heat shock factor from the inactive monomeric form to the active trimeric form that binds to the DNA and starts the transcription.[2] It has to be emphasized that this effect is not only seen upon heat shock but also upon exposure of cells to mitogens and toxic substances. In summary, not only the function of Hsp70 itself, but also its expression is strongly impaired in the aged organism.

MANY AGE-RELATED DISEASES MAY BE DUE TO NON-FUNCTIONAL OR PARTIALLY FUNCTIONAL CHAPERONE MACHINERY

This impaired ability to react to stress has severe consequences in the aged organism. Accumulation of misfolded proteins in an aged organism occurs especially in postmitotic cells, such as neurons. If the misfolded protein is protease resistant, the danger becomes even greater. The difficulties of chaperone expression in aging cells and an impaired ability to degrade the protein lead to a massive accumulation of proteins.[47] This effect can be observed in many neurodegenerative diseases.

Alzheimer's Disease and Parkinson's Disease

The best known example of such a folding-related neurodegenerative disease is Alzheimer's disease, which is characterized by senile plaques and neurofibrillary tangles that contain the beta-amyloid and hyperphosphorylated tau proteins, respectively. In this disease, both the small heat shock protein Hsp27, Hsp70, and ubiquitin (a 6 kDa heat-shock protein, which labels damaged proteins and directs them for proteolytic degradation) are induced in the targeted neurons and surrounding astrocytes. Neuronal chaperones can be found in neuritic plaques and neurofibrillary tangles.[48] Accumulated chaperones try to sequester the β-amyloid protein and other damaged proteins. However, this can also have an unwanted effect. By keeping the β-amyloid 1-40 peptide in a nonfibrillar form, they probably even enhance the neurotoxicity of the peptide and aggravate the disease.[49]

Also in the glial and astroglial cells of Parkinson's disease victims another Hsp, the heat shock protein αβ-crystallin, is expressed. In the brains of these patients abnormal structures in certain areas of the brain are found (Lewy bodies). As in the neuritic plaques and neurofibrillary tangles in Alzheimer's disease, these Lewy bodies also contain proteins that are entangled to aggregates together with various heat-shock proteins.[50] Obviously, in both diseases there is a disturbed protein homeostasis.

Huntington's Disease

Repeats of polyglutamine in a protein makes it vulnerable to aggregation. Diseases, such as Huntington's disease, a devastating, degenerative genetic brain disorder for which there is, at present, no effective treatment or cure, develops due to an expansion of polyglutamine segments in the respective mutated protein. Also here, chaperones are co-localized with the aggregates of theses polyglutamine containing proteins. Increase of chaperone levels, among them Hsp70 and the constitutive form Hsc70, are able to inhibit this protein aggregation and slow down the progress of the disease.[51]

In this field a very interesting review article by Nancy Bonini appeared.[52] She studied neurodegenerative diseases in *Drosophila*. *Drosophila* is an excellent model system for such studies because human genes causing neurogenerative diseases can be expressed in flies causing very similar phenotypes to those in the counterpart human disease. She inserted the mutant disease form of the gene encoding spinocerebellar ataxia type 3, a disease similar to Huntington's disease, where the flies are born with mildly to strongly degenerated eyes with loss of external and internal eye structure. If in these flies Hsp70 is co-expressed, the degeneration normally associated with the pathogenic polyglutamine protein is strongly suppressed, the external eye structure is fully restored to normal and the internal eye structure is strongly restored. An even stronger effect was obtained when, in addition to Hsp70, the cochaperone Hsp40 was co-expressed. Here, a complete restoration of the eye structure could be observed.

Prion-Related Diseases

Prion aggregation, which is the cause of Creutzfeld-Jakob disease, involves many chaperones. The causative agent is thought to be the prion protein, which normally does not form large aggregates and is sensitive to proteolysis. This protein can acquire a pathological isoform through a conformational change and becomes resistant to degradation and prone to excessive aggregation. The *in vivo* function of the prions is not entirely clear. Prion aggregation involves many chaperones. During aggregation they try to block the contact surfaces of the "sick and sticky" prion proteins, which tend to form large aggregates in the brain. In yeast, Hsp100 disassembles prion aggregates and generates the small prion seeds that initiate new rounds of prion propagation. In yeast, prions can be cured both by deletion or overexpression of Hsp100.[53] Thus, chaperones may both neutralize and activate aggregation processes depending on the exact chaperone–target ratio.

In summary, heat shock proteins seem to be involved in many neurogenerative diseases which are characterized by various deposits of large protein aggregates with associated chaperones in the brain. Although they are not responsible for the disease itself, they are implicated because of their role in protein aggregation prevention. From different studies in yeast and flies it could be shown that, by overexpressing or blocking different Hsps, the state of the disease could be improved or even completely reversed. This may lead to interesting therapeutic approaches to be considered in future.

Hsp70 FUNCTION AS A VACCINE AGAINST DIFFERENT TYPES OF CANCER

Another interesting aspect of heat shock proteins is their central role in the immune response and their role in fighting against various types of cancer. This knowledge is based in part from studies from the early to mid 1990s demonstrating that Hsps could function as powerful vaccines.[54] Hsps interact with specific receptors on antigen-presenting cells and are efficiently internalized by receptor-mediated endocytosis.[55] From different lines of research molecular chaperones have been identified as tumor-rejection antigens to which a protective antitumor immune response can be induced. The use of tumor-derived Hsps as vaccines takes advantage of the peptide binding properties of stress proteins. Purification of an Hsp is believed to copurify a specific peptide "fingerprint" of the cancer cells of origin containing many specific epitopes from melanoma antigens. To purify Hsp70-peptide complexes, ADP-affinity chromatography can be used.[56] These complexes are able to induce a protective antitumor immune response in both prophylactic and therapeutic manner[55] by secreting proinflammatory cytokines.[57] Interestingly, Hsps does not elicit an immune response to themselves. The practical procedure is to purify Hsps from a patient's surgical tumor specimen and to use these Hsp-antitumor peptide complexes as vaccines.

Based on the general properties of Hsps and on preclinical data available for murine tumors, Phase I/II studies have been conducted in melanoma patients with detectable tumor and in colorectal cancer patients rendered disease-free by complete resection of liver metastasis. In these studies, each patient was vaccinated with Hsp-peptide complexes isolated from his/her own tumor.[58,59] In both studies, *de novo* induction or an increase of an antitumor T-cell response could be achieved in the major part of the patients. For colorectal cancer, 17 out of 29 patients (59%) showed a post-vaccine immune response with production of interferon-γ. Although the number of patients was too small for any definitive conclusion on a possible clinical advantage of a vaccination with tumor-derived Hsp-peptide complexes, the results were promising. The disease-free rates and overall survival rates at 24 months of the 29 vaccinated patients were similar to those reported in the literature for patients who underwent similar curative surgery. However, there was a different clinical outcome between the patients who responded to the treatment and the nonresponders. Responders had statistically significant survival advantage at 24 months on both the overall survival (100% compared to 50% of nonresponding patients) and disease-free rate (51% compared to 8%) of nonresponding patients. These data suggest that the Hsp vaccination not only generated an antitumor response in about half of the treated colorectal patients but also may have actively contributed to improve their survival.

In the melanoma vaccination study, 11 of 23 patients (48%) showed an immunological response. Of the 28 patients, 2 had complete regression of the metastasis and 3 had their diseases remain stable for more than five months. Although this treatment seem to work only in about half of the patients, the results are promising, and further studies are awaited.

In this line of research there exists also other approaches that could be used.[54] One possibility would be to vaccinate cancer patients with an Hsp-peptide complex, in which the peptide consists of a well-known epitope against this type of cancer. Hsps are not only able to bind short peptides but also full-length proteins. Therefore, there would exist the possibility of vaccinating a patient with an Hsp in complex with a full-length tumor antigen. This would offer the advantage that such a vaccine could be applied to any patient with a tumor expressing that specific tumor antigen and no surgical specimen would be required.

Although this field of biomedical research is still in its infancy, many new and exciting clinical approaches can be expected in the near future.

REFERENCES

1. MORIMOTO, R.I., A. TISSIERES & C. GEORGOPOULOS. 1994. The biology of heat shock proteins and molecular chaperones. pp. 1–30. Cold Spring Harbor Laboratory Press. Cold Spring Harbor, NY.
2. HEYDARI, A.R., R. TAKAHASHI A. GUTSMANN, et al. 1994. Hsp70 and aging. Experientia 50: 1092–1098.
3. SÖTI, C. & P. CSERMELY. 2002. Chaperones and aging: role in neurodegeneration and in other civilizational diseases. Neurochem. Int. 41: 383–389.
4. HARTL, F.-U. 1996. Molecular chaperones in cellular protein folding. Nature 381: 571–580.
5. ELLIS, R.J. & S.M. VAN DER VIES. 1991. Molecular chaperones. Annu. Rev. Biochem. 60: 321–347.
6. BUCHNER, J. 1996. Supervising the fold: functional principles of molecular chaperones. FASEB J. 10: 10–19.
7. BECKER, J. & E.A. CRAIG. 1994. Heat-shock proteins as molecular chaperones. Eur. J. Biochem. 219: 11–23.
8. LANGER, T., C. LU, H. ECHOLS, et al. 1992. Successive action of DnaK, DnaJ, and GroEL along the pathway of chaperone-mediated protein folding. Nature 356: 683–689.
9. SCHRÖDER, H., T. LANGER, F-U. HARTL & B. BUKAU. 1993 DnaK, DnaJ, and GrpE form a cellular chaperone machinery capable of repairing heat-induced protein damage. EMBO J. 12: 4137–4144.
10. SZABO, A., T. LANGER, H. SCHRÖDER, et al. 1994 The ATP hydrolysis-dependent reaction cycle of the Escherichia coli Hsp70 system DnaK, DnaJ, and GrpE. Proc. Natl. Acad. Sci. USA 91: 10345–10349.
11. HAYES, S.A. & J.F. DICE. 1996 Roles of molecular chaperones in protein degradation. Cell Biol. 132: 255–258.

12. SCHATZ, G. & B. DOBBERSTEIN. 1996. Common principles of protein translocation across membranes. Science **271**: 1519–1526.
13. KNARR, G., M.J. GETHING, S. MODROW & J. BUCHNER. 1995. BiP binding sequences in antibodies. J. Biol. Chem. **270**: 27589–27594.
14. PALLEROS, D.R., K.L. REID, L. SHI & A.L. FINK. 1993. DnaK ATPase activity revisited. FEBS **336**: 124–128.
15. FLAHERTY, K.M., C. DE LUCA-FLAHERTY & D.B. MCKAY. 1190. Three-dimensional structure of the ATPase fragment of a 70K heat-shock cognate protein. Nature **346**: 623–628.
16. HARRISON, C.J., M. HAYER-HARTL, M. DI LIBERTO, et al. 1997. Crystal structure of the nucleotide exchange factor GrpE bound to the ATPase domain of the molecular chaperone DnaK. Science **276**: 431–435.
17. WILBANKS, S.M & D.B. MCKAY. 1995. How potassium affects the activity of the molecular chaperone Hsc70. II. Potassium binds specifically in the ATPase active site. J. Biol. Chem. **270**: 2251–2257.
18. WANG, T.-F., J. CHANG & C.J. WANG. 1993. Identification of the peptide binding domain of hsc70. 18-Kilodalton fragment located immediately after ATPase domain is sufficient for high affinity binding. Biol. Chem. **268**: 26049–26051.
19. GRAGEROV, A., L. ZENG, X. ZHAO, et al. 1994. Specificity of DnaK-peptide binding. J. Mol. Biol. **235**: 848–854.
20. ZHU, X., X. ZHAO, W.F. BURKHOLDER, et al. 1996. Structural analysis of substrate binding by the molecular chaperone DnaK. Science **272**: 1606–1614.
21. FLYNN, G.C., J. POHL, M.T. FLOCCO & J.E. ROTHMAN. 1991. Peptide-binding specificity of the molecular chaperone BiP. Nature **353**: 726–730.
22. RÜDIGER, S., L. GERMEROTH, J. SCHNEIDER-MERGENER & B. BUKAU. 1997. Substrate specificity of the DnaK chaperone determined by screening cellulose-bound peptide libraries. EMBO J. **16**: 1501–1507.
23. LANDRY, S.J., R. JORDAN, R. MCMACKEN & L.M. GIERASCH. 1992. Different conformations for the same polypeptide bound to chaperones DnaK and GroEL. Nature **355**: 455–457.
24. CASTELLI, C., L. RIVOLTINI, F. RINI, et al. 2004. Heat shock proteins: biological functions and clinical application as personalized vaccines for human cancer. Cancer Immunol. Immunother. **53**: 227–233.
25. MCCARTY, J.S., A. BUCHBERGER, J. REINSTEIN & B. BUKAU. 1995. The role of ATP in the functional cycle of the DnaK chaperone system. J. Mol. Biol. **249**: 126–137.
26. SCHMID, D., A. BAICI, H. GEHRING & P. CHRISTEN. 1994. Kinetics of molecular chaperone action. Science **263**: 971–973.
27. TAKEDA, S. & D.B. MCKAY. 1996. Kinetics of peptide binding to the bovine 70 kDa heat shock cognate protein, a molecular chaperone. Biochemistry **35**: 4636–4644.
28. WICKNER, S.H. 1990. Three *Escherichia coli* heat shock proteins are required for P1 plasmid DNA replication: formation of an active complex between *E. coli* DnaJ protein and the P1 initiator protein. Proc. Natl. Acad. Sci. USA **87**: 2690–2694.
29. GEORGOPOULOS, C., D. ANG, K. LIBEREK & M. ZYLICZ. 1990. Stress Proteins in Biology and Medicine. R. I. Morimoto, A. Tissieres & C. Georgopoulos. Eds.: 191–222. Cold Spring Harbor Laboratory Press. Cold Spring Harbor, NY.
30. BUKAU, B. & G.C. WALKER. 1990. Mutations altering heat shock specific subunit of RNA polymerase suppress major cellular defects of *E. coli* mutants lacking the DnaK chaperone. EMBO J. **9**: 4027–4036.
31. LIBEREK, K., J. MARSZALEK, D. ANG, et al. 1991. *Escherichia coli* DnaJ and GrpE heat shock proteins jointly stimulate ATPase activity of DnaK. Proc. Natl. Acad. Sci. USA **88**: 2874–2878.
32. CYR, D.M., T. LANGER & M.G. DOUGLAS. 1994. DnaJ-like proteins: molecular chaperones and specific regulators of Hsp70. Trends Biochem. Sci. **19**: 176–181.
33. GEORGOPOULOS, C. The emergence of the chaperone machines. 1992. Trends Biochem. Sci. **17**: 295–299.
34. HARTL, F.-U., R. HLODAN & T. LANGER. 1994. Molecular chaperones in protein folding: the art of avoiding sticky situations. Trends Biochem. Sci. **19**: 20–25.

35. Szabo, A., R. Korszun R, F.-U. Hartl & J. Flanagan. 1996. A zinc finger-like domain of the molecular chaperone DnaJ is involved in binding to denatured protein substrates. EMBO J. **15:** 408–417.
36. Packschies, L., H. Theyssen, A. Buchberger, *et al.* 1997. GrpE accelerates nucleotide exchange of the molecular chaperone DnaK with an associative displacement mechanism. Biochemistry **36:** 3417–3422.
37. Pierpaoli, E.V., E. Sandmeier, A. Baici, *et al.* 1997. The power stroke of the DnaK/ DnaJ/GrpE molecular chaperone system. J. Mol. Biol. **269:** 757–768.
38. Pierpaoli, E.V., E. Sandmeier, H.-J. Schönfeld & P. Christen. 1998. Control of the DnaK chaperone cycle by substoichiometric concentrations of the co-chaperones DnaJ and GrpE. J. Biol. Chem. **273:** 6643–6649.
39. Gisler, S.M., E.V. Pierpaoli & P. Christen. 1998. Catapult mechanism renders the chaperone action of Hsp70 unidirectional. J. Mol. Biol. **279:** 833–840.
40. Diamant, S., A.P. Ben-Zvi, B. Bukau B & P.J. Goloubinoff. 2000. Size-dependent disaggregation of stable protein aggregates by the DnaK chaperone machinery. Biol. Chem. **275:** 21107–21113.
41. Grimshaw, J.P.A., I. Jelesarov, H.-J. Schönfeld & P. Christen. 2001. Reversible thermal transition in GrpE, the nucleotide exchange factor of the DnaK heat-shock system. J. Biol. Chem. **276:** 6089–6104.
42. Grimshaw, J.P.A., I. Jelesarov, R.K. Siegenthaler & P. Christen. 2003. Thermosensor action of GrpE. The DnaK chaperone system at heat shock temperatures. J. Biol. Chem. **278:** 19048–19053.
43. Siegenthaler, R.K., J.P.A. Grimshaw & P. Christen. 2004. Immediate response of the DnaK molecular chaperone system to heat shock. FEBS Lett. **562:** 104–110.
44. Shpund, S. & D. Gershon. 1997. Alterations in the chaperone activity of HSP70 in aging organisms. Arch. Geront. Geriatr. **24:** 125–131.
45. Heydari, A.R., B. Wu, R. Takahashi, *et al.* 1993. Expression of heat shock protein 70 is altered by age and diet at the level of transcription. Mol. Cell Biol. **13:** 2909–2918.
46. Effros, R.B., X. Zhu & R.L. Walford. 1994. Stress response of senescent T lymphocytes: reduced hsp70 is independent of the proliferative block. J. Geront. **49:** B65–70.
47. Macario, A.J.L. & E. Conway de Macario. 2001. Molecular chaperones and age-related degenerative disorders. Adv. Cell. Aging Gerontol. **7:** 131–162.
48. Hamos, J.E., B. Oblas, D. Pulaski-Salo, *et al.* 1991. Expression of heat shock proteins in Alzheimer's disease. Neurology **41:** 345–350.
49. Stege, G.J., K. Renkawek, P.S. Overkamp, *et al.* 1992. The molecular chaperone alphaB-crystallin enhances amyloid beta neurotoxicity. Biochem. Biophys. Res. Commun. **262:** 152–156.
50. Jellinger, K.A. 2000. Cell death mechanisms in Parkinson's disease. J. Neural Transm. **107:** 1–29.
51. Hughes, R.E. & J.M. Olson. 2001. Therapeutic opportunities in polyglutamine disease. Nat. Med. **7:** 719–723.
52. Bonini, N. 2002. Chaperoning brain degeneration. Proc. Natl. Acad. Sci. USA **99:** 16407–16411.
53. Wegrzyn, R.D., K. Bapat, G.P. Newnam, *et al.* 2001. Mechanism of prion loss after Hsp104 inactivation in yeast. Mol. Cell Biol. **21:** 4656–4669.
54. Manjili, M.H., X.-Y. Wang, J. Park, *et al.* 2002. Immunotherapy of cancer using heat shock proteins. Front. Biosci. **7:** 43–52.
55. Srivastava, P. 2002. Interaction of heat shock proteins with peptides and antigen presenting cells: chaperoning of the innate and adaptive immune responses. Annu. Rev. Immunol. **20:** 395.
56. Peng, P., A. Menoret & P.K. Srivastava. 1997. Purification of immunogenic heat shock protein 70-peptide complexes by ADP-affinity chromatography. J. Immunol. Methods **204:** 13.
57. Asea, A., S.K. Kraeft, A.E. Kurt-Jones, *et al.* 2000. HSP70 stimulates cytokine production through a CD14-dependant pathway, demonstrating its dual role as a chaperone and cytokine. Nat. Med. **6:** 435–442.

58. BELLI, F., A. TESTORI, L. RIVOLTINI, *et al.* 2002. Vaccination of metastatic melanoma patients with autologous tumor-derived heat shock protein gp96-peptide complexes: clinical and immunologic findings. J. Clin. Oncol. **20:** 4169–4180.
59. MAZZAFERRO, V., J. COPPA, M. CARABBA, *et al.* 2003. Vaccination with autologous tumor-derived heat-shock protein gp96 after liver resection for metastatic colorectal cancer. Clin. Cancer Res. **9:** 3235–3245.

Central and Peripheral Effects of Insulin/IGF-1 Signaling in Aging and Cancer

Antidiabetic Drugs as Geroprotectors and Anticarcinogens

VLADIMIR N. ANISIMOV,[a] LEV M. BERSTEIN,[a] IRINA G. POPOVICH,[a] MARK A. ZABEZHINSKI,[a] PETER A. EGORMIN,[a] MARGARITA L. TYNDYK,[a] IVAN V. ANIKIN,[a] ANNA V. SEMENCHENKO,[a] AND ANATOLI I. YASHIN[b]

[a]N.N. Petrov Research Institute of Oncology, St. Petersburg 197758, Russia

[b]Center for Demographic Studies, Duke University, Durham, North Carolina 27708, USA

ABSTRACT: Studies in mammals have led to the suggestion that hyperglycemia and hyperinsulinemia are important factors both in aging and in the development of cancer. Insulin/insulin-like growth factor 1 (IGF-1) signaling molecules linked to longevity include DAF-2 and insulin receptor (InR) and their homologues in mammals and to inactivation of the corresponding genes followed by increased life span in nematodes, fruit flies, and mice. It is possible that the life-prolonging effect of caloric restriction are due to decreasing IGF-1 levels. A search of pharmacological modulators of life span–extending mutations in the insulin/IGF-1 signaling pathway and mimetics of effects of caloric restriction could be a direction in the regulation of longevity. Some literature and our own observations suggest that antidiabetic drugs could be promising candidates for both life span extension and prevention of cancer.

KEYWORDS: insulin; IGF-1; antidiabetic drugs; life span; cancer

INTRODUCTION

The potential link between aging and insulin/IGF-1 signaling has attracted substantial attention in recent years on the basis of evidence including an age-related increase in incidence of insulin resistance and type 2 diabetes in accelerated aging syndromes as well as life span extension by caloric restriction in rodents. Concomitant reduction in plasma insulin and plasma glucose levels, which implies increased sensitivity to insulin, emerges as a hallmark of increased longevity.[1,2] Hyperglycemia is an important aging factor involved in generation of advanced glycosylation end products (AGEs).[3,4] There is evidence that hyperinsulinemia favors accumulation of oxidized protein by reducing its degradation and facilitates protein oxidation by increasing the steady state level of oxidative stress.[5,6] Untreated diabetics with

Address for correspondence: Vladimir N. Anisimov, N.N. Petrov Research Institute of Oncology, St. Petersburg 197758, Russia. Voice: +7(812)596-8607; fax: +7(812)596-8947.

aging@mail.ru

Ann. N.Y. Acad. Sci. 1057: 220–234 (2005). © 2005 New York Academy of Sciences.

doi: 10.1196/annals.1356.017

elevated glucose levels suffer many manifestations of accelerated aging, such as impaired wound healing, obesity, cataracts, and vascular and microvascular damage.[7] It is worth stressing that hyperinsulinemia is an important factor not only in aging but also in the development of cancer.[7–11]

In organisms ranging from yeast to rodents, both caloric restriction and mutations in insulin/IGF-1 signaling pathway extend life span.[1,2,12–16] Both approaches have some unfavorable effects. For example, calorie restriction increases the level of serum glucocorticoids and decreases resistance to infection,[17–19] whereas genetic modifications on insulin/IGF-1 signaling pathway cause obesity, dwarfism, and cardiopulmonary lesions.[16] Reviewing the available data on the benefits and adverse effects of calorie restriction and genetic modifications, Longo and Finch[16] suggested three categories of drugs that may have potential to prevent or postpone age-related diseases and extend life span: drugs that (1) stimulate dwarf mutations and therefore decrease pituitary production of growth hormone; (2) prevent IGF-1 release from the liver, or (3) decrease IGF-1 signaling by the action on either extracellular or intracellular targets (one should remember though that age-associated or drug-induced decrease in IGF-1 production may lead to the lowering of cognitive function, thus making aging rather unsuccessful).[20]

The concept of caloric restriction mimetics is now being intensively explored.[21–23] Caloric restriction mimetics involve interventions that produce physiological and anti-aging effects similar to caloric restriction. The use of biguanide antidiabetics as a potential anti-aging treatment was suggested.[8,24–27] The antidiabetic drugs phenformin, buformin, and metformin were observed to reduce hyperglycemia and produce the following effects: improved glucose utilization; reduced free fatty acid utilization, gluconeogenesis, serum lipids, insulin and IGF-1, and reduced body weight both in humans and experimental animals.[7]

It was shown that hypoglycemic activity of Diabenol® (9-β-diethylaminoethyl-2,3-dihydroimidazo-(1,2-α) benzimidazol dihydrochloride) was 1.5 times higher than that of maninil (glibenclamide) and equal to the effect of glyclazide in rats, rabbits, and dogs.[28,29] It was shown that hypoglycemic effect of diabenol included both pancreatotropic and extrapancreatic pathways. Diabenol restores physiological profile of insulin secretion, decreases tissue resistance to insulin, prolongs hypoglycemic effect of insulin. It increases glucose utilization in glucose loading test in old obese rats. It was suggested that diabenol influences insulin receptors in peripheral tissues. Diabenol increases uptake of glucose by isolated rat diaphragm *in vitro* both without supplementation of insulin into the medium or with supplemented insulin. Diabenol also decreases platelet and erythrocyte aggregation and blood viscosity, inhibits mutagenic effect of 2-acetylaminofluorene, and has antioxidant activity.[28–32] Thus, these results suggest that like biguanides diabenol has a potential to increase the life span and inhibit carcinogenesis.

INSULIN/IGF-1 SIGNALING PATHWAY AND LONGEVITY IN MAMMALS

The intensive investigations in *C. elegans* since the 1990s, which have identified insulin signaling components including daf-2, age-1, and daf-16 as the genes whose mutations lead to life span extension, shed new light on molecular mechanisms underlying aging.[1,2,33–35] In *D. melanogaster*, the mutation modification of genes operating in signal transduction from insulin receptor to transcription factor *daf*-16 (*age*-1, *daf*-2, CHICO, InR) are strongly associated with longevity.[35–38] It was demonstrated that FKHR, FKHRL1, and AFX, mammalian homologues of *daf*-16 forkhead transcription factor, function downstream of insulin signaling and akt/PKB under cellular conditions.[39,40] However, it is an open question whether insulin signaling components, including forkhead transcriptional factors, play a critical role in aging and longevity in mammals as well as in *C. elegans* and *D. melanogaster*.

Daf-2 and InR are structural homologues of tyrosine kinase receptors in vertebrates and include the insulin receptor and the insulin-like growth factor type 1 receptor (IGF-1R). It was shown that in vertebrates the insulin receptor regulates energy metabolism, whereas IGF-1R promotes growth. Although decreased insulin-like signaling appears to increase longevity in *C. elegans* and *D. melanogaster*, whether the same is true in mammals is unclear. In at least three genes (*Pit1*[dw], *Prop1*[dw], *Ghr*) knockout leads to dwarfism with reduced levels of IGF-1 and insulin and to increased longevity.[41,42] In Snell and Ames dwarf mice, sexual maturation is delayed and few males are fertile, whereas females are invariably sterile.[43] These mice, as well as *Ghr*[−/−] mice, have significantly reduced glucose levels and fasting insulin levels, decreased tolerance to glucose, and increased sensitivity to insulin, which appears to be combined with reduced ability to release glucose in response to acute challenge.[1,43]

Strong support for the role of insulin/IGF-1 signaling pathway in the control of mammalian aging and for the involvement of this pathway in longevity of IGF-1–deficient mice was provided by Hsieh and colleagues.[44,45] It was shown that in the Snell dwarf mice, growth hormone deficiency leads to reduced insulin secretion and alterations in insulin signaling via InRβ, IRS-1, or IRS-2 and P13K affects genes involved in the control longevity. The authors concluded that the *Pit1* mutation may result in physiological homeostasis that favors longevity.

Reduction in glucose and insulin levels as well as an increase in the sensitivity to insulin are a well-documented response to caloric restriction in rodents and monkey.[46,47] It was shown that improved sensitivity to insulin in calorie-restricted animals is specifically related to reducing visceral fat.[48] It is worth noting that *Ghr*[−/−] mice have a major increase in the level of insulin receptors,[49] while Ames dwarf mice have a smaller increase in insulin receptor and substantially increased amount of insulin receptor substrates IRS-1 and IRS-2.[50] The development of tumors in Ames dwarf mice was postponed and the incidence was reduced as compared to the control.[51,52]

The crucial event of the effect of caloric restriction is lowered levels of insulin and IGF-1 and also increased insulin sensitivity in rodents[12,14] and monkeys.[53–55] Many characteristics of these long-lived mutants and growth hormone–receptor knockout mice resemble those of normal animals exposed to caloric restriction. These characteristics include reduced plasma levels of IGF-1, insulin, and glucose, with the consequent reductions in growth and body size, delayed puberty, and significantly increased sensitivity to insulin action.

In a bid to discover whether the IGF-1 receptor might control vertebrate longevity, Holzenberger and colleagues[56] inactivated the *Igf1r* gene by homologous recombination in mice. It was shown that *Igf1r*[−/−] mice died early in the life, whereas heterozygous *Igf1r*[+/−] mice live on average 26% longer than their wild-type littermates. It is worth noting that long-lived mice do not develop dwarfism and their energy metabolism was normal. Food intake, physical activity, fertility, and reproduction were also unaffected in $Igf1r^{+/-}$. The spontaneous tumor incidence in the aging cohort of $Igf1r^{+/-}$ mice was similar to that in wild-type controls. It is very important that these $Igf1r^{+/-}$ mice, and mouse embryonic fibroblasts derived from them, were more resistant to oxidative stress than controls. At the molecular level, insulin receptor substrate and the $p52$ and $p66$ isoforms of Shc, both main substrates of IGF-1 receptor, showed decreased tyrosine phosphorylation. $p66^{Shc}$ mediated cellular responses to oxidative stress. Two main pathways—the extracellular-signal regulated kinase (ERK)/mitogen-activated protein kinase (MAPK) pathway and the phosphatidylinositol 3-kinase (PI3K)-Akt pathway—were downregulated in $Igf1r^{+/-}$ mice.

The extension of longevity was observed in fat-specific insulin receptor knockout (FIRKO) mice.[57] These animals have reduced fat mass and were protected against age-related obesity and its subsequent metabolic abnormalities including deterioration in glucose tolerance, although their food intake was normal. Both male and female FIRKO mice were found to have an increase in mean life span (by 18%) with parallel increases in maximum life span. Extended longevity in FIRKO mice was associated with a shift in the age at which age-dependent increase in mortality risk becomes appreciable and a decreased rate of age-related mortality, especially after 36 months of age. In FIRKO mice, the resistance to obesity, despite normal food intake, suggests that metabolic rate is increased, rather than decreased.[58] The authors believe that decreased fat mass could lead to a decrease in oxidative stress in FIRKO mice. Another possibility is that the increased longevity in these mice is the direct result of altered insulin signaling.

Shimokawa and colleagues[59] designed a transgenic strain of rats whose GH gene was suppressed by an anti-sense GH transgene. Male rats homozygous for the transgene (*tg/tg*) had a reduced number of pituitary GH cells, a lower plasma concentration of IGF-1, and a dwarf phenotype. Heterozygous rats (*tg/−*) had an intermediate phenotype in plasma IGF-1, food intake, and body weight between *tg/tg* and control (*−/−*) rats. The life span of *tg/tg* rats was 5 to 10% shorter than *−/−* rats. In contrast, the life span of *tg/−* rats was 7 to 10% longer than *−/−* rats. It was found that tumors caused earlier death in *tg/tg* rats; in contrast, *tg/−* rats had reduced nonneoplastic diseases and a prolonged life span. Immunological analysis revealed a smaller population and lower activity of splenic natural killer cells in homozygous *tg/tg* rats. These results provided evidence that an optimal level of the GH-IGF-1 axis function is required for longevity in mammals.

It was shown that the incidence of mutations in the insulin regulatory region (IRE) of APO C-III T-455 C directly correlates with longevity in humans. This evidence shows that the mutation located downstream to *daf-16* in insulin signal transduction system is associated with longevity.[60] It is worth noting that centenarians display lower degree of resistance to insulin and lower degree of oxidative stress as compared with elderly persons before 90 years.[61,62] The authors suggest that centenarians may have been selected for appropriate insulin regulation as well as for the appropriate regulation of tyrosine hydroxylase (TH) gene, whose product is rate limiting in the synthesis of catecholamines, stress-response mediators. Catecholamine may increase free radical production through induction of the metabolic rate and autooxidation in diabetic animals.[63] Recent study on aging parameters of young (up to 39) and old (over 70) individuals having similar IGF-1 serum levels provides evidence of the important role of this peptide for life potential.[64] Roth and colleagues[65] analyzed data from the Baltimore Longitudinal Study of Aging and reported that survival was greater in men who maintained lower insulin levels.

EFFECT OF ANTIDIABETIC DRUGS ON LONGEVITY

Several years ago, it was suggested to use biguanide antidiabetics could be an anti-aging treatment.[8,24] The antidiabetic drugs, phenformin (1-phenylethylbiguanide), buformin (1-butylbiguanide hydrochloride), and metformin (N,N-dimethylbiguanide) were observed to reduce hyperglycemia, improve glucose utilization, reduce free fatty acid utilization, gluconeogenesis, serum lipids, insulin, somatomedin, reduce body weight, and decrease metabolic immunodepression both in humans and rodents.[7,8,66–70] Nowadays, phenformin is not used in clinical practice due to it side effects (mainly lactic acidosis) observed in patients with non-compensated diabetes. It is worth noting that during more than 10 years of administering phenformin to patients without advanced diabetes Dilman[7] observed no cases of lactic acidosis or any other side effects. Nevertheless we believe that the analysis of results of long-term administration of this drug as well as another antidiabetic biguanides (buformin and metformin) to non-diabetic animals seems very important for understanding links between insulin and longevity.

TABLE 1. Effect of antidiabetic biguanides on mortality rate in mice and rats

| Strain | Treatment | Animals (N) | Life span (days) | | | Ref. |
			Mean	Last decile of survivors	Maximum	
Mice						
C3H/Sn	control	30	450 ± 23.4	631 ± 11.4	643	25
	phenformin	24	545 ± 39.2 (+21.1%)	810 ± 0 ** (+28.4%)	810 (+26%)	
NMRI	control	50	346 ± 11.9	480 ± 9.2	511	71
	diabenol	50	369 ± 12.9	504 ± 6.4* (+5.9%)	518	
HER-2/ neu	dontrol	30	264 ± 3.5	297 ± 7.3	311	72
	metformin	24	285 ± 5.2 (+8%)	336 ± 2.7 (+17.9%)	340 (+9.3%)	
Rats						
LIO	control	41	652 ± 27.3	885 ± 11.3	919	73, 74
	phenformin	44	652 ± 28.7	974 ± 16.2** (+10.1%)	1009 (+9.8%)	
	control	74	687 ± 19.2	925 ± 22.5	1054	74, 75
	buformin	42	737 ± 26.4 (+7.3%)	1036± 38.9* (+12%)	1112 (+5.5%)	

NOTE: The difference with control is significant: *$P < 0.05$; **$P < 0.01$ (Student's t test).

The available data on the effect of antidiabetic biguanides on life span in mice and rats is summarized in TABLE 1. The treatment with phenformin prolonged the mean life span of female C3H/Sn mice by 21%, the mean life span of oldest decile of survivors by 28%, and the maximum life span by 5.5 months (by 26%) in comparison with the control. At the time of death of the last mice in the control group, 42% of phenformin-treated mice were alive.[25] Administration of phenformin failed influence the mean life span in LIO rats.[73,74] At the same time, the mean life span of the last decile of survivors was increased by 10%, and maximum life span was increased by 3 months (+10%) as compared with the controls. The treatment with phenformin slightly decreased the body weight of rats in comparison with the control. The age-related disturbances in estrus function were observed in 36% of 15- to 16-month-old rats of the control group and only in 7% of rats in phenformin-treated group.

The treatment with buformin slightly increased mean life span of rats (by 7%). The mean life span of the last decile of survivors increased by 12% and the maximum life span increased by 2 months (+5.5%) as compared with controls. The body weight of rats treated with buformin was slightly (5.2 to 9.4%) but statistically significantly decreased in comparison with the control from the age of 12 to 20 months. At the age of 16 to 18 months, 38% of control rats had disturbances in the estrus cycle (persistent estrus, repetitive pseudopregnancies, or anestrus), whereas in females treated with buformin these disturbances were observed only in 9% of rats.[75]

Recently it was found that metformin, like buformin and phenformin, significantly increases the life span of rats (G.S. Roth, personal communication). The chronic treatment of female transgenic HER-2/neu mice with metformin (100 mg/kg in drinking water) slightly decreased the food consumption but failed in reducing the body weight or temperature, slowed down the age-related rise in blood glucose and triglycerides level, as well as the age-related switch-off of estrous function, prolonged the mean life span by 8%, the mean life span of last decile of survivors by 13.1%, and the maximum life span by 1 months in comparison with control mice. The demographic aging rate represented by the estimate of respective Gompertz's parameter was decreased by 2.26 times. The metformin-treatment significantly decreased the incidence and size of mammary adenocarcinomas in mice and increased the mean latency of the tumors.[72]

There is limited data on the effects of other antidiabetic drugs on life span of animals. It is shown that treatment with diabenol failed to influence body weight gain dynamics, food and water consumption, and body temperature, slowed down age-related disturbances in estrous function, and increased life span of all and the most long-lived decile of NMRI mice. The treatment with diabenol inhibited spontaneous tumor incidence and increased mammary tumor latency in these mice. Diabenol treatment slowed down age-related changes in estrous function in HER-2/neu mice, failed to influence survival of these mice, and slightly inhibited the incidence and decreased the size of mammary adenocarcinoma metastases into the lung.[71]

It is worth noting that buformin was added as a supplement in various concentrations (from 1.0 to 0.00001 mg/mL) to nutrient medium during the larval stage and over the life span of *C. elegans*. The drug given at the concentration of 0.1 mg/mL increased the mean life span of the nematodes by 23.4% ($P < 0.05$) and the maximum life span by 26.1% as compared to the controls.[76]

Several other effects of treatment with antidiabetic biguanides, related to reproduction and aging, are known from earlier studies. For example, it decreased hypothalamic threshold of the sensitivity to feedback inhibition by estrogens,[77,78] which is one of the most important mechanisms regulating age-related decline and switch-off of the reproductive function.[40,78-80] It is worth noting that another antidiabetic biguanide, metformin, may improve menstrual regularity, leading to spontaneous ovulation, and enhance the induction of ovulation with clomiphene citrate in women with sclerocystic ovary syndrome.[81,82] The treatment with phenformin also decreased hypothalamic threshold of sensitivity to feedback regulation by glucocorticoids and by metabolic stimuli (glucose and insulin).[7] It was recently shown that elements involved in the insulin/IGF-1 signaling pathway are regulated at the expression and/or functional level in the central nervous system. This regulation may play a role in the brain's insulin resistance,[83] in the control of ovarian follicular development and ovulation,[84] and brain's control of life span.[14,55,54] Antidiabetic biguanides also alleviated age-related metabolic immunodepression.[7] These mechanisms can be involved in geroprotective effect of biguanides. Treatment with chromium picolinate, which increases insulin sensitivity in several tissues, including hypothalamus, significantly increased the mean life span and decreased the development of age-related pathology in rats.[85] It could be hypothesized that antidiabetic biguanides and possibly chromium picolinate, regulate thyrosine hydroxylase and insulin/IGF-1 signaling pathway genes—both associated with longevity.[35,86,87] It was shown that the polymorphism at the TH-INS locus affects non-insulin dependent type 2 diabetes,[88] and is associated with hypothalamic obesity,[89] sclerocystic ovary syndrome,[90] hypertriglyceridemia, and atherosclerosis.[91]

EFFECT OF ANTIDIABETIC DRUGS ON CARCINOGENESIS

Long-term treatment with phenformin significantly inhibited (by fourfold) the incidence of spontaneous mammary adenocarcinomas in female C3H/Sn mice.[25] The treatment with phenformin was followed by a 1.6-fold decrease in total spontaneous tumor incidence in rats,[73,74] whereas total tumor incidence was decreased by 49.5% in buformin-treated rats.[74,75]

The anticarcinogenic effect of antidiabetic biguanides has been demonstrated in several models of induced carcinogenesis (TABLE 2). Daily oral administration of phenformin or buformin suppressed DMBA-induced mammary tumor development in rats.[93,94,96] Phenformin-treated rats revealed a tendency toward a decrease in serum insulin level. The treatment with phenformin normalized the tolerance to glucose and serum insulin and IGF-1 level in rats exposed to intravenous injections of N-nitrosomethylurea (NMU) and inhibited mammary carcinogenesis in these animals.[95] Treatment of rats with 1,2-dimethylhydrazine (DMH) caused the decrease in the level of biogenic amines, particularly of dopamine in the hypothalamus, the decrease of glucose tolerance, and the increase of the blood level of insulin and triglycerides. The exposure to DMH also caused the inhibition of lymphocyte blastogenic response to phytohemagglutinin and lipopolysaccharide, the decrease in the level of antibody produced against sheep erythrocytes, and the decrease in phagocytic activity of macrophages.[100] Administration of phenformin started from the first injection of the carcinogen restored all the abovementioned immunological indices and inhibited DMH-induced colon carcinogenesis.[99,100] It is worth noting that colon 38 adenocarcinoma growth was significantly inhibited in liver-specific IGF-1–deficient mice

TABLE 2. Effect of antidiabetic biguanides on carcinogenesis in mice and rats.

Species	Drug	Carcinogen	Main target(s)	Effect of treatment	Ref.
Mouse	phenformin	spontaneous	mammary gland	inhibition	25
	metformin	spontaneous (HER-2/neu)	mammary gland	inhibition	72
	diabenol	spontaneous	mammary gland, lymphoma	iInhibition	71
	phenformin	MCA	subcutaneous tissue	inhibition	92
Rat	buformin	spontaneous	total incidence	inhibition	75
	phenformin	spontaneous	total incidence	inhibition	73
	phenformin	DMBA	mammary gland	inhibition	93, 94
	phenformin	NMU	mammary gland	inhibition	95
	buformin	DMBA	mammary gland	inhibition	96
	buformin	NMU transplacentally	nervous system	inhibition	97
	phenformin	NEU transplacentally	nervous system, kidney	inhibition	98
	phenformin	DMH	colon	inhibition	99, 100
	diabenol	DMH	colon	inhibition	71

ABBREVIATIONS: DMH, 1,2-dimethylhydrazine; MCA, 20-methylcholanthrene; NBOPA, N-nitrosobis-(2-oxopropyl)amine; NEU, N-nitrosoethylurea; NMU, N-nitrosomethylurea; X-rays, total-body X-ray irradiation.

whereas injections with recombinant human IGF-1 significantly promoted tumor growth and metastasis.[103]

In rats exposed to 1,2-dimethylhydrazine, treatment with diabenol significantly inhibited multiplicity of all colon tumors, decreased by 2.2 times the incidence of carcinomas in ascending colon and by 3.1 times their multiplicity. Treatment with diabenol was followed by higher incidence of exophytic and well-differentiated colon tumors as compared with the control rats exposed to the carcinogen alone (76.3% and 50%, and 47.4% and 14.7%, respectively).[71]

A decrease of glucose utilization in the oral glucose tolerance test was found in the 3-month-old female progeny of rats exposed to NMU on the 21st day of pregnancy.[97] The serum insulin level did not differ from the control, but the cholesterol level was higher in offspring of NMU-treated rats as compared with the control. Postnatal treatment with buformin started from the age of 2 months significantly inhibited the development of malignant neurogenic tumors in rats transplacentally exposed to NMU.[97] Similar results have been observed in rats exposed transplacentally to N-nitrosoethylurea (NEU) and postnatally to phenformin.[98] The authors observed the decrease of development of nervous system and renal tumors induced transplacentally with NEU. The treatment with phenformin also inhibited the carcinogenesis induced by a single total-body X-ray irradiation in rats.[101]

Vinnitski and Iakimenko[92] have shown that treatment with phenformin increased the immunological reactivity and inhibited carcinogenesis induced by subcutaneous administration of 20-methylcholanthrene in BALB/c mice. In high fat–fed hamsters, the treatment with N-nitrosobis-(2-oxopropyl)amine was followed by the development of pancreatic malignancies in 50% of cases, whereas no tumors were found in the hamsters treated with the carcinogen and metformin.[102]

Thus, anticarcinogenic effect of antidiabetic drugs has been demonstrated in relation to spontaneous carcinogenesis in mice and rats, in different models of chemical carcinogenesis in mice, rats, and hamsters, and in radiation carcinogenesis model in rats. Phenformin administered orally to mice potentiated the antitumor effect of cytostatic drug cyclophosphamide on transplantable squamous cell cervical carcinoma, hepatoma-22a, and Lewis lung carcinoma. Administration of phenformin to rats with transplanted Walker 256 carcinoma enhanced the antitumor effect of hydrazine sulfate.[103] It was observed that phenformin inhibits proliferation and induced enhanced and transient expression of the cell cycle inhibitor p21 and apoptosis in human tumor cells lines.[104,105]

The comparative study of 10 years of results of metabolic rehabilitation (including diets restricting fat and carbohydrates and treatment with antidiabetic biguanides) of cancer patients had shown significant increase in the survival of breast and colorectal cancer patients, the increase in the length of cancer-free period, and the decrease in the incidence of metastasis as compared with control patients.[70,106,107]

It was suggested that antidiabetic biguanides inhibit metabolic immunodepression developed in animals exposed to carcinogenic agents, which is similar to the immunodepression inherent to normal aging and specific age-related pathology.[7,8] If immunodepression is one of the important factors in carcinogenesis, then the elimination of metabolic immunodepression, which arises in the course of normal aging or under the influence of chemical carcinogens or ionizing radiation, can provide an anticarcinogenic prophylactic effect.[7,8,74,100]

Although it is known that free radicals are produced during metabolic reactions, it is largely unknown which factor(s) of physiological or pathophysiological significance, modulate their production in vivo. It has been suggested that hyperinsulinemia may increase free radicals and therefore promote aging independent of glycemia.[6,7,24,108,109] Plasma levels of lipid hydroperoxides are higher, and antioxidant factors are lower in individuals resistant to insulin-stimulated glucose disposal but otherwise glucose tolerant, nonobese, and normotensive.[108] This finding indicates that enhanced oxidative stress is present before diabetes ensues and therefore cannot simply be explained by overt hyperglycemia. There is substantial evidence supporting the hypothesis that selective resistance to insulin-stimulated (muscle) glucose disposal and the consequential compensatory hyperinsulinemia trigger a variety of metabolic effects, likely resulting in accelerated oxidative stress and aging.[6,7]

The biguanides inhibit fatty acid oxidation, gluconeogenesis in the liver, increase the availability of insulin receptors, decrease monoamine oxidase activity,[66,67] increase sensitivity of hypothalamo-pituitary complex to negative feedback inhibition, and reduce excretion of glucocorticoid metabolites and dehydroepiandrosterone sulfate.[7] These drugs have been proposed for the prevention of the age-related increase of cancer and atherosclerosis and for retardation of the aging process.[7,24] It has been shown that administration of biguanides to patients with hyperlipidemia lowers the level of blood cholesterol, triglycerides, and β-lipoproteins. Biguanides also inhibit the development of atherosclerosis and reduce hyperinsulinemia in men with coronary artery disease. Its increases hypothalamo-pituitary sensitivity to inhibition by dexamethasone and estrogens, causes restoration of estrous cycle in persistent-estrous old rats, improves cellular immunity in atherosclerotic and cancer patients, and lowers blood IGF-1 levels in cancer and atherosclerotic patients with Type IIb hyperlipoproteinemia.[7,67,68] Recently it was shown that metformin decreases platelet superoxide anion production in diabetic patients.[110] Similar effects have also been observed for diabenol.[30,71]

CONCLUSION

Striking similarities have been described in insulin/IGF-1 signaling pathways in yeast, worms, flies, and mice.[35] Many characteristics of mice that are long lived due to genetic modifications resemble effects of caloric restriction in wild-type (normal) animals.[1,26] Comparison of characteristics of animals exposed to these endogenous and exogenous influences shows a number of similarities but also some differences. The effects of antidiabetic drugs seem to be more effective than caloric restriction and genetic manipulation in both the prevention of age-related deterioration in glucose metabolism and insulin signaling pathway as well as in such important longevity parameters as infertility and resistance to oxidative stress and carcinogenesis.

ACKNOWLEDGMENTS

This work was supported in part by Grant 05-04-48110 from the Russian Foundation for Basic Research, and by Grant NSh-2293.2003.4 from the President of Russian Federation.

[*Competing interests*: The authors declare that they have no competing financial interests.]

REFERENCES

1. BARTKE, A., V. CHANDRASHEKAR, F. DOMINICI, *et al.* 2003. Insulin-like growth factor 1 (IGF-1) and aging: controversies and new insights. Biogerontology **4:** 1–8.
2. TATAR, M., A. BARTKE & A. ANTEBI. 2003. The endocrine regulation of aging by insulin-like signals. Science **299:** 1346–1351.
3. FACCHINI, F.S., N.W. HUA, G.M. REAVEN & R.A. STOOHS. 2000. Hyperinsulinemia: the missing link among oxidative stress and age-related diseases? Free Radicals Biol. Med. **29:** 1302–1306.
4. ELAHI, D., D.C. MULLER, J.M. EGAN, *et al.* 2002. Glucose tolerance, glucose utilization, and insulin secretion in aging. Novartis Found. Symp. **242:** 222–242.
5. XU, L. & M.Z. BARD. 1999. Enhanced potential for oxidative stress in hyperinsulinemic rats: imbalance between hepatic peroxisomal hydrogen peroxide production and decomposition due to hyperinsulinemia. Horm. Metab. Res. **31:** 278–282.
6. FACCHINI, F.S., N. HUA, F. ABBASI & G.M. REAVEN. 2001. Insulin resistance as a predictor of age-related diseases. J. Clin. Endocrinol. Metab. **86:** 3574–3578.
7. DILMAN, V.M. 1994. Development, Aging, and Disease: A New Rationale for an Intervention. Harwood Academic. Chur, Switzerland.

8. DILMAN, V.M. 1978. Aging, metabolic immunodepression, and carcinogenesis. Mech. Aging Dev. **8:** 153–173.
9. COLANGELO, L.A., S.M. GAPSTUR, P.H. GANN, et al. 2002. Colorectal cancer mortality and factors related to the insulin resistance syndrome. Cancer Epidemiol. Biomarkers Prev. **11:** 385–391.
10. GUPTA, K., G. KRISHNASWAMY, A. KARNAD & A.N. PEIRIS. 2002. Insulin: a novel factor in carcinogenesis. Am. J. Med. Sci. **323:** 140–145.
11. POLLAK, M.N., E.S. SCHERNHAMMER & S.E. HANKINSON. 2004. Insulin-like growth factors and neoplasia. Nature Rev. Cancer. **4:** 505–518.
12. WEINDRUCH, R. & R. WALFORD. 1988. The Retardation of Aging and Disease by Dietary Restriction. C.C. Thomas. Springfield, Ill.
13. ROTH, G.S., D.K. INGRAM & M.A. LANE. 1999. Calorie restriction in primates: Will it work and how will we know? J. Am. Geriatr. Soc. **46:** 869–903.
14. CHIBA, T., H. YAMAZA, Y. HIGAMI & I. SHIMOKAWA. 2002. Anti-aging effects of caloric restriction: involvement of neuroendocrine adaptation by peripheral signaling. Microsc. Res. Tech. **59:** 317–324.
15. KOUBOVA, J. & L. GUARENTE. 2003. How does calorie restriction work? Genes Dev. **17:** 313–321.
16. LONGO, V.D. & C.E. FINCH. 2003. Evolutionary medicine: from dwarf model systems to healthy centenarians. Science **299:** 1342–1346.
17. SUN, D., A.R. MUTHUKUMAR, R.A. LAWRENCE & G. FERNANDES. 2001. Effects of calorie restriction on polymicrobial peritonitis induced by cecum ligation and puncture in young C57BL/6 mice. Clin. Diagn. Lab. Immunol. **8:** 1003–1011.
18. MASORO, E.J. 2000. Caloric restriction and ageing: an update. Exp. Gerontol. **35:** 299–305.
19. MASORO, E.J. 2003. Subfield history: Caloric restriction, slowing aging, and extending life. Science's Sage, K.E. 2003, ns2 (26 February 2003). http://Sageke.sciencemag.org/cgi/content/full/ sageke;2003/8/re2.
20. DIK, M.G., S.M. PLUIJM, C. JONKER, et al. 2003. Insulin-like growth factor I (IGF-I) and cognitive decline in older persons. Neurobiol. Aging **24:** 573–581.
21. HADLEY, E.C., C. DUTTA, J. FINKELSTEIN, et al. 2001. Human implications of caloric restriction's effect on laboratory animals: an overview of opportunities for research. J. Gerontol. Ser. A **56:** 5–6.
22. MATTSON, M.P., W. DUAN, J. LEE, et al. 2001. Progress in the development of caloric restriction mimetic dietary supplements. J. Anti-Aging Med. **4:** 225–232.
23. WEINDRUCH, R., K.P. KEENAN, J.M. CARNEY, et al. 2001. Caloric restriction mimetics: metabolic intervention. J. Gerontol. Biol. Sci. **56:** 20–33.
24. DILMAN, V.M. 1971. Age-associated elevation of hypothalamic threshold to feedback control and its role in development, aging, and disease. Lancet **1:** 1211–1219.
25. DILMAN, V.M. & V.N. ANISIMOV. 1980. Effect of treatment with phenformin, dyphenylhydantoin, or L-DOPA on life span and tumor incidence in C3H/Sn mice. Gerontology **26:** 241–245.
26. ANISIMOV, V.N. 2003. Insulin/IGF-1 signaling pathway driving aging and cancer as a target for pharmacological intervention. Exp. Gerontol. **38:** 1041–1049.
27. ANISIMOV, V.N., A.V. SEMENCHENKO & A.I. YASHIN. 2003. Insulin and longevity: antidiabetic biguanides as geroprotectors. Biogerontology **4:** 297–307.
28. SPASOV, A.A., G.P DUDCHENKO & E.S. GAVRILOVA. 1997. Diabenol—new antidiabetic compounds with haemobiological activity. Vestn. Volgograd Med. Acad. **2:** 47–51.
29. ANISIMOVA, V.A., V.V. OSIPOVA & A.A. SPASOV. 2002. Synthesis and pharmacolgoical activity of N,N-disubstitutes 3-amino-1H-1,2-diazaohanalenes. Chem. Pharmaceut. J. **36:** 11–16.
30. SPASOV, A.A., O.V. OSTROVSKII, I.V. IVAKHNENKO, et al. 1999. The effect of compounds with antioxidant properties on thrombocyte functional activity. Eksp. Klin. Farmakol. **62:** 38–40.
31. MEZHERITSKI, V.V., A.L. PIKUS, A.A. SPASOV, et al. 1998. Synthesis and antioxidant activity N.N-substituted 3-amino-1H-1,2-diazophenalenes. Chem. Pharmaceut. J. **1:** 15–16.
32. ZINOVIEVA, V.N., O.V. OSTROVSKII, V.A. ANISIMOVA & A.A. SPASOV. 2003. Benzimidazole derivative inhibition of the mutagenic activity of 2-aminoanthracene. Gig. Sanit. **5:** 61–63.

33. KIMURA, K.D., H.A. TISSENBAUM, Y. LIU & G. RUVKUN. 1997. daf-2, an insulin receptor-like gene that regulates longevity and diapause in *Caenorhabditis elegans*. Science **277:** 942–946.
34. WOLKOV, C.A., K.D. KIMURA, M.S. LEE & G. RUVKUN. 2000. Regulation of *C. elegans* life span by insulin-like signaling in the nervous system. Nat. Genetics **290:** 147–150.
35. KENYON, C. 2001. A conserved regulatory system for aging. Cell **105:** 165–168.
36. CLANCY, D.J., D. GEMS, L.G. HARSHMAN, *et al.* 2001. Extension of life span by loss of CHICO, a Drosophila insulin receptor substrate protein. Science **292:** 104–106.
37. TATAR, M., A. KOPELMAN, D. EPSTEIN, *et al.* 2001. A mutant Drosophila insulin receptor homolog that extends life span and impairs neuroenodcrine function. Science **292:** 107–110.
38. DILLIN, A., D.K. CRAWFORD & C. KENYON. 2002. Timing requirements for insulin/IGF-1 signaling in *C. elegans*. Science **298:** 830–834.
39. RICHARDS, J.S., D.L. RUSSELL, S. OCHSNER, *et al.* 2002. Novel signaling pathways that control ovarian follicular development, ovulation, and luteinization. Recent Prog. Horm. Res. **57:** 195–220.
40. ROSSMANITH, W.G. 2001. Neuroendocrinology of aging in the reproductive system: gonadotropin secretion as an example. *In* Follicular Growth, Ovulation, and Fertilization: Molecular and Clinical Basis. A. Kumar & A.K. Mukhopadhayay, Eds.: 15–25. Narosa Publ. House. New Dehli.
41. FLURKEY, K., J. PAPACONSTANTINOU, R.A. MILLER & D.E. HARRISON. 2001. Life span extension and delayed immune and collagen aging in mutant mice with defects in growth hormone production. Proc. Natl. Acad. Sci. USA **98:** 6736–6741.
42. COSCHIGANO, K.T., D. CLEMMONS, L.L. BELLUSH & J.J. KOPCHICK. 2000. Assessment of growth parameters and life span of GHR/BP gene-disrupted mice. Endocrinology **141:** 2608–2613.
43. BARTKE, A. & D. TURYN. 2001. Mechanisms of prolonged longevity: mutants, knockouts, and caloric restriction. J. Anti-Aging Med. **4:** 197–203.
44. HSIEH, C.C., J.H. DeFORD, K. FLURKEY, *et al.* 2002. Implications for the insulin signaling pathway in Snell dwarf mouse longevity: a similarity with the *C. elegans* longevity paradigm. Mech. Ageing Dev. **123:** 1229–1244.
45. HSIEH, C.C., J.H. DeFORD, K. FLURKEY, *et al.* 2002. Effects of the Pit1 mutation on the insulin signaling pathway: implications on the longevity of the long-lived Snell dwarf mouse. Mech. Ageing Dev. **123:** 1245–1255.
46. WEINDRUCH, R. & R.S. SOHAL. 1997. Caloric intake and aging. N. Engl. J. Med. **337:** 986–994.
47. ROTH, G.S., D.K. INGRAM & M.A. LANE. 1999. Calorie restriction in primates: will it work and how will we know? J. Am. Geriatr. Soc. **46:** 869–903.
48. BARZILAI, N. & G. GUPTA. 1999. Interaction between aging and syndrome X: new insights on the pathophysiology of fat distribution. Ann. NY Acad. Sci. **892:** 58–72.
49. DOMINICI, F.P., G. AROSEGUI DIAZ, A. BARTKE, *et al.* 2000. Compensatory alterations of insulin signal transduction in liver of growth hormone receptor knockout mice. J. Endocrinol. **166:** 579–590.
50. DOMINICI, F.P., S. HAUCK, D.P. ARGENTION, *et al.* 2002. Increased insulin sensitivity and upregulation of insulin receptor, insulin receptor substrate (ISR)-1, and IRS-2 in liver of Ames dwarf mice. J. Endocrinol. **173:** 81–94.
51. MATTISON, J.M., C. WRIGHT, R.T. BRONSON, *et al.* 2000. Studies of aging in Ames dwarf mice: effects of caloric restriction. J. Am. Aging Assoc. **23:** 9–16.
52. IKENO, Y., R.T. BRONSON, G.B. HUBBARD, *et al.* 2003. Delayed occurrence of fatal neoplastic diseases in Ames dwarf mice: correlation to extended longevity. J. Gerontol. A Biol. Sci. Med. Sci. **58:** B291–B296.
53. LANE, M.A., E.M. TILMONT, H. DE ANGELIS, *et al.* 2000. Short-term calorie restriction improves disease-related markers in older male rhesus monkeys (*Macaca mulatta*). Mech. Ageing Dev. **112:** 185–196.
54. MATTSON, M.P., W. DUAN, J. LEE, *et al.* 2001. Progress in the development of caloric restriction mimetic dietary supplements. J. Anti-Aging Med. **4:** 225–232.

55. MATTSON, M.P., W. DUAN & N. MASWOOD. 2002. How does the brain control life span? Ageing Res. Rev. **1:** 155–165.
56. HOLZENBERGER, M., J. DUPOND, B. DUCOS, *et al.* 2003. IGF-1 receptor regulates life span and resistance to oxidative stress in mice. Nature **421:** 182–187.
57. BLUHER, M., B.B. KAHN & C.R. KAHN. 2003. Extended longevity in mice lacking the insulin receptor in adipose tissue. Science **299:** 572–574.
58. BLUHER, M., M.D. MICHAEL, O.D. PERONI, *et al.* 2002. Adipose tissue selective insulin receptor knockout protects against obesity and obesity-related glucose intolerance. Dev. Cell **3:** 25–38.
59. SHIMOKAWA, I., Y. HIGAMI, M. UTSUYAMA, *et al.* 2002. Life span extension by reducing growth hormone insulin-growth factor-1 axis in a transgenic rat model. Am. J. Pathol. **160:** 2259–2265.
60. ANISIMOV, S.V., M.V. VOLKOVA, L.V. LENSKAYA, *et al.* 2001. Age-associated accumulation of the apolipoprotein C-III gene T-455C polymorphism C allele in a Russian population. J. Gerontol. Biol. Sci. **56:** 27–32.
61. PAOLISSO, G., A. GAMBARDELLA, S. AMMENDOLA, *et al.* 1996. Glucose tolerance and insulin action in healthy centenarians. Am. J. Physiol. **270:** 890–896.
62. BARBIERI, M., M.R. RIZZO, D. MANZELLA, *et al.* 2003. Glucose regulation and oxidative stress in healthy centenarians. Exp. Gerontol. **38:** 137–143.
63. SINGAL, P.K., R.E. BEAMISH & N.S. DHALLA. 1983. Potential oxidative pathways of catecholamine in the formation of lipid peroxides and genesis of heart disease. Adv. Exp. Biol. Med. **161:** 391–401.
64. RUIZ-TORRES, A. & M. SOARES DE MELO KIRZNER. 2002. Ageing and longevity are related to growth hormone/insulin-like growth factor-1 secretion. Gerontology **48:** 401–407.
65. ROTH, G.S., M.A. LANE, D.K. INGRAM, *et al.* 2002. Biomarkers of caloric restriction may predict longevity in humans. Science **297:** 811.
66. MUNTONI, S. 1974. Inhibition of fatty acid oxidation by biguanides: implication for metabolic physiopathology. Adv. Lipid Res. **12:** 311–377.
67. MUNTONI, S. 1999. Metformin and fatty acids. Diabetes Care. **22:** 179–180.
68. DILMAN, V.M., L.M. BERSTEIN, M.N. OSTROUMOVA, *et al.* 1982. Metabolic immunodepression and metabolic immunotherapy. Oncology **39:** 13–19.
69. DILMAN, V.M., L.M. BERSTEIN, T.P. YEVTUSHENKO, *et al.* 1988. Preliminary evidence on metabolic rehabilitation in cancer patients. Arch. Geschwulstforsch. **58:** 175–183.
70. BERSTEIN, L.M., T.P. EVTUSHENKO, E.V. TSYRLINA, *et al.* 1992. Comparative study of five- and ten-year-long results of the metabolic rehabilitation of cancer patients. *In* Neuroendocrine System, Metabolism, Immunity, and Cancer: Clinical Aspects. K.P. Hanson & V.M. Dilman, Eds.: 102–112. N. N. Petrov Research Institute of Oncology. St. Petersburg.
71. POPOVICH, I.G., M.A. ZABEZHINSKI, P.A. EGORMIN, *et al.* 2005. Insulin in aging and cancer: new antidiabetic drug Diabenol as geroprotector and anticarcinogen. Int. J. Biochem. Cell Biol. **37:** 1117–1129.
72. ANISIMOV, V.N., P.A. EGORMIN, L.M. BERSTEIN, *et al.* 2005. Metformin slows down aging and development of mammary tumors in transgenic HER-2/neu mice. Bull. Exp. Biol. Med. **139:** 961–965.
73. ANISIMOV, V.N. 1982. Effect of phenformin on life span, estrus function, and spontaneous tumor incidence in rats. Farmakol. Toksikol. **45:** 127.
74. ANISIMOV, V.N. 1987. Carcinogenesis and Aging, Vol. 2. CRC Press. Boca Raton, Florida.
75. ANISIMOV, V.N. 1980. Effect of buformin and diphenylhydantoin on life span, estrus function, and spontaneous tumor incidence in female rats. Vopr. Onkol. **26:** 42–48.
76. BAKAEV, V.V. 2002. Effect of 1-butylbiguanide hydrochloride on the longevity in the nematoda *Caenorhabditis elegans*. Biogerontology **3:** 23–24.
77. ANISIMOV, V.N. & V.M. DILMAN. 1975. The increase of hypothalamic sensitivity to the inhibition by estrogens induced by the treatment with L-DOPA, diphenylhydantoin, or phenformin in old rats. Bull. Exp. Biol. Med. **80:** 96–98.
78. DILMAN, V.M. & V.N. ANISIMOV. 1979. Hypothalamic mechanisms of aging and of specific age pathology. I. Sensitivity threshold of hypothalamo-pituitary complex to homeostatic stimuli in the reproductive system. Exp. Gerontol. **14:** 161–174.

79. HUNG, A.J., M.G. STANBURY, M. SHANABROUGH, *et al.* 2003. Estrogen, synaptic plasticity, and hypothalamic reproductive aging. Exp. Gerontol. **38:** 53–59.
80. YAGHMAIE, F., S.A. GARAN, M. MASSARO & P.S. TIMIRAS. 2003. A comparison of estrogen receptor-alpha immunoreactivity in the arcuate hypothalamus of young and middle-aged C57BL6 female mice. Exp. Gerontol. **38:** 220.
81. AWARTANI, K.A. & A.P. CHEUNG. 2002. Metformin and polycystic ovary syndrome: a literature review. J. Obstet. Gynecol. Can. **24:** 393–401.
82. NESTLER, J.E., D. STOVALL, N. AKHTHER, *et al.* 2002. Strategies for the use of insulin-sensitizing drugs to treat infertility in women with polycystic ovary syndrome. Fertil. Steril. **77:** 209–215.
83. FERNANDES, M.L., M.J. SAAD & L.A. VELLOSO. 2001. Effect of age on elements of insulin-signaling pathway in central nervous system of rats. Endocrine **16:** 227–234.
84. RICHARDS, J.S., S.C. SHARMA, A.E. FALENDER & Y.H. LO. 2002. Expression of FKHR, FKHRL1, and AFX genes in the rodent ovary: evidence for regulation by IGF-I, estrogen, and the gonadotropins. Mol. Endocrinol. **16:** 590–599.
85. McCARTY, M.F. 1994. Longevity effect of chromium picolinate—rejuvenation of hypothalamic function. Med. Hypotheses **3:** 253–265.
86. DE BENEDICTIS, G., L. CAROTENUTO, G. CARRIERI, *et al.* 1998. Gene/longevity association studies at four autosomal loci (REN, THO, PARP, SOD2). Eur. J. Human Gen. **6:** 534–541.
87. DE BENEDICTIS, G., Q. TAN, B. JEUNE, *et al.* 2001. Recent advances in human gene-longevity association studies. Mech. Ageing Dev. **122:** 909–920.
88. HUXTABLE, S.J., P.J. SAKER, L. HADDAD, *et al.* 2000. Analysis of parent-offspring trios provides evidence for linkage and association between the insulin gene and type 2 diabetes mediated exclusively through paternally transmitted class III variable number tandem repeat alleles. Diabetes **49:** 126–130.
89. WEAVER, J.U., P.G. KOPELMAN & G.A. HITMAN. 1992. Central obesity and hyperinsulinemia in women are associated with polymorphism in the 5′ flanking region of the human insulin gene. Eur. J. Clin. Invest. **22:** 265–270.
90. WATERWORTH, D.M., S.T. BENNETT, N. GHARANI, *et al.* 1997. Linkage and association of insulin gene VNTR regulatory polymorphism with polycystic ovary syndrome. Lancet **349:** 986–990.
91. TYBAIERG-HANSEN, A., I.U. GERDES, K. OVERGAARD, *et al.* 1990. Polymorphism in 5′ flanking region of human insulin gene. Relationships with atherosclerosis, lipid levels, and age in three samples from Denmark. Arteriosclerosis **10:** 372–378.
92. VINNITSKI, V.B. & V.A. IAKIMENKO. 1981. Effect of phenformin, L-DOPA, and para-chlorophenylalanine on the immunological reactivity and chemical carcinogenesis in BALB/c mice. Vopr. Onkol. **27:** 45–50.
93. DILMAN V.M., L.M. BERSTEIN, M.A. ZABEZHINSKII & V.A. ALEXANDROV. 1974. On the effect of phenformin on the induction of mammary tumors in rats. Vopr. Oncol. Vol. **20:** 94–98.
94. DILMAN, V.M., L.M. BERSTEIN, M.A. ZABEZHINSKII, *et al.* 1978. Inhibition of DMBA-induced carcinogenesis by phenformin in the mammary gland of rats. Arch. Geschwulstforsch. **48:** 1–8.
95. ANISIMOV, V.N., N.M. BELOUS, I.A. VASILYEVA & V.M. DILMAN. 1980. Inhibitory effect of phenformin on the development of mammary tumors induced by N-nitrosomethylurea in rats. Exp. Onkol. **2:** 40–43.
96. ANISIMOV, V.N., M.N. OSTROUMOVA & V.M. DILMAN. 1980. Inhibition of blastomogenic effect of 7,12-dimethylbenz(a)anthracene in female rats by buformin, diphenylhydantoin, polypeptide pineal extract, and L-DOPA. Bull. Exp. Biol. Med. **89:** 723–725.
97. ALEXANDROV, V.A., V.N. ANISIMOV, N.M. BELOUS, *et al.* 1980. The inhibition of the transplacental blastomogenic effect of nitrosomethylurea by postnatal administration of buformin to rats. Carcinogenesis **1:** 975–978.
98. BESPALOV, V.G. & V.A. ALEXANDROV. 1985. Influence of anticarcinogenic agents on the transplacental carcinogenic effect of N-nitroso-N-ethylurea. Bull. Exp. Biol. Med. **100:** 73–76.
99. ANISIMOV, V.N., K.M. POZHARISSKII & V.M. DILMAN. 1980b. Effect of phenformin on the blastomogenic action of 1,2-dimethylhydrazine in rats. Vopr. Onkol. **26:** 54–58.

100. DILMAN, V.M., B.N. SOFRONOV, V.N. ANISIMOV, *et al.* 1977. Phenformin elimination of the immunodepression caused by 1,2-dimethylhydrazine in rats. Vopr. Onkol. **23:** 50–54.
101. ANISIMOV, V.N., N.M. BELOUS & E.A. PROKUDINA. 1982. Inhibition by phenformin of the radiation carcinogenesis in female rats. Exp. Onkol. **4:** 26–29.
102. SCHNEIDER, M.B., H. MATSUZAKI, J. HARORAH, *et al.* 2001. Prevention of pancreatic cancer induction in hamsters by metformin. Gastroenterology **120:** 1263–1270.
103. DILMAN, V.M. & V.N. ANISIMOV. 1979. Potentiation of antitumor effect of cyclophosphamide and hydrazine sulfate by treatment with the antidiabetic agent, 1-phenylethylbiguanide (phenformin). Cancer Lett. **7:** 357–361.
104. WU, Y., S. YAKAR, L. ZHAO, *et al.* 2002. Circulating insulin-like growth factor-1 levels regulate colon cancer growth and metastasis. Cancer Res. **62:** 1030–1035.
105. CARACI, F., M. CHISARI, G. FRASCA, *et al.* 2003. Effects of phenformin on the proliferation of human tumor cell lines. Life Sci. **74:** 643–650.
106. BERSTEIN, L.M., J.O. KVATCHEVSKAYA, T.E. POROSHINA, *et al.* 2004. Insulin resistance, its consequences for clinical course of the disease and possibilities of correction in endometrial cancer. J. Cancer Res. Clin. Oncol. **130:** 687–693.
107. BERSTEIN, L.M. 2005. Clinical usage of hypolipidemic and antidiabetic drugs in the prevention and treatment of cancer. Cancer Lett. **224:** 203–212.
108. FACCHINI, F.S., M.H. HUMPHREYS, F. ABBASI, *et al.* 2000. Relation between insulin resistance and plasma concentrations of lipid hydroperoxides, carotenoids, and tocopherols. Am. J. Clin. Nutr. **72:** 776–779.
109. FACCHINI, F.S., N. HUA, F. ABBASI & G.M. REAVEN. 2001. Insulin resistance as a predictor of age-related diseases. J. Clin. Endocrinol. Metab. **86:** 3574–3578.
110. GARGIULO, P., D. CACCESE, P. PIGNATELLI, *et al.* 2002. Metformin decreases platelet superoxide anion production in diabetic patients. Diabetes Metab. Res. Rev. **18:** 156–159.

The Phenomenon of the Switching of Estrogen Effects and Joker Function of Glucose

Similarities and Relation to Age-Associated Pathology and Approaches to Correction

LEV M. BERSTEIN, EVGENIA V. TSYRLINA, DMITRY A. VASILYEV, TATJANA E. POROSHINA, AND IRINA G. KOVALENKO

Laboratory of Oncoendocrinology, Professor N.N. Petrov Research Institute of Oncology, St. Petersburg, 197758, Russia

ABSTRACT: Estrogens and glucose are characterized by a myriad of functions that can be reduced to a small number of principal actions. In aging there is a simultaneous increase in the prevalence of diseases connected with estrogen deficiency as well as with estrogenic excess and associated with the phenomenon of the switching of estrogen effects (PSEE). Estrogens possess hormonal and genotoxic properties. An increase in genotoxic effect (isolated or combined with a decrease in hormonal effect) can influence the course of age-associated diseases that, contrary to the situation with adaptive hypersensitivity to estrogens, may become less favorable or more aggressive. Inductors of PSEE include smoking, irradiation, and aging. Yet with "glycemic load" and the endocrine effect of glucose (the stimulation of insulin secretion), reactive oxygen species are formed in multiple sites, including adipose tissue. The ratio between hormonal and genotoxic effects reflects a "joker" function of glucose and can be conditioned by endogenous (perhaps including genetic) and exogenous factors. The shift in this glucose-associated ratio may selectively encourage some chronic non-communicable diseases. Several groups of treatments can be distinguished including alleviators of PSEE and insulin resistance syndrome (biguanides, glitazones, statins, modifiers of adipocytokines secretion, etc.) as well as other compounds aimed to optimally orchestrate the balance between endocrine and DNA-damaging effects of estrogens and glucose.

KEYWORDS: estrogen and glucose effects; age-associated pathology; correction

Age-associated hormonal deviation has attracted the attention of many investigators, including Professor Vladimir Dilman.[1,2] The importance of an approach that "lays on the surface" of the problem is that it combines an evaluation of the possible consequences as well as causes of the aging. This approach makes the study of adult and pre-geriatric endocrinology two of the fundamentals of contemporary gerontology.

In this paper, we concentrate on two seemingly distant components of endocrine milieu: estrogens and glucose. Changes in estrogen production and glucose

Address for correspondence: Prof. Lev M. Berstein, M.D., Ph.D., Chief, Lab. Oncoendocrinology, N.N. Petrov Research Institute of Oncology, Pesochny, St. Petersburg 197758, Russia. Voice: 7-812-596-8654; fax: 7-812-596-8947.

levmb@endocrin.spb.ru

Ann. N.Y. Acad. Sci. 1057: 235–246 (2005). © 2005 New York Academy of Sciences.
doi: 10.1196/annals.1356.018

tolerance/insulin sensitivity are among most frequent endocrine shifts associated with aging and age-related pathology, including cardiovascular and neurodegenerative diseases and cancer of hormone-dependent tissues.[3–5] Particular properties of estrogens and glucose can explain some specific features of human physiology and pathology and reflect well-known interplay between endogenous and exogenous/environmental factors in an implementation of ontogenetic program.[2]

TWO TYPES OF HORMONAL CARCINOGENESIS AND BIVALENCE OF ESTROGENS

The hormonal environment and actions of hormones are responsible for more than 35–40% of all newly diagnosed neoplasia in men and women in developed countries.[6] It is likely that mitogenic as well as mutagenic effects of estrogens act in concert to initiate and promote the development of cancer.[7,8] The serious evidence for such a conclusion is provided, among other observations, by recent experiments in transgenic mice. Mice overexpressing the Wnt-1 gene develop mammary tumors with high incidence within a few months after birth. These mice were crossed with estrogen receptor-alpha knockout (ERKO) mice, and ER-beta was shown to be absent in the breast tissue of such animals. The ERKO/Wnt-1 breast extracts contained picomolar amounts of the 4-catechol estrogens (CE), but not their methoxy conjugates nor the 2-catechol estrogens or their methoxy conjugates. However, the formation of CE-3,4-quinones was detected in both tumors and hyperplastic mammary tissue. To assess the effect of estrogens in the absence of ER, half of the animals were oophorectomized on day 15 after birth. Castration reduced the incidence of breast tumors in aggregate, supporting the concept that DNA-damaging metabolites of estradiol may act in concert with ER-mediated mechanisms to induce breast cancer.[8] Additionally, recent studies demonstrated the ability of estradiol to activate the human CYP1B1 (estrogen-4-hydroxylase) gene via ER-alpha, thereby providing insights into the homeostasis of estrogen metabolism as well the interaction of potential pathways of estrogen-induced carcinogenesis.[9]

The dual role of estrogen in tumor development as both a hormone stimulating cell proliferation (mitogen) and as a procarcinogen that induces genetic damage (mutagen), confirms the possibility of the existence of two types of hormonal carcinogenesis: promotional (or, as we called it earlier, physiological) and genotoxic (which might be consequent as well as separated and independent stages of the one and same process).[7,10] Therefore, the search for conditions or factors modulating the genotoxic component in total estrogenic effect is essential because these factors may influence hormonal carcinogenesis, the biological properties of developing hormone-dependent tumors,[10,11] and the course of some physiological processes (such as senescence of the reproductive system based on free radical/"autosuffocation" mechanism[11]).

THE PHENOMENON OF SWITCHING OF ESTROGEN EFFECTS

As we demonstrated previously, tobacco smoke induces phased changes in the uterotrophic action of estrogens, resulting in attenuation of the hormonal component

TABLE 1. The phenomenon of switching of estrogen effects (PSEE) and its inductors

Agent/action	Variant of PSEE		
	Complete	Incomplete	Absent
Short-term treatment with tobacco smoke			+
Long-term treatment with tobacco smoke	+		
5% Ethanol		+	
15% Ethanol	+		
γ-Irradiation (0.2 gy)		+	
γ-Irradiation (2.0 gy)			+
Aging		+	

NOTE: See text and refs. 11 and 12 for terminology and detailed explanation.

of their effect (dynamics of uterine weight and proliferation index, induction of progesterone receptors, etc.) and in an increased rate of genotoxic damage (DNA unwinding or COMET assay). We adopted the term "phenomenon of switching of estrogen effects" (PSEE) to describe these changes[11,12] and sought to identify other factors inducing this phenomenon. The accumulated experimental data allowed for the assumption that PSEE can be described as complete, when simultaneous increase of the genotoxic component and decrease of the hormonal component in estrogenic effect is evident, and as incomplete, when only isolated an increase in DNA-damaging capacity is evident. The inductors of PSEE can be classified in a similar manner. In oophorectomized rats injected with estradiol, complete PSEE inductors included long-term "smoking" and drinking of 15% ethanol (i.e., levels equal to chronic alcoholism). Incomplete inductors included consumption of alcohol at more moderate concentrations, single whole-body γ-irradiation in the lesser of two investigated doses, and aging (TABLE 1).[11,12] Thus, "natural agents" (ones that are widespread and not excessive), when combined with abundant estrogenic stimulation, appear to be more dangerous cancer risk factors because incomplete PSEE results in an increased genotoxic effect of estrogens coupled with their retained hormonal influence.[11] Of note, the combined actions tobacco smoke and estrogen replacement therapy or tobacco smoke and increased levels of circulating estrogens in menopause lead to the above normal excretion of carcinogenic and genotoxic 4-hydroxyestrogens (TABLE 2)[13] and to increased risk of breast cancer.[14]

The practical significance of PSEE is also supported by its association with several other major human non-communicable diseases (e.g., atherosclerosis, hypertension, stroke, and osteoporosis), which simultaneously increase in number and frequency with advancing age until a certain period of life, notwithstanding the different and often opposite dependence of these illnesses on estrogen and other sex steroids.[11,15] Actually, the increase of cancer incidence during menopause and andropause is surprising and in contrast to the well-known parallel decrease in es-

TABLE 2. Urinary estrogen metabolites in nonsmoking and smoking postmenopausal women receiving estrogen replacement therapy

Estrogen fraction	Nonsmokers		Smokers	
	Before ERT	After ERT	Before ERT	After ERT
E1	9.70 ± 8.60	1232.6 ± 295.2	14.46 ± 13.24	951.5 ± 181.2
E2	2.87 ± 2.14	212.6 ± 42.2	3.65 ± 2.60	235.7 ± 40.5
E3	4.38 ± 2.37	178.3 ± 75.2	3.31 ± 2.49	131.2 ± 83.2
4-OHE1	2.29 ± 1.28^a	28.8 ± 10.6	1.09 ± 0.51^a	43.7 ± 18.2
4-OHE2	0.07 ± 0.08	23.5 ± 15.6^a	0.10 ± 0.13	60.8 ± 25.3^a
2-MOE1	2.49 ± 1.57	53.4 ± 15.1	2.36 ± 0.79	51.3 ± 21.9
2-MOE2	0.37 ± 0.26	3.10 ± 0.80	0.41 ± 0.25	3.20 ± 1.10

NOTE: Data are means and standard deviations (nmol/24 h).
ABBREVIATIONS: ERT, estrogen replacement therapy; E1, estrone; E2, estradiol; E3, estriol; 4-OHE1, 4-hydroxyestrone; 4-OHE2, 4-hydroxyestradiol; 2-MOE1, 2-methoxyestrone; 2-MOE2, 2-methoxyestradiol.
^{a}Difference between smokers and nonsmokers is significant ($P < 0.05$).

trogen and androgen production. Although this peculiar "scissors or swing rule" (increase of hormone-dependent cancer morbidity and decrease in mitogenic steroid levels) may be explained in different ways, one explanation is probably related to the mechanisms associated with adaptive hypersensitivity to estrogens.[8,16] Proof of this adaptive hypersensitivity is seen in observations of MCF-7 cells deprived of estrogen for more than 1.5–2 years. In these long-term estrogen-deprived (LTED) cells, enhanced responses to estradiol develop manifested by upregulation of membrane-associated ER-alpha, MAP-kinase, and phosphatidyl-inositol 3-kinase pathways as well as other non-genomic estrogen effects. Altogether these responses render cells more sensitive to the proliferative action of the hormone—the maximal proliferation index in LTED cells is registered at 10^{-14} M estradiol as compared to 10^{-10} M in wild-type cells.[8,16]

The concept of adaptive hypersensitivity to estrogens has important clinical implications and is used, in particular, to explain the superiority of aromatase inhibitors over antiestrogens for treatment of breast cancer.[16] Meanwhile, in atypical mammary epithelial hyperplasia, considered to be a premalignant breast lesion, a special mutant version of ER with hypersensitivity to estradiol was recently revealed.[17] These and other data make plausible the assumption that adaptive hypersensitivity to estrogen is associated not only with the mechanisms of hormonal therapy but also with the pathogenesis of hormone-related cancer. In comparison to the effect of PSEE, which drives the process of hormonal carcinogenesis to acquire genotoxic/mutagenic features and a receptor-negative state, adaptive hypersensitivity to estrogens presumably creates conditions for the development of less aggressive hormone-dependent tumors with more favorable biological properties.[11] It is important to remember that depending on the situation, estrogens can exhibit prooxidant[7] or antioxidant[18] features. This should be taken into consideration when analysis goes beyond estrogens and beyond malignancy.

INSULIN RESISTANCE AND MAJOR
NONCOMMUNICABLE DISEASES

An additional and rather broad topic considers the questions: How do age-associated changes in endocrine homeostasis increase the risk of hormone-dependent cancer? How do they create situations predisposing to simultaneous development of other common and ultimately lethal chronic human illnesses? What are the similarities and differences between the hormonal and metabolic pathways leading to the formation of these pathologies? Among the hormonal shifts already mentioned, according to available data the phenomenon of hyperinsulinemia/insulin resistance attracts special attention because it represents an important part of a so-called metabolic syndrome.[19,20]

Between the ages of 30–40, decreased sensitivity of the growth-hormone secretion regulatory centers to the inhibitory influence of glucose appear and progress. This results in temporary and transitory increases in the effects of growth hormone in peripheral tissues (including skeletal muscle and fat). Consequently, this leads, before growth hormone production is finally depleted with aging, to the lower utilization of glucose and to the decrease in insulin sensitivity.[2,21] This change is additionally supported by accumulation of fat not only in adipose tissue but also ectopically in muscles and the liver.[22] This chain of the events presents one of the possible scenarios for the development of age-associated insulin resistance syndrome. Typical associated characteristics include hyperinsulinemia, impaired carbohydrate tolerance, dyslipidemia, hypertension, and abdominal obesity[23] and closely resemble the state of cancrophilia described by Dilman.[2]

Under conditions of glucose intolerance the function of the principal energetic substrate, in accordance with the Randle cycle regulations, is acquired by free fatty acids. Their excessive oxidation together with age-dependent decrease in mitochondrial function,[24] dysfunction of receptors of peroxisome proliferators activators (in particular PPARγ), and the cytokine system (TNFα, IL-6, etc.) assist in furthering the progression of insulin resistance and the formation of a cluster of other metabolic pro-pathogenic factors. Over the past decade, insulin resistance has been recognized as a chronic, low-level, inflammatory state.[20] The role of a metabolic syndrome in mechanisms leading to the decrease in cellular immunity and disturbance of the co-agulation cascade was emphasized many years ago.[2] Additionally, hyperinsulinemia modifies steroid metabolism by stimulation of gonadal steroidogenesis and increasing free estradiol and testosterone fractions due to inhibition of sex hormones-binding globulin production.[25] In concert with IGF-I system, hyperinsulinemia creates a metabolic/mitogenic platform for the amplification of cellular proliferation.[26]

Prospective cohort studies confirmed age-associated hyperinsulinemia and resistance to insulin as a major underlying abnormality promoting cardiovascular disease, in particular atherosclerosis and hypertension.[27] The role of insulin resistance as a factor preceding the development of stroke, diabetes mellitus type 2, decrease of cognitive function, and other diseases whose frequency increases with aging is well known or presents a subject for consideration.[20] In contrast to the strong association between insulin resistance and cardiovascular pathology, cohort studies of the link of loss of sensitivity to insulin to risk of cancer have received less attention and support[19,28] although the idea of a probable connection was proposed some time ago.[2]

Meanwhile, there are several reasons to consider the suggestion that there is a common character of hormonal-metabolic factors predisposing to the development of major non-communicable diseases in humans including cancer. Due to the rather strong and often opposite influence of estrogens on age-associated chronic morbidity (osteoporosis and cardiovascular and neurodegenerative diseases on the one hand, and tumors of hormone-dependent tissues on the other[15,29]), the assumption was made that insulin resistance combined with estrogen deficiency might lead to the pathology of bone tissue and the cardiovascular and central nervous systems and, conversely, insulin resistance combined with estrogenic excess might lead to hormonal carcinogenesis.[11] Another example of the dissimilarity of the involved hormone-mediated mechanisms is associated with the distinctive role of the fetal programming phenomenon: the development of cancer (breast, endometrium, colorectal, etc.) is thought to be associated with newborn macrosomy,[30,31] while coronary heart disease, hypertension, stroke, and some signs of insulin resistance in elderly might occur more frequently among people who had low birth weight.[32] Certain differences exist also in regard to the predisposing role of dyslipidemia and glucose intolerance/diabetes mellitus; the latter subject deserves special discussion.

THE "JOKER" ROLE OF GLUCOSE

Links to Adipose Tissue, Genotoxicity, and PSEE

The association of obesity, especially its visceral type, with insulin resistance and increased cancer risk[2,6,33,34] quite naturally attracts attention to adipose tissue. In recent years, the simple notion that this tissue as merely a fat store is evolving rapidly into a complex paradigm of adipose tissue as multifunctional organ with strong endocrine function. The secretory compartment of adipose tissue consists of adipocytes, fibroblasts, and several other types of cells (including cells of inflammation/immunity, whose number increases with obesity[35]). These cells secrete multiple bioactive molecules, conceptualized as adipokines or adipocytokines.[36] Aromatase (estrogen synthetase) is found in adipose tissue too, and in a certain sense this tissue is reminiscent of the ovary because they are both capable of producing steroid and peptide factors. Peptide hormones of adipose tissue (leptin, resistin, adiponectin, TNFα, PAI-1, visfatin, etc.[5]) may influence tumor growth directly as well as through the reproductive system and other mechanisms.[11,37] Their pro-blastomogenic effects may differ considerably: i.e., leptin (probably via activation of MAP-kinase) increases aromatase activity and the proliferative index in mammary cancer cell lines, while a decrease of blood adiponectin concentration is described as a prospective risk factor for breast and endometrial cancer.[36,38]

Several adipose tissue hormones (say, PAI-1 and adiponectin) are associated in an opposite way also with glucose intolerance.[36] Glucose intolerance is an important component of the routes leading to insulin resistance and, according to some observations, increases risk of hormone-dependent cancer to a higher degree than does overt diabetes.[39] Glucose represents one of the basic components of the internal milieu and possesses two principal functions: (1) endocrine (in particular, the ability to induce insulin secretion) and (2) DNA-damaging related to formation of reactive oxygen species (ROS)[40] in different tissues, including adipose tissue.[41] It has been

TABLE 3. Comparative characteristics of healthy people distinguished by the rate of glucose-induced genotoxicity (GIGT)

Group	Age (years)	BMI	Chol (mM/L)	TG (mM/L)	MDA (nM/L)	TNF (pg/mL)	Glucose (120/0)	Insulin (120/0)	CML (120/0)	CML/Ins (120/0)
With increased GIGT ($N = 6$)	48.5±6.5	26.5±2.5	5.61±0.52	1.07±0.23	4.86±0.66	10.14±2.53	0.93±0.20	3.28±1.26	2.79±0.32	1.35±0.47
Without increased GIGT ($N = 8$)	51.1±3.2	31.2±2.9	6.69±0.37	1.32±0.17	4.38±0.97	4.36±1.07	1.19±0.09	4.85±0.79	1.04±0.10	0.27±0.05
P		0.09	0.07			0.05			0.02	0.02

ABBREVIATIONS: BMI, body mass index; Chol, cholesterol; TG, triglycerides; MDA, malonic dialdehyde; TNF, tumor necrosis factor-α; Glucose 120/0, ratio of blood glucose level 120 min after peroral glucose load, 40 g/m^2 body surface, to fasting glycemia; Insulin 120/0, ratio of blood insulin level 120 min after peroral glucose load, 40 g/m^2 body surface, to fasting insulinemia; CML 120/0, chemiluminescence data, 120 min after peroral glucose load, 40 g/m^2 body surface, to chemiluminescence data in blood mononuclears after fasting; CML/Ins 120/0, the ratio of the previous two parameters.

suggested that a shift in the ratio of these functions reflects a "joker" role of glucose as an important modifier of human pathology and its aggressiveness.[11,42] Therefore, we embarked on a study to investigate an individual effect of peroral glucose challenge on serum insulin level (ELISA) and ROS by mononuclears (luminol-dependent/latex-induced chemiluminescence) in a group of healthy people 28–66 years of age. Concentrations of glucose, blood lipids, carbonylated proteins, MDA and TNF as well as the percent of mononuclears with DNA-damage (COMET assay) were determined. From the study's data, we could distinguish two groups. In one group, glucose stimulation of ROS generation by mononuclears was increased and relatively prevailed over induction of insulin secretion (state of so-called glucose-induced genotoxicity, GIGT). In the second group, no signs of GIGT were evident (TABLE 3). People in the first group had lower body mass index and blood cholesterol and higher TNF concentration. The finding indicates that if the joker function of glucose is realized in "genotoxic mode," the phenotype (and perhaps genotype) of probands may be rather different from the one discovered in glucose-induced "endocrine prevalence." Thus a specific pro-endocrine or pro-mutagenic basis for different chronic diseases or for different features of the same disease can be created (such an assumption is also supported by the data on presence of DNA damage signs in some patients with atherosclerosis[43]). Although it is well known that increased TNF serum content may be associated with insulin resistance,[4,20] glucotoxins of different origin, including alpha-dicarbonyl methylglyoxal, are considered as an inductors of TNF[44]; therefore, combination of TNF excess with glucose-induced genotoxicity seems possible, perhaps reflecting a link in the chain of further pathological reactions.

The "glucose problem" is reminiscent of the abovementioned "estrogenic problem" based on the switching of estrogen effects in the direction of genotoxic predominance with aging and under the influence of some exogenic/environmental agents, such as smoking and irradiation.[10,12] In addition to age-related development of glucose intolerance,[1,2,4] the increased glycemic load registered during past decade in many countries is probably the most important diet-induced "natural" modifier of chronic diseases and hormone-associated cancer risk.[45] Thus, both these allied events (joker, endocrine or DNA-damaging, role of glucose, and PSEE) should be viewed when discussing mechanisms of the development of major noncommunicable human diseases, including cardiovascular pathology and hormone-dependent cancer, the possible existence of their two types (with endocrine or genotoxic prevalence), and the possible measures taken for prevention.

APPROACHES TO CORRECTION AND PREVENTION

Progress in understanding the mechanisms of age-associated pathology has shown that the aims of the prevention should include the aforementioned targets but should also go beyond them. Briefly stated, there are three main principles to consider for such an approach.

(1) When planning preventive measures, it is necessary to take into account the possible "benefits-risks ratio" of these measures in relation to the development of hormone-induced tumors and the whole group of non-communicable human diseases. For example, estrogen content and production in humans must be optimal.[10,11]

(2) The existence of the two different types of hormonal carcinogenesis demands the use of not only antimitogenic but also antigenotoxic/antioxidant actions to limit the formation of DNA-damaging hormonal metabolites or to increase their inactivation.[7,46] Correction of the phenomenon of switching of estrogen effects (PSEE) presents another step in the same direction.[12] The same principles may also apply the two basic variants to major chronic diseases besides cancer.

(3) Preventive measures need not be exclusively "anti-steroidal." Measures should also be aimed at components of peptidergic signaling (Ras-MAPK-PI3-kinase-system, IGF/IGF-binding proteins, etc.)[15,16,26] and at the deviations associated with formation of insulin resistance/metabolic syndrome. Besides modification of diet and increased physical activity, treatments may include antidiabetic biguanides, thiazolidinediones, and statins as well as TNFα antagonists, members of FGF-family, inhibitors of protein thyrosine phosphatase 1B, cytokine signaling (SOCS), COX-2 and glycation (AGE's production), and removal of visceral fat. Such recommendations, partly tested in clinical conditions,[2,47–51] justify the search for the correctors of "joker" function of glucose in the case of its optimal ratio shifts into abnormal endocrine or genotoxic region.[42] Of note, the antidiabetic insulin sensitizers glitazones are high affinity ligands for peroxisome proliferator-activated receptor, PPAR-γ, whose target gene is the UDP-glucuronosyltransferase (UGT) 1A9 enzyme. In its turn, an increased UGT1A9 expression is accompanied by an enhanced catecholestrogen glucuronidation, which is an important catabolic pathway for the inactivation/elimination of these genotoxic steroid metabolites.[52] On the other hand, estrogen-related receptor alpha is an effector of PPAR-γ coactivator PGC-1alpha and regulates the expression of genes involved in oxidative phosphorylation and mitochondrial biogenesis, which are reduced in insulin resistance.[24,53] Also, silent information regulators, or sirtuins (the mammalian orthologs of the life-extending gene SIR2 in yeast and nematodes), which recently attracted attention due to their mediating role in caloric restriction–related repression of PPARγ, in reduction of white adipose tissue, and in control of gluconeogenic/glycolytic pathways,[54] may also, as histone deacetylases, contribute to the modification of estrogen receptor-alpha activity.[55] These and other examples, additionally underscoring an intrinsic resemblance of glucose/estrogen pathways, demonstrate that the approaches focusing on different targets in the final end (and in accordance with thoughts of Professor V. Dilman[1,2]) may lead to the one main target—prevention or slowing down of age-associated human pathology.

ACKNOWLEDGMENTS

Supported in part by INTAS (01-434) and RFBR grants.

REFERENCES

1. DILMAN, V.M. 1971. Age-associated elevation of hypothalamic threshold to feedback control and its role in development, aging, and disease. Lancet **1:** 1211–1219.
2. DILMAN, V.M. 1994. Development, Aging and Disease: A New Rationale for an Intervention Strategy. Harwood Academy Publishers. Chur, Switzerland.

3. SIMPSON, E.R., M. MISSO, K.N. HEWITT, *et al.* 2005. Estrogen—the good, the bad, and the unexpected. Endocr. Rev. **26:** 322–330.
4. FERRANNINI, E., A. GASTALDELLI, Y. MIYAZAKI, *et al.* 2005. β-cell function in subjects spanning the range from normal glucose tolerance to overt diabetes: a new analysis. J. Clin. Endocrinol. Metabol. **90:** 493–500.
5. LAZAR, M.A. 2005. How obesity causes diabetes: not a tall tale. Science **307:** 373–375.
6. HENDERSON B.E. & H.S. FEIGELSON. 2000. Hormonal carcinogenesis. Carcinogenesis **21:** 427–433.
7. LIEHR, J.G. 1997. Dual role of oestrogens as hormones and pro-carcinogens: tumour initiation by metabolic activation of oestrogens. Eur. J. Cancer Prev. **6:** 3–10.
8. SANTEN, R.J. 2003. Endocrine-responsive cancer. *In* Williams textbook of endocrinology. P.R.Larsen *et al.*, Eds.: 1797–1833. W.B. Saunders. Philadelphia.
9. TSUCHIYA, Y., M. NAKAJIMA, S. KYO, *et al.* 2004. Human CYP1B1 is regulated by estradiol via estrogen receptor. Cancer Res. **64:** 3119–3125.
10. BERSTEIN, L.M. 2000. Hormonal carcinogenesis. Nauka Publishers. St. Petersburg.
11. BERSTEIN, L.M. 2004. Oncoendocrinology: Traditions, Present Situation and Perspectives. Nauka Publishers. St. Petersburg.
12. BERSTEIN, L., E. TSYRLINA, T. POROSHINA, *et al.* 2002. Switching (overtargeting) of estrogen effects and its potential role in hormonal carcinogenesis. Neoplasma **49:** 21–25.
13. BERSTEIN, L.M., E.V. TSYRLINA, O.S. KOLESNIK, *et al.* 2000. Catecholestrogens excretion in smoking and nonsmoking postmenopausal women receiving estrogen replacement therapy. J. Steroid Biochem. Mol. Biol. **72:** 143–147.
14. MANJER, J., R. JOHANSSON & P. LENNER. 2004. Smoking is associated with postmenopausal breast cancer in women with high levels of estrogen. Int. J. Cancer **112:** 324–328.
15. RAZANDI, M., P. OH, A. PEDRAM, *et al.* 2002. Estrogen receptors associate with and regulate the production of caveolin: implications for signaling, cellular actions, and human diseases. Mol. Endocrinol. **16:** 100–115.
16. SANTEN, R.J., R.X. SONG, Z. ZHANG, *et al.* 2003. Adaptive hypersensitivity to estrogen: mechanisms for superiority of aromatase inhibitors over selective estrogen receptor modulators for breast cancer treatment and prevention. Endocr. Relat. Cancer **10:** 111–130.
17. FUQUA, S.A.W., C. WILTSCHKE, Q.X. ZHANG, *et al.* 2000. A hypersensitive estrogen receptor-a mutation in premalignant breast lesions. Cancer Res. **60:** 4026–4029.
18. BORRAS, C., J. GAMBINI, M.C. GOMEZ-CABRERA, *et al.* 2005. 17beta-oestradiol upregulates longevity-related, antioxidant enzyme expression via the ERK1 and ERK2[MAPK]/NFkappaB cascade. Aging Cell **4:** 113–118.
19. FACCHINI, F.S., N. HUA, F. ABBASI & G.M. REAVEN. 2001. Insulin resistance as a predictor of age-related diseases. J. Clin. Endocrinol. Metab. **86:** 3574–3578.
20. FERNANDEZ-REAL, J.M. & W. RICART. 2003. Insulin resistance and chronic cardiovascular inflammatory syndrome. Endocrine Rev. **24:** 278–301.
21. VELDHUIS, J.D., J.T. PATRIE, K. FRICK, *et al.* 2004. Sustained growth hormone (GH) and insulin-like growth factor I responses to prolonged high-dose twice-daily GH-releasing hormone stimulation in middle-aged and older men. J. Clin. Endocrinol. Metab. **89:** 6325–6330.
22. YKI-JARVINEN, H. 2002. Ectopic fat accumulation: an important cause of insulin resistance. J. R. Soc. Med. **95:** 39–45.
23. REAVEN, G.M. 2003. The insulin resistance syndrome. Curr. Atheroscler. Rep. **5:** 364–371.
24. LOWELL, B.B.& G.I. SHULMAN. 2005. Mitochondrial dysfunction and type 2 diabetes. Science **307:** 384–387.
25. PORETSKY, L., D. SETO-YOUNG, A. SHRESTHA, *et al.* 2001. Insulin action pathway(s) in the human ovary. J. Clin. Endocrinol. Metab. **86:** 3115–3119.
26. LEROITH D. & C.T. ROBERTS. 2003. The insulin-like growth factor system and cancer. Cancer Lett. **195:** 127–137.
27. GINSBERG, H.N. 2000. Insulin resistance and cardiovascular disease. J. Clin. Invest. **106:** 453–458.

28. WEI, E.K., J. MA, M.N. POLLAK, *et al.* 2005. A prospective study of C-peptide, insulin-like growth factor-I, insulin-like growth factor–binding protein-1, and the risk of colorectal cancer in women. Cancer Epidemiol. Biomarkers Prev. **14:** 850–855.
29. DREYER, L. & J.H. OLSEN. 1999. Risk for nonsmoking-related cancer in atherosclerotic patients. Cancer Epidemiol. Biomarkers Prev. **8:** 915–918.
30. BERSTEIN, L.M. 1988. Newborn macrosomy and cancer. Adv. Cancer Res. **50:** 231–278.
31. MICHELS, K., D. TRICHOPOULOS, J.M. ROBINS, *et al.* 1996. Birthweight as a risk factor for breast cancer. Lancet **348:** 1542–1546.
32. BARKER, D.J.P. 1998. Mothers, babies, and health in later life. Churchill Livingstone. Edinburgh-London.
33. BERSTEIN, L.M. 1997. Macrosomy, obesity, and cancer. Nova Science. New York.
34. CALLE, E.E. & M.J. THUN. 2004. Obesity and cancer. Oncogene **23:** 6365–6378.
35. WEISBERG, S.P., D. MCCANN, D. DESAI, *et al.* 2003. Obesity is associated with macrophage accumulation in adipose tissue. J. Clin. Invest. **112:** 1796–1808.
36. MATSUZAWA, Y. 2005. Adipocytokines: emerging therapeutic targets. Curr. Atheroscler. Rep. **7:** 58–62.
37. BERSTEIN, L.M. 2005. Hormones of adipose tissue (adipocytokines): ontogenetic and oncologic aspects. Adv. Gerontol. **16:** 51–64.
38. MANTZOROS, C., E. PETRIDOU, N. DESSYPRIS, *et al.* 2004. Adiponectin and breast cancer risk. J. Clin. Endocrinol. Metab. **89:** 1102–1107.
39. SAYDAH, S.H., C.M. LORIA, M.S. EBERHARDT & F.L. BRANCATI. 2003. Abnormal glucose tolerance and the risk of cancer death in the United States. Am. J. Epidemiol. **157:** 1092–1100.
40. MOHANTY, P., W. HAMOUDA, R. GARG, *et al.* 2000. Glucose challenge stimulates reactive oxygen species (ROS) generation by leukocytes. J. Clin. Endocrinol. Metabol. **85:** 2970–2973.
41. LIN, Y., A.H. BERG, P. IYENGAR, *et al.* 2005. The hyperglycemia-induced inflammatory response in adipocytes: the role of reactive oxygen species. J. Biol. Chem. **280:** 4617–4626.
42. BERHSTEIN, L.M. 2005. The joker function of glucose in the development of main non-communicable human diseases. Vestn. Russian Acad. Med. Sci. **2:** 48–51.
43. ANDREASSI, M.G. & N. BOTTO. 2003. DNA damage as a new emerging risk factor in atherosclerosis. Trends Cardiovasc. Med. **13:** 270–275.
44. VLASSARA, H., W. CAI, J. CRANDALL, *et al.* 2002. Inflammatory mediators are induced by dietary glycotoxins, a major risk factor for diabetic angiopathy. Proc. Natl. Acad. Sci. USA **99:** 15596–15601.
45. BRAND-MILLER, J.C. 2003. Glycemic load and chronic disease. Nutr. Rev. **61:** 49–55.
46. BIANCO, N.R., G. PERRY, M.A. SMITH, *et al.* 2003. Functional implications of antiestrogen induction of quinone reductase: inhibition of estrogen-induced deoxyribonucleic acid damage. Mol. Endocrinol. **17:** 1344–1355.
47. CAULEY, J.A., J.M. ZMUDA, L.Y. LUI, *et al.* 2003. Lipid-lowering drug use and breast cancer in older women: a prospective study. J. Womens Health **12:** 749–756.
48. BERSTEIN, L.M. 2005. Clinical usage of hypolipidemic and antidiabetic drugs in the prevention and treatment of cancer. Cancer Lett. **224:** 203–212.
49. YAZDANI-BIUKI, B., H. STELZL, H.P. BREZINSCHEK, *et al.* 2004. Improvement of insulin sensitivity in insulin resistant subjects during prolonged treatment with the anti-TNF-alpha antibody infliximab. Eur. J. Clin. Invest. **34:** 641–642.
50. KHARITONENKOV, A., T.L. SHIYANOVA, A. KOESTER, *et al.* 2005. FGF-21 as a novel metabolic regulator. J. Clin. Invest. **115:** 1627–1635.
51. GABRIELY, I., X.H. MA, X.M. YANG, *et al.* 2002. Removal of visceral fat prevents insulin resistance and glucose intolerance of aging: an adipokine-mediated process? Diabetes **51:** 2951–2958.
52. BARBIER, O., L. VILLENEUVE, V. BOCHER, *et al.* 2003. The UDP-glucuronosyltransferase 1A9 enzyme is a peroxisome proliferator-activated receptor α and γ target gene. J. Biol. Chem. **278:** 13975–13983.
53. SCHREIBER, S.N., R. EMTER, M.B. HOCK, *et al.* 2004. The estrogen-related receptor alpha (ERRalpha) functions in PPARgamma coactivator 1alpha (PGC-1alpha)-induced mitochondrial biogenesis. Proc. Natl. Acad. Sci. USA **101:** 6472–6477.

54. PICARD, F. & L.GUARENTE. 2005. Molecular links between aging and adipose tissue. Int. J. Obes. Relat. Metab. Disord. **29**(Suppl 1): 36–39.
55. METIVIER, R., C.Y. LIN, S. DENGER, *et al.* 2005. Multiple mechanisms induce transcriptional silencing of a subset of genes, including oestrogen receptor alpha, in response to deacetylase inhibition by valproic acid and trichostatin. Oncogene Advance online publication, 2 May 2005; doi:10.1038/sj.onc.1208662.

Immunomodulation by Immunopeptides and Autoantibodies in Aging, Autoimmunity, and Infection

J. J. MARCHALONIS,[a] S. F. SCHLUTER,[a] R. T. SEPULVEDA,[a] R. R. WATSON,[b] AND D. F. LARSON[c]

[a]Department of Microbiology and Immunology, College of Medicine, University of Arizona, Tucson, Arizona 85724, USA

[b]Department of Health Promotion Sciences, Mel and Enid Zuckerman College of Public Health, Tucson, Arizona 85724, USA

[c]Carver Heart Center and Department of Surgery, College of Medicine, University of Arizona, Tucson, Arizona 85724, USA

ABSTRACT: The operation of the immune system is a complex orchestration of specific self and non-self–recognition capacities mediated by cells of the innate system acting in coordination with T and B lymphocytes in a series of processes modulated by cytokines. We provide evidence for a natural immunomodulatory system involving autoantibodies directed against a controlling segment of T cell receptor Vβ chains that downregulate production of stimulatory cytokines balanced by the peptides which in turn upregulate inflammatory activities mediated by TH1-type helper cells. TCR Vβ–derived peptides effective in retrovirally induced immunosupression could also reverse the effects of immunosenescence in aged mice by restoring the balance of TH1- and TH2-type immunity and the resistance of the animals to cardiac pathology caused by infection with coxsackievirus. An unexpected finding was an adaptive role of the T cells from peptide-treated mice in remodeling damaged hearts by increasing net collagen synthesis by cardiac fibroblasts.

KEYWORDS: T cell receptors; TH1; TH2; cytokines; retrovirus; coxsackievirus; immunosenescence; TCR Vβ; cardiopathology

INTRODUCTION

Suppression of many aspects of mammalian immunity, particularly activities of TH1-type helper T cells, is a consequence of retroviral infection[1–4] or the normal processes of aging.[5–7] However, this activity can be elevated in T cell–mediated autodestructive diseases.[8,9] Here we address the antagonistic roles of spontaneously arising autoantibodies to T cell receptor (TCR) variable domains[10,11] and TCR Vβ–derived

Address for correspondence: J. J. Marchalonis, Department of Microbiology and Immunology, College of Medicine, University of Arizona, 1501 N. Campbell Avenue, Tucson 85724, AZ. Voice: 520-626-6065; fax: 520-626-2100.
marchjj@email.arizona.edu

Ann. N.Y. Acad. Sci. 1057: 247–259 (2005). © 2005 New York Academy of Sciences.
doi: 10.1196/annals.1356.019

peptide epitopes[12,13] in maintaining the balance between TH1- and TH2-type immunities necessary for effective overall immunity.[14] We then illustrate the capacity of certain TCR VβCDR1 peptides to reverse negative effects of immunosenescence on the functional balance between TH1 and TH2 helper T cell subsets and on the restoration of resistance to cardiopathology following infection with coxsackievirus B in C57Bl/6 mice.[15]

NATURAL ANTIBODIES TO T CELL RECEPTOR VARIABLE DOMAINS

In studies designed to define and analyze the natural antibody repertoire common to vertebrate species, we investigated natural antibodies of man,[16–18] mouse,[19] and sharks.[20] Sharks are the lowest vertebrates to possess the genes essential for mounting the adaptive or combinatorial immune response[21] directed against a wide set of antigens including those involved in homeostasis, immunoregulation, and real-time response to infectious agents. Most notably, all three species possess spontaneously arising—that is, generated in the absence of purposeful immunization with the antigenic moieties—antibodies to shared idiotopes defined by the CDR1 segment of T cell receptor beta chain variable domains. The relative amounts and isotypes of these autoantibodies varied with physiological status, including aging[17,22] and pregnancy[23] as well as with allografts,[24] retroviral infections,[9,25,26] and autoimmune diseases—particularly rheumatoid arthritis and systemic lupus erythematosus.[11,16,17,22,27] On the basis of our findings in studies of rheumatoid arthritis and retroviral infections of humans and mice, we proposed that these antibodies directed against the VβCDR1 segments were part of an immunoregulatory process in which the body strove to downregulate T cells of an autodestructive character as defined by the overexpression of T cell subsets with varying particular Vβ gene products.[11] Studies by Jambeau and colleagues[28,29]—which reported that T cells of myasthenia gravis patients of the MHC class II DR3 haplotype showed a preferred restriction to Vβ5.1 and the levels of spontaneously arising autoantibodies to Vβ5.1 epitopes were inversely correlated with the severity of the disease—are also consistent with this hypothesis. Although these workers found autoantibodies to the CDR1 epitope, they found higher levels of autoantibodies to the CDR2 segment, and followed these in their investigations. Both the CDR1 and CDR2 segments are encoded by the Vβ gene and do not require rearrangement for their expression. In a rat autoimmunity model, Hashim and colleagues[30] reported that autoantibodies to the CDR2 segment of TCR Vβ8.2 arose spontaneously in animals injected with myelin basic protein in the generation of experimental allergic encephalomyelitis. Spontaneously arising T cell reactivity was also directed against the CDR2 peptide as well as to peptides corresponding to the third framework (FR3) and the CDR1 epitope.[31] It is noteworthy that naturally occuring human autoantibodies to VβCDr1 and FR3 epitopes were found in pooled normal IgG (IVIG) and in sera of patients with rheumatoid arthritis and systemic lupus erythematosus, although there was negligible reactivity to CDR2-associated peptides.[16]

Robey and colleagues determined that human monoclonals isolated from synovial and peripheral blood cells of rheumatoid arthritis patients reacted with the Vβ epitopes on the surface of D10-11 ovalbumin-specific T cells *in vitro*. This binding inhibited the production of the TH1-type cytokine IL2 following antigen stimulation of the cells.[10] Additional evidence for the capacity of natural antibodies to TCR Vβ CDR1 epitopes to interact with and influence the function of antigen-specific T cells came from experiments using immune-affinity purified human IgG (from IVIG) and murine monoclonal antibodies derived from splenic B cells of retrovirally infected mice. Cytotoxic murine T cells specific for the breast cancer–derived tumor EMT6 expressed predominantly Vβ1 or Vβ8 receptors.[32] Human natural antibodies specific for CDR1 peptide–defined epitopes of these Vβ gene products were isolated by immune affinity chromatography from IVIG and found to inhibit killing of the tumor cells *in vitro*. Virtually complete inhibition occurred when both purified antibody subsets were added simultaneously.[11] The murine monoclonal autoantibodies were able to synergize selectively with T cell superantigens,[33] thereby indicating binding to the combining site region of the TCR and modulating the signal given by the superantigen. The production and characterization of the monoclonal anti-TCRs have been described in detail elsewhere.[34,35] These results indicate that a likely function of these spontaneously arising anti TCRVβ antibodies is to downregulate TH1 type immunity.

IMMUNOMODULATION BY TCR Vβ–DERIVED PEPTIDES

We carried out experiments designed to determine whether the peptides themselves had any effect on the diminished immune system of C57Bl/6 mice infected with the LP-BM5 strain of murine leukemia virus. This infection produces a condition termed mouse AIDS (MAIDS), characterized by suppression of TH1-type immunity, enhancement of TH2 immunity, and an overall pattern of acquired immunodeficiency.[12] It was found (TABLE 1) that peptides corresponding to the CDR1 segment of human Vβ8.1 (now termed BV12s3) and Vβ5.2 reversed many of these negative effects induced by the retroviral infection, particularly restoring the balance between TH1 and TH2 activity, by upregulating TH1-type cytokines, such as IL-2 and INFγ, and downregulating TH2 cytokines, including IL4 and IL6.[12,13, 36–38] FIGURE 1 shows the sequences of the human peptides, their closest murine homologues, and a peptide derived from the CDR1 segment of the human lambda light chain MCG that served as a negative control in functional studies. Furthermore, the peptide-treated immunodeficient mice regained the capacity to resist infection by protozoan parasites such as *Cryptosporidium pavum*[38] and to regain resistance to the generation of cardiac pathology induced by secondary infection with coxsackievirus B3.[39] Healthy C57Bl/6 mice are not susceptible to infection by *C. parvum* or to the development of cardiac pathology following infection with coxsackie virus. How-

HuVβ8.1	CKPISGHNSLFWYRQT	biological activity
HuVβ5.2	CSPKSGTDTVSWYQQA	biological activity
MuBV13S1	CEPVSGHNDLFWYRQT	murine ortholog
MuBV12S1	CDPESGHDTLYWYQQP	murine ortholog
Mcg-CDR1	CTGTSSDVGGYNYVSWY	negative control

FIGURE 1. Comparative alignment of human and murine Vβ CDR1 segments illustrating homology between the two sets. The CDR1 segment of the human Vλ5 light chain is also shown. It was frequently used as a negative control in functional studies of peptide administration to MAIDS or aged mice.

TABLE 1. Summary of effects of treatment with pep Vβ8.1 CDR1 on immunological responses of C57Bl/6 mice infected with LP-BM5 strain of murine retrovirus (MAIDS)

Parameter	Murine AIDS	Effect of pep Vβ8.1
T cell mitogenesis	↓	↑
B cell miogenesis	↓	↑
NK cell activity	↓	↑
IL-2	↓	↑
INF-γ	↓	↑
INF-α	↑	↓
IL-4	↑	↓
IL-5	↑	↓
IL-6	↑	↓
IL-10	↑	↓
Ig synthesis	↑	↓
Splenomegaly	↑	↓
Resistance to parasites	↓	↑

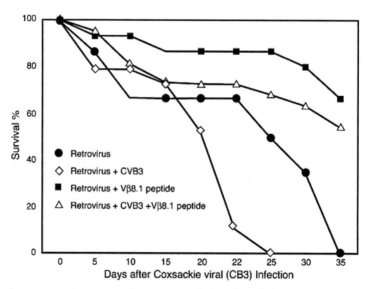

FIGURE 2. Survival curves for retrovirus-infected mice (LP-BM5) co-infected with coxsackie virus CVB3 (at 3 months following LP-BM5 infection) compared with animals given retrovirus alone, retrovirus plus TCR Vβ8.1 peptide, and retrovirus plus CVB3 plus Vβ8.1 peptide. Vβ8.1 peptide was administered as two injections within two weeks of the original retrovirus injection. Twenty mice were used per group. In addition, a control group of 10 non–retrovirally infected animals were followed during the course of the experiment. All survived the complete course of the trial (data not shown).

ever, immunosuppression either by retroviral infection or as a consequence of normal aging renders them susceptible to both deleterious consequences.

FIGURE 2 illustrates the enhancing effect of the Vβ8.1 peptide on survival of mice infected with the LP-BM5 retrovirus alone as well as those infected with coxsackievirus B3 in addition. Mice were infected with LP-BM5. The peptide-treated groups were given two intraperitoneal injections of peptide (200 µg per injection) within two weeks following the experimental infection. Sixteen animals were used per group. At three months following the LP-BM5 injection, two groups were infected with coxsackievirus. All of the mice given retrovirus and coxsackievirus, but not TCR Vβ8.1 peptide, died of cardiac failure by day 25. Unprotected animals given retrovirus alone survived approximately an additional 10 days. By contrast, approximately 60% of LP-BM5–infected mice given coxsackievirus plus TCR peptide were alive at day 35. More than 70% of peptide-treated LP-BM5–infected mice that were not given coxsackievirus survived until the end of the experiment at 35 days following initiation of the coxsackievirus infection. Coxsackievirus B–induced cardiopathology in aged mice will be considered below.

Analysis of the effects of the anti–TCR Vβ autoantibodies and the peptides themselves on the immune system, led us to propose the immunomodulatory scheme illustrated in FIGURE 3. Peptides generated from a control segment of endogenous T cell receptors upregulate TH1 immunity, and this is balanced by spontaneously arising autoantibodies to that segment. We consider the CDR1 loop, particularly of human Vβ8.1 (BV12s3), to be a "control segment" because it is the least variable of

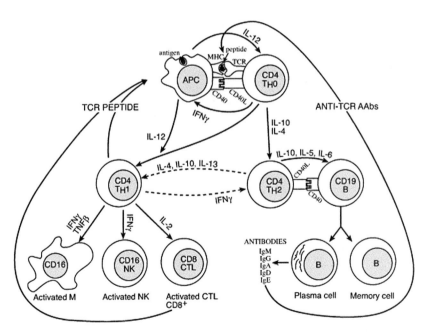

FIGURE 3. Scheme for the modulation of the immune response by TCR-derived peptides that upregulate TH1-type immunity and spontaneously arising anti-TCR Vβ autoantibodies that suppress this activity.

TCR Vβ hypervariable segments and shows 50% or more identity in comparison among the human Vβ chains. We were fortunate to be able to carry out functional studies *in vitro* and *in vivo* with mice using human peptides and antibodies because of the high degree of homology between T cell receptor Vβ segments of man and mouse.

APPLICATION TO IMMUNOSENESCENCE

Analysis of natural autoantibodies (NAABs) of humans ranging from approximately 20 to 90 years of age gave results consistent with general expectations that some autoantibodies would increase with age.[40,41] We found in particular that anti-TCR autoantibodies of the IgM isotype tended to decline with age, a result particularly noticeable after age 50, whereas autoantibodies of comparable specificity in the IgG isotype tended to increase in later years.[17] This phenomenon is illustrated in FIGURE 4 for spontaneously arising NAABs of healthy young (Y) individuals 20–40 years of age as compared with healthy elderly (E) individuals more than 70 years of age. This summary plot gives the frequency of positive sera per group to facilitate comparison. Statistical analysis of the original quantitative data using the Wilcoxon test supported the decrease in IgM NAABs with age (e.g., $P < 0.001$ for activity against pepB3) and the increase in IgG activity with aging (e.g., $P < 0.02$ for the pepVβ2.1 comparison). The fact that both groups had active immune systems is shown by their IgG antibodies to the common environmental antigen ovalbumin. Parallel results to those found to CDR1 peptides were also observed (not shown) with certain peptide epitopes corresponding to band 3 peptides associated[17] with the senescent cell antigen.[42]

The findings with the murine retrovirally immunosuppressed model suggested that comparable results may occur in the situation where immunosuppression is a natural consequence of aging. The thymus involutes substantially following puberty in humans.[43] In mice, the total number of thymocytes drops by more than 80% by 16 months of age.[44] This decline is associated with diminished lymphocyte prolifer-

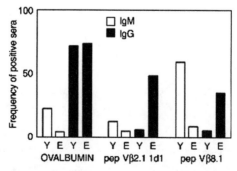

FIGURE 4. Comparison of IgM and IgG responses of healthy young (Y) and elderly (E) individuals to ovalbumin and to synthetic TCR CDR1 region peptides Vβ2.1 andVβ8.1. The data are given as frequency of positive set per group. The young group consisted of 28 individuals and the elderly group of 29. (Based on data of Marchalonis and colleagues.[17])

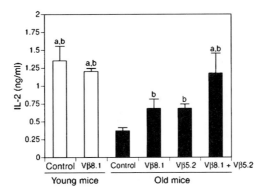

FIGURE 5. Effect of TCR peptide injection on IL-2 production by concanavalin A–stimulated splenocytes *in vitro* from young (4 weeks of age) and old (16 weeks of age) C57/B16 mice. Assays were performed in triplicate with data presented as mean ± standard deviation ($N = 6$). The letters indicate significant differences at $P < 0.05$: (**a**) compared with the old-mice group (with the exception of the group given both pep Vβ8.1 and pep Vβ5.2) and (**b**) compared with the control old-mouse group. (Based on the work of Liang and colleagues.[14])

ative capacity, decreased production of TH1 cytokines, and increased level of TH2 cytokines as well as increased levels of certain autoantibodies.[7,45] In addition, aged C57 BL/6 mice (16 months or older) resemble retrovirally immunosuppressed animals in becoming susceptibile to cardiopathology following infection with coxsackie virus B3.[15] FIGURE 5 illustrates a comparison in IL-2 production by concanavalin-A–stimulated splenocytes obtained from 4-week-old C57Bl/6 mice compared with comparable cells from animals 16 months of age. *In vitro* production of IL-2 by cells in the older animals was significantly ($P < 0.05$) lower than that produced by splenocytes from the young animals. Injection of the mice with Vβ 8.1 and Vβ 5.2 TCR–derived peptides, either individually, or in combination substantially increased the production of IL-2 by the mitogen-stimulated splenocytes of the older mice. FIGURE 6 illustrates the opposite situation, where the TH-2-type cytokine IL-4 is substantially increased in older animals relative to the young mice. However, this level is significantly decreased by the TCR Vβ–derived peptides. Control peptides, such as the homologue prepared from the CDR1 segment of human λ myeloma protein MCG, had no effect, and results were indistinguishable from those obtained when the animals were merely given saline instead of peptide solutions.

Neither young nor mature healthy C57BL\6 mice (3–10 months old) develop cardiopathology when challenged with coxsackievirus B3 (CVB3). To determine whether treatment with TCR Vβ 8.1 peptide could protect aged mice from the development of cardiopathology following CVB3 virus infection, we infected mice at 18 months of age with CVB3, sacrificing them at 12 days following infection to assess the extent of mononuclear cell infiltration and necrosis by histopathological examination of fixed, eosin-stained heart tissue. The results of this analysis are given in FIGURE 7. The severity of histopathology is scored as 1+, <10% heart tissue affected; 2+, moderate, with 10–25% tissue affected; and 3+, greater than 25% of the tissue showing evidence of necrosis and infiltration. Aged CVB3-infected mice

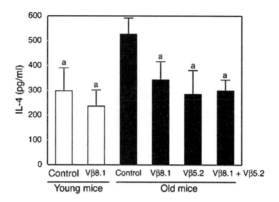

FIGURE 6. Effect of TCR peptide injection on IL-4 production *in vitro* by concanavalin A–stimulated splenocytes of young and old C57/ Bl6 mice. Values are mean ± standard deviation of triplicate assays on cells from 6 animals. (**a**) Significant difference at $P < 0.05$ by comparison with the old control group. (Based upon work by Liang and colleagues.[14])

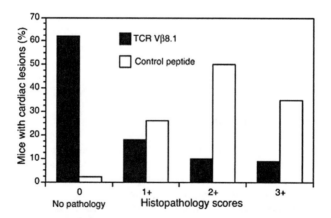

FIGURE 7. Histopathologic scores of aged C57/Bl6 mice infected with coxsackievirus B3 and treated with TCR peptide Vβ8.1 or peptide control. $N = 12$ samples per group. Pathologic score: 0, no lesions; 1+, mild multifocal nonsuppurative epicarditis to mild multifocal nonsuppurative myocarditis; 2+, mild focal to multifocal nonsuppurative myocarditis with myocardiocyte degeneration and necrosis; 3+, moderate focal to multifocal nonsuppurative myocarditis with myocardiocyte degeneration and necrosis; 4+, severe multifocal nonsuppurative myocarditis with myocardiocyte degeneration and necrosis. Mild damage is considered as <10% of heart tissue affected, moderate as 10–25%, and severe as >25%. (Based upon data of Sepulveda and colleagues.[15])

TABLE 2. Summary of effects of TCR peptide administration on parameters showing immunosenescence in aged mice

Cytokine or test	Old	Old + TCR peptide
B cell proliferation	↓	↑
T cell proliferation	↓	↑
IL-2	↓	↑
TNF-α	↑	↓
IL-6	↑	↓
IL-4	↑	↓

treated with the control peptide showed cardiopathology scores of 1+ (25%), 2+ (50%), and 3+ (25%). However, when identically infected mice were treated with TCR Vβ 8.1 peptide, 62% of the group showed no histopathology, with 18% showing 1+ scores, 10% showing 2+ scores, and only 9% showing 3+ scores. As would be expected from the enhanced TH1 activity of the aged mice following administration of Vβ 8.1 peptide, the burden of viruses infecting the heart, as judged by cardiac viral titer, was significantly diminished ($P<0.05$), with a decrease of approximately 80%.

The results with the aged mice documented the relative lack of histopathology in hearts of coxsackie-infected mice given the TCR Vβ peptide. Examples of phenomena correlated with immunosenescence and the effects of administration of TCR Vβ peptides are summarized in TABLE 2.

CONTRIBUTION OF T CELLS TO REMODELING HEARTS DAMAGED BY CVB INFECTION

Additional studies were carried out with the mice that were retrovirally immunosuppressed and given the TCR Vβ 8.1 peptide prior to CVB3 infection. They showed a restoration of cardiac function and collagen remodeling of the cardiac tissue.[46] Left ventricular stiffness was significantly increased in the treated mice, whereas it was decreased in those given retrovirus in the absence of TCR-derived peptide. Furthermore, the latter group also showed significantly increased end-diastolic and end-systolic volumes indicative of decreased functional pumping capacity. The results with the TCR peptide–treated animals indicative of a collagen remodeling within the heart were unexpected and surprising. In order to address possible explanations for the phenomomen, we performed experiments to determine whether splenic lymphocytes taken from uninfected animals, LP-BM5–infected animals, or LP-BM5–infected animals given a mixture of Vβ 8.1 and 5.2 peptides could interact *in vitro* with isolated cardiac fibroblasts to produce the collagen needed to remodel the coxsackievirus-damaged heart tissue. FIGURE 8 shows the effects of co-culturing these lymphocytes *in vitro* with cardiac fibroblasts. Lymphocytes from 3 mice in each group were pooled and added to the cardiac fibroblast cultures at a concentration of 10^5/mL. After 48 h of co-incubation the lymphocytes were removed and the cardiac fibroblast

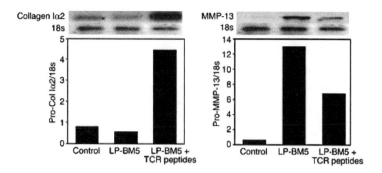

FIGURE 8. Modulation by lymphocytes of collagen (Iα2) and matrix metalloproteinase (MMP-13) synthesis by cardiac fibroblasts following *in vitro* co-culture. Splenic lympo-
cytes were harvested from the control, LP-BM5 only, and combined LP-BM5 plus TCR
peptide mice 12 weeks after treatment and co-cultured with naïve primary cardiac fibroblasts. Lymphocytes from three mice in each group were pooled and added to the cardiac
fibroblast cultures at a concentration of 5×10^5/mL. After 48 h of co-incubation, the lymphocytes were removed and the cardiac fibroblast RNA analyzed for pro-collagen Iα2 and
pro-MMP-13 with RT-PCR. The graphed data are the ratio of the candidate gene versus 18s
RNA.

mRNA analyzed for pro-collagen 1 α2 and pro-matrix metalloproteinases (MMP)-
13 using RT-polymerase chain reaction. The graphed data are the ratio of the candidate gene versus 18s RNA. LP-BM5 treatment alone caused a 30% decrease in synthesis of pro-collagen 1α2 and a 13-fold increase in pro-MMP-13 cardiac fibroblast
gene expression. However, the addition of the TCR peptides to the LP-BM 5 groups
resulted in a sixfold increase of pro-collagen and a twofold decrease in pro-MMP-13
gene expression. Thus, the effect of the treatment with the T cell peptides is to increase net collagen synthesis and decrease the level of matrix metalloproteinase that
would degrade the collagen. Cross-linking of the collagen to impart stiffness was also
increased. These results were confirmed for collagen by Western blot analysis showing an actual increase in the collagen as antigenic protein. The MMP results were
supported by assay of enzymatic activity. Supernatants isolated from *in vitro* cultured
lymphocytes of TCR peptide–treated animals likewise showed increased production
of collagen and decreased expression of MMP activity. Experiments are in progress
to determine what specific T cell–produced factors or cytokines are involved.

CONCLUSIONS AND DISCUSSION

The data presented here support the conclusion that the broad decline in TH1
function in aging can be reversed, at lease partially, by immunomodulatory peptides
derived from a "control region," the CDR1 segment, of TCRVβ. This process parallels the immunodulatory effects of the TCR peptides in murine retroviral infection,
where they restore the balance between TH1 and TH2 activities. We stress that this
is a restoration of balance required for normal immune function because we have not
observed overshoots in levels of TH1 cytokines relative to those produced by healthy
immunologically normal C57BL/6 mice. The increases we observed in the immuno-

logically deficient animals treated with appropriate TCR peptides approached the normal levels asymptotically. This strain of mouse generally gives a strong TH1-type response to microbial infection and it is necessary to use animals rendered immunodeficient by either retroviral infection or advanced age to demonstrate the restorative effects of the TCR Vβ–derived immunomodulatory peptides. The peptides found to be effective are derived from human Vβ sequence, but are extremely similar to homologues occurring in orthologous murine Vβ sequences. We have not observed any adverse effects following administration of high peptide doses to mice, and a Phase I safety trial carried out by Allergene, Inc. indicates that the Vβ8.1 peptide should be safe for human use at the doses tested. Efficacy trials have not yet been performed for human subjects (personal communication, W. Friedman, Allergene).

Spontaneously arising autoantibodies to the CDR1 segment downregulate TH1 activity, suggesting that a novel immunoregulatory mechanism acts in autoimmunity and normal immunization. Although we have focused on the VβCDR1 peptide in our studies, other workers have found spontaneously arising autoantibodies to the VβCDR2 segment in human autoimmune disease[28] and in rodent experimental autoimmunity.[30] The results of those studies are consistent with the model proposed here. This scheme resembles somewhat the idiotypic networks originally proposed by Jerne[47] for regulation of antibody production. However, it is quite distinct because the typical idiotypic models focus on the CDR3 segment formed by recombination of V, D, and J gene segments and is further diversified by junctional processes to form highly unique or individual markers. By contrast, the CDR1 segment used here is completely specified by the germline Vβ gene and shows a substantial degree of sequence identity (50% or greater) among all human Vβ genes reflected in antigenic crossreactions shown by monoclonal anti-Vβ autoantibodies.[35]

We found an unexpected role of TCR peptide–stimulated T cells in remodeling defective heart cells that had been infected with coxsackievirus. Net collagen synthesis was increased via downregulation of matrix metalloproteinase.[46] Experiments in which lymphocytes harvested from TCR peptide–treated mice were co-cultured *in vitro* with naïve primary cardiac fibroblasts showed that the lymphocytes exerted a direct control on fibroblast gene expression of pro-collagen, pro-MMP, and MMP enzymatic activity. These results raise the possibility that the helper T cell phenotype can differentially affect diastolic function and that modulation of TH activity could promote adaptive remodeling in heart failure and post-myocardial infarction. Traditionally, endocrine and mechanical factors are thought to be the major regulatory factors in these processes, but the immune system through TH polarization and subsequent generation of cytokines may play a substantive role as well.

ACKNOWLEDGMENTS

This work was supported in part by Arizona Disease Control Research Commission Grant 5-038 to J.J.M. and American Heart Grant 0455575Z to D.F.L.

REFERENCES

1. KLEIN, S.A. *et al.* 1997. Demonstration of the Th1 to Th2 cytokine shift during the course of HIV-1 infection using cytoplasmic cytokine detection on single cell level by flow cytometry. AIDS **11:** 1111–1118.

2. CLERICI, M. & G.M. SHEARER. 1993. A TH1→TH2 switch is a critical step in the etiology of HIV infection. Immunol. Today **3:** 107–111.
3. HUANG, D.S. *et al.* 1993. The kinetics of cytokine secretion and proliferation by mesenteric lymph node cells during the progression to murine AIDS, caused by LP-BM5 murine leukemia virus infection. Reg. Immunol. **5:** 325–331.
4. GAZZINELLI, R.T. *et al.* 1992. CD4+ subset regulation in viral infection. Preferential activation of the Th2 cells during progression of retrovirus-induced immunodeficiency in mice. J. Immunol. **148:** 182–188.
5. RAFI, A. *et al.* 2003. Immune dysfunction in the elderly and its reversal by antihistamines. Biomed. Pharmacother. **57:** 246–250.
6. DENG, Y. *et al.* 2004. Age-related impaired type 1 T cell responses to influenza: reduced activation *ex vivo*, decreased expansion in CTL culture *in vitro*, and blunted response to influenza vaccination *in vivo* in the elderly. J. Immunol. **172:** 3437–3446.
7. BOREN, E. *et al.* 2004. Inflamm-aging: autoimmunity and the immune-risk phenotype. Autoimmun. Rev. **3:** 401–406.
8. SKAPENKO, A. *et al.* 2005. The role of the T cell in autoimmune inflammation. Arthritis Res. Ther. **7:** S4–14.
9. LIBLAU, R.S. *et al.* 1996. Th1 and Th2 CD4+ T cells in the pathogenesis of organ-specific autoimmune diseases. Immunol. Today **16:** 34–38.
10. ROBEY, I.F. *et al.* 2002. Human monoclonal natural autoantibodies against the T cell receptor inhibit interleukin-2 production in murine T cells. Immunology **105:** 419–429.
11. SCHLUTER, S.F. *et al.* 2003. Natural autoantibodies to TCR public idiotopes: Potential roles in immunomodulation. Cell Mol. Biol. **49:** 193–207.
12. WATSON, R.R. *et al.* 1995. T cell receptor V beta complementarity-determining region 1 peptide administration moderates immune dysfunction and cytokine dysregulation induced by murine retrovirus infection. J. Immunol. **155:** 2282–2291.
13. MARCHALONIS, J.J. *et al.* 2001. T cell receptor-derived peptides in immunoregulation and therapy of retrovirally induced immunosuppression. Crit. Rev. Immunol. **21:** 57–74.
14. LIANG B. *et al.* 1998. Injection of T cell receptor peptide reduces immunosenescence in aged C57Bl/6 mice. Immunology **93:** 462–468.
15. SEPULVEDA, R.T., J.J. MARCHALONIS & R.R. WATSON. 2005. T cell receptor V beta 8.1 peptide reduces coxsackievirus-induced cardiopathology in aged mice. Cardiovasc. Toxicol. **5:** 21–28.
16. MARCHALONIS, J.J. *et al.* 1992. Human autoantibodies reactive with synthetic autoantigens from T cell receptor beta chain. Proc. Natl. Acad. Sci. USA **89:** 3325–3329.
17. MARCHALONIS, J.J. *et al.* 1993. Natural human antibodies to synthetic receptor autoantigens: correlations with age and autoimmune disease. Gerontology **39:** 65–79.
18. MARCHALONIS, J.J. *et al.* 2002. Natural recognition repertoire and the evolutionary emergence of the combinatorial immune system. FASEB J. **16:** 842–848.
19. DEHGHANPISHEH, K. *et al.* 1995. Production of IgG autoantibodies to TCRs in mice infected with the retrovirus LP-BM5. **7:** 31–36.
20. ADELMAN, M.K., S.F. SCHLUTER & J.J. MARCHALONIS. 2004. The natural antibody repertoire of sharks and humans recognizes the potential universe of antigens. Protein J. **23:** 103–118.
21. MARCHALONIS, J.J. *et al.* 1998. Phylogenetic emergence and molecular evolution of the immunoglobulin family. Adv. Immunol. **70:** 417–506.
22. MARCHALONIS, J.J. *et al.* 1993. Human autoantibodies to a synthetic putative T cell receptor beta-chain regulatory idiotype: expression in autoimmunity and aging. **10:** 1–15.
23. WANG, E. *et al.* 1994. IgG autoantibodies to "switch peptide" determinants of TCR alpha/beta in human pregnancy. Clin. Immunol. Immunopathol. **73:** 224–228.
24. MARCHALONIS, J.J. *et al.* 1996. Autoantibodies to T cell receptor beta chains in human heart transplantation: epitope and spectrotype analyses and kinetics of response. Exp. Clin. Immunogenet. **13:** 181–191.
25. LAKE, D.F. *et al.* 1994. Autoantibodies to the alpha/beta T cell receptors in human immunodeficiency virus infection: dysregulation and mimicry. Proc. Natl. Acad. Sci. USA **91:** 10849–10853.

26. MARCHALONIS, J.J. *et al.* 1997. Analysis of autoantibodies to T cell receptors among HIV-infected individuals: epitope analysis and time course. Clin. Immunol. Immunopathol. **82:** 174–189.
27. MARCHALONIS, J.J. *et al.* 1994. Synthetic autoantigens of immunoglobulins and T cell receptors: their recognition in aging, infection, and auto immunity. Proc. Soc. Exp. Biol. Med. **207:** 129–147.
28. JAMBOU, F. *et al.* 2003. Circulating regulatory anti-T cell receptor antibodies in patients with myasthenia gravis. J. Clin. Invest. **112:** 265–274.
29. JAMBOU, F. & S. COHEN-KAMINSKY. 2003, Immunoregulation by Vbeta specific antibodies in myasthenia gravis: mining physiological T cell homeostasis for TCR specific therapy. Cell Mol. Biol. **49:** 181–192.
30. HASHIM, G.A. *et al.* 1992. Spontaneous development of protective anti-T cell receptor autoimmunity targeted against a natural EAE-regulatory idiotope located within the 39-59 region of the TCR-V beta 8.2 chain. J. Immunol. **149:** 2803–2809.
31. VAINIENE M. *et al.* 1996. Natural immunodominant and experimental autoimmune encephalomyelitis-protective determinants within the Lewis rat V beta 8.2 sequence include CDR2 and framework 3 idiotopes. J. Neurosci. Res. **43:** 137–145.
32. KURT, R.A. *et al.* 1995. T lymphocytes infiltrating sites of tumor rejection and progression display identical V beta usage but different cytotoxic activities. J. Immunol. **154:** 3969–3974.
33. DEHGHANPISHEH, K. & J.J. MARCHALONIS. 1997. Retrovirally induced mouse anti-TCR monoclonals can synergize the *in vitro* proliferative T cell response to bacterial superantigens. Scand. J. Immunol. **45:** 645–654.
34. ROBEY, I.F. *et al.* 2000. Production and characterization of monoclonal IgM autoantibodies specific for the T-cell receptor. J. Protein Chem. **19:** 9–21.
35. ROBEY, I.F. *et al.* 2002. Specificity mapping of human anti-T cell receptor monoclonal natural antibodies: defining the properties of epitope recognition promiscuity. FASEB J. **16:** 642–652.
36. LIANG, B. *et al.* 1996. Effects of vaccination against different T cell receptors on maintenance of immune function during murine retrovirus infection. Cell Immunol. **172:** 126–134.
37. LIANG, B. *et al.* 1996. T cell-receptor dose and the time of treatment during murine retrovirus infection for maintenance of immune function. Immunology **87:** 198–204.
38. LIANG, B.L. *et al.* 1997. Prevention of immune dysfunction, vitamin E deficiency, and loss of Cryptosporidium resistance during murine retrovirus infection by T cell receptor peptide immunization. Nutr. Res. **17:** 677–692.
39. SEPULVEDA, R.T., J.J. MARCHALONIS & R.R. WATSON. 2003. T cell receptor V beta 8.1 peptide reduces coxsackievirus-induced cardiopathology during murine acquired immunodeficiency syndrome. J. Cardiovasc. Pharmacol. **41:** 489–497.
40. MAKINODAN T. & M.M.B. KAY. 1980. Age influence of the immune system. Adv. Immunol. **29:** 287–330.
41. NOBREGA, A. *et al.* 1996. The age-associated increase in autoreactive immunoglobulins reflects a quantitative increase in specificities detectable at lower concentration in young mice. Scand. J. Immunol. **44:** 437–443.
42. KAY, M.M. *et al.* 1990. Definition of a physiologic aging autoantigen by using synthetic peptides of membrane protein band 3: localization of the active antigenic sites. Proc. Natl. Acad. Sci. USA **87:** 5734–3738.
43. ARKING, R. 1998. Biology of Aging. Sinauer Associates, Inc. Sunderland, MA.
44. ORTMAN, C.L. *et al.* 2002. Molecular characterization of the mouse involuted thymus: aberrations in expression of transcription regulators in thymocyte and epithelial compartments. Int. Immunol. **14:** 813–822.
45. MILLER, R. 1990. Aging and the immune response. *In* Biology of Aging, 3rd edit. E.L. Schneider and J.W. Rowe, Eds.: 157–180. Academic Press, Inc. San Diego, CA.
46. YU, Q. *et al.* 2005. A role for T-lymphocytes in mediating cardiac diastolic function. Am. J. Physiol. In press.
47. JERNE, N.K. 1984. Idiotypic networks and other preconceived ideas. Immunol. Rev. **79:** 5–24.

Age at Onset

An Essential Variable for the Definition of Genetic Risk Factors for Sporadic Alzheimer's Disease

KATRIN BEYER,[a] JOSÉ I. LAO,[b] PILAR LATORRE,[c] AND AURELIO ARIZA[a]

Departments of [a]Pathology and [c]Neurology, Hospital Universitari Germans Trias i Pujol, Autonomous University of Barcelona, Badalona, Spain

[b]Department of Genetics and Molecular Medicine, Laboratorios Dr. Echevarne, Barcelona, Spain

ABSTRACT: The aim of our work was to detect minor loci acting as Alzheimer's disease (AD) genetic markers. We divided 206 AD patients and 186 individuals as controls into six age at onset/age-dependent groups. We studied polymorphisms of the genes of apolipoprotein E (APOE) and its promoter, cathepsin D, butyrylcholinesterase, cystatin C, methionine synthase, and cystathionine beta-synthase. Our results demonstrated that data analysis according to age at onset allows the detection of minor genetic risk factors for AD. Thus, the Th1/E47cs-G allele was an independent AD risk factor after 80 years, whereas the catD-T, BChE-K, CBS-844ins68, and CBS-VNTR 19 alleles are independent AD risk factors after 75 years. On the other hand, the CST3-A allele was an independent AD risk factor before 60 years while the CBS-VNTR allele 21 was an independent AD risk factor before 64 years. In contrast, the MS-AA genotype was an AD risk factor unrelated to age at onset. In conclusion, two main tasks remain to be accomplished to facilitate early detection of people at risk of developing AD: (1) the establishment of common criteria to carry out association studies for different genetic markers, including the introduction of AD age at onset as a crucial variable in each study, and (2) the definition of global and population-specific genetic markers for each age at onset AD subgroup.

KEYWORDS: Alzheimer's disease; apolipoprotein E; age at onset

INTRODUCTION

Alzheimer's disease (AD), the most common neurodegenerative disorder, is strongly associated with age and provides one of most illustrative examples of unsuccessful aging. AD affects about 10% of Americans over the age of 65 and 50% of those over the age of 85. More than 4 million Americans currently suffer from the disease and the number is projected to balloon to 10–15 million over the next several decades.[1]

Address for correspondence: Katrin Beyer, Department of Pathology, Hospital Germans Trias i Pujol, 08916 Badalona, Barcelona, Spain. Voice: 34-93-497 88 53; fax: 34-93-497 88 43.
jilaov@hotmail.com

Ann. N.Y. Acad. Sci. 1057: 260–278 (2005). © 2005 New York Academy of Sciences.
doi: 10.1196/annals.1322.021

Interruption of normal aging and accumulation of misfolded proteins are key features of this progressive neurodegenerative disorder. The earliest damage takes place in specialized brain structures, such as the entorrhinal cortex and hippocampus, that play critical roles in memory. Neuropathologically, the disease is characterized by extracellular amyloid plaques, with beta amyloid as their core peptide, and intraneuronal neurofibrillary tangles, with hyperphosphorylated tau as their main component.

Although the cause of AD remains a mystery, accumulating evidence suggests that AD encompasses a variety of diseases[2] whose various molecular mechanisms lead to the same neuropathological changes. AD may be divided into early-onset AD (EOAD), which accounts for 1–6% of all cases and has its onset before the age of 60 years, and late-onset AD (LOAD), which is responsible for the vast majority (94–99%) of cases and begins after 60 years of age. Only about 2–5% of cases are familial and most cases are thought to be of sporadic origin.

To date, mutations in three genes are known to cause three different types of familial EOAD. The first gene shown to be related to AD development was the amyloid precursor protein (APP) gene on chromosome 21.[3] EOAD resulting from APP mutations was named AD1 and accounts for 10–15% of familial AD. So far, 15 different APP mutations in 15 different families have been described. AD3 includes all familial AD cases caused by mutations of the presenilin 1 (PS1) gene on chromosome 14.[4] More than 80 PS1 mutations in 109 families are now known and they constitute the cause of up to 70% of familial EOAD cases. Finally AD4, the rarest AD type, develops as result of mutations of the presenilin 2 (PS2) gene on chromosome 1.[5]

The majority of AD cases are sporadic. Sporadic AD is genetically very heterogeneous and results from unknown multiple genetic and environmental influences.[2,6] The most studied and widely accepted AD genetic risk factor is the APOEɛ4 allele, whose role is now well documented.[7,8] The APOEɛ4 allele has been unanimously identified as an AD risk factor in almost all populations studied. Therefore, the APOE gene has been proposed to be a major locus, although its increased risk effect is mainly confined to AD patients between 60 and 80 years of age.[9,10] This suggests the existence of other genetic factors conferring AD risk in patients younger than 60 and older than 80 years of age. Nevertheless, most studies on this topic have provided controversial data in different AD populations and groups.[11–14] This lack of conclusive results has led to the proposal that the influence of these other genes (minor susceptibility loci) could be detected only when analyzing age at onset subgroups of AD patients in comparison with their respective age control subgroups.[15,16]

This claim seems reasonable if we consider that AD is related to aging, which is thought to be a genetically programmed process involving significant gene expression changes.[17–19] This process would result in the presence of quantitatively and qualitatively altered proteins at the various stages of aging, each of which would be characterized by distinctive protein expression profiles and protein–protein interactions.[20,21] Normal aging leads to longevity but, as happens in AD, genes conferring risk for early-life pathology can alter this process.[22]

The APOE gene, localized on chromosome 19, is considered to be an AD susceptibility gene because the APOEɛ4 allele is linked to an increased incidence of EOAD, both sporadic and familial.[23,24] It remains unclear whether this risk is secondary to diminished apoE4 function, lack of apoE3 and apoE2 protective effects, or the actual presence of apoE4.[24,25] The APOEɛ4 allele may exhibit gene dosage effects, so that individuals who are homozygous for APOEε4 demonstrate the most severe impairment in cognition and earliest age at onset of dementia.[26]

The mechanisms through which APOEε4 exerts its effect and contributes to AD remain unidentified, although it is known that APOEε4 enhances brain amyloid deposition in a dose-dependent manner[27,28] and transgenic mice lacking ApoE exhibit increased oxidative stress.[29]

A study of flanking markers has shown that polymorphisms of the APOE locus may confer risk for developing AD as well.[30] Specifically, three APOE promoter region polymorphisms have been found: –491AT polymorphism,[31] –427TC polymorphism,[32] and Th1/E47cs polymorphism at position –219.[33] In vitro expression studies have revealed a relationship between the presence of these polymorphisms and APOE gene transcription levels, so that the –491-T allele is associated with diminished transcription, the –427 polymorphism with unaltered transcription, and the Th1/E47-G allele with enhanced transcription.[32]

Another pathway recently seen to be involved in AD pathogenesis is that of homocysteine metabolism. Some association studies between homocysteine and cognitive function concluded that a strong, graded association exists between total plasma homocysteine levels and AD risk.[34] Other studies evinced that increased total homocysteine levels are a risk factor for pathologically confirmed AD[35,36] and are related to a more rapid disease progression.[37]

Homocysteine is a sulfur-containing amino acid that can induce apoptosis and cause increased neuronal vulnerability to excitotoxicity by mechanisms mainly involving DNA damage.[38]

Formed through methionine metabolic conversion, homocysteine is metabolized by remethylation and transsulfuration pathways. The remethylation pathway is controlled by vitamin B12–dependent methionine synthase (MS)[39] and methylenetetrahydrofolate reductase (MTHFR), whereas vitamin B6–dependent cystathionine beta synthase (CBS) takes part in the transsulfuration pathway. Reduced activity of any of these three enzymes would result in increased homocysteine plasma levels.[40]

Mutations of the MS gene, located on chromosome 1q43, lead to a methylcobalamin deficiency G (cblG) disorder.[41] Additionally, an A2765G polymorphism resulting in a glycine to aspartic acid substitution at codon 919 has been described; its effects on enzyme activity are currently unknown.[42]

On the other hand, inherited CBS deficiency is acknowledged to be the most frequent cause of homocystinuria in humans.[43] It is associated with more than 100 mutations of the CBS gene, which is located on chromosome 21q22.3.[44] The most frequent CBS gene mutations are a 68 base pair (bp) insertion 844ins68 on exon 8[45] and a 31 bp VNTR with 16–21 repeat units that spans the exon 13- intron 13 boundary of the gene. Both the 68 bp insertion and the increasing number of VNTR repeats are responsible for diminished CBS activity.[46]

An early role of cystatin C in AD amyloidogenesis has been suggested, since it colocalizes with Aβ in pyramidal cortical neurons.[47] Cystatin C, a secretory protein also found within the endosomal-lysosomal system,[48] is synthesized by neurons, astrocytes, and choroid plexus cells in the normal brain.[49] Its protease inhibitor function is regulated by either dimerization or cathepsin D–mediated endoproteolysis and, interestingly, both cystatin C and cathepsin D show increased neuronal levels in AD brains.[50] An Ala/Thr polymorphism on the CST3 gene corresponding to the penultimate amino acid of the cystatin C signal peptide has been identified as an AD risk factor.[15,48]

Dysfunction of the lysosomal system is closely tied to mechanisms of neurodegeneration and lysosomal proteases modify the vulnerability of neurons to degeneration.[51] Factors that are relevant to AD etiology (including Aβ peptide, APOEε4, cholesterol, and oxidative stress) decrease lysosome stability and promote lysosomal system dysfunction and cathepsin release.[52] Cathepsin D, a lysosomal hydrolase, is overexpressed in affected prefrontal cortex neurons of sporadic AD patients and is also markedly increased in patients with familial AD caused by PS mutations.[53]

A strong association between sporadic AD and the T allele of a catD gene polymorphism (catD-T), which consists of a C to T (alanine to valine) transition at position 224 in exon 2, has been reported.[54]

Finally, butyrylcholinesterase (BuChE), whose expression is increased in AD brains,[55] is associated with many aspects of AD pathology including compact β-amyloid plaques and neurofibrillary tangles[56] and it has been reported that BuChE may replace acetylcholinesterase (AChE) activity.[57] The K-variant of BuChE (BuChE-K), with a transition from G to A at codon 539 and a 30% reduction in enzymatic activity, has been claimed to act synergistically with APOEε4 in LOAD.[58] Furthermore, a BuChE-K association with a slower average rate of cognitive decline has been reported in AD patients.[59]

In the present work we summarize the results of various association studies between different genetic markers and AD. Our working hypothesis was based on the definition of major and minor loci. Major loci can be detected in a whole AD population without age at onset restrictions. Their influence on AD development may be strong (e.g., APOE) or weak, but it always applies to all AD age at onset subgroups. Minor loci, on the other hand, include those genes and allelic or genotypic variants that can be detected in only some AD age at onset subgroups.

Therefore, for successful association analyses, it would be of primary importance to subdivide the AD population into age at onset subgroups, such as patients with age at onset before 60 years, after 80 years, and between 60 and 80 years group according to intervals of at least 10 years (e.g., <60 years, 60–69 years, 70–79 years, ≥80 years). The analysis of these age at onset subgroups could define a specific genetic risk factor pattern for each subgroup and detect individuals at risk for the development of AD at different ages.

PATIENTS AND METHODS

Patients

The study included 206 patients with sporadic AD (age range, 48 to 85 years; mean age, 72.6 years; male-female ratio, 1:1.6) with clinical diagnosis of probable AD according to the DSM-IV and NINCDS-ADRDA criteria.[60] Patients were considered to have sporadic AD if they did not have a first-degree relative with either AD or progressive cognitive impairment. Age at onset was defined as the age when memory loss was first noticed by relatives. Informed consent was obtained from all subjects, either directly or from their legal guardians. Also investigated were 181 control subjects (age range, 45–84 years; mean age, 69.6 years; male-female ratio, 1:1.5) with no neurodegenerative disorder. As shown in TABLE 1, patients and controls were divided into six subgroups according to their age at onset/age.

TABLE 1. Age at onset/age groups for Alzheimer's disease patients and control subjects

Age at onset/age (years)	AD patients (N)	Control subjects (N)
<60	28	34
60–64	18	33
65–69	42	50
70–74	50	23
75–79	41	21
>80	27	20

TABLE 2. Polymorphisms analyzed in the present study

Polymorphism (references)	Primers (5′–3′)
APOE, exon 4 (61)	APOE-NI: TCCAAggAGCTgCAggCgg
	APOE-NII: gCTCgCggATggCgCTgA
	APOE491A: CACCATgTTggCCAggCTggTCTCAA
Th1/E47-TG (32,62)	TH1/E47L: ggAggAAggAggTggggCATAgAgg
	TH1/E47U: CAgAATggAggAgggTgTCTC
	TH1/E47L: ggAggTggggCATAgAggTCT
MS-AG (63)	MS-U: TgTTCCAgACAgTTAgATgAAAATC
	MS-L: gATCCAAAgCCTTTTACACTCCTC
CBS insertion (64)	CBSex8-U: gCTTTTgCTggCCTTgAgCC
	CBSex8-L: gAgggTgAgTTACAggCTgC
CBS-VNTR (64)	CBSvntr-U: TTCAAACAggTACCCAgTCACC
	CBSvntr-L: CCGAAgTgCCTgAgTACCTgT
CST3-AG (15,65)	CST3-U: gCgggTCCTCTCTATCTAgC
	CST3-L: ggTCTAgAACTCAgggCATTCCCggACA
catD-CT (66)	catD-U: gTgACAggCAggAgTTTggT
	catD-L: gggCTAAgACCTCATACTCACg
BuChE-WK (67)	BuChE-U: CTgTACTgTgTgTAgTTAgAgAAAATggC
	BuChE-K: ATggAATCCTgCTTTCCACTCCCATTCCgT
	BuChE-W: ATCATgTAATTgTTCCAgCgTAggAATCCT
	gCTTTCCACTCCCATTCTCC

DNA Analyses

Genomic DNA was extracted from peripheral blood cells according to standard procedures. One-step PCR was used to amplify regions containing polymorphisms of interest for APOE,[61] MS,[62] CBS,[63] CST3,[15,64] catD,[65] and BuChE.[66] The specific APOE promoter fragment was obtained by nested PCR.[32,67] Primers used for PCR amplifications are listed in TABLE 2. Digestion of obtained PCR products and/or electrophoresis was carried out as described.[61–67] Electrophoresis patterns were analyzed using the EDASystem 120 (Kodak).

Statistical Analysis

Allele and genotype frequencies were calculated by allele and genotype counting. The χ^2 statistic was used for preliminary analysis of allele and genotype distribution in case and control populations. Deviation of allele or genotype frequencies from Hardy-Weinberg equilibrium was calculated using the standard observed-expected chi-square goodness of fit test. Logistic regression was used to assess the main effects exerted on AD risk prediction by the different alleles or genotypes, gender, and age of controls or age at onset of AD cases (age/age at onset). Also analyzed were interactive effects among APOEε4 allele and all other analyzed alleles and genotypes. Odds ratios (ORs) and 95% confidence intervals (CIs) were calculated to estimate the relative risk conferred by the various polymorphisms. Significance levels were set at $P<0.05$ and statistical analyses were performed using SPSS version 10.0. Because, unlike APOEε2 or APOEε3, APOEε4 is a risk factor for AD, ε4-carrying genotypes (ε4/ε4 and ε4/ε3) were grouped against non-ε4–carrying genotypes to obtain sufficient interaction power in logistic regression analysis.

RESULTS

Major Loci

APOEε4 and Double Dose Effect

As expected, we found an accumulation of APOEε4 when the whole population was analyzed. As shown in TABLE 4, when the age at onset division was introduced, the highest APOEε4 frequency and association rate (0.42 vs. 0.10; OR = 14.9) was observed in the 60–64 years age subgroup. The APOEε4 frequency and association with AD diminished stepwise until it disappeared in AD patients older than 80 years of age.

In four of the six AD age at onset subgroups, we detected patients with the APOEε4/ε4 genotype (TABLE 3). Its frequency and odds ratios were markedly elevated in the 60–64 years age subgroup (0.25, OR = 17.8) and diminished stepwise in the 65–69 and 70–74 years age subgroups (TABLE 3).

Methionine Synthase

Hierarchical logistic regression analyses used to examine the main interactive effects exerted on AD risk by methionine synthase (MS) genotype, age, and gender revealed that the MS-AA genotype is a more important risk factor for AD than the MS-A allele alone. Additionally, to determine whether the MS-AA genotype acts as a re-

TABLE 3. Major loci associated with AD risk factors: APOEe4 allele, APOEe4/e4, and MS-AA genotypes

AD age at onset[a] groups (years)	APOEε4 AD[b]/C[c]	APOEε4/ε4 AD/C	MS-AA AD/C
All	0.46/0.22	0.07/0.0	0.84/0.64
	$P=0.002^d$, OR=3.8[e]	$P=0.013$, OR=2.1	$P<0.001$, OR=2.7
	CI[f]: 1.8–7.2	CI: 1.0–5.5	CI: 1.1–5.3
<60	0.32/0.25	0.04/0.0	0.87/0.69
	$P=0.096$, n.s.[g]	$P=0.024$, OR=1.8	$P<0.001$, OR=2.8
		CI: 0.9–3.7	CI: 1.5–5.4
60–64	0.65/0.17	0.25/0.0	0.85/0.68
	$P<0.001$, OR=7.2	$P<0.001$, OR=17.8	$P<0.001$, OR=2.6
	CI: 4.1–14.2	CI: 5.9–28.6	CI: 1.4–5.0
65–69	0.59/0.18	0.11/0.0	0.87/0.69
	$P<0.001$, OR=6.0	$P=0.008$, OR=6.2	$P<0.001$, OR=2.4
	CI: 3.8–13.1	CI: 4.1–12.8	CI: 1.3–4.7
70–74	0.50/0.12	0.07/0.0	0.81/0.62
	$P<0.001$, OR=4.9	$P=0.018$, OR=3.2	$P<0.001$, OR=2.2
	CI: 2.3–12.7	CI: 1.1–5.7	CI: 1.2–4.8
75–79	0.46/0.19	0.0/0.0	0.80/0.63
	$P<0.001$, OR=3.5	n.s.	$P=0.009$, OR=2.0
	CI: 1.2–11.9		CI: 1.1–4.5
≥80	0.27/0.42	0.0/0.0	0.82/0.55
	$P=0.0045$, OR=3.1	n.s.	$P<0.001$, OR=3.1
	CI: 1.4–8.3		CI: 1.4–7.2

[a]Age at onset subgroups of AD patients and their respective age subgroups of control subjects (C).
[b]Frequencies obtained for AD patients
[c]Frequencies obtained for control subjects.
[d]Value for χ^2 test of differences in APOE genotype frequencies between cases and controls. [e]Odds ratio (OR) estimates for the effect of the APOEε4 allele carrying genotypes on risk or protection for AD. [f]95% confidence interval.
[g]Nonsignificant.

stricted risk factor for AD depending on the age at onset, patients and controls were divided into subgroups according to their age at onset/age. Interestingly, accumulation of the AA genotype in AD patients could be detected within all age at onset subgroups, including the group younger than 60 years and the group ≥80 years of age (TABLE 3).

These results suggest that the MS gene can also be considered a major locus, although of much lesser influence than the APOE locus (OR=2.2–3.1 for MS-AA genotype vs. 7.2–14.2 for APOEε4-allele; TABLE 3).

Minor Loci

APOE Promoter: −491A/T and TH1/E47 Polymorphisms

The analysis of allelic and genotype frequencies of the two APOE promoter polymorphisms known to alter apoE expression levels revealed an accumulation of the Th1/E47-G allele in AD patients with age at onset ≥80 years (TABLE 4).

TABLE 4. Minor loci responsible for AD risk after 70 years: the Th1/E47-G allele from the APOE promoter, the CBS-ins, CBS-VNTR19, and catD-T alleles

AD age at onset[a] groups (years)	Th1/E47-G AD[b]/C[c]	CBS-ins AD[a]/C[b]	CBS-VNTR19 AD[a]/C[b]	catD-T AD[a]/C[b]
Total	0.45/0.42	0.16/0.10	0.16/0.12	0.08/0.08
	P^d=0.79, n.s.[e]	P=0.21, n.s	P=0.37, n.s.	P=1, n.s.
<60	0.41/0.44	0.25/0.12	0.14/0.13	0.10/0.10
	P=0.63, n.s.	P=0.002, OR[f]=3.7 CI[g]: 1.6–7.1	P=0.78, n.s.	P=1, n.s.
60–64	0.38/0.42	0.11/0.09	0.15/0.14	0.08/0.10
	P=0.46, n.s.	P=0.61, n.s.	P=0.69, n.s.	P=0.49, n.s.
65–69	0.42/0.52	0.14/0.16	0.17/0.14	0.11/0.12
	P=0.12, n.s.	P=0.68, n.s.	P=0.36, n.s.	P=0.91, n.s.
70–74	0.51/0.56	0.12/0.04	0.20/0.21	0.04/0.06
	P=0.24, n.s.	P=0.006, OR=3.4 CI: 1.6–8.0	P=0.78, n.s.	P=0.19, n.s.
75–79	0.41/0.34	0.17/0.20	0.31/0.0	0.17/0.06
	P=0.07, n.s.	P=0.16, n.s.	P<0.001, OR=6.3 CI: 2.6–9.5	P=0.19, OR=2.4 CI: 1.2–4.9
≥80	0.58/0.22	0.26/0.0	0.15/0.0	0.10/0.11
	P<0.001, OR=3.5 CI: 2.1–9.8	P<0.001, OR=5.1 CI: 3.4–9.7	P<0.001, OR=3.8 CI: 2.4–8.7	P=0.82, n.s.

[a]Age at onset subgroups of AD patients and their respective age subgroups of control subjects (C).
[b]Frequencies obtained for AD patients.
[c]Frequencies obtained for control subjects.
[d]Value for χ^2 test of differences in APOE genotype frequencies between cases and controls.
[e]Non-significant.
[f]Odds ratio (OR) estimates for the effect of the APOEε4-allele-carrying genotypes on risk or protection for AD.
[g]95% confidence interval.

CST3

Two main trends were observed in regard to the accumulation of CST3-A allele–carrying genotypes. The first trend was accumulation in the two AD patient subgroups with age at onset under 64 years (TABLE 5). Odds ratios indicated that these patients presented an almost threefold increase in AD risk as compared with GG-genotype carriers. The second trend consisted of the accumulation of CST3-A allele–carrying genotypes in control subjects of 75 years or older (TABLE 5). Our data show that control individuals of 75 years or older carrying at least one CST3-A allele enjoy a protective effect against AD about four times as high as that seen in GG-genotype carriers.

TABLE 5. Minor loci responsible for AD risk before 70 years (CST3-A and CBS-VNTR19 alleles) and at different ages (BuChE-K allele)

AD age at onset[a] groups (years)	CST3-A AD[b]/C[c]	CBS-VNTR21 AD/C	BuChE-K AD/C
Total	0.38/0.37 P^d=0.94, n.s.[g]	0.12/0.08 P=0.19, n.s.	0.16/0.10 P=0.11, n.s.
<60	0.47/0.31 P=0.003, OR[e]=2.8 CI[f]: 1.6–6.7	0.24/0.06 P<0.001, OR=4.7 CI: 2.4–7.7	0.23/0.13 P<0.001, OR=2.9 CI: 1.2–4.6
60–64	0.52/0.37 P=0.004, OR=2.1 CI: 1.3–5.7	0.17/0.09 P<0.001, OR=3.6 CI: 1.5–6.2	0.08/0.07 P=0.47, n.s.
65–69	0.33/0.28 P=0.34, n.s.[e]	0.14/0.08 P=0.06, n.s.	0.20/0.19 P=0.72, n.s.
70–74	0.30/0.26 P=0.76, n.s.[e]	0.10/0.08 P=0.33, n.s.	0.11/0.06 P=0.21, n.s
75–79	0.31/0.50 P=0.002, OR=3.4 CI: 1.8–8.7	0.05/0.07 P=0.48, n.s.	0.28/0.14 P<0.001, OR=2.6 CI: 1.2–4.9
≥80	0.37/0.58 P=0.003, OR=4.2 CI: 2.1–9.7	0.11/0.09 P=0.61, n.s.	0.18/0.16 P=0.78, n.s.

[a]Age at onset subgroups of AD patients and their respective age subgroups of control subjects (C).
[b]Frequencies obtained for AD patients.
[c]Frequencies obtained for control subjects.
[d]Value for χ^2 test of differences in APOE genotype frequencies between cases and controls.
[e]Nonsignificant.
[f]Odds ratio (OR) estimates for the effect of the APOEε4-allele-carrying genotypes on risk or protection for AD.
[g]95% confidence interval.

Cathepsin D (catD)

A significant accumulation of the catD-T allele was observed in the 75–79 years AD subgroup in comparison with their respective controls (TABLE 4). Odds ratios indicated a more than twofold increase of AD risk.

Butyrylcholinesterase (BuChE)

The BuChE-K allele was the only genetic marker that exhibited accumulation in different age at onset AD subgroups. On the one hand, the odds ratio of 2.1 in the patient subgroup with onset before 60 years indicated BuChE-K importance as an AD risk factor at these ages. On the other hand, BuChE-K was accumulated in the 75–79 year group, increasing almost fourfold the risk for AD development (OR = 3.9, TABLE 5).

Cystathionine Beta-Synthase (CBS)

CBS-844ins68 mutation. When undivided groups of cases and controls were studied, only a slight accumulation of the 844ins68 mutation was detected in AD patients. In contrast, division of cases and controls into age at onset/age subgroups revealed an important 844ins68 mutation accumulation in AD patients with disease onset at 75 years of age or later in comparison with the respective control subgroup. Odds ratio indicated an almost threefold increase in AD risk (TABLE 4).

CBS 31 bp VNTR. The most common genotype was VNTR18/18, which was present in more than 50% of all individuals. Determination of VNTR18/18 genotype, VNTR allele 19–carrying genotype, and VNTR allele 21–carrying genotype frequencies revealed a significant VNTR allele 19–carrying genotype accumulation in subgroups with AD onset at 75 years or later, in which the risk to develop AD showed a threefold increase (TABLE 4). On the contrary, VNTR allele 21–carrying genotype frequencies were elevated in patients with AD onset before 64 years of age, who showed an almost twofold AD risk increase (TABLE 5).

Distribution of Genetic Risk Factors among Age at Onset/Age Subgroups

On examination of the results of the various association studies carried out during the last six years, we detected characteristic patterns for the different age at onset/age subgroups. Taking into account that the APOEε4 allele is a major risk factor for AD onset between 60 and 80 years, the existence of additional genetic risk factors is expected. In complete agreement with this, we found additional risk factors mainly related with extreme AD onset ages (before 60 and in the late 70s). FIGURE 1 shows the AD marker distribution analysis of the six age at onset/age subgroups. Whereas the APOEε4 allele predominated in the three age subgroups ranging from between 60 and 74 years (FIG. 1b–d), AD patients with onset before 60 years were characterized by the accumulation of CST3-A, BuChE-K, and CBS-VNTR21 alleles (FIG. 1a). On the contrary, in the oldest age at onset subgroup there was accumulation of the APOE promoter Th1/E47-G, CBSins, and CBS-VNTR19 alleles (FIG. 1f). Interestingly, the 75–79 years subgroup exhibited accumulation of BuChE-K and catD-T in addition to CBSins and CBS-VNTR19 alleles (FIG. 1e).

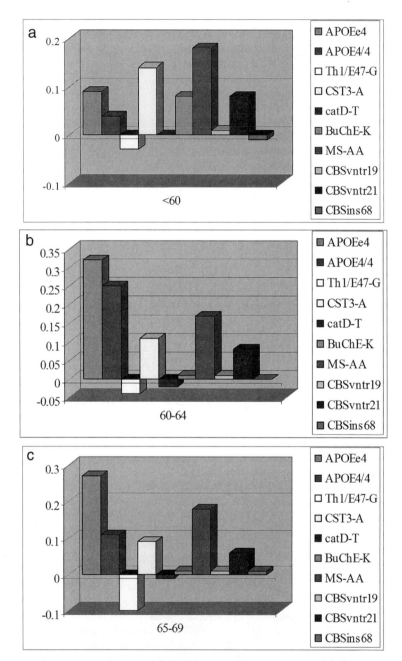

FIGURE 1. AD risk factor distribution within six age at onset/age-dependent AD groups. Results are represented after subtraction of control group frequencies from AD group frequencies: (**a**) AD onset before 60 years and controls <60 years old, (**b**) AD onset at 60–64 years and controls 60–64 years old, (**c**) AD onset at 65–69 years and controls 64–

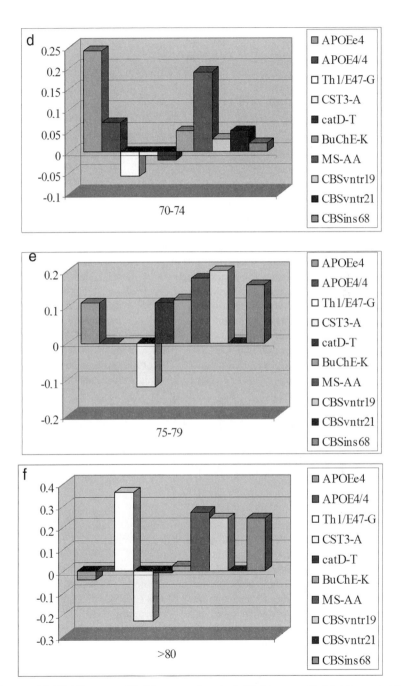

69 years old, **(d)** AD onset at 70–74 years and controls 70–74 years old, **(e)** AD onset at 75–79 years and controls 74–79 years old, and **(f)** AD onset at 80 years or later and controls >80 years old. The *key to the bars* (which is the same for all panels) applies left to right.

DISCUSSION

Cell metabolism comprises many different pathways and is characterized by the interaction of numerous proteins. Significant changes in gene expression have been observed with aging. Specifically, brain proteins involved in synaptic plasticity, vesicular transport, and mitochondrial function are downregulated, with induction of the stress response, antioxidants, and DNA repair genes.[19] Whereas many normal cell pathways have been well characterized, most mechanisms leading to neurodegeneration and related cell dysfunction are poorly understood.

The detection of age at onset–specific genetic influences suggests that dysfunctions leading to AD development occur at different levels of cell compartments, pathways, and mechanisms. On the one hand, recent data suggest that apoE, in conjunction with Aβ, plays an important role in the deposition of insoluble fibrils.[28] On the other hand, our results, in concordance with others, underlined that the well established risk factor APOEε4 plays an important role in AD development between 60 and 80 years. The APOEε4 dose effect was especially important for AD development between 60 and 64 years. The further detection of the APOE promoter Th1/E47-G allele as a risk factor for developing AD after 80 years showed that the APOE gene contains at least two loci with different impacts on AD risk, the common polymorphism at exon 4 being a major locus and the Th1/E47 polymorphism a minor locus.

The increased APOE gene expression caused by the Th1/E47cs-G allele may well result in apoE accumulation in the brain. Of interest, the majority of Th1/E47-G allele carriers did not present the APOEε4 allele. Therefore, our data suggest that only sustained apoE3 accumulation for many years would attain the effects shown by apoE4 in earlier AD stages.

AD patients exhibit increased levels of homocysteine, whose degradation is mediated by three enzymes (MS, MTHFR, and CBS). Our results show an accumulation of the MS-AA genotype in AD patients independent of age at onset of disease, including those subgroups comprising patients 60 years or younger and patients 80 years or older. Therefore, MS could be considered as an additional major locus, although of less influence than the APOE common polymorphism. Furthermore, if the MS-AG genotype results in enzyme activity reduction, DNA hypomethylation is to be expected.[62] Since methylation levels are of paramount importance in gene activation-deactivation mechanisms, hypomethylation associated with the MS-AG genotype could hamper AD development by delaying deactivation of protective key genes.

An inverse correlation between CBS enzyme activity and the number of tandem repeats defined by the 31 bp VNTR has been described in fibroblasts.[68] Thus, the presence of the VNTR allele 19 results in reduced CBS activity and the larger VNTR allele 21 causes an even greater reduction. The accumulation of CBS-VNTR allele 21–carrying genotypes in patients with an onset of disease before 65 years would be consistent with markedly reduced CBS activities in this subgroup. As a result, these patients would show higher homocysteine levels than non-VNTR allele 21 carriers and would experience homocysteine-mediated damage leading to EOAD.[38] On the other hand, the milder CBS activity reduction associated with the VNTR allele 19 would not elicit homocysteine-mediated damage until a later age and, consequently, its effect as an AD risk factor would be limited to ages greater than 75 years.

A 68-bp insertion (844ins68) has been identified within CBS gene exon 8 and different studies have shown that its frequency ranges from 7.5–18.8% in different populations.[69] Results of the present study show a significantly higher prevalence of the 844ins68 mutation in AD patients with an AD onset of 75 years or greater. Although the 844ins68 mutation apparently does not result in enzyme activity impairment or hyperhomocysteinemia,[45] low mRNA levels indicate that the insertion-carrying allele is poorly transcribed.[70]

Our results support the hypothesis that homocysteine metabolism genetics is directly involved in AD pathogenesis.[71] The accumulation of alleles that decrease the activity of the two main enzymes involved in homocysteine degradation was found to augment the risk of developing AD.

Although no functional effects of the cystatin C polymorphism have yet been reported, its location at the penultimate signal peptide amino acid suggests a capacity to impinge on the cystatin C secretory pathway. Normally, cystatin C is present intracellularly as monomers or inactive dimers. The latter convert to the active monomeric form before secretion. In this scenario, the signal peptide polymorphism would exert its effects on AD development by altering cystatin C aggregating properties and giving rise to the formation of dimers or complexes with other molecules such as Aβ.[10,72]

The catD-T allele may be associated with increased pro-cathepsin D secretion and altered intracellular maturation.[73] The cumulative effect of increased pro-cathepsin D levels could, at advanced age, contribute to lysosomal disturbances, which together with other alterations would enhance neurodegeneration.

BuChE participates in the hydrolysis of acetylcholine and patients carrying the APOEϵ4 allele present a severe cholinergic deficit. Therefore, it has been proposed that BuChE-K reduced enzymatic activity exerts a protective effect on cholinergic neurotransmission in subjects carrying the APOEϵ4 allele.[74] The fact that BuChE is further found within amyloid plaques[56] could suggest that BuChE-K conformation would increase interaction with some of plaque components.

Numerous previous studies have been carried out on almost every marker analyzed in the present work. At least three different reports showed deleterious effects of APOE promoter alleles,[31,33,75] but others could not confirm these findings.[76–78] With regard to homocysteine metabolism, very few studies have investigated the association between MTHFR polymorphisms and AD. Although some of them failed to detect allele frequency differences between AD cases and controls,[79–81] others did identify the MTHFR 677-T allele as a risk factor for AD.[82–84] The same can be said about the other markers investigated in our work. Two association studies identified CST3 as an AD risk factor,[15,48] whereas two others did not.[85,86] A recent meta-analysis of the catD A224V polymorphism included 14 independent studies that had probed the relation between the catD polymorphism and AD.[87] The main conclusion was that the catD-T allele is not a strong risk factor for AD, but differences of age at onset among studies were not taken into account in the meta-analysis. Finally, several studies on the association between the BChE-K allele and AD have yielded contradictory results.[66,88–90]

Leaving aside population variability, the controversial results obtained in the various association studies seem to be due to the use of different criteria for data analysis. The importance of considering age at onset/age subgroups for data analysis cannot be overemphasized. Only our subdivision of the AD patient group into smaller age at onset subgroups allowed identification of the specific influence exerted by minor loci. In this way, CBS-VNTR21 and CST3-A were shown to be risk factors for EOAD, whereas Th1/E47-G, CBS-VNTR19, CBSins, catD-T, and BuChE-K were seen to be risk factors for LOAD. Failure to implement this finer subdivision of age groups would have resulted in no detection of these minor loci as AD risk factors.

When our studies were started, the number of available AD patients and control individuals was just half the current figure. Close collaboration with the Department of Neurology of the "Germans Trias i Pujol" University Hospital has resulted in a steady increase in the number of patients and controls recruited. Interestingly, all the tendencies detected at the beginning of our investigation have been confirmed when studying larger numbers of patients.

Most importantly, our results strongly suggest the urgent need to establish well-defined criteria for genetic association studies. The introduction of age at onset/age subgroups should be the basis of each investigation of new genetic markers of AD risk. The vast majority of those markers seem to exert their influence as AD risk factors on certain age at onset subgroups only. Therefore, their detection is only possible when age at onset is taken into account. The establishment of common criteria would help to unify worldwide data, identify genetic risk differences among various populations, and obtain conclusive results. The latter could be used for the design of DNA chips including all the risk-conferring polymorphisms identified. Hopefully, this approach would make possible the identification of individuals at risk of developing AD at various ages. Finally, our observations lend credence to the claim that the morphologic hallmarks of AD are the shared end-result of a variety of molecular mechanisms that are relatively specific for each age at onset subgroup of AD patients.

REFERENCES

1. SMITH, M.A. 1998. Alzheimer disease. Int. Rev. Neurobiol. **42:** 1–54.
2. ROSES, A.D. 1998. Alzheimer disease: a model of gene mutations and susceptibility polymorphisms for complex psychiatric diseases. Am. J. Med. Genet. **81:** 49–57.
3. GOATE, A. *et al.* 1991. Segregation of a missense mutation in the amyloid precursor protein gene with familial Alzheimer's disease. Nature **349:** 704–706.
4. SHERRINGTON, R. *et al.* 1995. Cloning of a gene bearing missense mutations in early-onset familial Alzheimer's disease. Nature **375:** 754–760.
5. ROGAEV, E.I. *et al.* 1995. Familial Alzheimer's disease in kindreds with missense mutations in a gene on chromosome 1 related to the Alzheimer's disease type 3 gene. Nature **376:** 775–778.
6. ROSENBERG, R.N. 2000. The molecular and genetic basis of AD: The end of the beginning. Neurology **54:** 2045–2054.
7. FARRER, L.A. *et al.* 1997. Effects of age, sex, and ethnicity on the association between apolipoprotein E genotype and Alzheimer disease. A meta-analysis. APOE and Alzheimer Disease Meta Analysis Consortium. J. Am. Med. Assoc. **278:** 1349–1356.
8. SAUNDERS, A.M. *et al.* 1993. Association of apolipoprotein E allele e4 with late-onset familial and sporadic Alzheimer's disease. Neurology **43:** 1467–1472.
9. FRISONI, G.B. *et al.* 1998. The prevalence of apoE-epsilon4 in Alzheimer's disease is age dependent. J. Neurol. Neurosurg. Psychiatry **65:** 103–106.
10. TANG, M.X. *et al.* 1996. Relative risk of Alzheimer disease and age at onset distributions Caucasians and Hispanics in New York City. Am. J. Hum. Genet. **58:** 574–584.
11. COMBARROS, O. *et al.* 2002. Candidate gene association studies in sporadic Alzheimer's disease. Dement. Geriatr. Cogn. Disord. **14:** 41–54.
12. FARRER, L.A. *et al.* 2000. Association between angiotensin-converting enzyme and Alzheimer disease. Arch. Neurol. **57:** 210–214.

13. BLENNOW, K. *et al.* 2000. No association between the alpha2-macroglobulin (A2M) deletion and Alzheimer's disease, and no change in A2M mRNA, protein, or protein expression. J. Neural. Transm. **107:** 1065–1079.

14. KI, C.S. *et al.* 2001. Alpha-1 antichymotrypsin and alpha-2 macroglobulin gene polymorphisms are not associated with Korean late-onset Alzheimer's disease. Neurosci. Lett. **302:** 69–72.

15. CRAWFORD, F.C. *et al.* 2000. A polymorphism in the cystatin C gene is a novel risk factor for late-onset Alzheimer's disease. Neurology **55:** 763–768.

16. BEYER, K. *et al.* 2002. Identification of a protective allele against Alzheimer's disease in the APOE gene promoter. Neuroreport **13:** 1403–1405.

17. APELT, J., K. ACH & R. SCHLIEBS. 2003. Aging-related downregulation of neprilysin, a putative beta-amyloid-degrading enzyme, in transgenic Tg2576 Alzheimer-like mouse brain is accompanied by an astroglial upregulation in the vicinity of beta-amyloid plaques. Neurosci. Lett. **339:** 183–186.

18. BLALOCK, E.M. *et al.* 2003. Gene microarray in hippocampal aging: statistical profiling identifies novel processes correlated with cognitive impairment. J. Neurosci. **23:** 3807–3819.

19. LU, T. *et al.* 2004. Gene regulation and DNA damage in the ageing human brain. Nature **429:** 883–891.

20. JENKINS, G. 2002. Molecular mechanisms of skin ageing. Mech. Ageing Dev. **123:** 801–810.

21. ZHANG, Y. & B. HERMAN. 2002. Aging and apoptosis. Mech. Ageing Dev. **123:** 245–260.

22. LONGO, V.D. & C.E. FINCH. 2002. Genetics of aging and diseases: from rare mutations and model systems to disease prevention. Arch. Neurol. **59:** 1706–1708.

23. GROWDON, J. 2001. Incorporating biomoarkers into clinical drug trials in Alzheimer's disease. J. Alzheimer Dis. **3:** 287–292.

24. REBECK, G.W., M. KINDY & M.J. LADU. 2002. Apolipoprotein E and Alzheimer's disease: the protective effects of ApoE2 and E3. J. Alzheimer Dis. **4:** 145–154.

25. TETER, B. *et al.* 2002. The presence of apoE4, not the absence of apoE3, contributes to AD pathology. J. Alzheimer Dis. **4:** 155–163.

26. CASELLI, R.J. *et al.* 1999. Preclinical memory decline in cognitively normal Apolipoprotein E epsilon 4 homozygotes. Neurology **53:** 201–207.

27. POLVIKOSKI, T. *et al.* 1995. Apolipoprotein E, dementia, and cortical deposition of β-amyloid protein. N. Engl. J. Med. **333:** 1242–1247.

28. BALES, K.R. *et al.* 1997. Lack of apolipoprotein E dramatically reduces amyloid beta-peptide deposition. Nat. Genet. **17:** 263–264.

29. SHEA, T.B. & E. ROGERS. 2002. Folate quenches oxidative damage in brains of apolipoprotein E deficient mice: augmentation by vitamin E. Mol. Brain Res. **108:** 1–6.

30. CHARTIER-HARLIN, M.C. *et al.* 1994. Apolipoprotein E, epsilon 4 allele as a major risk factor for sporadic early and late-onset forms of Alzheimer's disease: analysis of the 19q13.2 chromosomal region. Hum. Mol. Genet. **3:** 569–574.

31. BULLIDO, M.J. *et al.* 1998. A polymorphism in the regulatory region of APOE associated with risk for Alzheimer's dementia. Nat. Genet. **18:** 69–71.

32. ARTIGA, M.J. *et al.* 1998. Allelic polymorphisms in the transcriptional regulatory region of apolipoprotein E gene. FEBS Lett. **421:** 105–108.

33. LAMBERT, J.C. *et al.* 1998. A new polymorphism in the APOE promoter associated with risk of developing Alzheimer's disease. Hum. Mol. Genet. **7:** 533–540.

34. SESHADRI, S. *et al.* 2002. Plasma homocysteine as a risk factor for dementia and Alzheimer's disease. N. Engl. J. Med. **346:** 476–483.

35. SMITH, A.D. 2002. Homocysteine, B vitamins, and cognitive deficit in the elderly. Am. J. Clin. Nutr. **75:** 785–786.

36. MCCADDON, A. *et al.* 1998. Total serum homocysteine in senile dementia of Alzheimer type. Int. J. Geriatr. Psychiatry **13:** 235–239.

37. CLARKE, R. *et al.* 1998. Folate, vitamin B12, and serum total homocysteine levels in confirmed Alzheimer's disease. Arch. Neurol. **55:** 1449–1455.

38. KRUMAN, I.I. *et al.* 2000. Homocysteine elicits a DNA damage response in neurons that promotes apoptosis and hypersensitivity to excitotoxicity. J. Neurosci. **20:** 6920–6926.

39. DRENNAN, C.L. *et al.* 1994. How a protein binds B_{12}: a 3.0 A X-ray structure of B_{12}-binding domains of methionine synthase. Science **266:** 1669–1674.
40. TSAI, M.Y. *et al.* 1999. Genetic causes of mild hyperhomocysteinemia in patients with premature occlusive coronary artery diseases. Atherosclerosis **143:** 163–170.
41. WATKINS, D. *et al.* 2002. Hyperhomocysteinemia due to methionine synthase deficiency, cblG: structure of the MTR gene, genotype diversity, and recognition of a common mutation, P1173L. Am. J. Hum. Genet. **71:** 143–153.
42. LECLERC, D. *et al.* 1996. Human methionine synthase: cDNA cloning and identification of mutations in patients of the cblG complementation group of folate/cobalamine disorders. Hum. Mol. Genet. **5:** 1867–1874.
43. MUDD, S.H., H.L. LEVY & F. SKOVBY. 2000. Disorders of transsulfuration. *In* The Metabolic and Molecular Basis of Inherited Disease. C.S. Scriver, A.L. Beaudet, W.S. Sly & D. Valle, Eds.: 1279–1327. McGraw-Hill. New York.
44. KRAUS, J.P. *et al.* 1998. The human cystathionine beta-synthase (CBS) gene: complete sequence, alternative splicing, and polymorphisms. Genomics **52:** 312–324.
45. TSAI, M.Y. *et al.* 1996. High prevalence of a mutation in the cystathionine beta-synthase gene. Am. J. Hum. Genet. **59:** 1262–1267.
46. LIEVERS, K.J. *et al.* 2003. Cystathionine beta-synthase polymorphisms and hyperhomocysteinaemia: an association study. Eur. J. Hum. Genet. **11:** 23–29.
47. LEVY, E. *et al.* 2001. Codeposition of cystatin C with amyloid-beta protein in the brain of Alzheimer's disease patients. J. Neuropathol. Exp. Neurol. **60:** 94–104.
48. FINCKH, U. *et al.* 2000. Genetic association of a cystatin C gene polymorphism with late-onset Alzheimer's disease. Arch. Neurol. **57:** 1579–1583.
49. LIGNELID, H., V.P. COLLINS & B. JACOBSSON. 1997. Cystatin C and transthyretin expression in normal and neoplastic tissues of the human brain and pituitary. Acta Neuropathol. (Berlin) **93:** 494–500.
50. CATALDO, A.M. *et al.* 1997. Increased neuronal endocytosis and protease delivery to early endosomes in sporadic Alzheimer's disease: neuropathologic evidence for a mechanism of increased beta-amyloidogenesis. J. Neurosci. **17:** 6142–6151.
51. BURSCH, W. 2001. The autophagosomal–lysosomal compartment in programmed cell death. Cell Death Diff. **8:** 569–581.
52. BAHR, B.A. & J. BENDISKE. 2002. The neuropathogenic contributions of lysosomal dysfunction. J. Neurochem. **83:** 481–489.
53. CATALDO, A.M. *et al.* 2004. Presenilin mutations in familial Alzheimer's disease and transgenic mouse models accelerate neuronal lysosomal pathology. J. Neuropathol. Exp. Neurol. **63:** 821–830.
54. PAPASSOTIROPOULOS, A. *et al.* 2000. A genetic variation of cathepsin D is a major risk factor for Alzheimer's disease. Ann. Neurol. **47:** 399–403.
55. GOMEZ-RAMOS, P., C. BOURAS & M. MORAN. 1994. Ultrastructural localization of butyrylcholinesterase on neurofibrillary degeneration sites in the brain of aged and Alzheimer's disease patients. Brain Res. **640:** 17–24.
56. GUILLOZET, A.L. *et al.* 1997. Butyrylcholinesterase in life cycle of amyloid plaques. Ann. Neurol. **42:** 909–918.
57. TASKER, A., E.K. PERRY & C.G. BALLARD. 2005. Butyrylcholinesterase: impact on symptoms and progression of cognitive impairment. Future Drugs **5:** 101–106.
58. LEHMANN, D.J., C. JOHNSTON & A.D. SMITH. 1997. Synergy between the genes for butyrylcholinesterase K variant and apolipoprotein E4 in late-onset confirmed Alzheimer's disease. Hum. Mol. Genet. **11:** 1933–1936.
59. HOLMES, C. *et al.* 2005. Rate of progression of cognitive decline in Alzheimer's disease: effect of butyrylcholinesterase K gene variation. J. Neurol. Neurosurg. Psychiatry **76:** 640–643.
60. MCKHANN, G., G. DRACHMAN & M. FOLSTEIN. 1984. Clinical diagnosis of Alzheimer's disease: report of the NINCDS-ADRDA Work Group under the auspices of the department of health and human services task force on Alzheimer's disease. Neurology **34:** 939–944.
61. BEYER, K. *et al.* 2001. Alzheimer's disease and the cystatin C gene polymorphism: an association study. Neurosci. Lett. **315:** 17–20.

62. MATSUO, K. *et al.* 2001. Association between polymorphisms of folate- and methionine-metabolizing enzymes and susceptibility to malignant lymphoma. Blood **97:** 3205–3209.

63. BEYER, K. *et al.* 2004. Cystathionine beta synthase as risk factor for Alzheimer's disease. Curr. Alzheimer Res. **1:** 127–133.

64. BALBÍN, M., A. GRUBB & M. ABRAHAMSON. 1993. An Ala/Thr variation in the coding region of the human cystatin C gene (CST3) detected as a SstII polymorphism. Hum. Genet. **92:** 206–207.

65. MCILROY, S.P. *et al.* 1999. Cathepsin D gene exon 2 polymorphism and sporadic Alzheimer's disease. Neurosci. Lett. **273:** 140–141.

66. WIEBUSCH, H. *et al.* 1999. Further evidence for a synergistic association between APOEε4 and BCHE-K in confirmed Alzheimer's disease. Hum. Genet. **104:** 158–163.

67. BEYER, K. *et al.* 2002. The Th1/E47cs-G APOE promoter allele is a risk factor for Alzheimer's disease of very late onset. Neurosci. Lett. **326:** 187–190.

68. LIEVERS, K.J. *et al.* 2001. A 31 bp VNTR in the cystathionine beta-synthase (CBS) gene is associated with reduced CBS activity and elevated post-load homocysteine levels. Eur. J. Hum. Genet. **9:** 583–589.

69. FRANCO, R. *et al.* 1998. The frequency of 844ins68 mutation in the cystathionine beta-synthase gene is not increased in patients with venous thrombosis. Haematologica **83:** 1006–1008.

70. GIUSTI, B. *et al.* 1997. Different distribution of the double mutant "T833C/68 bp insertion" in cystathionine beta-synthase gene in Northern and Southern Italian populations. Thromb. Haemost. **78:** 1293.

71. BEYER, K. *et al.* 2003. Methionine synthase polymorphism is a risk factor for Alzheimer's disease. Neuroreport **14:** 1391–1394.

72. MCCARRON, M.O. *et al.* 1999. The apolipoprotein epsilon 2 allele and the pathological features in cerebral amyloid angiopathy-related hemorrhage. J. Neuropathol. Exp. Neurol. **58:** 711–718.

73. TOUITOU, I. *et al.* 1994. Missense polymorphism (C/T244) in the human cathepsin D profragment determined by polymerase chain reaction single strand conformation polymorphism analysis and possible consequences in cancer cells. Eur. J. Cancer **30:** 390–394.

74. SOININEN, H. *et al.* 1995. A severe loss of choline acetyltransferase in the frontal cortex of Alzheimer patients carrying apolipoprotein ε4 allele. Neurosci. Lett. **187:** 79–82.

75. LAMBERT, J.C. *et al.* 1998. Pronounced impact of Th1/E47cs mutation compared with –491 AT mutation on neural APOE gene expression and risk of developing Alzheimer's disease. Hum. Mol. Genet. **7:** 1511–1516.

76. HALIMI, G. *et al.* 2000. Association of APOE promoter but not A2M polymorphisms with risk of developing Alzheimer's disease. Neuroreport **11:** 3599–3601.

77. REBECK, G.W. *et al.* 1999. Lack of independent associations of apolipoprotein E promoter and intron 1 polymorphisms with Alzheimer's disease. Neurosci. Lett. **272:** 155–158.

78. TOJI, H. *et al.* 1999. Apolipoprotein E promoter polymorphism and sporadic Alzheimer's disease in a Japanese population. Neurosci. Lett. **259:** 56–58.

79. MCILROY, S.P. *et al.* 2002. Moderately elevated plasma homocysteine, methylenetetrahydrofolate reductase genotype, and risk for stroke, vascular dementia, and Alzheimer's disease in Northern Ireland. Stroke **33:** 2351–2356.

80. BRUNELLI, T. *et al.* 2001. The C677T methylenetetrahydrofolate reductase mutation is not associated with Alzheimer's disease. Neurosci. Lett. **315:** 103–105.

81. SERIPA, D. *et al.* 2003. Methylenetetrahydrofolate reductase and angiotensin converting enzyme gene polymorphisms in two genetically and diagnostically distinct cohort of Alzheimer patients. Neurobiol. Aging **24:** 933–939.

82. POSTIGLIONE, A. *et al.* 2001. Plasma folate, vitamin B(12), and total homocysteine and homozygosity for the C677T mutation of the 5,10-methylene tetrahydrofolate reductase gene in patients with Alzheimer's dementia. A case-control study. Gerontology **47:** 324–329.

83. NISHIYAMA, M. *et al.* 2000. Apolipoprotein E, methylenetetrahydrofolate reductase (MTHFR) mutation and the risk of senile dementia: an epidemiological study using the polymerase chain reaction (PCR) method. J. Epidemiol. **10:** 163–172.
84. REGLAND, B. *et al.* 1999. The role of the polymorphic genes apolipoprotein E and methylene–tetrahydrofolate reductase in the development of dementia of the Alzheimer type. Dement. Geriatr. Cogn. Disord. **10:** 245–251.
85. MARUYAMA, H. *et al.* 2001. Lack of an association between cystatin C gene polymorphisms in Japanese patients with Alzheimer's disease. Neurology **57:** 337–339.
86. ROKS, G. *et al.* 2001. The cystatin C polymorphism is not associated with early onset Alzheimer's disease. Neurology **57:** 366–367.
87. NTAIS, C., A. POLYCARPOU & J.P.A. IOANNIDIS. 2004. Meta-analysis of the association of the cathepsin d ala224val gene polymorphism with the risk of Alzheimer's disease: a huge gene-disease association review. Am. J. Epidemiol. **159:** 527–536.
88. HILTUNEN, M. *et al.* 1998. Butyrylcholinesterase K variant and apolipoprotein E4 genes do not act in synergy in Finnish late-onset Alzheimer's disease patients. Neurosci. Lett. **250:** 69–71.
89. GRUBBER, J.M. *et al.* 1999. Analysis of associaton between Alzheimer disease and the variant of butyrylcholinesterase (BCHE-K). Neurosci. Lett. **269:** 115–119.
90. CRAWFORD, F. *et al.* 1998. The butyrylcholinesterase gene is neither independently nor synergistically associated with late-onset AD in clinic and community-based populations. Neurosci. Lett. **249:** 115–118.

The Neuronal and Immune Memory Systems as Supervisors of Neural Plasticity and Aging of the Brain

From Phenomenology to Coding of Information

KATICA D. JOVANOVA-NESIC AND BRANISLAV D. JANKOVIC[†]

Immunology Research Center "Branislav Jankovic," 11152 Belgrade, Serbia and Montenegro

ABSTRACT: The ultimate goal of this report is to learn how to manipulate the level of memory T cells for more effective treatment of such neurological diseases as multiple sclerosis (MS), where certain T cell subsets recognize self-antigens as opposed to pathogen antigens, and Alzheimer's disease (AD). Brain lesions (electrolitically, by kainic acid, with AlCl, and with 6-OHDA); stimulations (electrical, magnetic, or pharmacological); or restoration of some neurological functions (thermoregulatory and behavioral) by fetal graft allotransplantations in bilaterally lesioned anterior hypothalamic area (AHA-immune regulation) and nucleus basalis magnocellularis (NBM-experimental AD) in our studies were designed to reproduce immune and cognitive deficits induced by lesions of these brain structures. To localize memory traces in the immune system and in the brain we used ethanol and drugs such as kainic acid and 6-OHDA, which have been used very effectively to produce temporary lesions in the brain. Rats showed no learning and memory ability as well as inhibition of immune reactions.

KEYWORDS: rats; brain lesions; brain stimulation; fetal allotransplantation; behavior; experimental allergic encephalomyelitis; Alzheimer's disease

PREDICTING COGNITIVE IMPAIRMENT IN THE AGING BRAIN

The aging of the brain is a continual process starting in the embryo and can thus be analyzed with a development perspective based on data from multiple time points.[1,2] Alterations observed in the elderly could well be related to factors detectable early in life. For example, environmental manipulation early in life that reduces hypothalamo-hypophyseal-adrenal (HPA) axis dysfunction has been found to post-

[†]Deceased.

Address for correspondence: Katica D. Jovanova-Nesic, Immunology Research Center-Torlak, Vojvode Stepe, 458, 11152 Belgrade, Serbia and Montenegro. Voice: +381(11)397 66 74; fax: +381(11)2469 654.

jovanok@afrodita.rcup.bg.ac.yu
nesic@torlakinstitut.com

Ann. N.Y. Acad. Sci. 1057: 279–295 (2005). © 2005 New York Academy of Sciences.
doi: 10.1196/annals.1356.022

pone cognitive impairment in aged rats.[3] Thus, the effects of aging emerge before animals are generally considered old.[4]

Further research reinforced the growing understanding that many factors (extrinsic and intrinsic) besides age influence memory and cognitive ability.[5] Neuropathological studies have shown that neurons die in the entorhinal cortex (EC) and another brain regions involved in memory function even earlier than in the hippocampus.[6]

Also, biochemical investigations identified a panel of nine proteins that demonstrate altered expression in patients with Alzheimer's disease (AD).[7] A marker for oxidative stress in DNA may be identified persons with neurodegenerative disorders such as AD.[8] 8-OGH (8-hydroxy-ρ'-deoxyguanosine) detected in AD has four bases that make up DNA strands. Furthermore, prolonged administration of β-antibody decreased the accumulation of β-amyloid in the brain and rapidly increased in the blood.[9] Antibody therapy rapidly reversed the impairment in certain learning and memory tasks,[10] perhaps by blocking the nicotinic acetylcholine receptor, a key nerve cell signaling receptor in the hippocampus.[11] Elevated levels of C-reactive proteins, as a nonspecific inflammatory marker,[12] were seen in vascular dementia. The mainstay of current AD treatment is a drug that helps maintain the level of acetylcholine, a neurotransmitter crucial in the formation of memories[13] and depression.[14] Yet an NSAID (COX-2) inhibitor showed the rate of cognitive deterioration in people with mild to moderate AD.[15] A new therapeutic approach to postpone deteriorations associated with aging in the brain must incorporate these findings.

Behavioral and Cognitive Functions in an Experimental Alzheimer's Disease

Cholinergic deficit is highly correlated with cognitive decline in AD. By using of several learning paradigms in nucleus basalis magnocellularis (NBM)–lesioned rats as an experimental model of AD, a significant deficit in the learning and memory processes was found[16] (TABLE 1). Although it has been found that two-way active avoidance (AA) could be facilitated after ibotenic acid lesions of NBM,[17] our findings, similar to those of Kessler et al.,[18] LoConte et al.,[19] and Miyamoto et al.,[20] indicate that electrolytic lesions of NBM induce impairment of this task.[21] The most profound depression of acquisition of AA response[22] was found on days 4 and 5 as presented in TABLE 1. Although our data showed no significant changes in noncognitive (spontaneous motor activity, SMA) functions, there was significant emotional hyperactivity in open field test in NBM-lesioned rats. Additionally, the significant increase of number of inner squares entered and the decrease of defecation in these rats indicate the low level of fear response in this animal model of AD. Several other studies performed on animals with cholinergic deficit[23,24] support our findings. Results obtained in open field behavior in NBM-lesioned rats during four consecutive days indicate that there is no habituation—one of the well-defined open-field features. Adaptive status in NBM-lesioned rats was also disturbed. Escape latency in AA test was significantly dysregulated, no matter what the extrinsic environmental temperature (hot, cold, or optimal; TABLE 2). These results are in agreement with findings of others[25] that indicate that complex mechanism(s) are included in the regulation of cognitive and learning processes in the aging brain. It is possible that neurochemical and neuropathological changes in the hypothalamus of NBM-lesioned rats, in part, are related to those disturbances. It seems that NBM could play a key

TABLE 1. Influence of lesions of nucleus basalis magnocellularis (NBM) on behavioral performance in rats with electrolytically lesioned NBM versus sham-lesioned and intact control

5 Days	15 Min	4 Days	4 Days	4 Days	4 Days	4 Days
Two-way AA	SMA	defecation	ambulation	rearing	inner square	grooming
2nd $P<0.05$				no change		no change
3rd $P<0.01$	no change	↓	↑		↑	
5th $P<0.001$						
$P<0.001$	–	$P<0.05$	$P<0.001$	–	$P<0.05$	–

NOTE: There are no statistically significant differences between sham-lesioned and intact control group.

TABLE 2. Adaptive status in an experimental model of Alzheimer's disease

Core body temperature	Learned helplessness	Cold environment (4°C)	Hot environment (37°C)	Cold restraint stress (gastric lesions)
NBM-lesioned vs. sham-lesioned and intact group	↓	↑	↑	no change
Statistical significance	$P<0.001$	$P<0.001$	$P<0.001$	–

role in both (1) the filtering and bisecting of "limbic signals" carrying coded information from the CNS-external (objective–surrounding) and CNS-internal (cognitive–emotional) world, and (2) the processing and transmitting of these information together with sequential integration into circuitries of neocortical processing.[26]

In conclusion, the results of the present study support the view of a multipotent role of NBM in regulation of complex behavioral and neurophysiological processes, but there is no doubt that cognitive functions, as valid parameters for learning and memory, are strongly affected.

Immune Response in an Experimental Alzheimer's Disease

In contrast to detailed investigations of the immune response in AD patients, there are only a few short reports on the immune function in NBM-lesioned rats, an animal model of AD. The results[27] of our experiments (TABLE 3) showed that the relative weight of thymus, spleen, and adrenals compared with non-lesioned NBM were significantly lower in sheep red blood cell (SRBC)–immunized NBM-lesioned rats.[3]

Despite inconsistent data from clinical investigations of humoral immunity in AD patients,[28–30] the significant reduction of PFC response (TABLE 3), anti-SRBC antibody titer, Arthus skin reaction to BSA, and anti-BSA antibody titer indicate that humoral immune response is decreased in an animal model of AD. Also, the significant reduction of delayed hypersensitivity skin reaction to BSA after bilateral electrolytic

TABLE 3. Humoral and cell-mediated immune response in NBM-lesioned rats: Arthus reaction, delayed, and antibody production to protein antigens

Antigen	Assay	Effect	Statistical significance
SRBC	PFC response	suppression	$P<0.001$
SRBC	ELISA titer	suppression	$P<0.001$
BSA	antibody production	suppression	$P<0.001$
BSA	Arthus	suppression	$P<0.001$
BSA	delayed skin reaction	suppression	$P<0.01$ after 10 days $P<0.001$ after 21 days
BSA	thymus	involution, depletion (histological)	–
BSA	spleen	plasmocytic reaction	–
BSA	lymph nodes	no changes	–
BSA	adrenals	no changes	–
SRBC to stress (isolation)	spleen	relative body weight	–
SRBC to stress (isolation)	lymph nodes	relative body weight	–
SRBC to stress (isolation)	adrenals	relative body weight	$P<0.05$

lesions of NBM implies that cell-mediated immunity is also affected. Depression of cellular immunity is in agreement with the results obtained in patients with AD.[31–33]

Since several lines of investigation suggest the importance of studying the inter-relationship between the central nervous system (CNS) and the immune system as well as the endocrine and immune systems,[34,35] the negative influence of lesions of NBM on immune reactions in the rat could be related to significant immunomodulatory areas that are proposed to be located nearby. For example, it is well known that humoral and cell-mediated immune response could be suppressed by lesions of the hypothalamic area,[36] amygdaloid complex, locus coeruleus,[37,38] or neocortex.[39–41] Taken together, these facts suggest that the immune response is not under the control of a single region of the brain,[42] but rather is modulated by several brain structures that integrate and respond to external as well as internal stimuli.

IMMUNE MEMORY IN NEUROPROTECTION AND DEMYELINATION

Thymic involution leads to a diminished supply of fresh virgin T cells, so that the aging immune system comes to consist largely of memory cells that, probably by virtue of an over-responsive calcium extrusion pump, resist increases in calcium ion concentration upon encountering ConA or anti CD3 and CD4 molecules. The proportion of T cells that can enter the cell cycle and generate clones of proliferating cells or secrete IL-2 and other lymphokines whose secretion depends on Ca^{2+} signal generation therefore declines.[43]

Conventional antigen and alloantigen-responsive T cells are not separate lineages; the same cell may respond to both challenges. While the precursor cell frequency for recall antigen is much higher within the memory T cell subset than in the naive population, the same rule does not apply for alloantigens. The immune response against transplanted tissue is often so intense that it leads to graft rejection in the periphery, even in the presence of potent immunosuppression.

The unique nature of the communication between CNS and immune system can be observed, for example, in the dialog between CNS and T cells. Under normal conditions, activated T cells can cross the blood-brain barrier and enter the CNS parenchyma. However, only T cells capable of reacting with CNS antigens seem to persist there.[44] When new pathogens are encountered in the body, the immune system mounts a primary response during which affinity maturation is used to learn the structure of the pathogens. If the body is reinfected with a previously encountered pathogen, a rapid secondary response will be accomplished so fast and efficiently that we not aware we have been reinfected. An important characteristic of immune memory is that it is associative. That means B cells adapted to some type of pathogen can provide a rapid secondary response to some "structurally related" pathogens. This associative memory underlies the concept of inoculation.

Mechanisms of control might include (1) appropriate presentation of the antigen in a complex with MHC molecules, (2) the ability to evoke regulatory T cells, and (3) neuroendocrine effects on immune cells, such as regulation and activation.

Immune Response to Hypothalamic and Cortical Fetal Tissue Allografts

The dramatic effect of neonatal thymectomy on prolonging skin transplants, the long survival of grafts in children with thymic deficiency, and survival of xenografts in immunodeficient (nude) mice implicate T lymphocytes in these reactions. Also, it is known that specific cytotoxic T lymphocytes (CD8) and helper T cells (CD4), which represent an important arm of the rejection process, depend upon the expression in the donor's tissue of foreign major histocompatibility complex (MHC) class I and class II molecules.[45] Ideally, for successful allotransplantation, cells should be nonimmunogenic so as not to require immunosuppression in the recipient. Some investigators argue that autologous primary fibroblasts or epithelial cells from the brain may fulfill this criterion.[46–48] These are suggested as necessary but not sufficient criteria. In intact rat brain, weak MHC class I and II immunoreactivity was found only on the vascular endothelial cells within the chorioidal plexus and ependymal cells of the third ventricle. In our experiments, strong positive[49] staining with OX 18 (MHC-I) and OX 6 (Ia) on the cells in degenerative foci in anterior hypothalamus (AH)–lesioned (TABLE 4) and NBM-lesioned rats were detected, independent of the type of lesions (electrolytic or by kainic acid) as presented in TABLE 4. The expression of MHC molecules decreased in both AHA- and NBM-lesioned brain after 6 posttransplantation days. $CD4^+$ and $CD8^+$ T lymphocytes were seen 2–6 days after lesioning and disappeared 6–8 posttransplantation days. While some protection is offered to foreign antigen tissue grafts, delayed rejection can take place in the brain, particularly if there is wide genetic disparity between donor and host and because of disruption of blood-brain barrier during the grafting procedure. In our experiments, we showed that both fetal allografts in lesioned AHA and cortical tissue transplant in the kainic acid–lesioned NBM survive during the first posttransplantation month—that is, 28 days after transplantation. Other investigators demonstrated

TABLE 4. Immune response to hypothalamic (AH) and frontal cortical tissue allograft in AH-lesioned and NBM-lesioned rats, respectively

Antibody to molecules	Antigen expression	Days after transplantation	Substrate
MHC class I	↑	4–6	microglia
MHC class II (Ia)	↑	4–6	microglia
CD4+	↑	2–4	T helper lymphocytes
CD8+	↑	2–4	T cytotoxic
GFAP	↑	15–28	lymphocytes, astrocytes

that brain injury induces the release of neurotrophic factors that promote the survival of damaged neurons and enhance the viability and growth of transplanted tissue in the host brain.[50,51] Also the ability of grafted monoaminergic neurons in the case of anterior hypothalamus to establish synaptic contact with neuronal elements might be functionally very important. These "immunologically privileged sites,"[51] differ in terms of the degree of protection that they afford.[52] The mechanisms of brain graft protection are not complete understood. The presence of host leukocytes in the graft at the beginning is probably the result of locally increasing donor MHC expression. T cells are important players in the adaptive arm of the immune system. They respond to antigens through interactions of their specific antigen receptor with the antigen presented by MHC molecules and a group of costimulatory molecules.

The therapeutic approach of preventing or decreasing the secondary degeneration accompanying CNS trauma is termed neuroprotection.[53,54] T cells are important in the adaptive arm of the immune system. When activated, they can kill their target cells or produce cytokines that activate or suppress the growth or differentiation of other cells. Thus, T cells play a central role in the protection of tissues against foreign invaders, as well as in tissue maintenance.

The differences were found to be attributed to a beneficial T cell–dependent response spontaneously evoked after CNS insult in the resistant, but not in the susceptible, autoimmunity strains. Experimental autoimmune encephalomyelitis-resistant rats show better postinjury neuronal survival than susceptible rats because of their beneficial T cell–dependent response.[55,56] Because of this, a second encounter with the same antigen stimulates a higher level of immune reactivity, therefore providing life-long immunity to foreign molecules or long-term immunologic memory cells. The host response promotes host survival, rescues infected cells, and should establish lifelong immunity to reinfection (i.e., immunologic memory).

Besides, the beneficial T cell–dependent response to the insult was found to be independent of the type of the insult (mechanical or biochemical), the location (axon or cell body), or neural tissue structure of the primary insult. Yet, similarly Lewis and F344 rats were found to differ in their susceptibility to autoimmune diseases not associated with H-2. Although, these strains differ in their neuroendocrine interactions, difference in regulatory T cells could explain the difference in susceptibility.[55] Studies have attempted to link the outcome of neural trauma to genetic[56,57] or anatomic differences between strains, but no attention has been directed at the possibility that the response to trauma is controlled by the immune system.

Autoimmunity in the CNS: Experimental Autoimmune Encephalomyelitis

Neuronal survival after CNS insult is determined by a genetically encoded autoimmune response.[58] Experimental autoimmune encephalomyelitis (EAE) is a T cell–mediated, inflammatory demyelinating disease of the CNS. However, the failure of anti-brain antibodies to induce disease may be due to their inability to cross the blood-brain barrier. To avoid this barrier, DA rats (genetically susceptible for EAE) with cannulae inserted into the third ventricle of the brain were injected with myelin oligodendroglial (MOG) anti-goat antibodies that developed EAE. Immunocytochemical examination of the CNS of recipient animals revealed (our unpublished preliminary results) the presence of MMP-2 that exhibited all the characteristics of autoimmunity in the brain. Beneficial regulatory T cell response is present in resistant but not in susceptible autoimmunity strains. One of the major factors determining whether or not an autoimmune disease will develop in a particular individual appears to be the presence or absence, not of autoreactive T cells (found in multiple sclerosis patients or healthy individuals), but of mechanisms that regulate the proper functioning of the autoreactive T cells.[59] It may also explain the observed occurrence of multiple autoimmune diseases in a single, probably susceptible individual when the resistance breaks down.[60,61]

Deterioration of the brain tissue may lead, by "fooling" tissue into self-tolerance and by the activation of some brain-related autoimmune mechanisms, to the expression of some autoimmune phenomenon, but for the development of autoimmune disease the activation of genetically encoded autoimmune processes is necessary. The fact that autoimmune T cells can be both beneficial and destructive raises the question: Does the genetic predisposition to autoimmune disease affect the outcome to traumatic insult to the CNS? The source of different susceptibility to EAE has been located in both MHC class II antigens and elsewhere.[62,63] Further research is necessary for this question to be resolved.

BRAIN MEMORY AND NEURAL PLASTICITY IN AGING

Learning and Memory Mechanisms Underlying Age Differences in the Brain

Of particular interest have been experiments in which memory was affected by anti-brain antibodies.[64] The term "anti-brain antibodies" used here refers to immunoglobulin molecules that react with antigens of the neuron. Briefly, cats trained to respond by leg flexion to a "positive" tone (800 Hz) and a "negative" tone (700 Hz) becoming confused after injection of anti-brain antibodies.

In this experiment, marked disorganization and irregularity in hippocampal theta rhythm suggest that anti-brain antibodies could affect learning and memory processes in the brain. The highest incidence[65] of anti-S100 and NSE antibodies (TABLE 5) occurred in Alzheimer's disease and senile dementia, demonstrating the relationship between ELISA titer of anti-S100 and anti-NSE anti-brain antibodies and processes associated with dementia. For this purpose, the effects of chronic (6 months) ethanol (36%) consumption in animals on the expression of S100, brain protein 14-3-2, and NSE antibody in Wistar rats with EAE induced by guinea pig spinal cord in Complete Freund's Adjuvant were examined. Results[66] presented in TABLE 6 showed that

TABLE 5. Autoantibodies against human S100 and neuron-specific enolase (NSE) and guinea pig spinal cord in sera from neuropsychiatric patients and rats with induced experimental allergic encephalomyelitis (EAE)

Disease	Number of patients	Anti-S100 (%)	Anti-NSE (%)	Statistical significance
Alzheimer's disease	32	50	37.5	ND
Senile dementia	56	39.3	25.0	ND
Alcoholic syndrome	78	15.4	3.8	ND
Schizophrenia	466	13.3	6.6	ND
Parkinsonism	86	84.8	70.9	ND
Healthy blood donors	112	2.7	2.1	ND
EAE-Wistar rats	17	14.7	12.05	ND
Intact rats	20	3.1	1.7	$P<0.001$

TABLE 6. Autoantibodies against S100 and NSE in sera from rats with induced EAE after chronic ethanol consumption

Group	Number	Anti-S100	Anti-NSE
Ethanol-treated EAE-induced (E-EAE)	16	2.23	1.02
Ethanol-treated (Et)	19	7.40	3.34
Intact rats (IC)	20	2.52	1.32
Statistical significance	IC vs. E-EAE	–	–
	IC vs. Et	$P<0.001$	$P<0.001$

NSA and S100 protein were significantly increased in animals that consumed ethanol. In large measure, the autoimmune hypersensitivity response described here extend previous findings that showed a significant increase in the number of hypersensitivity cutaneous reactions to brain antigens, including S100 and NSE, in patients suffering from dementia and other psychiatric diseases. Alcohol[66] seems to influence most stages of the brain process to some degree. There is no doubt that alcohol affects the memory system of the brain, but much more work remains to be done exploring possible mechanism(s). Although molecular-genetic analysis may someday tell us the nature of the mechanisms of memory storage (i.e., in cerebellum or hippocampus or neocortex), such reductionistic analysis can never tell us what the memories are. Only a detailed characterization of neural circuits that code, store, and retrieve the memories can do this. Tonegawa (Picower Center for Learning and Memory) and collaborators[67] have now identified a critical molecular pathway that allows neurons to boost their production of new proteins rapidly during long-term memory formation and synaptic strengthening. To determine the mechanisms underlying the relevant plastic changes, Sherrington introduced the term *synapse*[68] (from the Greek *synapsis*, meaning junction), to refer to the specialized contact zone, described physiologically by Ramon y Cajal, where one neuron communicates with another.

Tonegawa supposed that there is a direct activational signal from the synapses to activational protein synthesis machinery, and the central component of this pathway is an enzyme called mitogen-activated protein kinase (MAPK). MAPK effectively provides a molecular switch that triggers long-term memory storage by mobilizing the protein synthesis machinery selectively inactivated in the adult brain.

The primary effect of alcohol appears to be on the transfer of information from-short-term to long-term storage. As the alcohol dose increases, the resulting memory impairments can become much more profound, sometimes culminating in blackout periods.[69]

Ethanol, when administered over a period of days or months, also interacts with opioid neurocircuits, such as those that regulate the secretion of β-endorphins and prodynorphin-derived peptides. The fact that opioid peptide secretion is influenced by ethanol consumption adds another level of complexity to the neural networks (e.g., GABA, glutamate, 5-HT, and DA) contributing to ethanol-seeking behavior.

The hypothesis adopted is that the magnitude of age-related decrements in memory function across different domains of memory can be accounted for by the amount of processing resource available or the mental effort required to encode and retrieve information.[70,71]

Measures of cognitive resource, all of which show evidence of substantial age-related decline are the speed of information processing, the amount of working memory capacity, the efficiency with which inhibition processes operate, and the neurobiological integrity (measured by sensory function that provide an index of "brain age" and are an assessment of neurobiological integrity).

Moreover, it has been demonstrated that tasks that are hypothesized to have more environmental support actually require less cognitive resource to perform.

Magnetic Field Effects on Memory and Aging of the Brain

Since treatment with anticholinergic drugs rarely, if at all, reverses the cognitive abnormalities of AD and may even exacerbate cognitive functions, the application of electromagnetic fields (EMFs) may be useful as an adjunctive modality and may even improve some cognitive functions in AD. Furthermore, it has been reported that physostigmine (a centrally acting cholinomimetic), and verapamil (a Ca^{2+} antagonist) could produce modest cognitive gains in some AD patients[72] and in an animal experimental model of AD.[73]

Brown proposed[74] the geophysical space-time continuum model, showing that through the use of subtle, pervasive electromagnetic cues an organism is able to maintain its orientation in both space and time. Interestingly, disorientation in space and time is one of the prominent features of AD. Also, the breakdown in AD might suggest involvement of the fine-tuning functions of pineal melatonin, which may occur early in the course of the disease.[75] Since the hypothalamus is the major target for melatonin action,[76] deregulation of pineal-hypothalamic work relationships might be a key pathogenic mechanism underlying not only depression[76] and immune dysfunctions,[77,78] but also the cognitive deficits in AD.[79]

Furthermore, it should be added that the behavioral effects of dementia induced by aluminum exposure are memory loss, loss of coordination, confusion, and disorientation.[80] Aluminum-induced[81] impairment of Na,K-ATPase (TABLE 7) occurs in a

TABLE 7. Effects of stationary magnetic fields on Na,K-ATPase in rats with lesioned by A1C13 nucleus basalis magnocellularis

Group	Na,K-ATPase (%)	V_{max} (nM)	K_m ($\times 10^{-5}$)
Lesioned-magnetic (AL-MF)	32.45	35.00	1.90
Aluminum (Al)	2.50	12.04	1.88
Intact control (IC)	6.78	24.01	2.70
P = AL vs. AL-MF	$P<0.001$	$P<0.001$	–
P = IC vs. AL	$P<0.001$	$P<0.001$	$P<0.001$

similar manner to age-related impairments.[82–84] The present results showed that $AlCl_3$ injected into the brain NBM significantly decrease ($P<.001$) Na,K-ATPase activity in rat erythrocyte membranes (from 6.78 ± 0.53 in intact animals to 2.57 ± 0.34 in the aluminium-treated group). In our experiments, magnetic stimulation of the brain with magnetic beads of 600 gauss flux density, permanent during the 10 days, restored the Na,K-ATPase activity inhibited by $AlCl_3$, and even significantly increased this activity more than few times (from 2.57 ± 0.34 in aluminum-treated group to 17.2 ± 1.02 in aluminium-treated magnetic-stimulated rats). Our study confirmed the well known findings that in the CNS aluminum blocks the action potential and electrical discharge of neurons[85,86] inhibits critical enzymes in the brain,[87] and reduces nervous system activity.[88–90] For Na,K-ATPase, the frequency maximum is about 60 Hz of pulsing EMFs,[91] suggesting that in this case EMFs coordinate with enzyme reaction.[92,93]

It is important to emphasize that while the forces acting on charge depend on the EMFs (the sum of endogenous and exogenous fields) and on the electrical characteristics of the particle, the dynamics of these particles as vibrations (e.g., electrons in atoms, atoms in molecules), displacement (mainly ions), and rotations (mainly molecules of water and neutral proteins) also depend on other properties (i.e., the mass, the kinds of particle atomic or molecular, the chemical bounds, etc.) and on neighboring particles.

CURRENT KNOWLEDGE OF CODING, RETRIEVAL, AND RECALL OF MEMORY IN THE AGING BRAIN: THEORETICAL APPROACH

The application of the laws of physics to living systems provides insight into their models of operation and suggests that this takes place in a controlled manner and close to the fundamental physical limits. Learning (engram) and memory (encoded information) can be classified as reflexive (repetitive/training located in hippocampus) and declarative on the basis of how information is stored and recalled. Declarative memory involves temporal lobe and associated structures of limbic system or the diencephalon and dramatically affect learning.

One of very simple form of short-term memory appears to be encoded by transient physical changes in the sensory receptor. A possible mechanism for encoding short-term memory is the storage of information in the form of ongoing neural activity maintained by excitatory feedback connections between neurons.

But how is long-term memory stored? How are changes maintained for years? Long-term memory is related to some plastic rather than dynamic changes—that is, to a permanent functional or structural change in the brain. Because of enduring nature of memory, it seems reasonable to postulate that in some way the changes must be reflected in long-term alterations of the connections between neurons. Memory traces are often localized in different place throughout the nervous system.[94] In AD, small soluble forms of amyloid-β appear to impair neuronal function and contribute to memory deficits *in vivo*.[95]

Also, many signaling pathways involve molecules called second messengers. These molecules may regulate short-term events (e.g., ion channels activity and neurotransmitter release) as well as long-term processes (e.g., synaptic plasticity, learning, and memory). Ca^{2+} itself can be a second messenger and it requires stimulating neuronal activities including the release of neurotransmitters. For memory to become long-term, it must go through a process known as consolidation. During consolidation, short-term memory is repeatedly activated. If during this repeated activation something interrupts the process—let's say, alcohol consumption—then short-term memory cannot be consolidated. It is believed that consolidation takes place in the hippocampus.

Neural networks are dynamic systems that depend both on the link matrix (matrices) and the dynamic levels and parameters of the neurons and synapses. To make a neural network that performs some specific task, we must choose how the units are connected to one another, and we must adjust the strengths of the influence[96] appropriately for one unit to influence another. Mental associations between ideas may derive in part from the overlap in their cell assemblies (the similarity factor) and in part from strong links between the assemblies (the contiguity factor). Erasure inhibition is necessary where the idea presented by a set of pyramidal neurons has had its turn on the mental stage, and it is time to activate other idea that was temporarily bypassed for processing and/or that is just now being excited. To reduce the noise level for idea recognition, it is desirable to clear what was already been completely processed "off the desk" so that it does not interfere with the processing of other ideas. While Landauer[97] doesn't measure everything, his estimate of memory capacity suggests that the capabilities of the human brain are more approachable than we had thought. While this might come as a blow to our egos, it suggests that we could build a device with the skills and abilities of a human being with little more hardware than we now have—if only we knew the correct way to organize that hardware!

Hypothesis of Chromosomal and Unsupervised Neural Network Application in the Aging Brain

Clearly, a behavior per se is not inherited: it is DNA that is inherited. The DNA of genes codes for proteins are important for the development, maintenance, and regulation of neural circuits that produce behavior.

The standard representation of genomic information by sequences of nucleotide symbols (triplets of nucleotides) or by sequences of amino acids in the corresponding polypeptide chains (for exons) has definite advantages for studying storage, search, and retrieval of genomic information.[98] Converting the DNA sequences into digital signals[99] opens the possibility of using signal processing methods in the analysis of genomic data.[99–101] The existence of large scale regularities, up to the scale of entire chromosomes, supports the view that extra-gene DNA sequences, which do not encode proteins, still play significant functional role. Such a view is in direct opposition to opinions prevalent at the time of publication of the first versions of the human genome sequence.[102,103] It was then believed that only the genes, containing information to synthesize proteins, were of real interest, while the remaining vast majority of the genome was simply "junk DNA." The genes have been considered the blueprints of any organism and the role of other endogenous and exogenous factors in the complex ontogeny, function, and dysfunction of the living has been diminished. This reductionist view, reminiscent of the old genetic concept "one gene–one trait," reformulated as "one gene–one protein," has also been motivated by the potential importance of genes for the pharmaceutical industry. The complexity of the proteome (the set of proteins existing in a cell) by far exceeds the complexity of its genome. The total number of genes in the human genome is only about 30,000, while the proteome comprises more that one million proteins. The protein coding is governed by the genetic code but processing of information depends on synaptic connections and transfer functions[104] and can be regulated by using a model of unsupervised neural networks.

Namely, processing neurons use logistic sigmoidal transfer functions in the autoassociator. To achieve an accurate reconstruction of an input example at its output, the autoassociator is implicitly forced to discover an appropriate nonlinear mapping of the original M-dimensional attribute space into a smaller n_2 dimensional space that captures the properties of the underlying distribution.[105] This requirement enforced specialization of the autoassociator for the distribution of training data, with the consequence of making large reconstruction errors on the examples represented in training data. Starting from a set of labeled sequences assigned to several classes of proteins and the distribution, the proposed learning approach consists of separate learning sequence distributions for each class.[106] Subsequently using the distribution models, a visualization step allows the evaluation of unlabeled sequences by measuring the extent to which they deviate from each training distribution model. Therefore, the selected sequences are distributionally[107] unusual with respect to available ordered and disordered proteins, but at the same time are more likely to have disordered properties. It is hoped that extended use of the proposed methodology by domain experts will prove its usefulness and provide more insight into taxonomy and nature of protein aging disorders, such as AD.

A greater challenge will be to assess the validity of a recall memory, recognizing, for example, that the telephone number or zipcode or name is almost correct, so that a new demand is made on the memory databanks (located in the neocortex) for information that can be recognized as correct. Thus, we have to consider the idea of two kinds of long-term memory: a databank memory that is stored in the neocortex, and a recognition memory that is in the conscious mind.

Finally, it can be concluded that stimulation (electrical or magnetic) acts as a mode of recall of past experience. We may regard this as a means for recovery of memories. It can be suggested that the storage of these memories is likely to be in cerebral areas close to the effective stimulation site. It is important to recognize that experimental recall is evoked from areas in the region of the disordered cerebral function displayed by the epileptic seizure.

EPILOGUE

This study has an interdisciplinary character par excellence and at the same time represents a great source of new ideas for young investigators (both experimenters and clinicians) and new questions regarding the phenomenological, cellular, and molecular levels that are involved in the encoding and transfer of information.

ACKNOWLEDGMENTS

We thank Professor Shoenfeld Yehuda, M.D., FCHR, Department of Medicine B and and Center for Autoimmune Disease, Sheba Medical Center, University of Tel Aviv, Tel Hashomer, Israel and Incumbent of Laura Schwartz-Kipp Chair for Research of Autoimmune Disease, for reading manuscript and his critical assistance. This work was supported by the Ministry of Science and Technology, Republic of Serbia.

[*Competing interests*: The authors declare that they have no competing financial interests.]

REFERENCES

1. COLEMAN, P., C. FINCH & J. JOSEPH. 1990. The need for multiple time points in aging studies. Neurobiol. Aging **11:** 1–2.
2. PIAZZA, P.V. *et al.* 1991. Dopamine activity is reduced in the prefrontal cortex to develop amphetamine self-administration. Brain Res. **567:** 169–174.
3. MEANEY, M.J. *et al.* 1988. Effects of neonatal handling on age-related impairments associated with the hippocampus. Science **239:** 766–768.
4. DELLU, F. *et al.* 1995. Reactivity to novelty during youth as a predictive factor of cognitive impairment in the elderly: a longitudinal study in rats. Brain Res. **653:** 51–56.
5. WEST, R. *et al.* 2002. Effects of time of day on age differences in working memory. J. Gerontol. B Psychol. Sci. Soc. Sci. **57:** 3–10.
6. KORDOWER, J.H. *et al.* 2001. Loss and atrophy of layer II entorhinal cortex neurons in elderly people with mild cognitive impairment. Ann. Neurol. **49:** 202–213.
7. CHOE, L.H. *et al.* 2002. Studies of potential cerebrospinal fluid molecular markers for Alzheimer's disease. Electrophoresis **23:** 2247–2251.
8. LOVELL, M.A. & W.R. MARKESBERY. 2001. Ratio of 8-hydroxyguanine in intact DNA to free 8-hydroxyguanine is increased in Alzheimer's disease ventricular cerebrospinal fluid. Arch. Neurol. **58:** 392–396.
9. DEMATTOS, R.B. *et al.* 2002. Brain to plasma amyloid-beta efflux: a measure of brain amyloid burden in mouse model of Alzheimer's disease. Science **295:** 2264–2267.
10. DODART, J.C. *et al.* 2002. Immunization reverses memory deficits without reducing brain a-beta burden in Alzheimer's disease model. Nature Neurosci. **5:** 452–457.
11. PETTIT, D.L., Z. SHAO & J.L. YAKEL. 2001. Beta-Amyloid (1-42) peptide directly modulates nicotinic receptors in the rat hippocampal slice. J. Neurosci. **21:** 120.
12. SCHMIDT, R. *et al.* 2002. Early inflammation and dementia: a 25-year follow-up of the Honolulu-Asia aging study. Ann. Neurol. **52:** 168–174.
13. NEWHOUSE, P.A. *et al.* 2001. Nicotinic treatment of Alzheimer's disease. Biol. Psychiatry **49:** 268–278.
14. NEUNDORFER, M.M. *et al.* 2001. A longitudinal study of the relationship between levels of depression among persons with Alzheimer's disease and level of depression among their family caregivers. J. Gerontol. B Psychol. Sci. Soc. Sci. **55:** 301–313.
15. AISEN, P.S. *et al.* 2003. Effects of rofecoxib or naproxen vs. placebo on Alzheimer's disease progression: a randomized controlled trial. J. Am. Med. Assoc. **289:** 2819–2826.

16. POPOVIC, M. *et al.* 1997. Open field behavior in nucleus basalis magnocellularis-lesioned rats treated with physostigmine and verapamil. Int. J. Neurosci. **91:** 181–188.
17. HEPLER, D. *et al.* 1985. Memory impairments following basal forebrain lesions. Brain Res. **346:** 8–14.
18. KESSLER, J., H. MARKOVITSCH & G. SIGG. 1986. Memory related role of the posterior cholinergic system. Int. J. Neurosci. **30:** 101–119.
19. LOCONTE, G. *et al.* 1982. Lesions of cholinergic forebrain nuclei: changes in avoidance behavior, and scopolamine actions. Pharmacol. Biochem. Behav. **17:** 933–937.
20. MIYAMOTO, M. *et al.* 1985. Lesioning of the rat basal forebrain leads to memory impairments in passive and active avoidance tasks. Brain Res. **328:** 97–104.
21. POPOVIC, M. *et al.* 1997. Effects of physostigmine and verapamile on active avoidance in an experimental model of Alzheimer's disease. Int. J. Neurosci. **90:** 87–97.
22. POPOVIC, M. *et al.* 1996. Behavioral and adaptive status in an experimental model of Alzheimer's disease in rats. Int. J. Neurosci. **86:** 281–299.
23. FISHER, A. *et al.* 1991. Cis-2 methyl-spirol (1,3-oxathiolanee-5-3') quinuclidine, an M1 selective cholinergic agonist, attenuates cognitive dysfunctions in an animal model of Alzheimer's disease. J. Pharmacol. Exp. Ther. **257:** 392–403.
24. OGISHMA, M. *et al.* 1992. Differential effects of i.c.v. AF64A injection, nucleus basalis of Meynert lesion, and scopolamine on place navigation and open-field behavior of rats. Yakubutsu Seishin Kodo **12:** 85–92.
25. WALSH, R.N. & R.A. CUMMINS. 1976. The open-field test: a critical review. Psychol. Bull. **83:** 482–504.
26. WENK, H. 1989. The nucleus basalis magnocellularis Meynert (NBmM) complex—a central integrator of coded 'limbic signals' linked to neocortical modular operation? A proposed (heuristic) model of function. J. Hirnforsch. **30:** 127–151.
27. POPOVIC, M. *et al.* 1997. Humoral and cell-mediated immune responses following lesions of the nucleus basalis magnocellularis in the rat. Int. J. Neurosci. **89:** 165.
28. MONSONEGO, A. & H.L. WEINER. 2003. Immunotherapeutic approaches to Alzheimer's disease. Science **302:** 834–838.
29. SCHORI, H. *et al.* 2002. Several immunodeficiency has opposite effect on neuronal survival in glutamate-susceptible and resistant mice: adverse effect of B cells. J. Immunol. **169:** 2861–2865.
30. COHEN, I.R. & M. SCHWARTZ. 1999. Autoimmune maintenance and neuroprotection of the central nervous system. J. Immunol. **100:** 111–114.
31. MILLER, R.A. 1991. Accumulation of hyporesponsive, calcium extruding memory T cells as a key feature of age-dependent immune dysfunction. Clin. Immunol. Immunopathol. **58:** 305–317.
32. ARAGA, S., H. KAGIMOTO & K. TAKAHASHI. 1991. Reduced natural killer activity in patients with dementia of the Alzheimer's type. Acta Neurol. Scand. **84:** 259–263.
33. SKIAS, D. *et al.* 1985. Senile dementia of Alzheimer's type (SDAT): reduced T8$^+$ cell-mediated suppressor activity. Neurology **35:** 1635–1638.
34. FAUMAN, M.A. 1982. The central nervous system and immune system. Biol. Psychiatry **17:** 1459–1482.
35. JANKOVIC, B.D. & N.H. SPECTOR. 1986. Effects on the immune system of lesioning and stimulation of the nervous system: neuroimmunomodulation. *In* Enkephalins and Endorphins. Stress and the Immune System. N.P. Plotnikoff, R.E. Faith, A.J. Murgo & R.A. Good, Eds.: 189–220. Plenum. New York.
36. JANKOVIC, B.D., K. JOVANOVA-NESIC & B.M. MARKOVIC. 1988. Neuroimmunomodulation: potentiation of delayed hypersensitivity and antibody production by chronic electrical stimulation of rat brain. Int. J. Neurosci. **39:** 153–164.
37. JANKOVIC, B.D. 1973. Structural correlates of immune microenvironment, *In* Microenvironmental Aspects of Immunity, B.D. Jankovic, Ed.: 1–4. Plenum. New York.
38. JOVANOVA-NESIC., V. NIKOLIC & B.D. JANKOVIC. 1993. Locus coeruleus and immunity. II. Suppression of experimental allergic encephalomyelitis and hypersensitivity skin reactions in rats with lesioned locus ceruleus. Int. J. Neurosci. **68:** 289–294.
39. KORNEVA, E.A. & V.M. KLIMENKO. 1976. Neuronale hypothalamus activitat und homeostatische reactionen. Ergeb. Exp. Med. **23:** 373–382.

40. ROSZMAN, T.L. *et al.* 1982. Hypothalamic-immune interactions. II. The effect of hypothalamic lesions on the ability of adherent spleen cells to limit lymphocyte blastogenesis. Immunology **45:** 737–742.
41. ABRAMSKY, O. *et al.* 1987. Effect of hypothalamic lesions on experimental autoimmune disease in rats. Ann. NY Acad. Sci. **496:** 360–365.
42. SPECTOR, N.H. 1980. The "central state" of the hypothalamus in health and disease: old and new concepts. *In* Physiology of the Hypothalamus. P. Morgane & J. Paksepp, Eds.: 453–517. Dekker. New York.
43. MILLER, R. 1991. Accumulation of hyporesponsive, calcium-extruding memory T cells as a key feature of age-dependent immune dysfunction. Clin. Immunol. Immunopathol. **58:** 305–317.
44. HICKEY, W.F., B.L. HSU & H. KIMURA. 1991. T-lymphocyte entry into the central nervous system. J. Neurosci. Res. **28:** 254–260.
45. SLOAN, D.J., M.J. WOOD & H.M. CHARLTON. 1991. The immune response to intracerebral neural graft. Trends Neurosci. **14:** 8–14.
46. GAGE, F.H., J. RAY & L.J. FISHER. 1995. Isolation, characterization and use of stem cells from the CNS. Annu. Rev. Neurosci. **18:** 159–192.
47. TUSZINSKY, M. & F.H. GAGE. 1995. Maintaining the neural phenotype after injury in the adult CNS. Mol. Neurobiol. **10:** 151–166.
48. DUAN, W.M., H. WINDER & P. BRUNDIN. 1995. Temporal pattern of host response against intrastriatal grafts of syngeneic, allogeneic or xenogeneic embryonic neural tissue in rats. Exp. Brain Res. **104:** 227–242.
49. JOVANOVA-NESIC, K. *et al.* 1995. Hypothalamic neural graft and immunity in the CNS. The 1995 International Co-Conference on Environmental Pollution an Neuroimmune Interaction an Environmental (ICON): 161.
50. NIETO-SAMPEDRO, M. *et al.* 1987. Effects of conditioning lesions on transplant survival. connectivity and function. Role of neurotrophic factors. Ann. NY Acad. Sci. **495:** 108–119.
51. SOARES, H.D., G.P. SINSON & T.K. MCINTOSH. 1995. Fetal hippocampal transplants attenuate CA3 pyramidal cell death resulting from fluid percussion brain injury in the rat. J. Neurotrauma **12:** 1059–1067.
52. STREILAIN, J.W. 1995. Unraveling immune privilege. Science **270:** 1158–1159.
53. FADEN, A.I. & S. SALZAM. 1992. Pharmacological strategies in CNS trauma. Trends Pharmacol. Sci. **3:** 29–35.
54. SMITH D.H., K. CASEY & T.K. MCINTOSH. 1995. Pharmacological therapy for traumatic brain injury: experimental approach. New Horiz. **3:** 562–572.
55. SUN, D., J.N. FHITAKER & B.D. WILSON. 1999. Regulatory T cells in experimental allergic encephalomyelitis. III. Comparison of disease resistance in Lewis and Fisher 344 rats. Eur. J. Immunol. **29:** 1101–1106.
56. FRIEDMAN, C. *et al.* 1999. Apolipoprotein E-epsilon4 genotype predicts a poor outcome in survivors of traumatic brain injury. Neurobiology **52:** 244–248.
57. MATTSON, M.P. *et al.* 2000. Presenilin-1 mutation increases neuronal vulnerability to focal ischemia *in vivo* and to hypoxia and glucose deprivation in cell culture: involvement of perturbed calcium homeostasis. J. Neurosci. **20:** 1358–1364.
58. KIPNIS, J. *et al.* 2001. Protective autoimmunity in acute and chronic CNS disorders: therapeutic vaccines. J. Neurosci. **21:** 4564–4571.
59. SHEVACH, E.M. 2000. Regulatory T cells in autoimmunity. Annu. Rev. Immunol. **18:** 423–449.
60. BELLONE, M. *et al.* 1999. Experimental myasthenia gravis in congenic mice. Sequence mapping and H-2 restriction of T helper epitopes on the alpha subunits of *Torpedo californica* and murine acetylcholine receptors. Eur. J. Immunol. **21:** 2303–2310.
61. GOEBELS, N. *et al.* 2000. Repertoire dynamics of autoreactive T cells in multiple sclerosis patients and healthy subjects: epitope spreading versus clonal persistence. Brain **123:** 508–518.
62. LIWINGSTON, K.D. *et al.* 1995. Susceptibility to actively-induced murine experimental allergic encephalomyelitis is not linked to genes of T cell receptor or CD3 complexes. Autoimmunity **21:** 195–201.

63. KUCHROO, V.K. & H.L. WEINER. 1998. Antigen-driven regulation of experimental autoimmune encephalomyelitis. Res. Immunol. **149:** 759–771.
64. JANKOVIC, B.D., L. RAKIC & M. SESTOVIC. 1969. Changes in electrical activity of the cockroach *Blatta orientalis* L. brain induced by anti-lobster brain antibody. Experientia **25:** 1049–1050.
65. JANKOVIC, B.D., J.S. JAKULIC & J. HORVAT. 1982. Delayed skin hypersensitivity reactions to human brain S-100 protein in psychiatric patients. Biol. Psychiatry **17:** 687–697.
66. JOVANOVA-NESIC, K. & L. RAKIC. 2004. Regulatory effects of S-100 protein in experimental allergic encephalomyelitis after ethanol consumption. 4th Int. Congress on Autoimmunity. Budapest. **11:** 43.
67. KELLER, R.J., A. GOVINDARAJAN, H-Y. JUNG, *et al.* 2004. Translational control by MAPK signaling in long-term synaptic plasticity and memory. Cell **115:** 467–469.
68. SHERRINGTON, C. 1906. The Integrative Action of the Central Nervous System. Yale University Press. New Haven.
69. WESTRICK, R.J. *et al.* 1988. Alcohol induced blackout. Psychiatry Res. **24:** 201–209.
70. PARKS, D.C. *et al.* 1997. Effects of age on event-based and time-based prospective memory. Psychol. Aging **12:** 314–327.
71. CABEZA, R. *et al.* 1997. Age-related differences in neural activity during memory encoding and retrieval: a positron emission tomography study. Neuroscience **17:** 391–400.
72. MESULAM, M.M., C. GEULA & M.A. MORAN. 1987. Anatomy of cholinesterase inhibition in Alzheimer's disease: Effect of physostigmine and tetrahydroaminoacridine on plaques and tangles. Ann. Neurol. **45:** 345–349.
73. POPOVIC, M. *et al.* 1995. Effect of verapamil on spontaneous motor activity and two-way avoidance learning in nucleus basalis-lesioned rats. Eur. J. Neurosci. **8:** 171.
74. BROWN, F.A. 1969. A hypothesis for extrinsic timing of circadian rhythm. Can. J. Biol. **47:** 287–298.
75. SANDYK, R. 1995. Improvement in short-term visual memory by weak electromagnetic fields in Parkinson's disease. Int. J. Neurosci. **81:** 67–82.
76. SANDYK, R., P.A. ANNINOS & N. TSAGAS. 1991. Magnetic fields and seasonality of affective illness: implication for therapy. Int. J. Neurosci. **58:** 261–267.
77. JOVANOVA-NESIC, K. & D. VUJIC. 1988. Magnetic brain stimulation and humoral immune response in the rat with lesioned anterior hypothalamus. Yugoslav Physiolog. Pharmacol. Acta. Proceedings of the 14th Congress of the Yugoslav Physiological Society, Belgrade, Sept. 20–24: 163–165.
78. JOVANOVA-NESIC, K. & A. SKOKLJEV. 1990. Magnetic brain stimulation and immune response in the rat with lesioned brain structure. Acupuncture Electro-Therapeutic Res. Int. J. **15:** 27–35.
79. SANDYK, R. 1994. Alzheimer's disease: improvement of visual memory and visuoconstructive performance by treatment with picotesla range magnetic fields. Int. J. Neurosci. **76:** 185–225.
80. STRUYS-PONSAR, C., O. GUILLARD & P. VAN DEN BOSCH DE AGUILAR. 2000. Effects of aluminum exposure on glutamate metabolism is a possible explanation for its toxicity. Exp. Neurol. **163:** 157–164.
81. JOVANOVA-NESIC, K. & L. RAKIC. 2004. Ethanol consumption affecting some brain proteins and cortical plasticity might disturb the neural network of long-term memory. The 7th Seminar on Neural Network Applications in Electrical Engineering, NEUREL-2004. IEEE **7:** 165–170.
82. CAUHAN, N.B., J.M. LEE & G.J. SIEGEL. 1997. Na,K-ATPase mRNA level and plaque load in Alzheimer's disease. J. Mol. Neurosci. **9:** 151–166.
83. SIEGEL, G.J., N.B. CAUHAN & J.M. LEE. 1998. Age-related Na,K,-ATPase mRNA expression and Alzheimer's disease. *In* Progress in Alzheimer's and Parkinson's Diseases. A. Fisher, I. Hanin & M. Yoshida, Eds.: 113–119. Plenum. New York.
84. HIKMET, K., O. PERNUR & O. BARIA. 2002. Comparison of activities of Na,K-ATPase in brains of rats. Gerontology **48:** 279–281.
85. XIE, T.D. *et al.* 1994. Recognition and processing of randomness fluctuating electric signals by Na^+, K^+-ATPase. Biophys. J. **67:** 1247–1251.

86. SAVALAINEN, K.M. *et al.* 1998. Interactions of excitatory neurotransmitters and xeno-biotics in excitatory and oxidative stress. Toxicol. Lett. **28:** 363–367.
87. HASPEL, P. *et al.* 1986. Erythrocyte cations and Na,K-ATPase pump activity in ath-letes and sedentary subjects. Eur. J. Appl. Physiol. Occup. Physiol. **55:** 249.
88. KVITNICKAYA-RUZHOVA, T.Y. & A.L. SHAPENKO. 1992. A comparative ultracytochem-ical and biochemical study of the choroid plexus ATPase in the aging brain. Tsitolo-gya **34:** 81–87.
89. LIGURI, G. *et al.* 1990. Changes in Na,K-ATPase, Ca^{2+}-ATPase and some soluble enzymes related to energy metabolism in brains of patients with Alzheimer's disease. Neurosci. Lett. **112:** 338–342.
90. FLICKER, C. *et al.* 1983. Behavioral and neurochemical effects following neurotoxic lesions of a major cholinergic input to the cerebral cortex in the rat. Pharmacol. Bio-chem. Behav. **18:** 973–981.
91. BLANK, M. & L. SOO. 2001. Optimal frequencies for magnetic acceleration of cyto-chrome oxidase and Na,K-ATPase reactions. Bioelectrochemistry **53:** 171–174.
92. BINDERA, C., T. SIMPLACEANU & S. POPESCU. 2001. The effect of magnetic fields on Na^+ transport through human erythrocyte membranes. Stud. Universitatis Babes Boliat, Physica Special issue: 1–12.
93. BLANK, M. & R. GOODMAN. 2003. Initial interactions in electromagnetic fields-induced biosynthesis. J. Cell. Physiol. **199:** 359–364.
94. MELENDES, R.I. *et al.* 2003. Alcohol stimulates the release of dopamine in the ventral pallidum but not in the globus pallidus: a dual probe microdialysis study. Neuropsy-chopharmacology **28:** 939–946.
95. ZERBINATTI, C.V. *et al.* 2004. Increased soluble amyloid-β peptide and memory defi-cits in amyloid model mice overexpressing the low density lipoprotein receptor-related protein. Proc. Natl. Acad. Sci. USA **101:** 1075.
96. SHALVI, D. & N.A. DECLARIS. 1997. An unsupervised neural networks approach to medical data mining. CIN: Computers, Informatics, Nursing **22:** 101–105.
97. LANDAUER, T.K. 1986. How much do people remember? Some estimates of the quan-tity of learned information in long-term memory. Cogn. Sci. **10:** 477–493.
98. CRISTEA, P.D. 2004. A model of chromosome longitudinal structure. The 7th Seminar on Neural Network Applications in Electrical Engineering, NEUREL-2004. IEEE **7:** 145–150.
99. CRISTEA, P.D. 2002. Conversation of nitrogenous base sequences into genomic sig-nals. J. Cell. Mol. Med. **6:** 279–303.
100. CRISTEA, P.D. 2004. Multiresolution phase analysis of genomic signals. 1st Int. Symp. on Control, Communication, and Signal Processing in Biological Science, 21–24 March, Tunisia, pp. 743–746.
101. CRISTEA, P.D. 2003. Large scale features in DNA genomic signals. Elsevier Sign. Proc. **83:** 871–888.
102. INTERNATIONAL HUMAN GENOME SEQUENCING CONSORTIUM. Initial sequencing and analysis of the human genome. Nature **409:** 860–869.
103. VENTER, J.C. 2001. Draft analysis of the human genome by Celera Genomics. Science **291:** 1304.
104. POKRAJAC, D. 2004. Application of unsupervised neural networks in data mining. The 7th Seminar on Neural Network Applications in Electrical Engineering, NEUREL-2004. IEEE **7:**17–20.
105. BISHOP, C.M. 1995. Network for Pattern Recognition. Oxford University Press. Oxford.
106. VUCETIC, S. 2003. Detection of unusual biological sequences using class-conditional distribution models. Proc. 3rd SIAM Data Mining Conference, pp. 279–283.
107. POKRAJAC, D. *et al.* 2003. Localized neural network based distributional learning for knowledge discovery in protein databases. Proc. IEEE/INNS Int. Conf. on Neural Networks, Budapest, July, 2004.

Hormones Are Key Actors in Gene X Environment Interactions Programming the Vulnerability to Parkinson's Disease

Glia as a Common Final Pathway

BIANCA MARCHETTI,[a,b] PIER ANDREA SERRA,[b] FRANCESCA L'EPISCOPO,[a] CATALDO TIROLO,[a] SALVO CANIGLIA,[a] NUCCIO TESTA,[a] SERENA CIONI,[c] FLORINDA GENNUSO,[a] GAIA ROCCHITTA,[b] MARIA SPERANZA DESOLE,[b] MARIA CLORINDA MAZZARINO,[c] EGIDIO MIELE,[b] AND MARIA CONCETTA MORALE[a]

[a]OASI Institute for Research and Care on Mental Retardation and Brain Aging (IRCCS), Neuropharmacology Section, 94018 Troina, Italy

[b]Department of Pharmacology, Faculty of Medicine, University of Sassari, 07100 Sassari, Italy

[c]Department of Biomedical Sciences, School of Pharmacy, University of Catania, Catania, Italy

ABSTRACT: Alterations in developmental programming of neuroendocrine and immune system function may critically modulate vulnerability to various diseases. In particular, genetic factors, including gender, may interact with early life events such as exposure to hormones, endotoxins, or neurotoxins, thereby influencing disease predisposition and/or severity, but little is known about the role of the astroglial cell compartment and its mediators in this phenomenon. Indeed, in the context of innate inflammatory mechanisms, a dysfunction of the astroglial cell compartment is believed to contribute to the selective degeneration of dopaminergic (DA) neurons in the substantia nigra pars compacta in Parkinson's disease (PD) and 1-methyl-4-phenyl-1,2,3,6-tetrahydropyridine (MPTP) model of PD. Hence, in response to brain injury the roles of astrocytes and microglia are very dynamic and cell type–dependent, in that they may exert the known proinflammatory (harmful) effects, but in certain circumstances they can turn into highly protective cells and exert anti-inflammatory (beneficial) functions, thereby facilitating neuronal recovery and repair. Here, we summarize our work suggesting a chief role of hormonal programming of glial response to inflammation and oxidative stress in MPTP-induced loss of DA neuron functionality and demonstrate that endogenous glucocorticoids and the female hormone estrogen (E_2) inhibit the aberrant neuroinflammatory cascade, protect astrocytes and microglia from programmed cell death, and stimulate recovery of DA neuron functionality, thereby triggering the repair process. The overall results highlight glia as a final common pathway directing

Address for correspondence: Bianca Marchetti, OASI Institute for Research and Care on Mental Retardation and Brain Aging (IRCCS), Section of Neuropharmacology, Via Conte Ruggero 73, 94018 Troina (EN), Italy. Voice: +39-935-936111; fax +39-935-653327.
bianca.marchetti@oasi.en.it

Ann. N.Y. Acad. Sci. 1057: 296–318 (2005). © 2005 New York Academy of Sciences.
doi: 10.1196/annals.1356.023

neuroprotection versus neurodegeneration. Such recognition of endogenous glial protective pathways may provide a new insight and may contribute to the development of novel therapeutic treatment strategies for PD and possibly other neurodegenerative disorders.

KEYWORDS: Parkinson's disease; glia; genes; environment; hormones; neuroinflammation; neurodegeneration; dopaminergic neurons; neuroprotection

INTRODUCTION

Experimental and epidemiological evidence strongly suggests that early events occurring in fetal life powerfully modulate the individual predisposition to different pathologic states in the adult life, including a certain number of neuromental and autoimmune disorders of the central nervous system (CNS).[1–5] In fact, there is a growing appreciation that pre-perinatal exposure to abnormal hormone levels or to pathogens may alter the developmental programming of neuroendocrine and immune reactions, which may produce dysfunctions later in life.[1–17] Bidirectional communication between the neuroendocrine and immune systems[18–20] represents a critical level modulating the susceptibility to certain inflammatory, autoimmune, and degenerative disorders of the CNS. For example, the hypothalamic-pituitary-adrenocortical (HPA) axis is subjected to programming by early life events, and early life exposure to endotoxin can alter HPA function and predisposition to inflammation.[2–5,11,21–25] Likewise, neonatal manipulations of hypothalamic-pituitary-gonadal (HPG) axis functionality during the first week of life result in long-lasting influences in the development of T cell– and B cell–dependent functions in adult animals, whereas immune system activation during neonatal life has long-lasting consequences for reproductive physiology.[7–10,13–16,26–31] Signals generated by the HPG axis, such as estrogens, and the hypothalamic decapeptide, luteinizing hormone–releasing hormone (LHRH) also remarkably interact at different levels of the immune axis.[17–20] Moreover, estrogens are important modulators of the stress (HPA) axis, thereby further modulating neuroendocrine-immune crosstalk in a gender-specific fashion.

Although several experimental studies have explored the mechanisms underlying long-term effects of pre-perinatal hormonal manipulations affecting CNS functions, little is known about the role of the astroglial cell compartment and its mediators in this phenomenon (FIG. 1).[32]

Indeed, accumulating evidence clearly suggests a pivotal role for glia and neuron–glial crosstalk both in the adult and aging brain. In the CNS, astroglial cells are recognized to play active roles in both health and disease states and to influence the development of the brain's response to a variety of insults. Astrocytes, which represent the major glial population, and microglial cells, like other tissue macrophages, are key components of the neuroendocrine–immune axis and, as such, are responsive to endocrine, neural, and immune factors.[32–39] In particular, astrocytes and macrophages/microglial cells express hormone receptors (i.e., glucocorticoid receptors [GRs] and estradiol receptors [ERs]) and are both a source and target of cytokine, growth, and neurotrophic factor activities in the brain.[26–36]

Although developmental and constitutive functions of the majority of these immunoregulatory molecules in the physiology of the normal, immune-privileged CNS are still not generally accepted, when CNS homeostasis is disturbed as a result of trauma, stroke, ischemia, infection, or degenerative processes, certain cytokines increase, as a result of blood–brain barrier disruption, or from local synthesis by invading immune cells. Then, most if not all neuropathologies are associated with glial cell activation.[37–41] Thus, activated astrocytes and microglia serve as endogenous sources of various cytokines involved in the orchestration of cellular responses aimed at rapid re-establishment of tissue integrity and subsequent repair.[40,41] Release of products of activated glia are thought to be important for initiating and guiding the infiltration of immune cells and for coordinating their activities in the brain tissue. Although glial responses may be considered beneficial mechanisms, when local production of cytokines exceeds the appropriate range, cytokine-mediated neurotoxicity may lead to severe neuronal damage.[32,37–39]

Besides others, one endogenous regulatory feedback mechanism may be represented by the hormonal background, and, in particular, glucocorticoids (GCs) and estrogens (E_2) are known regulators of astroglial cell function.[34,36,42,43] There is abundant evidence that cultured glia possess corticosteroid and estrogen receptors. Ligand binding studies have demonstrated the presence of a single population of GRs in both astrocytes, oligodendrocytes, and microglia. In glial cells, GCs and estrogens are known to modulate the expression of a variety of glial proteins, including

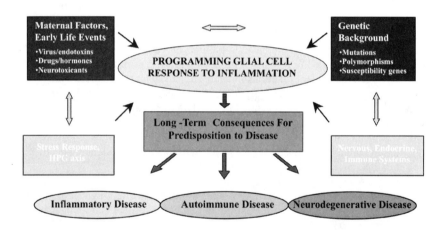

FIGURE 1. Schematic representation of the impact of perinatal genetic, hormonal, and environmental interactions on inflammatory glial cell response and individual resistance or susceptibility to inflammatory diseases during adult life. Genetic factors (e.g., sex, gene mutations, polymorphisms/susceptibility genes) can interact with maternal hormonal factors and external agents to which mother and fetus are exposed (drug treatments, bacteria, viruses, endotoxins, and/or environmental toxins such as heavy metals or pesticides) to alter the development of the neuroendocrine–immune system, in particular, the HPA and the HPG axis. The pivotal target of the overall interactions is glia, a key component of the neuroendocrine–immune system. Thus, an altered dialogue between the neuroendocrine and the immune system during development may irreversibly shape glial cells and "program" long-term effects in the mechanisms regulating immune responsiveness to inflammation, thereby contributing to individual vulnerability, propensity and predisposition to inflammatory, autoimmune, and neuromental disorders.[4,5,11,25,30,31,112]

glial fibrillary acidic protein (GFAP).[42,43] Of major interest, GCs and estrogens are potent inhibitors of inducible nitric oxide (iNOS)–derived NO in activated astrocytes and macrophage/microglial cells.[11,25] The inhibition of iNOS/NO by GC-bound GR is consistent with the ability of GR to inhibit NF-κB–mediated transcription of the iNOS gene by GC-induced increase in Ik-B protein levels or through protein–protein interactions between GR and NF-κB, thereby inhibiting interaction of the latter with iNOS promoter and preventing the induction of iNOS transcription.[11,25] Likewise, the inhibition of iNOS/NO by E_2 also appears to involve the NF-κB system (see next sections). Then, the major endogenous anti-inflammatory molecules, GCs, and the primary sex steroid hormone, E_2, may play active roles under *in vivo* conditions in which increased CNS levels of cytokines would have several adverse consequences. In fact, during inflammation, it seems highly possible that a sophisticate interplay between hormones of the stress and reproductive axes participated in the temporally and spatially correct expression of iNOS/NO in different organs and tissues, leading to the elimination of inflammatory agents with minimal tissue damage (see Marchetti *et al.*[25]).

In the present work, a prospective analysis of the impact of glucocorticoids and estrogens in modulating the vulnerability of nigrostriatal dopaminergic neurons to the neurotoxin 1-methyl-4-phenyl-1,2,3,6-tetrahydropyridine (MPTP)–induced dopaminergic (DA) neuron degeneration, a well-accepted animal model for Parkinson's disease (PD), is summarized. This work highlights the role of activated astrocytes, macrophages, and microglial cells as key elements of the neuroendocrine–immune dialogue involved in hormonal programming of CNS response to neuroinflammation. On the basis of our results, a hypothetical model is presented whereby the efficiency of the HPA–HPG immune interactions via steroid receptor–nitric oxide crosstalk represents a critical step in dictating susceptibility versus resistance to PD and possibly other degenerative CNS diseases.

PARKINSON'S DISEASE, GENES, AND HORMONAL PROGRAMMING OF VULNERABILITY: KEY ROLE FOR NEURON–GLIAL DIALOGUE

PD is a chronic progressive degenerative disorder characterized by the selective degeneration of mesencephalic DA neurons in the substantia nigra pars compacta (SN). The loss of dopaminergic afferents to the striatum and putamen results in extrapyramidal motor dysfunction, including tremor, rigidity, and bradykinesia (for extensive reviews, see Olanow *et al.*[44]). The mechanisms responsible for DA neuron degeneration in PD are still unknown, but oxidative stress, excitotoxicity, depletion of endogenous antioxidants, reduced expression of trophic factors, and dysfunction of protein degradation system are believed to participate in the cascade of events leading to DA neuron death.[44–48] Recent evidence clearly indicates that neuroinflammatory mechanisms may play an important role in the pathogenesis of PD and inhibition of inflammation proposed as a promising therapeutic intervention (FIG. 2).[48–58] Accordingly, the important role of glial-derived cytotoxic mediators is underlined by the ability of a number of genetic and pharmacological manipulations aimed at reducing different proinflammatory mediators, to significantly attenuate DA neurotoxicity in experimental models of PD. Conversely, a deficiency in major

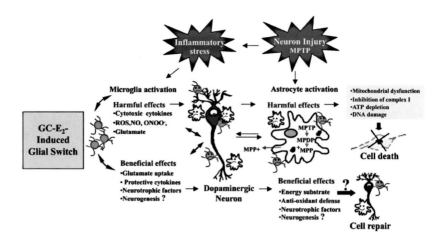

FIGURE 2. A schematic representation of detrimental and beneficial inflammatory pathways in PD. DA neurons in the SN represent a preferential target for inflammatory stressors, due to enzymatic and nonenzymatic autooxidation of DA generating H_2O_2, to the high toxicity of DA metabolites, and the interactions between iron (which is highly concentrated in SN) and H_2O_2 in the Fenton reaction, leading to highly toxic radicals. Injury of these neurons, as a result of specific chemical insults (e.g., MPTP), and/or a combination of genetic/ environmental factors (viruses, endotoxin, pesticides, susceptibility factors) leads to marked glial cell activation. MPTP is metabolized by astroglia to 1 methyl-4-phenylpyridinium (MPP+), which is concentrated in DA neurons. Here it inhibits cellular respiration, generates oxygen free radicals and NO, and initiates a cascade of cytotoxic events culminating in reduced DA release, locomotor deficits, and ultimately cell death by apoptosis. Astrocytes, which express glutathione at high levels, can protect neurons by scavenging radicals and glutamate, by harboring receptors for endogenous anti-inflammatory molecules (such as GCs), by providing energy support, trophic factors, "protective" cytokines, and possibly by stimulating repair processes. On the other hand, under conditions of chronic inflammatory stress, activated astrocytes and microglia may become dysfunctional and overexpress a variety of cytoyoxic mediators eventually resulting in DA neuron death.[25,37] The hypothetical role of endogenous GCs and estrogens (E_2) in "switching" the "harmful" into a "beneficial, protective" glial phenotype is illustrated.

endogenous anti-inflammatory signaling pathways in GR-deficient mice drastically exacerbates the vulnerability of nigral DA neurons to MPT neurotoxicity.[25]

Although several genes that cause certain forms of inherited PD (<10% cases) have been identified, most cases (>90%) appear to be sporadic and likely represent an interplay between both genetic and environmental influences.[59] More men than women develop PD, and aging and menopause in women (estrogen deficiency)[60] are recognized risk factors. Polymorphisms in candidate genes involved in dopamine metabolism, mitochondrial function, lipoprotein metabolism, and xenobiotic detoxification have been described. Polymorphisms of inflammatory genes have been also reported, including iNOS,[61] IL-6, and estrogen receptor beta (ER-beta) gene.[62] In addition, smoking and pesticides affect the probability of developing PD.[58,62] Rural living, dietary factors, exposure to metals, head injury, and exposure to infectious diseases during childhood have also been suggested to increase risk.[59] Therapeutic interventions generally consist in the management of motor symptoms with either L-DOPA or dopaminergic agents, but they do not delay progression. In humans, epidemiological evidence that chronic use of nonsteroidal anti-inflammatory drugs reduces risk by ~45%. Several anti-inflammatory agents and the tetracycline derivative minocycline are effective in experimental models of the disease.[32,37,52–58]

Within this context and of particular mention, genetic factors may interact with early life events such as exposure to hormones, endotoxins, or neurotoxins, thereby influencing disease predisposition and/or severity. In addition, developmental exposure to environmental toxins (such as pesticides/herbicides), either alone and/or in concert with other environmental (i.e., endotoxins, hormonal dysfunctions) and/or genetic "predisposing" factors, may synergistically increase dopaminergic neuron vulnerability.[11,25,59,63–66] Within this framework, we here develop the hypothesis that hormonal programming of glial response to inflammation dictates individual differences in disease susceptibility (FIG. 1).

In this respect, it seems particularly important to underline the common origin, organization, and functional integration of astrocytes in the brain and the thymic epithelial cell, in the thymus. Hence, brain glial cells and thymic cells arise from the same source (i.e., the neural crest).[67] Furthermore, the multipotential stem cells that form the glia of the brain, the enteric nervous system, and the thymus develop in the same region of the embryo.[67,68] Then, the neuropeptidergic, hormonal, growth factor, and transmitter network of astrocyte and thymocyte microenvironment is very similar.[69–72] Hence, it is not surprising that the HPA and HPG (and specifically glucocorticoids, LHRH, and estrogens) axes powerfully and dynamically interact with both thymocyte and astroglial cells, then realizing a further level of neuroendocrine–immune integration. In fact, thymocytes and astrocytes respond to a wide variety of mediators including cytokines and use similar transduction mechanisms.[19,20]

A key component is represented by the interactions between the neuroendocrine and immune systems during development and the response of the astroglial cell compartment during adult life (FIG. 1). In particular, the two master axes, which are responsible on one hand for the stress response and modulation of inflammation, and on the other, for reproductive hormone homeostasis, are involved: the HPA and the HPG axes. Then, GCs, the most potent anti-inflammatory and immunosuppressive agents known, via their cognate receptors (GRs), represent crucial vulnerability factors in experimentally induced parkinsonism via critical neuroendocrine–immune interactions.[11,25] Hence, early embryonic life exposure to GR antisense RNA in transgenic mice, leading to an abnormal response to stressful, inflammatory, and immune stimuli,[24,73–76] drastically increases DA neuron vulnerability to MPTP, via an exacerbation of the neuroinflammatory reaction (FIG. 3).[11,25] Conversely, environmental enrichment can confer resistance to MPTP,[64] thereby underlying a crucial role for the environment and the HPA axis, which is highly sensitive to environmental manipulations,[21–23] as critical factors involved in nigral DA neuron preservation. Thus, an altered dialogue between the neuroendocrine and the immune systems via the HPA axis, during development, may irreversibly shape glial cells and "program" long-term effects in the mechanisms regulating immune responsiveness to inflammation, thereby contributing to individual vulnerability, propensity, and/or predisposition to inflammatory, autoimmune, and neurodegenerative disorders (FIGS. 1 and 2).[25]

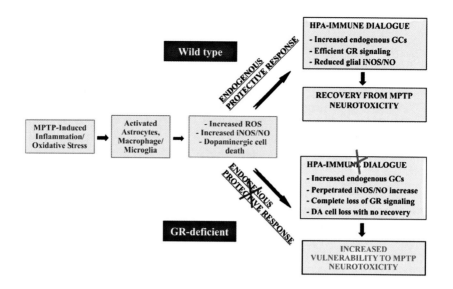

FIGURE 3. Schematic diagram depicting alteration of HPA–immune dialogue in GR-deficient mice responsible for increased vulnerability to MPTP neurotoxicity. In this model, the severity of MPTP-induced inflammation and oxidative stress as a result of activated astrocytes, macrophages, and microglia leading to impairment of dopaminergic neuron functioning is dependent on induction of an endogenous protective HPA–immune response. Thus, in Wt mice, the efficient GR signaling reduces glia neuroinflammatory reaction thereby stimulating the recovery process.[10] In Tg mice, however, the MPTP-induced increase in oxidative/nitrosative status of Tg mice further impairs GR signaling and exacerbates the GR deficiency at critical times of MPTP-induced dopaminergic neuron degeneration, suggesting GC-bound GRs as critical checkpoints in MPTP-induced neuroinflammatory reaction.[10]

Importantly enough, gender and sex steroid background also appear to strongly modulate vulnerability to PD, with mechanisms not completely elucidated. Several epidemiological studies have reported that the incidence and prevalence of PD is higher in men than in women.[59] Postmenopausal estrogen (E_2) deficiency has been reported to cause a worsening of Parkinson-related symptoms, whereas the severity of symptoms in women with early PD is diminished by the use of E_2.[60] Association between E_2 receptor gene polymorphism and PD disease with dementia and with age of onset of PD[62] has been reported.

These clinical results are supported by a body of experimental evidence indicating that the nigrostriatal DA system is subject to modulation by E_2 in rodents and nonhuman primates.[60,76–86] Then, the nigrostriatal DA system is exquisitely sensitive to gonadal hormone influence, and sexual differences are present in several parameters of the nigrostriatal DA neurons, as well as in the progression of diseases associated with this system.[60,76–86] Indeed, E_2 has been defined as a neuroprotectant for the nigrostriatal DA system.[60] The neuroprotective effects of E_2 have been reported against DA neurotoxicity induced by 6-hydroxidopamine, metamphetamine, and MPTP model of PD.[60,81–86] Both subtypes of the E_2 receptor, ER-alpha, and ER-beta mediate the effect of estrogen in the brain. ER-alpha is predominantly expressed in the hypothalamus and amygdala, indicating a main role of ER-alpha in estrogen modulation of neuroendocrine functions, autonomic events, and memory processing, whereas a higher level of expression of ER-beta has been described in regions such as the basal forebrain and cerebral cortex. Recent evidence of a relative abundance of ER-beta in the SN and ventral tegmental area suggest that this receptor may be important in E_2 modulation of midbrain DA systems.[87] In particular, studies conducted in ER-beta knockout mice indicate the importance of this receptor for neuronal survival, because specific degeneration of neuronal cell bodies particularly evident in SN was reported in aged ER-beta knockout mice.[88]

The mechanisms responsible for E_2 neuroprotection have been studied in a wide variety of neuronal systems including the DA neuronal system, both *in vivo* and *in vitro*, and have been shown to involve a multitude of direct genomic and nongenomic-mediated effects.[89–93] On the other hand, although the effects of endogenous and exogenous E_2 on the astroglial cell compartment are being extensively studied, and a glial involvement in E_2 neuroprotective effects has been clearly demonstrated, as recently and extensively reviewed by Garcia-Segura,[94,95] E_2 modulation of glial neuroinflammatory reaction in MPTP-induced experimental parkinsonism is not yet clear.

Collectively, this information appears to be of particular importance in view of the specific susceptibility of nigral DA neurons and for the crucial role that nigral astrocytes and microglia may play in DA neuron survival, recovery, and repair.

GLIAL ANTIOXIDANT AND ANTI-INFLAMMATORY NEUROPROTECTIVE MECHANISMS ARE DISRUPTED IN PD

Astrocytes are known to play a central role in the antioxidant defense of the brain and to exert a variety of neuroprotective functions (FIG. 2). Astrocytes scavenge reactive oxygen species released by neurons and remove glutamate from the extracellular space, thereby reducing the exposure of NMDA receptors to this excitatory amino acid; and they produce factors that induce antioxidant enzymes and express crucial neurotrophic molecules, regulating growth, differentiation, and survival of neurons, as part of the bidirectional, neuronal–glial interactions.[96–101] In particular, fetal mesencephalic neurons in culture are extremely vulnerable to serum deprivation or damaging compounds such as peroxynitrite, as compared with cultured glial cells, whereas interaction between astrocytes and neurons has been variously demonstrated to exert striking neurotrophic, differentiation, and neuroprotective effects.[25,96–101] Of particular mention, glia genotype has been shown to importantly contribute in determining mesencephalic neuron vulnerability.[11,25,102]

Under normal conditions, astrocytes exert a fundamental protective function against oxidative stress. This function appears to be of particular importance for nigral DA neurons. In fact, hydrogen peroxide (H_2O_2) is a product of dopamine conversion via monoamine oxidase (MAO) and could also be formed during dopamine autoxidation, and extraneuronal metabolism of dopamine by MAO B isozyme is performed by astrocytes (see Dimonte and Langston[103] for review). Then, changes in glial MAO-B activity as a result of astrocyte dysfunction can have a significant impact on DA functioning. Glutathione (GSH), the key molecule in the detoxification of H_2O_2, is present in high amounts in astrocytes as compared with neuronal cells.[99–102] Efflux of GSH from astroglial cells, which plays a vital role in defense of CNS against oxidative insults, is mediated by the ATP-dependent transporter and multidrug resistance–associated protein (Mrp1), and increased expression and membrane targeting of the Mrp1 export pump represents a further adaptive response involved in modulating vulnerability of astrocytes to various cytotoxic agents.[105] The presence of high GSH levels in astrocytes may protect them from oxidative injury, whereas reduced content of GSH may predispose DA neurons to be damaged by H_2O_2 formed in the process of DA turnover.[103]

Another important protective mechanism is represented by glial GRs. Endogenous GR signaling pathways in astrocytes and microglia have been shown to participate in the modulation of iNOS inducibility during neuroinflammation induced by MPTP or lipopolysaccharide, a specific activator of innate immunological responses.[11,25] Note that the ability of GC-bound GR to inhibit the transcriptional activity of NF-κB (p65-p50 heterodimers) and AP-1 (Jun-Fos heterodimers) is recognized to mediate most of the anti-inflammatory and immunosuppressive activity of GCs. In particular, the cytokine-dependent stimulation of iNOS, responsible for NO production, is mediated by NF-κB activation and is efficiently suppressed by GCs in most tissues and cell systems examined. Of major interest, we have recently shown that activation of innate immunity in wild-type (Wt) mice can trigger an autoinhibitory feedback mechanism responsible for inhibition of glial iNOS-derived NO, thereby attenuating glial overactivity and promoting the repair process.[25] In contrast, the GR deficiency of transgenic mice results in aberrant iNOS/NO expression and persistent NF-κB nuclear translocation[25] (FIG. 3).

GLUCOCORTICOIDS, NEURON–GLIAL DIALOGUE, AND VULNERABILITY TO PD

The HPA axis is of major interest for psychobiological and psychiatric research. The regulation of this neuroendocrine system has been shown to relate to prenatal factors, early life stress, several cognitive functions, brain reward pathway, and a variety of clinical disease states including chronic stress, post-traumatic stress disorder, depression, psychosis, inflammatory diseases, and metabolic syndrome. Impaired feedback regulation of relevant stress responses, especially immune activation/inflammation, may, in turn, contribute to stress-related pathologies. Several neurological diseases are accompanied by dysregulation of the HPA axis. Transgenic (Tg) mice expressing from early embryonic life antisense RNA directed against GRs were created to serve as animal models for the study of neuroendocrine changes occurring in stress-related disorders.[73–75] These mice show reduced GRs mRNA in the brain, pituitary, and immune organs, reduced GRs binding, and reduced HPA axis sensitivity to GCs.[24,73–75] As a consequence of GRs dysfunction, these Tg mice exhibit an aberrant response to stressful and immunogenic stimuli.[17,26,73–75,106] We recently reported that the GR deficiency renders these Tg mice resistant to myelin oligodendrocyte glycoprotein–induced autoimmune encephalomyelitis (EAE), via

NO-induced immunosuppression.[4] Of special interest, the GR deficiency of these Tg mice exacerbates the vulnerability of the nigrostriatal dopaminergic system to MPTP via hyperinduction of microglial-derived iNOS/NO.[11] The fact that the same GR deficiency that results in lack of glucocorticoid regulation of NO generation provides resistance to EAE via immunosuppression, but sensitivity to MPTP-induced parkinsonism via innate inflammatory mechanisms highlight the key role of the HPA-immune dialogue via GR–NO crosstalk in programming vulnerability to degenerative diseases of the CNS.[4,5,11,25]

GR DEFICIENCY INCREASES GLIAL HYPERTROPHY AND GLIAL iNOS IMMUNOREACTIVITY IN RESPONSE TO MPTP

Lesioning of the nigrostriatal DA system is known to cause reactive gliosis at the site of damage, the hallmark of which is an increase in GFAP immunoreactivity (IR). In addition, microglial cells (revealed by anti–Mac-1/CD11b antibody) and GFAP-IR cells express iNOS at high levels in striatum and SN, representing a recognized marker of astrocyte and microglia activation.[11,25]

To investigate whether differences in glial response to MPTP might contribute to the increased vulnerability of GR-deficient mice to experimental parkinsonism, we reacted striatal and nigral sections at various times after the lesion with antibodies against GFAP, the cytokine-inducible iNOS, responsible for generation of high levels of NO (a key molecule in the inflammatory response), the peroxynitrites footprint, nitrotyrosine, and the prototypic marker of activated macrophage/microglia, the membranolytic attack complex of complement (Mac-1/CD11b). Results of these analyses indicated an earlier and greater reactivity of GFAP-IR astrocytes and Mac-1-IR microglia exhibiting a strong iNOS immunofluorescent signal, as opposed to Wt counterparts, suggesting increased nitrosative stress as candidate mediator of early TH-IR neuron loss and impairment of dopaminergic functioning, as a result of the GR deficiency, and indicated glia as the key compartment involved in this phenomenon (FIGS. 3 and 4).[11,25] Indeed, when nigral sections were reacted with TUNEL reagent and the number of apoptotic cells were quantified, a drastic increase in dopaminergic neuronal cell death was observed in GR-deficient mice as compared with Wt mice. In addition, a sharp increase in the number of apoptotic astrocytes characterized the nigral response to MPTP in Tg mice.[10,24]

ESTROGEN AS NEUROIMMUNE MODULATOR: IMPLICATIONS FOR VULNERABILITY TO PD

In the CNS, estrogens promote neuron growth, prevent neuronal cell atrophy, and regulate synaptic plasticity. Within the reproductive system, estrogens are critical hormones that dictate both negative and positive feedback mechanisms on the LHRH neuronal system. In addition, estrogens lower plasma lipoproteins, influence the renin-angiotensin system, have antioxidative properties, and may act as calcium-blocking agents. Within the immune system, estrogens may act in both a proinflam-

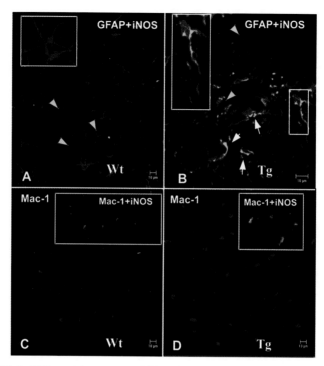

FIGURE 4. Differential response of iNOS in activated astrocytes and macrophages/
microglia in the SN of GR-deficient and Wt mice 1 day after MPTP exposure. One day after
MPTP exposure, nigral sections were processed for dual GFAP and iNOS, or Mac-1 and
iNOS immunocytochemistry.[11] (**A, B**) Confocal images of GFAP (revealed by CY3, in red
[*online*]) and iNOS (revealed by FITC, in green [*online*]) immunofluorescent reactions in
Wt (**A**) and Tg (**B**) mice. Note that GFAP-positive and GFAP-negative cells are faintly
stained by iNOS in the SN of Wt (**A**, *arrowhead*) at this time point. In Tg mice, a strong
iNOS immunofluorescent signal was localized in both GFAP-positive (**B**, *arrows*, **inset**) and
GFAP-negative round to oval (**B**, *arrowheads*) cells, identified as ameboid microglia. Fusion
confocal microscopic images reveal complete overlap of GFAP and iNOS in Tg, as revealed
by the bright yellow [*online*] immunofluorescence throughout the cell bodies, nuclei, and
processes of nigral Tg astrocytes (**B**, *arrow*, **inset**). (**C, D**) Numerous round Mac-1-IR (revealed by CY3, in red [*online*]) microglial cells are localized in the SN of GR-deficient (**D**)
as compared with Wt mice (**A**) on day 1 after MPTP discontinuance. Fusion confocal microscopic images reveal extensive overlap of Mac-1-IR and iNOS, appearing in yellow in Tg
(**D**, **inset**) as compared with Wt (**C**, **inset**) SN.

matory and antinflammatory manner. Much experimental literature has established
that the gonadal hormones are responsible not only for the sexual differentiation of
neural circuitry, which mediates a variety of reproductive behaviors and physiological mechanisms, but also for the generation of the sexually driven immunological
dimorphisms, as well as the sex-linked differential response of the HPA axis.[70] Thus,
as the female hormones, estrogens have received increased attention and consideration because of their neuroprotective properties. The precise underlying biochemical and molecular mechanisms are far from being completely disclosed. Besides

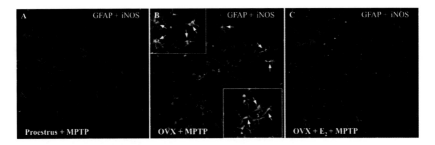

FIGURE 5. Effect of proestrous estrogen levels and estrogen deprivation in the response of iNOS in activated astrocytes and macrophages/microglia in the SN of MPTP-treated female mice.[112] Female mice were killed by quick decapitation, and the brain was processed for dual GFAP and iNOS, or Mac-1 and iNOS immunocytochemistry.[11] Confocal images of GFAP (revealed by CY3, in red) and iNOS (revealed by FITC, in green [*online*]) immunofluorescent reactions in proestrous (**A**), OVX (**B**), and OVX+E_2 (**C**) mice, 1 day after MPTP. Note that GFAP-positive and GFAP-negative cells are faintly stained by iNOS in the SN of proestrous (**A**) and in E_2-treated mice (**C**). In OVX mice, a strong iNOS immunofluorescent signal was localized in both GFAP-positive (**B**, *arrows*, **insets**) and GFAP-negative round to oval (**B**, *arrowheads*) cells, identified as ameboid microglia. Fusion confocal microscopic images reveal complete overlap of GFAP and iNOS in OVX, as revealed by the bright yellow [*online*] immunofluorescence throughout the cell bodies, nuclei, and processes of nigral OVX astrocytes (**B**, *arrow*, **inset**). Magnification × 20. *Scale bar* = 10 mm.

others, the crucial interaction of this class of hormones at the neuroendocrine–immune interface has been anticipated.[30,31]

Then, gender differences are known to play a central role in the regulation of innate, cell-mediated, and humoral immune responses outside the brain, and sex steroids are believed to play major roles in the development and/or severity of some autoimmune diseases such as lupus, multiple sclerosis, rheumatoid arthritis, and Graves's disease.[30,31] In particular, E_2 can exert direct effects on immune cells and/or modulate immune functions via neuroendocrine–immune interactions.[30,31,69] Recall that the effects of E_2 on immune cells are both dose- and cell type–specific, and, depending on the specific physiological/experimental condition, E_2 can act in both a proinflammatory and anti-inflammatory manner. In autoimmunity, Th1/Th2 balance and the activation of effector T cell subsets are often critical for the progression or the remission of disease, and E_2 might play a decisive role in changing the plasticity of these cells to drive the immune response toward the protection against disease. Thus, sex steroids shift T helper cells toward a Th2 phenotype, and cytokines produced by Th2 cells generally suppress EAE.[105] Indeed, ERs are present in T cells, macrophages, and dendritic cells at all stages of differentiation, and direct E_2 modulatory effects have been demonstrated. Of special mention, E_2 can potentiate the production of anti-inflammatory Th2 cytokines, IL-10, and TGF-beta, in antigen-specific T cells, resulting in full protection against EAE. Coupled with the recently reported effect of E_2 in driving the expansion of the "regulatory" T cell compartment,[105] these data clearly suggest the possibility that E_2 may dampen inflammation by directly influencing Th1/Th2 balance.[102] It is still not clear, however, whether a similar mechanism is operating within the CNS. In fact, signaling from activated cells in the periphery to the brain might also be influenced by E_2. Neuroimmune signaling involves, besides others, neuroendocrine–immune interactions (especially the stress axis), NO, and circulating cytokines, which interact with brain endothelium and macrophage populations in the brain; in turn, these signals are passed on to particular neuronal populations leading to mild and chronic brain cell inflammation. For some neurodegenerative diseases, a correlation between ongoing central and peripheral inflammation and increased susceptibility to neuronal death has been clearly demonstrated.[37]

Of particular mention, within the brain, E_2 via either ER-alpha and/or ER-beta has been shown to exert anti-inflammatory activity on activated macrophages and activated microglial cells *in vitro*, as revealed by the prevention of lipopolysaccharide-induced production of proinflammatory cytokines including TNF-alpha, iNOS-induced NO, COX-2, prostaglandin-E2 (PGE2), and metalloproteinase-9 (MMP-9).[108–110] Most importantly, E_2 can interact with the transcription factor NF-κB, a key regulator of inflammatory responses.[111] Moreover, E_2 has been shown to prevent the activation of brain macrophage, *in vivo*, by inhibiting proinflammatory gene expression.[110]

Collectively, this information points to E_2 neuroimmunomodulatory effects as potential gender-specific mechanisms underlying E_2 neuroprotection in PD.[112]

THE ESTROGENIC STATUS MODULATES THE RESPONSE OF THE GLIAL CELL COMPARTMENT TO MPTP

In normal cycling female rodents, fluctuations in plasma E_2 levels occur over a 4-day period. The cyclic changes in plasma E_2 have been reported to modulate structural parameters in various brain regions. In particular, region-specific changes of astrocyte and microglia morphology, cell number, and or immunoreactivity have been reported to occur, *in vivo*, as a function of gender and the estrous cycle, aging, gonadal hormone deprivation, and E_2 replacement.[94,95,113–116] In addition, sex steroids, in particular E_2, have been demonstrated to play a major role in modulating changes of glial reactivity, and/or proliferation occurring after brain injury.[94,95,117,118]

In female rodents, various parameters of nigrostriatal system functionality, including dopamine concentration, dopamine uptake sites, and DA1 receptor density, have been shown to vary according to the phases of the estrous cycle (see Refs. 76–79, 81, and the INTRODUCTION of this article). Based on this literature, the endogenous hormonal status at the time of injury may have a differential impact in MPTP-induced DA toxicity and/or in the ability of nigral neurons to recover. Hence, changes in astrocyte and microglial cell function may underlie the neurochemical and morphological alterations of nigrostriatal DA neurons as a result of estrogen cyclic variation, as well as after estrogen loss and estrogen replacement.

To address whether changes in estrogenic status might alter the astroglial response to MPTP underlying estrogen neuroprotection against nigrostriatal DA neurotoxicity, we assessed temporal changes in different indices of glial reactivity (immunocytochemistry and generation of iNOS-derived NO) and DA neuron functionality (striatal dopamine and its metabolites) after injection of the neurotoxin MPTP[110] at proestrus or estrus corresponding to high and low plasma E_2 levels, respectively, after bilateral ovariectomy (OVX, performed 2–3 weeks before MPTP treatment), both in the absence or the presence of a concomitant treatment with 17-beta (E_2, 1 μg/day) or 17-alpha (1 μg/day) estradiol.

ESTROGEN DEPRIVATION EXACERBATES ASTROCYTE AND MICROGLIAL REACTION: COUNTERACTION BY ENDOGENOUS AND EXOGENOUS E_2

Astrocyte Response

In accordance with our previous studies,[11] time-course analysis of the effect of MPTP on the astroglial cell compartment indicated that early (1 day) after MPTP treatment, immunolabeling for GFAP started to increase in striatum and midbrain of all MPTP-treated groups, but such an increase was different according to the estrogenic status. Then, lowest astrocyte reactivity was observed in intact female mice treated with MPTP at proestrus, corresponding to the phase of highest E_2 serum levels, or in OVX mice supplemented with E_2, whereas maximal astrocyte reaction was observed in E_2-deprived (OVX) mice, when plasma E_2 levels were almost undetectable.[110] In OVX mice, hypertrophic GFAP-IR astrocytes had enlarged cell bodies with longer and thicker processes, as compared with proestrous,[112] MPTP-treated counterparts. In contrast, treatment of OVX mice with E_2, but not 17-alpha estradiol, reduced astrocyte hypertrophy after MPTP.[112] Starting from 11 days on, numerous "healthy" astrocytes were localized in both striatum and SN in both intact and E_2-treated mice. In particular, double immunocytochemical labeling with GFAP and TH and confocal laser microscopy image analysis revealed a significant degree of recovery of TH immunoreactivity within nigral cell bodies of intact proestrus and E_2-treated mice, whereas in OVX females, a sharp reduction of GFAP-IR astrocytes and TH-IR neurons indicated the damage of astroglial cell compartment and absence of TH-IR neuron recovery.[112]

Microglia Response

Induction of a microglial reaction in the denervated striatum and SN of female mice was earlier and sharper in OVX as compared with intact MPTP-treated mice. Within the intact mice groups, proestrous mice exhibited a reduced reaction compared with estrous females. Hence, as early as 1 day after the lesion, when nigral sections were stained for the prototypic marker of activated macrophage/microglia, Mac-1/CD11b (revealed by CY3, in red), Mac-1 cells localized in the SN were far greater in OVX as compared with intact (not shown) female mice treated with MPTP on estrus, whereas a faint Mac-1 reaction was detected in mice treated with MPTP on the morning of proestrous. In our previous work,[11] we demonstrated that glial iNOS/NO is a determining factor triggering DA neurotoxicity in response to MPTP. The expression of iNOS and generation of reactive nitrites (RNI) thus were evaluated as a function of the estrogenic status, after E_2 deprivation and replacement with E_2, or 17-alpha estradiol. One day after MPTP injection in cycling females at proestrus, iNOS-IR was weaker (FIG. 5A), compared with estrous mice. In contrast, in OVX mice, a strong iNOS-IR signal (revealed by FITC in green) was localized in both the striatum and SN (FIG. 5B) in either GFAP-IR and Mac-1-IR cells. When microglial NO (measured by its decomposition product, nitrites) was determined at different time intervals during MPTP-induced neurotoxicity, results indicated a highly significant increase in E_2-deprived versus OVX females, further suggesting aberrant activation of microglial cytotoxic cascade as a result of gonadectomy.[112] In contrast, treatment with E_2, but not 17-alpha estradiol, sharply inhibited iNOS-IR in both striatum and SN (FIG. 5C).

Collectively, these data indicate that endogenous estrogenic status sharply modulates astrocyte and microglia functional iNOS/NO response to MPTP. Then, proestrous E_2 levels are associated with downregulation of MPTP-induced iNOS and reactive nitrite generation, whereas OVX exacerbates microglia reaction. In addition, they further show the ability of exogenous E_2 to stereospecifically counteract microglial generation of reactive nitrites at all time points tested.[112]

EXACERBATION OF ASTROGLIAL REACTION IN E_2-DEPRIVED MICE PRECEDES DA NEUROTOXICITY WITH NO RECOVERY

When different parameters of striatal DA functionality were measured at different time intervals after administration of MPTP, dopamine and its metabolites were shown to decrease; however, such reductions were inversely related to iNOS expression and the amount of nitrite generation. Compared with saline-injected mice, within the intact group, reductions of dopamine, DOPAC, and HVA were greater in females treated with MPTP at estrus compared with mice treated with MPTP in proestrus, 1 day after MPTP.[112] Interestingly, the protective effect of intact proestrous mice was reduced on day 7 after MPTP, whereas in estrus mice a somewhat further reduction was measured. Starting from 11 days on, however, neurochemical indices of DA functionality in both estrous and proestrous intact mice exhibited a certain degree of recovery. Withdrawal of gonadal hormones for 2–3 weeks, resulting in almost undetectable plasma E_2 concentration, drastically magnified MPTP-induced inhibition of striatal levels of dopamine, DOPAC, and HVA 1 day after MPTP. In addition, in OVX mice these levels further decreased at 7 and 11 days after MPTP, with no significant recovery until 28 days after MPTP, indicating a reduced ability of OVX females to recover.[112]

These findings clearly support a prominent protective role of circulating sex steroids against MPTP-induced DA neurotoxicity and further show an important effect of E_2 endogenous proestrous levels at the time of MPTP injection, whereas gonadal hormone withdrawal drastically increases MPTP-induced DA neurotoxicity with no recovery. In contrast, treatment of OVX mice with E_2 resulted in a complete counteraction of MPTP-induced decrease of DA and its metabolites at all time points, indicating the ability of both endogenous and exogenous E_2 to protect DA neurons and to stimulate the recovery/repair process.[112]

Given that endogenous and exogenous E_2 (1) powerfully counteracts astrocyte cell death, (2) inhibits the aberrant macrophage/microglial generation of peroxynitrite in response to MPTP, (3) inhibits nitrosative stress and oxidative stress, (4) significantly counteracts the loss of nigral TH neurons, and (5) stimulates the repair process, it seems tempting to speculate that a causal relationship exists between the exacerbation of glia reactivity of E_2-deprived females and increased vulnerability of nigral DA neurons to MPTP, thereby implicating E_2-induced switch of proinflammatory astroglial phenotype as determining factor in DA neuron protection (FIG. 2).[112]

CONCLUSION

Interaction between astrocytes and neurons has been proposed as an important factor limiting neuronal death from excitotoxins, serum deprivation or growth factor inhibition, oxidants, and nitric oxide toxicity. Decreased function of astroglial cells (e.g., as a result of oxidative/nitrosative stress) may lead to reduced expression of growth and neurotrophic factors, reduced scavenging properties, and reduced antioxidant enzymes production, with consequent damage to neurons and inhibition of repair processes. Given the crucial role of glia–neuron interactions in neuronal growth, survival differentiation and synapse formation, alterations in glia–neuron crosstalk as observed in GR-deficient mice and E-deprived mice suggest a pivotal role for endogenous glucocorticoids and E_2 in restraining harmful innate inflammatory reactions, thereby contributing to enhance astroglial neuroprotective functions and possibly helping the repair process (FIG. 6). This mechanism might represent an important compensatory response to bring about the resolution of the injury, to restore homeostasis, and to stimulate the repair process.[11,25,112]

The findings herein reviewed in the MPTP model of PD suggest that endogenous GCs and E_2 participate in the modulation of astrocyte and microglia reactivity and iNOS inducibility during neuroinflammation induced by MPTP. In particular, increased levels of GCs as a result of MPTP injection as well as peak E_2 levels appear

FIGURE 6. Schematic representation of potential ER/GR–NO crosstalk in glial neuroprotective functions. Stimulation of innate immunity has bidirectional effects on activated astroglial cells, resulting in proinflammatory and anti-inflammatory cascades with positive and negative influences on neuronal survival/protection/rescue. Activation of glial GR–NO crosstalk[25]) and ER–NO crosstalk[112] may provide a further endogenous mechanism reducing neuroinflammatory detrimental effects, while promoting astroglial "proregenerative" functions.[11,25,112]

to contribute to restrain astroglial cell response to nitrosative stress and consequently protect dopaminergic neurons against the described cytotoxic cascades, whereas partial GR knockout or estrogen deprivation in 2–3 week OVX mice sharply up-regulates various indices of astrocyte and microglia activation, resulting in exacerbation of DA neurotoxicity. Note that intact female mice appear more resistant when MPTP is administrated at proestrous, as compared with females treated at estrous. This assumption seems corroborated by counteraction of such effect by administration of the ER antagonist. On the other hand, the hormonal background of estrus is able to significantly protect striatal indices of DA functionality against MPTP neurotoxicity as compared with OVX mice. Furthermore, intact female mice exhibit a significant degree of recovery from DA neurotoxicity, both at morphological and functional levels, thereby supporting a protective role of circulating ovarian hormones.

According to the findings presented it seems tempting to speculate that endogenous GCs and E_2, by limiting endogenous generation of iNOS-derived NO, are likely to protect both astroglial cells and nigral DA neurons against the toxic effects of excessive stimulation of innate immunity (FIGS. 2 and 6). Within this context, recall that whereas an acute brain insult can trigger a compensatory endogenous glial response, which includes the activation of neurogenesis in specialized brain regions,[118] a chronic neuroinflammatory condition can have detrimental effects for neuronal repair, rescue, and/or regeneration,[120–124] and further studies are clearly needed to elucidate the glucocorticoid and gonadal hormone modulation of neurogenesis in the MPTP model of parkinsonism.

Pre-existing inflammatory conditions, such as giant cell arteritis and systemic lupus erythematosus, cause predisposition to neurodegenerative diseases, as do a range of acute and chronic infections, principally respiratory. For example, in multiple sclerosis, systemic respiratory infections have been estimated to account for up to 30% of recurrences. Consistent with these findings, aging (the life period characterized by the higher incidence of neurodegenerative diseases) is accompanied by a two- to four-fold increase in plasma/serum levels of inflammatory mediators, including cytokines and acute-phase proteins. Besides concomitant chronic diseases, several factors may contribute to this low-grade inflammation, such as an increased amount of fat tissue (overweight) and decreased production (and loss of protective effect) of sex steroids.[37]

In summary, the herein-described glucocorticoid and estrogenic activation of glial anti-inflammatory and "protective" functions may provide a further mechanism reducing the detrimental effects of neuroinflammation, while promoting cytokine activation of astroglial "proregenerative" functions. This mechanism might represent a compensatory/adaptive response to stimulate the repair process (FIG. 6). Note that recent results of the Women's Health Initiative Memory Study clinical trial have caused a reconsideration of the efficacy of estrogen hormone therapy as a strategy to prevent age-related cognitive decline and dementia, and several important issues still remain unresolved, such as the timing and the duration of estrogen therapy as well as the combined presence of progestins. Development of selective receptor modulators for the brain (SERMs), also called NeuroSERMs, certainly represent one major pharmacological challenge to target E_2-specific subtype receptors and to prevent brain-related climateric symptoms and neurodegenerative diseases.[125] The complementary action of estrogen on astrocyte and microglia herein reviewed in the MPTP model of PD may provide a potential pharmacological target and a new insight into the therapeutic potential of these hormones in PD and a further means to design "bifunctional" molecules aimed, on the one hand, at preserving astrocyte functionality, while, on the other, restraining the aberrant stimulation of innate immunity.

ACKNOWLEDGMENTS

We thank the Italian Ministry of Health (Strategic Research Project Contract No. 189), Italian Ministry of Research (MURST), and OASI (IRCCS) Institution for Research and Care on Mental Retardation and Brain Aging, Troina (EN) Italy.

REFERENCES

1. AHBON, E. *et al.* 2000. Prenatal exposure to high levels of glucocorticoids increases the susceptibility of cerebellar granule cells to oxidative stress-induced cell death. Proc. Natl. Acad. Sci. USA **97:** 14726–14730.
2. SHANKS, N., S. LAROCQUE & M. MEANEY. 1995. Neonatal endotoxin exposure alters the development of the hypothalamic-pituitary adrenal axis: early illness and later responsivity to stress. J. Neurosci. **15:** 376–384.
3. SHANKS, N. *et al.* 2000. Early-life exposure to endotoxin alters hypothalamic-pituitary-adrenal function and predisposition to inflammation. Proc. Natl. Acad. Sci. USA **97:** 5645–5650.
4. MARCHETTI, B. *et al.* 2002. Exposure to a dysfunctional glucocorticoid receptor from early embryonic life programs the resistance to experimental allergic encephalomyelitis via nitric oxide-induced immunosuppression. J. Immunol. **168:** 5848–5859.
5. MARCHETTI, B. *et al.* 2001. Stress, the immune system and vulnerability to degenerative disorders of the central nervous system in transgenic mice expressing glucocorticoid receptor antisense RNA. Brain Res. Rev. **36:** 259–272.
6. PIERPAOLI, W. & H.O. BESEDOVSKY. 1975. Interdependence of the thymus in programming neuroendocrine functions. Clin. Exp. Immunol. **20:** 323–329.
7. PIERPAOLI, W. & E. SORKIN. 1972. Hormones, thymus, and lymphocyte function. Experentia **28:** 1385–1389.
8. PIERPAOLI, W. & H.O. BESEDOVSKY. 1975. Interdependece of the thymus in programming neuroendocrine functions. Clin. Exp. Immunol. **20:** 323–329.
9. PIERPAOLI, W., H.G. KOOP, J. MULLER & M. KELLER. 1977. Interdependence between neuroendocrine programming and the generation of immune recognition in ontogeny. Cell. Immunol. **29:** 16–26.
10. MARCHETTI, B., F. GALLO, Z. FARINELLA & M.C. MORALE. 1996. Unique neuroendocrine-immune (NEI) interactions during pregnancy. *In* The Physiology of Immunity. M. Kendal & J. Marsh, Eds.: 297–328. CRC Press. London.
11. MORALE, M.C. *et al.* 2003. Glucocorticoid receptor deficiency increases vulnerability of the nigrostriatal dopaminergic system: critical role of glial nitric oxide. FASEB J. **18:** 164–166.
12. McCANN, S.M. *et al.* 1993. Endocrine aspects of neuroimmunomodulation: methods and overview. *In* Neurobiology of Cytokines, Part A. Vol. 16, Methods in Neurosciences. E.B. de Souza, Ed.: 187–210. Academic Press. San Diego.
13. PIERPAOLI, W. 1994. The pineal gland as ontogenetic scanner of reproduction, immunity, and aging. The aging clock. Ann. N.Y. Acad. Sci. **741:** 46–49.
14. MARCHETTI, B. *et al.* 1990.Crosstalk communication in the neuroendocrine-reproductive axis: age-dependent alterations in the common communication networks. Ann. N.Y. Acad. Sci. **594:** 309–325.
15. MARCHETTI, B., F. GALLO, Z. FARINELLA & M.C. MORALE. 1995. Neuroendocrineimmunology (NEI) at the turn of the century: towards a molecular understanding of basic mechanisms and implications for reproductive physiopathology. Endocrine **3:** 845–861.

16. PIERPAOLI, W. *et al.* 1997. Circadian melatonin and young-to-old pineal grafting postpone aging and maintain juvenile conditions of reproductive functions in mice and rats. Exp. Gerontol. **32:** 587–602.

17. MARCHETTI, B. *et al.* 1997. Developmental consequences of hypothalamic-pituitary-adrenocortical system disruption: impact on thymus gland maturation and the susceptibility to develop neuroimmune diseases. *In* Developmental Correlates of Stress System Dysfunction. B. Marchetti, Ed. Dev. Brain Dysfunct. **10:** 503–527.

18. MARCHETTI, B. *et al.* 1990. The neuroendocrine-immune connections in the control of reproductive functions. *In* Major Advances in Human Female Reproduction E.Y. Adashi and S. Mancuso, Eds.: Serono Symposia Publications from Raven Press. **73:** 279–289.

19. BLALOCK, J.E. 1992. Neuroimmunoendocrinology. J.E. Blalock, Ed. Chem. Immunol. **52:** Karger. Basel.

20. BLALOCK, J.E. 1994. Shared ligands and receptors as a molecular mechanism for communication between the immune and neuroendocrine systems. Ann. N.Y. Acad. Sci. **741:** 292–298.

21. MASON, D. 1991. Genetic variation in the stress response: susceptibility to experimental and implications for human inflammatory disease. Immunol. Today **12:** 57–60.

22. MEANEY, M.J. *et al.* 1996. Early environmental regulation of forebrain glucocorticoid receptor gene expression: implications for the adrenocortical response to stress. Dev. Neurosci. **18:** 49–72.

23. MEANEY, M.J. *et al.* 1993. Basis for the development of individual differences in the hypothalamic-pituitary-adrenal stress response. Cell. Mol. Neurobiol. **13:** 321–347.

24. MORALE, M.C. *et al.* 1995. Disruption of hypothalamic-pituitary adrenocortical system in transgenic animals expressing type II glucocorticoid receptor antisense ribonucleic acid permanently impairs T cell function: effects on T cell trafficking and T cell responsiveness during postnatal development. Endocrinology **136:** 3949–3960.

25. MARCHETTI, B. *et al.* 2005. Glucocorticoid receptor-nitric oxide crosstalk and vulnerability to experimental parkinsonism: pivotal role for glia-neuron interactions. Brain Res. Rev. **48:** 302–321.

26. MORALE, M.C. *et al.* 1991. Blockade of central and peripheral luteinizing hormone-releasing hormone (LHRH) receptors in neonatal rats with a potent LHRH-antagonist inhibits the morphofunctional development of the thymus and maturation of cell-mediated and humoral immune responses. Endocrinology **128:** 1073–1085.

27. MARCHETTI, B. *et al.* 2000. Gender, neuroendocrine-immune interactions, and neuron-glial plasticity: the role of luteinizing hormone releasing hormone (LHRH). Ann. N.Y. Acad. Sci. **917:** 678–709.

28. MARCHETTI, B. 1996. The LHRH-astroglial network of signals as a model to study neuroimmune interaction: assessment of messenger systems and transduction mechanisms at cellular and molecular levels. Neuroimmunomodulation **3:** 1–27.

29. MARCHETTI, B. 1997. Cross-talk signals in the CNS: the role of neurotrophic and hormonal factors, adhesion molecules, and intercellular signaling agents in a luteinizing hormone-releasing hormone (LHRH)-astroglia interactive network. Trends Biosci. **2:** 1–32.

30. MARCHETTI, B. *et al.* 2001. The hypothalamic-pituitary-gonadal axis and the immune system. *In* Psychoneuroimmunology, 3rd ed., Vol. 1. R. Ader, D.L. Felton & N. Cohen, Eds.: 363–387. Raven Press. New York.

31. MORALE, M.C. *et al.* 2001. Special Feature: Neuroendocrine-immune (NEI) circuitry from neuron-glial interactions to function: focus on gender and HPA–HPG interactions on early programming of the NEI system. Immunol. Cell Biol. **79:** 400–417.

32. MARCHETTI, B., H. KETTENMANN & W.J. STREIT. 2005. Glia-neuron crosstalk in neuroinflammation, neurodegeneration, and neuroprotection. Brain Res. Rev. **48:** 129–489.

33. PEARCE, B. & G.P. WILKIN. 1995. Eicosanoids, purine, and hormone receptors. *In* Neuroglia. H. Kattenmann & B.R. Ransom, Eds.: 377–386. Oxford University Press. New York.

34. YUNG-TESTAS, I., J.M. RENOIR, G.L. BRUGNARD & E.E. GREENE. 1992. Demonstration of steroid hormones receptor and steroid action in primary cultures of rat glial cells. J. Steroid Biochem. Mol. Biol. **41:** 3–8.

35. YUNG-TESTAS, I., J.M. RENOIR, J.M. GSE & E.E. BEAULIEU. 1991. Estrogen inducible progesterone receptors in primary cultures of rat glial cells. Exp. Cell Res. **193:** 12–19.
36. CADEPOND, C. *et al.* 2002. Steroid receptors in various glial cell lines expression and functional studies. Ann. N.Y. Acad. Sci. **973:** 484–487.
37. MARCHETTI, B. & M.P. ABBRACCHIO. 2005. To be or not to be (inflammed) is that the question in anti-inflammatory drug therapy of neurodegenerative diseases? Trends Pharmacol. Sci. **26:** 517–525.
38. ALOISI, F. 2001. Immune function of microglia. Glia **36:** 165–179.
39. BANATI, R., B.J. GEHRMANN, P. SCHUBERT & G.W. KREUTZBERG. 1993. Cytotoxicity of microglia. Glia **7:** 111–118.
40. STREIT, W.J. 2002. Physiology and pathophysiology of microglial cell function. *In* Microglia in the Regenerating and Degenerating Central Nervous System. W.J. Streit, Ed.: 1–15. Springer. New York.
41. STREIT, W.J. 2002. Microglia as neuroprotective: immunocompetent cells of the CNS. Glia **40:** 133–139.
42. NICHOLS, N.R. 1999. Glial responses to steroids as markers of brain aging. J. Neurobiol. **40:** 485–601.
43. NICHOLS, N.R. *et al.* 1990. Messenger RNA for glia fibrillary acidic protein is decreased in the rat brain following acute and chronic corticosterone treatment. Mol. Brain Res. **7:** 1–7.
44. OLANOW, C.W., A.H. SHAPIRA & Y. AGID. 2003. Neuroprotection for Parkinson's disease: prospects and promises. Ann. Neurol. **53** (Suppl. 3): 51–52.
45. LANGSTON, J.W. *et al.* 1999. Evidence of active nerve cell degeneration in the substantia nigra of humans years after 1-methyl-4-phenyl-1,2,3,6-tetrahydropyridine exposure. Ann. Neurol. **46:** 598–605.
46. JENNER, P. 2003. Oxidative stress in Parkinson's disease. Ann. Neurol. **53:** 26–38.
47. JENNER, P. & C.W. OLANOW. 1996. Oxidative stress and the pathogenesis of Parkinson's disease. Neurology **47:** 161–170.
48. MCNAUGHT, K.S.P. & P. JENNER. 2000. Extracellular accumulation of nitric oxide, hydrogen peroxide, and glutamate in astrocytic cultures following glutathione depletion, complex I inhibition, and/or lipopolysaccharide-induced activation. Biochem. Pharm. **60:** 979–988.
49. HUNOT, S. *et al.* 1999. Fc epsilonRII/CD23 is expressed in Parkinson's disease and induces, *in vitro*, the production of nitric oxide and tumor necrosis factor-alpha in glial cells. J. Neurosci. **19:** 3440–3447.
50. HERRERA, A.J. *et al.* 2000. The single intranigral injection of LPS as a new model for studying the selective effects of inflammatory reactions on an opaminergic system. Neurobiol. Dis. **7:** 429–447.
51. HIRSCH, E.C., S. HUNOT, P. DAMIER & B. FAUCHEUX. 1998. Glial cells and inflammation in Parkinson's disease: a role in neurodegeneration? Ann. Neurol. **44** (Suppl. 1): 5115–5120.
52. IRAVANI, M.M. *et al.* 2002. Involvement of inducible nitric oxide synthase in inflammation-induced dopaminergic neurodegeneration. Neuroscience **110:** 49–58.
53. GAO, H.M., B. LIU, W. ZHANG & J.S. HONG. 2003. Novel anti-inflammatory therapy for Parkinson's disease. Trends Pharmacol. Sci. Rev. **24:** 395–401.
54. GAO, H.M. *et al.* 2002. Microglial activation–mediated delayed and progressive degeneration of rat nigral dopaminergic neurons: relevance to Parkinson's disease. J. Neurochem. **81:** 1285–1297.
55. MCGEER, P.L. & E.G. MCGEER. 2004. Inflammation and neurodegeneration in Parkinson's disease. Parkinsonism Relat. Disord. **10** (Suppl. 1): 3–7.
56. HARTMANN, A. *et al.* 2003. Inflammation and dopaminergic neuronal loss in Parkinson's disease: a complex matter. Exp. Neurol. **184:** 561–564.
57. SCHIESS, M. 2003. Nonsteroidal anti-inflammatory drugs protect against Parkinson neurodegeneration: can an NSAID a day keep Parkinson disease away? Arch. Neurol. **60:** 1043–1044.
58. CHEN, H. *et al.* 2003. Nonsteroidal anti-inflammatory drugs and the risk of Parkinson's disease. Arch. Neurol. **60:** 1059–1064.

59. WARNER, T.T. & A.H.V. SCHAPIRA. 2003. Genetic and environmental factors in the cause of Parkinson's disease. Ann. Neurol. **53:** 16–25.
60. DLUZEN, D.E. & M.W.I. HORSTINK. 2003. Estrogen as a neuroprotectant of nigrostriatal dopaminergic system. Endocrine **21:** 67–75.
61. LEVECQUE, C. *et al.* 2003. Association between Parkinson's disease and polymorphisms in the nNOS and iNOS genes in a community-based case-control study. Hum. Mol. Genet. **12:** 79–86.
62. WESTBERG, L. *et al.* 2004. Association between the estrogen receptor beta gene and age onset of Parkinson's disease. Psychoneuroendocrinology **29:** 993–998.
63. BETARBET, R. *et al.* 2000. Chronic systemic pesticide exposure reproduces features of Parkinson's disease. Nat. Neurosci. **3:** 1301–1306.
64. BEZARD, E. *et al.* 2003. Enriched environment confers resistance to 1-methyl-4-phenyl-1,2,3,6-tetrahydropyridine and cocaine: involvement of dopamine transporter and trophic factors. J. Neurosci. **23:** 10999–11007.
65. BEZARD, E. *et al.* 2000. Adaptive changes in the nigrostriatal pathway in response to increased 1-methyl-4-phenyl-1,2,3,6-tetrahydropyridine-induced neurodegeneration in the mouse. Eur. J. Neurosci. **12:** 2892–2900.
66. LING, Z.D. *et al.* 2004. Combined toxicity of prenatal bacterial endotoxin exposure and post-natal 6-hydroxydopamine in the adult rat midbrain. Neuroscience **124:** 619-628.
67. BOCKMAN, D.E. & M.L. KIRBY. 1984. Dependence of thymus development on derivatives of neural crest. Science **223:** 498–500.
68. BODEY, B., B. BODEY, JR., S.E. SIEGEL, *et al.* 1996. Identification of neural crest derived cells within the cellular microenvironment of the thymus employing a library of monoclonal antibodies raised against neuronal tissue. In Vivo **10:** 39–48.
69. MAIER, C.C. *et al.* 1992. Thymocytes express a mRNA that is identical to hypothalamic luteinizing hormone-releasing hormone mRNA. Cell. Mol. Neurobiol. **12:** 447–454.
70. PEIFFER, A., M.C. MORALE, N. BARDEN & B. MARCHETTI. 1994. Modulation of glucocorticoid receptor gene expression in the thymus by the sex steroid hormone milieu and correlation with sexual dimorphism of immune response. Endocrine **2:** 181–191.
71. MORALE, M.C., F. GALLO, N. BATTICANE & B. MARCHETTI. 1992. The immune response evokes up-and-down modulation of a beta$_2$-adrenergic receptor messenger RNA concentration in the male rat thymus. Mol. Endocrinol. **6:** 1513–1522.
72. MARCHETTI, B., M.C. MORALE, P. PARADIS & M. BOUVIER. 1994. Characterization, expression, and hormonal control of a thymic ß$_2$-adrenergic receptor. Am. J. Physiol. **267:** 718–731.
73. PÉPIN, M.C., F. POTHIER & N. BARDEN. 1992. Impaired type II glucocorticoid-receptor function in mice bearing antisense RNA transgene. Nature **355:** 725–728.
74. STEC, I., N. BARDEN, J.M.H.M. REUL & F. HOLSBOER. 1993. Dexamethazone nonsuppression in transgenic mice expressing antisense RNA to the glucocorticoid receptor. J. Psychiatric Res. **28:** 1–5.
75. BARDEN, N. *et al.* 1997. Endocrine profile and neuroendocrine challenge tests in transgenic mice expressing antisense-RNA against the glucocorticoid receptor. Neuroendocrinology **3:** 212–220.
76. DI PAOLO, T., P. BÉDARD & A. DUPONT. 1982. Effects of estradiol on intact and denervated striatal dopamine receptors and on dopamine levels: a biochemical and behavioral study. Can. Physiol. Pharmacol. **60:** 350–357.
77. LEVESQUE, D., S. GAGNON & T. DI PAOLO. 1989. Striatal dopamine receptor density fluctuates during the rat estrous cycle. Neurosci. Lett. **10:** 345–350.
78. BECKER, J.B. 1990. Direct effect of 17beta estradiol on striatum: sex differences in dopamine release. Synapse **5:** 157–164.
79. MORISSETTE, M. & T. DI PAOLO. 1993. Effect of chronic estradiol and progesterone treatments of ovariectomized rats on brain dopamine uptake sites. J. Neurochem. **60:** 1876–1886.
80. MORISSETTE, M. & T. DI PAOLO. 1993. Sex and estrous cycle variations on rat striatal dopamine uptake sites. Neuroendocrinology **58:** 16–22.
81. DLUZEN, D.E. & J.L. MCDERMOT. 2002. Estrogen, anti-estrogen, and gender differences in methamphetamine neurotoxicity. Ann. N.Y. Acad. Sci. **965:** 136–156.

82. GRANDBOIS, M., M. MORISSETTE, S. CALLIER & T. DI PAOLO. 2000. Ovarian steroids and raloxifene prevent MPTP-induced dopamine depletion in mice. Neuroreport **11:** 343–346.

83. CALLIER, S., M. MORISSETTE, M. GRANDBOIS & T. DI PAOLO. 2000. Stereospecific prevention by 17-beta estradiol of MPTP-induced dopamine depletion. Synapse **37:** 245–251.

84. DLUZEN, D.E. & J.L. MCDERMOTT. 2004. Developmental and genetic influences upon gender differences in methamphetamine-induced nigrostriatal dopaminergic neurotoxicity. Ann. N.Y. Acad. Sci. **1025:** 205–220.

85. D'ASTOUS, M., M. MORISSETTE & T. DI PAOLO. 2004. Effect of estrogen receptor agonists treatment in MPTP mice: evidence of neuroprotection by ER alpha agonist. Neuropharmacology **47:** 1180–1188.

86. D'ASTOUS, M., T.M. GAJJAR, D.E. DLUZEN, & T. DI PAOLO. 2004. Dopamine transporter as a marker of neuroprotection in methamphetamine-lesioned mice treated acutely with estradiol. Neuroendocrinology **79:** 296–304.

87. CREUTZ, L.M. & M.F. KRITZER. 2002. Estrogen receptor beta immunoreactivity in the midbrain of adult rats: regional, subregional, and cellular localization in the A10, A9, and A8 dopamine cell groups. J. Comp. Neurol. **446:** 288–300.

88. WANG, L., S. ANDERSSON, M. WARNER & J.A. GUSTAFSSON. 2001. Morphological abnormalities in the brain of estrogen receptor beta knockout mice. Proc. Natl. Acad. Sci. USA **98:** 2792–2796.

89. BEHL, C. 2002. Estrogen as a neuroprotective hormone. Nat. Rev. Neurosci. **3:** 433–442.

90. DUBAL, D.B. *et al.* 1999. Estradiol modulates bcl-2 in cerebral ischemia: a potential role for estrogen receptors. J. Neurosci. **19:** 6385-6393.

91. DUBAL, D.B. & P.M. WISE. 2001. Neuroprotective effects of estradiol in middle-aged female rats. Endocrinology **142:** 43–48.

92. SAWADA, H., M. IBI & T. KIHARA. 1998. Estradiol protects mesencephalic dopaminergic neurons from oxidative stress-induced neuronal death. J. Neurosci. Res. **54:** 707–719.

93. SAWADA, H. *et al.* 2002. Estradiol protects dopminergic neurons in a MPP+Parkinson's disease model. Neuropharmacology **42:** 1056–1064.

94. AZCOITIA, D. *et al.* 2001. Astroglia play a key role in the neuroprotective actions of estrogen. Prog. Brain Res. **132:** 469–478.

95. GARCIA-OVEJERO, D. *et al.* 2005. Glia-neuron crosstalk in the neuroprotective mechanisms of sex steroid hormones. Brain Res. Rev. **48:** 2130–2142.

96. TAKESHIMA, T., J.M. JOHNSTON & J.W. COMMISSIONG. 1994. Mesencephalic type 1 astrocytes rescue dopaminergic neurons from death induced by serum deprivation. J. Neurosci. **14:** 4769–4779.

97. GALLO, M.C. *et al.* 2000. Basic fibroblast growth factor (bFGF) acts on both neurons and glia to mediate the neurotrophic effects of astrocytes on LHRH neurons in culture. Synapse **36:** 233–253.

98. GALLO, M.C. *et al.* 2000. Basic fibroblast growth factor (bFGF) priming increases the response of LHRH neurons to neurotrophic factors. J. Neuroendocrinol. **12:** 941–959.

99. DRINGEN, R., J. GUTTERER & J. HIRRLINGER. 2000. Glutathione metabolism in brain. Metabolic interaction between astrocytes and neurons in the defense against reactive oxygen species. Eur. J. Biochem. **267:** 4912–4916.

100. PARK, C., H. ZHANG & G.E. GIBSON. 2001. Co-culture with astrocytes or microglia protects metabolically impaired neurons. Mech. Ageing Dev. **123:** 21–27.

101. MCNAUGHT, K.S.P. & P. JENNER. 1999. Altered glial function causes neuronal death and increases neuronal susceptibility to 1-methyl-4-phenylpyridinium- and 6-hydroxydopamine-induced toxicity in astrocytic/ventral mesencephalic co-cultures. J. Neurochem. **73:** 2469.

102. SMEYNE, O., O. GOLOUBEVA & R.J. SMEYNE. 2001. Strain-dependent susceptibility to MPTP and MPTP(+)-induced parkinsonism is determined by glia. Glia **34:** 73-80.

103. DI MONTE, D.A. & J.W. LANGSTON. 1995. Idiopathic and 1-methyl-4phenyl-1,2,3,6-tetrahydropyridine (MPTP)-induced parkinsonism. *In* Neuroglia. H. Kettenmann & B.R. Ransom, Eds.: 997–989. New York. Oxford University Press.

104. GEGG, M.E. *et al.* 2003. Differential effect of nitric oxide on glutathione metabolism and mitochondrial function of astrocytes and neurones: implications for neuroprotection/neurodegeneration? J. Neurochem. **86:** 228–237.
105. GENNUSO, F. *et al.* 2004. Bilirubin protects astrocytes from its own toxicity inducing upregulation and translocation of multigrug resistance-associated protein 1 (Mrp 1). Proc. Natl. Acad. Sci. USA **101:** 2470–2475.
106. SACEDON, R. *et al.* 1999. Partial blockade of T cell differentiation and marked alteration of the thymic microenvironment in transgenic mice with a dysfunctional glucocorticoid receptor function. J. Neuroimmunol. **98:** 157–167.
107. POLANCZYK, M.J. *et al.* 2004. Cutting edge: estrogen drives expansion of the CD4$^+$CD25$^+$ regulatory T cell compartment. J. Immunol. **173:** 2227–2230.
108. BRUCE-KELLER, A.J. *et al.* 2000. Anti-inflammatory effects of estrogen on microglial activation. Endocrinology **141:** 3646–3656.
109. DREW, P.D. & J.A. CHAVIS. 2000. Female sex steroids: effects upon microglial cell activation. J. Neuroimmunol. **111:** 77–85.
110. VEGETO, E. *et al.* 2003. Estrogen receptor-alpha mediates the brain anti-inflammatory activity of estradiol. Proc. Natl. Acad. Sci. USA **100:** 9614–9619.
111. DODEL, R.C., Y. DU & K.R. BALES. 1999. Sodium salicylate and 17 beta estradiol attenuate nuclera transcription factor NF-κB translocation in cultured rat astroglia cultures following exposure to amyloid beta (1-40) and lipopolisaccharide. J. Neurochem. **73:** 1453–1460.
112. MORALE, M.C. *et al.* 2005. Estrogens, neurinflammation, and neuroprotection in Parkinson's disease: glia dictates resistance versus vulnerability to neurodegeneration. Neuroscience. In press.
113. LONG, J.M. *et al.* 1998. Stereological analysis of astrocyte and microglia in the aging mouse hippocampus. Neurobiol. Aging **19:** 497–503.
114. LUQUIN, S., F. NAFTOLIN & L.M. GARCIA-SEGURA. 1993. Natural fluctuation and gonadal hormone regulation of astrocyte immunoreactivity in dentate gyrus. J. Neurobiol. **24:** 913–924.
115. LEI, D.-L. *et al.* 2003. Effects of estrogen and raloxifene on neuroglia number and morphology in the hippocampus of aged female mice. Neuroscience **121:** 659–666.
116. MOUTON, P.R. 2002. Age and gender affect microglia and astrocyte number in the brain of mice. Brain Res. **956:** 30–36.
117. GARCIA-OVEJERO, D., S. VEIGA, L.M. GARCIA-SEGURA & L.L. DONCARLOS. 2002. Glial expression of estrogen and androgen receptors after rat brain injury. J. Comp. Neurol. **450:** 256–271.
118. MILLER, D.B., S.F. ALI, J.P. O'CALLAGHAN & S.C. LAW. 1998. The impact of gender and estrogen on striatal dopamine neurotoxicity. Ann. N.Y. Acad. Sci. **844:** 153–165.
119. GARCIA-SEGURA, L.M. *et al.* 1999. Role of astroglia in estrogen regulation of synaptic plasticty and brain repair. J. Neurobiol. **40:** 574–584.
120. LIBERTO, C.M. *et al.* 2004. Pro-regenerative properties of cytokine-activated astrocytes. J. Neurochem. **89:** 1092–1100.
121. EKDAL, J.H. *et al.* 2003. Inflammation is detrimental for neurogenesis in the adult brain. Proc. Natl. Acad. Sci. USA **100:** 13632–13637.
122. FAULKNER, J.R. 2004. Reactive astrocytes protect tissue and preserve function after spinal cord injury. J. Neurosci. **24:** 2143–2155.
123. MONJE, M.L., H. TODA & T.D. PALMER. 2003. Inflammatory blockade restores adult hippocampal neurogenesis. Science **302:** 1760–1765.
124. LEHNARDT, S. *et al.* 2003. Activation of innate immunity in the CNS triggers neurodegeneration through a Toll-like receptor 4-dependent pathway. Proc. Natl. Acad. Sci. USA **100:** 8514–8519.
125. LONARD, D.M. & C.L. SMITH. 2002. Molecular perspectives on selective estrogen receptor modulators (SERMs): progress in understanding their tissue-specific agonist and antagonist actions. Steroids **67:** 15–24.

Neurodegenerative Diseases

A Common Etiology and a Common Therapy

WALTER PIERPAOLI

Walter Pierpaoli Foundation of Life Sciences, Orvieto, Italy

Jean Choay Institute for Neuroimmunomodulation, Riva San Vitale, Switzerland

ABSTRACT: The variety of names of neurodegenerative diseases (NDDs) does not indicate that there is a wide variety of causes and a multiple number of cures. In fact NDDs derive from a common and repetitive, almost monotonous multicausal origin. NDDs are initiated invariably by a sudden or silent insidious decrease in immunologic resistance of the T cell–dependent or delayed type, produced by a large variety of psychological-emotional and/or environmental "stressors" (e.g., social, family-domestic, economic, alimentary, traumatic, and professional). These stressors increase the vulnerability of tissues (in this case, a section of the central or peripheral nervous system) to attack by a common virus (e.g., adenoviruses and herpesviruses). This attack creates a vicious circle leading to emergence of virus-generated tissue autoantigens and then to formation of autoantibodies. Use of corticosteroids and immunosuppressive drugs dramatically worsen and "eternalize" the diseases with further immunosuppression. Invariably, onset of NDDs is anticipated by a clear-cut alteration of the hormonal cyclicity, which closely controls immunity. My experience with patients in the last five years indicates a new approach to prevent and cure NDDs, based on a system totally divergent from present therapies. In fact "resetting the hormonal cyclicity clock" results in restoration of hormone-dependent antiviral immunity, arrest of disease progression, and at least partial recovery of neural functions, whatever the origin, anatomic location, and course of pathology.

KEYWORDS: neurodegenerative diseases; Parkinson's disease; multiple sclerosis; Alzheimer's disease; immunity; viral infections; stress

INTRODUCTION

Long ago, I suspected that neither corticosteroids nor toxic immunosuppressive drugs bring durable benefits for the cure of neurodegenerative diseases (NDDs).[1] Having been involved for many years in the "art" of neuroimmunomodulation (NIM), namely in the studies linking the brain to the immune system,[2–6] it seemed obvious to me that "hiding" the symptoms of NDDs would never reveal to us the primary cause of these diseases. For that reason I was induced to write an article in 1987 (adapted in the APPENDIX to this article), which was obviously totally neglected. Now I am exhuming it for a number of reasons: (1) I discovered how to better modify

Address for correspondence: Dr. Walter Pierpaoli, Via San Gottardo 77, CH-6596 Gordola, Switzerland. Voice: +41 91 7451940; fax: +41 91 7451046.

pierpaoli.fnd@bluewin.ch

Ann. N.Y. Acad. Sci. 1057: 319–326 (2005). © 2005 New York Academy of Sciences.
doi: 10.1196/annals.1356.024

and maintain the hormonal control of immunity, thanks to the "clock" that maintains daily rhythmicity.[7–10] (2) I was able to arrest the progress of autoimmune processes by returning the immune system to its own self-control, and finally (3) I could observe the effects of my interventions in patients suffering from multiple sclerosis, Parkinson's disease, Alzheimer's disease, and similar diseases with many names, but similar or even identical etiology. Differences depend on the localization of the neural damage.[11] Obviously, I do not mean those pathologic disorders stemming from genetic anomalies, neoplastic processes, or vascular diseases that result in neural damage. However, myelopathies, neurodegenerative processes, and vasculitis are often intrinsically identical, closely linked, and propelled by the same cause—a self-promoting autoimmune process.[11] I am now happy to see that this article is still valid.

COMMENTS

I wish to summarize here the conditions leading to the initiation and progression of NDDs.

- Psychological, environmental, traumatic situations produce "di-stress"—anguish, and anxiety—in particular in relation to family or working situations with stresses, such as quarrels, litigations, mob rule, economic pressure, etc. The "di-stress" condition can be acute and highly traumatic (as would occur, e.g., from the loss of a close person in the family) or prolonged and not even perceived by the subject (or simply repressed).
- The mental-psychological conditions of uneasiness, depression, and anguish result inevitably in silent or overt immunosuppression.The person gets used to repeated and often prolonged infections (laryngotracheal, bronchial, urinary), which are "cured" with antibiotics that, obviously, are themselves immunosuppressive and have no effect on viruses.
- A common virus of the most varied type, strain, or category—often adenoviruses or herpesviruses—is activated by a self-maintained condition of immunosuppression. The dire circle begins!
- The virus combines with cell components, often in the membrane, and generates autoantigens that promote the formation of autoantibodies. The syndrome is now typically autoimmune. Generally the pathology produces a further immunosuppression and a spreading of more infectious diseases, "cured" further with toxic antibiotics that destroy natural immunity and the gut flora.
- Doctors intervene with massive acute or chronic administration of corticosteroids and quite often with extremely poisonous cytotoxic and antimitotic drugs. The idea underlying this treatment is to abrogate synthesis of auto-antibodies and to eliminate pain. In fact, the outcome is a "hiding" of the inflammatory process and a deeper immunodepression.
- Depending on the location of the viral aggression,the combination of viruses and autoantibodies produces typical inflammatory and granulomatous reactions (plaques, infiltrates, demyelinization) that "eternalize" and propagate the process.
- The typical expression of the NDD becomes evident depending on the neural "district" involved (such as in multiple sclerosis, chorea, Alzheimer's disease, and Parkinson's disease).

As suggested many years ago in the earlier article, I propose here a totally different approach to the prevention and to the cure of NDDs, based on the fact that the conditions produce massive alterations of the neuroendocrine-immune system. My interventions have been applied to a variety of patients and are based on the concept that to arrest or to cure NDDs we must visualize first the hormonal damage produced to the patients by the causes mentioned and then repair centrally the neurohormonal change. In other words, we must bring back *self*-control to immunity by resetting the hormonal cyclicity that controls the expression of immunity. This can be done by measuring peripheral levels of hormones. Very often, the thyroid gland is involved—in fact, we see a huge variety of thyroid diseases that are in many cases autoimmune diseases. The intervention based on correction of central (hypothalamic-pituitary-pineal-peripheral glands) regulatory functions results in a cascade of positive events that restore immunity and return the body to self-control of immunity. Obviously, arrest and restoration of motile functions depend on the degree and progression of damage produced to the central (the brain) or the peripheral nervous tissues, often involving vascularization (vasculitis). My interventions are based on the prolonged administration of an association of endogenous, even if synthetic, molecules suitable to adjust day-night hormonal rhythms to their original and physiological circadian periodicity and consequently to boost the immune system, in particular cell-mediated antiviral immunity. The results will be reported in specialized journals.

[*Competing interests*: The author declares that he has no competing financial interests.]

REFERENCES

1. PIERPAOLI, W. 1987. Neuroimmunomodulation: an approach to the therapy of neurological diseases. *In* Clinical Neuroimmunology. WJ.A. Aarli, W.M.H. Behan & P.O. Behan, Eds.: 500–505. Blackwell Scientific Publications, Oxford.
2. PIERPAOLI, W. & E. SORKIN. 1972. Hormones, thymus and lymphocyte functions. Experientia **28**: 1385–1389.
3. PIERPAOLI, W. & H.O. BESEDOVSKY. 1975. Role of the thymus in programming of neuroendocrine functions. Clin. Exp. Immunol. **20**: 323–338.
4. PIERPAOLI, W., H.G. KOPP & E. BIANCHI. 1976. Interdependence of thymic and neuroendocrine functions in ontogeny. Clin. Exp. Immunol. **24**: 501–506.
5. PIERPAOLI, W. & G. MAESTRONI. 1977. Pharmacological control of the immune response by blockade of the early hormonal changes following antigen injection. Cell. Immunol. **31**: 355–363.
6. PIERPAOLI, W. & N.H. SPECTOR, EDS. 1988. Neuroimmunomodulation: Interventions in Aging and Cancer. Ann. N.Y. Acad. Sci. Vol. 521. New York.
7. PIERPAOLI, W. & G. MAESTRONI. 1988. Neuroimmunomodulation: some recent views and findings. Intern. J. Neurosci. **39**: 165–175.
8. PIERPAOLI, W. & W. REGELSON. 1994. Pineal control of aging: effect of melatonin and pineal grafting on aging mice. Proc. Natl. Acad. Sci. USA **91**: 787–791.
9. PIERPAOLI, W., D. BULIAN, A. DALL'ARA, *et al.* 1997. Circadian melatonin and young-to-old pineal grafting postpone aging and maintain juvenile conditions of reproductive functions in mice and rats. Exp. Gerontol. **32**: 587–602.
10. PIERPAOLI, W. & V.A. LESNIKOV. 1997. Theoretical considerations on the nature of the pineal "aging clock." Gerontology **43**: 20–25.
11. GOETZ, C.G. 2003. Textbook of Clinical Neurology. Saunders, Elsevier Science. Philadelphia.

APPENDIX

Neuroimmunomodulation: An Approach to the Therapy of Neurological Diseases

By W. PIERPAOLI

Introduction: Neuroimmunomodulation (NIM) has assumed its name and significance and received acceptance only recently,[1,2] but it is not a new medical discipline. The term "psychoneuroimmunology" (PNI) has also been proposed,[3] but this is in fact more restrictive, because the prefix *neuro* can include psychological and behavioral aspects, while that of *psycho* induces one to believe that the subject concerns only the relationship between strictly psychological and behavioral disorders and immunological abnormalities. In addition, the orientation and approach of investigators working in this interdisciplinary field of research differ profoundly, depending on their original scientific training (as scientist or clinician in immunology, biochemistry, endocrinology, neurophysiology, etc.). Thus the task of defining NIM is difficult. In the last analysis, the recent acceptance of NIM as a valid integrative science is based on clear-cut evidence of a bilateral connection between hormones and factors of the central and peripheral neuroendocrine system, and those tissues and cells of the immune system that lead to humoral and cell-mediated immune responses. For an overall view on NIM, the reader is referred to recent publications and reports.[3,4]

The Immunologic Dogma of Specificity: An immunologist's view of NIM can be quite different from that of an endocrinologist or neurologist. The accepted, fundamental concepts in immunogenetics and cellular immunology have led to an obvious scepticism towards NIM. In fact, while hormones and "factors" are certainly needed for amplification and the final expression of an immune reaction (e.g., the proliferation to T and B cells, the blastic transformation of lymphocytes, the secretion of lymphokines, etc.), they do not induce, according to an immunologist's view, the appearance or formation of sterically and/or chemically defined "receptors" on antigen-reactive cells. The appearance of antigenic determinants on specific antigen-sensitive cells is genetically programmed in a germline fashion. Hormones, on the other hand, simply modulate or condition the second-stage phase (differentiation, proliferation) of the hormone-independent, genetically determined, specific immunocytes.[5,6] In the labyrinth of this semantic problem, the immunologist's view is that a primary immune reaction may, in fact, take place *in vitro* without any connection with the central nervous system (CNS).[5] This, however, is a deceptively simplistic view which leads to the conclusion that, if a lymphocyte can react to an antigen, mature to a plasma cell *in vitro,* and produce immunoglobulins, NIM is unnecessary for the expression of an immune response. Unfortunately, sensitization to antigens *in vitro* requires the presence of hundreds of "factors" or "hormones" which are in the medium in which the immune cells are cultured. Those "factors" are just those preformed products of the neuroendocrine system that are supplied "ready for use" in the calf serum added to the culture medium, and which allow a "primary" immune response to take place.[7] In addition, it is most likely that emergence of germline antigenic deter-

Adapted by permission from: *Clinical Neuroimmunology,* 1987. J.A. Aarli, W.M.H. Behan & P.O. Behan, Eds.: 500–505. Blackwell Scientific Publications. Oxford.

minants on precursors of immune cells also needs hormones. It may be, although this is not certain, that only binding of antigens to specific receptors on immune cells does not require the presence of hormones.[7,6]

These considerations automatically lead the non-immunologist to conclude that the immune response is completely generated, expressed, and amplified by the neuroendocrine microenvironment and the assumption that hormones constitute "non-specific" participants of an immune response is a paradoxical assertion because, in fact, it has been known for many years that an immune response generated *in vitro* (primary immunization, inductive phase) cannot be evoked at all without the presence of at least some of the elements of the neuroendocrine system, the number of which is increasing daily. It is thus evident that the immune and neuroendocrine systems cannot co-exist, from conception to death, without constant interdependence.[8,9] This is demonstrated by the fact that antigenic confrontation with genetically different immune cells during ontogeny irreversibly changes neuroendocrine and immune recognition and regulation.[10] Thus the assumption that there is dissociation between "specific" and "non-specific" elements in the development and performance of the immune system, is blatantly naive.

An Immunologist's View of Neuropathology: One of the questions we can ask is whether or not, in those syndromes or diseases of the central or peripheral nervous system in which immune reactions or autoimmune disorders are recognized, the immune or autoimmune manifestations preceded or followed the onset of the symptoms and the overt clinical expression of the disease. From the NIM point of view, this is a crucial point for understanding the etiology and pathogenesis of multiple sclerosis (MS), systemic lupus erythematosus (SLE), and many other diseases in which an immune or autoimmune component is recognized or suspected. Of course, the presence of active immunologic processes is proof that the disorder is, in fact, an autoimmune disease, but it does not reveal anything about its etiology. Therefore, we must intervene with immunosuppressant, anti-inflammatory drugs and corticosteroids. Generally, the immunologist confines his attention to the actual immunologic symptoms and does not consider those derangements and alterations which are combined with the neurological and immune disorders or may even have produced them. This is not to blame the immunologist, but it is hardly useful in the cure or prevention of a disease. In fact, the "immunologic cure" would be, not the elimination of symptoms, but rather the eradication of the cause by, for instance, the induction of specific tolerance to autoantigens, a kind of "reversal" of the cause of the disease. Thus, we must try to understand whether the "immune" disease is indeed a side-effect or a byproduct of another, or several, concomitant acute or chronic derangements of other organs, tissues, or cells.

Neuropathology from the Point of View of Neuroimmunomodulation: If we hypothesize that some or most neurologic diseases may be derived from other than primary disorders of the immune system, we might be able to evaluate those diseases from a new viewpoint. How far are pathologic affections of the nervous system with immunologic disorders only a consequence of primary disorders of other systems? Very often, in neurologic diseases there is an enormous variety of symptoms and alterations that affect other organs. This constellation of diseases with a huge variety of symptoms can hardly be attributed to a primary immune disorder. Thus, the best confirmation for the validity of the NIM approach to neurologic diseases is the pathologic picture. The fact that an immune or autoimmune process is recognized as such is not proof for classification of the disease as "autoimmune." This applies to most neurologic afflictions like MS and to diseases of the skin like SLE or psoriasis. Strangely enough, in most cases immunologic alterations have been viewed as the cause and not the consequence of a neurological disease. In fact we think that, as is commonly the case in other systemic diseases with immune alterations that are cured by immunosuppressive drugs and corticosteroids (e.g. SLE and rheumatoid arthritis), many neurologic diseases may be initiated by a primary derangement of neuroendocrine regulation which is unobserved and yet determines the onset of the immune disorder. Most patients go to the doctor when they already have symptoms and the symptoms are then analyzed and treated, mainly with drugs aiming at eliminating the actual pains and the local or general manifestations of disease. Put differently, however, a clinician could not act otherwise. The same is true when we cure cancer, arteriosclerosis, or even infectious diseases.

The NIM view of neurologic diseases is different. Why should we cure only the symptoms? What prevents us from trying new therapies on the basis of observations of animal models or accompanying family alterations, and of all those anomalies that precede onset of the disease? The timing of appearance of a neurologic disease among the members of the same family may help us to recognize those symptoms that precede the manifestation of the disease. Why do we not consider more carefully the fact that the immune system is totally dependent on neuroendocrine regulation? Why do we not analyze systematically the possible derangements of hypothalamic, pituitary, adrenal, gonadal or thyroid function? Again, the way to a "physiological therapy" is deceptively simple, to the point of being naive.

Possible NIM Intervention in Neurohormonal Derangements and Autoimmune Processes in Neuropathology: It is now firmly established that bi-directional connections and signals link the neuroendocrine and immune systems. This reciprocal relationship has been demonstrated by a large variety of models in developing and mature organisms.[6,8-11] It is especially noteworthy that such bidirectional links are established very early in ontogeny and constitute a basic aspect of the final brain programming for neuroendocrine and immune functions in the adult.[9] Mature cells of the immune system, carrying on their surface antigenic determinants, which are different from those of the host allotype (alloantigens), when injected into immunologically and endocrinologically immature recipients produce an irreversible change (tolerance) to the immune-recognition apparatus of the recipient (allotolerance to tissues and cells of the donor) and, at the same time, permanent changes of neuroendocrine regulation as expressed by hormone levels in blood.[10] Thus the two-way circuit seems to be established very early in ontogeny and to constitute an integral part of the system by which the organism distinguishes "self" from "non-self." Construction of immune responsiveness is thus a part of the early brain programming for immune responsiveness and endocrine regulation.[10]

On the other hand, at the anatomic, morphologic, and ultrastructural levels also, an intimate connection exists between central and peripheral nervous tissues and the cells of the lymphohemopoietic organs like the thymus, the spleen, the bone marrow and lymph nodes.[12,13] Immunologic signals are detected immediately by the neuroendocrine system, in particular by the gonadal adrenal system.[14] This suggests strongly a close link between those cells, tissues, and hormones involved in reproduction physiology and in maintenance of self (immunity, resistance to infectious agents, or an "intruder"). We have proposed, however, that maintenance of "self" by immune resistance became subordinate to maintenance of self by reproduction on an evolutionary scale.[9]

There is an extraordinary variety of neurologic, behavioral, and endocrine alterations and disorders in animal models, or in humans, before the onset or during the course of autoimmune diseases, including those of the central and peripheral nervous system in which an immune or autoimmune component is recognized. As mentioned in the previous sections, however, primary attention has been devoted to the presence, expression, detection, measurements and cure of the immunologic symptoms. Most typically, the use of corticosteroids is generally accepted as a means of curing symptoms in many autoimmune diseases. Unfortunately, until recent discoveries in neuroendocrinology and pharmacology, and the development of new psychomimetic drugs, a NIM approach to the cure of autoimmune diseases affecting the neural system, and based on possible elimination or diminution of persisting neuroendocrine derangements rather than relief of immune symptoms, was not feasible. Nowadays, one can aim at early detection of possible endocrine derangements in the initial phase of neurologic diseases, by systematic measurements of basal levels of hormones in blood, analysis of the function of endocrine glands, and detection of specific deficiencies or increases in production of hormones. These findings would indicate the necessity for intervention with replacement therapy, the use of promoters or inhibitors of hormone synthesis or release, the use of antagonists to hormone receptors, or of some of the increasing variety of drugs acting on the CNS and affecting the production or function of neurotransmitters or hormones.

It may be that one cannot cure an immune deficiency or an autoimmune disease once it is established, or reverse its course. We may succeed in diminishing the intensity of the symptoms, however, by affecting immunologic (lymphomonokines, interferons, etc.) or humoral factors (neurotransmitters, hormone-releasing, and inhibitory agents, polypeptide hormones, and so on), rather than controlling their targets, namely the immune cells whose secretory products, in the last analysis, produce the morphologic alterations in neurologic diseases. Nowadays, the distinction between the cells and factors of the immune and neural systems is becoming more and more artificial: cells of the immune system are apparently able to produce hormones which were formerly considered to be produced only by certain endocrine glands,[15] while cells of the neural system can all produce those factors, like interleukins, which are secreted by lymphocytes.[16]

To summarize, an approach to the cure of neurologic diseases via neuroimmunomodulation aims at specific intervention on those primary neuroendocrine derangements which undoubtedly promote and/or affect the onset and maintenance of immune disorders linked to a number of neurologic diseases, such as MS, myasthenia gravis, idiopathic polymyositis, polyneuritis. and other immune-related neuropathies.

REFERENCES

1. SPECTOR, N.H. 1980. The central state of the hypothalamus in health and disease. Old and new Concepts. *In* Physiology of the Hypothalamus. P. J. Morgan & J. Panksepp, Eds.: 453–503. Marcel Dekker, Inc. New York.
2. SPECTOR, N.H. & A. KORNEVAE. 1981. Neurophysiology, immunophysiology, and neuroimmunomodulation. *In* Psychoneuroimmunology R. Ader, Ed.: 449–473. Academic Press. New York.
3. ADER, R., ED. 1981. Psychoneuroimmunology. Academic Press. New York.
4. GUILLEMIN, R., M. COHN & T. MELNECHUK, Eds. 1985. Neural Modulation of Immunity. Raven Press. New York.
5. COHN, M. 1985. What are the "must" elements of immune responsiveness? *In* Neural Modulation of Immunity R. Guillemin, M. Cohn & T. Melnechuk, Eds.: 3–25. Raven Press. New York.
6. PIERPAOLI, W., N. FABRIS & E. SORKIN. 1970. Developmental hormones and immunological maturation. *In* Hormones and the Immune Response. G.E.W. Wolstenholme & J. Knight, Eds.: 126–53. Ciba Foundation Study Group No. 36. Churchill Livingstone. London.
7. AMBROSE, C.T. 1970. The essential role of corticosteroids in the induction of the immune response *in vitro*. *In* Hormones and the Immune Response. G.E.W. Wolstenholme & J. Knight, Eds.: 100–125. Ciba Foundation Study Group No. 36. Churchill Livingstone. London.

8. JANKOVIC, B.D., K. ISAKOVIC, K. MICIC & Z. KNEZEVIC. 1981. The embryonic lympho-neuro-endocrine relationship. Clin. Immunol. Immunopathol. **18:** 108–120.
9. PIERPAOLI, W. 1981. Integrated phylogenetic and ontogenetic evolution of neuroendo-crine and identity-defence, immune functions. *In* Psychoneuroimmunology. R. Ader, Ed.: 575–606. Academic Press. New York.
10. PIERPAOLI, W., H.G. KOPP, J. MUELLER & M. KELLER. 1977. Interdependence between neuroendocrine programming and the generation of immune recognition in ontogeny. Cell. Immunol. **29:** 16–27.
11. PIERPAOLI, W. & H.O. BESEDOVSKY. 1975. Role of the thymus in the programming of neuroendocrine functions. Clin. Exp. Immunol. **20:** 323–338.
12. BULLOCH, K. 1985. Neuroanatomy of lymphoid tissue: a review. *In* Neural Modulation of Immunity. R. Guillemin, M. Cohn & T. Melnechuk, Eds.: 111–141. Raven Press. New York.
13. ZETTERSTROEM, B.E.M., T. HOEKFELT, K.A. NORBERG & P. OLSSON. 1973. Possibilities of a direct adrenergic influence on blood elements in the spleen. Acta Chir. Scand. **139:** 117–122.
14. MAESTRONI, G.J.M. & W. PIERPAOLI. 1981. Pharmacological control of the hormonally mediated immune response. *In* Psychoneuroimmunology. R. Ader, Ed.: 405–428. Academic Press. New York.
15. SMITH, E.M., W.J. MEYER & J.E. BLALOCK. 1982. Virus-induced corticosterone in hypophysectomized mice: a possibie lymphoid adrenal axis. Science **218:** 1311–1312.
16. FONTANA, A., F. KRISTENSEN, R. DUBS, *et al.* 1982. Production of prostaglandin E and an interleukin-I like factor by cultured astrocytes and C6 Glioma cells. J. Immunol. **129:** 2413–2419.

Clinical Perspectives for the Use of Melatonin as a Chronobiotic and Cytoprotective Agent

DANIEL P. CARDINALI, ANALÍA M. FURIO, AND MARÍA P. REYES

Department of Physiology, Faculty of Medicine, University of Buenos Aires, Buenos Aires, Argentina

ABSTRACT: The circadian time system involves periodic gene expression at the cellular level, synchronized by a hierarchically superior structure located in the hypothalamic suprachiasmatic nuclei. Treatment of circadian rhythm disorders has led to the development of a new type of agent called "chronobiotics," among which melatonin is the prototype. In elderly insomniacs, melatonin treatment decreased sleep latency and increased sleep efficiency, particularly slow-wave sleep. The effect of melatonin on sleep is the consequence of increasing sleep propensity (by augmenting the amplitude of circadian clock oscillation via MT_1 receptors) and of synchronizing the circadian clock via MT_2 receptors. Daily melatonin production decreases with age and in several pathologies, attaining its lowest values in Alzheimer's disease (AD) patients. About 45% of AD patients have disruptions in their sleep and "sundowning" agitation. Generally, melatonin treatment decreases sundowning in AD patients and reduced variability of sleep onset time. Both open and controlled studies have indicated a significant decrease of cognitive deterioration in AD patients treated with melatonin. The mechanisms accounting for the possible therapeutic effect of melatonin in AD patients may be manifold. On one hand, melatonin treatment promotes slow-wave sleep in the elderly and could be beneficial by augmenting the restorative phases of sleep. On the other hand, melatonin protects neurons against β-amyloid toxicity. By its combined chronobiotic and cytoprotective properties melatonin provides an innovative neuroprotective strategy to reduce the cost of lifetime treatment of some neuropsychiatric disorders.

KEYWORDS: melatonin; aging; insomnia; sleep; Alzheimer's disease; circadian; neuroprotection; chronobiotics

SYNCHRONIZATION OF THE CIRCADIAN CLOCK

Many biological functions wax and wane in cycles that repeat each day, month, or year. Such patterns do not just reflect an organism's passive response to environmental changes. Rather, they reflect the organism's biological rhythms—that is, its ability to keep track of time and to direct changes in function accordingly.[1]

Research in animals and humans has shown that only a few such environmental cues, like light–dark cycles, are effective entraining agents ("zeitgebers") for the cir-

Address for correspondence: Dr. D.P. Cardinali, Departamento de Fisiología, Facultad de Medicina, UBA, Paraguay 2155, 1121 Buenos Aires, Argentina. Voice/fax: 54-11-59509611.
dcardinali@fmed.uba.ar

Ann. N.Y. Acad. Sci. 1057: 327–336 (2005). © 2005 New York Academy of Sciences.
doi: 10.1196/annals.1356.025

cadian oscillator. An entraining agent can actually reset, or phase shift, the internal clock. Depending on when an organism is exposed to such an entraining agent, circadian rhythms can be advanced, delayed, or not shifted at all. Therefore, adjusting the daily activity pattern to the appropriate time of day involves a rhythmic variation in the influence of the zeitgeber as a resetting factor.

In mammals, a hierarchically major circadian oscillator is located in the suprachiasmatic nuclei (SCN) of the hypothalamus. This circadian master clock acts like a multifunctional timer to adjust the homeostatic system, including sleep and wakefulness, hormonal secretions, and various other bodily functions, to the 24-h cycle. Lesions of the SCN eliminate all circadian-driven rhythms. Inversely, SCN transplants to animals whose own SCN had been ablated, can restore circadian activity rhythms. Every single SCN cell exerts a waxing and waning of the firing rate with a predictable circadian rhythm. Synchronized by paracrine signals, the SCN produces an output signal that "drives" endogenously generated daily oscillations in sleep—wakefulness, alertness, performance, hormone secretion, and many other physiological functions.[2]

The current view of the physiological regulation of the circadian rhythm of sleep/wakefulness (the major circadian rhythm in the body) holds that it is regulated by two components, namely, a circadian (~24 h) component and a homeostatic component.[3] Melatonin plays a major role in the circadian component that regulates the timing of sleep. The circadian rhythm in synthesis and secretion of pineal melatonin is closely associated with sleep rhythm in both normal and blind subjects. The onset of nighttime melatonin secretion is initiated 2 h in advance to individual's habitual bedtime and has been shown to correlate with the onset evening sleepiness.[4,5] Several studies implicate endogenous melatonin in the physiological regulation of sleep propensity.

Aging has been associated with a significant reduction in sleep efficiency and continuity[6] and coincides with a significant reduction in amplitude of the melatonin rhythm.[7-13] Increase in early morning awakenings and difficulty in falling sleep have been frequent reported in the elderly. Impaired melatonin secretion has been shown to be associated with the sleep disorders that are encountered in elderly insomniacs.[14] Indeed aging may be a process resulting in part from relative melatonin deficiency and melatonin can be effective for improving the quality of life in the elderly.[15,16]

CHRONOBIOTICS

Drugs that influence the circadian apparatus are referred as chronobiotics.[17] Melatonin is the prototypic chronobiotic. Melatonin plays a major function in the coordination of circadian rhythmicity.[18,19] Melatonin secretion is an "arm" of the biologic clock in the sense that it responds to signals from the SCN and that the timing of the melatonin rhythm indicates the status of the clock, both in terms of phase (i.e., internal clock time relative to external clock time) and amplitude. From another point of view, melatonin is also a chemical code of night: the longer the night, the longer the duration of its secretion. In many species, this pattern of secretion serves as a time cue for seasonal rhythms.[20] In addition to controlling seasonal reproduction, melatonin is involved in several physiological functions such as sleep regulation,[21,22] immune function,[23,24] inhibition of tumor growth,[25] blood pressure regulation,[26] retinal physiology,[27] control of circadian rhythms,[28] control of human mood and behavior,[29] and free radical scavenging.[30] Most of these functions are relevant for normal and pathological aging.

Like the effects induced by the external zeitgeber light, effects by the internal zeitgeber melatonin on the circadian clock are time-dependent.[28] Entraining free-running circadian rhythms by administering melatonin is only possible if the SCN is intact. Daily timed administration of melatonin to rats shifts the phase of the circadian clock, and this phase shifting may partly explain the effect of melatonin on sleep in humans. Indirect support for such a physiological role derived from clinical studies on blind subjects that showed free running of their circadian rhythms and treated with melatonin. More direct support for this hypothesis was provided by the demonstration that the phase response curve for melatonin was opposite (i.e., 180 degrees out of phase) to that of light.[19,31]

Once formed in the pineal gland, melatonin is released into capillaries and can reach all tissues of the body within a very short period.[18,32] In the liver, melatonin is first hydroxylated and then conjugated with sulfate and glucuronide. In human urine, 6-sulfatoxymelatonin has been identified as the main metabolite. In the brain melatonin is metabolized into kynurenine derivatives. It is of interest that the well-documented antioxidant properties of melatonin are shared by some of its metabolites including cyclic 3-hydroxymelatonin, N^1-acetyl-N^2-formyl-5-methoxykynuramine (AFMK)[33] and, with highest potency, N^1-acetyl-5-methoxykynuramine (AMK).[34] Thus an "antioxidant cascade" seems to be triggered by melatonin administration to experimental animals and humans.

Melatonin action involves interaction with specific receptors in the cell membrane,[35] with nuclear receptors[36] and with intracellular proteins such as calmodulin,[37] dihydronicotinamide riboside:quinone reductase 2[38] or tubulin-associated proteins.[39] As already mentioned, melatonin is also a potent antioxidant acting as a free radical scavenger as well as via induction of antioxidant enzymes.[30,40]

Within the SCN, melatonin reduces neuronal activity in a time-dependent manner. The effects of melatonin on SCN activity are mediated by two different receptors that are insensitive during the day, but sensitive at dusk and dawn (MT_2; causes phase shifts) and during early night period (MT_1; decreases neuronal firing rate).[41] The evening increase in melatonin secretion is associated with an increase in the propensity for sleep via interaction with SCN MT_1 receptors.[42] Melatonin (in a dose of 3–5 mg daily, timed to advance the phase of the internal clock by interaction with MT_2 receptors in the SCN) can maintain synchronization of the circadian rhythm to a 24-h cycle in sighted persons who are living in conditions likely to induce a free-running rhythm, and it synchronizes the rhythm in some persons after a short period of free-running. In blind subjects with free-running rhythms, it has been possible to stabilize, or entrain, the sleep/wake cycle to a 24-h period by giving melatonin, with resulting improvements in sleep and mood.[43] In normal aged subjects[44] and in demented patients with desynchronization of the sleep/wake cycle[45] melatonin administration is helpful to reduce the variation of onset time of sleep. The phase shifting effects of melatonin are also sufficient to explain its effectiveness as a treatment for circadian-related sleep disorders, such as jet lag or the delayed phase sleep syndrome.[46–48]

CHRONOBIOTIC AND CYTOPROTECTIVE PROPERTIES
OF MELATONIN IN DEMENTED PATIENTS

Alzheimer's disease (AD) is an age-associated neurodegenerative disease with severe disruption of the sleep/wake rhythm.[49] A chronobiological sleep disturbance known as "sundowning" has been reported in many AD patients. Symptoms of sundowning agitation include circadian disorganization, a reduced ability to maintain attention to external stimuli, disorganized thinking and speech, a variety of motor disturbances including agitation, wandering and repetitive physical behaviors, and perceptual and emotional disturbances. This sleep/wake disturbance becomes increasingly more marked with progression of the disease. These clinical findings strongly argue in favor of the disruption of the circadian time-keeping system in AD.[50,51] Indeed, lesioned SCN neurons in AD patients exhibits neurofilament tangles,[52] the damage of SCN neurons being suggested as one of the major causes for circadian rhythm disturbances in demented patients.[53]

The disturbed sleep/wake rhythm seen in AD patients is associated with high degree of irregularity of melatonin secretion.[54] Studies undertaken in AD patients by a number of investigators have shown that AD melatonin levels are significantly lower than age-matched control subjects.[55-58] The decreased melatonin levels in AD patients correlated with a disturbed sleep-wake rhythm and disrupted circadian rhythmicity.[59] It is of interest that melatonin levels in the cerebrospinal fluid are decreased during the progression of AD neuropathology (as determined by the Braak stages), even in cognitive-intact subjects with the earliest AD neuropathology (preclinical AD). The circadian melatonin rhythm disappeared in both pre-symptomatic and diagnosed AD and AD patients.[60] The melatonin decrease was specifically profound in AD patients carrying the apolipoprotein E-epsilon 4/4 genotype, which is a predictor of early onset AD.[56] Furthermore, a reduced hippocampal MT_2 melatonin receptor expression has been reported in AD patients.[61] The decrease in melatonin production seen during preclinical AD stages (Braak stages I–II) has been suggested to be due to dysfunctional regulation of pineal gland by SCN. These changes in melatonin secretion could contribute to the frequent initial symptoms in "AD-like" patients, including sleep disruption and nightly restlessness.[59]

In view of the altered melatonin secretion and disrupted circadian rhythmicity found in AD it was logical to treat AD patients with melatonin. The efficacy of melatonin for treatment of AD patients has been supported by several studies. In a first examination of the sleep-promoting action of melatonin in a small non-homogeneous group of elderly demented patients, 7 of 10 dementia patients having sleep disorders and treated with melatonin (3 mg orally at bed time) showed decreased sundowning and reduced variability of sleep onset time.[62] Melatonin reduced the variability in sleep onset time in demented patients exhibiting sundowning.[62] In another study, 10 individuals with mild cognitive impairment were given 6 mg melatonin before bedtime. Improvement was found in sleep, mood, and memory.[63] Further studies reported the efficacy of melatonin in improving sleep and alleviating sundowning in elderly AD patients.[64] The retrospective account of AD patients after a 2–3-year period of treatment with 6–9 mg melatonin/day indicated that all experienced improved sleep quality.[65,66] A significant observation in these studies was the halted evolution of the cognitive and amnesic alterations expected in comparable populations of patients not receiving melatonin. Asayama and colleagues[67] reported, in a double-blinded, placebo-controlled study that administration of 3 mg melatonin for 4 weeks significantly prolonged actigraphy-evaluated sleep time and decreased activity in the night. As in the study of Brusco and colleagues,[65] significant improvement in cognitive performance was detected.[67] Another study using actigraphic evaluation of the sleep/wake cycle noted remarkable effects of melatonin (3 mg/day) in improving nocturnal sleep and reducing sundowning in AD.[68] A beneficial effect of melatonin in improving sleep quality was also noted in 45 AD patients treated for a period of 4 months using 6–9 mg/day.[59]

In a multicenter, randomized, placebo-controlled clinical trial of two dose formulations of oral melatonin, 157 subjects with AD and nighttime sleep disturbance were randomly assigned to one of three treatment groups: placebo, 2.5-mg slow-release melatonin, or 10-mg melatonin given daily for 2 months.[69] When sleep was defined by actigraphy, trends for increased nocturnal total sleep time and decreased wake after sleep onset in the melatonin groups were observed. On subjective measures, caregiver ratings of sleep quality showed significant improvement in the 2.5-mg sustained-release melatonin group relative to placebo.[69]

The mechanisms accounting for the therapeutic effect of melatonin in AD patients remain unknown. On the one hand, melatonin treatment promotes mainly non-REM sleep in the elderly[22] and can thus be beneficial in AD by augmenting the restorative phases of sleep, culminating in the augmented secretion of growth hormone[70] and neurotrophins. On the other hand, melatonin has very strong cytoprotective properties in a number of neurodegenerative disorders via its antioxidant and antiapoptotic effects.[30] In addition, other effects of melatonin, like interference with the phosphorylation system may occur. In AD patients, the neurofibrillary tangles are composed of abnormally bundled cytoskeletal fibers, due to hyperphosphorylation of tau (a microtubule-associated protein) and of neurofilament H/M subunits, processes that lead to misfolding and accumulation of these proteins, along with disruption of microtubules. *In vitro*, melatonin prevented cell death and the decline in mitochondrial activity induced by okadaic acid, a potent protein phosphatase inhibitor, in SH-SY5Y[71] and N1E-115[72] neuroblastoma cells. Melatonin also inhibited the phosphorylation and accumulation of neurofilaments. Similar data were obtained in neuroblastoma N2a cells with calyculin A, an inhibitor of protein phosphatases 2A and 1.[73] Melatonin effect was not only based on antioxidant properties, but also on interference with the phosphorylation system, especially stress kinases.

Animal studies strongly indicate the ability of melatonin to prevent β-amyloid (Aβ) toxicity. For example, melatonin had the ability to protect against the changes in circadian rhythmicity induced by microinjection of Aβ peptide[25–35] into the SCN of golden hamsters.[74] Melatonin prevents the death of neuroblastoma cells exposed to Aβ polypeptide.[75–78] Animal models of AD have also been used to study the possible antioxidant and antiapoptotic actions of melatonin in arresting the neuronal lesions. Melatonin inhibited Aβ deposition in APP 695 transgenic mice.[79] In the APP695 transgenic mouse, senile plaques appear in the cortex as early as at 8 months of age. These mice also display behavioral impairments and memory deficits.[80] Evaluation of long term administration of melatonin revealed that melatonin (10 mg/kg) alleviated learning and memory deficits. It also reduced the number of apoptotic neurons.[79] In another model of transgenic mice, the senile plaques consisting of deposited Aβ are able to induce the secretion of interleukins IL-6 and IL-1β[81] In this study, melatonin attenuated the Aβ-induced secretion of IL-1β and IL-6, again a beneficial antiinflammatory effect related to antioxidative protection.

Recent studies show that apoptosis of astrocytes may contribute to the pathogenesis of AD. Astrocytes exhibit tau hyperphosphorylation and accumulation as well as activation of stress kinases in AD pathology. Moreover, astrocytes produce apoE4,[82] which aggravates Aβ effects. The Aβ protein and astrocyte-neuron interactions seem to mutually potentiate neurodegeneration in AD.[83] Feng and Zhang found in C6 astroglioma cells that melatonin was able to prevent increases in NO formation, as induced by A1-42.[84] This study suggested that melatonin's ability to rescue C6 cells from apoptosis is associated with its ability to scavenge peroxynitrite. From all these studies it is clear that melatonin treatment conveyed neuroprotection against oxidative injury by maintaining survival of both neuronal cells and glial cells.

CONCLUDING REMARKS

Melatonin may provide an innovative neuroprotective strategy in normal and pathological aging. Melatonin protects against several mechanisms of neuronal death, including oxyradical-mediated damage and apoptosis. Melatonin also has very strong cytoprotective activity in a number of situations including osteoporosis,[85] ischemia–reperfusion,[30] and diabetic microangiopathy.[86,87] Through restoration of slow-wave sleep melatonin treatment can result in better regulation of neuronal metabolism. Indeed better sleep must be considered as a neuroprotective strategy that can potentially improve the course and outcome of several brain disorders associated with aging, and thus the quality of life of the affected individuals and their family members. To promote and protect an appropriate sleep can substantially reduce the costs of treatment and management—in particular, the enormous costs of lifetime treatment of some neuropsychiatric disorders.

ACKNOWLEDGMENTS

Studies in authors' laboratory were supported by grants from the Agencia Nacional de Promoción Científica y Tecnológica, Argentina (PICT 14087) and the University of Buenos Aires (ME 075). D.P.C. is a Research Career Awardee from the Argentine Research Council (CONICET).

[*Competing interests*: The authors declare that they have no competing financial interests.]

REFERENCES

1. GACHON, F. *et al.* 2004. The mammalian circadian timing system: from gene expression to physiology. Chromosoma **113:** 103–112.
2. SAPER, C.B. *et al.* 2005. The hypothalamic integrator for circadian rhythms. Trends Neurosci. **28:** 152–157.
3. DIJK, D.J. & J.F. DUFFY. 1999. Circadian regulation of human sleep and age-related changes in its timing, consolidation, and EEG characteristics. Ann. Med. **31:** 130–140.

4. TZISCHINSKY, O. *et al.* 1993. The association between the nocturnal sleep gate and nocturnal onset of urinary 6-sulfatoxymelatonin. J. Biol. Rhythms **8:** 199–209.
5. ZHDANOVA, I.V. *et al.* 1996. Effects of low oral doses of melatonin, given 2–4 hours before habitual bedtime, on sleep in normal young humans. Sleep **19:** 423–431.
6. CARSKADON, M.A. *et al.* 1982. Sleep fragmentation in the elderly: relationship to daytime sleep tendency. Neurobiol. Aging **3:** 321–327.
7. IGUCHI, H. *et al.* 1982. Age-dependent reduction in serum melatonin concentrations in healthy human subjects. J. Clin. Endocrinol. Metab. **55:** 27–29.
8. DORI, D. *et al.* 1994. Chrono-neuroendocrinological aspects of physiological aging and senile dementia. Chronobiologia **21:** 121–126.
9. GIROTTI, L. *et al.* 2000. Low urinary 6-sulphatoxymelatonin levels in patients with coronary artery disease. J. Pineal Res. **29:** 138–142.
10. SIEGRIST, C. *et al.* 2001. Lack of changes in serum prolactin, FSH, TSH, and estradiol after melatonin treatment in doses that improve sleep and reduce benzodiazepine consumption in sleep-disturbed, middle-aged, and elderly patients. J. Pineal. Res. **30:** 34–42.
11. MISHIMA, K. *et al.* 2000. Supplementary administration of artificial bright light and melatonin as potent treatment for disorganized circadian rest-activity and dysfunctional autonomic and neuroendocrine systems in institutionalized demented elderly persons. Chronobiol. Int. **17:** 419–432.
12. LUBOSHITZKY, R. *et al.* 2001. Actigraphic sleep-wake patterns and urinary 6-sulfatoxymelatonin excretion in patients with Alzheimer's disease. Chronobiol. Int. **18:** 513–524.
13. MISHIMA, K. *et al.* 2001. Diminished melatonin secretion in the elderly caused by insufficient environmental illumination. J. Clin. Endocrinol. Metab. **86:** 129–134.
14. HAIMOV, I. *et al.* 1994. Sleep disorders and melatonin rhythms in elderly people. Br. Med. J. **309:** 167.
15. KARASEK, M. *et al.* 2002. Future of melatonin use in the therapy. Neuroendocrinol. Lett. **23:** 118–121.
16. KARASEK, M. 2004. Melatonin, human aging, and age-related diseases. Exp. Gerontol. **39:** 1723–1729.
17. DAWSON, D. & S.M. ARMSTRONG. 1996. Chronobiotics—drugs that shift rhythms. Pharmacol. Ther. **69:** 15–36.
18. CARDINALI, D.P. & P. PÉVET. 1998. Basic aspects of melatonin action. Sleep Med. Rev. **2:** 175–190.
19. KENNAWAY, D.J. & H. WRIGHT. 2002. Melatonin and circadian rhythms. Curr. Top. Med. Chem. **2:** 199–209.
20. REITER, R.J. 1980. The pineal and its hormones in the control of reproduction in mammals. Endocr. Rev. **1:** 109–131.
21. WURTMAN, R.J. & I. ZHDANOVA. 1995. Improvement of sleep quality by melatonin. Lancet **346:** 1491.
22. MONTI, J.M. *et al.* 1999. Polysomnographic study of the effect of melatonin on sleep in elderly patients with chronic primary insomnia. Arch. Gerontol. Geriatr. **28:** 85–98.
23. GUERRERO, J.M. & R.J. REITER. 2002. Melatonin-immune system relationships. Curr. Top. Med. Chem. **2:** 167–179.
24. ESQUIFINO, A.I. *et al.* 2004. Circadian organization of the immune response: a role for melatonin. Clin. Appl. Immunol. Rev. **4:** 423–433.
25. BLASK, D.E. *et al.* 2002. Melatonin as a chronobiotic/anticancer agent: cellular, biochemical, and molecular mechanisms of action and their implications for circadian-based cancer therapy. Curr. Top. Med. Chem. **2:** 113–132.
26. SCHEER, F.A. *et al.* 2004. Daily night time melatonin reduces blood pressure in male patients with essential hypertension. Hypertension **43:** 192–197.
27. DUBOCOVICH, M.L. *et al.* 1999. Molecular pharmacology and function of melatonin receptor subtypes. Adv. Exp. Med. Biol. **460:** 181–190.
28. KUNZ, D. 2004. Chronobiotic protocol and circadian sleep propensity index: new tools for clinical routine and research on melatonin and sleep. Pharmacopsychiatry **37:** 139–146.
29. SRINIVASAN, V. 1997. Melatonin, biological rhythm disorders, and phototherapy. Indian J. Physiol. Pharmacol. **41:** 309–328.

30. REITER, R.J. *et al.* 2005. When melatonin gets on your nerves: its beneficial actions in experimental models of stroke. Exp. Biol. Med. **230:** 104–117.
31. LEWY, A.J. *et al.* 1998. The human phase response curve (PRC) to melatonin is about 12 hours out of phase with the PRC to light. Chronobiol. Int. **15:** 71–83.
32. MACCHI, M.M. & J.N. BRUCE. 2004. Human pineal physiology and functional significance of melatonin. Front. Neuroendocrinol. **25:** 177–195.
33. ALLEGRA, M. *et al.* 2003. The chemistry of melatonin's interaction with reactive species. J. Pineal Res. **34:** 1–10.
34. RESSMEYER, A.R. *et al.* 2003. Antioxidant properties of the melatonin metabolite N^1-acetyl-5-methoxykynuramine (AMK): scavenging of free radicals and prevention of protein destruction. Redox. Rep. **8:** 205–213.
35. DUBOCOVICH, M.L. *et al.* 2000. Melatonin receptors. *In* The IUPHAR Compendium of Receptor Characterization and Classification, 2nd Edition. IUPHAR, Ed.: 271–277. IUPHAR Media. London.
36. WIESENBERG, I. *et al.* 1998. The potential role of the transcription factor RZR/ROR as a mediator of nuclear melatonin signaling. Rest. Neurol. Neurosci. **12:** 143–150.
37. ANTON-TAY, F. *et al.* 1998. Modulation of the subcellular distribution of calmodulin by melatonin in MDCK cells. J. Pineal Res. **24:** 35–42.
38. MAILLIET, F. *et al.* 2004. Organs from mice deleted for NRH:quinone oxidoreductase 2 are deprived of the melatonin binding site MT3. FEBS Let. **578:** 116–120.
39. CARDINALI, D.P. *et al.* 1997. Melatonin site and mechanism of action: single or multiple? J. Pineal Res. **23:** 32–39.
40. LEON, J. *et al.* 2004. Melatonin and mitochondrial function. Life Sci. **75:** 765–790.
41. DUBOCOVICH, M.L. *et al.* 2003. Molecular pharmacology, regulation, and function of mammalian melatonin receptors. Front. Biosci. **8:** 1093–1108.
42. LAVIE, P. 2001. Sleep-wake as a biological rhythm. Annu. Rev. Psychol. **52:** 277–303.
43. SKENE, D.J. *et al.* 1999. Melatonin in circadian sleep disorders in the blind. Biol. Signals Recept. **8:** 90–95.
44. CARDINALI, D.P. *et al.* 2002. A double blind-placebo controlled study on melatonin efficacy to reduce anxiolytic benzodiazepine use in the elderly. Neuroendocrinol. Lett. **23:** 55–60.
45. CARDINALI, D.P. *et al.* 2002. The use of melatonin in Alzheimer's disease. Neuroendocrinol. Lett. **23:** 20–23.
46. CARDINALI, D.P. *et al.* 2002. A multifactorial approach employing melatonin to accelerate resynchronization of sleep-wake cycle after a 12 time-zone westerly transmeridian flight in elite soccer athletes. J. Pineal Res. **32:** 41–46.
47. ARENDT, J. 2003. Importance and relevance of melatonin to human biological rhythms. J. Neuroendocrinol. **15:** 427–431.
48. BEAUMONT, M. *et al.* 2004. Caffeine or melatonin effects on sleep and sleepiness after rapid eastward transmeridian travel. J. Appl. Physiol. **96:** 50–58.
49. MCCURRY, S.M. *et al.* 2000. Treatment of sleep disturbance in Alzheimer's disease. Sleep Med. Rev. **4:** 603–628.
50. GIUBILEI, F. *et al.* 2001. Altered circadian cortisol secretion in Alzheimer's disease: clinical and neuroradiological aspects. J. Neurosci. Res. **66:** 262–265.
51. HARPER, D.G. *et al.* 2001. Differential circadian rhythm disturbances in men with Alzheimer disease and frontotemporal degeneration. Arch. Gen. Psych. **58:** 353–360.
52. STOPA, E.G. *et al.* 1999. Pathologic evaluation of the human suprachiasmatic nucleus in severe dementia. J. Neuropathol. Exp. Neurol. **58:** 29–39.
53. SKENE, D.J. & D.F. SWAAB. 2003. Melatonin rhythmicity: effect of age and Alzheimer's disease. Exp. Gerontol. **38:** 199–206.
54. MISHIMA, K. *et al.* 1999. Melatonin secretion rhythm disorders in patients with senile dementia of Alzheimer's type with disturbed sleep-waking. Biol. Psych. **45:** 417–421.
55. UCHIDA, K. *et al.* 1996. Daily rhythm of serum melatonin in patients with dementia of the degenerate type. Brain Res. **717:** 154–159.
56. LIU, R.Y. *et al.* 1999. Decreased melatonin levels in postmortem cerebrospinal fluid in relation to aging, Alzheimer's disease, and apolipoprotein E-epsilon 4/4 genotype. J. Clin. Endocrinol. Metab. **84:** 323–327.

57. OHASHI, Y. *et al.* 1999. Daily rhythm of serum melatonin levels and effect of light exposure in patients with dementia of the Alzheimer's type. Biol. Psych. **45:** 1646–1652.
58. FERRARI, E. *et al.* 2000. Pineal and pituitary-adrenocortical function in physiological aging and in senile dementia. Exp. Gerontol **35:** 1239–1250.
59. CARDINALI, D.P. *et al.* 2002. The use of melatonin in Alzheimer's disease. Neuroendocrinol. Lett. **23:** 20–23.
60. WU, Y.H. *et al.* 2003. Molecular changes underlying reduced pineal melatonin levels in Alzheimer's disease: alterations in preclinical and clinical stages. J. Clin. Endocrinol. Metab. **88:** 5898–5906.
61. SAVASKAN, E. *et al.* 2005. Reduced hippocampal MT2 melatonin receptor expression in Alzheimer's disease. J. Pineal Res. **38:** 10–16.
62. FAINSTEIN, I. *et al.* 1997. Effects of melatonin in elderly patients with sleep disturbance. A pilot study. Curr. Ther. Res. **58:** 990–1000.
63. JEAN-LOUIS, G. *et al.* 1998. Melatonin effects on sleep, mood, and cognition in the elderly with mild cognitive impairment. J. Pineal Res. **25:** 177–183.
64. COHEN-MANSFIELD, J. *et al.* 2000. Melatonin for treatment of sundowning in elderly persons with dementia. Arch. Gerontol. Geriatr. **31:** 65–76.
65. BRUSCO, L.I. *et al.* 1998. Melatonin treatment stabilizes chronobiologic and cognitive symptoms in Alzheimer's disease. Neuroendocrinol. Lett. **19:** 111–115.
66. BRUSCO, L.I. *et al.* 1998. Monozygotic twins with Alzheimer's disease treated with melatonin: case report. J. Pineal Res. **25:** 260–263.
67. ASAYAMA, K. *et al.* 2003. Double blind study of melatonin effects on the sleep-wake rhythm, cognitive, and noncognitive functions in Alzheimer-type dementia. J. Nippon Med. Sch. **70:** 334–341.
68. MAHLBERG, R. *et al.* 2004. Melatonin treatment of day-night rhythm disturbances and sundowning in Alzheimer's disease: an open-label pilot study using actigraphy. J. Clin. Psychopharmacol. **24:** 456–459.
69. SINGER, C. *et al.* 2003. A multicenter, placebo-controlled trial of melatonin for sleep disturbance in Alzheimer's disease. Sleep **26:** 893–901.
70. VAN COEVORDEN, A. *et al.* 1991. Neuroendocrine rhythms and sleep in aging men. Am. J. Physiol. **260:** 651–661.
71. WANG, Y.P. *et al.* 2004. Melatonin ameliorated okadaic-acid induced Alzheimer-like lesions. Acta Pharmacol. Sin. **25:** 276–280.
72. BENITEZ-KING, G. *et al.* 2003. Melatonin prevents cytoskeletal alterations and oxidative stress induced by okadaic acid in N1E-115 cells. Exp. Neurol. **182:** 151–159.
73. LI, X.C. *et al.* 2005. Effect of melatonin on calyculin A-induced tau hyperphosphorylation. Eur. J. Pharmacol. **510:** 25–30.
74. FURIO, A.M. *et al.* 2002. Effect of melatonin on changes in locomotor activity rhythm of Syrian hamsters injected with beta amyloid peptide 25-35 in the suprachiasmatic nuclei. Cell. Mol. Neurobiol. **22:** 699–709.
75. PAPPOLLA, M.A. *et al.* 1997. Melatonin prevents death of neuroblastoma cells exposed to the Alzheimer amyloid peptide. J. Neurosci. **17:** 1683–1690.
76. PAPPOLLA, M.A. *et al.* 1999. Alzheimer beta protein–mediated oxidative damage of mitochondrial DNA: prevention by melatonin. J. Pineal Res. **27:** 226–229.
77. PAPPOLLA, M.A. *et al.* 2000. An assessment of the antioxidant and the antiamyloidogenic properties of melatonin: implications for Alzheimer's disease. J. Neural Transm. **107:** 203–231.
78. PAPPOLLA, M.A. *et al.* 2002. The neuroprotective activities of melatonin against the Alzheimer beta-protein are not mediated by melatonin membrane receptors. J. Pineal Res. **32:** 135–142.
79. FENG, Z. *et al.* 2004. Melatonin alleviates behavioral deficits associated with apoptosis and cholinergic system dysfunction in the APP 695 transgenic mouse model of Alzheimer's disease. J. Pineal Res. **37:** 129–136.
80. KAWARABAYASHI, T. *et al.* 2001. Age-dependent changes in brain, CSF, and plasma amyloid (beta) protein in the Tg2576 transgenic mouse model of Alzheimer's disease. J. Neurosci. **21:** 372–381.
81. CLAPP-LILLY, K.L. *et al.* 2001. Melatonin reduces interleukin secretion in amyloid-beta stressed mouse brain slices. Chem. Biol. Interact. **134:** 101–107.

82. HARRIS, F.M. *et al*. 2004. Astroglial regulation of apolipoprotein E expression in neuronal cells. Implications for Alzheimer's disease. J. Biol. Chem. **279:** 3862–3868.
83. MALCHIODI-ALBEDI, F. *et al*. 2001. Astrocytes contribute to neuronal impairment in beta A toxicity increasing apoptosis in rat hippocampal neurons. Glia **34:** 68–72.
84. FENG, Z. & J.T. ZHANG. 2004. Protective effect of melatonin on beta-amyloid-induced apoptosis in rat astroglioma C6 cells and its mechanism. Free Radic. Biol. Med. **37:** 1790–1801.
85. CARDINALI, D.P. *et al*. 2003. Melatonin effects on bone: experimental facts and clinical perspectives. J. Pineal Res. **34:** 81–87.
86. REYES TOSO, C. *et al*. 2002. Vascular reactivity in diabetic rats: effect of melatonin. J. Pineal Res. **33:** 81–86.
87. REYES TOSO, C. *et al*. 2004. Effect of melatonin on vascular reactivity in pancreatectomized rats. Life Sci. **74:** 3085–3092.

Melatonin Influences on the Neuroendocrine– Reproductive Axis

BEATRIZ DÍAZ LÓPEZ, E. DÍAZ RODRÍGUEZ, COLMENERO URQUIJO, AND C. FERNÁNDEZ ÁLVAREZ

Departamento de Biología Funcional, Área Fisiología, Facultad de Medicina, Universidad de Oviedo, C /Julián Clavería, 6. 33006 Oviedo, Spain

ABSTRACT: The neuroendocrine–reproductive axis designates the functional activity of the hypothalamus–pituitary–gonadal axis. A delicate synchronization of many inputs at these three different levels is vital for normal reproductive function. From the median basal hypothalamus, the median eminence releases gonadotrophin releasing hormone into the portal circulation to reach the anterior pituitary gland. Gonadotrophin releasing hormone is obviously a key hormone for the regulation of the secretion of gonadotrophins LH and FSH.

KEYWORDS: melatonin; aging; pituitary gland; median eminence; LH; FSH; prolactin

SEXUAL MATURATION

The process of sexual maturation is a very complex phenomenon mediated by the ontogeny of the hypothalamus–pituitary–gonadal axis, which begins during intrauterine life. Administering melatonin to rats during pregnancy produced a delayed onset of puberty and altered hormonal developmental patterns in their female offspring. The ovarian oocyte developmental pattern showed an increase in cellular volume from 25 to 30 days of age in control offspring. In the offspring of pinealectomized or melatonin-treated mothers, however, such an increase was not observed, and nuclear and nucleolar volumes significantly decreased ($P < 0.05$). These data indicate that the maternal pineal gland and melatonin are necessary for normal sexual maturation. Thus, melatonin could also be involved in the decline of the reproductive system.

AGING OF THE FEMALE REPRODUCTIVE SYSTEM AND MELATONIN

Aging of the female reproductive system is marked by discrete stages in the disappearance of regular sexual cyclicity. A deficient GnRH drive or reduced respon-

Address for correspondence: Beatriz Díaz López, Departamento de Biología Funcional, Área Fisiología, Facultad de Medicina, Universidad de Oviedo. C /Julián Clavería, 6. 33006 Oviedo, Spain. Voice: (34) 985 102713; fax: (34) 985103534.

beatrizd@uniovi.es

Ann. N.Y. Acad. Sci. 1057: 337–364 (2005). © 2005 New York Academy of Sciences. doi: 10.1196/annals.1356.026

siveness to the GnRH signal accounts for the age-related decline in reproductive function as reflected by an attenuated pro-estrus LH surge in middle-aged rats. The 24-h rhythm of melatonin production is very robust in young animals, as in humans, but it deteriorates with age. Thus in old animals, the amount secreted is lower than in young individuals, and the administration of melatonin during aging may be beneficial in delaying age-related physiological changes. The action of melatonin on the sexual maturation processes has been thoroughly investigated, but its action during aging of the reproductive system has not been previously studied. The sexual cycle of the rats, the estrus cycle, exhibits a regularity of 4 to 5 days. Before ovulation takes place, in the estrus phase, preovulatory surges of gonadotropins, prolactin, and estradiol levels are observed in young rats. However, the profile of LH, FSH, prolactin, and estradiol, studied throughout the different stages of the estrus cycle in middle-aged rats, showed no preovulatory surges of these hormones. Melatonin treatment during aging of the female reproductive system enhanced the amount of hormones released during the surge in the proestrus and may have improved synchronization. The increased gonadotropin concentrations observed in aged rats were decreased by melatonin treatment.

The chronological pattern of the gonadotropins LH and FSH showed significantly increased values ($P < 0.01$) in aged rats (25 months old), as well as an increased magnitude of the response to a bolus of luteinizing hormone releasing hormone (LHRH) (50 ng LHRH/100 g body weight). Melatonin treatment during aging significantly decreased ($P < 0.01$) gonadotropin concentrations to levels similar to those observed in young rats and restored pituitary responsiveness to LHRH to levels similar to those encountered in young rats. A similar effect upon increased hypothalamic neurokinin A (NKA) and substance P (SP) concentrations in aged rats was observed after melatonin treatment.

In Vitro *Studies*

In young and old female rats, pituitary FSH content increased to values similar to FSH released from *in vitro* pituitaries to incubation medium. Pituitary LH and prolactin contents were significantly higher ($P < 0.01$) than the amounts released to incubation medium. Melatonin was found to exert an inhibitory influence on the secretory processes, rather than on the biosynthetic processes. This functional activity of melatonin reduced the secretory activity of young rats to levels demonstrated by old rats.

INFLUENCE OF MATERNAL PINEAL GLAND AND PRENATAL MELATONIN ON POSTNATAL OFFSPRING DEVELOPMENT

Female Rat Sexual Maturation

The process of sexual maturation is a very complex phenomenon mediated by the ontogeny of the hypothalamus–pituitary–gonadal axis that commences during intrauterine life.[1] The pineal hormone melatonin crosses the placenta, passes from the mother rat to the fetus.[2] Melatonin can be also be transferred to rat pups through maternal milk.[3] Thus, melatonin can act directly on the fetus and on the newborn to influence subsequent somatic and sexual development. We performed new experiments in order to develop a better knowledge of this matter, given the scarce data available at the beginning of the 1990s.

Female offspring of control group mothers (control offspring), pinealectomized mothers (PIN-X offspring), or melatonin-treated mothers (250 μg/100 g body weight) during pregnancy (MEL offspring) were studied. The results show a delayed ($P < 0.05$) onset puberty, a decreased percentage of proestrus, and significantly lower ($P < 0.01$) LH concentrations in MEL offspring. However, LH values of PIN-X offspring were similar to those of control offspring. For lower LH values, higher melatonin concentrations could be expected, but no differences in melatonin concentrations were observed 5.5 h after dark, which is the time peak of the nocturnal rhythm in the rat.[4]

Pituitary Reproductive Hormonal Development

As prenatal melatonin treatment produced delayed sexual maturation, the next step was to investigate the developmental pattern of the gonadotropins luteinizing hormone (LH), follicle stimulating hormone (FSH), and prolactin in the same offspring groups.[5]

Female Offspring

To study female offspring, we followed the classification and protocols of Ojeda[6] concerning postnatal maturation; neonatal period between birth and day 7 of life, animals being examined on day 5; infantile period, between 8 and 21 days, animals being studied on day 15; juvenile or prepubertal period, from 21 to 32 days of age, animals being studied at 25 and 30 days; pubertal period from day 32 to day of vaginal opening, animals being examined at 35 days of age.

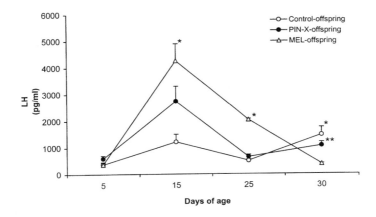

FIGURE 1. Effect of pinealectomy or melatonin treatment (250 μg/100 g body weight) during pregnancy on plasma LH levels of female rat offspring. Values are mean ± SEM. 15 days: *$P < 0.01$ vs. control. 25 days: *$P < 0.01$ vs. control and PIN-X. 30 days: *$P < 0.01$ and **$P < 0.05$ vs. MEL. (Reproduced from Diaz López *et al.*[44] with permission.)

Plasma LH Levels

LH (FIG. 1) exhibited great variability throughout postnatal development with low concentrations during the neonatal period and increased basal secretion during the infantile period (15 days) in the three offspring. In the control and PIN-X off-spring, LH values returned to lower values at the end of the prepubertal period. However, a different pattern of development was observed in MEL offspring. In these, the peak in the infantile period (15 days) was significantly higher ($P < 0.01$) than that of control offspring, and values remained elevated during the juvenile period (25 days), suggesting that prenatal melatonin treatment affects the normal development of the neuroendocrine–reproductive axis. At the end of the prepubertal period (30 days), higher plasma LH values were observed in control and PIN-X offspring. This is probably the result of the disappearance of direct central nervous inhibition of LH secretion.[7] However, in MEL offspring at 30 days of age such an increase of LH levels was not observed. Thus, the decrease of LH appeared to be delayed until the end of the juvenile phase (30 days), when significantly lower values than in control and PIN-X ($P < 0.01$, $P < 0.05$, respectively) offspring were observed. This, accompanied by hyperprolactinemia, which is usually associated with amenorrhea and/or infertility, could be indicators of the delayed sexual maturation.

PLASMA FSH AND PROLACTIN

Prenatal melatonin treatment or pinealectomization of the mothers did not change postnatal FSH levels in female offspring, with peak values at 15 days of age. Prenatal melatonin treatment produced hyperprolactinemia in 30-day-old offspring.

Male Offspring

Sexual development in male control rats progressed rapidly, as shown by the significantly increased LH and FSH levels at 25 and 30 days of age, as well as by the higher prolactin levels at 30 days of age. These results (FIG. 2) confirm previous data on male pubertal development, which begins during the juvenile period and extends from 21 to 35 days (the so-called "FSH period"). Indeed, FSH secretion is increased markedly, with peak plasma values observed between 30 and 35 days.[8] However, in MEL offspring FSH values were inhibited during the neonatal period and at 25, 30, or 55 days as compared to control offspring. This would affect the sexual maturation of the rat by reducing FSH stimulation of Leydig cells. In PIN-X offspring a decrease in LH levels during the neonatal and infantile periods and at 30 days of age as well as a decrease of FSH levels at 15 and 25 days as compared to control rats were observed. It is known that the pineal gland begins to function around 15 days of age in rats.[9] This functioning may be responsible for the normalization of hormone levels in PIN-X offspring earlier than in MEL offspring: at 25 or 30 days of age when the secretory melatonin rhythm is established. Our results suggest that melatonin treatment during pregnancy influences the ontogeny of the hypothalamus–pituitary–gonadal axis that begins during intrauterine life. These changes resulted in alterations in gonadotropin and prolactin secretion of both female and male rats during sexual development.

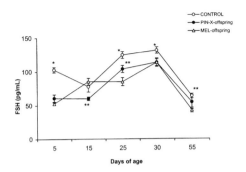

FIGURE 2. Effect of pinealectomy or melatonin treatment (250 µg/100 g body weight) during pregnancy on plasma FSH levels of male rat offspring. Values are expressed as mean ± SEM. 5 days: *$P < 0.01$ vs. PIN-X offspring and PIN-X offspring+MEL. 15 days: **$P < 0.05$ vs. PIN-X offspring + MEL. 25 days: *$P < 0.01$ vs. PIN-X offspring + MEL and **$P < 0.05$ vs. control. 30 days: *$P < 0.01$ vs. PIN-X offspring + MEL. 55 days: **$P < 0.05$ vs. PIN-X offspring + MEL. (Reproduced from Diaz López *et al.*[44] with permission.)

PITUITARY HORMONAL FEEDBACK SYSTEM

Female Offspring

In female rats, physiological levels of estradiol are markedly effective in inhibiting gonadotropin release in prepubertal as compared to postpubertal sexual stages. Thus, sensitivity of the hypothalamic–pituitary axis results in a negative LH feedback response to estradiol benzoate (EB) during the infantile period. After 19 days of age, however, the animals showed a biphasic pattern, with a negative feedback in the first hours followed by a positive effect at 31 or 55 hours after EB injection. Basal LH secretion is already pulsatile before the first preovulatory surge.[10]

As there were no references about the possible influence of prenatal melatonin treatment on the gonadotropin-steroid feedback during postnatal life, we examined in detail, throughout sexual development, the changes in sensitivity of the gonadotrophins and prolactin response to estrogen feedback in female offspring of control and melatonin-treated rats (150 µg/100 g body weight) during pregnancy. To study female offspring we followed the classification of Ojeda[6] concerning postnatal maturation as previously described, with animals being studied on days 15 and 18, 21, 23, 25, and 30, 33, and 35 days of age. Female offspring received 50 µg EB/rat by subcutaneous injection at the specified ages. Basal blood samples were taken at 10:00 AM, and blood samples 24, 31, 48, and 55 h after EB.

LH Feedback Mechanism

The results (FIG. 3) show that a negative feedback or no LH feedback effect developed after EB administration in control offspring up to 30 days of age and that positive feedback was established from day 33 on. This response was altered in MEL offspring, during the prepubertal phase, when a significantly increased ($P < 0.05$) LH secretion was found on day 25, 31 h after EB injection. Puberty is initiated at the

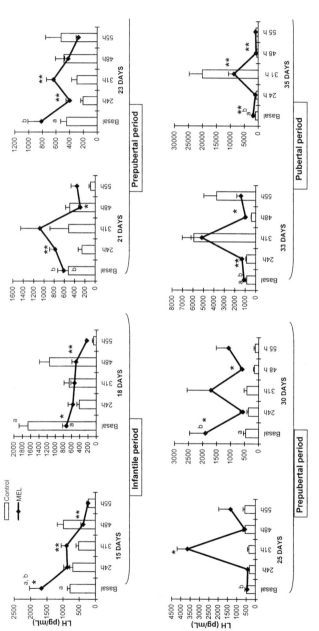

FIGURE 3. Plasma LH response to estradiol benzoate (50 μg) subcutaneous administration on 15-, 18-, 21-, 23-, 25-, 30-, 33-, 35-day-old female offspring of control and melatonin-treated (150 μg/100 g body weight) mother rats. Values are mean ± SEM. *$P < 0.01$; **$P < 0.05$ vs. control group. Longitudinal study. Basal value vs. post EB injection. (*Infantile period*) 15-day-old offspring: control offspring, [a]$P < 0.01$ vs. 55 h; MEL offspring, [a]$P < 0.01$ vs. 55 h, [b]$P < 0.05$ vs. 55 h. 18-day-old offspring: control offspring, [a]$P < 0.01$ vs. 24, 31, and 55 h; MEL offspring: [a]$P < 0.01$ vs. 55 h. (*Prepubertal period*) 21-day-old offspring: control offspring, [b]$P < 0.05$ vs. 55 h; MEL offspring: [b]$P < 0.05$ vs. 48 and 55 h. 23-day-old offspring: control offspring, [a]$P < 0.01$ vs. 24 h; MEL offspring: [b]$P < 0.05$ vs. 24 and 55 h. 25-day-old offspring: MEL offspring: [b]$P < 0.05$ vs. 31 h. 30-day-old offspring: control-offspring, [a]$P < 0.01$ vs. 48 and 55 h, MEL offspring: [b]$P < 0.05$ vs. 48 h. (*Pubertal period*) 33-day-old offspring: control offspring, [a]$P < 0.01$ vs. 31 and 48 h, [b]$P < 0.05$ vs. 55 h; MEL offspring: [a]$P < 0.01$ vs. 31 h, MEL offspring: [b]$P < 0.05$ vs. 31 and 48 h. 35-day-old offspring: control offspring, [a]$P < 0.01$ vs. 31 h, MEL offspring: [b]$P < 0.05$ vs. 31 and 48 h. (Reproduced from Díaz *et al.*[12] with permission.)

point when levels of endogenous estrogens present in the system are no longer an effective inhibitor of LH and FSH secretion.[11] We can conclude that melatonin administration to the mothers induced precocious initiation of puberty at 25 days of age, when a single exogenous dose of 50 μg of EB was administered. However, at 30 days of age the positive response was lost in MEL offspring, which again showed a negative feedback effect 48 h after EB and in controls a negative feedback was found during the second postinjection day. At 33 days of age, when 56.25–64.28% of animals in both groups studied showed vaginal opening, an LH positive feedback response was observed 31 h after the EB injection. In control offspring this positive response continued to 35 days of age, whereas in MEL offspring it remained at similar levels to 33 days of age and was subsequently highly depressed, showing a two-fold lower increase as compared to control offspring.

FSH and the Prolactin Feedback Mechanism

Similarly, the EB-FSH feedback system was altered in MEL offspring: negative feedback lasted for a longer time, and during the pubertal phase of sexual development the positive feedback effect was observed earlier in MEL offspring (at 33 days of age, while in the control offspring it was apparent at 35 days of age). At this last age, in MEL offspring the positive-feedback response did not persist, a finding similar to what was found in LH response to EB.

A positive prolactin response to EB at all ages studied in control offspring was observed. The typical pulsatility with higher values in the afternoon appeared for the first time by 30 days of age. However, in MEL offspring no pulsatile response was observed at any age. All these data point to the existence of differential sensitivity of the gonadotropin-releasing system to estrogen negative feedback from offspring of control and MEL-treated rats. The data also suggest that prenatal melatonin treatment affects the postnatal development of gonadotropins and prolactin feedback mechanisms to EB as early as in intrauterine life.[1]

Male Offspring

The modification in the sensitivity of the negative feedback effect of gonadal steroids on gonadotrophin secretion is one of the principal events involved in the onset of puberty. Much less testosterone is needed to suppress LH levels in prepubertal than in pubertal males, which reflect a "shift" at puberty in the response of gonadotropins to androgens.[13] At puberty, sensitivity to gonadal steroid negative feedback declines sharply.[14]

The purpose of this investigation was to examine the effect of prenatal melatonin administration on the shift in the steroid feedback process of gonadotrophins and prolactin response throughout sexual development. To study male offspring, we followed the classification system of Ojeda.[6] Infantile-period animals (between 8 and 21 days) were studied at 17 and 21 days of age. Rats in the juvenile or prepubertal period (extending from 21 to 35 days of age) were studied at 30 and 35 days of age. Pubertal-period animals (from day 35 to 60 days) were studied at 40 and 60 days. At selected ages, all male rats received a single dose of testosterone propionate (TP) 100 μg/100 g body weight by subcutaneous injection. Basal blood samples were taken at 10 AM. Blood samples were obtained 8 and 24 h after TP administration.

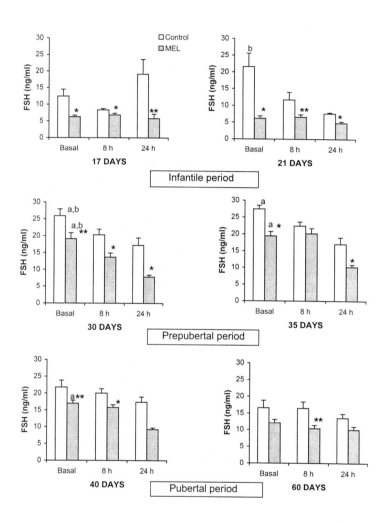

FIGURE 4. Plasma FSH response to testosterone propionate (TP) (100 µg/100 g body weight) administration on 17-, 21-, 30-, 35-, 40-, and 60-day-old male offspring of control and melatonin-treated (150 µg/100 g body weight) mother rats. Values are mean ± SEM. *P < 0.01 vs. control group; **P < 0.05 vs. control group. Basal values vs. post-TP injection from longitudinal study. (*Infantile period*) 21-day-old offspring: control offspring, [b]P < 0.05 vs. 24 h. (*Prepubertal period*) 30-day-old offspring: control offspring, [a]P < 0.01 vs. 24 h; [b]P < 0.05 vs. 8 h; MEL offspring, [a]P < 0.01 vs. 24 h, [b]P < 0.05 vs. 8 h. 35-day-old offspring: control offspring, [a]P < 0.01 vs. 8 and 24 h; MEL offspring: [a]P < 0.01 vs. 24 h. (*Pubertal period*) 40-day-old offspring: MEL offspring, [a]P < 0.01 vs. 48 h. (Reproduced from Díaz et al.[15] with permission.)

LH Feedback Mechanism

In control offspring, LH levels in response to TP showed the classical negative feedback effect as early as 8 h after TP injection, from the infantile period to the pubertal period. The highest magnitude of negative response from infantile to pubertal periods was observed at 30 days of age. However, 5 days later, at 35 days of age, the negative response was attenuated. This indicates that at 35 days of age, which is a transitional stage from the prepubertal to the pubertal period, the neural mechanisms involved in feedback response are at a more advanced phase of sexual development. In MEL offspring, the negative response to TP was observed as early as 17 days of age. The highest magnitude of response to TP was observed 5 days later than in control offspring, at 35 days of age.

FSH Feedback Mechanism

FSH secretion (FIG. 4) showed a negative feedback response to supraphysiological androgen levels in immature rats. In MEL offspring, decreased FSH concentrations at most time points studied and delayed negative feedback effect were observed. In MEL offspring, negative feedback was observed for the first time at 30 days of age, whereas in control offspring it was observed at 21 days of age. This feedback effect disappears at 35 days of age in control offspring, whereas in MEL offspring it remained until 40 days of age.

Prolactin Feedback Mechanism

Plasma prolactin levels increase from infantile to pubertal periods. TP produced a stimulatory action on the mechanism involved in prolactin release in control offspring at all ages studied. In MEL offspring the increased prolactin values after TP injection were blunted at 17 and 35 days of age in control offspring. During the pubertal period, although the MEL offspring group showed positive feedback, the magnitude of the response was lower than in control offspring.[15]

The maturation processes of pituitary responsiveness to both LHRH and steroid feedback action are probably the most important steps in the regulation of sexual maturation in the rat. All these data suggest that modification of the fetal endocrine environment, caused by prenatal melatonin administration, induced changes in the sensitivity of gonadotrophins and the prolactin-feedback response to exogenous androgens indicative of a delayed sexual development of male offspring.

Ovarian Oocyte Development

We investigated the maturational stage of oocytes of prepubertal female rats when mother rats were pinealectomized or treated with melatonin (250 µg/100 g body weight) during pregnancy. Female offspring were studied at the juvenile or prepubertal period (days 21–35). It is known that growth of primordial follicles to primary, secondary and tertiary stages entails enlargement of the oocyte. Ovaries were obtained from 25-, 30-, and 34-day-old prepubertal rats for cytometric studies of cellular volume (FIG. 5).

This 9-day chronological progression study shows the influence of the maternal pineal gland on oocyte development and also shows that the maternal pineal gland is necessary for a normal postnatal oocyte development. Enlargement of oocyte volume from 25 to 30 days was observed in the follicle oocytes that were examined in control offspring. However, in MEL offspring, oocyte volumes showed an opposite developmental pattern with a tendency toward reduced oocyte volume from 25 to 34 days of age and with a significantly lower volume at 34 days of age than at 25 days. The oocyte and nuclear volumes are correlated in control offspring. They have sig-

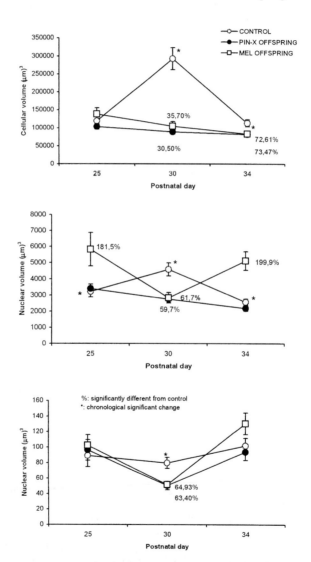

FIGURE 5. Differential proportionality changes of cellular, nuclei, and nucleoli volumes (μm^3) in the experimental groups from control offspring. All values are mean ± SEM. % = significantly different from control offspring ($P < 0.05$). * = chronological significant changes ($P < 0.01$). (Reproduced from Fernández et al.[16] with permission.)

nificantly higher values at 30 days of age than at 25 and 34 days. In MEL offspring an opposite pattern of development was again observed, with lower nuclear volumes at 30 days of age than at 25 or 34 days. Pinealectomy of the mother resulted in an intermediate level of oocyte development, without significant alterations in cellular volume from 25 to 34 days and with higher nuclear volume values at 25 days of age than at 30 or 34 days. Also, the decrease of the nucleole observed at 30 days of age in offspring of either PIN-X or MEL offspring suggests a temporary arrest of follicular development, which was not observed in control offspring. Relatively small cellular volumes could indicate an arrest of the growth of the oocytes. FIGURE 5 shows that the differential proportionality between cellular and nuclear volumes must indicate highly significant regulatory mechanisms.[16]

DEVELOPMENT OF GONADAL STEROID HORMONES

Estradiol

Investigation of the influence of the maternal pineal gland and melatonin on oocyte development was accompanied by a study of the estradiol levels from the neonatal period (day 7), to the infantile period (day 15), and up to the prepubertal period

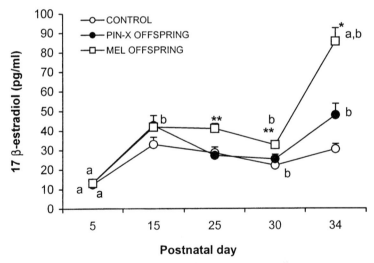

FIGURE 6. Effect of pinealectomy or melatonin treatment (250 μg/100 g body weight) during pregnancy on 17β-estradiol concentrations (pg/mL). All values are mean ± SEM. $N = 6–11$. Comparisons among experimental groups: 25-day-old offspring, $**P < 0.05$ vs. control and PIN-X offspring; 30-day-old offspring, $**P < 0.05$ vs. control offspring; 34-day-old offspring, $*P < 0.01$ vs. control and PIN-X offspring. Chronological study of control offspring, $^{a}P < 0.01$ vs. 15, 25, 30, and 34 days old and $^{b}P < 0.05$ vs. 15 and 34 days old. Chronological study of PIN-X offspring, $^{a}P < 0.01$ vs. 15, 25, 30, and 34 days old and $^{b}P < 0.05$ vs. 25 and 30 days old. Chronological study of MEL offspring, $^{a}P < 0.01$ 5 days old vs. 15, 25, 30, and 34 days old, and 34 days old vs. 30 and 15 days old and $^{b}P < 0.05$ 30 days old vs. 15 and 25 days old, and 34 days old vs. 25 days old.

(days 25 and 30). Plasma estradiol levels were determined in control, PIN-X, and MEL offspring (250 μg/100 g body weight) (FIG. 6). Results show that plasma estradiol levels significantly increased from 5 to 15 days of age in all groups. These increased levels are maintained at 25 days of age in control and MEL offspring. Estradiol levels of MEL offspring were significantly higher ($P < 0.05$) than in the other two groups. This difference is still observed at 30 days of age between MEL offspring and control offspring. In the PIN-X offspring, from the highest estradiol levels at 15 days of age, values significantly decreased in 25- and 30-day-old female rats. However, in MEL offspring, significantly lower values were not observed until day 30.[16]

Testosterone

In the present investigation, plasma testosterone levels in male offspring of control and melatonin-treated mother rats (150 μg/100 g body weight) were determined during sexual development (FIG. 7). Testosterone propionate (TP) was administered by subcutaneous injection at a single dose of 100 μg/100 g body weight, The animals that received TP were killed one day after injection.

In control offspring + TP and MEL offspring+TP, testosterone values were detectable through all phases of sexual development. In control offspring + placebo, testosterone values were observed at 17–18 days of age, but then non detectable values were observed until the end of prepubertal period (35–36 days of age). Testosterone concentrations increased throughout the pubertal period. In MEL offspring+placebo, detectable testosterone values were found only at 17–18 days of age and then delayed until 60–61 days of age, such values being significantly higher ($P < 0.05$) than in MEL offspring + TP. This indicates that increased exposure to melatonin during intrauterine life results in an inhibitory effect on postnatal androgen biosynthesis.[17]

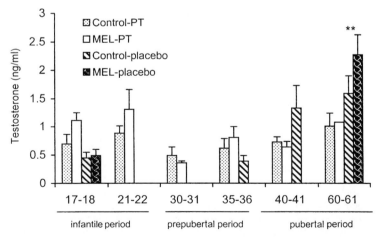

FIGURE 7. Developmental changes in plasma testosterone levels of male offspring of control and melatonin-treated mother rats (150 μg/100 g body weight) from infantile to pubertal period. Values are mean ± SEM. **$P < 0.05$ vs. MEL offspring + MEL. (Reproduced from Díaz et al.[17] with permission.)

MELATONIN AND AGING

Melatonin Action on Female Reproductive System during Aging

Different factors are involved in the decline of the reproductive system with age, such as a decrease in ovarian function and increased resistance to gonadotropins.[16,17] The aging of the female reproductive system in rats is marked by discrete stages in the disappearance of regular estrus cyclicity. Transition to acyclicity is characterized by increased LH pulse duration but decreased amplitude. Changes in the pulse generator function may play a role in the age-related transition to acyclicity. Studies in young and middle-aged female rats have indicated that reduced pituitary responsiveness to LHRH may contribute to the delayed and attenuated proestrus LH surge that precedes reproductive senescence.[20]

Bearing in mind that melatonin secretion is reduced in senescence[21] and that it has been found to inhibit aging processes[22] the possibility that nocturnal supplementation of melatonin may prevent or delay some of these changes was tested in the following investigations. On the other hand, as we have previously found that melatonin was necessary for normal sexual maturation to take place, we thought that melatonin could also be involved in the decline of the sexual reproductive functions.

Effect of Melatonin on Middle-Aged Female Sexual Cycle Hormones

Middle-aged female rats show alterations of their reproductive system such as reduced LH surges.[23] This supports the hypothesis that a decrease in the excitatory inputs to LHRH neurons may be involved directly in the reduction of the activity of hypothalamic–pituitary–ovarian axis observed during aging. However, it is not known to what extent the hypothalamus and pituitary are involved in this process, nor is the role of melatonin understood. To investigate this question, 11-month-old female rats, showing irregular sexual cycles (estrus cycle) indicative of a pre-acyclic stage equivalent to perimenopause in women, were used. The profile of the reproductive hormones LH, FSH, prolactin, and 17-B-estradiol were studied throughout the different stages of their estrus cycle (FIG. 8). The results showed that control middle-aged rats do not have the preovulatory peaks of the gonadotrophins LH and FSH, prolactin, or estradiol; instead, values were similar in all stages of the estrus cycle. Melatonin administration (150 μg/100 g body weight) for two months, enhanced the amount of LH, FSH, and prolactin released during the surge at the proestrus day, and may have improved synchronization.

The circadian cycle (FIG. 9) of gonadotropins showed that melatonin administration resulted in significantly lower ($P < 0.01$, $P < 0.05$) LH values from 07.00 to 14.00 h; meanwhile, at 17.00 h, the hour of the afternoon preovulatory surge of proestrus, a peak value equivalent to that found in young rats was again observed. The circadian cycle of FSH concentrations during the day of proestrus in control middle-aged rats increased progressively from 07.00 to 01.00 h. Melatonin administration resulted in a clear peak value at 17.00 h. Surges in the secretion of LH, FSH, and prolactin are preceded and stimulated by the increasing concentrations of follicular estrogens. The results of the present study indicate that melatonin treatment in middle-aged rats had a positive influence on this mechanism; estradiol concentrations were significantly increased ($P < 0.05$) on the morning of proestrus, which pre-

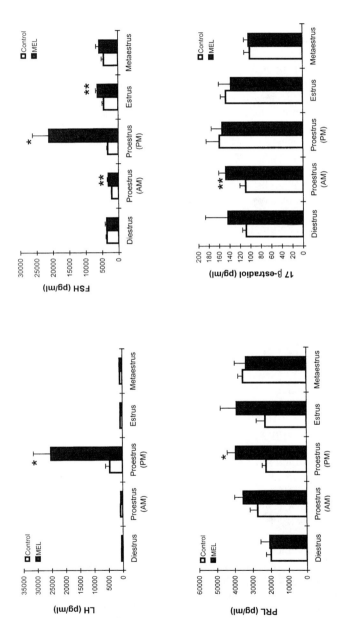

FIGURE 8. Concentrations of plasma LH, FSH, prolactin, and estradiol in all stages of the estrus cycle in control and melatonin-treated 11–13-month-old female rats. Rats were treated with 150 μg melatonin per 100 g body weight for two months. *$P < 0.01$, **$P < 0.05$ vs. control group. (Reproduced from Díaz et al.[24] with permission.)

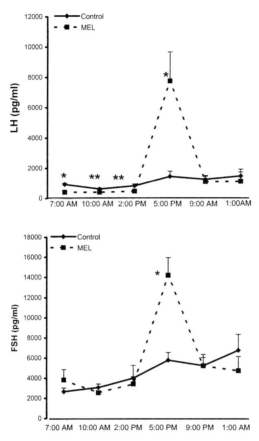

FIGURE 9. Plasma LH levels (pg/mL), and FSH, on proestrus in 11–15-month-old control and melatonin-treated rats (150 μg/100 g body weight) for two months. *$P < 0.01$; **$P < 0.05$ vs. control group. (Reproduced from Díaz et al.[24] with permission.)

ceded the gonadotropins and prolactin surges observed in the afternoon of proestrus.[24]

Chronological Pattern

The chronological pattern of gonadotropins and prolactin indicate that an important change in secretory processes occurs from 23- to 25-month-old rats, with gonadotrophins increasing while prolactin decreases (FIG. 10). Melatonin treatment (150 μg/100 g body weight) for two months significantly reduced ($P < 0.01$) gonadotropins LH and FSH concentrations in 25-month-old acyclic rats to levels similar to those encountered in 5-month-old cyclic control rats at the diestrus phase. However, melatonin treatment significantly decreased ($P < 0.05$) prolactin concentrations in the young cyclic rats, but significantly increased ($P < 0.05$) prolactin concentrations in the old acyclic rats.

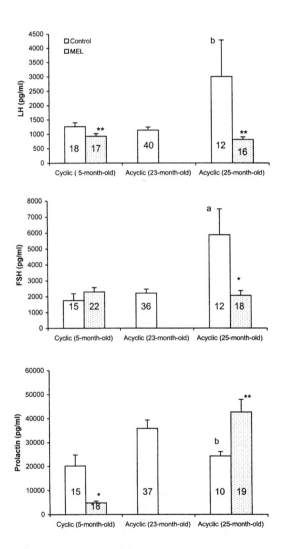

FIGURE 10. Concentrations of LH, FSH, and prolactin (PRL) in cyclic and acyclic control and melatonin-treated rats (150 μg/100 g body weight for two months). *Numbers within histograms* represent the number of animals per group. *$P < 0.01$, **$P < 0.05$ vs. control group. LH and FSH: [a]$P < 0.01$, [b]$P < 0.05$ vs. cyclic control and acyclic control (23 months). Prolactin: [b]$P < 0.05$ vs. acyclic control (23 months). (Reproduced from Díaz *et al.*[25] with permission.)

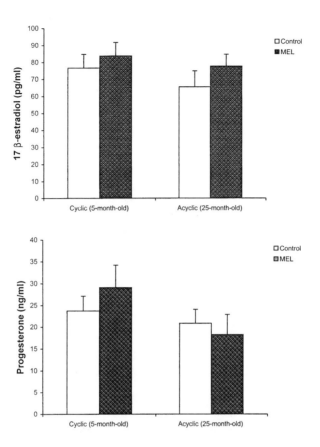

FIGURE 11. Plasma 17β-estradiol and progesterone levels of cyclic and acyclic female rats, control and melatonin-treated (150 μg/100 g body weight). Data are mean ± SEM. (Reproduced from Díaz *et al.*[25] with permission.)

There were no differences in estradiol or progesterone concentrations between the young cyclic and old acyclic control rats (FIG. 11). This lack of different concentrations may be due to the phase in which cyclic rats were studied, the diestrus phase, which is the quiescent phase of the estrus cycle and is characterized by low steroid levels. No effect of melatonin treatment was observed on the steroid hormones studied either in cyclic rats at diestrus phase, or in acyclic rats.[25]

Pituitary Gland Responsiveness to LHRH

The responsiveness to a bolus injection of LHRH (50 ng LHRH per 100 g body weight) in middle-aged (12-month-old) female rats showing irregular duration of the estrus cycles was examined (FIG. 12). The results showed a stimulatory effect of melatonin treatment (150 μg/100 g body weight) for one month on the gonadotropins

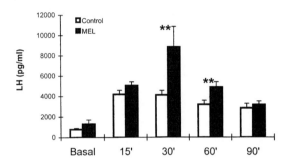

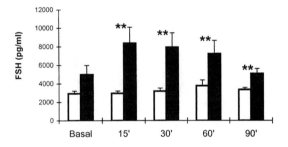

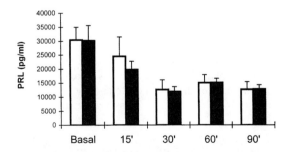

FIGURE 12. Concentrations of plasma LH, FSH, and prolactin (PRL) at diestrus after injection of 50 ng LHRH/100 g body weight in control and melatonin-treated (150 µg/100 g body weight for one month) 12-month-old female rats. **$P < 0.05$ vs. control group. (Reproduced from Díaz et al.[24] with permission.)

response to LHRH as treated rats showed a significantly greater ($P < 0.05$) LH secretion surge than did control rats. In the control group FSH showed no response to LHRH; however, a positive FSH response was observed in melatonin-treated rats.[24]

Pituitary responsiveness was studied in young (4-month-old) rats showing regular repetitive (4–5 day) estrus cycles and in aged (24-month-old) female rats showing persistent estrus (PE) or repetitive diestrus (RD) pseudopregnancy, indicative of an acyclic stage. The magnitude of the LH and FSH response to LHRH was higher in acyclic 24-month-old rats than it was in cyclic 4-month-old control rats, which indicates altered pituitary responsiveness to LHRH in aged control rats. Melatonin treatment (150 μg/100 g body weight) for one month during aging restored the pituitary responsiveness, effecting a similar response to that in young cyclic rats.[26]

IN VITRO HORMONE SECRETION FROM THE PITUITARY GLAND AND MEDIAN EMINENCE

To define more critically the possible role of melatonin treatment (150 μg/100 g body weight daily for two months) in neuroendocrine–reproductive axis functional activity during aging, we investigated the *in vitro* secretory activity of the anterior pituitary gland in young cyclic (3-month-old) and old acyclic (23-month-old) female rats. In the same study, the secretory activity of the median eminence was also investigated, given the importance that this region of median basal hypothalamus exerts on the functions of the hypothalamo-pituitary system.[27]

Pituitary Hormones Release and Content

LH release at the basal period (I-1) from cyclic-MEL and acyclic control groups was significantly reduced ($P < 0.01$; $P < 0.05$) as compared to the cyclic control (FIG. 13). No significant differences in LH release and hemipituitary content due to age or melatonin treatment were observed in the second incubation period (I-2). However, hemipituitary LH content was significantly higher ($P < 0.01$) in all groups studied than LH released from the respective groups in the I-2 incubation.

FSH release in both incubation periods (I-1 and I-2) and FSH hemipituitary content were significantly lower ($P < 0.01$) in cyclic MEL and acyclic control groups than in the cyclic control group. Only in cyclic MEL was hemipituitary content significantly higher ($P < 0.05$) than FSH released from the respective group.

Prolactin release in the I-1 period was significantly decreased ($P < 0.06$) in cyclic MEL and acyclic control as compared to cyclic control. In the I-2 period, prolactin release was significantly lower ($P < 0.05$) in acyclic control than in cyclic control. Hemipituitary I-2 content did not show significant differences among groups studied; however, concentrations were significantly higher ($P < 0.01$) than prolactin released from the respective groups at the I-2 incubation period.[27]

The results show that melatonin treatment of young rats significantly decreased basal LH, FSH, and prolactin secretory processes, with values similar to those seen in old acyclic rats. However, melatonin did not affect the pituitary LH content, and values 10- to 30-fold higher than LH secreted to incubation media were found. The significance of the biphasic course of pituitary LH content and release may represent the necessary pituitary LH stores available to induce an ovulatory discharge of LH[20] in cyclic rats. This suggests that at these two different sexual stages a selective LH release from the pituitary must be functionally active. On the other hand, the neuroendocrine mechanisms responsible for pituitary LH stores in acyclic rats could be mediated by an altered gonadal steroid environment in the aged animals[28] or by changes in age-dependent hypothalamo–pituitary axis function.[29] Pituitary FSH

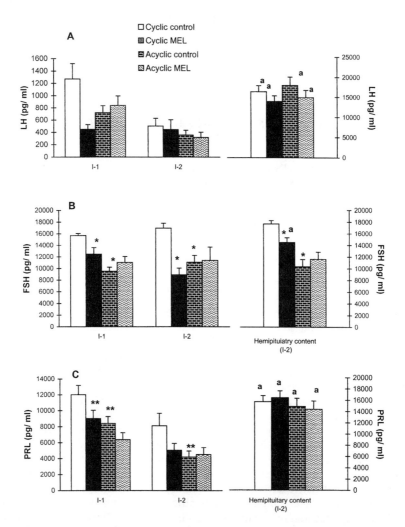

FIGURE 13. *In vitro* hemipituitary LH, FSH, and prolactin (PRL) release and content from cyclic and acyclic control and melatonin-treated (150 μg/100 g body weight for two months) female rats. For sample sizes, see FIGURE 13. (**A**) I-1, *$P<0.01$ vs. cyclic control and acyclic MEL. **$P < 0.05$ vs. cyclic control. Hemipituitary content (I-2), ** $P < 0.05$ vs. cyclic control. Hemipituitary content (I-2), *$P < 0.01$ vs. I-2 respective groups. (**B**) I-1, *$P < 0.01$ vs. cyclic control, I-2 cyclic MEL. (**C**) I-1 and I-2, **< 0.05 cyclic control. Hemipituitary content (I-2), ᵃ$P < 0.01$ vs. I-2 respective groups. (Reproduced from Fernández et al.[27] with permission.)

content rose to levels similar to FSH released from *in vitro* pituitaries. Pituitary gland prolactin content was significantly higher than prolactin released to incubation media. This indicates that the inhibitory influence of melatonin on the pituitary hormones is mainly exerted on secretory processes, rather than on the biosynthetic process.

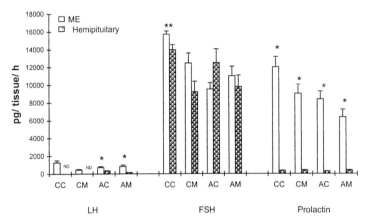

FIGURE 14. Basal rate release of LH, FSH, and prolactin (PRL) by *in vitro* hemipituitaries and median eminences (ME) after 1-h incubation period. *Bars* indicate the mean ± SEM. Cyclic-control (CC, N = 15), cyclic MEL (CM, N = 15), acyclic control (AC, N = 13), acyclic-MEL (AM, N = 18). *P < 0.01 vs. ME; **P < 0.05 vs. ME. ND = no detectable values. (Reproduced from Fernández *et al.*[27] with permission.)

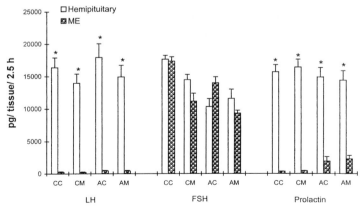

FIGURE 15. Tissue content of LH, FSH, and prolactin (PRL) from *in vitro* hemipituitaries and median eminences (ME) after 2.5-h incubation period. *Bars* indicate the mean± SEM. For sample sizes, see FIGURE 13. CC: cyclic control, CM: cyclic MEL, AC: acyclic control, AM: acyclic MEL. *P < 0.01 vs. ME. (Reproduced from Fernández *et al.*[27] with permission.)

Comparisons between Pituitary Gland and Median Eminence Hormones Release and Content

LH release from the pituitary was observed in all groups studied, but from the median eminence only in acyclic control and acyclic MEL groups, with significantly lower ($P < 0.01$) values than those found in the pituitary (FIG. 14). LH content was significantly higher ($P < 0.01$) in pituitary than in median eminence in all groups studied (FIG. 15).

FSH release and content showed similar values in the pituitary and median eminence in all groups studied, with the exception that significantly lower values ($P < 0.05$) were observed in FSH released from the median eminence in the cyclic control group. LH and prolactin release and content showed significantly higher values ($P < 0.01$) in the pituitary than in the median eminence.

Our results clearly demonstrate for the first time that LH was released from pituitary in an amount much higher than from the median eminence. This coincides with previous findings indicating that LH is present in the hypothalamus of rats.[30] However, the results obtained for LH secreted from the median eminence clearly demonstrate a secretory process different from the one observed in the pituitary. The fact that similar values were obtained for the LH content in the median eminence of young cyclic and old acyclic rats, but opposite results for amounts of LH secreted between cyclic control and acyclic control rats, suggests that there exists a different neuroendocrine mechanism in the median eminence at both reproductive stages, with increased secretory activity during aging.

In contrast, FSH released by the median eminence was not affected by age, and similarities between median eminence and pituitary FSH neuroendocrine mechanisms could be observed, contrary to what was found regarding LH. Finally, the 40- to 75-fold lower median eminence LH content as compared with the pituitary clearly indicates a different secretory mechanism of LH from FSH.

Prolactin release from the median eminence in amounts 10- to 30-fold lower than from the pituitary were observed in all groups studied. Similarly, prolactin tissue content was 40-fold lower in the median eminence than the pituitary in cyclic rats and sevenfold lower in acyclic rats. This result points to the different biosynthetic process in the median eminence of old acyclic as compared with young cyclic rats.

In conclusion, our results indicate that pituitary LH and prolactin release and content show significantly higher levels than those observed in the median eminence. This activity is maintained during senescence. FSH showed a different profile, and showed similar levels for pituitary and for median eminence.[27]

MELATONIN, TACHYKININS, AND AGING

The mammalian tachykinins (TKs) neurokinin A (NKA) and substance P (SP) are neuropeptides widely distributed in the central nervous system (CNS) and other organs. NKA is very often present with SP, and some of its effects are similar to those of SP on neurons and peripheral tissues. The presence of TKs in the hypothalamus, pituitary gland, and other brain tissues has been previously described.[31,32] There is evidence that TKs may be involved in the control of the secretion of anterior pituitary hormones.[33] Castration and administration of gonadal steroids have been shown to alter TKs levels in the hypothalamus and anterior pituitary.[34,35] The primary function of these peptides is not completely understood, but it is known that they are potential regulators of basal blood flow and therefore they may control the blood flow of many organs and tissues.[36]

Recent results show that aging of the CNS is associated with changes in tachykinin expression. The numbers of SP-like immunoreactivity (LI) perikarya in the hypothalamic suprachiasmatic nucleus of aged hamsters at day and night time were augmented three- to fourfold when compared to adult animals.[37] It is important to know the influence of age on brain TKs, as they may be implicated in the physiopathology of some common diseases in elderly patients.[38]

It is known that in old animals the amount of melatonin secreted is lower than in young animals. Consequently, melatonin treatment may be beneficial in delaying age-related degenerative conditions. Administration of pineal extracts prevents cell degeneration, delays onset of age-related diseases, and prevents reproductive senescence in old animals.[22] Our previous results[24,25] showed that melatonin administration during aging regulates the activity of the hypothalamo-pituitary unit and particularly improves gonadotropin secretion in response to LHRH in middle-aged female rats.

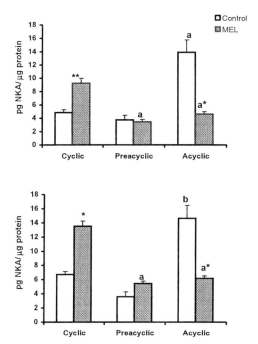

FIGURE 16. NKA and SP immunoreactive substances in the hypothalamus of cyclic (five-month-old), preacyclic (15-month-old), and acyclic (25-month-old) control and melatonin-treated (150 μg/100 g body weight) female rats. Data: mean ± SEM. Control group (N = 15-24), MEL group (N = 16–27): **$P < 0.05$ vs. cyclic control. *$P < 0.01$ vs. acyclic control. ${}^{a}P < 0.01$ preacyclic MEL and acyclic MEL vs. cyclic MEL. ${}^{a}P < 0.01$, ${}^{b}P < 0.05$ acyclic control vs. preacyclic-control and cyclic-control. (Reproduced from Fernández *et al.*[39] with permission.)

NKA AND SP IN BRAIN AND BODY ORGANS

The aim of this investigation was to study the effects of melatonin treatment during aging on the NKA and SP stores in neural structures related to the reproductive functions, such as hypothalamus, pituitary, and pineal glands, and in brain areas implicated in neurodegenerative age-dependent diseases, such as the striatum. In a second phase, we investigated the possibility that melatonin may also influence TK concentrations in different body organs during aging. For this purpose three ages were studied: young or cyclic (3 months old), middle-aged or preacyclic (13 months old), and old or acyclic (23 months old) rats. Melatonin treatment was a subcutaneous shot of 150 µg melatonin/100 g body weight for two months. These were the first studies in which the effect of melatonin during aging of the rat were reported in brain or body organs.[39,40]

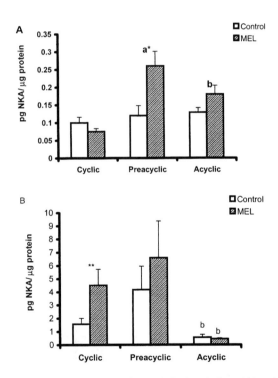

FIGURE 17. NKA immunoreactive substances in the pituitary (**A**) and pineal (**B**) of cyclic (five-month-old), precyclic (15-month-old), and acyclic (25-month-old) control and melatonin-treated (150 µg/100 g body weight) female rats. Data are presented as the mean ± SEM at each age. The sample sizes were between 15 and 28 in control groups and between 15 to 30 in the melatonin-treated groups. (**A**) *$P < 0.01$ vs. preacyclic control. [a]$P < 0.01$ preacyclic MEL vs. cyclic MEL. [b]$P < 0.05$ acyclic MEL vs. cyclic MEL. (**B**) **$P < 0.05$ vs. cyclic control. [b]$P < 0.05$ vs. acyclic control vs. preacyclic control and acyclic MEL vs. preacyclic MEL. (Reproduced from Fernández et al.[39] with permission.)

Hypothalamus

The results (FIG. 16) obtained in female control rats showed that in aged female rats NKA and SP levels significantly increased ($P < 0.01$, $P < 0.05$) as compared with young and middle-aged rats. This suggests that such an increase may have some relationship with the loss of sexual cyclicity. In addition, melatonin treatment significantly reduced NKA and SP concentrations observed in aged rats, with values in the range of those found in young rats.

Pituitary and Pineal Glands

Pituitary NKA stores in the anterior pituitary (FIG. 17) of female control rats were not affected by age. Melatonin treatment resulted in increased concentrations of NKA in middle-aged and aged-rats.

Significantly lower NKA values ($P < 0.05$) were found in the pineal glands of aged control and melatonin-treated rats as compared to the two other ages studied. Melatonin treatment of young cyclic rats resulted in significantly increased ($P < 0.05$) pineal NKA values as compared to cyclic control rats.

Ovary

SP concentrations (FIG. 18) in the aged control rats showed significantly increased ($P < 0.01$) values as compared to young- and middle-aged rats. In the melatonin-treated groups differences among the three ages studied disappeared. There are few reports on the role of TKs in the ovary, but some indicate that these peptides are present in the ovary and they may affect the secretion of ovarian steroids.[41]

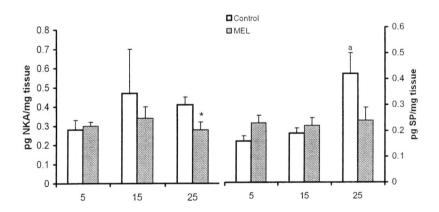

FIGURE 18. Developmental pattern of NKA and SP immunoreactive substances in the ovary of control and melatonin-treated (150 μg/100 g body weight) female rats. Data are mean ± SEM at each age. For NKA, samples size were $N = 7, 7, 7$ for the respective control age groups, $N = 7, 8, 7$ for the respective melatonin age groups. For SP, samples size were $N = 11, 11, 8$ for the respective control age groups, $N = 11, 12, 10$ for the respective melatonin age groups. *$P < 0.01$ vs. control group. For SP, [a]$P < 0.01$ vs. young and middle-aged control groups. (Reproduced from Fernández *et al.*[40] with permission.)

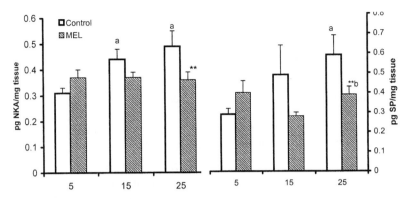

FIGURE 19. Developmental pattern of NKA and SP immunoreactive substances in the pancreas. Data are mean ± SEM at each age. For NKA, sample size were N = 15, 12, 12 for the respective control age groups, N = 20, 13, 15 for the respective melatonin age groups. For SP, sample size were N = 15, 11, 12 for the respective control age groups, N = 20, 12, 16 for the respective melatonin age groups. **P < 0.05 vs. control group. For NKA, [a]P < 0.01 vs. young-control. For SP, [a]P < 0.01 vs. young control; [b]P < 0.05 vs. middle-aged MEL. (Reproduced from Fernández *et al.*[40] with permission.)

Pancreas

Our results (FIG. 19) show significantly higher (P < 0.01) NKA and SP concentrations in old female control rats. This increase was blunted by melatonin treatment and was lowered to values similar to those observed in young rats. This indicates a positive and beneficial effect of melatonin on pancreatic TK concentrations. Previous data have also reported that the removal of the endogenous source of melatonin in rodents—that is, pinealectomy, induced hypoglycemia, and hyperinsulinemia[42] and reduced glucose-induced insulin release.[43]

Liver

In our study, no modifications in liver TKs concentrations throughout the ages studied were found. Melatonin had no influence on these concentrations. This indicates that as no fluctuations of the values assessed are observed, melatonin fails to affect them.

The current results are relevant to the understanding of the interactions between TKs and melatonin, and the possible influence of melatonin on changes occurring during the aging of different organs. The effects of melatonin were observed mainly on the hypothalamus, ovary, and pancreas in old rats, with a reduction in the concentrations of TK as compared with those observed in the young groups.

ACKNOWLEDGMENTS

The authors wish to express their gratitude to M. F. Suarez, S.J., for his help in the English edition. This work has been supported by the Spanish Ministry of Health, Fondo de Investigación Sanitaria. FIS No. 97/0988.

[*Competing interests*: The authors declare that they have no competing financial interests.]

REFERENCES

1. DUBOIS, P.M. 1985. Ontogeny of hypothalamic luteinizing hormone releasing-hormone (GnRH) and pituitary receptors in fetal and neonatal rats. Endocrinology **116:** 1565–1576.
2. KLEIN, D.C. 1972. Evidence for the placental transfer of ^{3}H-N-acetyl-melatonin. Nature **237:** 118.
3. REPPERT, S.M. & D.C. KLEIN. 1979. Transport of maternal ^{3}H-melatonin in the neonatal rat. Endocrinology **102:** 582–588.
4. COLMENERO, M.D. *et al.* 1991. Melatonin administration during pregnancy retards sexual maturation of female offspring in the rat. J. Pineal Res. **11:** 23–27.
5. DÍAZ, E. *et al.* 1995. Effect of mother pineal gland on offspring's somatic and ovarian maturation. Neuroendocrinol. Lett. **17:** 163–170.
6. OJEDA, S.R. *et al.* 1980. Recent advances in endocrinology of puberty. Endocrinol. Rev. **1:** 228–257.
7. DOCKE, F. *et al.* 1980. Evidence for a direct central nervous inhibition of LH secretion during sexual maturation of female rats. Endocrinology **75:** 1–7.
8. KETELSLEGER, J.M. *et al.* 1978. Developmental changes in testicular gonadotropin receptors: plasma gonadotropins and testosterone in the rat. J. Endocrinol. **47:** 391–392.
9. TANG, P.L. & S.F. PANG. 1988. The ontogeny of pineal and serum melatonin in male rats at mid-light and mid-dark. J. Neural. Transm. **72:** 43–53.
10. ANDREWS, W.W. *et al.* 1981. The maturation of estradiol-negative feedback in female rats: evidence that the resetting of the hypothalamic "gonadostat" does not precede the first preovulatory surge of gonadotropins. Endocrinology **109:** 2022–2031.
11. DLUZEN, D. & V. D. RAMÍREZ. 1979. Escape from estrogen negative feedback of LH and FSH release during puberty in the female rat. Presented at the Sixty-First Annual Meeting of the Endocrine Society. Anaheim, CA.
12. DÍAZ, E. *et al.* 1999. Prenatal melatonin exposure influences the maturation of gonadotropin and prolactin estradiol-benzoate feedback system. J. Steroid Biochem. Mol. Biol. **9:** 81–82.
13. BLOCH, G.J. *et al.* 1974. Effect of testosterone on plasma LH in male rats of various ages. Endocrinology **94:** 947–951.
14. RAMIREZ, V.D. & S.M. MCCANN. 1963. A comparison of the regulation of luteinizing hormone (LH) secretion in immature and adult rats. Endocrinology **72:** 452–464.
15. DÍAZ, E. *et al.* 2000. Effect of prenatal melatonin on the gonadotropin and prolactin response to the feedback effect of testosterone in male offspring. J. Steroid Biochem. Mol. Biol. **72:** 61–69.
16. FERNÁNDEZ, B. *et al.* 1995. Maternal pineal gland participates in prepubertal rats' ovarian oocyte development. Anat. Rec. **243:** 461–465.
17. DIAZ, E. *et al.* 1999. Developmental changes of hypothalamic, pituitary, and striatal tachykinins in response to testosterone: influence of prenatal melatonin. Peptides **20:** 501–508.
18. NAVOT, D. *et al.* 1994. Age-related decline in female fertility is not due to diminished capacity of the uterus to sustain embryo implantation. Fertil. Steril. **61:** 97–101.
19. HOFFMAN, G.E. *et al.* 1995. Evaluation of the reproductive performance of women with elevated day 10 progesterone levels during ovarian reserve screening. Fertil. Steril. **63:** 979–983.
20. SMITH, W.A. *et al.* 1982. Altered pituitary responsiveness to gonadotropin-releasing hormone in middle-aged rats with 4-day estrous cycles. Endocrinology **111:** 1843–1848.
21. WALDHAUSER, F. & H. STEGER. 1986. Changes in melatonin secretion with age and pubescence. J. Neural Transm. **21:** 183–197.
22. PIERPAOLI, W. *et al.* 1997. Circadian melatonin and young-to-old pineal grafting postpone aging and maintain juvenile conditions of reproductive functions in mice and rats. Exp. Geront. **32:** 587–602.

23. ARIAS, P. *et al.* 1996. Effects of ageing on N-methyl-D-aspartate (NMDA)-induced GnRH and LH release in female rats. Brain Res. **740:** 234–238.
24. DÍAZ, E. *et al.* 1999. Effect of exogenous melatonin on neuroendocrine-reproductive function of middle-aged female rats. J. Reprod. Fertil. **117:** 331–337.
25. DÍAZ, E. *et al.* 2000. Effects of aging and exogenous melatonin on pituitary responsiveness to GnRH in rats. J. Reprod. Fertil. **117:** 331–337.
26. FERNÁNDEZ, C. & E. DÍAZ. 1999. *In vitro* pituitary responsiveness to LHRH in young and old female rats. Influence of melatonin. Mech. Ageing Dev. **112:** 75–83.
27. FERNÁNDEZ, C. *et al.* 2000. Aging and melatonin influence on *in vitro* gonadotropins and prolactin secretion from pituitary and median eminence. Mech. Ageing Dev. **114:** 173–183.
28. STEGER, R.W. & J.J. PELUSO. 1982. Effects of age on hormone levels and *in vitro* steroidogenesis by rat ovary and adrenal. Exp. Aging Res. **8:** 203–208.
29. HUANG, H.H. *et al.* 1978. Patterns of sex steroid and gonadotropin secretion in aging female rats. Endocrinology **103:** 1855–1859.
30. EMANUELE, N.V. *et al.* 1981. Hypothalamic luteinizing hormone (LH): characteristics and response to hypophysectomy. Biol. Reprod. **25:** 321–326.
31. DEBELJUK, L. & A. BARTKE. 1994. Immunoreactivity substance P and neurokinin A in the hypothalamus and anterior pituitary gland of the Siberian and Syrian hamster and of the rat. J. Reprod. Fertil. **101:** 427–434.
32. DEBELJUK, L. *et al.* 1995. Developmental changes of tachykinins in the hypothalamus and anterior pituitary of female Siberian hamster from prepuberty to adulthood. Peptides **16:** 827–831.
33. DEBELJUK, L. *et al.* 1987. Effect of anti-substance P serum on prolactin and gonadotropins in hyperprolactinemic rats. Regul. Pept. **19:** 91–98.
34. DEBELJUK, L. *et al.* 1990. Neurokinin A levels in the hypothalamus of rats and mice: effects of castration, gonadal steroids, and expression of heterologous growth hormone genes. Brain Res. Bull. **25:** 717–721.
35. VILLANÚA, M.A. *et al.* 1992. Effects of neonatal administration of monosodium glutamate and castration on neurokinin A levels in the hypothalamus and anterior pituitary of rats. Peptides **13:** 377–381.
36. SEVERINI, C. *et al.* 2002. The tachykinin peptide family. Pharmacol. Rev. **54:** 285–322.
37. REUSS, S. & K. BURGER. 1994. Substance P-like immunoreactivity in the hypothalamic suprachiasmatic nucleus of *Phodopus sungorus*—relation to day-time, photoperiod, sex, and age. Brain Res. **638:** 189–195.
38. POMPEI, P. & R. SEVERINI. 1999. *In situ* hybridization analysis of protachykinin-A mRNA levels in young and old rats. Mol. Brain Res. **64:** 132–136.
39. FERNÁNDEZ, C. *et al.* 2002. Age differences in neurokinin A and substance P from the hypothalamus, pituitary, pineal gland, and striatum of the rat. Effect of exogenous melatonin. Peptides **23:** 941–945.
40. FERNÁNDEZ, C. *et al.* 2002. Developmental pattern of tachykinins during aging in several organs: effect of exogenous melatonin. Peptides **23:** 1617–1623.
41. DEBELJUK, L. & M. LASAGA. 1999. Modulation of the hypothalamo-pituitary-gonadal axis and the pineal gland by neurokinin A, neuropeptide K, and neuropeptide gamma. Peptides **20:** 285–299.
42. MILCU, S.M. *et al.* 1971. The effect of pinealectomy on the plasma insulin in rats. *In* The Pineal Gland. G.E.W. Woltensholme & J. Knight, Eds.: 345–357. Churchill Livingstone. Edinburgh.
43. DÍAZ, B. & E. BLÁZQUEZ. 1986. Effect of pinealectomy on plasma glucose, insulin, and glucagon levels in the rat. Horm. Metab. Res. **18:** 225–229.
44. DÍAZ LÓPEZ, B. M.D. COLMENERO URQUIJO, M.E. DÍAZ RODRÍGUEZ, *et al.* 1995. Effect of pinealectomy and melatonin treatment during pregnancy on the sexual development of the female and male rat offspring. Eur. J. Endocrinol. **132:** 765–770.

Caloric Restriction Mimetics

The Next Phase

GEORGE S. ROTH,[a] MARK A. LANE,[b,c] AND DONALD K. INGRAM[b]

[a]*GeroScience, Inc., Pylesville, Maryland 21132, USA*

[b]*Laboratory of Experimental Gerontology, Intramural Research Program, National Institute on Aging, National Institutes of Health, Baltimore, Maryland 21224, USA*

ABSTRACT: Calorie restriction (CR) mimetics are agents or strategies that can mimic the beneficial health-promoting and anti-aging effects of CR, the only intervention conclusively shown to slow aging and maintain health and vitality across the phylogenetic spectrum. Our lead compound, developed at the National Institute on Aging, was 2-deoxyglucose, an analogue of the native sugar, that acted as a glycolytic inhibitor, having limited metabolism and actually reducing overall energy flow—analogous to CR. This agent reduced insulin levels and body temperature of rats, similar to the physiological effects of CR, but toxicity was noted in long-term studies, which apparently prevented life-span extension. We previously demonstrated that lower insulin and body temperature (as well as maintenance of dehydroepiandrosterone levels) correlate with longevity in non-CR humans. The recent work of other investigators shows that humans subjected to short-term CR also have lower insulin and body temperature. Obviously, longer-term CR is extremely difficult to maintain; hence, the need for CR mimetics. The next phase of calorie restriction studies includes basic investigations as well as possible clinical trials of a number of candidate CR mimetics, ranging from glycolytic inhibitors to lipid-regulating agents to antioxidants and specific gene modulators. The scope of these ongoing studies in various laboratories, as well as their practical implications, are reviewed and analyzed here.

KEYWORDS: nutrition; longevity; glucose; glycolysis; insulin; 2-deoxyglucose

CALORIC RESTRICTION AND AGING

One of the major tenets of experimental gerontology is that caloric restriction (CR) is the most robust and reproducible strategy for both maintaining and extending health, function, and longevity.[1] This intervention has now been investigated systematically for seven decades, in a wide range of species/models, under various experimental conditions, and as such has become the gold standard against which to

Address for correspondence: Donald K. Ingram, Laboratory of Experimental Gerontology, Gerontology Research Center, National Institute on Aging, 5600 Nathan Shock Drive, Baltimore, MD 21224. Voice: 410-558-8180; fax: 410-558-8302.

geor@iximed.com

[c]Currently serving as a Special Volunteer to the National Institute on Aging.

Ann. N.Y. Acad. Sci. 1057: 365–371 (2005). © 2005 New York Academy of Sciences.
doi: 10.1196/annals.1356.027

compare other "anti-aging" manipulations.[2] An important distinction is that CR, unlike many interventions, extends both average and maximal (genetically determined) life span, suggesting alteration of fundamental biological processes that control aging processes. Many other interventions alter only average life span and may therefore be targeting specific disease processes rather than processes that affect vulnerability to a wide range of chronic diseases and loss of function.[2] For example, improvements in medical care, nutrition, and overall health have resulted in significant increases in average life span in developed countries over the last 100 years, while maximal human life span has remained essentially unchanged. The beneficial effects of CR on health, function, and longevity have been reviewed extensively elsewhere,[1–4] and the reader is referred to the literature for more specific details.

For present purposes, a few examples of the beneficial effects will be mentioned with special relevance to CR mimetics. First, CR lowers circulating insulin levels and increases sensitivity to this hormone.[5] This effect reduces predisposition to diabetes, as well as hypertension, and other complications of the metabolic syndrome, and based on experiments in animal models, may also be related to longevity extension.[6] Second, CR lowers core body temperature in homeotherms.[7,8] This is consistent with reduced and more efficient energy expenditure and is undoubtedly related to increased life span. Reducing body temperature in poikilotherms is a reliable means of life span extension. However, thermodynamically, the modest temperature reduction observed in mammals under most CR conditions is insufficient to account for all of its beneficial effects, and this phenomenon is probably more symptomatic of a fundamental metabolic shift from growth and reproduction to survival maintenance.[2,7] Finally, CR delays the onset and reduces the incidence of diseases ranging from neoplasia to cardiovascular and cognitive deterioration,[1–4] in parallel with delaying and/or decreasing the rate of "normal" age-associated reductions in physiological, behavioral, and adaptive and stress-protective performance and responsiveness.[1–4]

Various hypotheses have been proposed to elucidate the mechanism(s) by which CR produces the above benefits. These have been reviewed comprehensively elsewhere.[1–4,7] For present purposes, a thermodynamic strategy whereby an organism's perception of a reduced energy supply elicits a coordinated set of transcriptional regulatory events leading to increased protection against a variety of stressors and the damage they cause as well as the maintenance of reserve capacity will serve as an adequate working hypothesis.

RELEVANCE OF CALORIC RESTRICTION TO HUMANS

We first reviewed the potential relevance of CR to humans in 1999[9] in response to the obvious challenge that its health, function, and longevity promoting effects might be either a laboratory curiosity or restricted to short-lived species. Such reservations and related ones continue to be expressed,[10] and the debate will no doubt continue until definitive human data are obtained (see below). However, a number of observations are at least consistent with the notion that CR may indeed work in people.

First, several studies in nonhuman primates (some still ongoing) have reported beneficial effects of CR on health and function, if not yet survival.[11,12] Since almost all of the previous work on dietary manipulation of aging was restricted to models with maximal life spans of only a few years or less—dogs are a recent notable exception[13]—monkeys provide a species much closer in both life span and genetics to humans. Essentially all the studies of CR in nonhuman primates attempted thus far have provided data consistent with the idea that anti-disease, if not anti-aging, effects accrue.[11,2] Since rhesus macaques, the most commonly employed monkey species, have maximal life spans on the order of forty years,[12] current findings would argue against CR only benefiting short-lived species.

We have also previously reviewed the literature regarding anecdotal and short-term human CR studies.[2,9] Again, whether CR is a consequence of cultural and/or geographical conditions (such as Muslims fasting during Ramadan and Okinawans eating less than mainland Japanese, or limited scientific endeavors like Biosphere 2), various health benefits, including increased protection against cardiovascular and neoplastic disease, have been well documented.[2] We recently obtained data from a non-CR population of men in the Baltimore Longitudinal Study of Aging that indicated that three robust physiological markers of the CR state in animal models predicted survival in humans as well.[14] These data indicated lower body temperature and circulating insulin levels and maintenance of higher levels of dehydroepiandrosterone sulfate (DHEAS). Although the factors that contribute to a CR phenotype in non-restricted individuals remain unclear (e.g., genetics, lifestyle, etc.), the very fact that these biomarkers were associated with survival enhancement offers a possible strategy for evaluating anti-aging interventions, such as CR.

Most recently, the National Institute on Aging has funded short-term human CR studies at Washington University, Louisiana State University, and Tufts University.[15] Although much work remains to be completed, preliminary data already confirm the reductions in plasma insulin levels and body temperature observed repeatedly in animal studies.[12]

Taken together, these findings bode well for the possibility that humans may indeed benefit from CR, at least in terms of disease protection, if not from a generalized metabolic reprogramming toward increased survival.[2] Clearly more data will be necessary to draw definitive conclusions, but full life span studies remain unrealistic for practical reasons.

CALORIC RESTRICTION MIMETICS

Despite the very suggestive evidence presented that CR might indeed be beneficial for humans, it is our contention that, for various (some obvious) reasons, most people would be unwilling or unable to adopt such a strategy. Consequently, we proposed the idea of CR mimetics in 1998.[16] These agents or interventions are defined as manipulation of energy metabolism creating the metabolic shift seen in CR without the requirement for reduced food intake to obtain the health, function, and longevity-promoting effects of CR. Some have more broadly defined CR mimetics to include any intervention that increases either average or maximal life span or both. We believe that the link to energy metabolism is critical since energy intake is well established as the fundamental manipulation responsible for the varied beneficial effects of CR on aging. Our first study validated the concept of CR mimetics by employing the glucose analogue, 2-deoxyglucose, to elicit reductions in plasma insulin and body temperature of rats without reducing food consumption over a 6-month period.[16]

Subsequently, a National Institutes of Health consensus workshop reviewed the status of mimetics in conjunction with other aspects of CR, leading to increased interest.[17] The study of CR mimetics was further defined in 2002[18] and the term *CR mimetic* is increasingly found in gerontological literature.

CALORIC RESTRICTION MIMETICS: THE NEXT PHASE

With the background information firmly established, the CR mimetics field has evolved into a vibrant intellectual and commercial enterprise. At the time of this writing at least a dozen companies and an equal number of government and academic laboratories are engaged in research in this new field. Because of the proprietary nature of much of the private work, only limited data are available, although some information regarding general approaches toward CR mimetic strategies can be obtained publicly. For purposes of the present review, the focus will be restricted to the latter.

CR mimetic strategies can be loosely grouped into categories of glycolytic inhibitors, antioxidants, sirtuin regulators, lipid regulators, autophagic enhancers, insulin sensitizers, and miscellaneous. These categories are not mutually exclusive.

As mentioned above, the first compound to be tested, 2-deoxyglucose, is a glycolytic inhibitor, which acts primarily on the enzyme phosphohexose isomerase, thereby reducing the glucose flux through cells.[19] Our initial studies validated the concept that 2-deoxyglucose could duplicate important physiological indices of the CR phenotype without significantly reducing food intake in rats.[16] Unfortunately, this compound exhibits an apparently very narrow window between efficacy and toxicity. Although there is some disagreement as to the possible extent and species specificity of this toxicity,[20] other shorter-term studies have at least suggested some beneficial effects for neuroprotection.[21] Similar positive results have been obtained with iodoacetate[22] and phenformin.[23]

The category of antioxidants has been reviewed extensively by others in many contexts[24] and will not be considered further here, except for two points. First, with few exceptions, survival experiments with antioxidants generally reveal an increase in median, but not maximal life span, probably due primarily to anti-disease effects.[24] Second, some antioxidants, such as alpha lipoic acid and resveratrol are currently being promoted as CR mimetics.[2] Clearly, protection against oxygen radical damage is enhanced by CR and is at least partially involved in CR mechanism(s).[1,3]

Resveratrol also falls into the category of sirtuin regulators. Sirtuins are histone deacetylases, which serve as gene silencers.[25] A considerable body of literature regarding their role in the aging of short-lived organisms and possible role in mediating the effects of CR already exists.[25] It is important to note here, however, that at least one experiment in mammals suggests that sirtuin effects may be secondary to those of insulin following CR.[26] Serum from CR rats can stimulate these genes in cultured cells. However, the effect is partially blocked by insulin, suggesting some primary mediation by the reductions in levels of this hormone.[26] Thus, their relevance may not be as direct as that of glycolytic inhibitors. Another link between sirtuins and CR suggests that they may act on PPAR gamma, a member of the nuclear receptor superfamily,[27] to inhibit fat deposition.[28] Masoro has recently pointed out some of the difficulties of extrapolating this observation to actual CR.[29]

Nevertheless, the concept of lipid regulation itself may be important in developing CR mimetics, since long chain fatty acids may elicit some of the benefits of CR, such as tumor suppression, enhanced immune response, and protection against oxyradical damage.[1,3] In addition, carnitine, which is currently being promoted (in combination with alpha lipoic acid) as a CR mimetic, facilitates fatty acid oxidation in the mitochondrion.[2] Moreover, Olbetam® (active ingredient, Acipimox), a hypolipidemic drug, has been shown to enhance autophagy, the removal of damaged cellular components.[30] The latter process is itself a legitimate CR mimetic target, since it slows with age, leading to a number of decrements associated with the presence of abnormal macromolecules.[31]

Agents that increase sensitivity to insulin provide another category of mimetics. As mentioned above, lower circulating insulin levels, coupled with increased sensitivity to this hormone, are key biomarkers of the CR state.[1] PPAR gamma, also a potential target for sirtuins, is additionally a target for thiazolidinedione antidiabetic drugs.[27] Biguanides, such as phenformin and metformin, have achieved some limited success in mimicking segmental effects of CR, including gene activation, but unfortunately, longevity effects remain somewhat equivocal.[32]

Finally, a number of other miscellaneous agents might also fall tangentially into this category,[2,17] but will not be discussed to avoid further dilution of our definition of CR mimetics.

Ultimately, if commercial/proprietary barriers can be breached, it may be possible to develop a CR mimetic "cocktail," containing ingredients from multiple categories. Some preliminary work already supports the feasibility of this approach. Wald and Law have proposed a "polypill," containing a cholesterol-lowering statin, several hypertensive drugs, aspirin, and folic acid to avoid the cardiovascular risks of increasing age.[33] While the polypill might not conform to the strictest definition of CR mimetics, this strategy is at least consistent with anti-disease effects of CR. An even more ambitious and direct approach has been attempted by Rollo and colleagues, in which mice have been fed a diet composed of more than thirty vitamins, minerals, herbs, antioxidants, and other nutritionally active compounds, some already discussed here, in an attempt to enhance both function and longevity in mice.[34] An initial report[34] demonstrated cognitive enhancement in treated mice, and most recently survivorship data suggested a modest extension of median life span.[35]

Thus, it is conceivable that future cocktails, containing various combinations of candidate "segmental" CR mimetics might be devised. Since no single mimetic would appear to have the potential for completely duplicating the effects of CR, this strategy would provide maximal mimicry, and the best option for eventual product applications.

SUMMARY AND CONCLUSIONS

CR remains the most robust intervention for slowing aging and maintaining health and vitality. Recent experiments with both monkeys and humans suggest that limiting food intake may indeed be a relevant anti-aging strategy for human application. Unfortunately, the 30–40% reduction in calories over much of the life span necessary to achieve maximal benefit is probably beyond the capability of most people. For this reason, CR mimetics could provide a practical anti-aging strategy. Subsequent to early experiments with glycolytic inhibitors, antioxidants, and other selected agents, several academic, government, and private sector laboratories have broadened the search for true CR mimetics to include a number of other compounds.[36,37] Although no single candidate mimetic appears to produce all the beneficial effects of actual CR, cocktails containing combinations of these substances appear to be feasible and may ultimately provide the most successful approach.

[*Competing interests*: The authors have the following potential competing financial interests: stock in GeroScience, Inc.]

REFERENCES

1. WEINDRUCH, R. & R.L WALFORD. 1988. The Retardation of Aging and Disease by Dietary Restriction. Charles C Thomas. Springfield, IL.
2. ROTH, G.S. 2005. The Truth About Aging: Can We Really Live Longer and Healthier? Windstorm Creative. Port Orchard, WA.
3. YU, B.P., ED. 1994. Modulation of Aging Processes by Dietary Restriction. CRC Press. Boca Raton, FL.
4. FISHBEIN, L., ED. 1991. Biological Effects of Dietary Restriction. Springer-Verlag. Berlin.
5. LANE, M.A., S.S. BALL, D.K. INGRAM, et al. 1995. Diet restriction in rhesus monkeys lowers fasting glucose stimulated and glucoregulatory endpoints during an intravenous glucose challenge. Am. J. Physiol. 268: E941–948.
6. KATIC, M. & C.R. KAHN. 2005. The role of insulin and IGF-1 signaling in longevity. Cell. Mol. Life Sci. 62: 320–323.
7. ROTH, G.S., D.K. INGRAM & M.A. LANE. 1995. Slowing aging by caloric restriction. Nature Med. 1: 414–415.
8. LANE, M.A., D.J. BAER, W.V. RUMPLER, et al. 1996. Dietary restriction lowers body temperature in rhesus monkeys, consistent with a postulated anti-aging mechanism in rodents. Proc. Natl. Acad. Sci. USA 93: 4159–4164.
9. ROTH, G.S., D.K. INGRAM & M.A. LANE. 1999. Caloric restriction in primates: Will it work and how will we know? J. Am. Geriatrics Soc. 47: 896–903.
10. DEGREY, A.D.N.J. 2005. The unfortunate influence of weather on the rate of aging: why human caloric restriction or its emulation may only extend life expectancy by 2–3 years. Gerontology 51: 73–82.
11. ROBERTS, S.B., X. PI-SUNYER, L. KULLER, et al. 2001. Physiologic effects of lowering caloric intake in nonhuman primates and nonobese humans. J. Gerontol. Biol. Sci. Med. Sci. 56: 66–75.
12. ROTH, G.S., J.A. MATTISON, M.A. OTTINGER, et al. 2004. Aging in rhesus monkeys: relevance to human health interventions. Science 305: 1423–1426.
13. KEALY, R.D., D.F. LAWLER, J.M. BALLAM, et al. 2002. Effects of diet restriction on life span and age-related changes in dogs. J. Am. Vet. Med. Assoc. 220: 1315–1320.
14. ROTH, G.S., M.A. LANE, D.K. INGRAM, et al. 2002. Biomarkers of caloric restriction may predict longevity in humans. Science 207: 811.
15. HADLEY, E.C., E.G. LAKATTA, H.R. WARNER & R.J. HODES. 2005. The future of aging therapies. Cell 120: 557–567.
16. LANE, M.A., D.K. INGRAM & G.S.ROTH. 1998. 2-Deoxy-D-glucose feeding in rats mimics physiological effects of caloric restriction. J. Anti-Aging Med. 1: 327–337.
17. WEINDRUCH, R., K.P. KEENAN, J.M. CARNEY, et al. 2001. Caloric restriction mimetics: metabolic interventions. J. Gerontol. Biol. Sci. 56: B1–10.
18. LANE, M.A., D.K. INGRAM & G.S. ROTH. 2002. The serious search for an anti-aging pill. Sci. Am. 287: 36–41.
19. REZEK, M. & E.A. KROEGER. 1972. Glucose antimetabolites and hunger. J. Nutr. 106: 143–157.
20. WAN, R., S. CAMANDOLA & M.P. MATTSON. 2004. Dietary supplementation with 2-deoxy-D-glucose improves cardiovascular and neuroendocrine stress adaptation in rats. Am. J. Physiol. Heart Circ. Physiol. 287: 186–193.

21. LEE, J., A.J. BRUCE-KELLER, Y. KRUMAN, *et al.* 1999. 2-Deoxy-D-glucose protects hippocampal neurons against excitotoxic and oxidative injury: evidence for the involvement of stress proteins. J. Neurosci. Res. **57:** 48–61.
22. GUO, Z., J. LEE, M. LANE & M. MATTSON. 2001. Iodoacetate protects hippocampal neurons against excitotoxic and oxidative injury: involvement of heat-shock proteins and Bcl-2. J. Neurochem. **79:** 361–370.
23. Lee J., S.L. Chan, C. Lu, *et al.* 2002. Phenformin suppresses calcium responses to glutamate and protects hippocampal neurons against excitotoxicity. Exp. Neurol. **175:** 161–167.
24. CUTLER, R.G., L. PACKER, J. BERTRAM & A. MON, EDS. 1995. Oxidative Stress and Aging. Birkhauser. Basel, Switzerland.
25. GUARENTE, L. & F. PICARD. 2005. Calorie restriction—the SIR2 connection. Cell **120:** 473–482.
26. COHEN H.Y., C. MILLER, K.J. BITTERMAN, *et al.* 2004. Calorie restriction promotes mammalian cell survival by inducing the SIRT1 deacetylase. Science **305:** 390–392.
27. LAZAR, M.A. 2005. PPAR gamma, 10 years later. Biochimie **87:** 9–13.
28. PICARD, F., M. KURTEV, N. CHUNG, *et al.* 2004. Sirt1 promotes fat mobilization in white adipocytes by repressing PPAR gamma. Nature **429:** 771–776.
29. MASORO, E.J. 2004. Role of sirtuin proteins in life extension by caloric restriction. Mech. Ageing Dev. **125:** 591–594.
30. BERGAMINI, E. & H.L. SEGAL. 1987. Effects of antilipolytic drugs on hepatic peroxisomes. *In* Peroxisomes in Biology and Medicine. H.D. Fahimi, Ed.: 295–303. Springer-Verlag. Berlin.
31. WARD, W.F. 2002. Protein degradation in the aging organism. Prog. Molec. Subcell. Biol. **29:** 35–42.
32. DILMAN, V.M. & V.N. ANISIMOV. 1980. Effect of treatment with phenformin, diphenylhydantoin, or L-dopa on lifespan and tumor incidence in C3H/Sn mice. Gerontology **26:** 241–246.
33. WALD, N.J. & M.R. LAW. 2003. A strategy to reduce cardiovascular disease by more than 80%. Br. Med. J. **326:** 1423–1427.
34. LEMON, J.A., D.R. BOREHAM & C.D. ROLLO. 2003. A dietary supplement abolishes age-related cognitive decline in transgenic mice expressing elevated free radical processes. Exp. Biol. Med. **228:** 800–810.
35. LEMON, J.A., D.R. BOREHAM & C.D. ROLLO. 2005. A complex dietary supplement extends longevity of mice. J. Gerontol. Biol. Sci. **60:** 275–279.
36. HURSTING S.D., J.A. LAVIGNE, D. BERRIGAN, *et al.* 2003. Calorie restriction, aging, and cancer prevention: mechanisms of action and applicability to humans. Annu. Rev. Med. **54:** 131–152.
37. SINCLAIR, D.A. 2005. Toward a unified theory of caloric restriction and longevity regulation. Mech Ageing Dev. **126:** 987–1002.

Mechanisms Regulating Melatonin Synthesis in the Mammalian Pineal Organ

CHRISTOF SCHOMERUS AND HORST-WERNER KORF

Dr. Senckenbergische Anatomie, Institut für Anatomie II, Johann Wolfgang Goethe-Universität Frankfurt, Theodor-Stern-Kai 7, 60590 Frankfurt/Main, Germany

ABSTRACT: The day/night rhythm in melatonin production is a characteristic feature in vertebrate physiology. This hormonal signal reliably reflects the environmental light conditions and is independent of behavioral aspects. In all mammalian species, melatonin production is regulated by norepinephrine, which is released from sympathetic nerve fibers exclusively at night. Norepinephrine elevates the intracellular cAMP concentration via β-adrenergic receptors and activates the cAMP-dependent protein kinase A. This pathway is crucial for regulation of the penultimate enzyme in melatonin biosynthesis, the arylalkylamine N-acetyltransferase (AANAT); cAMP/protein kinase A may, however, act in different ways. In ungulates and primates, pinealocytes constantly synthesize AANAT protein from continually available $Aanat$ mRNA. During the day—in the absence of noradrenergic stimulation—the protein is immediately destroyed by proteasomal proteolysis. At nighttime, elevated cAMP levels cause phosphorylation of AANAT by protein kinase A. This posttranslational modification leads to interaction of phosphorylated AANAT with regulatory 14-3-3 proteins, which protect AANAT from degradation. Increases in AANAT protein are paralleled by increases in enzyme activity. Stimulation of the cAMP/protein kinase A pathway may also activate pineal gene expression. In rodents, transcriptional activation of the $Aanat$ gene is the primary mechanism for the induction of melatonin biosynthesis and results in marked day/night fluctuations in $Aanat$ mRNA. It involves protein kinase A–dependent phosphorylation of the transcription factor cyclic AMP response element-binding protein (CREB) and binding of phosphorylated CREB in the promoter region of the $Aanat$ gene. In conclusion, a common neuroendocrine principle, the nocturnal rise in melatonin, is controlled by strikingly diverse regulatory mechanisms. This diversity has emerged in the course of evolution and reflects the high adaptive plasticity of the melatonin-generating pineal organ.

KEYWORDS: AANAT; biological rhythms; circadian; norepinephrine; pCREB; pineal gland; protein kinase C

Address for correspondence: Dr. C. Schomerus, Dr. Senckenbergische Anatomie, Institut für Anatomie II, Johann Wolfgang Goethe-Universität Frankfurt, Theodor-Stern-Kai 7, 6059 Frankfurt/Main, Germany. Voice: (49) 69 6301 6059; fax: (49) 69 6301 6017.
schomerus@em.uni-frankfurt.de

Ann. N.Y. Acad. Sci. 1057: 372–383 (2005). © 2005 New York Academy of Sciences.
doi: 10.1196/annals.1356.028

INTRODUCTION

The daily changes in environmental light conditions have profound impact on all aspects of life. As a response, a wide panoply of processes in physiology and behavior in all kinds of living organisms is rhythmic.[1] Some of these processes are merely reactive; others are anticipatory and strictly controlled by an internal clock. In vertebrates, the master clock is located in the anterior hypothalamus above the optic chiasm, in the suprachiasmatic nucleus (SCN) and generates an endogenous "circadian" rhythm with a period very close to 24 hours.[2–4] The SCN is the core of a complex neuronal circuit that entrains body functions to the ambient photoperiod: the photoneuroendocrine system. Apart from the circadian rhythm generator, it comprises two additional components to which the SCN is tightly linked. One component is the photodetector(s) that continuously convey light information to set the clock and compensate for seasonal changes in the photoperiod. In vertebrates, the most important photodetector is the retina.[5,6]The other component is effectors that translate SCN rhythmicity into rhythmic output messages propagated within the organism through humoral and/or neural routes.[7,8]

LET'S MAKE MELATONIN! COMMON PRINCIPLES

The major endocrine effector of the photoneuroendocrine system is the pineal gland located in the roof of the diencephalon. Each night, the pineal gland generates a message encoding darkness, the neurohormone melatonin, that is spread throughout the body via the circulation.[9–11] The day/night rhythm in melatonin reliably reflects the environmental light conditions and is independent of behavioral aspects. Changes in the night length, such as seasonal changes, are translated into changes in the duration and/or amplitude of the melatonin surge. Melatonin is a lipophilic compound that is not stored in the pinealocytes, but immediately released upon its formation from the cell into blood capillaries. Thus, the daily rhythm in circulating melatonin is caused by changes in the rate of the melatonin production. The source of melatonin production is the amino acid tryptophan. It is converted into serotonin. Serotonin is metabolized into melatonin in two steps, which are catalyzed by arylalkylamine N-acetyltransferase (AANAT) and hydroxyindole-O-methyltransferase (HIOMT). The first enzyme N-acetylates serotonin to form N-acetylserotonin; the second O-methylates N-acetylserotonin to form melatonin. In all vertebrates, AANAT is considered the key enzyme in melatonin production because all regulatory mechanisms converge at the control of AANAT enzyme activity. Hence it is often called the melatonin rhythm enzyme.[10]

Remarkably, the regulatory cues that control AANAT and melatonin biosynthesis vary among vertebrate species. Mammals differ from non-mammalian species, such as like fish and birds, in that their pinealocytes have in the course of evolution lost direct photosensitivity and the capacity of circadian rhythm generation.[11] Rather, melatonin production is solely driven by the internal clock in the SCN. A multisynaptic pathway connects the SCN with the pineal gland. The final part of this circuit consists of sympathetic nerve fibers that invade the pineal parenchyma and release norepinephrine (NE) in a circadian manner exclusively at night.[7,12,13]

NE is the dominant neurotransmitter for regulation of AANAT and melatonin synthesis in mammals. NE stimulation of pinealocytes initiates a sequence of molecular processes that has been found in pinealocytes from all mammalian species investigated so far. This standard program of pineal signaling is switched on by NE activation of two subtypes of adrenergic receptors. First, activation of β_1-adrenergic receptors leads to increases in the intracellular cAMP concentration. The rise in cAMP levels is closely and inevitably followed by activation of the cAMP-dependent protein kinase A (PKA). Both elevated cAMP levels and PKA activation are indispensable for stimulation of AANAT and melatonin production in all mammalian species. Second, activation of α_1-adrenergic receptors leads to increases in the intracellular calcium concentration ($[Ca^{2+}]_i$). The rise in $[Ca^{2+}]_i$ is caused by release of calcium ions from intracellular stores followed by calcium influx into the pinealocyte.[14,15] The functional role of $[Ca^{2+}]_i$ for regulation of melatonin synthesis is less clear than that of cAMP and appears to vary with respect to the species investigated.

The NE-dependent activation of the β_1-adrenergic/cAMP/PKA and α_1-adrenergic/$[Ca^{2+}]_i$ pathways is basically conserved in mammalian physiology. However, the downstream mechanisms that link these signaling cascades with AANAT activation and melatonin production exhibit marked species-to-species variations. In particular, these mechanisms differ between rodents and nonrodent species—for example, ungulates and primates.

RODENTS LIKE IT TRANSCRIPTIONAL

As in other mammalian species, an increase in the intracellular cAMP concentration is essential for melatonin production in rodents. In rat pinealocytes, cAMP levels are elevated approximately 100-fold upon NE stimulation.[16] The marked increase in the cAMP concentration is controlled through a combinatorial mechanism involving both α_1- and β_1-adrenergic receptors; activation of both receptors is required for the full effect to occur.[16,17] NE-dependent activation of β_1-adrenergic receptors elevates the intracellular cAMP concentration by increasing adenylate cyclase activity. Concomitant NE activation of α_1-adrenergic receptors potentiates the β_1-adrenergic effects on intracellular cAMP levels, AANAT activity, and melatonin production at a postreceptor site through increases in $[Ca^{2+}]_i$ and activation of protein kinase C (PKC). Notably, α_1-adrenergic activation of $[Ca^{2+}]_i$ or direct activation of PKC by phorbol esters alone does not increase intracellular cAMP levels or AANAT activity in rat pinealocytes.[17,18] It is not clear whether such an α_1/β_1-adrenergic combinatorial mechanism occurs in rodents other than the rat.

In the rat, and apparently also in other rodents, the cAMP/PKA pathway controls transcriptional mechanisms to regulate melatonin synthesis.[10,11] The transcription factor cyclic AMP response element (CRE)–binding protein (CREB) is positioned in the center of these processes. As in other systems, CREB is constitutively expressed in the pineal gland, and its transcriptional impact depends on phosphorylation at serine 133, but not on de novo protein synthesis.[19] β_1-Adrenergic stimulation of the cAMP/PKA pathway leads to CREB phosphorylation in the nuclei of basically all rat pinealocytes; in contrast, α_1-adrenergic stimulation is without effect.[20,21] Phosphorylated (p) CREB enhances transcription of genes bearing a CRE in their promoter regions. Among these is the *Aanat* gene.[22] Enhanced transcription of the *Aanat* gene elicits a dramatic rise in *Aanat* mRNA in the pinealocyte, from nearly undetectable levels at daytime to maximal levels in the middle of the night.[23,24] With a temporal lag of approximately two hours, the increase in *Aanat* mRNA levels is followed by similar increases in AANAT protein levels and activity and melatonin production. Other targets of pCREB in the pineal gland are the genes for the β_1-adrenergic receptor[25] and for the transcription factor ICER (inducible cAMP early repressor).[26] ICER is a potent inhibitor of cAMP-inducible genes that is involved in the inhibition of *Aanat* mRNA accumulation at the end of the night.[19] It should be mentioned that CREB phosphorylation is not only elicited by β_1-adrenergic stimulation, but can also be caused by various other cAMP-elevating agents, including vasoactive intestinal peptide (VIP) and pituitary adenylate cyclase-activating polypeptide (PACAP), which may have modulatory roles in regulation of melatonin synthesis.[27] These findings emphasize the central integrative role of pCREB for control of melatonin production in rodents.

During the last years, the importance of transcriptional mechanisms to regulate pineal functions in rodents has been further emphasized because the list of genes that are rhythmically expressed in the rat pineal gland has grown. It includes genes encoding transcription factors,[28] enzymes,[29] transporters,[30] and receptors.[31,32] This list continues to grow as a result of cDNA microarray analysis studies.[33–38] From these investigations it is becoming clear that there is a global increase in gene expression at night in the rodent pineal gland mediated through pCREB-dependent, but apparently also through pCREB-independent transcriptional processes.

Although the indispensable role of transcriptional AANAT activation in rodents is generally acknowledged, the transcriptional mechanisms are apparently supplemented by posttranslational processes that control proteolytic degradation in a cAMP-dependent manner.[39] This mechanism allows for a more rapid switch-off of AANAT protein and melatonin production than transcriptional processes—for example, upon unexpected light at night. Under these conditions, cAMP levels immediately drop and terminate the protection of AANAT from degradation whereas AANAT mRNA is still elevated. It is not yet clear whether or to which extent this mechanism also contributes to the NE-induced activation of AANAT early in the night by stabilizing the enzyme; analysis of this issue in rodents is confounded by the essential role of transcription in this group of species.

Characteristically, melatonin synthesis is terminated already before the end of the night. In the rat, downregulation of melatonin production involves precise interaction of several regulatory processes. First, the release of NE from sympathetic nerve fibres is attenuated as the night progresses.[40] A second mechanism to terminate melatonin production is the dephosphorylation of pCREB. Experiments on cultured rat pinealocytes showed that withdrawal of NE causes rapid pCREB dephosphorylation, which is followed by decreases in *Aanat* mRNA and AANAT protein levels as well as in melatonin synthesis. Inhibitor studies suggest that the sharp decline in pCREB levels reflects the activation of protein serine/threonine phosphatase PSP1.[41] Finally, the pCREB-dependent transcriptional activation of the inhibitory transcription factor ICER appears to be involved in terminating melatonin synthesis. It is possible that, in the rodent pineal gland, the rhythmic transcription of CRE-bearing genes like the *Aanat* gene is influenced by a shift in balance between the stimulatory transcription factor CREB and the inhibitory transcription factor ICER, which is elicited by the sharp decrease in pCREB levels at the end of the night and the moderate increase in ICER protein levels in the second part of the night.[19,26,42]

UNGULATES LIKE IT POSTTRANSLATIONAL

Several comparative analyses have shown that models of pineal signal transduction based solely on the rat (FIG. 1) may not apply to other mammalian species. Primary cell cultures of bovine pinealocytes have proven to be an excellent model for investigating the melatonin-generating system in nonrodent species. In a couple of studies, both similarities and differences in pineal signaling between rat and bovine were detected. It was found that notable features of regulation of melatonin production in the rat pinealocyte are also evident in the bovine pinealocyte—for example, that stimulation of β_1-adrenergic receptors leads to cAMP accumulation and that increased cAMP levels are mandatory for the onset of melatonin synthesis.[43] However, the NE-induced rise in cAMP levels is distinctly less prominent in the cow (twofold) than in the rat (100-fold[16]). The reason for this difference is not clear.

Another common principle in rat and bovine pineal signaling is that α_1-adrenergic stimulation increases $[Ca^{2+}]_i$. As is true for rat pinealocytes, α_1-adrenergic stimulation of bovine pinealocytes by itself does not stimulate AANAT or melatonin production, but it does potentiate the β_1-adrenergically induced increase in AANAT activity. In contrast to the rat pinealocyte, potentiation of AANAT activity is not accompanied by potentiating effects on cAMP levels in the bovine pinealocyte. This suggests that the α_1-adrenergic potentiation of AANAT activity in bovine pinealocytes occurs downstream to the formation of cAMP by adenylate cyclase. Similar downstream effects have been described in rat pinealocytes.[44] It should be mentioned that α_1-adrenergic potentiation of β_1-adrenergically induced AANAT activity in the bovine pinealocyte is also not accompanied by a potentiating effect on melatonin production. The molecular basis for this finding is unclear.

With respect to pineal calcium signaling, it is remarkable that, in both rat and bovine pinealocytes, $[Ca^{2+}]_i$ is elevated not only by α_1-adrenergic, but also by cholinergic stimulation.[15,43,45,46] The functional role of cholinergic signal transduction in the mammalian pineal gland is debated. In neonatal pinealocytes, acetylcholine (ACh) elevates $[Ca^{2+}]_i$ via muscarinic rather than nicotinic acetylcholine receptors.[46] In the second postnatal week, pinealocytes gain responsiveness to nicotine and gradually lose responsiveness to muscarinic cholinergic stimuli. Since ACh-evoked cellular events may be different depending on the functional subclass of receptor that is present, the transient existence of muscarinic acetylcholine receptors would permit ACh to elicit temporary effects in early pineal development. In pinealocytes from adult rats, acetylcholine exclusively activates nicotinic ACh receptors via influx of calcium ions into the cells through nifedipine-sensitive voltage-gated calcium channels.[45] The increase in $[Ca^{2+}]_i$ triggers L-glutamate exocytosis from pineal microvesicles, which reportedly inhibits NE-induced melatonin synthesis through paracrine and/or autocrine mechanisms.[47,48] In the adult bovine pinealocyte, ACh induces increases in $[Ca^{2+}]_i$ through both nicotinic ACh receptors and muscarinic ACh receptors.[43] However, cholinergic and glutamatergic stimuli do not inhibit NE-induced increases in AANAT activity or in melatonin production in bovine pinealocytes. Thus, the functional significance of ACh receptors and also that of voltage-gated calcium channels in bovine pinealocytes remains to be elucidated.

While the basic mechanisms in regulation of the second messengers cAMP and Ca^{2+} and also their functional significance are similar in rat and bovine pinealocytes,

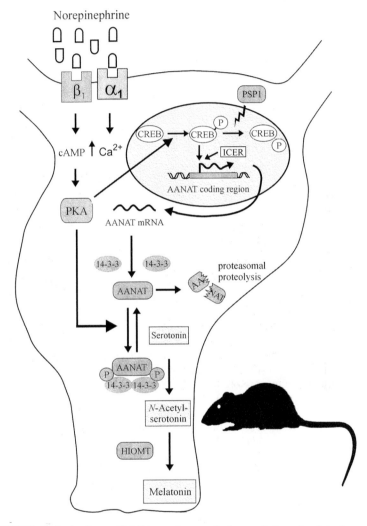

FIGURE 1. Activating and inhibitory elements in the regulation of melatonin synthesis in the rat pinealocyte. Activating elements: cAMP increase; activation of PKA; phosphorylation of CREB; phosphorylation of AANAT. Inhibitory elements: PSP1-mediated pCREB dephosphorylation; ICER-mediated attenuation of AANAT transcription; proteasomal proteolysis of AANAT. *Abbreviations*: α_1-adrenergic receptor; β_1,β_1-adrenergic receptor; PKA, cAMP-dependent protein kinase A; CREB, cyclic AMP responsive element-binding protein; ICER, inducible cAMP early repressor; PSP1, protein serine/threonine phosphatase 1; AANAT, arylalkylamine *N*-acetyltransferase; 14-3-3, 14-3-3 protein; HIOMT, hydroxyindole *O*-methyltransferase; P, phosphate group.

major differences between both species are evident concerning the downstream mechanisms that control AANAT. Investigations on ovine and bovine pinealocytes showed that the *Aanat* gene is constitutively expressed and that *Aanat* mRNA levels do not fluctuate on a day/night basis.[10,49,50] Accordingly, the cAMP/PKA pathway does not control melatonin production via transcriptional activation of AANAT. In these species, melatonin synthesis is regulated by controlling the rate at which AANAT protein is destroyed by proteolytic degradation.[10,12,50] In unstimulated pinealocytes, cAMP levels are low, and AANAT protein, which is constantly synthesized from continually available *Aanat* mRNA, is immediately degraded by the proteasome. Upon NE stimulation, cAMP levels are elevated, and newly synthesized AANAT protein is protected from proteasomal proteolysis through cAMP/PKA-dependent phosphorylation of the enzyme. This posttranslational mechanism was first detected in rat pinealocytes in which protection of AANAT proteolysis supplements the transcriptional activation by stabilizing the enzyme and is responsible for the rapid degradation of AANAT upon unexpected light at night.[39] With transcriptional regulation of AANAT absent in ungulates, control of AANAT proteolysis is the dominant regulatory mechanism in these species (FIG. 2). Several observations indicate that this may be also valid for primates. In the rhesus monkey, day and night levels of pineal *Aanat* mRNA levels are similar.[51] Moreover, in both primates[52,53] and ungulates,[54] AANAT activity and melatonin synthesis start to increase immediately at the onset of night, presumably because *Aanat* mRNA is continually available and no extra time is required for AANAT mRNA to accumulate, as is the case in rodents. These common motifs in pineal physiology in ungulates and primates document the particular importance of the bovine pinealocyte as a model system.

AANAT is phosphorylated at two PKA consensus sites located in regulatory regions of the enzyme that flank a central catalytic core with binding domains for arylalkylamines and acetyl-CoA. Phosphorylation of the amino acids threonine-31 and serine-205 switches the fate of AANAT from destruction to protection and activation because it promotes binding of AANAT to 14-3-3 proteins.[12,55,56] The 14-3-3 proteins represent a versatile family of ubiquitous regulatory proteins that act on a variety of target proteins through binding, mostly in a phosphorylation-dependent manner.[57–59] The effect of pAANAT/14-3-3 complex formation is at least twofold. First, 14-3-3 acts like a "magic hood" and shields AANAT from metabolic processes that lead to proteolysis. Presumably, this is accomplished by blocking access of phosphatases and/or macromolecules required for proteasomal degradation. Second, 14-3-3 increases the affinity of AANAT for its substrate, serotonin, thus facilitating the conversion of serotonin to *N*-acetylserotonin at nighttime when the serotonin concentration is low.

Remarkably, cAMP/PKA seems to activate also transcriptional processes in bovine pinealocytes because activation of the cAMP/PKA pathway leads to CREB phosphorylation.[60] Apparently, CREB phosphorylation represents a very conserved element in pineal physiology, irrespective of whether or not AANAT is controlled via transcriptional mechanisms. However, the target genes for pCREB in bovine pinealocytes remain to be elucidated.

Although the cAMP/PKA pathway has a crucial role for AANAT activation in bovine pinealocytes, it should be emphasized that other signaling mechanisms may contribute to regulation of the enzyme. Recently, it was shown that protein kinase C activation by phorbol esters induces marked increases in AANAT protein levels,

AANAT activity, and melatonin production.[61] The effects were of similar magnitude as compared to those induced by NE. A direct stimulatory effect of phorbol esters on AANAT activity has also been described in CHO cells stably transfected with human AANAT.[62] These data strikingly differ from data from rat pinealocytes showing that, in this species, phorbol ester have no effect on any of these parameters.[17,18] No-

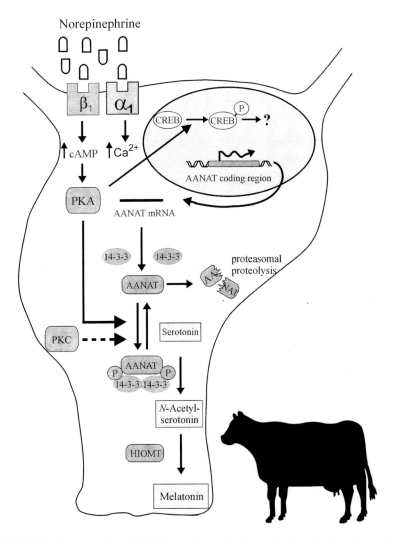

FIGURE 2. Activating and inhibitory elements in the regulation of melatonin synthesis in the bovine pinealocyte. Activating elements: cAMP increase; activation of PKA; activation of PKC; phosphorylation of AANAT. Inhibitory elements: proteasomal proteolysis of AANAT. *Abbreviations*: α_1-adrenergic receptor; β_1,β_1-adrenergic receptor; PKA, cAMP-dependent protein kinase A; CREB, cyclic AMP responsive element-binding protein; AANAT, arylalkylamine *N*-acetyltransferase; 14-3-3, 14-3-3 protein; PKC, protein kinase C; HIOMT, hydroxyindole *O*-methyltransferase; P, phosphate group.

tably, *Aanat* mRNA levels did not change upon phorbol ester treatment in bovine pinealocytes, indicating that phorbol esters control AANAT at the posttranscriptional level. The most interesting aspect of the bovine study is that responses to phorbol esters were not accompanied by increases in cAMP levels. These data lead to the provocative conclusion that in bovine pinealocytes melatonin synthesis may also be influenced by a mechanism that is cAMP-independent and involves protein kinase C. Future studies need to focus on the identification of putative physiological activator(s) of PKC in bovine pinealocytes. First results indicate that α_1-adrenergic agonists,[43] ATP and neuropeptide Y (Schomerus, unpublished observations) did not elicit any changes in AANAT protein and activity in bovine pinealocytes. Another interesting aspect to be addressed relates to the mechanisms that link PKC with AANAT activation.

CONCLUSIONS

In conclusion, a common neuroendocrine principle, the nocturnal rise in melatonin, is controlled by strikingly diverse regulatory mechanisms. This diversity has emerged in the course of evolution and reflects the highly adaptive plasticity of the melatonin-generating pineal organ. The findings from bovine pinealocytes emphasize the need for comparative studies to understand the regulation of melatonin biosynthesis in the pineal organ of each species of interest and suggest that additional, not-yet-identified mechanisms may contribute to regulation of the melatonin-generating system in the cow as well as in primates.

An obvious question is what is the functional significance of the observed interspecies differences in the regulatory mechanisms of melatonin biosynthesis. It is tempting to speculate that this may be related to behavioral aspects—for example to diurnal (ovine, bovine) versus nocturnal activity (most rodents). However, this appears to be not the case since it was shown that in a diurnally active rodent, the grass rat (*Arvicanthis ansorgei*), AANAT is also controlled through transcriptional activation.[63]

Irrespective of the answer to this question, it is clear that the nocturnal increase in melatonin is an essential feature for mammalian physiology because the integrity of this signal is preserved throughout evolution while diverse regulatory mechanisms have developed to ensure rhythmic melatonin production.

ACKNOWLEDGMENT

This study was supported by the Deutsche Forschungsgemeinschaft (DFG).

[*Competing interests*: The authors declare that they have no competing financial interests.]

REFERENCES

1. HASTINGS, M.H., A.B. REDDY & E.S. MAYWOOD. 2003. A clockwork web: circadian timing in brain and periphery, in health and disease. Nat. Rev. Neurosci. **4:** 649–661.

2. MOORE, R.Y. 1993. Organization of the primate circadian system. J. Biol. Rhythms **8:** 3–9.
3. WEAVER, D.R. 1998. The suprachiasmatic nucleus: a 25-year retrospective. J. Biol. Rhythms **13:** 100–112.
4. REPPERT, S.M. & D.R. WEAVER. 2002. Coordination of circadian timing in mammals. Nature **418:** 935–941.
5. HANNIBAL, J. 2002. Neurotransmitters of the retino-hypothalamic tract. Cell Tissue Res. **309:** 73–88.
6. FOSTER, R.G. 2005. Neurobiology: bright blue times. Nature **433:** 698–699.
7. KALSBEEK, A. & R.M. BUIJS. 2002. Output pathways of the mammalian suprachiasmatic nucleus: coding circadian time by transmitter selection and specific targeting. Cell Tissue Res. **309:** 109–118.
8. PANDA, S. & J.B. HOGENESCH. 2004. It's all in the timing: many clocks, many outputs. J. Biol. Rhythms **19:** 374–387.
9. KLEIN, D.C. 1985. Photoneural regulation of the mammalian pineal gland. *In* Photoperiodism, melatonin and the pineal gland. D. Evered & S. Clark, Eds.: 38–56. Ciba Foundation Symposium 117. Pitman. London.
10. KLEIN, D.C., S.L. COON, P.H. ROSEBOOM, *et al.* 1997. The melatonin rhythm generating enzyme: molecular regulation of serotonin *N*-acetyltransferase in the pineal gland. Rec. Progr. Hormone Res. **52:** 307–358.
11. KORF, H.W., C. SCHOMERUS & J.H. STEHLE. 1998. The pineal organ, its hormone melatonin, and the photoneuroendocrine system. Adv. Anat. Embryol. Cell Biol. **146:** 1–100.
12. GANGULY, S., J.A. GASTEL, J.L. WELLER, *et al.* 2001. Role of a pineal cAMP-operated arylalkylamine *N*-acetyltransferase/14-3-3 binding switch in melatonin synthesis. Proc. Natl. Acad. Sci. USA **98:** 8083–8088.
13. MØLLER, M. & F.M.M. BAERES. 2002. The anatomy and innervation of the pineal gland. Cell Tissue Res. **309:** 139–150.
14. SUGDEN, L.A., D. SUGDEN & D.C. KLEIN. 1987. Alpha 1-adrenoceptor activation elevates cytosolic calcium in rat pinealocytes by increasing net influx. J. Biol. Chem. **262:** 741–745.
15. SCHOMERUS, C., E. LAEDTKE & H.W. KORF. 1995. Calcium responses of isolated, immunocytochemically identified rat pinealocytes to noradrenergic, cholinergic, and vasopressinergic stimulations. Neurochem. Int. **27:** 163–175.
16. VANECEK, J., D. SUGDEN, J.L. WELLER & D.C. KLEIN. 1985. Atypical synergistic α_1- and β-adrenergic regulation of adenosine 3′,5′-monophosphate and guanosine 3′,5′-monophosphate in rat pinealocytes. Endocrinology **116:** 2167–2173.
17. SUGDEN, D., J. VANECEK, D.C. KLEIN, *et al.* 1985. Activation of protein kinase C potentiates isoprenaline-induced cyclic AMP accumulation in rat pinealocytes. Nature **314:** 359–360.
18. ZATZ, M. 1985. Phorbol esters mimic alpha-adrenergic potentiation of serotonin *N*-acetyltransferase induction in the rat pineal. J. Neurochem. **45:** 637–639.
19. MARONDE, E., M. PFEFFER, J. OLCESE, *et al.* 1999. Transcription factors in neuroendocrine regulation: rhythmic changes in pCREB and ICER levels frame melatonin synthesis. J. Neurosci. **19:** 3226–3336.
20. ROSEBOOM, P.H. & D.C. KLEIN. 1995. Norepinephrine stimulation of pineal cyclic AMP response element-binding protein phosphorylation: involvement of a β-adrenergic/cyclic AMP mechanism. Mol. Pharmacol. **47:** 439–449.
21. TAMOTSU, S., C. SCHOMERUS, J.H. STEHLE, *et al.* 1995. Norepinephrine-induced phosphorylation of the transcription factor CREB in isolated rat pinealocytes: an immunocytochemical study. Cell Tissue Res. **282:** 219–226.
22. BALER, R., S. COVINGTON & D.C. KLEIN. 1997. The rat arylalkylamine *N*-acetyltransferase gene promotor. J. Biol. Chem. **272:** 6979–6985.
23. BORJIGIN, J., M.M. WANG & S.H. SNYDER. 1995. Diurnal variation in mRNA encoding serotonin *N*-acetyltransferase in the pineal gland. Nature **378:** 783–785.
24. ROSEBOOM, P.H., S.L. COON, R. BALER, *et al.* 1996. Melatonin synthesis: analysis of the more than 150-fold nocturnal increase in serotonin *N*-acetyltransferase messenger ribonucleic acid in the rat pineal gland. Endocrinology **137:** 3033–3044.

25. PFEFFER, M., E. MARONDE, C.A. MOLINA, et al. 1999. Inducible cyclic AMP early repressor protein in rat pinealocytes: a highly sensitive natural reporter for regulated gene transcription. Mol. Pharmacol. **56:** 279–289.

26. STEHLE, J.H., N.S. FOULKES, C.A. MOLINA, et al. 1993. Adrenergic signals direct rhythmic expression of transcriptional repressor CREM in the pineal gland. Nature **356:** 314–320.

27. SCHOMERUS, C., E. MARONDE, E. LAEDTKE & H.W. KORF. 1996. Vasoactive intestinal peptide (VIP) and pituitary adenylate cyclase-activating polypeptide (PACAP) induce phosphorylation of the transcriptional factor CREB in subpopulations of rat pinealocytes: immunocytochemical and immunochemical evidence. Cell Tissue Res. **286:** 305–313.

28. BALER, R. & D.C. KLEIN. 1995. Circadian expression of transcription factor Fra-2 in the rat pineal gland. J. Biol. Chem. **270:** 27319–27325.

29. MURAKAMI, M., Y. HOSOI, T. NEGISHI, et al. 1997. Expression and nocturnal increase of type II iodothyronine deiodinase mRNA in rat pineal glands. Neurosci. Lett. **227:** 65–67.

30. BORJIGIN, J., A.S. PAYNE, J. DENG, et al. 1999. A novel pineal night-specific ATPase encoded by the Wilson disease gene. J. Neurosci. **19:** 1018–1026.

31. BALER, R., S.L. COON & D.C. KLEIN. 1996. Orphan nuclear receptor RZRbeta: cyclic AMP regulates expression in the pineal gland. Biochem. Biophys. Res. Commun. **220:** 975–978.

32. COON, S.L., S.K. MCCUNE, D. SUGDEN & D.C. KLEIN. 1997. Regulation of pineal alpha$_{1B}$-adrenergic receptor mRNA: day/night rhythm and beta-adrenergic receptor/cyclic AMP control. Mol. Pharmacol. **51:** 551–557.

33. HUMPHRIES, A., D.C. KLEIN, R. BALER & D.A. CARTER. 2002. cDNA array analysis of pineal gene expression reveals circadian rhythmicity of the dominant negative helix-loop-helix protein-encoding gene, Id-1. J. Neuroendocrinology **14:** 101–108.

34. HUMPHRIES, A., J.L. WELLER, D.C. KLEIN, et al. 2004. NGFI-B (Nurr77/Nr4a1) orphan nuclear receptor in rat pinealocytes: circadian expression involves an adrenergic-cyclic AMP mechanism. J. Neurochem. **91:** 946–955.

35. FUKUHARA, C., J.C. DIRDEN & G. TOSINI. 2003. Analysis of gene expression following norepinephrine stimulation in the rat pineal gland using DNA microarray technique. J. Pineal Res. **35:** 196–203.

36. KIM, J.S., S.L. COON, S. BLACKSHAW, et al. 2005. Methionine adenosyltransferase:adrenergic-cAMP mechanism regulates a daily rhythm in pineal expression. J. Biol. Chem. **280:** 677–684.

37. PRICE, D.M., C.L. CHIK, D. TERRIFF, et al. 2004. Mitogen-activated protein kinase phosphatase-1 (MKP-1): >100-fold nocturnal and norepinephrine-induced changes in the rat pineal gland. FEBS Lett. **577:** 220–226.

38. GAILDRAT P., M. MØLLER, S. MUKDA, et al. 2005. A novel pineal-specific product of the oligopeptide transporter PepT1 gene: circadian expression mediated by cAMP activation of an intronic promoter. J. Biol. Chem. **280:**16851–16860.

39. GASTEL, J.A., P.H. ROSEBOOM, P.A. RINALDI, et al. 1998. Melatonin production: proteosomal proteolysis in serotonin N-acetyltransferase regulation. Science **279:** 1358–1360.

40. DRIJFHOUT, W.A., S. VAN DER LINDE, C. KOOI, et al. 1996. Norepinephrine release in the rat pineal gland: the input from the biological clock measured by in vivo microdialysis. J. Neurochem. **66:** 748–755.

41. KOCH, M., V. MAUHIN, J.H. STEHLE, et al. 2003. Dephosphorylation of pCREB by protein serine/threonine phosphatases is involved in inactivation of Aanat gene transcription in rat pineal gland. J. Neurochem. **85:** 170–179.

42. VON GALL, C., A. LEWY, C. SCHOMERUS, et al. 2000. Transcription factor dynamics and neuroendocrine signalling in the mouse pineal gland: a comparative analysis of melatonin-deficient C57BL mice and melatonin-proficient C3H mice. Eur. J. Neurosci. **12:** 964–972.

43. SCHOMERUS, C., E. LAEDTKE, J. OLCESE, et al. 2002. Signal transduction and regulation of melatonin synthesis in bovine pinealocytes: impact of adrenergic, peptidergic and cholinergic stimuli. Cell Tissue Res. **309:** 417–428.

44. YU, L., N.C. SCHAAD & D.C. KLEIN. 1993. Calcium potentiates cyclic AMP stimulation of pineal arylalkylamine N-acetyltransferase. J. Neurochem. **60:** 1436–1443.
45. LETZ B., C. SCHOMERUS, E. MARONDE, et al. 1997. Stimulation of a nicotinic ACh receptor causes depolarization and activation of L-type Ca^{2+} channels in rat pinealocytes. J. Physiol. **499:** 329–340.
46. SCHOMERUS, C., E. LAEDTKE & H.W. KORF. 1999. Analyses of signal transduction cascades in rat pinealocytes reveal a switch in cholinergic signaling during postnatal development. Brain Res. **833:** 39–50.
47. YAMADA, H., A. OGURA, S. KOIZUMI, et al. 1998a. Acetylcholine triggers L-glutamate exocytosis via nicotinic receptors and inhibits melatonin synthesis in rat pinealocytes. J. Neurosci. **18:** 4946–4952.
48. YAMADA, H., S. YASTUSHIRO, S. ISHIO, et al. 1998b. Metabotropic glutamate receptors negatively regulate melatonin synthesis in rat pinealocytes. J. Neurosci. **18:** 2056–2062.
49. COON, S.L., P.H. ROSEBOOM, R. BALER, et al. 1995. Pineal serotonin N-acetyltransferase: expression cloning and molecular analysis. Science **270:** 1681–1683.
50. SCHOMERUS, C., H.W. KORF, E. LAEDTKE, et al. 2000. Selective adrenergic/cyclic AMP-dependent switch-off of proteasomal proteolysis alone switches on neural signal transduction: an example from the pineal gland. J. Neurochem. **75:** 2123–2132.
51. COON, S.L., E. DEL OLMO, W.S. YOUNG 3RD & D.C. KLEIN. 2002. Melatonin synthesis enzymes in *Macaca mulatta*: focus on arylalkylamine N-acetyltransferase (EC 2.3.1.87). J. Clin. Endocrinol. Metab. **87:** 4699–4706.
52. ARENDT, J. 1998. Melatonin and the pineal gland: influence on mammalian seasonal and circadian physiology. Rev. Reprod. **3:** 13–22.
53. REPPERT, S.M., M.J. PERLOW, L. TAMARKIN & D.C. KLEIN. 1979. A diurnal melatonin rhythm in primate cerebrospinal fluid. Endocrinology **104:** 295–301.
54. HEDLUND, L., M.M. LISCHKO, M.D. ROLLAG & G.D. NISWENDER. 1977. Melatonin: daily cycle in plasma and cerebrospinal fluid of calves. Science **195:** 686–687.
55. OBSIL, T., R. GHIRLANDO, D.C. KLEIN, et al. 2001. Crystal structure of the serotonin N-acetyltransferase complex: a role for scaffolding in enzyme regulation. Cell **105:** 257–267.
56. ZHENG, W., D. SCHWARZER, A. LEBEAU, et al. 2005. Cellular stability of serotonin N-acetyltransferase conferred by phosphonodifluoromethylene alanine (Pfa) substitution for Ser-205. J. Biol. Chem. **280:** 10462–10467.
57. FU, H., R.R. SUBRAMANIAN & S.C. MASTERS. 2000. 14-3-3 proteins: structure, function, and regulation. Annu. Rev. Pharmacol. Toxicol. **40:** 617–647.
58. AITKEN, A., H. BAXTER, T. DUBOIS, et al. 2002. Specificity of 14-3-3 isoform dimer interactions and phosphorylation. Biochem. Soc. Trans. **30:** 351–360.
59. KLEIN, D.C., S. GANGULY, S.L. COON, et al. 2003. 14-3-3 proteins in pineal photoneuroendocrine transduction: how many roles? J. Neuroendocrinol. **15:** 370–377.
60. SCHOMERUS, C., E. LAEDTKE E & H.W. KORF. 2003. Norepinephrine-dependent phosphorylation of the transcription factor cyclic adenosine monophosphate responsive element binding protein in bovine pinealocytes. J. Pineal Res. **34:** 103–109.
61. SCHOMERUS, C., E. LAEDTKE & H.W. KORF. 2004. Activation of arylalkylamine N-acetyltransferase by phorbol esters in bovine pinealocytes suggests a novel regulatory pathway in melatonin synthesis. J. Neuroendocrinol. **16:** 741–749.
62. FERRY, G., J. MOZO, C. UBEAUD, et al. 2002. Characterization and regulation of a CHO cell line stably expressing human serotonin N-acetyltransferase (EC 2.3.1.87). Cell. Mol. Life Sci. **59:** 1395–1405.
63. GARIDOU, M.L., F. GAUER, B. VIVIEN-ROELS, et al. 2002. Pineal arylalkylamine N-acetyltransferase gene expression is highly stimulated at night in the diurnal rodent, *Arvicanthis ansorgei*. Eur. J. Neurosci. **15:** 1632–1640.

Effects of Melatonin in Age-Related Macular Degeneration

CHANGXIAN YI,[a] XIAOYAN PAN,[a] HONG YAN,[a] MENGXIANG GUO,[a]
AND WALTER PIERPAOLI[b]

[a]Department of Fundus Diseases, Zhongshan Ophthalmic Center, Sun Yat-Sen University,
510060 Guangzhou, China

[b]Jean Choay Institute for Neuroimmunomodulation, Riva San Vitale, Switzerland

ABSTRACT: Age-related macular degeneration (AMD) is the leading cause of
severe visual loss in aged people. Melatonin has been shown to have the capacity to control eye pigmentation and thereby regulate the amount of light reaching the photoreceptors, to scavenge hydroxyradicals and to protect retinal
pigment epithelium (RPE) cells from oxidative damage. Therefore, it is reasonable to think that the physiological decrease of melatonin in aged people may
be an important factor in RPE dysfunction, which is a well known cause for initiation of AMD. Our purpose is to explore a new approach to prevent or treat
AMD. We began case control study with a follow-up of 6 to 24 months. One
hundred patients with AMD were diagnosed and 3 mg melatonin was given
orally each night at bedtime for at least 3 months. Both dry and wet forms of
AMD were included. Fifty-five patients were followed for more than 6 months.
At 6 months of treatment, the visual acuity had been kept stable in general.
Though the follow up time is not long, this result is already better than the otherwise estimated natural course.[1,2] The change of the fundus picture was remarkable. Only 8 eyes showed more retinal bleeding and 6 eyes more retinal
exudates. The majority had reduced pathologic macular changes. We conclude
that the daily use of 3 mg melatonin seems to protect the retina and to delay
macular degeneration. No significant side effects were observed.

KEYWORDS: aging; melatonin; fundus disease; retina; age-related macular degeneration; AMD; macula

INTRODUCTION

Age-related macular degeneration (AMD) is the leading cause of severe visual
loss in people over the age of 50 years in the western world and accounts for approximately 50% of all cases of registered blindness.[3] The rising prevalence of this disease in Asia seems parallel to the same trend in the developed countries. The
etiology of AMD has yet to be clarified, though intensive study has strongly suggested that aging, heredity, and vascular diseases, smoking, UV light exposure, and mal-

Address for correspondence: Prof. Dr. Changxian Yi, Department of Medical Retina, Zhongshan Ophthalmic Center, Sun Yat-Sen University, Guangzhou 510060 China. Voice: +86-20-
87331541; fax: +86-20-87333271.
yichang@public.guangzhou.gd.cn

Ann. N.Y. Acad. Sci. 1057: 384–392 (2005). © 2005 New York Academy of Sciences.
doi: 10.1196/annals.1356.029

nutrition (such as lack of trace elements like zinc) could be the possible causes.[4] Therefore, there is no definitely proven method for preventing or curing the disease or even significantly slowing its pathogenetic course. Until recently, the most modern therapies are practically focused on surgical intervention, such as PDT (photosensitivity dynamic therapy).[5] They are aimed at destroying or halting the proliferation of the subretinal neovascularization (SRNV). The entire ophthalmology community is eagerly trying various methods to treat AMD.

Melatonin is a small molecule that exists in the human body. Its function remains largely a mystery to us. However, its multiple functions on human beings are getting more and more attention. We know now some of its function in relation to eye structure. For examples, melatonin can control eye pigmentation and thereby to regulate the amount of light reaching the photoreceptors.[6] Melatonin has been demonstrated to have the capacity to scavenge the hydroxyl radical.[7] Therefore, it is reasonable to think that the physiological decrease of melatonin in aged people may be an important factor in the dysfunction of retinal pigment epithelial cells, which is a well known reason for initiation of AMD. It is logical to study whether melatonin could have a effect on the progress of AMD.

METHODS AND CLINICAL DATA

Patients

One hundred patients with AMD were diagnosed and given 3 mg melatonin per night for several months from October 1998 to May 2005 in the outpatient department of the Zhongshan Ophthalmic Center. In this group, only 55 patients were followed from 6 to 24 months. The patients group had average age of 71 years (range from 55 to 82 years), 36 were male and 19 female. All together, 110 eyes were checked and recorded in this study. Both dry and wet forms of AMD were included in the study. For the dry form of AMD, the patients must have druses, pigment change, and decreased vision unrelated with other ocular diseases of the macula. For the wet form AMD (exudative AMD), the patients must have no vitreous hemorrhage or other end-stage features of AMD such as macular atrophy and scar.

Due to practical limitations, the melatonin given in this study was manufactured by various companies. Half the patients were given "T-melatonin" prepared by Chronolife (Riva San Vitale, Switzerland). However, the dosage of melatonin for all patients was reported by the manufacturers to be the same and accurately formulated. The basic difference between T-melatonin and other brands of melatonin available is that T-melatonin contains zinc, which is a key trace element for more than 200 important enzymes in the human body. Zinc has been suggested to have positive effect on the treatment of AMD.[8] Oral zinc doses was associated with a smaller decrease in frequency of progressive visual loss in patients with druses and atrophic changes, which is the typical early change of AMD.[9]

General and Ophthalmic Conditions

Eighty-two (82%) patients have some kind of hypertension and have been taking some kind of antihypertension medicine. Fifteen patients had various forms of diabetes and 2 of the 110 eyes had had vitrectomy operations performed as a result of diabetic retinopathy. Eleven eyes had had cataract operations performed before beginning of the melatonin treatment. Thirty-two patients had some degree of senile cataract, but had not had cataract operations. Three eyes had laser coagulation because of the peripheral retinal degeneration or retinal tears. Three eyes had received laser treatment because of diabetic retinopathy or complications of retinal vein occlusion. None of the patients had ever taken any kind of photocoagulation on the macula including TTT (trans-pupil thermal therapy) and PDT (photodynamic treatment) for the treatment of AMD.

The patients were examined approximately every two months during the treatment, and every 3 months after finishing the melatonin. The Snellen visual chart and slit lamp examination were performed. Macula were examined with a +90D pre-set lens. The peripheral fundus was examined with a binocular indirect ophthalmoscope. The fundus photo was taken at least three times for each patient at the beginning of the study and at the end of the treatment as well as during the sixth month after the treatment started. All patients had at least once fundus fluorescein angiography (FFA) to confirm the diagnosis of age-related macular degeneration. Electroretinogram (ERG) and visual electronic potential (VEP) were not part of the basic information collected in this study.

Typical Fundus Improvement

Typical fundus improvement is shown in FIGURES 1–5 below.

RESULTS

One hundred patients were given melatonin, and their visual change was recorded during the follow-up period. However, only 55 patients were able to be followed for more than 6 months. At the sixth month after the treatment, there was no significant

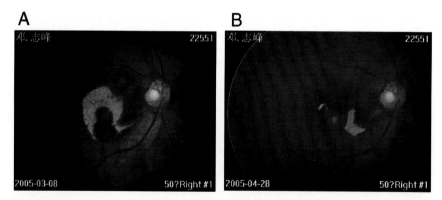

FIGURE 1. (**A**) A 67-year-old male retired teacher. Visual acuity had been decreasing for 2 years. (**B**) The same patient with a stable visual acuity at 0.3 with remarkable improvement in subretinal macular hemorrhage.

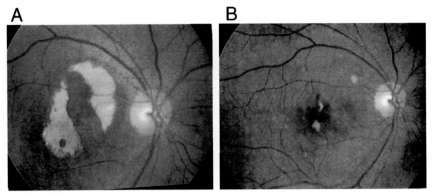

FIGURE 2. Female, 71 years old with OD wet AMD, visual acuity improved from 0.2 (**A**) to 0.4 (**B**).

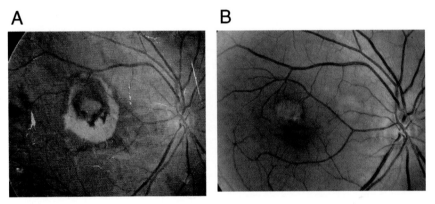

FIGURE 3. Male, 58 years old, visual acuity increased from 0.2 (**A**) to 0.6 (**B**) one year ater. The subretinal hemorrhage and exudate was remarkably absorbed.

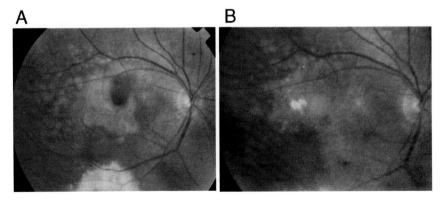

FIGURE 4. Man, 65 years old, with 0.1 visual acuity with no change after 3 months of treatment, but fundus improved. Macular edema reduced.

A   B

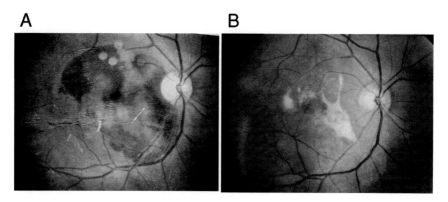

FIGURE 5. Visual acuity improved from 0.2 (**A**) to 0.3 (**B**) after 6 month, wet AMD had been largely changed to dry AMD. Subretinal hemorrage absorbed, some subretinal fibrosis developed, the lesion become much stable.

TABLE 1. Change of visual acuity after administration of melatonin[a]

Time visual acuity checked	Visual acuity 0.05–0.1*	Visual acuity 0.1–0.2	Visual acuity 0.2–0.4	Visual acuity 0.4 or more
Beginning	6	54	38	12
6 months	8	54	38	10

NOTE: Visual acuity range indicates Snellen chart visual acuity at the beginning of the study.
[a] Numbers of eyes with the visual acuity at the beginning time point and after several months. The totals represent initial measurements and those at 6-month follow up. Some patients were lost to the study.

TABLE 2. Fundus change after six months[a]

Fundus change	Retinal blood	Retinal exudates	Atrophy
Reduced or same	102	96	0
Increased	8	6	22

[a]Number of eyes that changed macular presentations at the sixth month.

change in allocation of the patients (TABLE 1). It has been difficult to summarize as completely as we would have liked because of the decrease in patient numbers. However, this result is already better than the otherwise estimated natural course, which has been demonstrated deterioration as the natural course.[1,2] The change in the fundus picture is also promising (TABLE 2). Among the 55 patients, only 8 eyes showed an increase in retinal blood and 6 eyes showed an increase in retinal exudates. The majority had a reduction compared to pathologic changes before treatment in these patients, though atrophy was slightly increased. However, the atrophy could be viewed as a positive change because of the relative stability for the vision.

TABLE 3. Other main ophthalmic alterations after a 6-month melatonin treatment[a]

Ocular change	Lens opacity	Intra-ocular pressure	New vitreous hemorrhage	Subjective ocular feeling
Better	0	slightly 32	4	40
Same	110	78	106	58
Worse	0	0	4	12

[a]Number of eyes with ocular change at the sixth month after beginning treatment.

TABLE 4. Subjective report about general condition of eyes after melatonin[a]

General change	Sleeping	General energy and health
Better	22	43
Same	33	12
Worse	0	0

[a]Number of patients with change in general condition after 6 months of melatonin treatment.

Other major ophthalmic alterations besides the visual change at the 6-month exam after taking melatonin are summarized in TABLE 3. In the course of aging, the lens opacity (cataract) is usually increased; however, this process can be very slow. It is just a clinical judgment, not a practical objective measurement; we did not observe any obvious increase in cataract development. The intra-ocular pressure was slightly better in 16 patients (32 eyes), but the difference in the values did not reach statistical significance. The general ocular experience, such as flare, dryness, clarity, comfort appeared to have improved, although at this time this impression has not been measured.

We also recorded the general condition of this group of patients (TABLE 4), and improvements in sleeping and appetite were demonstrated. A generally healthier feeling is quite commonly reported among the melatonin users.

DISCUSSION

The most important pathologic change that threatens visual acuity in AMD is subretinal neovascularization (SRNV), which can lead to hemorrhage, retinal edema, exudates, RPE tears, detachment of the RPE, and so forth. All forms of subretinal neovascularization, whether extrafoveal, juxtafoveal, or subfoveal, possess great potential for causing visual damage. Within three years, about 44% of the patients may lose six lines or more of their vision.[1,2] However, the patients treated with T-melatonin or melatonin without zinc have shown practically no deterioration in visual acuity in this study. Although the follow-up time is relatively short, it is suggestive of some positive influence on AMD. At this time, it is still too early for us to conclude the definite therapeutic function of melatonin on AMD, as further large multi-center trials may be warranted. It is also worthwhile here to review some of the function of melatonin, which may support our results.

The Relationship of AMD, Aging, and Melatonin

Though we can list many possible causes of AMD, from genetic to accumulations of free-radical damage to cardiovascular diseases, aging is still the most convincing explanation for the development of AMD. Melatonin does apparently have some anti-aging functions, which have been discuss and demonstrated by many researchers.[10–14] The best and most complete layman's description can be found in the book *The Melatonin Miracle*.[15]

Influence of Melatonin on the Eye

The earliest change in the pathologic progression of AMD is the dysfunction of retinal pigment epithelium (RPE). It was reported that melatonin can control eye pigmentation and thereby regulate the amount of light reaching the photoreceptors. It has also been shown that melatonin can induce light-dependent photoreceptor disc shedding in cell cultures.[6,16] One of the main mechanisms in damaging the RPE in AMD is through the accumulation of various free radicals. Melatonin has been demonstrated to have the capacity to scavenge the hydroxyl radical.[7,17,18] Therefore, it is reasonable to think that the physiological decrease of melatonin in aged people may be an important factor in dysfunction of retinal pigment epithelial cells. Recent studies have also shown that melatonin is related to many eye diseases besides AMD,[19,20] especially in uveitis.[21] There is always some inflammation in the pineal gland in experimental uveitis in the animal model.

Melatonin and Ocular Physiology

Many researchers have demonstrated that eye and pineal gland are closely connected in certain pathologic processes.[22–25] The pineal gland and eyes also have some common points in morphology and embryology. There are also some common biochemical features.[26] The pinealocyte and the retinal photoreceptor appear to use the same biosynthetic pathway for converting 5-HT (serotonin) to melatonin.[27] Recent studies show that melatonin may play an important role in maintaining the normal function of the retina,[4,16,28,29] perhaps via its control of dopamine systems in the inner retina. Melatonin plays an important role in dark adaptation and in various retinal processes that exhibit a circadian rhythm. Melatonin and glutamate may represent "co-transmitters," that provide the visual pathway with two types of signals, with melatonin providing widespread modulatory influences on the discrete visual information conveyed via glutamatergic circuits.[28] There are also reports indicating that melatonin enhances GABAergic inhibition of light-evoked dopamine release. This mechanism may underlie the light/dark difference in dopamine release in vertebrate retina.[30] Researchers also have found various melatonin receptors in both eye and brain.[31,32] Melatonin synthesis is stimulated within minutes after exposure to darkness, and may reflect an increase in N-acetyl transferase activity. Melatonin is not stored in the eyes, rather it diffuses freely throughout the retina immediately after it is synthesized; and the dark-induced increase in retinal melatonin release is a synthesis-coupled response and does not involve separate secretion mechanisms. Houston researchers believe that melatonin is a local paracrine effector of dark-adaptive responses in the retina.[33] Several lines of evidence indicate that retinal photoreceptors produce melatonin. However, there are other potential melatonin sources in the retina, and melatonin synthesis can be regulated by feedback from the inner retina.[34,35]

CONCLUSION

Melatonin supplementation among the aged population may be beneficial in preventing, relieving, or reducing the severity of AMD, which is one of the leading causes for blindness in the elder. It may help vision through its improvement of general health or through direct effect on the ocular tissues such as RPE. However, further studies involving a bigger randomized sample are needed before its effect can be proven. The basic mechanism also remains to be understood in the near future.

Considering what we know about the interaction of melatonin, RPE, and the retina and the fact that we still have no reliable pharmaceutical treatment for AMD, it would be useful to study possible effects of melatonin on AMD. Even though this report is not based on a randomized controlled study, we have compared our results with data available in the literature and conclude that the potential for melatonin to be used as a treatment for AMD deserves further study.

[*Competing interests*: The authors declare that they have no competing financial interests.]

REFERENCES

1. SCUPOLA, A., G. COSCAS, G. SOUBRANE, *et al.* 1999. Natural history of macular subretinal hemorrhage in age-related macular degeneration. Ophthalmologica **213:** 97–102.
2. AVERY, R.L., S. TEKRAT, B.S. HAWKINS, *et al.* 1996. Natural history of subfoveal subretinal hemorrhage in age-related macular degeneration. Retina **16:** 183–189.
3. SLAKTER, J.S. & M. STUR. 2005. Quality of life in patients with age-related macular degeneration: impact of the condition and benefits of treatment. Surv. Ophthalmol. **50:** 263–273.
4. VAN LEEUWEN, R., C.C. KLAVER, J.R. VINGERLING, *et al.* 2003. Epidemiology of age-related maculopathy. Eur. J. Epidemiol. **18:** 845–854.
5. HOOPER, C.Y. & R.H. GUYMER. 2003. New treatments in age-related macular degeneration. Clin. Exp. Ophthalmol. **31:** 376–391.
6. PANG, S.F. & D.T. YEW. 1979. Pigment aggregation by melatonin in the retinal pigment epithelium and choroid of guinea pigs *Cavia porcellus.* Experientia **35:** 213–233.
7. TAN, D.X., L.C. MANCHESTER, R.J. REITER, *et al.* 2000. Significance of melatonin in antioxidative defense system: reactions and products. Biol. Sign. Recept. **9:** 137–159.
8. COMER, G.M., T.A. CIULLA, M.H. CRISWELL, *et al.* 2004. Current and future treatment options for nonexudative and exudative age-related macular degeneration. Drugs Aging **21:** 967–992.
9. NEWSOME, D.A., M. SWARTZ, N.C. LEONE, *et al.* 1988. Oral zinc in macular degeneration. Arch. Ophthalmol. **106:** 192–198.
10. PIERPAOLI, W., A. DALL'ARA & E. PEDRINIS. 1991. The pineal control of aging: the effects of melatonin and pineal grafting on the survival of old mice. Ann. N.Y. Acad. Sci. **620:** 291–313.
11. REITER, R.J. 1992. The aging pineal gland and its physiological consequence. BioEssay **14:** 169–175.

12. MAESTRONI, G.J.M,. A. CONTI & W. PIERPAOLI. 1988. Pineal melatonin: its fundamental immunoregulatory role in aging and cancer. Ann. N.Y. Acad. Sci. **521:** 140–148.
13. LEHRER, S. 1979. Pineal effect on longevity. J. Chron. Dis. **32:** 411–412.
14. PIERPAOLI, W. & W. REGELSON. 1994. Pineal control of aging: effect of melatonin and pineal grafting on aging mice. Proc. Natl. Acad. Sci. USA **94:** 787–791.
15. PIERPAOLI, W. & W. REGELSON. 1995. The Melatonin Miracle. Simon & Schuster. New York.
16. BESHARSE, J.C. & D.A. DUNIS. 1983. Methoxyindoles and photreceptor metablism--activation of rod shedding. Science **219:** 1341–1343.
17. POEGGELER, B., R.J. REITER, D.X. TAN, et al. 1993. Melatonin, hydrohyl radical-mediated oxidative damage, and aging: a hypothesis. J. Pineal Res. **14:** 151–158.
18. LIANG, F.Q., L. GREEN, C. WANG, et al. 2004. Melatonin protects human retinal pigment epithelial (RPE) cells against oxidative stress. Exp. Eye Res. **78:** 1069–1075.
19. WURTMAN, R.J. 1985. Melatonin secretion as a mediator of circadian variations in sleep and sleepiness. J. Pineal Res. **2:** 301–303.
20. NAIR, N.P.V., N. HARIHARASUBRAMANIAN, C. PILAPIL, et al. 1986. Plasma melatonin: an index of brain aging in humans? Biol. Psychiatry **21:** 141–150.
21. TOUITOU, Y., P. LE HOANG, B. CLAUSTRAT, et al. 1986. Decreased nocturnal plasma melatonin peak in patients with a functional alteration of the retina in relation with uveitis. **70:** 170–174.
22. REITER, R.J. 1981. The mammalian pineal gland: structure and function. Am. J. Anat. **162:** 287–313.
23. KALSOW, C.M. & W.B. WACKER. 1978. Pineal gland involvement in retina-reduced experimental allergic uveitis. Invest. Ophthalmol. Visual Sci. **17:** 775–783.
24. BROEKHUISE, R.M., H.J. WINKENS & E.D. KUHLMANN. 1986. Induction of experimental autoimmune uveoretinitis and pinealitis by IRBP. Comparision to uveoretinitis induced by S-antigen and opsin. Curr. Eye Res. **5:** 231–240.
25. KORF, H.W., R.G. FOSTER, P. EKSTROEM, et al. 1985. Opsin-like immunoreatction in the retinae and pineal organs of four mammalian species. Cell Tissue Res. **242:** 645–648.
26. FOLEY, P.B., K.D. CAIRNCROSS & A. FOLDES. 1986. Neurosci. Biobehav. Rev. **10:** 273–293.
27. PIERCE, M.E., D. BARKER, J. HARRINGTON, et al. 1989. Cyclic AMP-dependent melatonin production in Y79 human retinoblastoma cells. J. Neurochem. **53:** 307–310.
28. REDBURN, D.A. 1988. Neurotransmitter systems in the outer plexiform layer of mammalian retina. Neurosci. Res. **8:** 127–136.
29. SCHER, J., E. WANKIEWICZ, G.M. BROWN, et al. 2003. AII amacrine cells express the MT1 melatonin receptor in human and macaque retina. Exp. Eye Res. **77:** 375–382.
30. BOATRIGHT, J.H., N.M. RUBIM & P.M. IUVONE. 1994. Regulation of endogenous dopamine release in amphibian retina by melatonin: the role of GABA. Visual Neurosci. **11:** 1013–1018.
31. REPPERT, S.M., C. GODSON, C.D. MAHLE, et al. 1995. Molecular characterization of a second melatonin receptor expressed in human retina and brain: the Mel1b melatonin receptor. Proc. Natl. Acad. Sci. USA **92:** 8734–8738.
32. WIECHMANN, A.F. & C.R. WIRSIG-WIECHMANN. 1994. Melatonin receptor distribution in the brain and retina of a lizard, *Anolis carolinensis*. Brain Behav. Evol. **43:** 26–33.
33. REDBURN, D.A. & C.K. MITCHELL. 1989. Darkness stimulates rapid synthesis and release of melatonin in rat retina. Visual Neurosci. **3:** 391–403.
34. CAHILL, G.M. & J.C. BESHARSE. 1992. Light-sensitive melatonin synthesis by Xenopus photoreceptors after destruction of the inner retina. Visual Neurosci. **8:** 487–490.
35. BANAS, I., B. BUNTNER & T. NIEBROJ. 1995. Levels of melatonin in the serum of patients with retinitis pigmentosa. Klin. Oczna **97:** 321–323.

Effects of Melatonin in Perimenopausal and Menopausal Women

Our Personal Experience

G. BELLIPANNI, F. DI MARZO, F. BLASI, AND A. DI MARZO

Menopause Center, Madonna delle Grazie Health Institute, Velletri, Rome, Italy

ABSTRACT: The purpose of this clinical trial on possible effects of nocturnal MEL administration in perimenopausal women was to find if MEL by itself modifies levels of hormones and produces changes of any kind, independently of age (42–62 years of age) and the stage of the menstrual cycle. It is accepted that a close link exists between the pineal gland, MEL, and human reproduction and that a relationship exists between adenohypophyseal and steroid hormones and MEL during the ovarian cycle, perimenopause, and menopause. Subjects took a daily dose of 3 mg synthetic melatonin or a placebo for 6 months. Levels of melatonin were determined from five daily saliva samples taken at fixed times. Hormone levels were determined from blood samples three times over the 6-month period. Our results indicate that a cause-effect relationship between the decline of nocturnal levels of MEL and onset of menopause may exist. The follow up controls show that MEL abrogates hormonal, menopause-related neurovegetative disturbances and restores menstrual cyclicity and fertility in perimenopausal or menopausal women. At present we assert that the six-month treatment with MEL produced a remarkable and highly significant improvement of thyroid function, positive changes of gonadotropins towards more juvenile levels, and abrogation of menopause-related depression.

KEYWORDS: melatonin; menopause; aging

A progressive reduction of nocturnal serum melatonin (MEL) concentration is observed in aging healthy humans.[1] It has been demonstrated that the pineal gland is responsible for the development and maintenance of neuroendocrine and sexual functions. In fact, the identification of the key role of the pineal gland in the control of reproductive biology emerged from the observation of precocious puberty in children carrying pineal tumors.[2] In nonhuman mammals, such as sheep, seasonal fer-

Address for correspondence: Giulio Bellipanni, M.D., Ph.D., Medical Director, "Madonna della Grazie" Health Institute, Via Salvo D'Acquisto 67, 00049 Velletri (Rome), Italy. Voice: +39-06-96441671.

giuliobellipanni@yahoo.it

Ann. N.Y. Acad. Sci. 1057: 393–402 (2005). © 2005 New York Academy of Sciences.
doi: 10.1196/annals.1356.030

tility and reproduction during young, reproductive age, are strictly dependent on night levels of MEL, which vary according to the seasons (temperature and photoperiod). In this case MEL exerts an inhibitory action on gonadotropins and thyroid hormones.[3–5] The effect of MEL seems to be related, at least to a large extent, to regulation of high-affinity MEL receptors in the neurohypophysis and on modulation of leutotropic hormone-releasing hormone (LH-RH) receptor density in the hippocampus and in the brain in general (anterior hypothalamus, suprachiasmatic nuclei, preoptic area).[6–8] MEL administration at high, pharmacological doses, inhibits gonadal function in male hamsters and this effect is especially evident when MEL is given at daytime rather than with nocturnal periodicity.[9,10] MEL at extremely high doses and for periods of years has also been given to thousands of young women in the attempt to develop a new kind of contraceptive pill without early or late side effects.[11] Also, MEL thwarts the growth of hormone-sensitive tumors, such as certain breast cancers, and can increase the density of hormone receptors on cancer cells.[12]

In the course of aging, age-relate illnesses and disabilities are clearly linked to a progressive decline of reproductive-sexual functions in men (andropause) and women (menopause). The progressive decline of this basic function leads to severe side-effects psychosomatic in nature and to a chain of negative events that stem from the close relationship between neuroendocrine and immune capacity and functions. In fact, it has been demonstrated that LH-RH is a powerful immuno-regulating and immuno-enhancing hormone both in ontogeny and in adult life.[13]

Findings from our laboratory have demonstrated that administration of MEL in the night hours and transplantation of pineal glands from young into older mice and rats maintain juvenile thyroid function and significantly delay their aging and/or prolongs their life.[14,15] A clear-cut activity of MEL and the grafting of young-to-old pineal glands is the maintenance of gonadal and sexual functions. A most remarkable maintenance of juvenile levels of LH-RH receptors both in the brain (hippocampus) and in the gonads is achieved by transplantation of the pineal gland from a young donor rat into older recipients.[16] This effect becomes more pronounced in the course of months after grafting, which demonstrates that, at least in rats, there is more than just a simple return of sexual functions to more juvenile levels.[16] The general conclusion derived from others' and our own work is that circadian, nocturnal administration of MEL may postpone endocrine aging and maintain or reconstitute more juvenile sexual functions at a time of life (e.g., between 40 and 60 years of age in women) when changes of ovarian cyclicity become evident (premenopausal, perimenopausal, and menopausal age). Along with the complete absence of noxious side-effects in MEL treatment, there are many somatic and psychological benefits for women of all ages. Many women, just before and after menopause, undergo a hormone replacement therapy (HRT) to correct levels of ovarian steroids, such as estrogens and progesterone, to premenopause values. HRT also protects against the psychosomatic side-effects of menopause and against heart disease and osteoporosis. Even if MEL could not, by itself, prevent completely menopause and maintain ovarian cyclicity, it may enhance and improve the efficiency of HRT and thus add more years of treatment until there is an opportunity for prevention, delay, or mitigation of menopause and its related psychosomatic problems. An early intervention with night administration of MEL before the initiation of ovarian dysfunction in relatively young women may indicate to us whether or not MEL can modify the onset or the course of menopause in women.

PATIENTS AND METHODS

Patients and Treatment

A formal and fully documented application for the conduction of the clinical trial was delivered to the Ministry of Public Health, Rome, Italy. All women joining the clinical trial were asked to sign an informed consent. They were given detailed information on the character, duration, and aims of the research. Women were carefully selected who had no relevant pathologies, no history of use of drugs, hormones, or herbal preparations, were living a normal lifestyle with typical Mediterranean diet rich of carbohydrates and fresh vegetables, and had the normal sleep habits of the population of the region south of Rome. All women did not smoke and did not abuse alcohol.

All women completed a questionnaire that included questions about life habits, physiological data, data concerning previous pathologies, treatment-related side-effects, treatment-induced alterations, effects on perimenopausal symptoms (neurovegetative, sleep, psychological) were recorded at time 0, and at 3, 6, 12, 24 months of MEL or placebo treatment. The questions concerned duration and character of their menstrual cycle and/or psychosomatic, symptoms associated with perimenopause and menopause (such as irritability, morning mood, depression, insomnia, night sweats, headache, stypsis, amnesia, hot flushes, palpitations, and body weight increase).

The recruited premenopausal (45 women), perimenopausal (56 women), and postmenopausal (38 women) women age ranged from 42 to 62 years of age. They were divided in two age groups (42–49 and 50–62) of similar age and peri- or postmenopausal symptoms. In fact, in most countries and latitudes perimenopausal symptoms and onset of menopause do not depend precisely on the age but rather on individual, environmental, and genetical variability. The two groups of women were given plastic bottles with 60 MEL capsules of 3 mg (MEL, synthetic, 100% purity) or placebo at the beginning of each two-month period. The bottles with MEL or placebo were certified as a GMP (Good Manufacturing Practice) product and were a gift from Eurochem Ltd, Munich, Germany. They were only marked with a code number. The opaque white capsules of MEL and placebo were identical. It was unknown to the women and to the doctor which of the two groups received MEL or placebo. In fact the key of the code number (MEL or placebo) was unknown to the patients and to the doctor and was opened only at the end of the first six-month trial.

All women were asked to take the capsules at bedtime, between 10 and 11 p.m. For obvious ethical reasons, administration of placebo was restricted to the initial six months of this ongoing clinical study.

At the end of the six-month trial, final results were collected from a total of 139 women, with small number variations in the group studied depending on minor failure of some determinations of MEL in saliva, T3, T4, LH, and follicle-stimulating hormone (FSH).

Methods

A correlation exists between blood and salivary levels of MEL.[17] In order to evaluate the basal levels of MEL in the recruited women, MEL was measured in the saliva of all women immediately before the initiation of the trial (late summer 1997).

All women in the study were given a commercial kit for the quantitative determination of MEL in saliva (Diagnos-Techs Inc., Kent, WA). Samples of saliva for measurement of MEL were taken at 2 P.M., 10 P.M., 2 A.M., 8 A.M., and 10 A.M. For testing MEL in saliva, 500 mL of standards, controls, and unknown patient saliva samples were dispensed into labeled tubes. In a second step, fixed amounts of assay buffer, MEL antiserum and ^{125}I-labeled MEL tracer were added to the tubes and incubated for 36–48 h at room temperature. In this reaction the unlabeled MEL of standards, controls, and patients samples and the ^{125}I-labeled MEL compete for the binding sites of the MEL antibody. In a final step, the antibody-bound fraction is precipitated by the addition of a second antibody. Samples were incubated for another 30 min and centrifuged. The supernatant was decanted and the pellet was counted in a Gamma counter. The concentration of MEL of the unknown patient was read from the calibration curve. The sensitivity of the assay is approximately 1.5 pg/mL.

A negative correlation was found between levels of MEL in saliva and age (see below). The correlation between basal MEL levels and menopause was not evaluated.

For determination of hormones in blood, heparinized blood was taken from all fasting women between 8 and 10 A.M., at time 0, 3, and 6 months, irrespective of menstrual cyclicity. The following hormones were assayed by standard laboratory techniques: thyroid hormones (TSH, total T3 and T4), LH, FSH, prolactin (PRL), estrone, estradiol, and progesterone.

Statistics

Results are expressed as mean ± SD. The significance between the means was assessed using paired Student's t test. ANOVA test (one-way) was used where appropriate. Correlations were determined by linear regression by the least square method. The differences between the various regression lines were evaluated by analysis of covariance. Chi-square test was used to establish significance of LH increment in MEL-treated women. Differences were considered statistically significant when $P < 0.05$.

RESULTS

Basal Levels of MEL in Saliva

As mentioned above, basal levels of MEL in the saliva were measured in all women before beginning the clinical trial. The women were divided into three groups according to their average nocturnal MEL levels between 10 p.m. and 2 a.m. Low levels were considered those below 20 pg/mL, medium levels were those higher than 20 pg/mL and lower than 300 pg/mL, and high levels those around 300 pg/mL. The age range of the women in our trial was relatively narrow (42–62 years of age), a negative correlation was clearly present between age of the women and basal night levels of MEL ($N = 139$; $r = -0.263$; $P < 0.05$).

Effects of MEL on Blood Levels of Thyroid Hormones

At time 0, before initiation of the trial, a positive correlation was found between basal levels of total T3 and MEL in all women pooled together ($N = 139$; $r = 0.350$; $P < 0.05$). This correlation was absent between MEL and total T4 ($N = 139$;

$r = 0.076; P = 0.594$). In the course of the six-month clinical trial, significant changes of thyroid hormone levels were observed in the MEL-treated women. Evening administration of MEL produced a significant increase of total T3 and T4 after 3 and 6 months of MEL administration. When compared to placebo-treated women, the increment of T3 was significant after 3 and 6 months of treatment while T4 only after 6 months.

There were no differences in TSH levels. However, a slightly significant enhancement of T3 was measurable also in placebo-treated women after 6 months of treatment. When thyroid hormone levels were evaluated according to the basal levels of MEL, it was found that the effects of MEL on the increase of T3 and T4 were particularly significant in women with low (below 20 pg/mL) night MEL levels group when T3 and T4 were measured at 3 and 6 months and for the group with the medium (above 20 and below 300 pg/mL) night levels of MEL after six months of treatment. Only T3 was positively modified after six months of treatment in the group of women with high (300 pg/mL) basal levels of MEL. An increase of T3 was also observed in placebo-treatment women after six months of treatment, which was significant in women with high basal levels of MEL. No effects were observed on levels of TSH. MEL administration also produced a significant increment of total T4 in women with low basal level of MEL, when compared to placebo-treated women.

Effects of MEL on Blood Levels

LH

As is well known, progression of aging in women leads to increased blood levels of LH.[18] A consistent increase of basal LH in relation to the increasing age of women was in fact visible at time 0. Also a clear-cut negative correlation between basal levels of MEL and LH ($N = 76; r = -0.314; P < 0.05$) was found in all women at time 0. When the levels of LH were evaluated in the younger or older age groups (42–49 and 50–62 years of age) in relation to the percentage of women displaying an increment of LH, it was seen that MEL produces a significant decrease in plasma LH only in the younger and not in the older women (see TABLE 1). In fact, in the pla-

TABLE 1. Melatonin effect on LH in perimenopausal and menopausal women

Treatment	Age (years)	N1	N2	%
Melatonin	43–49	38	10	28.6
Placebo	42–49	34	20	61
Melatonin	50–62	30	22	75
Placebo	50–69	38	18	47.6
Melatonin	43–62	71	34	48.6
Placebo	42–59	75	40	53.8

NOTE: Effects of circadian, evening administration of melatonin on LH levels in perimenopausal and menopausal women (42–62 years old) grouped according age. Melatonin produces a more pronounced decrement of LH in the younger women (N1 = number of women taken into consideration; N2 = number of women where the increment of LH after six months is ≥ 10%; %, percent of women where LH increases.

cebo-treated women, no correlation was found between the increment ($\Delta\%$) of LH level and the age of the women after six months of treatment, while in MEL-treated women a significant and positive correlation was found between age and LH increment. This demonstrates that the effects of MEL in controlling and maintaining low levels of LH are much more pronounced in the younger women (43–49 years of age) than in the older women (50–62 years of age).

FSH

FSH was measured in all women before initiation of placebo or MEL treatment. As predicted[18] basal levels of FSH increase with age. Similar to LH, a negative correlation exists between basal levels of MEL and FSH at time 0 in all women ($N = 72$; $r = -0.322$; $P < 0.05$). Treatment with MEL for six months produced a significant decrease ($\Delta\%$) of FSH especially in women with low basal MEL levels ($N = 70$; $r = 0.468$; $P < 0.05$). After six months of treatment, no correlation was seen between increment of FSH ($\Delta\%$) and basal levels of MEL in the placebo-treated women ($N = 74$; $r = -0.057$; $P = 0.795$).

Estrogens, Progesterone, and PRL

The large variability of the values measured in blood samples taken at different times of the menstrual cycle and in menopausal women with age varying from 42 to 62 years of age, prevented the evaluation for significant differences in these hormones between MEL or placebo-treated women within the six-month period of MEL treatment.

Menstrual Cyclicity

Within the relatively short period of MEL administration, only episodical improvement of regularity and duration of menstrual cycles were reported. However, 12 menopausal women (at 1 and 2 years after total cessation of the menses) reported a re-acquisition of normal (bleeding and duration) menstrual cycles.

Other Psychosomatic and Neurovegetative Changes

From the analysis of the completed questionnaire after the initial six months of MEL or placebo treatment, the evaluation of different typical perimenopause or menopause-related symptoms or alterations did not disclose a clear-cut difference between the two groups, with the notable exception of a very significant improvement of mood and complete disappearance of morning depression in the MEL-treated women. Only 6.7% of MEL-treated women reported the continuation of morning depression compared to 21% of placebo-treated women ($P < 0.05$, χ^2-test). Although not significant, many women reported a tendency to amelioration of hot flushes, palpitations, and improvement of quality and duration of sleep.

DISCUSSION

The purpose of this initial trial on possible effects of nocturnal MEL administration in perimenopausal women was to find if MEL by itself modifies levels of hormones and produces changes of any kind, independently of age (42–62 years of age) and the stage of the menstrual cycle. It is undisputed that a close link exists between the pineal gland, MEL, and human reproduction[19] and that a relationship exists between adenohypophyseal and steroid hormones and MEL during the ovarian cycle, perimenopause and menopause.[20]

The preliminary findings emerging from this ongoing clinical trial help to focus and to restrict our attention on the significant changes of thyroid, and of some hypophyseal hormones, namely LH and FSH, after evening administration of exogenous, oral MEL in the course of six months. In fact, no statistically significant changes were observed in the levels gonadal steroids and PRL. This may depend on the different age of women and on the short treatment period (6 months) and because the blood samples were taken with no regard to menstrual cyclicity.

The initial measurement of MEL in the saliva of all women established a criterion for dividing the women in low, medium, and high MEL subjects to see whether or not MEL can produce endocrine changes depending on the basal individual levels of MEL. It was also important to divide all women in two age groups (42–49 and 50–62 years of age) to verify if MEL can affect neuroendocrine functions in relation to the age of the subject. In fact, these separations allowed us to differentiate MEL-reactive and MEL-unreactive women and to disclose an higher sensitivity of younger women to MEL treatment. This is extremely relevant in view of the fact that prevention of perimenopausal and menopause-related endocrine changes may be initiated in women.

The most important and largely unexpected finding was the clear-cut effect of MEL on thyroid function (see TABLE 2). Our trial was initiated in late summer and was concluded in late winter, a season when physiological levels of thyroid hormones are higher.[21] In spite of the obvious variability of T3 and T4 values, when the effect of MEL was evaluated in relation to the basal levels of MEL in the saliva it can be seen that MEL produced a significant increase of T3 and T4 in the low and medium MEL groups, but only T3 increased in the high MEL group. Apparently this effect of MEL on thyroid function can be exerted only when the pineal gland produces less MEL. In the placebo-treated women, endogenous levels of MEL in the group with high MEL seem to affect thyroid function positively. The interpretation of this placebo-dependent change could also be attributed, in the high MEL group, to the physiological changes of thyroid function in relation to temperature and season.[21,22] It is thus clear that low MEL levels can be suggestive of low thyroid function and that administration of MEL may prevent and cure the decline of thyroid function in perimenopausal women. Our findings are confirmed by the positive correlation existing between basal levels of MEL and T3 at time 0 in all women. These results seem also to suggest that thyroid deficiency may be a common and apparently latent endocrine disorder for initiation and progression of menopause in women, and also an unsuspected cause for at least some of the psychosomatic and psychic, neurovegetative symptoms. That MEL can restore deranged thyroid function in perimenopausal women is not a surprise if we consider that MEL modulates the 5′ thyroid deiodinase, that MEL or pineal grafting improve thyroid function in old rodents, and that the pineal gland contains relatively large amounts of TRH. MEL was found to decline significantly from premenopause to postmenopause and to correlate negatively with serum FSH, suggesting that MEL may be determinant for the initiation of menopause.[23]

TABLE 2. Effects of melatonin on thyroid function in perimenopausal and menopausal women

Melatonin concentration in saliva	T3 (ng/mL)			T4 (ng/mL)			TSH (µg/mL)		
	0 Months	3 Months	6 Months	0 Months	3 Months	6 Months	0 Months	3 Months	6 Months
MEL (N=76)	1.44±0.19	1.58± 0.21	1.65±0.22	85.2±8.3	90.2±8.3	94.1±8.6	1.05±0.48	1.03±0.56	1.11±0.65
High (N=25)	1.41±0.23	1.56±0.28	1.67±0.33	86.9±9.7	88.9±11	90.2±10.3	1.02±0.39	1.01±0.37	1.24±0.53
Medium (N=23)	1.43±0.13	1.44±0.13	1.56±0.08	86.4±6.6	90.5±7.2	95.6±8.7	0.95±0.45	1.04±0.68	1.13±0.64
Low (N=25)	1.39±0.20	1.61±0.17	1.60±0.15	84.8±7.1	92.8±4.6	95.4±6.2	1.06±0.39	1.00±0.42	1.14±0.42
Placebo (N=78)	1.43±0.20	1.48±0.22	1.54±0.19	85.5±9.2	86.1±11.3	86.4±9.6	1.08±0.54	1.10±0.66	1.16±0.69
High (N=24)	1.40±0.14	1.47±0.19	1.71±0.11	85.3±9.1	85.1±9.1	93±9.2	1.09±0.52	1.10±0.30	1.21±0.39
Medium (N=28)	1.47±0.22	1.49±0.25	1.55±0.25	86.8±7.4	90.3±11.2	88.2±10.9	0.97±0.38	1.06±0.44	0.99±0.46
Low (N=26)	1.39±0.07	1.53±0.23	1.52±0.18	86.4±5.8	85.7±13.7	85.4±8.9	1.05±0.36	1.20±0.45	1.19±0.36

NOTE: Effects of evening administration of melatonin on thyroid function in perimenopausal and menopausal women (42–62 years of age) groups according to their basal melatonin levels at night. High = 300 pg/mL, Medium = more than 20 pg/mL and less than 300 pg/mL, Low = less than 20 pg/mL.

FSH may also represent an early endocrine marker of reproductive aging. In our trial the administration of MEL produces significant changes in LH and FSH levels, which seem to be related to the original basal levels of MEL and to the age of the women. MEL depresses the production of FSH only in women with low MEL levels. The apparent partial recovery of more juvenile gonadotropin function in the MEL-treated women is consonant with the notion that progressive increase of LH and FSH in aging women signals an increased central hypothalamic resistance and insensitivity to feedback regulation, with increasing number of LHRH receptors in the hippocampus and compensatory increase of LH and FSH secretion. This effect of MEL in the restoration of reproductive functions is evident in aging experimental animals. Our results indicate that a cause-effect relationship between the decline of nocturnal levels of MEL and onset of menopause may in fact exist. The follow up controls show that MEL abrogates hormonal, menopause-related neurovegetative disturbances and restores menstrual cyclicity and fertility in perimenopausal or menopausal women. At present we can assert that the six-month treatment with MEL produced a remarkable and highly significant improvement of thyroid function, positive changes of gonadotropins towards more juvenile levels, and abrogation of menopause-related depression.

[*Competing interests*: The authors declare that they have no competing financial interests.]

REFERENCES

1. IGUCHI, H., K. KATO & H. IBAYASHI. 1982. Age-dependent reduction in serum melatonin concentrations in healthy human subjects. J. Clin. Endocrinol. Metab. **55:** 27–29.
2. ZRENNER, C. 1985. Theories of pineal function from classical antiquity to 1900: a history. Pineal Res. Rev. **3:** 1–40.
3. TAMARKIN, L., C.J. BAIRD & O.F.X. ALMEIDA. 1985. Melatonin: a coordinating signal for mammalian reproduction. Science **227:** 714–720.
4. ARENDT, J. 1995. Melatonin and the Mammalian Pineal Gland. Chapman and Hall. London.
5. VRIEND, J. & M. STEINER. 1988. Melatonin and thyroid function. *In* Melatonin: Clinical Perspectives. A. Miles, D.R.S. Philbrick & C. Thompson, Eds.: 92–117. Oxford University Press. Oxford.
6. REPPERT, S.M., D.R. WEAVER, S.A. RIVKEES & E.G. STOPA. 1988. Putative melatonin receptors in a human biological clock. Science **227:** 714–720.
7. MORGAN, P.J., P. BARRETT, G. DAVIDSON, *et al.* 1992. Melatonin regulates the synthesis and secretion of several proteins by pars tuberalis cells of the ovine pituitary. J. Neuroendocrinol. **4:** 557–563.
8. WEAVER, D.R., J.H. STEHLE, E.G. STOPA & S.M. REPPERT. 1993. Melatonin receptors in human hypothalamus and pituitary: implications for circadian and reproductive responses to melatonin. J. Clin. Endocrinol. Metab. **76:** 295–301.
9. REITER, R.J. 1980. The pineal and its hormones in the control of reproduction in mammals. Endocr. Rev. **1:** 109–131.
10. ARENDT, J. 1986. Role of pineal gland and melatonin in seasonal reproductive function in mammals. Oxford Rev. Reprod. Biol. **8:** 266–320.
11. COHEN, M., J. JOSIMOVICH & BRZEZINSKI. 1996. Melatonin: From Contraception to Breast Cancer Prevention. Sheba Press. Potomac, MD.
12. REGELSON, W. & W. PIERPAOLI. 1987. Melatonin, a rediscovered antitumor hormone? Its relation to surface receptors, steroid metabolism, immunologic response, and chronobiologic factors in tumor growth and therapy. Cancer Invest. **5:** 379–385.

13. MARCHETTI, B., F. GALLO, Z. FARINELLA, et al. 1998. LHRH is a primary signaling molecule in the neuroimmune network. Ann. N.Y. Acad. Sci. **840:** 205–248.
14. PIERPAOLI, W., A. DALL'ARA, E. PEDRINIS & W. REGELSON. 1991. The pineal control of aging. The effects of melatonin and pineal grafting on the survival of older mice. Ann. N.Y. Acad. Sci. **621:** 291–313.
15. PIERPAOLI, W. & W. REGELSON. 1994. The pineal control of aging. The effects of melatonin and pineal grafting on aging mice. Proc. Natl. Acad. Sci. USA **91:** 787–791.
16. PIERPAOLI, W., A. DALL'ARA, B. MARCHETTI, et al. 1997. Circadian melatonin and young-to-old pineal grafting postpone aging and maintain juvenile conditions of reproductive functions in mice and rats. Exp. Gerontol. **32:** 587–602.
17. NOWAK, R., I.C. MCMILLEN, J. REDMAN & R.V. SHORT. 1987. The correlation between serum and salivary melatonin concentrations and urinary 6-hydroxymelatonin sulphate excretion rates: two noninvasive techniques for monitoring human circadian rhythmicity. Clin. Endocrinol. **27:** 445–452.
18. AHMED-EBBIARY, N.A., E.A. LENTON & I.D. COOKE. 1994. Hypothalamic–pituitary aging: progressive increase in FSH and LH concentrations throughout the reproductive life in regularly menstruating women: Clin. Endocrinol. **41:** 199–206.
19. REITER, R.J. 1998. Melatonin and human reproduction. Ann. Med. **30:** 103–108.
20. FERNANDEZ, B., J.L. MALDE, A. MONTERO & D. ACUNA. 1990. Relationship between adenohypophyseal and steroid hormones and varias in serum and urinary melatonin levels during the ovarian cycle, perimenopause and menopause in healthy women. J. Steroid Biochem. **35:** 257–262.
21. NICOLAU, G.Y., E. HAUS, L. PLINGA, et al. 1992. Chronobiology of pituitary-thyroid functions. Rom. J. Endocrinol. **30:** 125–148.
22. MAES, M., K. MOMMEN, D. HENDRICKX, et al. 1997. Components of biological variation, including seasonality, in blood concentrations of TSH, FT3, FT4, PRL, cortisol, and testosterone in healthy volunteers. Clin. Endocrinol. **46:** 587–598.
23. BELLIPANNI, G. et al. 2001. Effects of melatonin in perimenopausal and menopausal women: a randomized and placebo controlled study. Exp. Gerontol. **36:** 297–310.

Modulatory Factors of Circadian Phagocytic Activity

MONICA L. HRISCU

Institute of Public Health, 6 L. Pasteur St., 400349 Cluj-Napoca, Romania

ABSTRACT: Basal phagocytosis of neutrophils, a crucial component of non-specific immunity, is an eminently circadian parameter. In mice and rats, rates of phagocytosis peak in the second half of the dark span. A lipopolysaccharide-induced phagocytic response challenged in this period appears the most significant in amplitude and duration. Neonatal administration in rats of the neurotoxic agent monosodium glutamate, which induces massive destruction of the arcuate nucleus, suppresses the phagocytic response and moderately inhibits basal phagocytosis. Physiological phagocytosis *in vivo* appears to depend on the presence of the nocturnal melatonin surge. Functional pinealectomy, achieved in rats by a 2-week exposure to constant light, lowered the average circadian level of phagocytosis, damped the characteristic rhythm of neutrophil adherence, and decreased the neutrophil count and the amplitude of its circadian oscillation. In an *in vivo* study, adult rats were given alcohol intraperitoneally (0.1 mL ethanol/kg body weight, 1:10 in saline solution), alone or co-administered with melatonin (1 mg/kg body weight), for 16 days, once a day at 20:00 h. Alcohol-treated animals displayed a drastically depressed and flattened phagocytic curve, a marked decline of the adherence ability, and a reduction of the mean circadian neutrophil and lymphocyte count. Addition of melatonin significantly increased circadian values of phagocytosis, restored adherence, and prevented the numeric depletion of lymphocytes induced by alcohol. In correlation with other experimental evidence, our data speak for a physiological role of melatonin in upkeep of neutrophil phagocytosis and hint towards the existence of several pineal-hypothalamic pathways regulating different components of phagocytosis *in vivo*.

KEYWORDS: neutrophils; phagocytosis; phagocytic response; adherence; circadian rhythms; monosodium glutamate; constant light; melatonin; SCN; AHA; arcuate nucleus

INTRODUCTION

In the mammalian circadian system, the main external stimulus is light. Photic information, received by the retina, is transmitted via the retino-hypothalamic tract to the central oscillator of the system, the suprachiasmatic nucleus (SCN).[1] Several effector pathways originating in the internal clock head towards the neighboring hypothalamic nuclei.[2,3] The dorso-caudal periventricular efferents reaches the

Address for correspondence: Monica L. Hriscu, Institute of Public Health, 6 L. Pasteur St., 400349 Cluj-Napoca, Romania. Voice: +40-727-305103; fax: +40-264-593112.
mlhriscu@yahoo.com

Ann. N.Y. Acad. Sci. 1057: 403–430 (2005). © 2005 New York Academy of Sciences.
doi: 10.1196/annals.1356.032

paraventricular nucleus and, from here, via a monosynaptic pathway, the sympathetic preganglionic neurons in the spinal intermediolateral nucleus, which innervates the pineal gland through the superior cervical ganglion.[4] The environmental light-dark cycle is transduced to an internal, neuroendocrine signal, represented by the rhythmic secretion of pineal melatonin, which occurs with no exceptions during the dark period. Melatonin acts upon specific effector areas of the brain, such as the hypophyseal pars tuberalis (PT). It also acts on the retina, modulating its photosensitivity, and upon the SCN as well. The density of the melatonin receptors, both in the PT and in the SCN, exhibits a circadian rhythm that is downregulated by the melatonin concentration.[5,6]

Melatonin synthesis exhibits a prominent circadian rhythm, low during day and high at night. The rhythm is endogenous, persisting in animals kept under constant dark. The fact that melatonin secretion is linked in a mandatory manner to the dark span makes "night" a meaningful factor with respect to all immune, endocrine, and other rhythms that might depend on the endogenous melatonin surge.

Melatonin has well-documented immunostimulatory properties in many compartments of the immune function. Nevertheless, its effects may differ in relation to variables such as species, sex, and age of the organism; maturation and physiological status of the immune system; activation pathway; assessed parameter; dose and administration route; circadian timing; season.

Classic research has proven that melatonin enhances specific immune activity. Administered in the evening in mice, melatonin increases the number of plaque-forming cells during the primary response against sheep red blood cells[7,8] and enhances the formation of IgM and IgG antibodies.[9] It stimulates the responsivity of NK cells to γ-IFN and IL-2 in human subjects,[10] while chronic administration of melatonin increases both the number of phenotypically identifiable NK cells and their spontaneous activity.[11,12] Melatonin activates human monocytes in a dose-dependent manner, being able to induce cytotoxicity, IL-1, IL-6, and TNF secretion.[13,14]

Regarding the effect of melatonin upon microphagocytes, a fair amount of data suggests that it would generally have a stimulatory role. Indolic precursors, such as tryptophan and serotonin, can activate phagocytosis in unicellular organisms[15] and in mammalian macrophages.[16] A number of *in vitro* studies revealed that pharmacological doses of melatonin exert a concentration-dependent activation effect on ring dove (*Streptopelia risoria*) heterophils,[17] with enhanced phagocytosis for latex beads. Also, in ring dove heterophils, phagocytosis appears to be the highest during the nocturnal melatonin surge.

FACTS, PREMISES, HYPOTHESES

While just three decades ago the scientific world had not been introduced to the concept of neuroimmunomodulation,[18] an impressively large body of experimental data has been reported recently that confirm the existence of interactions at multiple levels among the nervous, immune, and endocrine systems. The existence of such interactions can explain how emotional status, stress, environmental conditions, and other factors can alter individual reactivity in organisms confronted with infections, autoimmune diseases, neoplasia, or pathophysiological changes associated with aging.

TABLE 1. Premises of and questions addressed by experiments described here

Area of knowledge	Premises	Selected references	Questions addressed
Phagocytic activity of neutrophils (PHAG)	Basal PHAG is a periodic function, with highest values after dark, in guinea pigs, humans, ring doves	19, 20, 21, 22	Does phagocytosis oscillate on a circadian basis in nocturnal species, too? According to what pattern?
	The phagocytic response displays marked circadian variability	Personal observation	Is the phagocytic response related in a meaningful manner with the circadian moment of antigen encounter?
Melatonin (MLT)	The pineal gland and/or MLT have a general stimulatory effect on various immune compartments, including PHAG, *in vivo* and *in vitro*, physiologically and pharmacologically	23, 24, 25, 26	Is phagocytosis dependent (as in level and/or variation pattern) on the endogenous melatonin secretion?
			Is exogenous melatonin able to support neutrophil activity in conditions of concomitant low-dose alcohol exposure?
Hypothalamus	The anterior hypothalamic area (AHA) controls the basal level of PHAG	19, 27	Is hypothalamic modulation of PHAG under pineal/melatonin control?
	The tubero-mammillary area (TM) controls the circadian rhythm of PHAG		
	TM and the arcuate nucleus (ARC) are involved in triggering the phagocytic response	19, 28	
	SCN, AHA, ARC, *pars tuberalis* are areas rich in melatonin receptors	29	
	SCN receives afferences involved in rhythm entrainment, via the RHT	1, 30	
	SCN and ARC are linked bidirectionally	31, 32	
	Monosodium glutamate (MSG) is neurotoxic for both ARC and RHT	33, 34	

The hypothalamus appears to play a modulatory and integrative role in most neuroendocrine-immune interactions. Several hypothalamic areas and more than one downstream system have been proven to be involved in controlling neutrophil functions; however, many aspects of *in vivo* phagocytosis in rodent neutrophils have not yet been addressed. TABLE 1 summarizes the questions experiments described in the present review have sought to answer, as well as some of the scientific facts they were based on.

GENERAL METHODS

Animals and General Experimental Conditions

All animals (mice and rats) employed in the experiments described herein have been purchased from the laboratory rodent breeding facility of the University of Medicine and Pharmacy in Cluj-Napoca, Romania. They were allowed to adapt to the new environment for at least two weeks before beginning the experiment. Unless otherwise specified, the animals were kept under natural lighting conditions and controlled temperature ($24 \pm 1°C$). Food and water were given once a day at $12:00 \pm 1$ h. During the experiments, manipulations in the dark span were carried out under dim red light. All assessments were made from venous blood obtained from the tail. Blood sampling and microscopic assessments were always performed by the same operators.

Basal Phagocytic Activity

The micromethod used, described in detail elsewhere,[35] employs only 40 μL of whole blood. Briefly, heparinized whole blood was mixed with a living *Escherichia coli* suspension and incubated for 30 min at 37°C, under continuous slow stirring. The blood-bacteria mixture was spread, dried, and fixed immediately. Neutrophils on the Giemsa-stained smears were examined microscopically. Phagocytic activity was expressed as the number of bacteria engulfed by 100 neutrophils (phagocytic index, PI) and also as the percentage rate of neutrophils having engulfed at least one bacterium (phagocytic percentage, PP).

The Phagocytic Response

The phagocytic response represents the increase as an absolute number and differential count of the blood neutrophils, as well as the augmentation of their phagocytic ability *in vitro* (increase of the phagocytic index and of the percentage rate of phagocytosing cells), following an *in vivo* antigenic challenge.

In our experiment (carried out in mice), the phagocytic response was induced by the intraperitoneal injection of 10 μg *E. coli* lipopolysaccharide per 25 g body weight mouse (1 mg lyophilized *E. coli* LPS 055:B5, Sigma Chemical Co., was reconstituted with 5 mL sterile saline. The injected volume was 0.05 mL). In the experiment assessing the effect of monosodium glutamate, the phagocytic response in mouse pups was induced by the intraperitoneal injection of 0.05 mL of diluted bacterial suspension (*E. coli* in HBSS, 0.1×10^8 germs/mL).

Adherence of Neutrophils

Neutrophil adherence to nylon fibers was assessed by means of a microassay system, employing fresh whole blood.[36] After a first neutrophil count, capillary microhematocrit-type tubes containing a standardized amount of nylon fiber were filled with heparinized blood and incubated for 30 min at 37°C, under permanent slow axial rotation. The percentage adherence rate was given by the mean difference in the neutrophil count in the blood sample before and after incubation, all assessments being performed in duplicate. According to MacGregor and colleagues,[37] the process of neutrophil adhesion to nylon fibers is similar to that of neutrophil adhesion to the vascular wall.

Statistics

Results were expressed as average per group ± standard deviation. Means per group or means at different time points were compared using the Student's t test (statistical significance threshold $P<.05$). For every assessed rhythm, a single cosinor analysis[38] was employed to determine rhythm-adjusted mean (MESOR) and amplitude (A, difference between average and highest/lowest value) and to define acrophase (Φ) and statistical validity of the rhythms (P).

CIRCADIAN RHYTHM OF BASAL PHAGOCYTOSIS IN MICE

Several immune and hematological parameters display highly reproducible circadian oscillations. One example is the number of cells in the peripheral blood, in humans and animals.[39–43] A large number of biological rhythms are correlated with the activity type of the species, diurnal or nocturnal, respectively.[44] Many immunohematological rhythms peak during the resting period—at night in humans and during the day in mice. Such parameters are the total leukocyte number, the peripheral lymphocyte and eosinophil count and the T helper:T suppressor ratio.[44–46] Other rhythms display a chronostructure that appears to be different in humans and mice. For instance, circulating neutrophils and monocytes reach their acrophase in humans in the second half of the active span, while in mice in the first half of the resting period.[45]

The phagocytic activity also undergoes circadian variations. The reticuloendothelial system in CBA mice also displays rhythmical activity, with maximal values at the end of the dark span.[47,48] The first proof that phagocytosis of neutrophils also oscillates according to a circadian pattern was noted in guinea pigs (*Cavia porcellus*), where the phagocytic acrophase occurs soon after the light-dark transition.[19,20] A similar rhythm in humans was reported.[21] As both humans and guinea pigs are diurnal, respectively equivocal species, their pattern could not be simply extrapolated in nocturnal laboratory rodents, such as mice and rats. We attempted therefore to study the basal circadian rhythm of neutrophil phagocytosis in mice.[49]

Animals and Experimental Design

The study was carried out on NMRI mice, young healthy males of 20–25 g each. Nine animals, housed three per container, were kept under a natural light-dark alternation of 12:12 ± 0.5 h, in a period of the year close to the autumnal equinox. Assessment of the phagocytic activity was made over a few days, covering the following time points: 06:00, 09:00, 12:00, 15:00, 18:00, 21:00, 24:00, and 03:00. Blood was drawn only once a day from each animal to avoid excessive strains caused by blood loss and manipulation.

Results

The phagocytic index (FIG. 1) displayed a clear-cut circadian curve. Phagocytosis began to rise significantly after dark onset and attained its highest values at about 03:00 (3.56 h). A clear-cut decline was noticed after 06:00, with low values during the morning. Trough values were met at 15:00. Rhythm parameters, as given by the single cosinor analysis, were: MESOR = 153.40 ± 3.40; A = 46.04 ± 4.81; $\Phi = 3.56$ h ± 0.03; $P = .001$.

T-test values for each time point vs. all the other points appeared to be significant only for the 03:00 h value: $P_{(24:00 \text{ vs. } 03:00)} = .0004$; $P_{(03:00 \text{ vs. } 06:00)} = .0072$. Other statistically significant differences were registered between 09:00–12:00 h (decrease) and 21:00–24:00 (increase). These data suggest that the acrophase has a rather peaked shape, with both the after-dark increase and the decrease during the morning hours being fairly abrupt.

The percentage rate of phagocytosing neutrophils (PP) and the phagocytic efficiency (PE, calculated as the PI:PP ratio) displayed temporal structures similar to that of the phagocytic index, only with earlier acrophases (data not shown).

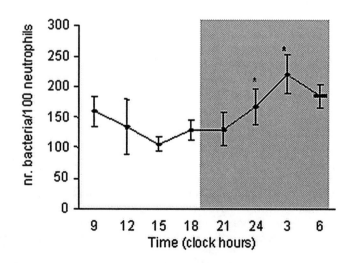

FIGURE 1. The phagocytic index in mice over a 24-h period. Each value is the mean ± SD of nine determinations. The *shaded band* corresponds to the hours of darkness. *$P < 0.05$ against values at all other time points. (From Hriscu.[55] Modified and reproduced by permission.)

EVOLUTION OF THE PHAGOCYTIC RESPONSE CHALLENGED AT DIFFERENT CIRCADIAN TIMES

Early studies concerning the non-specific immune response[50] showed that direct humoral stimulation of the hypothalamus by microbial antigens induces a phagocytic response, more intense in immunized than in naïve animals, which peaks 4–6 hours after the antigen encounter. In immunized animals, the response fades out in 24 hours, while nonimmunized animals retain increased phagocytic activity values for a longer period of time.

Bilateral electrolytic destruction of the anterior endocrinotropic hypothalamus in rats is followed by a diminution of the primary specific immune response,[51] while triggering of the secondary immune response is under control of the tubero-mammillary area, at the same time depending on the antigen's nature (erythrocytic, bacterial or viral). Bilateral lesions in this area induce effects that range from attenuation of the response against bacteria and bacterial products to a complete suppression of the phagocytic response in animals immunized against the influenza virus. In conclusion, the nonspecific immune response, as well as the specific one, appears to undergo hypothalamic modulation.

A large body of experimental data suggests the existence of circadian variation of the immune reactivity. Mice are far more susceptible to the *E. coli* endotoxin in the middle of the light span (12:00) than at midnight.[52] The rate of both nonspecific and antigen-induced leukocyte migration differs in relation with the time of the day when stimulation occurs;[44] the same is true for the inflammation caused by the oxazolone-induced immune response,[53] for the infectious and noninfectious chronic granulomatous inflammation,[54] and others. In the experiment described herein,[55] we attempted to verify whether a relationship could be established between the circadian moment of antigenic stimulation and the evolution of the subsequent phagocytic response.

Animals and Experimental Design

NMRI mice, young males weighing 20–25 g, were kept for 14 days in a controlled light-dark 12:12 environment (lights on between 09:00–21:00). The phagocytic response was assessed in five animals at each of the following time points: 10:00, 16:00, 22:00, and 04:00. In order to induce a primary response, different (non-immunized) animals were used each time. The phagocytic index was tested immediately before injecting the bacterial antigen (baseline value) and 4, 8, and 24 h after. The experiment took place in July.

Results

The LPS injection delivered at 10:00 or at 16:00 induced a phagocytic response whose amplitude at 4 h did not exceed 10% of the initial value. This early increase was followed by a depression at 8 h, which at the 16:00 time point went below the baseline value. At 24 h, phagocytosis had recovered, the PI exceeding slightly the 4-h level. All these variations lacked statistical meaning.

The antigenic challenge at 22:00 caused an increase of about 30% ($P = 0.049$) of the PI at 4 h. A marked depression ($P = 0.011$) occurred 8 hours after the challenge,

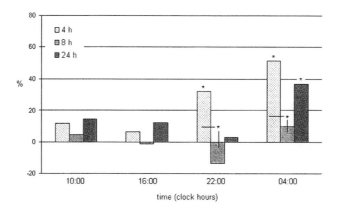

FIGURE 2. Evolution of the phagocytic response challenged at different time points. Percentage difference of average values 4, 8, and 24 hours after antigenic stimulation are shown against the phagocytic index before stimulation (*abscissae*). *$P < 0.05$. (From Hriscu.[55] Modified and reproduced by permission.)

the PI falling below the initial value. At 24 h, the phagocytic activity had risen again to a level slightly above baseline.

The LPS injection at 04:00 induced more than a 50% increase of the initial value (51.8%; $P = 0.044$ at 4 h). After a depression at 8 h, during which the level still retained an almost 10% elevation, the PI rose again, at 24 h exceeding the initial value by 36.8% ($P = 0.019$). FIGURE 2 illustrates the evolution of the phagocytic response at the four different experimental points.

EFFECTS OF NEONATAL ADMINISTRATION OF MONOSODIUM GLUTAMATE ON THE PHAGOCYTIC ACTIVITY IN MICE

The endogenous circadian clock is entrained by photic signals received by the retina and conveyed through the direct pathway of the retino-hypothalamic tract (RHT).[30] Experimental disruption of this neural pathway blocks the entrainment to the light-dark cycle.[56–59]

Glutamate (MSG) acts as a neurotransmitter in the main excitatory pathways of the isocortex, including the retino-hypothalamic tract (RHT). Light impulses received by the retina determine glutamate release from the synaptic terminals in the RHT, which influence the neurons of the circadian clock.[60] MSG is neurotoxic if administered in early developmental stages. One of its reported effects is a massive destruction of the arcuate nucleus, affecting as much as 80–90% of the neurons in this structure.[33] It also severely damages the primary visual pathway and the RHT.[34] On the other hand, it was shown that electrolytic or MSG-induced destruction of the arcuate nucleus inhibits NK activity and markedly reduces the activity of suppressor cell in spleen and bone marrow,[33,61,62] while a partial electrolytic lesion of the arcuate nucleus depresses the phagocytic response.[28] According to Belluardo and colleagues,[33] immune activity in MSG-treated mouse pups tends to recover after the first month of life.

The present study[35] was designed to investigate the influence of MSG administration on basal phagocytosis and on the phagocytic response, in order to elucidate the role played by the arcuate nucleus in the neural control of these parameters.

Animals and Experimental Design

The study was carried out on 23 newborn NMRI mice. The pups originated from three different nests of 5, 10, and 8 animals, respectively. Each mother with her litter was housed separately and fed *ad libitum*, the pups being reared naturally.

MSG-treated pups ($N = 15$) received 2 mg/g body weight MSG (Carl Roth GmbH, Karlsruhe, Germany, 6% solution in normal saline). Control animals ($N = 8$) received an equivalent volume (30 µL/g body weight) of vehicle. Both were administered once a day in intramuscular injection, for 7 days, starting on the day after birth. On day 32, half of the pups within each group were sacrificed by carotid section and basal phagocytic activity was assessed at 11.00 h. The phagocytic response (see METHODS) was tested the following day in the rest of the animals. The bacterial suspension was delivered at 14:00 h. The phagocytic activity (PI and PP) was assessed before stimulation and 4 hours after. A differential neutrophil and lymphocyte count was also performed.

Results

Following the neonatal MSG treatment, basal phagocytic activity decreased by 27% ($P < 0.1$) (FIG. 3). The percentage rate of phagocytosing neutrophils (PP) also showed a 7% diminution as compared to control animals. The lymphocyte count appeared significantly lower in MSG-treated animals (39% as compared to 59% in controls).

In vehicle-receiving animals, the antigenic challenge induced a marked increase of all parameters that characterize the phagocytic response. The neutrophil count increased from a baseline value of 38% to 65% ($P < 0.5$); both the PI and the PP rose significantly over the value before stimulation (70% and 34%, respectively, $P < .01$).

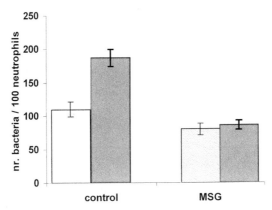

FIGURE 3. The phagocytic index during basal phagocytosis (*light rectangles*) and during the phagocytic response (*dark rectangles*), in controls and in MSG-treated mouse pups.

Injection of the bacterial suspension in MSG-treated animals elicited almost no response: the PI rose by 7.5% and the PP by 3.8% (FIG. 3), while the neutrophil count declined from 59% to 49%. It is noteworthy that only three of the seven MSG-treated pups that received bacterial suspension survived, while in the control group the survival rate was of 100%. Also, the average weight of MSG-treated pups constantly exceeded that of control animals (10.2 g vs. 7.2 g, respectively, in the 32nd day).

PHAGOCYTOSIS AND ADHERENCE OF NEUTROPHILS IN RATS SUBMITTED TO CONSTANT LIGHT

Nocturnal melatonin secretion can be suppressed by exposure to constant light, which induces a functional inhibition of the pineal gland.[5,6] In rats, one minute of fluorescent white light of 150 lux intensity is enough to induce the rapid decline in the activity of the limiting enzyme N-acetyl transferase (NAT), thus decreasing plasma melatonin to low day levels.[63] Studies carried out on different species proved that, following suppression of the daily dark span, circulating melatonin remains at almost undetectable concentrations over the whole circadian cycle.[64-66]

We attempted to find out whether functional pinealectomy accomplished through constant light exposure affects neutrophils (cell count, phagocytosis, adherence, and their circadian dynamics, respectively) and to discern to what extent their rhythmic oscillation is attributable to a normal light-dark cycle and consequently to the presence of nocturnal melatonin.[36,67,68]

Animals and Experimental Design

The experiment employed young male Wistar rats of 180–200 g. The LD group ($N = 4$) was maintained under an artificial light-dark 12:12 regimen (1500–2000 lux white fluorescent light, on between 07:00-19:00; 5–15 lux in the dark phase). The LL group ($N = 4$) was submitted to constant light with the same characteristics as in the light phase of the LD group.

The experimental conditions were applied for 15 days before and ongoing during the testing period. Blood was collected at four time points of the 24-h cycle (10:00, 16:00, 22:00, and 04:00). Assessed parameters were phagocytic index (PI), neutrophil adherence, total leukocyte count, absolute and differential neutrophil, and lymphocyte count. The study was carried out in December.

Results

In the LD group, the PI exhibited a circadian rhythm with the acrophase at about 04:00 and the nadir at about 10:00. The PI remained low over the whole light period and in the first dark hours; a marked increase ($P = 0.03$) occurred between 22:00 and 04:00. Characteristics of the rhythm were MESOR = 329.6 ± 15.00; amplitude 56.74 ± 21.2; acrophase 2.39 h; $P = 0.058$.

In the LL group, the circadian structure (timing of acrophase and nadir) of the PI was largely preserved. The amplitude of the oscillation increased by 25%; nevertheless, the phagocytic level appeared depressed, with the MESOR of only 60% of that

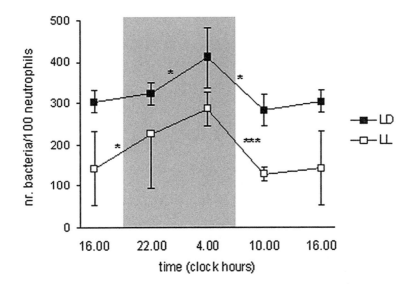

FIGURE 4. The circadian phagocytic index in rats kept under 12:12 light-dark conditions (LD) and under constant light (LL). *$P < 0.05$; ***$P < 0.001$ between adjacent values. (From Hriscu.[55] Modified and reproduced by permission.)

in LD. In LL animals, the circadian curve assumed an ascending slope earlier in the afternoon: a significant increase of the phagocytic activity occurred between 16:00 and 22:00 and the acrophase was advanced (1.43 h) (FIG. 4).

The ability of neutrophils to adhere appeared to be a rhythmic function with two circadian maxima: a main peak at 10:00 and a second lower one (of half the amplitude) at 22:00. The rhythm had a MESOR of 64.7±8.1 and a double amplitude of the higher peak of 37.8±16.5. Calculated acrophase was at 11.1 h.

Constant light induced a slight decrease of the adherence ability (MESOR about 10% lower than in LD animals). The adherence was significantly depressed both during the circadian period of high values, at 10:00 ($P = 0.01$), and in the evening at 22:00 ($P = 0.009$). The acrophase occurred earlier (9.5 h) and the amplitude diminished in animals exposed to constant light. The secondary increase at 22:00 almost disappeared, the rhythm losing its characteristic shape (FIG. 5).

In LD rats, the peripheral leukocyte and neutrophil count peaked in the morning hours (10:00), remained high over the light period, decreased towards evening, reaching the nadir at 22.00, and began to rise again in the following dark hours. Continuous light exposure lowered the amplitude and the MESOR of both rhythms by about 25%. In both, however, the timing of overall high values remained unaffected (10:00) (FIG. 6). The percentage of circulating neutrophils did not suffer marked alterations, except for a general decrease of about 10% (data not shown).

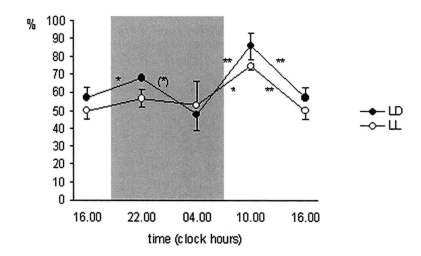

FIGURE 5. The percentage adherence rate of neutrophils over a 24 h-cycle, in rats kept under 12:12 light-dark conditions (LD) and under constant light (LL). *$P < 0.1$; *$P < 0.05$; **$P < 0.01$ between adjacent values. (From Hriscu.[55] Modified and reproduced by permission.)

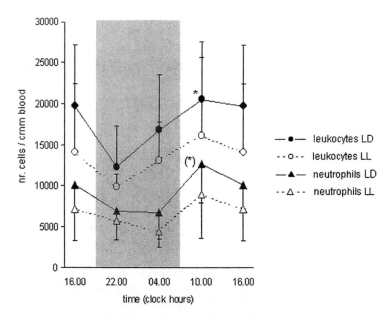

FIGURE 6. Circadian variation of the leukocyte and neutrophil count in rats submitted to 12:12 light-dark alternation (LD) and to constant light (LL). *$P < 0.1$; *$P < 0.05$ against the lowest value. (From Hriscu.[55] Modified and reproduced by permission.)

EFFECTS OF MELATONIN ADDITION ON THE NEUTROPHIL
ACTIVITY IN ALCOHOL-TREATED RATS

Chronic ethanol consumption in humans is frequently associated with an altered immune status.[69] Besides other pathological changes, alcohol inhibits certain neutrophil functions, such as chemotaxis and reactive oxygen species production,[70,71] mobilization of membrane receptors from the granules,[72] as well as leukotriene production and elastin aggregation and release *in vitro*.[73] It has been shown that four months of alcohol exposure in rats induce cytochemical and ultrastructural changes of the neutrophil granules, including their reduction, redistribution, and atypical accumulation, as well as appearance of autophagic vacuoles in the cytoplasm of neutrophil leukocytes.[74] These findings might thus provide a morpho-biochemical basis for alcohol-induced suppression of neutrophil functions. A decreased phagocytic ability of granulocytes from human subjects with chronic ethanol consumption was also reported.[75]

Many of the pathologic effects of alcohol exposure were accounted for by free radical formation and/or by depletion of the antioxidant systems. *In vitro* data indicate a significant cytotoxic effect in granulocytes incubated with alcohol.[76] Melatonin acts as a potent free radical scavenger,[77,78] thus displaying a protective effect against cellular and tissue lesions induced via this mechanism.[79,80] The experiment allowed us to investigate the effects *in vivo* on neutrophil parameters of a low-dose ethanol exposure, without or with co-administration of a pharmacological dose of melatonin in rats.[81]

Animals and Experimental Design

The experiment was carried out on Wistar rats, young males weighing 180–220 g. All animals, housed four per cage, were maintained under an 12:12 artificial LD regimen during the whole experiment (1500–2000 lux white fluorescent light, on between 09.00 and 21.00; 5–15 lux in the dark phase). The study was carried out in April.

Treatment Groups

In the ALC group ($N = 4$), animals received intraperitoneally 0.1 mL/100 g body weight ethanolic saline 1:10, representing an ethanol dose of 0.1 ml/kg body weight. In the ALC + MLT group ($N = 4$), melatonin (*N*-acetyl 5-methoxytryptamine, Sigma Chemical Co.) was dissolved (1 mg/mL) in the same ethanolic saline administered to the ALC group. The rats were given the same volume of the final solution, representing a dose of 100 µg/100 g body weight melatonin in addition to the 0.1 mL/kg body weight ethanol. The treatment was administered under identical conditions and at the same time as for the ALC group. Results in the ALC- and ALC-MLT–treated groups were compared against those in LD (untreated) animals in the previous experiment, which served as control.

The treatment was administered daily at 20.00 h ± 10 min (1 h before light offset), 16 days before and ongoing during the 4 days of sampling. Blood was drawn at 10:00, 16:00, 22:00, and 04:00 for all groups, over four consecutive days, one time point per day. The amount of blood drawn once from one animal was of approximately 100 microliters.

Results

Ethanol administration induced a severe depression of phagocytosis over the whole circadian span (MESOR decreased by more than 60%) and especially at night, in the hours when physiologic acrophase is noticed in control animals. The nocturnal peak of PI disappeared, the highest values occurring in ethanol-treated animals at 16:00 (calculated acrophase at 18.1 h). The circadian rhythm of phagocytosis in rats exposed to ethanol appeared thus deeply altered, losing statistical validity, as proven by single cosinor rhythmometry ($P = 0.1$).

Addition of melatonin was followed by an almost 50% increase of the mean level of phagocytosis (MESOR) as against the alcohol-treated group. The circadian pattern of the rhythm tended to recover ($P = 0.02$), the amplitude was restored and even increased, and the acrophase shifted closer to its physiologic timing (20.34 h ± 0.4). The highest value ($P = 0.005$ against the lowest) occurred 2 hours after injection of the alcohol-melatonin solution and was significantly higher ($P = 0.02$) than the corresponding value in ALC animals. Nevertheless, the phagocytic activity remained low when compared to untreated animals (MESOR reaching only 55% of that in controls) (FIG. 7).

Alcohol induced a marked decline in adherence (MESOR 56.5% of that in controls). The mean peak lost amplitude, becoming lower than the one at 22:00; but the characteristic circadian pattern was preserved.

In ALC+MLT animals, adherence was largely restored, recovering to values very close to those in controls and exceeding them significantly at 04:00 ($P = 0.009$). The biphasic pattern changed thus to a continuous curve with high values in the hours

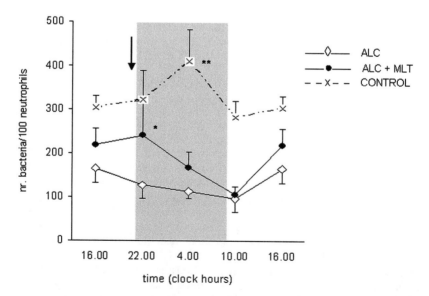

FIGURE 7. Circadian PI in untreated rats and after 16 days ethanol and ethanol+melatonin treatment. The *arrow* indicates the time of injection. *$P < 0.05$; **$P < 0.01$ against the lowest value.

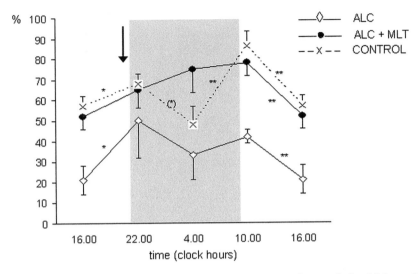

FIGURE 8. Circadian adherence of neutrophils in untreated rats and after 16 days ethanol and ethanol+melatonin treatment. The *arrow* indicates the time of injection. *$P < 0.1$; *$P < 0.05$; **$P < 0.01$ between adjacent values.

following the alcohol-melatonin injection and during the whole dark span, when control animals experience a physiologic decrease (FIG. 8).

Ethanol treatment induced a depression of about 40% of the mean circadian leukocyte count. A marked, though transitory, leukocytosis ($P = 0.03$ and $P = .04$ against values at 16:00 and at 10:00, respectively) occurred 2 hours after alcohol injection, the leukocyte count remaining low over the rest of the nycthemere.

Co-administration of melatonin was followed by an increase of the trough values during the light phase (which nevertheless remained below the corresponding levels in controls) and by a massive and prolonged nocturnal leukocytosis accounting for a reversed circadian rhythm, with highest values at night (FIG. 9).

In control animals, neutrophils and lymphocytes (both their absolute counts and the differential distribution) oscillated in reverse phase: neutrophils reached the maximum in the first half of the light span, while lymphocytes rose in the afternoon and again at night. Alcohol induced an important numeric and also proportional depression of neutrophils, the absolute count going as low as 50% of that in controls. Two hours after alcohol administration, a short-lived neutrophilia was noticed, during which the number of blood neutrophils became twice as high as the average value at the rest of the time points.

Association of melatonin was followed by a striking increase of the neutrophils at 22:00. As with total leukocytes, that reaction showed a marked individual variability ranging between 100–300% of the initial value, and thus statistically less powerful ($P = 0.008$ and $P = 0.02$ against the preceding, respectively the following time point) than suggested by the amplitude of the peak. Except for this transient surge, melatonin did not alter the neutrophil count, which remained as low as in the ALC group (FIG. 10 A, B).

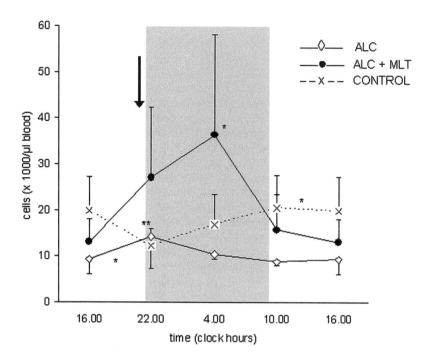

FIGURE 9. Circadian rhythm of the leukocyte count in untreated rats and after 16 days ethanol or ethanol+melatonin treatment. The *arrow* indicates the time of injection. *$P < 0.05$; **$P < 0.01$.

Not only neutrophils were affected by the alcohol treatment, but lymphocytes as well. Their mean circadian value decreased by an average of 30% in ALC rats, a decline that reached 45% during the light hours. At 22:00, following the alcohol injection, a slight lymphocytosis could also be noticed.

Co-administered melatonin caused a marked overall increase of the lymphocyte count, trough values (16:00) recovering to those in controls and the rest going far beyond. The lymphocyte percentage also exceeded that in controls, except for the 22:00 time point (FIG. 10 C, D).

DISCUSSION

As in previously studied species, the phagocytic activity of neutrophils in mice and rats appears to be a circadian function. In guinea pigs[19,20] and humans,[21] it attains maximal values soon after the light-dark transition. Rodríguez and colleagues[22] described in ring dove (*Streptopelia risoria*) heterophils a phagocytic rhythm with the acrophase at about 04:00, concurrent with the maximum of plasma melatonin. In mice and rats, the acrophase also occurred in the second half of the dark span (between 02:00 and 04:00). Nocturnal activation of phagocytosis begins

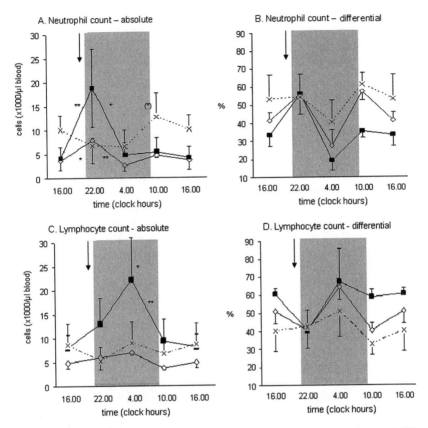

FIGURE 10. Circadian rhythm of neutrophils and lymphocytes (absolute and differential count) in untreated rats and after 16 days ethanol or ethanol+melatonin treatment. The *arrow* indicates the time of injection. *$P < 0.1$; *$P < 0.05$; **$P < 0.01$ between adjacent values. *Dotted line*, control; *open rhombs*, ALC; *solid squares*, ALC + MLT).

after dark with an increase of the percentage of neutrophils displaying phagocytic activity and is followed in the next hours by an increase in the average number of bacteria engulfed by a neutrophil.

Both in humans and guinea pigs, on one hand, and in mice and rats, on the other hand, phagocytosis increases in the last hours of the active period. In humans, the maxima of the neutrophil count and of their phagocytic activity are concurrent, while in mice and rats the two peaks are about 6 hours apart: phagocytosis peaks in the second half of the dark span, while the highest level of circulating neutrophils occurs in the morning hours. It appears that phagocytosis in rodents and perhaps in other groups is linked to the type of activity of a specific species (see the equivocal guinea pig and the diurnal human versus the nocturnal mouse and rat), integrated with the circadian oscillations of the immune system.

Many experimental studies provide evidence that basal phagocytosis is under the control of the anterior endocrinotropic hypothalamus. Electrolytic damage to the

AHA is followed by a severe decrease of the neutrophil enzymes,[27] while its destruction in guinea pigs induces changes of phagocytosis (average decrease by about 30%, circadian variation pattern preserved) that are very similar to the ones we obtained in animals exposed to constant light, thus deprived of endogenous melatonin. Experimental thymectomy was also followed by a comparable decrease of basal phagocytosis.[82]

It is known that AHA exerts its action via the thymus. Electrolytic lesions to the anterior hypothalamus in mice determine an increase in the plasma glucocorticoids, a rapid and severe involution of the thymus, and the decrease of peripheral leukocytes.[83–85] All these effects can be counteracted by melatonin.[86–88] On the other hand, AHA is one of the cerebral areas with relatively abundant melatonin receptors.[29] Our results obtained in LL rats corroborated these data and allow us to hypothesize that pineal melatonin exerts its stimulatory effects on the basal phagocytic activity via AHA, which acts mainly by enhancing the thymic release of phagocytosis-enhancing peptides. Further support for the idea that the effects of light are AHA-mediated is the finding that long-lasting constant light produces morphologic alterations of the neurons in several hypothalamic regions, including the preoptic area of the anterior hypothalamus.[89]

The LPS-induced phagocytic response appears to be different in shape and amplitude in correlation with the time of the day when the antigen encounter takes place, therefore with the phase of basal phagocytosis at the time of challenge. Antigenic stimulation in the light period, corresponding to a decreasing (10:00) or close to minimal (16:00) basal activity, induces a long-lasting and mild enhancement of the phagocytic activity. Whether this low amplitude response is biologically meaningful remains a question. If the challenge is superimposed on the ascendant slope of the basal curve (22:00), the induced response is relatively high, though it turns off soon, moreover being followed by an interval of marked depression. When the challenge coincides with the peak of basal phagocytosis, there follows an immediate and intense phagocytic response that lasts for at least the next 24 hours. The two-phase pattern of the phagocytic response (with values at 8 hours always lower than values at 4 and at 24 hours, respectively) appears to be characteristic.

The results of this study, as well as other experimental data,[36] suggest that the amplitude of the phagocytic response is conditioned simultaneously by the employed antigen dose and by the time of day when the antigenic challenge takes place. A small antigen dose (5 µg LPS/25 g body weight) elicits a phagocytic response only during the high reactivity period, at night (data not shown). Injection of a doubled LPS dose is followed by enhanced phagocytosis at any time of the day, but the induced response is much higher in the dark span than in the light hours. From a practical perspective, these results draw attention to the necessity for careful circadian planning of immunomodulatory interventions, either experimental or therapeutic, as the physiologic oscillations of the immune system may induce very different effects depending on the circadian moment of the intervention.

Attempts to localize superior modulatory centers of phagocytosis[19] have shown that bilateral electrolytic lesions of the tubero-mammillary area (TM) are followed not only by a drastic depression of phagocytosis, but also by the loss of its circadian oscillations, as well as by a suppression of the phagocytic response against bacterial antigens, inhibition that lasts longer if the stimulation occurs during the light period. On the other hand, it was proven more than 50 years ago[90] that electrical stimulation of the TM area induces release from the bone marrow of a phagocytosis-enhancing peptidic factor. The functional relationship that exists between TM and immunity was later confirmed by Forni and colleagues,[61] who reported that radio frequency destruction of the hypothalamic tubero-infundibular region permanently abrogates NK cell activity in mice.

Felegean and colleagues[28] suggested that the arcuate nucleus, located in the TM area, would be involved in a particular manner in triggering the phagocytic response. They showed that limited electrolytic lesions of this structure in rats severely diminished the amplitude of the phagocytic response. Our experimental results corroborate this finding: the extensive damage to the arcuate nucleus inflicted by the neonatal MSG treatment practically blocked the phagocytic response and rendered the organism unable to cope with an antigenic challenge, up to the point where less than 50% of the pups survived LPS administration. Initiation of the phagocytic response appears to depend on the integrity of the arcuate nucleus and, to a lesser extent, on that of other adjacent hypothalamic structures.

The arcuate nucleus, whose chemical damage affects basal phagocytic activity only moderately, might, however, control its circadian rhythm. This hypothesis is supported by the existence of bidirectional connections between neuronal subpopulations in the SCN and in the arcuate nucleus, respectively, which express a variable tonus over the light-dark cycle.[31,32] Also, the arcuate nucleus includes certain neuronal groups exhibiting intrinsic rhythmic activity, synchronized with, but not generated in, the SCN.[91]

We hypothesize that the circadian variation of basal phagocytosis and of the phagocytic response is due to a rhythmically oscillating tonus of the hypothalamic centers (the anterior and preoptic hypothalamus and the tubero-mammillary area, respectively). A fair body of experimental evidence supports this idea. The depression of phagocytosis that occurs in guinea pigs with bilateral lesions of the anterior hypothalamic (AHA) and preoptic area is deeper in the late light and early dark hours, at a time when basal phagocytic activity in control animals attains its circadian trough.[19] In guinea pigs exposed to a 6:6 light-dark alternation, the phagocytic rhythm tends to become synchronized with the lighting regimen, peaking during both dark spans over a 24-h cycle.[19] Not only exposure to constant light, but also olfactory bulbectomy depresses phagocytosis and alters its circadian dynamics.[92] The above data suggest that the intrinsic activity of hypothalamic centers that control phagocytosis is subjected to photic (the light-dark cycle) and olfactory modulatory influences.

There are several indirect and also direct indications that endogenous melatonin would play a role in the neutrophil function, yet the existing information is not always concordant. Melatonin is directly involved in the neutrophil metabolism *in vitro*.[93,94] *In vitro* studies employing pharmacological doses of melatonin (5–100 μM) revealed a dose-dependent activation of the phagocytic function of ring dove heterophils.[17] However, such doses are far above the physiologically available range. It was later shown that melatonin concentrations corresponding to the nocturnal physiological level would also enhance phagocytosis in *Streptopelia* heterophils and decrease the superoxide anion levels.[95] Barriga and colleagues[96] reported similar results in mice. Our data, showing that phagocytosis in rats is higher during the nocturnal melatonin surge and lower in animals kept under constant light, also suggest that melatonin would be physiologically involved in the phagocytic function.

However, the hypothesis that circadian dynamics of basal phagocytic activity depends on nocturnal melatonin secretion is refuted by the finding that the oscillations persist and even show increased amplitude in rats exposed to constant light. In the absence of the time cue represented by dark onset (and not by the melatonin surge, which in rats occurs several hours later), the rhythm tends to lose synchronization, what leads to an advance of the acrophase in LL rats and to enhanced variability of the individual evolutions.

Neutrophil adherence appears to be synchronous not with the rhythm of phagocytosis, but with that of other components of the inflammatory process, such as vascular permeability (higher at midday than at midnight[54]) and chemotactic migration of granulocytes, which is also higher in the morning hours.[97,98] These data suggest that the mechanisms controlling basal phagocytosis *in vivo* are different from those that regulate the inflammatory reaction, adherence included, and point towards the complex functional role of the latter, which exceeds its role in the phagocytic process.

There is little experimental work regarding the adherence of neutrophils, and data concerning the influence of melatonin are particularly scarce. Rodríguez and colleagues[17] found no modification of the nylon fiber adherence of *Streptopelia* heterophils incubated with melatonin, irrespective of the used dose. *In vivo*, Lotufo and colleagues[99] showed that melatonin concentrations within the physiologic nocturnal range exert an inhibitory action upon rolling and adhesion of rat leukocytes. Generally speaking, melatonin is considered to play an anti-inflammatory role, enhancing apoptosis of neutrophils,[100] inhibiting their adhesion to endothelia, and reducing neutrophil infiltration into damaged tissues.[101,102] These effects are mediated by specific receptors.[103] However, it should be considered that various pharmacologically active chemicals, such as sodium salicylate and hydrocortisone succinate, have a different effect on neutrophil adhesion *in vitro* and *in vivo*, being able to inhibit adhesion *in vivo*, but not *in vitro*.[104] Our data show that constant light induces a certain degree of inhibition to neutrophil adhesion, resulting both in a decreased amplitude and in a lower circadian average of the rhythm. It appears thus that endogenous melatonin is needed to maintain the characteristics of basal neutrophil adherence. The inhibition of the adherence capacity of neutrophils in LL rats is probably also attributable, at least in part, to the increased level of plasma cortisol. Constant light activates the hypothalamic-hypophyseal-adrenal axis, stimulating cortisol secretion. Cortisol, prednisone, as well as other steroid and nonsteroid anti-inflammatory drugs, possess a proven inhibitory effect upon neutrophil adhesion *in vivo*, part of their pharmacological action residing probably on this effect.[104]

Saeb-Parsy and Dyball[31] described the existence of a specific neuronal subpopulation in the SCN, with projections to the arcuate nucleus and/or to the supraoptic region. The activity of these neurons reportedly has two maxima, close to the light-dark and dark-light transitions, respectively. While a phase relation between this rhythm and that of neutrophil adherence does by no means imply a causal relationship, it might however hint at a possible control mechanism of adherence involving this specific neuronal group. It is known that the SCN mediates the functions of melatonin related to the circadian regulation of various rhythmic activities, while seasonal changes related to the physiology of reproduction are mediated by receptors in pars tuberalis.[105] The ML1 receptors mRNA expression within the SCN exhibits under LD conditions a robust circadian rhythm with two maxima, having a time disposition very similar to the ones described before.[106] At the same time, melatonin receptor density in the SCN is under control of the light-dark cycle and is specifically increased at day, even in pinealectomized animals.[107] These data might account for the existence of a complex system, melatonin-SCN-arcuate nucleus, responsible for regulating circadian dynamics of neutrophil adherence.

In the experiment regarding the effects of exogenous melatonin in alcohol-exposed rats, there must be stated that the initial purpose of the experiment was to investigate the effect of melatonin in normal rats. Probably the most commonly used vehicle for injectable melatonin, as indicated by many studies, is 1:10 ethanolic saline. This is the vehicle we used, injecting it to animals meant to serve as controls. The obtained data presented a striking difference as in activity levels and circadian patterns of the assessed parameters when compared with untreated animals in previous experiments. It was thus reasonable to assume that the low-dose alcohol administration had affected leukocyte functions. Therefore, we employed as controls untreated rats kept in a controlled light-dark 12:12 environment (the LD group in the previous study). Conditions in the two experiments had been similar with regard to animals, bacterial strain used for assessment of phagocytosis, methods, and environmental (lighting) conditions. One identifiable difference was the time of the year when the two experiments were carried out (December for treatment groups, April for the control group). Circannual variability of the investigated parameters, to the extent to which it is known, has to be taken into account when interpreting the results. However, while changes in number of cells and in activity levels can be attributed to a certain extent to the physiological circannual oscillations, it is unlikely that circadian patterns of the rhythms would have been significantly biased by this factor.

Obviously, in the absence of a perfect sham-treated group, we cannot attribute the entire range of changes occurring in ALC rats to the alcohol itself. For instance, the transiently increased leukocyte number following administration of the ethanolic solution is most probably a nonspecific effect of the injecting procedure. Also, the extremely low phagocytic level at 22:00 could be a summative effect of acute alcohol administration and of handling. Yet, the fact that all parameters display consistently decreased levels over the whole 24-h span time suggests an actual effect of the administered alcohol.

There are certain experimental data indicating functional inhibition of neutrophils by alcohol. A 100 mM dose of ethanol suppresses *in vitro* phagocytosis of *Macacus rhesus* neutrophils, an effect that the granulocyte colony-stimulating factor is able to attenuate.[108] In our experiment, rats receiving 0.1 mL/kg body weight of alcohol for 16 days, without or with melatonin addition, displayed a marked inhibition of the phagocytic activity over the whole circadian span and especially in the 2–8 hours following the injection. The physiologic acrophase, as seen in untreated rats, has been replaced in alcohol-exposed animals by an interval of minimal phagocytic activity, the circadian rhythm of phagocytosis appearing completely disrupted. It appears thus that alcohol-induced inhibition of neutrophil activity may also take place *in vivo*.

Several studies suggested also that alcohol might inhibit neutrophil adherence. MacGregor and colleagues[104] reported that neutrophils in whole blood incubated with ethanol lose their ability to adhere in a direct relationship with the alcohol concentration, and also that 20 mL/kg of a 20% ethanolic solution injected to rabbits induces an acute inhibition of adherence. In our experimental model, 16 days of low-dose alcohol exposure in rats was also followed by a significant around-the-clock depression of the adherence percentage rate, affecting especially the high values.

Aside from its direct action on the phagocytes, alcohol effects are also mediated by suppression of the mechanism involved in granulocyte colony-stimulating factor production.[109–111] Rich experimental evidence attests that melatonin acts in a synergic manner with colony-stimulating factor, intervening in CFU-GM proliferation.[112] Recent data suggest that the hematogenic bone marrow would be the site of a *de novo* melatonin secretion, with intracellular and/or paracrine functions.[113] Whether melatonin exerts its protective and growth-stimulating functions on the myeloid line directly and/or via granulocyte colony-stimulating factor remains subject to further investigation.

It was shown that some of the effects of chronic alcohol administration are thymus and spleen atrophy, alteration of the leukocyte populations in these organs and in periphery[114] and enhanced apoptosis of thymocytes.[115] We here bring evidence for the fact that pharmacologic melatonin doses are able to prevent the peripheral leukopenia induced by alcohol exposure. One of the mechanisms possibly involved

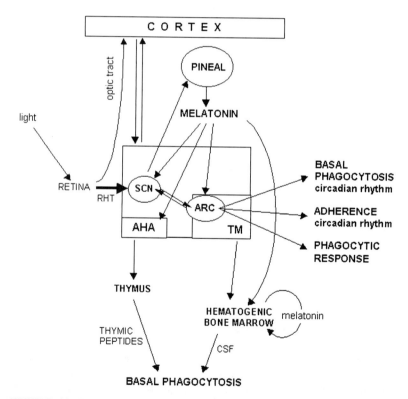

FIGURE 11. Integrative scheme of the central mechanisms modulating circadian neutrophil activity. RHT, retino-hypothalamic tract; SCN, suprachiasmatic nucleus; AHA, anterior hyopthalamic area; TM, tubero-mammillary area; ARC, arcuate nucleus; CSF, colony-stimulating factor. (From Hriscu.[55] Modified and reproduced by permission.)

is inhibition of thymocyte apoptosis, demonstrated *in vivo* and *in vitro*[116] and supported by the receptor (ML2)-mediated antiapoptotic role of melatonin, as in the case of cells subjected to thermal shock.[117]

It is very likely that melatonin's protective role on the leukocytes also involves its antioxidative properties. Also, melatonin was shown to prevent membrane fluidity changes associated with lipid peroxidation in different experimental models.[101,118–120] This might be also an explanation for the higher degree of inhibition exerted by alcohol upon phagocytosis, which depends on membrane rheology and as such would be affected by a decrease of the membrane fluidity more so than adherence. Moreover, ethanol was shown to impair release of oxygen metabolites from neutrophils, but not the delivery of cells at the inflammatory site,[121] a finding that is in accordance with our data indicating that adherence is less affected by alcohol exposure.

CONCLUSION

Taking together the above, we may conclude that basal and reactive phagocytic activity and neutrophil adherence are periodic functions. Their innate rhythms, part of the circadian oscillation of the whole immune system, are regulated by hypothalamic centers and are susceptible to circadian modulation under the impact of endogenous and exogenous factors, among which are melatonin and the light-dark cycle. Based on the presented evidence, we propose a functional scheme integrating the central mechanisms with photic afferences that modulate neutrophil activity (FIG. 11). In this diagrammatic representation, some links have been certified, while other are still hypothetical, awaiting their experimental confirmation.

ACKNOWLEDGMENTS

The author gratefully acknowledges the contribution of Germaine Cornélissen-Guillaume, M.D., Ph.D., and F. Halberg, M.D. (Halberg Chronobiology Center of the University of Minnesota, USA), who performed the single cosinor rhythmometry of the data series showed herein.

[*Competing interests*: The author declares that she has no competing financial interests.]

REFERENCES

1. PICKARD, G.E. 1982. The efferent connections of the suprachiasmatic nucleus of the golden hamster with emphasis on the retinohypothalamic projection. J. Comp. Neurol. **211:** 65–83.
2. BERK, M.L. & J.A. FINKELSTEIN. 1981. An autoradiographic determination of the efferent projections of the suprachiasmatic nucleus of the hypothalamus. Brain Res. **226:** 1–13.
3. SWANSON, L.W. & W.M. COWAN. 1975. The efferent connections of the suprachiasmatic nucleus of the hypothalamus. J. Comp. Neurol. **160:** 1–12.

4. PICKARD, G.E. & F.W. TUREK. 1983. The hypothalamic paraventricular nucleus mediates the photoperiodic control of reproduction but not the effects of light on the circadian rhythm of activity. Neurosci. Lett. **43:** 67–72.
5. ARENDT, J. 1995. Melatonin and the Mammalian Pineal Gland. Chapman & Hall. London.
6. VANECEK, J. 1998. Cellular mechanisms of melatonin action. Phys. Rev. **78:** 687–721.
7. MAESTRONI, G.J.M., A. CONTI & W. PIERPAOLI. 1986. Role of the pineal gland in immunity: circadian synthesis and release of melatonin modulates the antibody response and antagonizes the immunosuppressive effect of corticosterone. J. Neuroimmunol. **13:** 19–30.
8. MAESTRONI, G.J.M., A. CONTI & W. PIERPAOLI. 1988. Role of the pineal gland in immunity. III. Melatonin antagonizes the immunosuppressive effect of acute stress via an opiatergic mechanism. Immunology **63:** 465–469.
9. MAESTRONI, G.J.M., A. CONTI & W. PIERPAOLI. 1987. Role of the pineal gland in immunity. II. Melatonin enhances the antibody response via an opiatergic mechanism. Clin. Exp. Immunol. **68:** 384–391.
10. ANGELI, A. et al. 1994. Circadian rhythms and multihormonal control of human NK cell. In Advances in Pineal Research. G.J.M. Maestroni, A. Conti & R.J. Reiter, Eds.: 274–282.
11. ANGELI, A. et al. 1988. Effects of exogenous melatonin on human natural killer NK cells activity. An approach to the immunomodulatory role of the pineal gland. In The Pineal Gland and Cancer. D. Gupta, A. Attanasio & R.J. Reiter, Eds.: 145–147.
12. GATTI, G. et al. 1990. Circadian stage-dependent enhancement of human natural killer cell activity by melatonin. J. Immunol. Res. **2:** 108–116.
13. MORREY, K.M. et al. 1994. Activation of human monocytes by the pineal hormone melatonin. J. Immunol. **153:** 2671–2680.
14. BARJAVEL, M.J. et al. 1998. Differential expression of the melatonin receptor in human monocytes. J. Immunol. **160:** 1191–1197.
15. CSABA, G. 1993. Presence in and effects of pineal indoleamines at very low level of phylogeny. Experientia **498:** 627–634.
16. SÁNCHEZ S. et al. 2004. Effect of tryptophan administration on circulating levels of melatonin and phagocytic activity. J. Appl. Biomed. **2:** 169–177.
17. RODRÍGUEZ, A.B. et al. 1997. Melatonin and the phagocytic process of heterophils from the ring dove Streptopelia risoria. Molec. Cell. Biochem. **168:** 185–190.
18. SPECTOR, N.H. 1990. Neuroimmunomodulation takes off. Immunol. Today **11:** 381–383.
19. BACIU, I. et al. 1988. Changes of phagocytic biological rhythm by reduction of circadian times and by influences upon hypothalamus. Intern. J. Neuroscience. **41:** 143–153.
20. BACIU, I. et al. 1994. Chrono-meta-analysis of circadian phagocytosis rhythms in blood of guinea pigs on two different lighting regimens. Chronobiologia **21:** 299–302.
21. BONGRAND, P. et al. 1988. Are there circadian variations of polymorphonuclear phagocytosis in man? Chronobiol. Int. **5:** 81–83.
22. RODRÍGUEZ, A.B. et al. 1999. Correlation between the circadian rhythm of melatonin, phagocytosis, and superoxide anion levels in ring dove heterophils. J. Pineal Res. **26:** 35–42.
23. MAESTRONI, G.J.M., A. CONTI & W. PIERPAOLI. 1987. The pineal neurohormone melatonin and its physiologic, opiatergic immunoregulatory role. Mem. Inst. Oswaldo Cruz **82:** 75–79.
24. MAESTRONI, G.J.M. & W. PIERPAOLI. 1981. Pharmacologic control of the hormonally mediated immune system. In Psychoneuroimmunology. R. Ader, Ed.: 405–425. Academic Press. New York.
25. RADOSEVIC-STASIC, B. et al. 1983. Immune response of rats after pharmacologic pinealectomy. Period. Biol. **85:** 119–121.
26. RODRÍGUEZ, A.B. & R.W. LEA. 1994. Effect of pinealectomy upon the nonspecific immune response of the ring dove (Streptopelia risoria). J. Pineal Res. **163:** 159–166.

27. SAULEA, G. 1999. The olfactory bulb and immunity. Ph.D. thesis. University of Medicine and Pharmacy Cluj-Napoca (Romanian).
28. FELEGEAN A., M. JUNIE & I. BACIU. 1997. Effects of the electrolytic damage of the arcuate nucleus on the blood neutrophils phagocytic activity and phagocytic response. Rom. J. Physiol. **34:** 75–82.
29. WILLIAMS, L.M. et al. 1995. Melatonin receptors in the rat brain and pituitary. J. Pineal Res. **194:** 173–177.
30. CARD, J.P. & R.Y. MOORE. 1991. The organization of visual circuits influencing the circadian activity of the suprachiasmatic nucleus. In Suprachiasmatic Nucleus: The Mind's Clock. D.C. Klein, R.Y. Moore & S.M. Reppert, Eds.: 51–76. Oxford University Press. New York.
31. SAEB-PARSY, K. & R.E. DYBALL. 2003. Defined cell groups in the rat suprachiasmatic nucleus have different day/night rhythms of single-unit activity in vivo. J. Biol. Rhythms **181:** 26–42.
32. SAEB-PARSY, K. & R.E. DYBALL. 2003. Responses of cells in the rat suprachiasmatic nucleus in vivo to stimulation of afferent pathways are different at different times of the light/dark cycle. J. Neuroendocrinol. **159:** 895–903.
33. BELLUARDO, N., G. MUDO & M. BINDONI. 1990. Effects of early destruction of the mouse arcuate nucleus by monosodium glutamate on age-dependent natural killer activity. Brain Res. **534:** 225–233.
34. CHAMBILLE, I. & J. SERVIÈRE. 1993. Neurotoxic effects of neonatal injections of monosodium L-glutamate (L-MSG) on the retinal ganglion cell layer of the golden hamster: anatomical and functional consequences on the circadian system. J. Comp. Neurol. **338:** 67–82.
35. HRISCU, M. et al. 1997. Effects of monosodium glutamate on blood neutrophils phagocytic activity and phagocytic response in mice. Rom. J. Physiol. **34:** 95–101.
36. HRISCU, M. 2003. Modulatory factors of circadian phagocytic activity. PhD thesis. Babes-Bolyai University Cluj-Napoca.
37. MACGREGOR R.R., E. MACARAK & N. KEFALIDES. 1978. Comparative adherence of granulocytes to endothelial monolayer and nylon fiber. J. Clin. Invest. **61:** 697–702.
38. CORNÉLISSEN, G. & F. HALBERG. 1994. Introduction to chronobiology. Medtronic chronobiology Seminar No. 7. Library of Congress: Catalog Card No. 94-060580.
39. HAUS, E. et al. 1983. Chronobiology in hematology and immunology. Am. J. Anat. **168:** 467–517.
40. LEVI, F. et al. 1983. Large amplitude circadian rhythm in helper suppressor ratio of peripheral blood lymphocytes. Lancet **20:** 462–463.
41. LEVI, F. et al. 1985. Circadian and/or circahemedian rhythms in nine lymphocyte-related variables from peripheral blood of healthy subjects. J. Immunol. **134:** 217–222.
42. REINBERG, A. et al. 1977. Rythmes circadiens et circannuels de leucocytes, protéines totales, immunoglobulines A, G, M. Étude chez 9 adultes jeunes et sains. Nouv. Presse Méd. **6:** 819.
43. RITCHIE, A.W. et al. 1983. Circadian variation of lymphocyte subpopulations: a study with monoclonal antibodies. Br. Med. J. **288:** 1773–1775.
44. BUREAU, J.P. et al. 1987. Chronobiologie de l'inflammation. Path. Biol. **356:** 942–950.
45. SWOYER J., E. HAUS & L. SACKETT-LUNDEEN. 1987. Circadian reference values for hematologic parameters in several strains of mice. Prog. Clin. Biol. Res. **227:** 281–296.
46. KUREPA, Z., S. RABATIC & D. DEKARIS. 1992. Influence of circadian light–dark alternations on macrophages and lymphocytes of CBA mouse. Chronobiol. Int. **95:** 327–340.
47. SZABO, I., T.G. KOVATS & F. HALBERG. 1978. Circadian rhythm in murine reticuloendothelial function. Chronobiologia **5:** 137–143.
48. SZABO, I., T.G. KOVATS & F. HALBERG. 1981. Circadian rhythm in phagocytic index of CBA mice, replicated in two studies. In Chronobiology, Proc. XIII Int. Conf. Int. Soc. Chronobiol. Pavia, Italy, 1977. F. Halberg, L. Scheving, E.W. Powell, D.K. Hayes, Eds.: 197–201. Il Ponte, Milan.
49. HRISCU, M. et al. 1998. Circadian rhythm of phagocytosis in mice. Rom. J. Physiol. **35:** 315–319.

50. BENETATO, G.R., C. OPRISIU & I. BACIU. 1947. Système nerveux central et phagocytose contribution expérimentale grâce à la méthode de la "tête isolée." J. Physiol. Paris **39:** 191–197.
51. BACIU, I. & A. IVANOF. 1984. The role of hypothalamic centers in the immune specific response. Rev. Rom. Physiol. **21:** 251–259.
52. HALBERG, F. 1960. Temporal coordination of physiological function. Cold Spring Harb. Symp. Quant. Biol. **25:** 289–310.
53. POWNALL, R., P.A. KABLER & M.S. KNAPP. 1979. The time of day of antigen encounter influences the magnitude of the immune response. Clin. Exp. Immunol. **36:** 347–354.
54. LOPEZ, C. *et al.* 1997. Circadian rhythm in experimental granulomatous inflammation is modulated by melatonin. J. Pineal Res. **23:** 72–78.
55. HRISCU, M. 2004. Circadian phagocytic activity of neutrophils and its modulation by light. J. Appl. Biomed. **2:** 199–211.
56. DARK, J.G. & D. ASDOURIAN. 1975. Entrainment of the rat's activity rhythm by cyclic light following lateral geniculate nucleus lesions. Physiol. Behav. **15:** 295–301.
57. PICKARD, G.E., M.R. RALPH & M. MENAKER. 1987. The intergeniculate leaflet partially mediates effects of light on circadian rhythms. J. Biol. Rhythms **2:** 35–56.
58. JOHNSON, R.F., R.Y. MOORE & L.P. MORIN. 1988. Loss of entrainment and anatomical plasticity after lesions of the hamster retino-hypothalamic tract. Brain Res. **460:** 297–313.
59. JOHNSON, R.F., R.Y. MOORE & L.P. MORIN. 1989. Lateral geniculate lesions alter circadian activity rhythms in the hamster. Brain Res. Bull. **22:** 411–422.
60. EBLING, F.J. 1996. The role of glutamate in the photic regulation of the suprachiasmatic nucleus. Prog. Neurobiol. **50:** 109–132.
61. FORNI G. *et al.* 1983. Radio frequency destruction of the tubero-infundibular region of hypothalamus permanently abrogates NK cell activity in mice. Nature **306:** 181–185.
62. MUDO, G., M. BINDONI & N. BELLUARDO. 1994. Neuroregulation of natural suppressor cell activity. Intern. J. Neurosci. **75:** 129–137.
63. ILLNEROVA, H. & J. VANECEK. 1979. Response of rat pineal serotonin N-acetyltransferase to one minute light pulse at different night times. Brain Res. **167:** 431–434.
64. ANISIMOV, V.N. 2002. The light–dark regimen and cancer development. Neuroendocrinol. Lett. **23c** 28–36.
65. BLASK D.E. *et al.* 2002. Light during darkness, melatonin suppression, and cancer progression. Neuroendocrinol. Lett. **23:** 52–56.
66. SAMEJIMA, M. *et al.* 2000. Light- and temperature-dependence of the melatonin secretion rhythm in the pineal organ of the lamprey, *Lampetra japonica.* Jpn. J. Physiol. **50:** 437–442.
67. HRISCU, M. *et al.* 2002–2003. Circadian phagocytic activity in rats under light–dark and constant light regimens. Rom. J. Phys. **39–40:** 17–26.
68. HRISCU, M. *et al.* 2002–2003. Neutrophil adherence in rats submitted to light–dark alternance and to constant light. Rom. J. Phys. **39–40:** 27–33.
69. SZABO, G. 1999. Consequences of alcohol consumption on host defence. Alcohol **34:** 830–834.
70. KAWAKAMI, M. *et al.* 1989. Immunologic consequences of acute ethanol ingestion in rats. J. Surg. Res. **47:** 412–417.
71. PATEL, M. *et al.* 1996. Human neutrophil functions are inhibited *in vitro* by clinically relevant ethanol concentrations. Alcohol Clin. Exp. Res. **2:** 275–283.
72. NILSSON, E. *et al.* 1996. *In vitro* effects of ethanol on polymorphonuclear leukocyte membrane receptor expression and mobility. Biochem. Pharmacol. **51:** 225–231.
73. NILSSON, E., C. EDENIUS & J.A. LINDGREN. 1995. Ethanol affects leukotriene generation and leukotriene-induced functional responses in human polymorphonuclear granulocytes. Scand. J. Clin. Lab. Invest. **7:** 589–596.
74. TODOROVIC, V. *et al.* 1999. Cytochemical and ultrastructural alterations of cytoplasmic granules of rat peripheral blood neutrophils induced by chronic alcoholism and malnutrition. Indian J. Med. Res. **109:** 105–114.
75. SCHLEIFER, S.J. *et al.* 1999. Immune changes in alcohol dependent patients without medical disorders. Alcohol. Clin. Exp. Res. **237:** 1199–1206.

76. COLOME PAVON, J.A. *et al.* 2003. Free radicals and cytotoxicity of ethanol over human leucocytes in peripheral blood. An. Med. Interna **20**: 396–398.
77. SAHNA, E. *et al.* 2003. Melatonin protects against myocardial doxorubicin toxicity in rats: role of physiological concentrations. J. Pineal Res. **35**: 257–261.
78. DZIGIEL, P. *et al.* 2003. Melatonin stimulates the activity of protective antioxidative enzymes in myocardial cells of rats in the course of doxorubicin intoxication. J. Pineal Res. **35**: 183–187.
79. MELCHIORRI, D. *et al.* 1997. Suppressive effect of melatonin administration on ethanol-induced gastroduodenal injury in rats *in vivo*. Br. J. Pharmacol. **121**: 264–270.
80. BILICI, D. *et al.* 2002. Melatonin prevents ethanol-induced gastric mucosal damage possibly due to its antioxidant effect. Dig. Dis. Sci. **47**: 856–861.
81. HRISCU, M. *et al.* 2003. Alcohol and melatonin influence circadian rhythmic phagocytic activity and adherence of neutrophils in rats [abstract]. The Fourth Symposium on Chronomics, Tokyo. November, 2003.
82. BACIU I. *et al.* 1996. Effects of thymectomy on blood neutrophils phagocytic activity and phagocytic response in mice. Rom. J. Physiol. **33**: 75–81.
83. PIERPAOLI, W. & C.X. YI. 1990. The involvement of pineal gland in immunity and aging. I. Thymus-mediated, immunoreconstituting, and antiviral activity of thyrotropin-releasing hormone. J. Neuroimmunol. **27**: 99–109.
84. LESNIKOV, V.A. & I.P. TSVETKOVA. 1985. Stereotaxic coordinates of the hypothalamus in mice. Fiziol. Zhurn. **71**: 798–804 (Russian).
85. LESNIKOV, V.A., S.B. ADJIEVA & E.A. KORNEVA. 1989. Endogenous splenic colony-formation in mice in experimental pathology of the CNS. Patol. Fiziol. **6**: 14–17 (Russian).
86. MAESTRONI, G.J.M. & A. CONTI. 1990. The pineal neurohormone melatonin stimulates activated CD4+, Thy-1+ cells to release opioid agonists with immunoenhancing and anti-stress properties. J. Neuroimmunomodul. **28**: 167–176.
87. MAESTRONI, G.J.M. & A. CONTI. 1996. Melatonin and the immuno-hematopoietic system: therapeutic and adverse pharmacological correlates. Neuroimmunomodulation **3**: 325–332.
88. LESNIKOV, V.A. *et al.* 1992. The involvement of pineal gland and melatonin in immunity and aging. II. Thyrotropin-releasing hormone and melatonin forestall involution and promote reconstitution of the thymus in anterior hypothalamic area AHA-lesioned mice. Intern. J. Neurosci. **62**: 141–153.
89. LIN, Y.L., M.Y. TAI & Y.F. TSAI. 1990. Morphological changes in the hypothalamic neurons of female rats exposed to continuous illumination. Chin. J. Physiol. **33**: 291–298.
90. BENETATO, G.R., I. BACIU & V. VITEBSKI. 1951. Mecanismul de activare al fagocitozei de catre sistemul nervos central. II. Despre acþiunea fagocitostimulatoare a nervilor micsti si rolul maduvei osoase în activarea fagocitozei. Commun. Acad. Rom. **l**: 707–709 (Romanian).
91. ABE, M. *et al.* 2002. Circadian rhythms in isolated brain regions. J. Neurosci. **22**: 350–356.
92. SAULEA, G. *et al.* 1998. Influence of bilateral olfactory bulbectomy on the circadian rhythm of phagocytic activity in mice. Rom. J. Physiol. **35**: 309–314.
93. RECCHIONI, R. *et al.* 1998. Melatonin increases the intensity of respiratory burst and prevents L-selectin shedding in human neutrophils *in vitro*. Biochem. Biophys. Res. Commun. **252**: 20–24.
94. OLSEN, L.F. *et al.* 2003. A model of the oscillatory metabolism of activated neutrophils. Biophys. J. **84**: 69–81.
95. RODRÍGUEZ, A.B. *et al.* 2001. Physiological concentrations of melatonin and corticosterone affect phagocytosis and oxidative metabolism of ring dove heterophils. J. Pineal Res. **31**: 31–38.
96. BARRIGA, C. *et al.* 2001. Circadian rhythm of melatonin corticosterone and phagocytosis: effect of stress. J. Pineal Res. **30**: 180–187.
97. BUREAU, J.P., L. GARELLY & P. VAGO. 1991. Nycthemeral variations on LPS- and BCG-induced PMN migration in normal mice. Int. J. Tissue React. **134**: 203–206.
98. GARELLY, L., J.P. BUREAU & G. LABREQUE. 1991. Temporal study of carrageenan-induced PMN migration in mice. Agents Actions **33**: 225–228.
99. LOTUFO, C.M. *et al.* 2001. Melatonin and N-acetylserotonin inhibit leukocyte rolling and adhesion to rat microcirculation. Eur. J. Pharmacol. **430**: 351–357.

100. CHEN, J.C. *et al.* 2003. Altered neutrophil apoptosis activity is reversed by melatonin in liver ischemia-reperfusion. J. Pineal Res. **34:** 260–264.
101. REITER, R.J., J.J. GARCIA & J. PIE. 1998. Oxidative toxicity in models of neurodegeneration: responses to melatonin. Restor. Neurol. Neurosci. **12:** 135–142.
102. CUZZOCREA, S. *et al.* 1997. Protective effect of melatonin in carrageenan-induced models of local inflammation: relationship to its inhibitory effect on nitric oxide production and its peroxynitrite scavenging activity. J. Pineal Res. **23:** 106–116.
103. DUBOCOVICH, M.L. *et al.* 2003. Molecular pharmacology, regulation and function of mammalian melatonin receptors. Front. Biosci. **8:** 1093–1108
104. MACGREGOR R.R., P. SPAGNUOLO & A. LENTNEK. 1974. Inhibition of granulocyte adherence by ethanol, prednisone, and aspirin, measured with an assay system. N. Engl. J. Med. **291:** 642–645.
105. REITER, R.J. 1993. The melatonin rhythm: both a clock and a calendar. Experientia **49:** 654–664.
106. POIREL, V.J. *et al.* 2002. MT1 melatonin receptor mRNA expression exhibits a circadian variation in the rat suprachiasmatic nuclei. Brain Res. **946:** 64–71.
107. GAUER, F. *et al.* 1994. Daily variations in melatonin receptor density of rat pars tuberalis and suprachiasmatic nuclei are distinctly regulated. Brain Res. **641:** 92–98.
108. STOLTZ, D.A. *et al.* 1999. Ethanol suppression of the functional state of polymorphonuclear leukocytes obtained from uninfected and simian immunodeficiency virus infected rhesus macaques. Alcohol Clin. Exp. Res. **23:** 878–884.
109. BAGBY, G.J. *et al.* 1998. Suppression of the granulocyte colony-stimulating factor response to Escherichia coli challenge by alcohol intoxication. Alcohol Clin. Exp. Res. **22:** 1740–1745.
110. STOLTZ, D.A. *et al.* 2000. *In vitro* ethanol suppresses alveolar macrophage TNF-alpha during simian immunodeficiency virus infection. Am. J. Respir. Crit. Care Med. **161:** 135–140.
111. STOLTZ, D.A. *et al.* 2002. Effects of *in vitro* ethanol on tumor necrosis factor-alpha production by blood obtained from simian immunodeficiency virus-infected rhesus macaques. Alcohol Clin. Exp. Res. **26:** 527–534.
112. HALDAR, C., D. HAUSSLER & D. GUPTA. 1992. Effect of the pineal gland on circadian rhythmicity of colony forming units for granulocytes and macrophages CFU-GM from rat bone marrow cell cultures. J. Pineal Res. **12:** 79–83.
113. CONTI, A. *et al.* 2000. Evidence for melatonin synthesis in mouse and human bone marrow cells. J. Pineal Res. **28:** 193–202.
114. HELM, R.M. *et al.* 1996. Flow cytometric analysis of lymphocytes from rats following chronic ethanol treatment. Alcohol **13:** 467–471.
115. WANG, J.F. & J.J. SPITZER. 1997. Alcohol-induced thymocyte apoptosis is accompanied by impaired mitochondrial function. Alcohol **14:** 99–105.
116. SAINZ, R.M. *et al.* 1995. The pineal hormone melatonin prevents *in vivo* and *in vitro* apoptosis in thymocytes. J. Pineal Res. **19:** 178–188.
117. CABRERA, J. *et al.* 2003. Melatonin prevents apoptosis and enhances HSP27 mRNA expression induced by heat shock in HL-60 cells: possible involvement of the MT2 receptor. J. Pineal Res. **35:** 231–238.
118. OCHOA, J.J. *et al.* 2003. Melatonin protects against lipid peroxidation and membrane rigidity in erythrocytes from patients undergoing cardiopulmonary bypass surgery. J. Pineal Res. **35:** 104–108.
119. KARBOWNIK, M. & R.J. REITER. 2002. Melatonin protects against oxidative stress caused by delta-aminolevulinic acid: implications for cancer reduction. Cancer Invest. **20:** 276–286.
120. KARBOWNIK, M. *et al.* 2000 Protective effects of melatonin against oxidation of guanine bases in DNA and decreased microsomal membrane fluidity in rat liver induced by whole body ionizing radiation. Mol. Cell. Biochem. **211:** 137–144.
121. NILSSON, E. *et al.* 1995. Rabbit polymorphonuclear granulocyte function during ethanol administration: migration and oxidative responses in a joint with immune complex synovitis. Clin. Exp. Immunol. **102:** 137–143.

The Role of NF-κB in Protein Breakdown in Immobilization, Aging, and Exercise

From Basic Processes to Promotion of Health

MARINA BAR-SHAI,[a] ELI CARMELI,[b] AND ABRAHAM Z. REZNICK[a]

[a]Department of Anatomy and Cell Biology, Rappaport Faculty of Medicine, Technion–Israel Institute of Technology, Haifa 31096, Israel

[b]Department of Physical Therapy, Sackler Faculty of Medicine, Tel Aviv University, Ramat Aviv, Tel-Aviv, Israel

ABSTRACT: Between the ages of 20 and 80 years, humans lose about 20–30% of their skeletal muscle weight. This phenomenon has been termed sarcopenia of old age and is directly involved in the well-being of the aged. With aging, people tend to be less mobile and are frequently bedridden, which exacerbates the muscle weight loss. The molecular mechanisms responsible for the muscle protein breakdown during immobilization in aging have been studied in our laboratory in a model of 24-month-old Wistar rats, immobilized for 4 weeks. Subsequently we investigated the activation of the intracellular and extracellular proteolytic systems in the immobilized muscles. A similar group of young (6-month-old) rats was examined and compared to the older rats. The involvement of NF-κB transcription factor in muscle atrophy was assessed in immobilized muscles of young and old animals. There were marked differences in the kinetics and the pattern of NF-κB activation in young versus old muscles. It seems that in both young and old animals in the early stages of limb immobilization, an alternative pathway of NF-κB activation can be observed. However, in late stages of immobilization, the canonic pathway of NF-κB activation (p65/p50 complex with I-κB α degradation) is predominant. Interestingly, the canonic activation pathway is more prominent in muscles from old animals compared to young ones. The activation of NF-κB has been observed also in muscles subjected to acute and intense exercise, implying that inflammatory processes may take place under the conditions of intense exercise. This may cause muscle damage and protein breakdown. Therefore, using NF-κB pathway inhibitors may prove beneficial in attenuating NF-κB–associated muscle damage in both disuse atrophy and strenuous exercise.

KEYWORDS: NF-κB; skeletal muscle; immobilization; exercise; proteolysis

Address for correspondence: A.Z. Reznick, Department of Anatomy and Cell Biology, Rappaport Faculty of Medicine, Technion–Israel Institute of Technology, P.O. Box 9649, Haifa 31096, Israel. Voice: 972-48295388; fax: 972-48295403.
reznick@tx.technion.ac.il

Ann. N.Y. Acad. Sci. 1057: 431–447 (2005). © 2005 New York Academy of Sciences.
doi: 10.1196/annals.1356.034

INTRODUCTION

The Aging of Skeletal Muscle

Skeletal muscle wasting, common in elderly humans and animals, is often referred to as "sarcopenia of old age."[1] *Sarcopenia* is a generic term for the overall loss of muscle mass, strength, and quality (structural composition, innervation, contractility, capillary density, fatiguability, and glucose metabolism). In old age, there is marked decline of 20–30% in muscle mass in comparison to muscle mass in young individuals.[2–5] Sarcopenia results in muscle weakness leading to an increased prevalence for falls, greater morbidity, and loss of functional autonomy.[6] There have been several proposals regarding the underlying biochemical mechanisms for age-related sarcopenia.[3] These include reduction in mediating factors involved in activation of progenitor myoblasts;[7] decreased muscle protein synthesis;[8] role of reactive oxygen species;[9,10] metabolic consequences of alteration in enzyme activities, nitrogen imbalance, impaired glucose metabolism;[11] and imbalance between degradation and removal of "old" damaged muscle proteins.[12] Of late, the role of mitochondria in muscle degeneration and sarcopenia has been proposed.[13,14]

Factors that lead to skeletal muscle aging are complex and numerous. They include biochemical and metabolic changes in muscle tissue, alterations in muscle fiber size and composition, and loss of muscle mass.[15–18] However, it is not clear whether this phenomenon results from the decline in muscle use in old age, or, conversely, it is a direct result of the aging process.[4,19,20] The above alterations are even more pronounced in aged humans and animals that remain immobilized for extended periods. In these individuals, skeletal muscles, especially those of the lower limbs, undergo progressive atrophy and loss of both mass and function. This process is termed "disuse muscle atrophy," and it may result in deleterious outcomes, such as weakness, instability, and frequent falls, that often aggravate the lives of the aged people.

Muscle Proteolytic Systems

Proteolytic systems in skeletal muscles can be divided into extracellular and intracellular systems.

Extracellular proteolytic systems exist outside muscle fibers in the surrounding connective tissue and include:

- Matrix metalloproteinases (MMPs). MMPs are produced and secreted by both muscle fibers and either resident or invading phagocytes. The most prevalent and principal MMPs in skeletal muscle are MMP-2 and MMP-9.[21–23]

- Lysosomal proteases (cathepsins) and hydrolases (acid phosphatase). These are also found in resident or invading phagocytes in the connective tissue surrounding muscle fibers.

Intracellular proteolytic systems exist within muscle fibers and include:

- Lysosomal proteases and hydrolases (these are found in lysosomes inside the muscle fibers[24]);

- Calcium-dependent proteases (calpains) associated mainly with muscle Z-lines and sarcomeric proteins;[24]

- The ubiquitin-proteasome pathway;[24]
- Miscellaneous cytosolic and membranous proteases;
- Mitochondrial proteases.

It has been shown that the first three intracellular proteolytic systems outlined above are the major systems that take part in muscle breakdown in disuse atrophy, limb immobilization, and cachexia, and act in concert.[24] The calcium-dependent calpains are activated in the early stages of muscle breakdown[25] and are responsible for the release of myofibrillar proteins from the sarcomeric structure. Subsequently, myofibrillar proteins are polyubiquitinated and degraded by the ubiquitin–proteasome system. Conversely, membrane protein components, such as receptors and protein channels, are monoubiquitinated and shuttled to lysosomes for eventual degradation.[24]

The Involvement of TNF-α and NF-κB in Muscle Protein Breakdown

In the last few years, the connection between inflammatory processes, NF- κB activation, and protein degradation has been described. In 1997, Sen and colleagues[26] showed that in L-6 myoblasts, TNF-α could activate NF-κB, and this activation was redox-regulated.[26] In subsequent studies, Li and colleagues demonstrated that in C2C12 myocyte cell line, TNF-α induced protein loss, which was mediated by oxidative stress and NF-κB activation.[27] In other studies, it has been shown that the exposure to either hydrogen peroxide or TNF-α led to NF-κB–dependent protein ubiquitination and the expression and activity of ubiquitin ligases of the E3 family.[28–30]

In the early 1990s, similar studies showed that TNF-α was capable of increasing both ubiquitin gene expression and the ubiquitination of rat skeletal muscle proteins.[31,32] Finally, these researchers demonstrated that TNF-α could directly induce the expression of the ubiquitin-proteasome pathway components in rat soleus muscles.[33]

NF-κB Signaling Pathways and Muscle Breakdown

All NF-κB family members are expressed in skeletal muscle.[25] The NF-κB transcription factor complex has been implicated in muscle atrophy attributed to both disuse and cachexia, but the specific family members involved in the two types of atrophy are distinct. This is important, because it indicates that there are differences in the molecular signaling for these two types of atrophy and, therefore, that there may be more specific molecules to target in the development of therapies.

In cachexia the major upstream targets are circulating cytokines, especially TNF-α, which induces the phosphorylation of NF-κB inhibitor protein, I-κB α, by activating the specific kinase IKK. Upon phosphorylation, I-κB α is polyubiquitinated and degraded by the proteasome, enabling NF-κB nuclear translocation and transcriptional activity. The most ubiquitous NF-κB form, p65/p50 heterodimer, is activated through the above pathway, which is considered the classic or canonic mode of activation.[34]

In muscle disuse, an alternative pathway of NF-κB activation has been recently elucidated. It was first demonstrated[35] that muscle disuse leads to increased transcriptional activity of NF-κB. The luciferase activity from an NF-κB–dependent reporter plasmid injected into soleus muscles from rats that were subjected to unloading for 7 days was increased compared with that in weight-bearing control rats, whereas AP-1– or NFAT (nuclear factor of activated T cells)–dependent reporter plasmids did not increase with unloading. Moreover, while the prototypical NF-κB family member p65 did not show increased nuclear levels, p50 and Bcl-3 (a nuclear I-κB family member) were markedly increased. Moreover, injection of an NF-κB reporter plasmid led to an increase in luciferase activity in wild-type soleus muscles as the result of unloading but was unaffected by unloading in the p50 knockout mice. The same results were found using bcl-3 knockout mice, suggesting that p50 and Bcl-3 are required for unloading-induced muscle atrophy.[36] Because p65 does not undergo nuclear translocation during unloading atrophy,[35] the NF-κB pathway activated during unloading is different from the canonic (p65/ p50-dependent) pathway.

Corresponding findings were described recently in muscles from old rats that had been immobilized for four weeks. In the first two weeks of immobilization, the classic p65/p50 dimer of NF-κB was downregulated in the immobilized muscles, while in the third and the fourth weeks of immobilization, there was an upregulation of that dimer in quadriceps muscles of old immobilized rats.[21] This finding may point to the presence of an alternative NF-κB pathway activation in the early stages of muscle immobilization, probably similar to the one described by the Kandarian group.[35]

In the present manuscript, the mode of NF-κB activation under the conditions of immobilization, aging and physical exercise are further described and analyzed.

MATERIALS AND METHODS

Animals

Pathogen-free young (6 months old, 250–280 g body weight) and old (24 months old, 300–340 g body weight) female Wistar rats were used in this study. The animals were kept at room temperature with a natural night-day cycle and sustained with standard rat chow and water ad libitum. The rats were randomly divided into five groups of six animals that were immobilized for one, two, three, and four weeks, or alternatively, were not immobilized (untreated controls). Right hindlimbs of the animals were externally fixed by passing metal immobilization appliances through femur and tibia of the anesthetized animals as previously described,[35] and subsequently securing the entire construct with an external immobilization frame. Contralateral (left) hindlimbs served as controls. Additional animals remained non-immobilized, and served as untreated controls. Rats were sacrificed under ether anesthesia at 7, 14, 21, and 28 days following limb immobilization. Hindlimb muscles (gastrocnemius, soleus, plantaris, and quadriceps) were removed for further study. The protocols for animal maintenance and experimental procedures were approved by the Animal Ethics and Welfare Committee of the Technion-Israel Institute of Technology.

Muscle Processing

Both right and left quadriceps muscles were dissected out and weighed (wet weight). All muscle samples were taken from the mid-belly region in order to avoid the sites immediately adjacent to the external fixation pins. One part of each muscle was immediately frozen in liquid nitrogen and stored at $-70°C$ for further biochemistry and molecular biology studies. The other part of each muscle was rapidly frozen in isopentane chilled by liquid nitrogen, for frozen sections and histochemistry. Muscles were thawed and minced on ice in 7 volumes of homogenization buffer containing 100 mM Trizma-HCl, 0.3 M KCl, 2.5 mM $MgCl_2$, 1 mM DTT, 0.5 mM PMSF, 1 μg/mL leupeptin, and 0.1% Triton-100. The muscles were homogenized in a Polytron homogenizer (Kinematica, Lucerne, Switzerland) three times for 20 seconds. Subsequently, the homogenates were centrifuged at 24,000g for 30 minutes. Protein content in supernatants was determined using Bradford reagent (Bio-Rad, Hercules, CA) and bovine serum albumin was used as standard. The supernatants were used for Western blotting.

Western Blotting

For Western blotting, 10 μg of proteins of either cytoplasmatic or nuclear extracts were boiled in Laemmli buffer, separated by sodium dodecyl sulfate-polyacrylamide gel electrophoresis (SDS-PAGE) on 10% acrylamide gels, and electroblotted onto nitrocellulose membranes (Biological Industries Inc., Kibbutz Bet Haemek, Israel). The equality of sample loading in each lane was confirmed by Coomassie Brilliant Blue gel staining and by Ponceau Red membrane staining following the blotting. Membranes were blocked in 1% dry milk powder in Tris-buffered saline (TBS) and probed with suitable primary antibodies. Subsequently, the membranes were washed in TBS with 0.5% Tween 100, and incubated with secondary horseradish peroxidase–conjugated antibodies (Santa Cruz Biotechnology Inc., Santa Cruz, CA). The blots were incubated with the enhanced chemiluminescence reagent (Santa Cruz Biotechnology Inc., Santa Cruz, CA) according to the manufacturer's instructions. Chemiluminescence was recorded by exposing the membranes to Fuji films. Primary antibodies used were: mouse anti ubiquitin antibody, rabbit anti p65 NF-κB subunit antibody, rabbit anti I-κB α antibody, mouse anti phosphorylated I-κB α antibody (recognizes phosphorylated serine 32).

Extraction of Nuclear Proteins and Analysis of NF-κB Activation

Quadriceps muscles were thawed, weighed, and minced on ice in 7 volumes of homogenization buffer containing 100 mM Trizma-HCl, 0.3 M KCl, 2.5 mM $MgCl_2$, 1 mM DTT, 0.5 mM PMSF, 1 μg/mL leupeptin and 0.1% Triton 100. The muscles were homogenized in a Polytron homogenizer (Kinematica, Lucerne, Switzerland) three times for 20 seconds. Subsequently, the homogenates were centrifuged at 24,000g for 30 minutes. After separation of the supernatant containing the cytosolic proteins, the pellet containing the nuclei was incubated with nuclear extraction buffer: 20 mM HEPES, 1.5 mM $MgCl_2$, 420 mM NaCl, 0.2 mM EDTA, 1 mM DTT, 0.5 mM PMSF, 1 μg/mL leupeptin, for 30 minutes on ice with strong agitation. Subsequently, the mixture was centrifuged at 24,000g for 30 minutes. The supernatant containing nuclear proteins was separated, and protein concentration was determined according to the method of Bradford. NF-κB activation kinetics in nuclear extracts from control and immobilized hindlimb quadriceps muscles was measured by Western blotting.

Densitometric Analysis

Densitometric quantitative analysis of the protein bands detected by Western blot was carried out using Bio1D software (Vilber Lourmat, Torcy, France).

Statistical Analysis

Comparisons of the means of densitometric analyses and specific immobilization groups were made using the unpaired Student's t-test with significant values set at $P < 0.05$. The results of the experiments were expressed as mean fold increase versus the control values ± SEM.

RESULTS

NF-κB Activation and I-κB α in Control and Immobilized Quadriceps Muscles from Young Rats

In order to assess the mode and the kinetics of NF-κB activation in control and immobilized quadriceps muscles of young rats, NF-κB p65 subunit nuclear translocation and cytoplasmatic I-κB α degradation were measured by Western immunoblot as a function of the four weeks of immobilization.

In control quadriceps muscles of young rats there was slow and steady increase in NF-κB p65 activation, peaking in the fourth week of hindlimb immobilization (FIG. 1A). This was not accompanied by a decrease in cytoplasmatic I-κB α levels until the fourth week, when peak p65 nuclear translocation was observed (FIG. 1B).

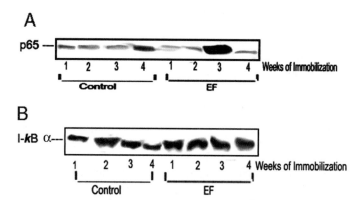

FIGURE 1. The effect of hindlimb immobilization on NF-κB activation in control and immobilized quadriceps muscles from young rats. (**A**) The effect of hindlimb immobilization by external fixation (EF) on NF-κB p65 nuclear translocation in control and immobilized quadriceps muscles from young rats. Control (untreated) and immobilized quadriceps muscles from young rats underwent nuclear protein extraction. p65 nuclear translocation was analyzed by Western immunoblot with anti-p65 antibody as a function of the duration of immobilization. The results are representative of one of three assays in muscles from three different animals after 1, 2, 3, and 4 weeks of immobilization. (**B**) The effect of hindlimb immobilization by EF on I-κB α cytoplasmatic levels in control and immobilized quad-

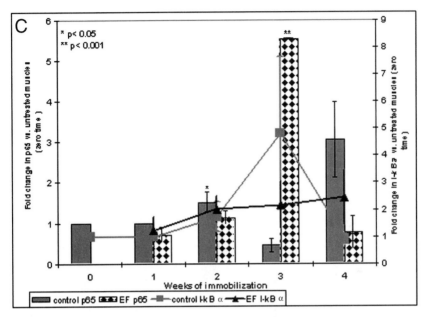

riceps muscles from young rats. Control (untreated) and immobilized quadriceps muscles from young rats underwent cytoplasmatic proteins extraction. I-κB α levels were analyzed by Western immunoblot with anti-I-κB α antibody as a function of the duration of immobilization. The results are representative of one of three assays in muscles from three different animals after 1, 2, 3, and 4 weeks of immobilization. (C) Densitometric analysis of p65 nuclear translocation plotted against I-κB α levels, in control and immobilized quadriceps muscles from young rats, as a function of the duration of immobilization. Data are expressed as average-fold change of NF-κB activation and I-κB α levels ± SEM at every time point in comparison to the untreated control muscles. *$P < 0.05$, **$P < 0.001$ for p65 values versus control values of the untreated muscles. The statistics of relative intensities of the bands were derived from two to three independent experiments.

In immobilized muscles of young rats, slowly increasing levels of p65 nuclear translocation were observed in the muscle nuclei, starting with the second week following immobilization and reaching a maximal level during the third week of immobilization (FIG. 1A and C). This was not accompanied by I-κB α degradation (FIG. 1B and C).

NF-κB Activation and I-κB α Levels in Control and Immobilized Quadriceps Muscles from Old Rats

In order to assess the mode and the kinetics of NF-κB activation in control and immobilized quadriceps muscles of old rats, NF-κB p65 subunit nuclear translocation and cytoplasmatic I-κB α degradation were measured by Western immunoblot as a function of the four weeks of immobilization.

It has been found that in control quadriceps muscles of old rats, there was little significant change in NF-κB p65 activation during the entire period of immobiliza-

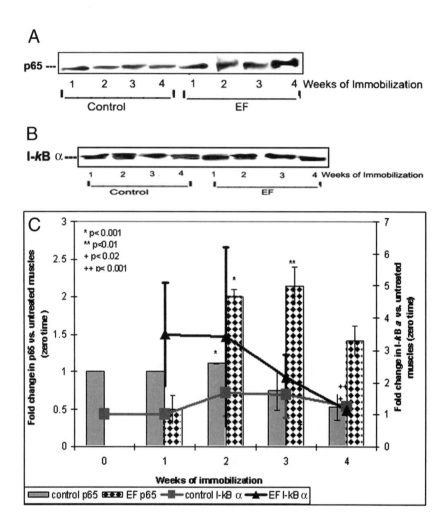

FIGURE 2. The effect of hindlimb immobilization on NF-κB activation in control and immobilized quadriceps muscles from old rats. (**A**) The effect of hindlimb immobilization by external fixation (EF) on NF-κB p65 nuclear translocation in control and immobilized quadriceps muscles from old rats. Control (untreated) and immobilized quadriceps muscles from old rats underwent nuclear proteins extraction. p65 Nuclear translocation was analyzed by Western immunoblot with anti-p65 antibody as a function of the duration of immobilization. The results are representative of one of three assays in muscles from three different animals after 1, 2, 3, and 4 weeks of immobilization. (**B**) The effect of hindlimb immobilization by EF on I-κB α cytoplasmatic levels in control and immobilized quadriceps muscles from old rats. Control (untreated) and immobilized quadriceps muscles from old rats underwent cytoplasmatic proteins extraction. I-κB a degradation was analyzed by Western immunoblot with anti-I-κB α antibody as a function of four weeks of immobilization. The results are representative of one of three assays in muscles from three different animals after 1, 2, 3, and 4 weeks of immobilization. (**C**) Densitometric analysis of p65 nuclear translocation plotted against I-κB α levels, in control and immobilized quadriceps muscles from old rats,

tion, with a slight tendency for decrease (FIG. 2A and C). This was accompanied by relatively unchanged levels of cytoplasmatic I-κB α (FIG. 2B and C). However, the extent of p65 nuclear translocation was markedly lower in control than in the immobilized quadriceps muscles. In immobilized muscles of old rats, significant increase in the levels of p65 nuclear translocation was observed in the muscle cell nuclei, starting with the second week following immobilization (FIG. 2A and C). This was paralleled by I-κB α degradation, becoming apparent after 3 to 4 weeks of immobilization (FIG. 2B and C). Moreover, the changes in p65 nuclear translocation and I-κB α levels in control muscles were less statistically significant that those in the immobilized muscles (FIG. 2C).

I-κB α Serine Phosphorylation in Control and Immobilized Quadriceps Muscles from Old Rats

In order to attempt to elucidate the mechanism of NF-κB activation in control and immobilized quadriceps muscles of old rats, I-κB α serine phosphorylation, which is a prerequisite for the canonic NF-κB activation, has been measured by Western immunoblot as a function of four weeks of immobilization.

It has been found that in control quadriceps muscles, there was no significant change in the levels of I-κB α serine phosphorylation during the entire period of immobilization (FIG. 3A and B). This was consistent with the low and steady NF-κB activation levels in the control old muscles during the entire immobilization period (FIG. 2A). In contrast, in immobilized quadriceps muscles from old animals, there was a gradual decrease in I-κB α serine phosphorylation starting with the first week following immobilization, with an increase in the fourth week of immobilization (FIG. 3A and B). This was in agreement with NF-κB activation in the immobilized muscles during the second and third weeks of immobilization, because the phosphorylated I-κB α is rapidly degraded in the proteasome (FIG. 1A and C).

Myosin Heavy-Chain Ubiquitination in Control and Immobilized Quadriceps Muscles from Old Rats

In order to assess myosin heavy-chain (MHC) ubiquitination, which is a prerequisite for MHC degradation, in control and immobilized muscles from old rats, MHC ubiquitination was measured by Western immunoblot.

---

as a function of the duration of immobilization. Data are expressed as average-fold change of NF-κB activation and I-κB α levels ± SEM at every time point in comparison to the untreated control muscles. *$P < 0.001$, **$P < 0.01$ for p65 values versus control values of the untreated muscles; [+]$P < 0.02$, [++]$P < 0.001$ for I-κB α values versus control values of the untreated muscles. The statistics of relative intensities of the bands were derived from three independent experiments.

A

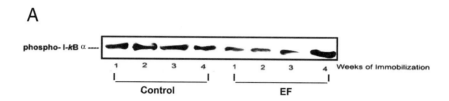

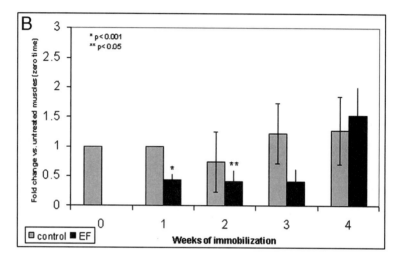

FIGURE 3. The effect of hindlimb immobilization on I-κB α serine phosphorylation in control and immobilized quadriceps muscles from old rats. (**A**) The effect of hindlimb immobilization by external fixation (EF) on I-κB α serine phosphorylation in control and immobilized quadriceps muscles from old rats. Control (untreated) and immobilized quadriceps muscles from old rats underwent cytoplasmic proteins extraction. I-κB α serine phosphorylation was analyzed by Western immunoblot with anti-phospho-I-κB α antibody as a function of the duration of immobilization. The results are representative of one of three assays in muscles from three different animals after 1, 2, 3, and 4 weeks of immobilization. (**B**) Densitometric analysis of I-κB α serine phosphorylation in control and immobilized quadriceps muscles from old rats, as a function of the duration of immobilization. Data are expressed as average-fold change of I-κB α serine phosphorylation ± SEM at every time point in comparison to the untreated control muscles. *$P < 0.001$, **$P < 0.05$ for I-κB α serine phosphorylation values versus control values of the untreated muscles. The statistics of relative intensities of the bands were derived from three independent experiments.

It has been found that in control muscles, there was a gradual decrease in MHC ubiquitination starting with the first week of immobilization. In the fourth week of immobilization, this decrease was the most apparent (FIG. 4A and B). In contrast to that, in immobilized muscles, there was a decrease in MHC ubiquitination in the first week of immobilization, followed by an increase in MHC ubiquitination during the second and third weeks of immobilization, with subsequent significant decrease in the ubiquitinated MHC levels in the fourth week of immobilization (FIG. 4A and B). Thus it seems that there were transient levels of MHC ubiquitination with a peak at the second and the third weeks of immobilization.

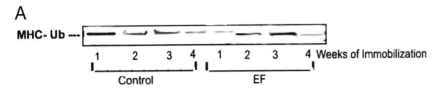

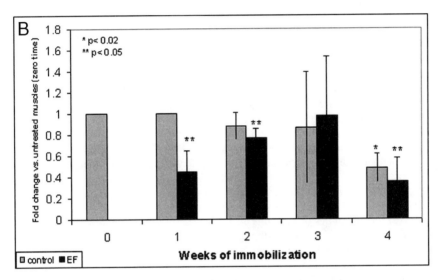

FIGURE 4. The effect of hindlimb immobilization on MHC ubiquitination in control and immobilized quadriceps muscles from old rats. (**A**) The effect of hindlimb immobilization by external fixation (EF) on MHC ubiquitination in control and immobilized quadriceps muscles from old rats. Control (untreated) and immobilized quadriceps muscles from old rats underwent cytoplasmatic proteins extraction. MHC ubiquitination was analyzed by Western immunoblot with anti-ubiquitin antibody as a function of the weeks of immobilization. The results are representative of one of three assays in muscles from three different animals after 1, 2, 3, and 4 weeks of immobilization. (**B**) Densitometric analysis of MHC ubiquitination in control and immobilized quadriceps muscles from old rats, as a function of the duration of immobilization. Data are expressed as average-fold change of MHC ubiquitination ± SEM at every time point in comparison to the untreated control muscles. *$P <$ 0.02, **$P <$ 0.05 for MHC ubiquitination values versus control values of the untreated muscles. The statistics of relative intensities of the bands were derived from three independent experiments.

DISCUSSION

The activation of proteolytic systems in muscle immobilization in aged animals and humans is a complex multi-step phenomenon. In the last few years it became apparent that several extracellular and intracellular protein degradation systems are involved in the above phenomenon. Their triggering and the kinetics of activation may be divergent in muscles of young compared to old animals. For example, it was shown that in young rats a marked increase in the activity of MMPs, acid phosphatase, and the ubiquitin-proteasome was observed in the late stages of immobilization.[22] It seems that this activation of the proteolytic systems was synchronized in young animals, but was desynchronized in immobilized muscles of old animals. Indeed, the kinetics of acid phosphatase activation in old animals showed a different pattern during the four-weeks immobilization as compared to young rats.[21,22] Similarly, MMP-2 and MMP-9 activity as a function of the four weeks of immobilization was also different between young and old animals.[21,22]

NF-κB Activation in Muscle Disuse and Immobilization

The role of the various intracellular proteolytic systems in muscle immobilization has been suggested in a previous publication.[38] In this report, a distinction was made between the fast stage of muscle protein breakdown (mediated by calcium-dependent proteases, calpains) and the slow, prolonged phase of muscle wasting (mediated by the lysosomes and the ubiquitin-proteasome systems).[38] In the aforementioned report, a pivotal role in muscle breakdown was attributed to oxidative stress and NF-κB activation. It seems that the upregulation of NF-κB transcriptional activity is a common phenomenon in various types of muscle wasting. Subsequently, it has been demonstrated that in the first 10 days of immobilization, usually characterized by little to no infiltration of inflammatory cells, NF-κB is activated through a non-canonic pathway that involves the transcriptional activity of a p50/p50/Bcl3 complex.[35] However, in our studies of aged rats whose hindlimbs were immobilized by external fixation, in the third and the fourth weeks of immobilization, the increase in the nuclear translocation of the canonic p65/p50 dimer was observed,[21] which is also supported by the data presented in this manuscript. Moreover, the contribution of cytokines, especially TNF-α, is similar to the conditions reported for cachexia and the canonic activation of NF-κB.[24] Finally, the pattern of p65 activation and I-κB α degradation presented in FIGURES 1 and 2 points to the possibility that the activation of the canonic NF-κB pathway in the later stages of immobilization is more dominant in old compared to young animals. This may reflect a more pronounced and earlier inflammatory response in muscles of old animals compared to young ones. In immobilized muscles of young animals, a peak of p65 nuclear translocation is observed three weeks after immobilization (FIG. 1A and C), and is accompanied by I-κB α degradation, however statistically insignificant.

In old immobilized muscles, there is an increase in p65 nuclear translocation starting with the second week following immobilization (FIG. 2A and C). This is accompanied by low levels of I-κB α degradation (FIG. 2B and C) and by I-κB α dephosphorylation (FIG. 4A and B). In contrast to that, in control muscles of old animals there is a slight increase in p65 nuclear translocation during the second week of immobilization, after which p65 nuclear levels seem to decrease during the third and the fourth weeks of immobilization (FIG. 2A and C). No substantial I-κB α phosphorylation (FIG. 4A and B) or degradation (FIG. 2B and C) is observed in control muscles. It is, therefore, possible to conclude that in immobilized muscles of old rats, p65 is persistently activated with only insignificant I-κB α degradation. This may point to a coexistence of a noncanonic activation that begins prior to the infiltration of monocytes into muscle (and, therefore, may originate in muscle-dependent factors) with the classic p65 activation pathway in the third to fourth weeks of immobilization. In contrast to that, there is transient p65 activation in the control muscles during the second week of immobilization, which does not involve I-κB α phosphorylation and degradation. This is quite similar to p65 activation in immobilized muscles during the third week of immobilization in young animals.

In general, it is possible to conclude that p65 is activated through divergent kinetics and patterns in control and immobilized muscles of old animals. While in untreated muscles of old animals, p65 is only slightly activated, in immobilized muscles the activation is more persistent and prominent and begins as early as the second week after immobilization. The outcome in old animals is markedly different from p65 activation in control and immobilized muscles of young animals, as described above. It is important to emphasize that the persistent NF-κB activation during most of the immobilization period in old animals may lead to deleterious outcomes such as greater muscle wasting and lesser capacity for recovery than in young animals.[37,38]

One possible hint at the effects of the persistent NF-κB activation in old immobilized muscles may be the ubiquitination of MHC. MHC is a major myofibrillar protein that is released from sarcomeres by calpains in the early stages of muscle breakdown. Subsequently, MHC undergoes ubiquitination, which marks it for proteasomal degradation during the slow stage of muscle wasting. Therefore, the levels of MHC ubiquitination may serve as an indication for the rates of its release from sarcomeres and its turnover in the atrophying muscles. While in untreated muscles of old rats a decrease in MHC ubiquitination could be observed only in the fourth week of immobilization, two distinct events of MHC de-ubiquitination were observable in immobilized muscles of old animals—one event after one to two weeks of immobilization, the other after four weeks of immobilization (FIG. 4A and B). As MHC de-ubiquitination may indicate MHC degradation in the proteasome (thus leading to the disappearance of the marked MHC), it seems that in control muscles of old rats some degree of MHC degradation occurs after four weeks of immobilization, which follows transient NF-κB activation (as detailed in this report). However, in immobilized muscles of old rats, there are two possible peaks of MHC degradation, one occurring early (during the first and the second week of immobilization), and the other one late (during the fourth week of immobilization). This may be attributed to the initial wave of MHC degradation after the early (fast) stage of muscle proteins breakdown, followed by the slower and more massive MHC degradation during the slow stage of muscle wasting. This observation in old animals may hint at the nature of the significant loss of muscle mass after immobilization in old animals. Also, it is important to emphasize the correlation between the alleged MHC degradation and NF-κB activation in immobilized muscles of old animals, raising the possibility of either direct or circumstantial link between these two events.

The Effect of Exercise of NF-κB Activation

There have been conflicting reports on the effect of exercise on NF-κB activation in muscle cells and other tissues. If indeed certain modes of training may cause muscle damage, one would expect that NF-κB might participate in the processes of muscle damage as manifested by protein breakdown because of damaging exercise regimens. Assuming that in most exercises muscle protein synthesis is stimulated, it is plausible to expect downregulation of NF-κB activation under these conditions.

A number of reports in recent years indicated that intense physical exercise led to the increase in the activation of NF-κB signaling pathway. Thus, Ji and colleagues[39] have found that in rats exercised to exhaustion, there were high levels of NF-κB activation in muscles as compared to unexercised rats. Muscle cytosolic IKK and I-κB α content was concomitantly decreased. Treatment with antioxidants, such as pyrrolidine dithiocarbamate (PDTC), almost completely abolished the activation of NF-κB signaling cascade.[39] Similarly, more recent work by Ho and colleagues reported that NF-κB activity increased by 50% in rat gastrocnemius muscles 1–3 hours after 60 minutes of treadmill running.[40] Inhibition of p38 and ERK MAP kinases pathways resulted in as much as 76% inhibition of IKK phosphorylation. This suggested that these kinases may influence the activation of IKK and NF-κB during exercise.[40]

Sen has suggested a mechanism for the possible mode of NF-κB activation due to exercise.[41] According to this monograph, intense acute exercise may cause the oxidation of glutathione in muscle cells, resulting in increased oxidative stress and NF-κB activation.[41] However, regular exercise may increase glutathione levels in muscle tissue, resulting in downregulation of NF-κB activity. Indeed, other reports of studies in humans have suggested that acute fatiguing exercise appears to decrease NF-κB activity in muscles.[42]

Another work on the effect of aging and exercise showed that with age, there was an increase in NF-κB content in rat livers. This was significantly attenuated by regular exercise.[43] Therefore, regular exercise may reduce the increase in inflammatory response observed with aging.[43]

Immobilization and Exercise—From Basic Studies to Promotion of Health

Recent studies on the role of NF-κB in immobilization and disuse atrophy have elucidated several molecular events that connect NF-κB to muscle protein breakdown. The specific activation of some particular E3 ubiquitin ligases in muscle disuse and atrophy, established NF-κB as one of the central regulators of muscle breakdown.[28,44] Moreover, genetic deletion of these E3 ligases, such as atrogin-1/MAFbx or MURF1, slowed down muscle atrophy.[44] In a more comprehensive study, transgenic mice that overexpressed muscle-specific IKK (MIKK mice) were used. In these mice a profound increase in NF-κB activity and muscle wasting was observed.[44] Another strain of transgenic mice expressing muscle I-κB a super-repressor (MISR mice) exhibited no overt phenotype, where the basal activity of NF-κB was decreased by 80%. Furthermore, the expression of E3 ligase MURF-1 was increased in MIKK mice, and both genetic and pharmacological inhibition of the NF-κB/ MURF-1 pathway attenuated muscle atrophy.[45] Therefore, the use of NF-κB pathway inhibitors may hint to the possible approach of downregulating NF-κB activity in muscles and thus decreasing the extent of muscle atrophy. In addition, several *in vivo*-compatible approaches, such as oligo decoy nucleotides or proteasome inhibitors[45] may also prove beneficial in diminishing muscle atrophy in aging, disuse, and immobilization.

Exercise and muscle activity are essential to the well-being of humans and animals. With age, the capacity to engage in training and exercise is reduced. Thus, mice of various ages were exercised for short periods of 6 weeks and for long periods throughout life. While young and middle-aged animals improved their muscle capacities biochemically and morphologically, 25-month-old mice were affected negatively by exercise, exhibiting reduced muscle mass and increase muscle damage. Consequently, the concept of threshold of age in exercise was established, according to which—depending on the exercise regimen—physical exercise may be detrimental to the aged organism. The conclusion is that training and exercise may be beneficial to aging people as long as they do not pass the threshold of age.[46] As muscle damage inflicted by exercise in aged individuals may be associated in part with inflammatory processes, it is logical to assume that NF-κB may be at least partially responsible for the detrimental effects of physical exercise in old age. In the event of the upregulation of NF-κB by exercise, it may also be possible to use NF-κB pathway inhibitors to attenuate the damage caused by NF-κB–dependent inflammatory processes. Thus, using NF-κB pathway modulators may also shed more light on the role of NF-κB in immobilization, exercise, and aging.

ACKNOWLEDGMENTS

This study was supported by the Krol Foundation, Lakewood, New Jersey and by a grant from the vice-president of research of the Technion–Israel Institute of Technology. This study was also supported by a grant from the Myers-JDC-Brookdale Institute of Gerontology and Human Development, and Eshel—the Association for the Planning and Development of Services for the Aged in Israel.

[*Competing interests*: The authors declare that they have no competing financial interests.]

REFERENCES

1. MORLEY, J.E., R.N. BAUMGARTNER, R. ROUBENOFF, *et al.* 2001. Sarcopenia. J. Lab. Clin. Med. **137**: 231–243.
2. EVANS, W.J. 1995. What is sarcopenia? J. Gerontol. A Biol. Sci. Med. Sci. **50**: 5–8.
3. SHORT, K.R. & K.S. NAIR. 1999. Mechanisms of sarcopenia of aging. J. Endocrinol. Invest. **22**: 95–105.
4. SHORT, K.R. & K.S. NAIR. 2001. Muscle protein metabolism and the sarcopenia of aging. Int. J. Sport Nutr. Exerc. Metab. **11**: S119—127.
5. DESCHENES, M.R. 2004. Effects of aging on muscle fibre type and size. Sports Med. **34**: 809–824.
6. VANDERVOOT, A.A. & T.B. SYMONS. 2001. Functional and metabolic consequences of sarcopenia. Can. J. Appl. Physiol. **26**: 90–101.
7. CRISONA, N.J, K.D. ALLEN & R.C. STROHMAN. 1998. Muscle satellite cells from dystrophic (mdx) mice have elevated levels of heparan sulphate proteoglycan receptors for fibroblast growth factor. J. Muscle Res. Cell. Motil. **19**: 43–51.
8. SCHONEICH, C., R.I. VINER, D.A. FERRINGTON & D.J. BIGELOW. 1999. Age-related chemical modification of the skeletal muscle sarcoplasmic reticulum Ca-ATPase of the rat. Mech. Ageing Dev. **107**: 221–231.
9. SELSBY, J.T., A.R. JUDGE, T. YIMLAMAI, *et al.* 2005. Life long calorie restriction increases heat shock proteins and proteasome activity in soleus muscles of Fisher 344 rats. Exp. Gerontol. **40**: 37–42.
10. LEEUWENBURGH, C., C.M. GURLEY, B.A. STROTMAN & E.E. DUPONT-VERSTEEGDEN. 2005. Age-related differences in apoptosis with disuse atrophy in soleus muscle. Am. J. Physiol. Regul. Integr. Comp. Physiol. **288**: R1288–1296.
11. RANALLETTA, M., H. JIANG, J. LI, *et al.* 2005. Altered hepatic and muscle substrate utilization provoked by GLUT4 ablation. Diabetes **54**: 935–943.
12. NAIR, K.S. 1995. Muscle protein turnover: methodological issues and the effect of aging. J. Gerontol. A. Biol. Sci. Med. Sci. **50**:107–112.

13. Bua, E.A., S.H. McKiernan, J. Wanagat, *et al.* 2002. Mitochondrial abnormalities are more frequent in muscles undergoing sarcopenia. J. Appl. Physiol. **92:** 2617–2624.

14. McKenzie, D., E. Bua, S. McKiernan, *et al.* 2002. Mitochondrial DNA deletion mutations: a causal role in sarcopenia. Eur. J. Biochem. **269:** 2010–2015.

15. Volpi, E., R. Nazemi & S. Fujita. 2004. Muscle tissue changes with aging. Curr. Opin. Clin. Nutr. Metab. Care. **7:** 405–410.

16. Russ, D.W. & J.A. Kent-Braun. 2004. Is skeletal muscle oxidative capacity decreased in old age? Sports Med. **34:** 221–229.

17. Carmeli, E., R. Coleman & A.Z. Reznick. 2002. The biochemistry of aging muscle. Exp. Gerontol. **37:** 477–489.

18. Carmeli, E. & A.Z. Reznick. 1994. The physiology and biochemistry of skeletal muscle atrophy as a function of age. Proc. Soc. Exp. Biol. Med. **206:** 103–113.

19. Dorrens, J. & M.J. Rennie. 2003. Effects of ageing and human whole body and muscle protein turnover. Scand. J. Med. Sci. Sports. **13:** 26–33.

20. Tipton, K.D. 2001. Muscle protein metabolism in the elderly: influence of exercise and nutrition. Can. J. Appl. Physiol. **26:** 588–606.

21. Bar-Shai, M., E. Carmeli, R. Coleman, *et al.* 2005. The effect of hindlimb immobilization on acid phosphatase, metalloproteinases, and nuclear factor-kappaB in muscles of young and old rats. Mech. Ageing Dev. **126:** 289–297.

22. Reznick, A.Z., O. Menashe, M. Bar-Shai, *et al.* 2003. Expression of matrix metaloproteinases, inhibitor, and acid phosphatase in immobilized hindlimbs of rats. Muscle Nerve **26:** 51–59.

23. Carmeli, E., M. Moas, A.Z. Reznick & R. Coleman. 2004. Matrix metalloproteinases and skeletal muscle: a brief review. Muscle Nerve. **29:** 191–197.

24. Jackman, R.W. & S.C. Kandarian. 2004. The molecular basis of skeletal muscle atrophy. Am. J. Physiol. Cell. Physiol. **287:** C834–C843.

25. Zarzhevsky, N., R. Coleman, G. Volpin, *et al.* 1999. Muscle recovery after immobilisation by external fixation. J. Bone Joint Surg. Br. **81:** 896–901.

26. Sen, C.K., S. Khanna, A.Z. Reznick, *et al.* 1997. Glutathione regulation of tumor necrosis factor-alpha–induced NF-kappa B activation in skeletal muscle–derived L6 cells. Biochem. Biophys. Res. Commun. **237:** 645–649.

27. Li, Y.P, R.J. Schwartz, I.D. Waddell, *et al.* 1998. Skeletal muscle myocytes undergo protein loss and reactive oxygen-mediated NF-kappaB activation in response to tumor necrosis factor alpha. FASEB J. **12:** 871–880.

28. Li, Y.P., Y. Chen, J. John, *et al.* 2005. TNF-alpha acts via p38 MAPK to stimulate expression of the ubiquitin ligase atrogin1/MAFbx in skeletal muscle. FASEB J. **19:** 362–370.

29. Li, Y.P., S.H. Lecker, Y. Chen, *et al.* 2003. TNF-alpha increases ubiquitin-conjugating activity in skeletal muscle by upregulating UbcH2/E220k. FASEB J. **17:** 1048–1057.

30. Li, Y.P. & M.B. Reid. 2000. NF-kappaB mediates the protein loss induced by TNF-alpha in differentiated skeletal muscle myotubes. Am. J. Physiol. Regul. Integr. Comp. Physiol. **279:** R1165–R1170.

31. Garcia-Martinez, C., M. Llovera, N. Agell, *et al.* 1994. Ubiquitin gene expression in skeletal muscle is increased by tumor necrosis factor-alpha. Biochem. Biophys. Res. Commun. **201:** 682–686.

32. Llovera, M., C. Garcia-Martinez, N. Agell, *et al.* 1994. Ubiquitin gene expression is increased in skeletal muscle of tumor-bearing rats. FEBS Lett. **3:** 311–318.

33. Llovera, M., C. Garcia-Martinez, N. Agell, *et al.* 1997. TNF can directly induce the expression of ubiquitin-dependent proteolytic system in rat soleus muscles. Biochem. Biophys. Res. Commun. **230:** 238–241.

34. Janssen-Heininger, Y.M., M.E. Poynter & P.A. Baeuerle. 2000. Recent advances towards understanding redox mechanisms in the activation of nuclear factor kappaB. Free Radic. Biol. Med. **28:** 1317–1327.

35. Hunter, R.B., E. Stevenson, A. Koncarevic, *et al.* 2002. Activation of an alternative NF-kappaB pathway in skeletal muscle during disuse atrophy. FASEB J. **16:** 529–538.

36. Hunter, R.B. & S.C. Kandarian. 2004. Disruption of either the Nfkb1 or the Bcl3 gene inhibits skeletal muscle atrophy. J. Clin. Invest. **114:** 1504– 1511.

37. ZARZHEVSKY, N., O. MENASHE, R. NAGLER, *et al.* 2000. Capacity for recovery and possible mechanisms in immobilization atrophy of young and old animals. Ann. N.Y. Acad. Sci. **928:** 212–225.
38. JI, L.L., M.C. GOMEZ-CABRERA, N. STEINHAFEL & J. VINA. 2004. Acute exercise activates nuclear factor (NF)-kappaB signaling pathway in rat skeletal muscle. FASEB J. **18:** 1499–1506.
39. HO, R.C., M.F. HIRSHMAN, Y. LI, *et al.* 2005. Regulation of I{kappa} B kinase and NF-{kappa}B in contracting adult rat skeletal muscle. Am. J. Physiol. Cell. Physiol. In press.
40. SEN, C.K. 1999. Glutathione homeostasis in response to exercise training and nutritional supplements. Mol. Cell. Biochem. **196:** 31–42.
41. DURHAN, W.J., Y.P. LI, E. GERKEN, *et al.* 2004. Fatiguing exercise reduces DNA binding activity of NF-kappaB in skeletal muscle nuclei. J. Appl. Physiol. **97:** 1740–1745.
42. RADAK, Z., H.Y. CHUNG, H. NAITO, *et al.* 2004. Age-associated increase in oxidative stress and nuclear factor kappaB activation are attenuated in rat liver by regular exercise. FASEB J. **18:** 749–750.
43. GLASS, D.J. 2003. Signalling pathways that mediate skeletal muscle hypertrophy and atrophy. Nat. Cell. Biol. **5:** 87–90.
44. CAI, D., J.D. FRANTZ, N.E. TAWA, JR., *et al.* 2004. IKKbeta/NF-kappaB activation causes severe muscle wasting in mice. Cell **119:** 285–298.
45. EPINAT, J.-C. & T.D. GILMORE. 1999. Diverse agents act at multiple levels to inhibit the REL/NF-κB signal transduction pathway. Oncogene **18:** 6896–6909.
46. REZNICK, A.Z., E.H. WITT, M. SILBERMANN & L. PACKER. 1992. The threshold of age in exercise and antioxidant action. *In* Free Radicals and Aging. I. Emerit & B. Chance, Eds. **62:** 423–427. Berkhauser Verlag. Basel.

The "Multiple Hormone Deficiency" Theory of Aging

Is Human Senescence Caused Mainly by Multiple Hormone Deficiencies?

T. HERTOGHE

University Centre of Charleroi, Belgium

ABSTRACT: In the human body, the productions, levels and cell receptors of most hormones progressively decline with age, gradually putting the body into various states of endocrine deficiency. The circadian cycles of these hormones also change, sometimes profoundly, with time. In aging individuals, the well-balanced endocrine system can fall into a chaotic condition with losses, phase-advancements, phase delays, unpredictable irregularities of nycthemeral hormone cycles, in particular in very old or sick individuals. The desynchronization makes hormone activities peak at the wrong times and become inefficient, and in certain cases health threatening. The occurrence of multiple hormone deficits and spilling through desynchronization may constitute the major causes of human senescence, and they are treatable causes. Several arguments can be put forward to support the view that senescence is mainly a multiple hormone deficiency syndrome: First, many if not most of the signs, symptoms and diseases (including cardiovascular diseases, cancer, obesity, diabetes, osteoporosis, dementia) of senescence are similar to physical consequences of hormone deficiencies and may be caused by hormone deficiencies. Second, most of the presumed causes of senescence such as excessive free radical formation, glycation, cross-linking of proteins, imbalanced apoptosis system, accumulation of waste products, failure of repair systems, deficient immune system, may be caused or favored by hormone deficiencies. Even genetic causes such as limits to cell proliferation (such as the Hayflick limit of cell division), poor gene polymorphisms, premature telomere shortening and activation of possible genetic "dead programs" may have links with hormone deficiencies, being either the consequence, the cause, or the major favoring factor of hormone deficiencies. Third, well-dosed and -balanced hormone supplements may slow down or stop the progression of signs, symptoms, or diseases of senescence and may often reverse or even cure them. If hormone deficiencies and imbalances are the major causes of senescence, what then is the treatment? Crucial for the treatment of senescent persons is to make a correct diagnosis by making up an anamnesis of all symptoms related to hormone disturbances, conducting a thorough physical examination, and getting laboratory tests done such as serum and 24-hour urine analyses. The physician should look not only for hormone deficiencies, including the mildest ones, but also for any alterations in hormone circadian cycles, and for the presence of any factors—nutritional, dietary, behavioral, lifestyle, environmental (including illumination and

Address for correspondence: T. Hertoghe, University Centre of Charleroi, Belgium.
thierry.hertoghe@skynet.be

Ann. N.Y. Acad. Sci. 1057: 448–465 (2004). © 2004 New York Academy of Sciences.
doi: 10.1196/annals.1322.035

indoor, outdoor, or dietary pollutants)—that cause or aggravate hormone deficiencies. After completion of the detailed diagnostic phase and obtaining and analyzing the results of the tests, treatment can start. In general, before supplying hormones, all other factors that contribute to senescence should be eliminated. After that, supplements of the missing hormones can then be administered, carefully respecting an appropriate timing of their intake, and eventually recommending measures such as lifestyle changes to restore circadian rhythmicity.

KEYWORDS: senescence; hormone deficiency; endocrine system; circadian cycles; hormone replacement therapy

INTRODUCTION

The search for the cause and remedy of human senescence (pathological aging), is a very ancient quest. It probably dates as far back as the dawn of mankind several million years ago. A multitude of scientific observations accumulating since the beginning of modern endocrinology, more than hundred years ago (modern endocrinology started in 1891 with the first efficient endocrine hormonal supplementation, namely the first thyroid treatment),[1] may possibly help to solve these enigmas.

SENESCENCE IS ASSOCIATED WITH A DECLINE OF MOST HORMONE LEVELS.

Elderly persons have significantly lower levels and productions of most important hormones compared to young adults. The levels of growth hormone (GH) and insulin-like growth factor-1 (IGF-1),[2–4] (nocturnal) melatonin,[5–7] TSH,[8] thyroid hormones (in particular the most active one, triiodothyronine),[9–14] calcitonin,[15,16] DHEA (in particular the sulphated form in the serum and its urinary 17-keto-metabolites),[17–20] aldosterone,[21] estrogens,[22,23] testosterone in men[24–26] and women[27] progressively decrease with age in adults. Even the secretion of insulin weakens with age, in particular when the needs are higher, in response to glucose intake, for example.[28]

Cortisol seems to be the exception among the major hormones. Its serum level has been found to be higher in the evening and at night in elderly persons compared to younger subjects. This is only the case in some studies,[29,30] but not all. In other studies, no significant differences in serum cortisol were observed in aging adults,[31–33] while in another investigation very variable levels were reported following the subgroup of elderly persons[34] (in some elderly subjects lower cortisol levels were found, in others normal or higher levels). In one study a lower serum cortisol was reported.[35] In fact, the serum level of cortisol is just one piece of the entire endocrine axis, an axis, which, in parallel with what happens to other endocrine axes, slowly deteriorates in other parts. Indeed, researchers have observed in persons of advanced age a lower daily cortisol secretion[36,37] and a lower cortisol metabolic clearance,[38,39] which means that cortisol diffuses slower and less from the serum into the target cells. Moreover, the cells of many tissues (such as the brain, including the hypothalamus, the liver, the kidneys, the muscles) of elderly persons may fail using cortisol adequately because they have a lower number of glucocorticoid receptors compared to young people,[40–44] Besides, the lower urinary excretion of the 17-

hydroxysteroids, the metabolites of cortisol[45,46] observed in aging persons constitutes a proof of the decrease in metabolic activity. The crucial problem in most elderly persons is not that they have too much cortisol, in fact they produce and use smaller amounts, but that they are considerably more depleted in anabolic hormones than in cortisol, the body's major catabolic hormone (see FIG. 1). The resulting imbalance with predominance of catabolism upon anabolism may trigger or accelerate the development of pathological aging.[47–49]

But not all hormone levels decline with age. Some hormones such as, for instance, parathormone[50] and somatostatin[51] increase.

Most studies on hormone levels are cross-sectional. A few longitudinal (prospective) studies are found. They too confirm the global age-related decline of the endocrine system.[18,20,52]

At what age start hormone levels to decline? For most hormones, the levels begin decreasing around age 25–35. The decline is generally slow and progressive, not sudden. The abrupt drop of female hormones at menopause constitutes an exception, although even menopause is preceded by many years of progressive weakening of ovarian function, recognizable by the appearance of irregular menstrual cycles and sometimes profound (female) hormone imbalances. What is the speed of decline in hormones? Always too quickly. The serum levels of DHEA[17–19] and GH[53,54] are often the first and the quickest to decrease by 1 to 5 % every year in both genders.[53,54] Serum testosterone in women declines more sharply. The levels in women of 40 years of age, for instance, have been reported to be more than 50 % lower than those found in younger women of 21 years of age. The fact that in sharp contrast to men, nearly all (more than 90%) of the serum testosterone of a woman derives from DHEA , whose level quickly declines, explains the discrepancy.[27,55]

Similar to cortisol,[38,39] the supply of many other hormones to the target cells is additionally reduced in elderly persons by a gradual decrease in the hormones' metabolic clearance. Hormones remain a longer time in the blood, and, consequently, penetrate less into the target cells. This reduction has been confirmed for thyroid hormones,[56,57], estrogens[58] and testosterone.[59,60] The metabolic activity of thyroid hormones is further decreased by an age-related reduction of the conversion of poorly active thyroxine or T_4 into the much more active triiodothyrone ot T_3)[14] But even if hormones do penetrate into the target cells, they may not easily be used because the number of receptors inside of the target cells for them often also decreases with age (easily more than 50 % less in old individuals). This has been reported for critical hormones such as thyroid hormones,[61] glucocorticoids[40–44] and insulin.[62,63]

SENESCENCE IS ASSOCIATED WITH DISTURBANCES, SOMETIMES PROFOUND, OF MOST CIRCADIAN ENDOCRINE CYCLES

In young adults, most hormones are secreted according to specific circadian patterns. The persistence of a well-balanced endocrine circadian rhythmicity is essential to health. Higher concentrations of hormones are necessary at certain times of the day or the night when the needs are higher.

Human senescence alters slowly but progressively many of the hormone rhythms.[64–74] It decreases the amplitude of the circadian rhythms in the serum of hormones such as GH,[64] melatonin,[65,66] TSH,[67,68] DHEA sulphate,[71] aldosterone,[72] and testosterone.[73–74] For some hormones, such as testosterone, the circadian rhythm even disappears in elderly men.[73,74]

More bothersome is that senescence tends to advance in time (phase advance) circadian cycles of vital hormones such as growth hormone,[64] melatonin, TSH and cortisol,[65] an effect comparable to the shift created by an eastward flight from Paris to Moscow or Dallas to New York, bringing a person to a time zone one to two hours earlier. All these shifts would not upset us very much if they were of equal importance and in the same direction, but they often are not. While some endocrine cycles remain relatively stable, many others phase advance, and some even undergo phase delays. The changes in phase are in general different for each hormone. The unequal changes progressively worsen and desynchronises the hormone rhythms in elderly persons. The desynchronisation of the endocrine rhythms confuses the aging person. It is a shift to hormone chaos. It affects the quality of life and the health in the same way as the jet lag or shift work affects health. At moments, when a person should sleep, signals are send—through inappropriate secretion of hormones—that it is time to be active. At other times of the day, when the person wants to be active, sleep-inducing hormones such as melatonin may be secreted, while the levels of hormones that keep a person awake and provide energy such as cortisol, may desperately be low.

MOST AGING THEORIES CAN BE EXPLAINED AND SUPPORTED BY THE "MULTIPLE HORMONE DEFICIENCY" THEORY OF AGING

Most "aging theories", which are explanations of proven or presumed mechanisms leading to senescence, can themselves be explained and supported by the concept that senescence is mainly caused by the age-related decline of hormones levels, resumed in the "multiple hormone deficiency" theory of aging.

Let us first consider the "free radical" theory of aging, the actually best accepted explanation for senescence. Its concept is that pathological aging results mainly from increased free radical damage, itself a consequence of the lower antioxidant activity in the serum of elderly persons that allows higher concentrations of free radical products.[75,76]

A Decline in Hormone Levels May Be at the Origin of this Aggression. How?

First, lower endocrine states such as hypothyroidism may alter the quality of the intestinal cells and impair the absorption of indispensable foods, including antioxidant vitamins and trace elements, in the digestive tract of elderly persons.[77] Next, lower endocrine states bring in a hypometabolic state that slows down the endogenous production of antioxidant enzymes. Many hormones are potent stimulators of the formation of antioxidant systems. Third, and possibly the most important factor, is that lower endocrine states automatically decrease the antioxidant activity because many hormones are themselves potent antioxidants! Melatonin, for example, is about twice a more powerful antioxidant than vitamin E, and four times more than glutathione and ascorbic acid (vitamin C). In addition, melatonin works faster and more effectively than these other antioxidants.[78] It neutralizes efficiently the excessive free radical formation caused by GSM phones,[79] alcohol,[80] contusion head trauma,[81] mycotoxins that are contaminants in food,[82] oxidative injury induced by ischemia,[83] and other factors.

Peptide hormones such as thyroid hormones,[84–89] parathyroid hormone,[90] and insulin[91] tend to stimulate the production of antioxidant systems (superoxyde dismutase, catalase, glutathione peroxidase, general antioxydant activity, etc.), while lipid (steroid) hormones such as DHEA (dehydroepiandrosterone),[92–94] hydrocortisone,[95] estrogens,[75,96–103] testosterone[101–105] may have direct antioxidant effects.

Hormone deficiencies such as found in aging individuals can also be at the source of other mechanisms that favor senescence such as glycation,[106] cross-linking of proteins,[107] imbalanced apoptosis system,[102,108–112] failure of (macroscopic and microscopic) repair systems,[113–125] including the repair systems of damaged DNA[115,123] as various observations suggest. DNA repair systems, for example, may fail easier in the presence of a hormone deficiency, but normalize again with addition of melatonin[115] and insulin[115] for example.

The subtle hormone deficits that are found in elderly persons can similarly be the major cause of the age-related decline in immune function,[126–211] which is believed to be another chief mechanism of senescence. Growth hormone (GH) and IGF-1 (which contributes to many effects of GH), for example, can stimulate the immune system at all levels. On one hand, they stimulate the cellular immunity. For instance, GH and/or IGF-1 promote in animals the growth, the regeneration, the cellularity and the thymosine-beta-4 production of the thymus,[126–128] the production of lymphocytes T,[129] and the reactivity of lymphocytes to antigens.[130] In (by head injury) immunodepressed humans, IGF-1 therapy beneficially increased the ratio of CD4 (helper) on CD8 (suppressor cells).[131] In other studies, GH was reported to restore immunity by increasing the counts of leucocytes, lymphocytes and thymocytes, by boosting natural killer cell activity, and elevating the serum IgM levels in hypophysectomized rats and the serum IgG levels in normal rats.[132] GH also stimulates phagocytic activity of neutrophils and monocytes in humans.[133]

On the other hand, GH and IGF-1 stimulate also the humoral immunity. IGF-1 has been reported to stimulate the production of lymphocytes B.[134] GH-therapy elevates the production of immunoglobulins in (by surgery) immunodepressed humans,[135] and stimulates the resistance against bacterial and viral infections in rats and humans.[136–137] The existence of all these effects help understand why, when GH and IGF-1 levels are low, the immune system suffers. But GH and IGF-1 are not the main immune-stimulating hormones. Thyroid hormones, melatonin and if judiciously used—at not too high doses—cortisol, are more potent immunity boosters. Any decrease in the levels of these hormones will weaken the immune system even more.

Moreover, the progressive decline in endocrine function with aging may also be a contributing factor to or at the root of "genetic" causes of aging such as genetic limits to cell proliferation[114,212–221](the Hayflick limit for fibroblast proliferation, for example), poor gene polymorphisms,[226–244] progressive telomere shortening,[245–246] and possibly also the activation of genetic "dead programs." Dead programs are considered by certain prominent researchers to be the prime causes of aging (Pierpaoli's aging clock theory for example).[247] Hormone deficiencies in parents may favor the occurrence of genetic inequalities in their offspring. It is also possible that genetic factors can be the cause or major contributing factor to hormone deficiencies.

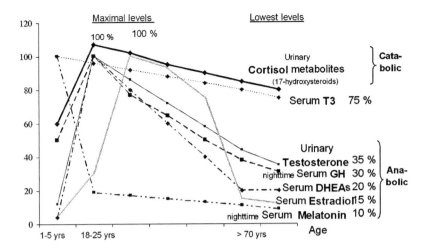

FIGURE 1. Schematic representation of the decline of hormone (metabolite) levels with aging. The much sharper decline in the levels of anabolic hormones compared to that of cortisol (metabolites) and other catabolic hormones, may accelerate senescence. DHEAs = DHEA sulfate; urinary testosterone values valid for men, serum estradiol values for women. (Adapted from Refs. 2, 3, 5–8, 13, 17, 18, 22, 23, 26, 45, 46.)

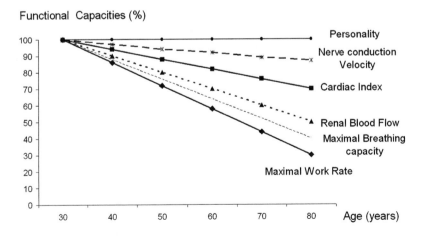

FIGURE 2. Representation of the progressive decline of the functional capacities with aging, a decline that bewgins for most around age 30. Cross-sectional data for male partic-ipan ts from the Baltimore Longitudinal Study. (G.T. Baker III & G.R. Martin, in *Geriatric Medicine* 1(1): 4, Fig. 1.1, as adapted from Refs. 248–250.)

SENESCENCE RESEMBLES HORMONE DEFICIENCIES

During human senescence, many functional capacities progressively decline in parallel to the general decline in hormone levels. FIGURES 1 and 2 are schematic representations of such progressive losses.[248–250]

Many, if not most, of the signs, symptoms and diseases of senescence, including a lower quality of life and fatigue,[251–323] depression,[278,311,324–377] memory loss and Alzheimer's disease,[378–400] insomnia,[401–408] progressive loss of sexual desire, sensitivity and potency,[409–421] cardiovascular diseases (hypercholesterolemia,[422–436] atherosclerosis,[437–452] arterial hypertension,[423,424,453–472] coronary heart insufficiency and other forms of cardiac disease,[439–441,460,473–487] stroke and various cerebrovascular deficits,[488–498] obesity,[251,402,422,499–521] diabetes,[432,522–536] rheumatism,[537–552] osteoporosis,[22,252,253,499,553–589] cancer,[590-658] increased mortality,[192,478,658-667] are similar to those of hormone deficiencies, and their appearance can be caused or favored by hormone deficiencies.

When we see and examine elderly persons, are we looking at the marks of pathological aging or in fact of hormone deficiencies? Put to the extreme: does a separate disease or physiological state called senescence, distinctive from any known endocrine deficiency or imbalance, exist or not? The question remains open: It is not impossible that when we see aging signs in people, we are in fact looking at signs of hormone deficiencies and that senescence is mainly a mixing of multiple hormone deficiencies, a multiple hormone deficiency syndrome.

HORMONE TREATMENTS THAT CURE OR ATTENUATE SENESCENCE CONSEQUENCES

If hormone deficiencies and imbalances are the major causes of senescence, then logically the correction of these endocrine disorders with hormone supplements should cure or stop the progression of signs, complaints and diseases related to senescence, and prolong life. What is the evidence here for?

For each symptom or disease closely related to aging, there exist a multitude of scientific experiments with various types of hormone (replacement) therapy that have given significant improvements.

For example, hormone administration to patients or animals has been reported to boost the quality of life and to decrease fatigue,[251-258,273,274,279,311,554,669–748] alleviate depression,[278,311,358,679,680,701,702,749–781] improve the failing memory and slow down Alzheimer's disease,[396,397,749,782–813] cure or help insomnia,[682,683,749,812–826] improve low sexual desire, sensitivity, and potency[358,359,411,413,415,416,711–714,735,736,829–851] ameliorate cardiovascular diseases (including hypercholesterolemia,[427,499,553,850–873] atherosclerosis,[436,441,449,856,876–901] arterial hypertension,[455,456,466,881,882,902–926] coronary heart insufficiency and other forms of cardiac disease,[736,856,861,887,927–948] stroke and various cerebrovascular deficits,[489,490,856,950–961] reduce obesity and poor body compostion[258,402,514,553,669–671,852,962–1008] counter diabetes,[514,523,529,860,983,1003–1005,1007–1032] calm down rheumatism,[543–545,1035–1052] correct osteoporosis,[253,556–558,568,960,961,997,1007,1051–1086] cancer,[138,139,685–689,938,1089–1154]   and   seem   to   prolong survival.[202,689,1115,1128,1129,1150,1151,1155–1164] The beneficial effects only appear or are generally greater when there is an endocrine deficiency prior to the treatment.

Much of the available data of the beneficial effects of hormone therapies comes from a growing number of animal studies, mostly short-lived animals such as rodents. Controlling senescence in species with a short lifespan may not always give the adequate information. Humans belong to a long-living species. It is likely that they possess better physiological mechanisms, in particular of endocrine nature, that make them live longer. Much of the data on humans comes from patients with severe disease such as AIDS, cardiovascular disease, septic shock, where hormones therapies often decrease the severity of disease and sometimes prolonged survival. Yet, many elderly persons who are affected by senescence are not necessarily overtly sick patients. The beneficial effects of hormone therapies on sick persons, may not fully inform on what happens to persons without overt disease when they preventively take hormone treatments to correct mild hormone deficiencies. Another point to highlight is that the evidence on hormone treatments applied to human subjects, although abundant, is usually based on investigations with small numbers of persons individually very heterogeneous and different from each other.

Aside from these objections, it cannot be denied that solid scientific evidence does exist in support of the beneficial, if not prominent, role of hormone replacement against several illnesses of senescence.

ABOUT THE "NEGATIVE" STUDIES ON HORMONES

Some adverse impacts of hormone replacement have been reported in a minority of studies, but the data is worth analyzing.

In one publication, for example, GH treatment was found to double the death rate in critically ill patients.[1165,1166] However, to draw from this finding the conclusion that GH replacement should not be given to adults, would be premature, as important biases are found in the study. First, the study subjects were not healthy or moderately ill individuals, but severely sick, critically ill patients in intensive care units (ICU). The poor status of ICU-patients is not representative of the healthier state of the large number of elderly persons.

Second, it is probable that the treatment was ill-balanced. It is possible that GH may be given only to critically ill patients on the condition they receive simultaneously glucocorticoids. Why? Because GH lowers by −20 to −40 % cortisol's production, serum levels and metabolic activity (reflected by a drop in the urinary excretion of 17-hydroxysteroids),[1167–1169] This is disturbing when taking into account that a good adrenal function with a high (stimulatable) cortisol secretion is absolutely essential for the survival of critically ill patients, especially when they suffer from severe infections.[192] A deficiency in cortisol—caused or aggravated by GH therapy—may be the explanation for the excessive mortality rate found in the critically ill patients who received GH. A greater number of deaths by infections and multiple organ failure were found in these patients, conditions where cortisol treatment—at least when given early—greatly helps surviving.[193–202,1159,1171] Third, supraphysiological doses, 10 to 75 times higher than necessary for the correction of a partial or even complete

GH deficiency, were given to the patients in a critical condition.[1170] Imagine that dying persons would receive 100 liters of water per day in the place of the optimal 1.5 to 2 liters a day. Would the excess liquid not kill the patients?

The observations should be an incentive more for physicians to use safe physiological doses of hormones and avoid overdosing, and to better learn to balance the endocrine system with perhaps multiple hormone replacement, rather than using one hormone treatment alone.

In two very large studies an increased risk of breast cancer with female hormone replacement was reported.[1172,1173] Media coverage was impressive and installed fear in women in need of sex hormones. The major bias here is that all of the participants in the study took structurally modified derivatives of the safer endogenously produced progesterone. The women took in general also derivatives of the human estrogens. Modified molecules are harder to metabolize by the body and have different, often undesirable, metabolic effects. A report of a 20 to 50 % reduction in breast cancer incidence in women with breast cysts who took during years percutaneous or oral bio-identical progesterone, is on this point enlightening,[1116] as it is in sharp contrast with what happens with the non bio-identical progestogens in the two studies. These derivatives of natural progesterone were the greatest wrongdoers as they are the ones that seem to increase the most the breast cancer risk. More instructive is the fact that the combined use of synthetic progestogen medroxypreogesterone and horse—not human—conjugated estrogens, for example, highly stimulate the proliferation of epithelial cells of the human breast glands in vitro, while the addition of bio-identical estradiol and progesterone to epithelial cell cultures does it 20 times less![1174,1175] Thirdly, the oral route of intake of estrogen that nearly all women took during the two studies, is the worst. In the British study, only 30 women on more than a million did take the form that is actually considered as the safest, transdermal estradiol gel, for a short period of time, a maximum of two of the four years of the study, as the product appeared on the U.K. market only two years after the start of the 4-year study. The transdermal estradiol gel is a better form than the other transdermal form, the estradiol patch, as it provides—thanks to a daily application - more stable estrogen levels, and is, for this reason, also easier to be correctly balanced with progesterone treatment. The conclusions of the two big studies can, therefore, not be automatically applied to the use of safer forms.[496,497] Interestingly, the use of the same toxic, foreign-to-the-human-body female hormone replacement therapies was shown in an Australian study not to increase the risk of breast cancer if testosterone—also a hormone that decreases with age in women—is added to the treatment.[1149] This normalization of the risk is probably due to a better balanced treatment. Further, in studies on women with previous breast cancer who take female hormone replacement after the cancer, the survival is generally better (up to a 80 % decrease in 5-year mortality!) and breast cancer recurrence is often less frequent![1116,1163]

Here again, such reports should be an incentive more for physicians to use bio-identical hormones, forms that are the best adapted to the body, and not derivatives, to use the better routes of intake (oral route for estrogens in not advisable), to use physiological doses and not supraphysiological, and to balance better the endocrine system, and in particular to add small doses of androgens to female hormone replacement therapy. A well-balanced multiple hormone replacement therapy of aging adults provides greater efficacy and safety.

MULTIPLE HORMONE DEFICIENCIES

If senescence is mainly due to hormone deficiencies, it is certainly not due to one hormone deficiency alone as no single hormone therapy has on its own been able to totally halt or completely reverse senescence. It is much more likely that senescence is due to a combination of many hormone deficiencies as many hormone levels are lower in aging persons. Differences in rate and severity of senescence between people can be caused by individual differences in the amount of hormones left in the body and in the degree of desynchronisation of the circadian cycles of these hormones.

The view that senescence may be due to one or more hormone deficiencies was already suggested by Brown-Sequard, one of the pioneers of modern hormone replacement, more than 100 years ago at the end of the 19th century, when he injected himself with extracts of monkey testicles to treat his failing energy, including sexual energy; and senescent looks.[1176] The view that it may be due to a series of hormone deficiencies had been suggested by Pincus, the inventor of the first birth control pill, some 50 years ago. Jacques Hertoghe, my father, in the late 1970s proposed the name *multiple deficiency syndrome* to all pathologies that are difficult to treat by a single treatment, meaning that most of these ailments from chronic fatigue and depression to diabetes, obesity, osteoporosis, cardiovascular diseases and cancer, are in fact due to multiple hormone and nutritional, including vitamin, trace element and mineral deficiencies.[1177] A first argument to support the concept that pathological aging (senescence) is mainly due to a multitude of hormone deficiencies, a multiple hormone deficiency syndrome, rather than to a multitude of hormone, vitamin, mineral, trace element deficiencies is that aging is associated with a polyhormonal decline in all men and women, while nutrients do not necessarily decrease in well-nourished elderly persons. A second argument is that the adverse effects of hormone deficiencies are generally far greater than the consequences of nutritional deficiencies. A last, but not least, argument is that many physical signs of senescence can be reversed generally with far greater efficacy by corrective hormone therapy than what is usually seen with the use of nutritional—nonhormonal—supplements.

IS SENESCENCE TOTALLY CAUSED BY ENDOCRINE DEFICIENCIES?

Senescence is due to multiple causes that damage tissues such as trauma, infections, toxics, stress, and not only endocrine dysfunction. Nevertheless, hormone deficiencies may profoundly influence the frequency and impact of these aggressors. Poor repair after trauma may be due to hormone deficiencies, as may be proneness to infections and increased vulnerability to toxics and stress, providing again a central role to endocrine dysfunction in the development of senescence.

TREATING SENESCENCE: A MULTIPLE HORMONE REPLACEMENT THERAPY

If the progressive decrease in hormone levels, the gradual alterations in hormone rhythms and the increasingly greater hormone imbalances cause senescence, then supplying the missing hormones, a multiple hormone replacement therapy, at the right time of the day and in an adequate balance, should form the best treatment of senescence. The treatment should also include the correction of any factor that decrease or increase hormone levels or disturb their circadian rhythms. Toxic foods must be avoided, healthy foods increased. The lifestyle should be optimized, including getting enough sleep and minimizing stress.

Which Hormones?

Which hormones are implicated in the multiple hormone deficiency syndrome?

In principle, all hormones whose levels decline with advancing age, from the sex hormones to growth hormone, from melatonin, DHEA to thyroid hormones. In underweight elderly persons even a too low serum (fasting) level of the major anabolic hormone insulin may contribute to catabolism, and require a supplementation in very small doses.

Must We Wait Longer before Treating?

If the amount of available scientific evidence on hormones and diseases, is today considerable and our experience with some hormone therapies important (thyroid hormones have been administered since more than 110 years, sex hormones for approximately 70 years), further studies are nevertheless recommended.

Some hormones may not be as efficient as initially thought. A treatment of young adults for example with androstenedione, an androgen hormone produced by the adrenal glands and the ovaries and available over the counter in health food stores, was reported to be inefficient to increase exercise resistance in young men, even at high doses.[1178]

Other unexpected risks may be discovered. DHEA for example, is a hormone that has been reported to be among the safest of all hormones and a potent anti-cancer hormone, including against breast cancer in rodent studies. In humans too, it may help prevent breast cancer as a lower level of its sulphated form is found in breast cancer patients (post- and premenopausal women confounded), for example. The higher the serum levels of DHEA sulphate, the lower the risk.[1179–1183] Nevertheless, in two reports[1184,1185] (on 3 studies)[1186] a higher breast cancer risk was observed in postmenopausal women with high serum DHEA sulphate. The information should incite us to be cautious against the use of DHEA therapy alone in postmenopausal women. It is possible that a polytherapy associating DHEA to female hormone replacement is safer in postmenopausal women. Other intriguing facts such as the association of an increased risk of diabetes for women with high serum testosterone levels, while high levels are linked with a reduced risk in men, incite us to be prudent and limit therapy doses of testosterone in women to the physiological range,[532,533,1186] and take be more careful with women with diabetes risk.

The more and better information we get, the better insight we will have in endocrine treatments, and better know how to increase the safety and efficacy of these treatments. However, the constant need for more information, basic to most medical therapies, should not constitute an excuse for refusing hormone replacement treatments to elderly persons who desperately need them to increase their quality of life and counter the devastating effects of senescence. If physicians wait too long before deciding on endocrine treatment, their elderly patients may not be there anymore to profit from the decision.

The only reasons for not treating are special high-risk pathologies, a lack of experience and competency or some philosophical opposition. The evidence is rather predominantly in favor of providing prudent hormone replacement therapies to aging adults in the presence of regular check-ups and cancer screening, and firmly putting into practice all the above-mentioned precautions for hormone use.

If no solid scientific grounds exist to justify waiting longer to correct the age-related endocrine deficiencies prudently with small and personalized doses of bio-identical hormones, there is, however, solid evidence to avoid *long-term* use of some structurally-foreign-to-the-human- body hormones (such as medroxyprogesterone acetate and conjugated estrogens in postmenopausal women) and using supraphysiological doses and inadequate routes of administration (oral route for estrogens, for example) .

CONCLUSION

Is Senescence Greatly Evitable by Restoring the Hormone Levels to the Youthful Ones?

In the light of the vast scientific literature, of which only a fraction is presented here in the references, it seems likely that the subtle and numerous deficits in hormones with advancing age, is the primary cause of human senescence (pathological aging). The correction of all these deficits by an adequate multiple hormone replacement therapy constitute then the best solution we have for the moment to prevent and treat most of pathological aging. Today, we are not yet able to completely stop senescence by hormone treatments as not all the hormones that decline with age are available as medications, but the future is promising.

REFERENCES (Keyed to Sections)

INTRODUCTION

1. MURRAY, G. 1891. First treatment of a case of myxoedema with thyroid gland.

SENESCENCE IS ASSOCIATED WITH A DECLINE OF MOST HORMONE LEVELS

GH
2. RUDMAN, D. *et al.* 1981. Impaired growth hormone secretion in the adult population: relation to age and adiposity. J. Clin. Invest. **67:** 1361–1369.
3. BANDO, H. *et al.* 1991. Impaired secretion of growth hormone-releasing hormone, growth hormone and IGF-I in elderly men. Acta Endocrinol. **124:** 31–33.
4. IRANMANESH, A., G. LIZARRALDE & J.D. VELDHUIS. 1991. Age and relative adiposity are specific negative determinants of the frequency and amplitude of growth hormone (GH) secretory bursts and the half-life of endogenous GH in healthy men. J. Clin. Endocrinol. Metab. **73:** 1081–1088.

Melatonin
5. WALDHAUSER, F., G. WEISZENBACHER, E. TATZER, *et al.* 1988. Alterations in nocturnal serum melatonin levels in humans with growth and aging. J. Clin. Endocrinol. Metab. **66:** 648–652.

6. WALDHAUSER, F., J. KOVACS & E. REITER. 1998. Age-related changes in melatonin levels in humans and its potential consequences for sleep disorders. Exp. Gerontol. **33:** 759–772.
7. LANGER, M., J. HARTMANN, H. TURKOF & F. WALDHAUSER. 1997. Melatonin in the human: an overview. Wien Klein. Wochenschr. **109:** 707–713.

TSH

8. WIENER, R., R.D. UTIGER, R. LEW & C.H. EMERSON. 1991. Age, sex, and serum thyrotropin concentrations in primary hypothyroidism. Acta Endocrinol. **124:** 364–369.

Thyroid Hormones

9. BERMUDEZ, F., M.I. SURKS & J.H. OPPENHEIMER. 1975. High incidence of decreased serum triiodothyronine concentration in patients with nonthyroidal disease. J. Clin. Endocrinol. Metab. **41:** 27–40.
10. HESCH, R.D., J. GATZ, H. JUPPNER & P. STUBBE. 1977. TBG-dependency of age-related variations of thyroxine and triiodothyronine. Horm. Metab. Res. **9:** 141–146.
11. HERRMANN, J. *et al.* 1981. Thyroid function and thyroid hormone metabolism in elderly people. Low T3-syndrome in old age? Klin. Wochenschr. **59:** 315–323.
12. DJORDJEVIC, M.Z., N.D. PAUNKOVIC, V.B. DJORDJEVIC-LALOSEVIC & D.S. PAUNKOVIC. 1990. The effect of age on *in vitro* thyroid function tests in adult patients on a chronic hemodialysis program. Srp. Arh. Celok. Lek. **118:** 291–293.
13. SPAULDING, S.W. 1987. Age and the thyroid. Endocrinol. Metab. Clin. North Am. **16:** 1013–1025.
14. SMEULERS, J. *et al.* 1979. Decreased triiodothyronine (T3) production in constant reverse T3 production in advanced age. Ned. Tijdschr. Geneeskd. **123:** 12–15.

Calcitonin

15. BUCHT, E. *et al.* 1995. Serum calcitonin forms and concentrations in young and elderly healthy females. Calcif. Tissue Int. **56:** 32–37.
16. PEDRAZZONI, M. *et al.* 1988. Calcitonin levels in normal women of various ages evaluated with a new sensitive radioimmunoassay. Horm. Metab. Res. **20:** 118–119.

DHEA

17. ORENTREICH, N., J.L. BRIND, R.L. RIZER & J.H. VOGELMAN. 1984. Age changes and sex differences in serum dehydroepiandrosterone sulfate concentrations throughout adulthood. J. Clin. Endocrinol. Metab. **59:** 551–555.
18. ORENTREICH, N. *et al.* 1992. Long-term longitudinal measurements of plasma dehydroepiandrosterone sulfate in normal men. J. Clin. Endocrinol. Metab. **75:** 1002–1004.
19. LABRIE, F. *et al.* 1997. Marked decline in serum concentrations of adrenal C19 sex steroid precursors and conjugated androgen metabolites during aging. J. Clin. Endocrinol. Metab. **82:** 2396–2402.
20. NAFZIGER, A.N., S.J. BOWLIN, P.L. JENKINS & T.A. PEARSON. 1998. Longitudinal changes in dehydroepiandrosterone concentrations in men and women. J. Lab. Clin. Med. **131:** 316–323.

Aldosterone

21. HEGSTAD, R. *et al.* 1983. Aging and aldosterone. Am. J. Med. **74:** 442–448.

Estrogens

22. KHOSLA, S. *et al.* 1998. Relationship of serum sex steroid levels and bone turnover markers with bone mineral density in men and women: a key role for bioavailable estrogen. J. Clin. Endocrinol. Metab. **83:** 2266–2274.
23. SHERMAN, B.M., J.H. WEST & S.G. KORENMAN. 1976. The menopausal transition: analysis of LH, FSH, estradiol, and progesterone concentrations during menstrual cycles of older women. J. Clin. Endocrinol. Metab. **42:** 629–636.

Testosterone

24. DESLYPERE, J.P. & A. VERMEULEN. 1985. Influence of age on steroid concentrations in skin and striated muscle in women and in cardiac muscle and lung tissue in men. J. Clin. Endocrinol. Metab. **61:** 648–653.
25. DESLYPERE, J.P. & A. VERMEULEN. 1984. Leydig cell function in normal men: effect of age, life-style, residence, diet, and activity. J. Clin. Endocrinol. Metab. **59:** 955–962.
26. MORER-FARGAS, F. & H. NOWAKOWSKI. 1965. The urinary excretion of testosterone in males. Acta Endocrinol. **49:** 443–452.
27. ZUMOFF, B., G.W. STRAIN, L.K. MILLER & W. ROSNER. 1995. Twenty-four-hour mean plasma testosterone concentration declines with age in normal premenopausal women. J. Clin. Endocrinol. Metab. **80:** 1429–1430.

Insulin: Reduced Acute Responses to Glucose in Older Subjects

28. DECHENES, C.J., C.B. VERCHERE, S. ANDRIKOPOULOS & S.E. KAHN. 1998. Human aging is associated with parallel reductions in insulin and amylin release. Am. J. Physiol. **275:** E785–791.

Cortisol

29. FERRARI, E. *et al.* 2000. Pineal and pituitary–adrenocortical function in physiological aging and in senile dementia. Exp. Gerontol. **35:** 1239–1250.
30. MAGRI, F. *et al.* 2000. Association between changes in adrenal secretion and cerebral morphometric correlates in normal aging and senile dementia. Dement. Geriatr. Cogn. Disord. **11:** 90–99.
31. JENSEN, H.K. & M. BLICHERT-TOFT. 1971. Serum corticotrophin, plasma cortisol, and urinary secretion of 17 -ketogenic steroids in the elderly (age group 66–94 years). Acta Endocrinol. **66:** 25.
32. GRAY, A., H.A. FELDMAN, J.B. MCKINLAY & C. LONGCOPE. 1991. Age, disease, and changing sex hormone levels in middle-aged men: results of the Massachusetts Male Aging Study. J. Clin. Endocrinol. Metab. **73:** 1016–1025.
33. FERRARI, E. *et al.* 2001. Age-related changes of the hypothalamic–pituitary–adrenal axis: pathophysiological correlates. Eur. J. Endocrinol. **144:** 319–329.
34. LUPIEN, S. *et al.* 1996. Longitudinal study of basal cortisol levels in healthy elderly subjects: evidence for subgroups. Neurobiol. Aging **17:** 95–105.
35. ZIETZ, B., S. HRACH, J. SCHOLMERICH & R.H. STRAUB. 2001. Differential age-related changes of hypothalamus–pituitary–adrenal axis hormones in healthy women and men: role of interleukin 6. Exp. Clin. Endocrinol. Diabetes **109:** 93–101.
36. SHARMA, M. *et al.* 1989. Circadian rhythms of melatonin and cortisol in aging. Biol. Psychiatry **25:** 305–319.
37. ROMANOFF, L.P. *et al.* 1961. The metabolism of cortisol 4-C14 in young and elderly men. 1. Secretion rate of cortisol and daily excretion of tetrahydrocortisol, allotetrahydrocortisol, tetrahydrocortisone, and cortalone (20 alpha and 20 beta). J. Clin. Endocrinol. Metab. **21:** 1413.
38. SAMUELS, L.T. 1957. Factors affecting the metabolism and distribution of cortisol as measured by levels of 17 -hydroxycorticosteroids in blood. Cancer **10:** 7465.
39. WEST, C.D. *et al.* 1961. Adrenocortical function and cortisol metabolism in old age. J. Clin. Endocrinol. Metab. **21:** 1197–1207.
40. TIMIRAS, P.S. 1995. Steroid-secreting endocrines: adrenal, ovary, testis.: III. The adrenal cortex. *In* Hormones and Aging. P. S. Timiras, W. D. Quay, A. Vernadakis, Eds.: 28–35. CRC Press. Boca Raton-New York.
41. SHARMA, R. & P.S. TIMIRAS. 1988. Glucocorticoid receptors, stress, and aging. *In* Regulation of Neuroendocrine Aging. A. V. Everit & J. R. Walton, Eds.: 98. Karger. Basel.
42. ROTH, G.S. 1974. Age-related changes in specific glucocorticoid binding by steroid-responsive tissues of rats. Endocrinology **94:** 82–90.
43. LATHAM, K.R. & C.E. FINCH. 1976. Hepatic glucocorticoid binders in mature and senescent C57BL/6J male mice. Endocrinology **98:** 1480–1489.

44. SHARMA, R. & P.S. TIMIRAS. 1987. Age-dependent regulation of glucocorticoid receptors in the liver of male rats. Biochim. Biophys. Acta **930:** 237–243.
45. WEYKAMP, C.W. *et al.* 1989. Steroid profile for urine: reference values. Clin. Chem. **35:** 2281–2284.
46. RASMUSON, S. *et al.* 2001. Increased glucocorticoid production and altered cortisol metabolism in women with mild to moderate Alzheimer's disease. Biol. Psychiatry **49:** 547–552.

Imbalaced Anabolic/Catabolic Hormonal Balance with Age

47. SAPOLSKY, R.M., L.C. KREY & B.S. MCEWEN. 1986. The neuroendocrinology of stress and aging: the glucocorticoid cascade hypothesis. Endocr. Rev. **7:** 284–301.
48. VALENTI, G. 2004. Neuroendocrine hypothesis of aging: the role of corticoadrenal steroids. J. Endocrinol. Invest. **27**(Suppl. 6): 62–63.
49. DEBIGARE, R. *et al.* 2003. Catabolic/anabolic balance and muscle wasting in patients with COPD. Chest **124:** 83–89.

Parathormone: Increased Serum Level with Age

50. MCKANE, W.R. *et al.* 1996. Role of calcium intake in modulating age-related increases in parathyroid function and bone resorption. J. Clin. Endocrinol. Metab. **81:** 1699–1703.

Somatostatin: Increased Serum Level with Age

51. ROLANDI, E. *et al.* 1987. Twenty-four-hour beta-endorphin secretory pattern in the elderly. Acta Endocrinol. **115:** 441–446.

Longitudinal Prospective Studies

52. GAPSTUR, S.M. *et al.* 2002. Serum androgen concentrations in young men: a longitudinal analysis of associations with age, obesity, and race. The CARDIA male hormone study. Cancer Epidemiol. Biomarkers Prev. **11:** 1041–1047.

Quick Decline of GH, DHEA in Men and Women, and Testosterone in Women with Age

53. RUDMAN, D. & U.M.P. RAO. 1992. The hypothalami–growth hormone–somatomedin C axis: the effect of Aging. *In* Endocrinology and Metabolism in the Elderly. J. C. Morley & S. O. Korenman, Eds.: 28–35. Blackwell Science. Boston.
54. VERMEULEN, A. 1976. Plasma levels and secretion rate of steroids with anabolic activity in man. Environ. Qual. Saf. Suppl. **5:** 171–180.

Testosterone Derives in Women for More Than 90% from the Much More Quickly Declining Serum DHEA[77]

55. LABRIE, F. *et al.* 1998. DHEA and the intracrine formation of androgens and estrogens in peripheral target tissues: its role during aging. Steroids **63:** 322–328.

Decreases of Metabolic Clearance of Hormones with Age

56. GREGERMAN, R.I., G.W. GAFFNEY, N.W. SHOCK, S.E. CROWDER. 1962. Thyroxine turnover in euthyroid men with special reference to changes with age. J. Clin. Invest. **41:** 2065–2074.
57. KATZEFF, H.L. 1990. Increasing age impairs the thyroid hormone response to overfeeding. Proc. Soc. Exp. Biol. Med. **194:** 198–203.
58. LONGCOPE, C. 1971. Metabolic clearance and blood production rates of estrogens in postmenopausal women. Am. J. Obstet. Gynecol. **111:** 778–781.
59. WANG, C. *et al.* 2004. Testosterone metabolic clearance and production rates determined by stable isotope dilution/tandem mass spectrometry in normal men: influence of ethnicity and age. J. Clin. Endocrinol. Metab. **89:** 2936–2941.
60. BAKER, H.W. *et al.* 1976. Changes in the pituitary–testicular system with age. Clin. Endocrinol. **5:** 349–372.

Reductions of the Number of Hormone Receptors with Age

61. KVETNY, J. 1985. Nuclear thyroxine and triiodothyronine binding in mononuclear cells in dependence of age. Horm. Metab. Res. **17:** 35–38.
62. PISU, E. *et al.* 1980. Diurnal variations in insulin secretion and insulin sensitivity in aged subjects. Acta Diabetol. Lat. **17:** 153–160.
63. BORG, L.A., N. DAHL & I. SWENNE. 1995. Age-dependent differences in insulin secretion and intracellular handling of insulin in isolated pancreatic islets of the rat. Diabetes Metab. **21:** 408–414.

SENESCENCE IS ASSOCIATED WITH DISTURBANCES, SOMETIMES PROFOUND, OF MOST CIRCADIAN ENDOCRINE CYCLES

Growth Hormone (Gh): Reduced Amplitude and Phase Advance with Age

64. MAZZOCCOLI, G. *et al.* 1997. Age-related changes of neuroendocrine-immune interactions in healthy humans. J. Biol. Regul. Homeost. Agents **11:** 143–147.

Melatonin: Reduced Amplitude and Phase Advance with Age

65. VAN COEVORDEN, A. *et al.* 1991. Neuroendocrine rhythms and sleep in aging men. Am. J. Physiol. **260:** E651–661.
66. ZHOU, J.N. *et al.* 2003. Alterations in the circadian rhythm of salivary melatonin begins during middle age. J. Pineal Res. **34:** 11–16.

TSH: Reduced Amplitude and Phase Advance with Age

67. GREENSPAN, S.L., A. KLIBANSKI, J.W. ROWE & D. ELAHI. 1991. Age-related alterations in pulsatile secretion of TSH: role of dopaminergic regulation. Am. J. Physiol. **260:** E486–491.
68. BARRECA, T. *et al.* 1985. Twenty-four-hour thyroid-stimulating hormone secretory pattern in elderly men. Gerontology **31:** 119–123.

Cortisol: Higher Serum Cortisol in the Evening and at Night, and Phase Advance of Cortisol Rhythm

69. VGONTZAS, A.N. *et al.* 2003. Impaired nighttime sleep in healthy old versus young adults is associated with elevated plasma interleukin-6 and cortisol levels: physiologic and therapeutic implications. J. Clin. Endocrinol. Metab. **88:** 2087–2095.
70. SHERMAN, B., C. WYSHAM & B. PFOHL. 1985. Age-related changes in the circadian rhythm of plasma cortisol in men. J. Clin. Endocrinol. Metab. **61:** 439–443.

DHEA Sulphate: Reduced Amplitude up to Disappearance of its Circadian Rhythm with Age

71. DEL PONTE, A. *et al.* 1990. Changes in plasma DHEAS circadian rhythm in elderly men. Prog. Clin. Biol. Res. **341:** 791–796.

Aldosterone: Reduced Amplitude but Maintenance with Age

72. CUGINI, P. *et al.* 1992. Effect of aging on circadian rhythm of atrial natriuretic peptide, plasma rennin activity, and plasma aldosterone. J. Gerontol. **47:** B214–219.

Testosterone: Reduced Amplitude and Desynchonisation of Its Circadian Rhythm with Age

73. BREMNER, W.J., M.V. VITIELLO & P.N. PRINZ. 1983. Loss of circadian rhythmicity in blood testosterone levels with aging in normal men. J. Clin. Endocrinol. Metab. **56:** 1278–1281.

Starts in Middle Age

74. LUBOSHITZKY, R., Z. SHEN-ORR & P. HERER. 2003. Middle-aged men secrete less testosterone at night than young healthy men. J. Clin. Endocrinol. Metab. **88:** 3160–3166.

MOST AGING THEORIES CAN BE EXPLAINED AND SUPPORTED BY THE MULTIPLE HORMONE DEFICIENCY THEORY OF AGING

Excessive Free Radical Formation

75. BEDNAREK-TUPIKOWSKA, G. *et al.* 2001. Serum lipid peroxide levels and erythrocyte glutathione peroxidase and superoxide dismutase activity in premenopausal and postmenopausal women. Gynecol. Endocrinol. **15:** 298–303.
76. BHAGWAT, V.R. 1997. Relationship of erythrocyte superoxide dismutase, serum lipid peroxides, and age. Indian J. Med. Sci. **51:** 45–51.
77. MISRA, G.C., S.L. BOSE & A.K. SAMAL. 1991. Malabsorption in thyroid dysfunctions. J. Indian Med. Assoc. **89:** 195–197.

Melatonin

78. SOFIC, E. *et al.* 2005. Antioxidant capacity of the neurohormone melatonin. J. Neural Transm. **112:** 349–358.
79. AYATA, A. *et al.* 2004. Oxidative stress-mediated skin damage in an experimental mobile phone model can be prevented by melatonin. J. Dermatol. **31:** 878–883.
80. BAYDAS, G. & M. TUZCU. 2005. Protective effects of melatonin against ethanol-induced reactive gliosis in the hippocampus and cortex of young and aged rats. Exp. Neurol. **194:** 175–181.
81. KERMAN, M. *et al.* 2005. Does melatonin protect or treat brain damage from traumatic oxidative stress? Exp. Brain Res. **163:** 406–410.
82. ABDEL-WAHHAB, M.A., M.M. ABDEL-GALIL & M. EL-LITHEY. 2005. Melatonin counteracts oxidative stress in rats fed an ochratoxin A contaminated diet. J. Pineal Res. **38:** 130–135.
83. OZACMAK, V.H. *et al.* 2005. Protective effect of melatonin on contractile activity and oxidative injury induced by ischemia and reperfusion of rat ileum. Life Sci. **76:** 1575–1588.

Thyroid Hormones

84. ANTIPENKO, A.Y. & Y.N. ANTIPENKO. 1994. Thyroid hormones and regulation of cell reliability systems. Adv. Enzyme Regul. **34:** 173–198.
85. TSENG, Y.L. & K.R. LATHAM. 1984. Iodothyronines: oxidative deiodination by hemoglobin and inhibition of lipid peroxidation. Lipids **19:** 96–102.
86. BOZHKO, A.P. & I.V. GORODETSKAIA. 1998. The role of thyroid hormones in the prevention of disorders of myocardial contractile function and antioxidant activity during heat stress. Ross. Fiziol. Zh Im I M Sechenova **84:** 226–232.
87. FAURE, P., L. OZIOL, Y. ARTUR & P. CHOMARD. 2004. Thyroid hormone (T3) and its acetic derivative (TA3) protect low-density lipoproteins from oxidation by different mechanisms. Biochimie **86:** 411–418.
88. BRZEZINSKA-SLEBODZINSKA, E. 2003. Influence of hypothyroidism on lipid peroxidation, erythrocyte resistance and antioxidant plasma properties in rabbits. Acta Vet. Hung. **51:** 343–351.
89. OZIOL, L., P. FAURE, N. BERTRAND & P. CHOMARD. 2003. Inhibition of *in vitro* macrophage-induced low density lipoprotein oxidation by thyroid compounds. J. Endocrinol. **177:** 137–146.

Parathyroid Hormone

90. CHEN, S.M. & T.K. YOUNG. 1996. Effect of parathyroid hormone on antioxidant enzyme activity and lipid peroxidation of erythrocytes in five-sixths nephrectomized rats. Nephron **73:** 235–242.

Insulin: Oxidative Stress in Diabetes

91. ABOU-SEIF, M.A. & A.A. YOUSSEF. 2001. Oxidative stress and male IGF-1, gonadotropin, and related hormones in diabetic patients. Clin. Chem. Lab. Med. **39:** 618–623.

DHEA

92. IWASAKI, Y. *et al.* 2004. Dehydroepiandrosterone-sulfate inhibits nuclear factor-kappaB-dependent transcription in hepatocytes, possibly through antioxidant effect. J. Clin. Endocrinol. Metab. **89:** 3449–3454.
93. BEKESI, G. *et al.* 2000. *In vitro* effects of different steroid hormones on superoxide anion production of human neutrophil granulocytes. Steroids **65:** 889–894.
94. BEDNAREK-TUPIKOWSKA, G. *et al.* 2000. Influence of dehydroepiandrosterone on platelet aggregation, superoxide dismutase activity, and serum lipid peroxide concentrations in rabbits with induced hypercholesterolemia. Med. Sci. Monit. **6:** 40–45.

Cortisol

95. GAVAN, N. & H. MAIBACH. 1997. Effect of topical corticosteroids on the activity of superoxide dismutase in human skin *in vitro*. Skin Pharmacol. **10:** 309–313.

Estrogens

96. SUGISHITA, K., F. LI, Z. SU & W.H. BARRY. 2003. Antioxidant effects of estrogen reduce [Ca2+]i during metabolic inhibition. J. Mol. Cell Cardiol. **35:** 331–336.
97. HAN, H.J. *et al.* 2001. Effects of sex hormones on Na+/glucose cotransporter of renal proximal tubular cells following oxidant injury. Kidney Blood Press. Res. **24:** 159–165.
98. BARP, J. *et al.* 2002. Myocardial antioxidant and oxidative stress changes due to sex hormones. Braz. J. Med. Biol. Res. **35:** 1075–1081.
99. AZEVEDO, R.B. *et al.* 2001. Regulation of antioxidant enzyme activities in male and female rat macrophages by sex steroids. Braz. J. Med. Biol. Res. **34:** 683–687.
100. MASSAFRA, C. *et al.* 2000. Effects of estrogens and androgens on erythrocyte antioxidant superoxide dismutase, catalase, and glutathione peroxidase activities during the menstrual cycle. J. Endocrinol. **167:** 447–452.

NOTE

References 101–1186 may be found as a data supplement to the online version of this paper at Annals Online <www.annalsonline.org>.

Stress-Induced Hypocortisolemia Diagnosed as Psychiatric Disorders Responsive to Hydrocortisone Replacement

SUZIE E. SCHUDER

Hoag Memorial Hospital Presbyterian, Department of Neurology-Psychiatry, Newport Beach, California 92660 USA

ABSTRACT: In patients of all ages, many disorders labeled as psychiatric may actually be due to hormonal insufficiencies. For example, cortisol deficiency is rarely taken into account in a medical or psychiatric work-up, so persons with mild to moderate cortisol insufficiency are for the most part relegated to receiving a psychiatric diagnosis when, in fact, the same disorder is represented. However, the symptoms of cortisol insufficiency appear to closely parallel such psychiatric disorders as post traumatic stress disorder (PTSD) and addictions. There has been some question of whether substance abuse *causes* a hypocortisolemic state. In reviewing the literature and obtaining detailed histories of addicted patients, it appears that childhood trauma, also known as "early life stress" (ELS), instead may elicit a hypocortisolemic state. This leads some to self-medicate with an addictive substance to quell the pain of a cortisol insufficiency, both physical and emotional. In fact, the literature supports the concept that addictive substances increase cortisol in predisposed patients. Patients with a variety of psychiatric disorders including addictions were found to have signs and symptoms of mild or moderate hypocortisolemia. Generally, an appropriate comprehensive examination supported a diagnosis of cortisol insuffiency. For the most part, these patients were succesfully treated with physiologic doses of bio-equivalent hydrocortisone, along with replacement of any other deficient hormone. By correcting underlying hormonal insufficiencies, many patients improved, with some patients having a total reversal of psychiatric symptoms. It is therefore reasonable to evaluate and treat hormonal insufficiencies with hormones prior to using psychotropic medication.

KEYWORDS: stress; trauma; cortisol; PTSD; hormones; substance abuse; aging; paranoia; irritability; hostility; startle; alcoholism; psychotropic; DSM

Exponentially expanding medical knowledge, intensified by the advance of computer-oriented technologies, has furthered the development of medical specialization, which in turn has spawned even more highly focused subspecialties. This is advantageous in that complex medical and especially surgical procedures require a higher level of expertise and therefore more intense educational training. Unfortunately, this level of expertise has been at the cost of a cohesive, holistic perspective. The problem is complicated by the relative lack of communication between the dif-

Address for correspondence: Suzie E. Schuder, M.D., 901 Dover Drive, Suite 204, Newport Beach, CA, 92660 USA. Voice: 949-722-9884; fax: 949-722-9885.
sschuder@yahoo.com

Ann. N.Y. Acad. Sci. 1057: 466–478 (2005). © 2005 New York Academy of Sciences.
doi: 10.1196/annals.1356.036

ferent disciplines.[1] The same medical problems appear to have discipline-specific diagnoses with various treatment methods based on specialty-driven approaches.

The problem of this fragmented approach to medicine was further complicated by a subtle transformation to symptom-based treatment. This shift in part is due to a decrease in time allocated for patient visits. Limits on third-party reimbursements to physicians impose time constraints on patient care. In some studies, primary care physicians average about ten minutes per patient.[2] In one study the average time for a patient interview was 6.7 minutes.[3] Clearly these brief visits are not conducive to a comprehensive approach to patient care. However, an expeditious approach using symptom relief, is preferable to the costly and time-consuming explorations of a possible underlying problem. This quick, symptom-centered approach to medicine has been described by some as "McMedicine."

The imposition of time constraints created a need for rapid resolutions to patients' problems. This need was met by a pharmaceutical industry that produced and continues to produce medications providing symptom relief. The patient feels better and a minimal amount of time is required of the doctor. Whatever the problem, there is a pill offering an "antidote." For example, patients with insomnia are given a soporific medication. If the patient feels depressed, there is an antidepressant for him. In fact, in the United States, medications offering symptom relief are advertised directly to the public so that patients can ask doctors for a specific brand of medication to relieve their symptoms.[4,5]

Psychiatry fits well into this symptom-based paradigm because psychiatric diagnoses are based on symptom clusters. Psychiatry has developed unique diagnoses based on the Diagnostic and Statistical Manual (DSM), the current version being DSM IV-TR. The original purpose of developing such a manual was to standardize international research parameters. With a world-wide consensus about the exact identity of a mental problem, shared research information multiplies the usefulness of the data and provides greater validity to the knowledge gained. The ultimate purpose of developing research parameters was to discover the etiology, appropriate treatment, and prevention of mental problems.[6]

On the basis of a variety of statistically derived symptom clusters and the array of behavioral responses to emotional discomfort, the manual sorted them into specific diagnostic groupings. These classifications are decided on by a DSM committee that continually revises the criteria by including or excluding various symptoms. The classifications were not meant to replace an etiologically based diagnosis. In fact, it is clearly stated in the DSM that a psychiatric diagnosis is made only if the symptoms are not caused by a medical disorder or by the effects of alcohol or drugs.[6] Therefore, a psychiatric diagnosis is, by definition, one of exclusion.

In reality, however, if the primary complaint is emotional, with the presenting symptoms consistent with a DSM-defined disorder, then the initial diagnosis can be considered a psychiatric disorder. This may mistakenly eliminate a search for any known medical problems that can produce the same symptoms. Diagnostic codes give the psychiatric determinations some validity. It then appears quite appropriate to give the patients psychotropic medications to treat diagnostically indicated and coded psychiatric problems. Pharmaceutical company studies of drugs specifically designed to treat these psychiatric disorders, based on DSM criteria, legitimize the diagnosis. These double-blind placebo-controlled trials of drugs become the basis of what has been called "evidence-based treatment."

TABLE 1. Comparing post-traumatic stress disorder and cortisol insufficiency

Post-traumatic stress syndrome	Low cortisol
Intense psychological stress to external cues	Anxious, nervous
Intense response to external cues of trauma	Poor stress tolerance
	Hypersensitivity to environment
	Stress-induced fatigue
Avoidance	
Avoids thoughts of trauma	Absent minded
Avoids activities, places, etc.	Forgetful
Decreased interest and participation in activities	Feeling spacey, confused
Feeling detached from others	Depression
Feeling future foreshortened	Depression
Increased symptoms of arousal	
Trouble falling/staying asleep	Hypersensitivity of all senses
Hypervigilant	Paranoid feelings
Irritable or outburses of anger	Irritable/hostile
Trouble concentrating	Concentration problems
Exaggerated startle response	Poor stress tolerance

However, because psychiatrically diagnosed disorders are not etiologically derived, the designated therapies are based on treating symptoms. Additionally, patients may not be aware of any other indicators of a medical problem possibly associated with the emotional symptoms and would not mention them to the physician during their brief visit. Additionally, the diagnosis was already made. This negates the necessity of any further investigation into a cause.

Moreover, primary care physicians are encouraged to look for psychiatric disorders, especially depression. When the symptoms fit the diagnosis of depression they are urged to treat these patients with antidepressants.[5] If the patients' depressive symptoms respond to an antidepressant, it further reduces the likelihood of finding a physiologic problem that can be treated. Unfortunately, using a psychiatric diagnosis without further investigation into a possible cause is not necessarily a benign treatment choice. An example of this is the depression that is a manifestation of hypothyroidism. If patients with this particular depression remain untreated for hypothyroidism, a number of morbid conditions can develop, the most damaging of which is dilated cardiomyopathy or congestive heart failure.[7,8] Clearly, hormones have functions beyond the effects on emotions.

With aging, hormone deficiencies are a fact, as most of the body's hormones diminish. However, hormone deficiencies, not merely as a product of aging, may cause symptoms categorized as psychiatric disorders. Hormone deficiencies can generate emotional symptoms, create perceptual distortions, and affect the ability to think. Even so, patients are given a psychiatric diagnosis without an adequate investigation into a treatable medical cause. A side-by-side comparison of a hormone deficiency and a psychiatric diagnosis shows that they have same symptoms[9] (TABLE 1).

It is apparent that the psychiatric diagnoses of post-traumatic stress syndrome (PTSD) and hypocortisolemia have the same symptoms.[6,9] Of the symptoms listed for PTSD in DSM IV-TR, not all are not required to make the diagnosis. For a diagnosis of PTSD, two requirements are essential. The person must have been exposed to a traumatic event during which he or she witnessed the death of another or experienced physical harm, or even the threat of physical harm, to self or others. The second absolute requirement is that the person's response to the experience be intense fear, helplessness, or horror. In children, this manifests as disorganized or agitated behavior and can occur any time a child feels unsafe and incapable of dealing with intense fear. Basically, it can be any traumatic event or even a chaotic environment that creates intense fear. In the category of "re-experiencing the traumas" only one of the following conditions is required: intrusive thoughts of the event, dreams or nightmares, flashbacks, an intense response to the external cues of the trauma, or intense psychological stress to external cues. Of these choices, the last two items easily fit most cases of hypocortisolemia. In the category of "avoidance or numbing of feelings," only three of five of the following conditions is required: avoids thoughts of trauma; avoids activities, places, etc.; decreased interest and participation in activities; feeling detached from others; or feeling that the future is foreshortened. In the category of "increased symptoms of arousal" only two of the following items are necessary for the diagnosis of PTSD: trouble falling or staying asleep, irritability or outbursts of anger, trouble concentrating, hypervigilant, or an exaggerated startle response. Additionally, for the diagnosis of PTSD, the symptoms have to last more than a month and impair functioning in social, occupational, and other areas.[6]

Patients with hypocortisolemia can easily meet diagnostic criteria for PTSD. They can feel anxious or nervous, have poor stress tolerance, be hypersensitive to the environment, with a hypersensitivity to sound, light, taste, smell, and touch. Additionally, many patients experience extreme fatigue after a stressful situation. They can often feel depressed, spacey, absent-minded, forgetful or confused. All these symptoms can be interpreted as being avoidant and thus consistent with PTSD symptoms. The hypocortisolemic symptoms of sensory hypersensitivity, paranoia, irritability, hostility, problems concentrating, and an intolerance to stress can readily correspond to the PTSD subcategory of increased arousal.[9]

Additionally, the symptoms of hypocortisolemia may be interpreted as meeting the diagnostic criteria for numerous other psychiatric disorders.[6] For example, children who present with disorganized or agitated behavior and meet diagnostic criteria for PTSD can also be viewed as having the mood fluctuations of a bipolar II disorder. Concentration difficulties are given the designation of attention deficit disorders (ADD). The psychiatric journals abound with discussions of bipolar II disorders misdiagnosed as ADD.[10] At times they are categorized as comorbid disorders.

Even a mild to moderate cortisol insufficiency can produce certain emotional states that are common to a variety of psychiatric disorders. Emotional states can generate distinctive patterns of behavior. Furthermore, behavior is influenced by individualized interpretations of stress, the severity of the insufficiency, and cultural differences. Given these differences, it is not surprising that the manifestations of hypocortisolemic symptoms can also be assigned to a multiplicity of psychiatric diagnoses such as PTSD, ADD, depression, bipolar disorder, anxiety disorders, and personality disorders.

Poor stress tolerance, a symptom associated with hypocortisolemia, may be due to compensatory surges of adrenalin creating a fearful emotional state that leaves patients feeling overwhelmed. They have difficulty tolerating "one more thing" and thus can feel trapped by others or by life's circumstances. Life feels unmanageable, out of control, giving patients the perception that they are victims, at the mercy of everyone and everything. Taken to the extreme it can be seen as paranoia. In addition to inducing fear, adrenalin excesses intensify the senses to the point of sensory overload. In fact, patients may be so sensitive that they can even feel the pain of others. This hypersensitivity coupled with an undercurrent of fear may help create a pessimistic attitude, perhaps to the point of being classified as depression.[1]

A lack of sufficient cortisol with a compensatory adrenalin release may cause some patients to feel irritable or hostile. Emotional extremes because of an adrenergic overload may result in a behavioral response of anxious outbursts or explosions of anger.[12] Accusatory expressions, for some, may be a way to dispel, through projection, intolerable emotions, such as fear or guilt. All these behaviors can be seen as personality features when they may be no more than a response to hypocortisolemia.[6] Not all people who are hypocortisolemic choose to behave in this manner. There is a choice, or free will, whereby people may change their perceptions and alter their behavior. The option to choose behavior alters physiologic response, emotional reactions, and even the balance of hormones.[13–16]

Numerous studies have shown that people with PTSD have low cortisol.[17–20] Although traumatized, other people with an adequate amount of cortisol did not go on to develop PTSD.[21] In fact, cortisol normally might even increase ten-fold during a traumatic event and then return to normal levels when the trauma ends. This is a healthy response to a physical or emotional stress. A healthy stress response triggers a release of adrenalin in addition to a cortisol increase.[9] Contrary to the general public's erroneous concept that cortisol creates stress, the role of cortisol is to facilitate a healthy adaptation to a particular stressor. Cortisol mobilizes glucose to provide the fuel necessary for surviving a fearful event. Cortisol also counters the damaging physiologic response to stress by limiting inflammation and suppressing hyperimmunity.[9,22]

The term "adrenal burnout" has been used to describe an abnormal physiologic response to stress. At first, usually in childhood, the adrenal gland is thought to release enough cortisol in a crisis, but with ongoing trauma and a continuous outpouring of cortisol there is a point when the adrenal gland can no longer produce enough cortisol resulting an inadequate response to stress. In these people, even though cortisol levels rise, the amount may not be enough to meet their needs.

A negative feedback loop regulates the release of cortisol from the adrenal cortex. Without pathology, the adrenal gland is well regulated by hypothalamic and pituitary mechanisms. Insufficient quantities of cortisol stimulate an increase in the hypothalamic production of cortisol-releasing hormone (CRH). CRH then increases the pituitary release of adrenocorticotropic hormone (ACTH). The larger amounts of ACTH released by the pituitary gland would normally induce the adrenal glands to release the necessary amount of cortisol. In people with a relative insufficiency, even though there is an increase in the amount of cortisol, it is still inadequate to support an optimal response to stress.

The effect of an inadequate cortisol level can be debilitating, especially to the brain. Cortisol facilitates neuronal utilization of glucose for energy to allow the cell to function and live. Neurons cannot survive without glucose. Without energy, brain cell function is severely limited.[23] In essence, the brain begins to shut down. Brain cells go into survival mode and diminish or cease making neurotransmitters. This decrease in brain activity may be perceived as a loss of pleasure or depression. To coin a phrase, "a screaming boredom" seems to be a fitting description for the feeling evoked by the brain's need to survive. This state, viewed from a DSM perspective, may fill the requirements for a variety of diagnoses, depression, ADD, with or without hyperactivity, and the anxiety spectrum disorders. This partially functioning state invoked by low brain glucose can also fit the diagnosis of chronic fatigue syndrome.[24] It is not surprising that with a limited ability to utilize brain cells that stressful situations evoke intolerance and a limited ability to cope. In this case, challenging situations become crises.[23]

A limited supply of glucose to the brain causes an endogenous response in at least three ways. A perceived hypoglycemic state induces glucose hyperphagia or an overwhelming drive to eat sweets or carbohydrates, which are readily converted to glucose.[25] There may also be an intense drive to have something that increases cortisol for that particular person, such as coffee, alcohol, or drugs. This response to a particular substance is not universal and appears to be mediated by a genetic predisposition.

Another response to cortisol insufficiency is the rapid conversion of the prohormone thyroxine (T4) to the activated form of triiodothyronine (T3).[26,27] This also increases the brain's ability to utilize glucose. Adrenalin has a similar effect. The lack of adequate fuel for brain cells is a danger signal to the adrenal medulla to respond with surges of adrenalin.[28] Though compensatory measures are essential to allow the brain to utilize glucose, they can cause a multitude of problems. The combination of T3 and adrenalin surges produces intense physiologic effects.[26] The emotional interpretation of these bodily processes and resultant behavior is determined by each person based on individual attitudes and cultural biases.

As specified in the DSM, psychiatric diagnoses are defined by specific groupings of symptoms and categories of behavior. However, the responses to the same physiologic responses, surges of T3 and adrenalin, are sweating, shaking, an increase in heart rate, palpitations, and the feelings associated with a "fight or flight" reaction. These biologic events induce the subjective feelings of anxiety, fear, and paranoia along with the possible behavioral effects of an increased startle response, irritability, hostility, or anxious outbursts. The variety of emotional and behavioral responses correlates with an array of psychiatric diagnoses such as PTSD, panic disorder, numerous anxiety disorders, substance abuse disorders, and even personality disorders.[6]

A relative hypocortisolemic state may underlie or at least exacerbate stress-related medical problems. Chronic fatigue syndrome may be due to relative hypocortisolemia that would lead to an inadequate supply of glucose to cells, especially in the brain cells, resulting in low energy.[29] Fibromyalgia may be the result of an unchecked inflammatory response in the muscles and skin due to inadequate cortisol.[24] Irritable bowel syndrome (IBS) may also be linked to a dysregulation of the HPA axis.[30] An inadequate release of cortisol in response to a stress can, in the genetically predisposed, lead to inflammation and swelling of the bowel walls that would result in the symptoms associated with IBS. Other disorders connected to a relative insufficiency of cortisol are the autoimmune disorders, such as rheumatoid arthritis or lupus. Because cortisol modulates the immune system, in those prone to autoimmune disorders, an inadequate amount might exacerbate or perhaps even trigger a hyperimmune state.

Because many of the medical problems caused by a cortisol insufficiency respond, at least in part, to antidepressants, and in particular serotonin reuptake inhibitors (SSRIs), it may be possible that SSRIs induce an increase in cortisol. In spite of a lack of financial incentive to study antidepressant effects on the production of cortisol, there are a few studies, with paroxetine and sertraline, that do show an increase in cortisol in cortisol-deficient patients.[31–33]

A hormone imbalance as a cause of depression has been postulated since the 1970s. At times patients with psychiatrically diagnosed disorders were noted to have high levels of cortisol. Attempting to find a laboratory marker for depression, there were a number of studies that used a low evening dose of dexamethasone to check for the effect on cortisol the following morning. A normal response results in cortisol suppression. The hypothesis that depression is linked to nonsuppression was not supported. Using dexamethasone as a diagnostic tool was abandoned because only about half of the depressed patients showed cortisol suppression with dexamethasone.[21]

Some depressed patients have high cortisol levels and an excess of CRH. The hypothalamus continues to release CRH, an apparent malfunction in the negative feedback loop. A proposed mechanism is that cortisol, although excessive in blood, has limited access to the brain because of the blood–brain barrier, the capillary network supplying the brain. The blood–brain barrier also restricts entry to excessive or harmful substances. Lining the capillary endothelium, a molecular sentry, a phosphorylated glycoprotein (P-gp), acts to limit the amount of cortisol allowed to cross into the brain. Pariente and others found that the antidepressants they studied appeared to block P-gp at the blood–brain barrier, allowing more cortisol to enter the brain and hypothalamus, thus shutting down excessive CRH production.[33] Perhaps this may be, at least in part, the mechanism of antidepressant action. If this is correct, then P-gp may limit cortisol entry into the brain regardless of blood cortisol levels. Antidepressants may still allow for cortisol to pass through to the brain, enabling glucose to provide energy to brain cells. Theoretically, P-gp's limiting cortisol entry into the brain may be the reason why a cortisol-deficient patient does not respond to cortisol replacement.

It is not surprising to find that a condition of low cortisol may be one of the driving forces that lead people to alcoholism or substance abuse. The patients with PTSD, prone to developing an addiction, seem to do so because the particular addicting substance increases their cortisol.[34] This hypothesis seems to be validated by the studies done with alcoholics and naltrexone. Studies at Rockefeller University and Yale showed that alcohol increases cortisol in alcoholics and also that naltrexone appears to increase and stabilize cortisol levels through HPA activation.[35] This naltrexone-mediated increase in cortisol may also be the mechanism for reducing alcohol craving.[34,35]

People with addictions appear to have stress-related hypocortisolemia with a genetic predisposition to have cortisol levels rise with their particular substance of abuse.[12] Because caffeine is known to increase cortisol,[36] it is interesting to note that Alcoholics Anonymous meetings often include coffee, sweets, and drama. This ubiquitous availability of sweets at meetings may be due to a sugar craving from

TABLE 2. Other symptoms of hypocortisolemia

Diarrhea with stress	Cravings for:
Irritable bowel syndrome	sweets
Sore throat	pasta, bread
Flu-like symptoms	salty food
Achy skin	pickles, vinegar
Headaches	lemon
Trouble staying hydrated	spicy food
Fatigue more in afternoon	Autoimmune disorders
Stress-induced fatigue	arthritis
Low blood pressure	lupus
Need to keep moving	obsessive–compulsive disorders?
Allergies	multiple sclerosis?
Sinusitis	schizophrenia?

hypocortisolemia and the drama may be helpful in providing more adrenalin, another compensatory mechanism.[28,37,38]

In addition to emotional and behavioral symptoms, many somatic symptoms are associated with hypocortisolemia, although most patients do not manifest all the symptoms[9] (TABLE 2). The digestive tract problems associated with hypocortisolemia, particularly under stressful circumstances, consist of diarrhea, abdominal cramping, irritable bowel symptoms, and nausea with or without vomiting. Other symptoms commonly linked with inflammation are flu-like symptoms, achy skin, a sore throat, sinusitis, headaches, and painful joints and muscles. With a decrease in cortisol, patients who are genetically predisposed will express a hyperimmune state. Allergies flare and autoimmune conditions, such as rheumatoid arthritis, lupus, or multiple sclerosis, worsen with stress, indicating a possible association to a hypocortisolemic state. In that cortisol supports the circulatory system, lowered levels of cortisol can result in low blood pressure with orthostatic hypotension, with patients reporting momentary visual loss or a dizziness when standing up abruptly. If the levels of cortisol become extremely low, the result can be circulatory collapse or shock. Some unusual food preferences or cravings, in various combinations, seem to be a component of this condition and can consist of a need for sugar or carbohydrates such as bread or pasta, salty food, pickles, vinegar, lemon, and or spicy foods.[9] Fatigue, often but not necessarily in the afternoon, is a common feature of hypocortisolemia. Frequently, stressful conditions induce fatigue[9] (TABLE 2). At times some patients will have excessive amounts of nervous energy and seem to be "highly charged." This may be do to the compensatory release of adrenalin in hypocortisolemia. In fact, high-risk activities or novelty-seeking behaviors are associated with PTSD and therefore with hypocortisolemia.[17,38,39]

In addition to the emotional, behavioral, and somatic symptoms, many patients with hypocortisolemia may have physical signs as well. Some patients with autoimmune disorders may also have vitiligo, a spotty depigmentation of the skin. Many patients have hyperpigmentation at pressure points such as joints, elbows, knees, knuckles, and palmar creases. A brown pigmentation also can be seen below the eye in some of the patients. Scars acquired after the onset of a cortisol insufficiency will be pigmented, while earlier scars are not pigmented. Clothing pressure can also create pigmented areas such as pigment under bra straps or at the waistline from the pressure of belts. Pigmented areas are not seen in all patients with cortisol insufficiency. The presence of hyperpigmentation requires melanocytes that are able to produce melanin and presupposes an intact pituitary gland able to produce an excess of adrenocorticotropic hormone (ACTH) in response to hypocortisolemia.[40]

Laboratory evaluations, though necessary, may not always corroborate even obvious clinical signs and symptoms. This may be due to laboratory testing limitations. One problem is the evanescent property of cortisol. In that cortisol release and utilization is an adaptive mechanism to any stress, fluctuations in levels would be expected. Blood drawing traumatizes some patients. Other patients may be harried by time constraints with the need to rush to work or their next scheduled activity. In these patients, the blood levels will be higher than in patients who are relatively calm. The amount of a hormone is not the only constituent that determines an adequate hormonal response. Other factors influence the efficacy of any hormone.[41]

Hormones exist in the blood in two states, bound to a protein and in the "free" state, or unbound. Total cortisol is a level of both bound and unbound factions, and often appears on laboratory reports as merely "cortisol." If the amount of total cortisol is within the reference range or even appears to be above the reference range, the actual amount of useful cortisol might still be very low. When a hormone is bound to a protein carrier it is slated for elimination and not available for use by the body. For cortisol, the principle protein carrier is transcortin, also called cortisol binding globulin (CBG). When CBG is present in excessive amounts, as seen in women on oral contraceptives or even those who take oral estrogen, the amount of CBG increases dramatically, greatly diminishing the amount of available cortisol in spite of a high total cortisol level.[42,43]

Another issue is the lack of clinical laboratory standardization. This is an international problem. Each laboratory uses different equipment, different reagents, and a variety of individual clinical pathologists who interpret the results.[44,45] The samples obtained to determine reference ranges are collected from the first three thousand or so results of a test ordered by doctors. These results are then used to determine the standard range by applying the concept of a bell-shaped curve. The statistical model maintains that ninety-five percent of the samples fall within two standard deviations from the mean and thus determines the reference range. Unfortunately, the samples used to make this determination are from patients who have the specific abnormalities being investigated.[46,47]

A 24-hour collection of urine for cortisol and cortisol metabolites is also helpful in that it is not at merely one point in time like a blood test. Instead, it is an average value for the day. Even though this may add some knowledge of the patient's physical state, not all patients who present with signs and symptoms have a low 24-hour urinary free cortisol, although they may have abnormally low levels of cortisol metabolites, suggesting that even though cortisol is available it is not being utilized. The symptoms of hypocortisolemia, in these cases, are postulated to be due to a defect in the cellular receptors for cortisol. However, these patients still respond to physiologic doses of hydrocortisone.[41] It is essential to remember that laboratory testing is merely an adjunct to the evaluative process and is not intended to replace good medical judgment based on a comprehensive history and an examination of patients for signs and symptoms of a disorder.

Another test frequently done in patients presenting with hypocortisolemia is the stimulation of the adrenal gland using an ACTH analogue, Cortrosyn®. This is a method to determine whether the source of hypocortisolemia is due to primary adrenal insufficiency or from a secondary adrenal insufficiency due to inadequate amounts of pituitary ACTH or a hypothalamic insufficiency of cortisol-releasing hormone (CRH). The test has been performed by injecting 250 micrograms of Cortrosyn®, although this amount is considered excessive by some studies, in that all but the patients with the most severe forms of hypocortisolemia respond to this dose. Several studies suggest that a 1-microgram dose can uncover the mild to moderate cases of low cortisol.[48] Again, a laboratory evaluation is not intended to replace good medical judgment.

In spite of a variety of laboratory testing results, patients demonstrating the signs and symptoms of mild to moderate hypocortisolemia appear to respond to low physiologic doses of hydrocortisone in a number of studies. Patients with symptoms of hypocortisolemia who were diagnosed with chronic fatigue syndrome, responded to low doses of hydrocortisone.[50] Additionally, patients identified as having PTSD who are given low doses of cortisol at the time of the trauma do not seem to develop symptoms of PTSD.[49,51]

In the aging population, hormone insufficiencies and hormone treatments are commonplace. However, a cortisol insufficiency is often overlooked in spite of the presence of signs and symptoms. The cause may not be due to aging, but to a long-standing deficiency exacerbated by aging. Over time, most patients develop a variety of coping skills for mild to moderate cortisol insufficiency, at times known as a low adrenal reserve. If fatigue, pain, memory deficits, and concentration difficulties worsen, they are then attributed to "getting old."

Some studies describe an increase in serum cortisol in the elderly. However, they may still manifest the signs and symptoms of a hypocortisolemia. As in some younger people, cellular cortisol receptors may be less sensitive. This limits cortisol function in spite of seemingly adequate levels. Testing for cortisol and its metabolites in a 24-hour urine sample will show high or normal urinary free cortisol with below normal metabolites, the total amount of hydroxycorticosteroids.[41] Other studies of older patients actually demonstrate a decrease in serum cortisol. This is thought to be due to an age-related decrease in adrenal cortex sensitivity to ACTH.[52] Additionally, the signs and symptoms of a relative cortisol deficiency correspond to the symptoms normally associated with aging.

Moreover, when elderly patients have symptoms that are primarily emotional, or demonstrate behavioral dysfunction, they are readily diagnosed with a psychiatric disorder and are treated with psychotropic medications. As physicians, we have an obligation to include a thorough evaluation of each patient despite age, prior to making a psychiatric diagnosis, which by definition is a diagnosis of exclusion. Furthermore, in addition to medication side effects, treating symptoms with psychotropic medications may mask any underlying medical problems, which can result in continued, perhaps progressive morbidity and possibly mortality. The most important element of providing good patient care is the ability to make an accurate diagnosis by obtaining a good history and performing a physical exam. Treating the underlying problem is a better approach to patient care. In the case of a relative cortisol insufficiency, if patients are treated with physiologic doses of hydrocortisone, the symptoms for the most part abate.[41,51] Once a medical diagnosis can be either ruled out or confirmed and treated, then a psychiatric diagnosis and treatment with psychotropic medications may be appropriate and perhaps even necessary.

[*Competing interests*: The author declares that she has no competing financial interests.]

REFERENCES

1. NEAME, R. 2000. Creating an infrastructure for the productive sharing of clinical information.Top. Health Inf. Manage. **20:** 85–91.
2. STANGE, K.C. *et al.* 1998. Illuminating the "black box." A description of 4454 patient visits to 138 family physicians. J. Fam. Pract. **46:** 377–389.
3. OGANDO, D.B. *et al.* 1995. How much time does the physician dedicate to his patients? A study of contents of office visits and their length. Ateneo Primaria **15:** 290–296.
4. OLFSON, M. *et al.* 2002. National trends in the outpatient treatment of depression. J. Am. Med. Assoc. **287:** 203–209.
5. TINSLEY, J.A. *et al.* 1998. A survey of family physicians and psychiatrists. Psychotropic prescribing practices and educational needs. Gen. Hosp. Psychiatry **20:** 360–367.
6. AMERICAN PSYCHIATRIC ASSOCIATION. 2000. Diagnostic and Statistical Manual of Mental Disorders, 4th ed., text revision. American Psychiatric Press. Washington, DC.
7. KOZDAG, G. *et al.* 2005. Relation between free triiodothyronine/free thyroxine ratio, echocardiographic parameters and mortality in dilated cardiomyopathy. Eur. J. Heart Fail. **7:** 113–118.
8. BEZDAH, L. *et al.* 2004. Hypothyroid dilated cardiomyopathy. Ann. Cardiol. Angeiol. **53:** 217–220.
9. OTH, D.N. *et al.* 1998. The adrenal cortex. Effects of glucocorticoids on diseases of the adrenal cortex. Hypofunction. *In* Williams' Endocrinology, 9th edit., **12:** 544–563. W. B. Saunders Company. Philadelphia.
10. BREMNER, J.D. *et al.* 2003. Assessment of the hypothalamic-pituitary-adrenal axis over a 24-hour diurnal period and in response to neuroendocrine challenges in women with and without childhood sexual abuse and post-traumatic stress disorder. Biol. Psychiatry **54:** 710–718.
11. IWATA, M. *et al.* 2004. A case of Addison's disease presented with depression as a first symptom. Seishin Shinkeigaku Zasshi **106:** 1110–1116.
12. GERRA, G. *et al.* 2004. Aggressive responding in abstinent heroin addicts: neuroendocrine and personality correlates. Prog. Neuropsychopharmacol. Biol. Psychiatry **28:** 129–139.
13. ALLEN, J.J. *et al.* 2001. Manipulation of frontal EEG asymmetry through biofeedback alters self-reported emotional responses and facial EMG. Psychophysiology **38:** 685–693.
14. HARDMAN, E. *et al.* 1997. Frontal interhemispheric asymmetry: self regulation and individual differences in humans. Neurosci. Lett. **221:** 117–120.
15. TAMAI, H. *et al.* 1993. Changes in plasma cholecystokinin concentrations after oral glucose tolerance test in anorexia nervosa before and after therapy. Metabolism **42:** 581–584.
16. ABELSON, J.L. *et al.* 2005. Cognitive modulation of the endocrine stress response to a pharmacological challenge in normal and panic disorder subjects. Arch. Gen. Psychiatry **62:** 668–675.
17. YEHUDA, R. *et al.* 1991. Low urinary cortisol excretion in PTSD. J. Nerv. Ment. Dis. **178:** 366–369.
18. KANTER, E.D. *et al.* 2001. Glucocorticoid feedback sensitivity and adrenocortical responsiveness in post-traumatic stress disorder. Biol. Psychiatry **50:** 238–245.

19. YEHUDA, R. *et al.* 1998. Predicting the development of post-traumatic stress disorder from the acute responses to a traumatic event. Biol. Psychiatry **44:** 1305–1313.
20. YEHUDA, R. *et al.* 1993. Enhanced suppression of cortisol following dexamethasone administration in post-traumatic stress disorder. Am. J. Psychiatry **150:** 83–86.
21. FOUNTOULAKIS, K. *et al.* 2004. Relationship among dexamethasone suppression test, personality disorders, and stressful life events in clinical subtypes of major depression: an exploratory study. Ann. Gen. Hosp. Psychiatry **3:** 15.
22. FAUCI, A.S. 1979. Immunosuppressive and anti-inflammatory effects of glucocorticoids. *In* Glucocorticoid Hormone Action. J.D. Baxter *et al.*, Eds.: 449–465. Springer-Verlag. New York.
23. MARTIN, A.D. *et al.* 1984. The stimulation of mitochondrial pyruvate carboxy after dexamethosone treatment of rats. Biochem. J. **219:** 107–115.
24. GRIEP, E.N. *et al.* 1998. Function of the hypothalamic-pituitary-adrenal axis in patients with fibromyalgia and low back pain. J. Rheumatol. **25:** 1374–1381.
25. SCHULTES, B. *et al.* 2005. Processing of food stimuli is selectively enhanced during insulin-induced hypoglycemia in healthy men. Psychoneuroendocrinology **30:** 496–504.
26. LARSEN, P.R. *et al.* 1998. The Thyroid Gland. *In* Williams' Endocrinology, 9th edit., **12:** 408–409. W.B. Saunders Company. Philadelphia.
27. KARLOVIC, D. *et al.* 2004. Increase of serum triiodothyronine concentration in soldiers with combat-related chronic post-traumatic stress disorder with or without alcohol dependence. Wien Klin. Wochenschr. **116:** 385–390.
28. BORG, WP. *et al.* 1995. Local ventromedial hypothalamus glucopenia triggers counter-regulatory hormone release. Diabetes **44:** 180–184.
29. DEMITRACK, M.A. *et al.* 1991. Evidence for impaired activation of the hypothalamic-pituitary-adrenal axis in patients with chronic fatigue syndrome. J. Clin. Endocrinol. Metab. **73:** 1224–1234.
30. PATACCHIOLI, F.R. *et al.* 2001. Actual stress, psychopathology and salivary cortisol levels in the irritable bowel syndrome (IBS). J. Endocrinol. Invest. **24:** 173–177.
31. SCHLOSSER, R. *et al.* 2000. Effects of subchronic paroxetine administration on nighttime endocrinological profiles in healthy male volunteers. Psychoneuroendocrinology **25:** 377–388.
32. SAGUD, M. *et al.* 2002. Effects of sertraline treatment on plasma cortisol, prolactin and thyroid hormones in female depressed patients. Neuropsychobiology **45:** 139–143.
33. PARIANTE, C.M. *et al.* 2004. Do antidepressants regulate how cortisol affects the brain? Psychoneuroendocrinology **29:** 423–447.
34. ANTON, R.F. *et al.* 2004. Naltrexone effects on alcohol consumption in a clinical laboratory paradigm: temporal effects of drinking. Psychopharmacology **173:** 32–40.
35. O'MALLEY, S.S. *et al.* 2002. Naltrexone decreases craving and alcohol self-administration in alcohol-dependent subjects and activates the hypothalamo-pituitary-adrenocortical axis. Psychopharmacology **160:** 19–29.
36. LOVALLO, W.R. *et al.* 1996. Stress-like adrenocorticotropin responses to caffeine in young healthy men. Pharmacol. Biochem. Behav. **55:** 365–369.
37. GALEN, L.W. *et al.* 1997. The utility of novelty seeking, harm avoidance, and expectancy in the prediction of drinking. Addict. Behav. **22:** 93–106.
38. WANG, S. *et al.* 1997. Relationships between hormonal profile and novelty seeking in combat-related post-traumatic stress disorder. Biol. Psychiatry **41:** 145–151.
39. YEHUDA, R. 2002. Post-traumatic stress disorder. N. Engl. J. Med. **346:** 108–114.
40. ZATOUROFF, M. 1976. A Colour Atlas of Physical Signs in General Medicine. Wolfe Medical Publications Ltd. London, England.
41. JEFFERIES, W.M. 1996. Safe Uses of Cortisol. Charles C Thomas. Springfield, IL.
42. WIEGRATZ, I. *et al.* 2003. Effect of four different oral contraceptives on various sex hormones and serum-binding globulins. Contraception **67:** 25–32.
43. JUNG-HOFFMAN, C. *et al.* 1992. Serum concentrations of ethinylestradiol, 3-keto-desogestrel, SHBG, CBG, and gonadotropins during treatment with a biphasic oral contraceptive containing desogestrel. Horm. Res. **38:** 184–189.
44. STEELE, B.W. *et al.* 2005. Total long-term within-laboratory precision of cortisol, ferritin, thyroxine, free thyroxine, and thyroid stimulation hormone assays based on a

College of American Pathologists fresh frozen serum study: do available methods meet medical needs for precision? Arch. Pathol. Lab. Med. **129:** 318–322.
45. FRANZINI, C. 1991. The reference method value. Ann. Ist Super Sanita. **27:** 359–363.
46. WRIGHT, E.M. *et al.* 1999. Calculating reference intervals for laboratory measurements. Stat. Methods Med. Res. **8:** 93–112.
47. HENNY, J. *et al.* 2000. Need for revisiting the concept of reference values. Clin. Chem. Lab. Med. **38:** 589–595.
48. THALER, L.M. *et al.* 1998. The low dose (1-microg) adrenocorticotropin stimulation test in the evaluation of patients with suspected central adrenal insufficiency. J. Clin. Endocrinol. Metab. **83:** 2726–2729.
49. SCHELLING, G. *et al.* 2004. Can post-traumatic stress disorder be prevented with glucocorticoids? Ann. N.Y. Acad. Sci. **1032:** 158–166.
50. CLEARE, A.J. *et al.* 2001. Hypothalamo-pituitary-adrenal axis dysfunction in chronic fatigue syndrome, and the effects of low-dose hydrocortisone therapy. J. Clin. Endocrinol. Metab. **86:** 3545–3554.
51. AERNI, A. *et al.* 2004. Low-dose cortisol for symptoms of post-traumatic stress disorder. Am. J. Psychiatry **161:** 1488–1490.
52. GIORDANO, R. *et al.* 2001. Elderly subjects show severe impairment of dehydroepiandrosterone sulphate and reduced sensitivity of cortisol and aldosterone response to the stimulatory effect of ACTH. Clin. Endocrinol. **55:** 259–265.

Use of Telomerase to Create Bioengineered Tissues

JERRY W. SHAY AND WOODRING E. WRIGHT

Department of Cell Biology, University of Texas Southwestern Medical Center at Dallas, Dallas, Texas 75390-9039, USA

ABSTRACT: Telomeres are repetitive DNA (TTAGGG) elements at the ends of chromosomes. Telomerase is a ribonucleoprotein complex that catalyzes the addition of telomeric sequences to the ends of chromosomes. The catalytic protein component of telomerase (hTERT) is expressed only in specific germ line cells, proliferative stem cells of renewal tissues, and cancer cells. The expression of hTERT in normal cells reconstitutes telomerase activity and circumvents the induction of senescence. Telomeres shorten with each cell division, eventually leading to senescence (aging), due to incomplete lagging DNA strand synthesis and end-processing events, and because telomerase activity is not detected in most somatic tissues. There are specific tissues and locations in which replicative senescence likely contributes to the decline in human physiological function with increased age and with chronic illnesses. While expressing hTERT in cells results in the maintenance of telomere length and greatly extended life span, blocking replicative aging systemically would be predicted to increase the potential for tumor formation. However, there are many situations in which the transient rejuvenation of cells could be beneficial. Ectopic expression of hTERT has been shown to immortalize human skin keratinocytes, dermal fibroblasts, muscle satellite (stem), and vascular endothelial, myometrial, retinal-pigmented, and breast epithelial cells. In addition, human bronchial, corneal and skin cells expressing hTERT can be used to form organotypic (3D) cultures (bioengineered tissues) that express differentiation-specific proteins, demonstrating that hTERT by itself does not alter normal physiology. The production of hTERT-engineered tissues offers the possibility of producing tissues to treat a variety of chronic diseases and age-related medical conditions that are due to telomere-based replicative senescence.

KEYWORDS: telomeres; bioengineered tissue; telomerase; aging; cell proliferation; cell division; senescence

Telomeres are specialized DNA–protein complexes at the ends of linear chromosomes consisting of long tandem arrays of double-stranded TTAGGG repeats, a G-rich 3′ single-strand overhang (that can form a special loop structure), and telomere-associated proteins. The work of Muller[1] and McClintock[2] in the 1930s led to the concept that telomeres function to "cap" chromosomal termini to prevent degrada-

Address for correspondence: Jerry W. Shay, University of Texas Southwestern Medical Center at Dallas, Department of Cell Biology, 5323 Harry Hines Boulevard, Dallas, Texas 75390-9039. Voice: 214-648-3282; fax: 214-648-8694.
Jerry.Shay@UTSouthwestern.edu

Ann. N.Y. Acad. Sci. 1057: 479–491 (2005). © 2005 New York Academy of Sciences.
doi: 10.1196/annals.1356.037

TABLE 1. Telomere hypothesis of aging and cancer

Telomeres: repetitive DNA (TTAGG)n
"cap" chromosome ends
hide telomeres from the DNA damage response
promote proper chromosome segregation
Telomeres: progressively lost with each cell division
"end-replication" problem and end-processing events
senescence (aging) occurs when few telomeres are short
Pre-malignant tissues have very short telomeres
telomere attrition may be a tumor-supporessor pathway
Cancer cells are almost always immortal
must engage a mechanism for stabilizing telomere length for the growth of the advanced tumor

tion, end-to-end fusions, and recombination, thereby maintaining chromosomal integrity (TABLE 1).[3–12] Subsequent work on telomeres has substantiated this model and advanced our understanding of the pathogenesis of complex human diseases associated with aging, including cancer.[13–23] Evidence from human tissues and model organisms has established that telomere maintenance and the cellular response to telomere dysfunction are crucial in the maintenance of chromosomal stability. These advances in the basic understanding of telomere dynamics are now being translated into clinically relevant cancer therapeutic applications. In addition, knowledge of the relationship of telomeres and telomerase to stem cell biology may in the near future have an impact on the management of patients with chronic diseases associated with increased age.[24–33]

Telomerase is a ribonucleoprotein cellular reverse transcriptase that maintains chromosome ends.[34] Telomerase uses its internal RNA component (complementary to the telomeric single-stranded overhang) as a template in order to synthesize telomeric repeats (TTAGGG)n directly onto the ends of chromosomes. Telomerase is present in most fetal tissues, normal adult male proliferative germ cells, inflammatory cells, in rapidly dividing cells of renewal tissues, and in most tumor cells.[35–38] After adding six bases, the enzyme pauses while it repositions (translocates) the template RNA for the synthesis of the next six base pair repeat.[39] This extension of the 3′ DNA template end compensates for the end-replication problem and other end-processing events.

In normal somatic cells progressive telomere shortening is observed, eventually leading to greatly shortened telomeres and to a limited ability to continue to divide (FIG. 1). Telomere shortening has been demonstrated to be a molecular clock mechanism that counts the number of times a cell has divided (a cellular replication monitor), and, when a few telomeres are short enough, cellular senescence (growth arrest) occurs.[40–41] It has been proposed that shortened telomeres in mitotic (dividing) cells may be responsible for some of the changes we associate with normal aging.[4,9,10,33] The strongest correlation of the role of short telomeres in human physiology involves patients with aplastic anemia[42] (with mutations in the hTERT

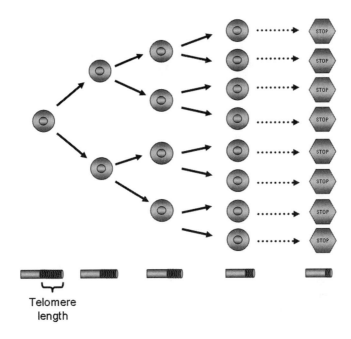

FIGURE 1. Aging of normal cells: a model illustrating that almost all human somatic cells can divide only a limited number of times. Even stem-like cells that are telomerase-competent show gradual telomere erosion eventually leading to replicative senescence.

gene) and patients with dyskeratosis congenita (with mutations in the hTR gene or a protein associated with hTR, dyskerin).[43-48] In both of these human disorders, there is progressive bone marrow failure that directly correlates with short telomeres. While short telomere dynamics correlate with the onset of these diseases, resulting in a shortened patient life span, it is not yet proven that telomere shortening limits normal human longevity. Although. in the elderly, relatively shortened white blood cell telomere length may correlate with premature death[4,33] and disease onset, the causality of this relation is not established. Thus, the putative role of telomere length as one of probably many different mechanisms that contribute to human aging remains a matter of conjecture and speculation.

It is fair to ask how we might go from correlations to more direct evidence that changes in cellular activity owing to short telomeres make humans more susceptible to some of the diseases and dysfunction associated with increased age. Specifically, how would one determine whether a subset (threshold) of cells in specific organs that ceased to divide contributes to tissue deterioration, and indeed, some aspects of aging itself? While the ultimate answers regarding the cellular basis of *in vivo* aging remain elusive, if one could "rejuvenate" shortened telomeres in such tissues and demonstrate that tissue function was improved, then a cause-and-effect relationship would be more tenable. At the present time ongoing research efforts are allowing

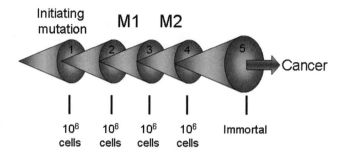

FIGURE 2. The clonal model of cancer progression. For a cancer cell to continue to grow, several tumor-suppressor genes and oncogenes must be altered within the same cell. If replicative senescence limits the maximal number of replications, then short telomeres would limit cell growth and be a potent tumor-suppressor mechanism.

biologists to accrue evidence that points the way toward a better understanding of how cells, tissues, and organisms change with age.

A major disease associated with aging is cancer. Since most cases of cancer occur in the 65-year and older segment of the population, it is important to understand the underlying mechanisms at the cellular level. For a normal cell to become a malignant cancer cell, multiple genetic and epigenetic changes must occur within a single clonal population (FIG. 2). The initial mutant cell probably uses up at least 20 divisions expanding to a population size large enough to have a reasonable chance of a second event occurring. As most mutations are recessive, an additional 20 divisions may be needed to expand again following elimination of the remaining wild-type allele. If replicative aging limited the number of available divisions, for example to 80–100 doublings, then most pre-malignant cells would come up against this barrier to proliferation after accumulating only a few mutations and would be prevented from progressing to cancer.[10,20,23] Indeed, most precancerous tissue lesions such a prostatic intraepithelial neoplasia, ductal carcinoma *in situ*, Barrett's esophagus, liver cirrhosis, and inflammatory bowel disease have critically short telomeres.[24–27] One key question is if these short telomeres are a tumor-suppressor pathway preventing cancer or an oncogenic pathway that leads to genomic instability and an increased susceptibility to cancer. However, the need to overcome this limitation, usually by upregulating the expression of telomerase in order to maintain telomere length, would likely explain why ~90% of all tumors express high levels of telomerase.[36–37]

The best protection against cancer would be to severely restrict the number of available divisions, but a person's cells must have sufficient proliferative capacity to provide for normal growth, maintenance, and repair functions. Within the context of organismal aging, evolution would select for the number of divisions required during the expected average human life span, enough for both routine cell turnover and to respond to a variety of sporadic stresses (trauma, disease), without modern medical interventions (such as improved sanitation, vaccines, antibiotics, anithypertensives, laproscopic surgery, etc). Evolutionarily processes should select against an excessive number of divisions (for example, sufficient for a 200-year human life span), since the extra divisions would increase the risk of cancer without any benefit if most of us died before we were 40. As humans now live well beyond 40, we are observing an increase in chronic diseases in which replicative aging may reduce the efficiency of cell turnover and tissue maintenance and may contribute to the declining physiology. However, it is important to note that replicative aging would be expected to be only one of a large number of maintenance/repair functions contributing to aging. Even if replicative aging contributed only to 10% of what we consider aging, and assuming we understood this well enough to slow down or reverse the telomere losses, it would potentially be an important advance.

While there has been much speculation in the last decade about the relationship between telomeres and aging, up until recently no one had looked for an association between human mortality and telomere length. To probe that issue, Cawthon and coworkers[4] examined blood samples previously collected from 60- to 97-year-olds. They measured the length of telomeres in DNA purified from white blood cells. People with short telomeres died from infectious diseases more than eight times as often as did those with longer telomeres, and they died from heart disease more than three times as frequently. One interpretation is that short telomeres in immune cells with shortened telomeres might repel invaders poorly. It is important to note that while this study is interesting, it is still only a correlation and does not prove a relationship between mortality and telomere length. However, there are many additional studies showing correlations of short telomeres with a variety of age-associated diseases.[24-33]

The process by which cells regulate the onset of cellular senescence represents an important question for understanding replicative aging and the mechanisms by which tumor cells escape these limits and become immortal. So what is cellular senescence? In contrast to tumor cells, which can divide forever (are "immortal"), normal human cells have a limited capacity to proliferate (are "mortal").[7,8] In general, cells cultured from a fetus divide more times in culture than those from a child, which in turn divide more times than those from an adult. The length of the telomeres decreases both as a function of donor age and with the number of times a cell has divided in culture.[9]

There are two telomere-related mechanisms responsible for the proliferative failure of normal cells (FIG. 3). The first, M1 (Mortality Stage 1), occurs when there are still at least several thousand base pairs of telomeric sequences left at the end of most of the chromosomes. M1 appears to be induced by a telomere "uncapping," leading to a DNA damage signal produced by one or a few of the 92 telomeres that have particularly short telomeres.[40] We and others have previously demonstrated that the timing of M1 is dependent on the shortest telomeres,[49-51] but these studies did not distinguish between one or a few short sentinel telomeres or a more general monitoring of short telomeres (FIG. 3). Our results show that approximately 10% of the telomeres are responsible for 90% of the telomeric end-association events at senescence, and thus a subset of telomeres is being used to time the onset of replicative arrest.[40,52] A reasonable hypothesis is that a few of the 10 shortest telomeres induces a sufficient DNA damage signal (FIG. 3) that leads to a growth arrest of cells for years. As such, this initial M1 growth arrest is likely to be a potent tumor suppressor pathway.

However, since we are living longer and are exposed to a variety of environmental insults (carcinogens leading to oxidative damage, misdirected DNA repair, etc.), an

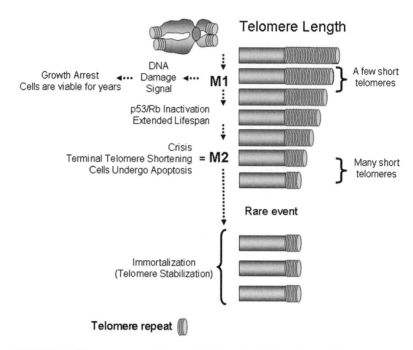

FIGURE 3. There are two major steps limiting the ability of normal human cells to divide indefinitely. They are called mortality stage I and mortality stage 2 (M1, and M2). It has been shown that shortened telomeres are important in both of these stages, because introduction of telomerase (hTERT) either before M1 or between M1 and M2 results in direct cell immortalization.

occasional alteration in a critical cell cycle checkpoint pathway (e.g., p53 or pRB mutations) can lead to an extended life span of cells. In this situation the DNA damage signal from the shortened telomeres fails to produce the checkpoint arrest and cells continue to divide without a mechanism to maintain telomeres (FIG. 3). This results in continuing shortening of telomeres until many telomeres have become critically shortened leading to crisis, M2, or what we call "terminal telomere shortening" (FIG. 3). While the M1 mechanism causes a stable growth arrest mediated by the tumor suppressor genes such as *p53,* if the actions of *p53* are blocked, either by genetic mutation or by epigenetic processes such as methylation, then cells can continue to divide and telomeres continue to shorten until the M2 (Mortality Stage 2) mechanism is induced. M2 represents the physiological result of critically short telomeres, when cells are no longer able to protect the ends of the chromosomes, so that end-degradation and end-to-end fusions occurs, leading to genomic instability and cell death (apoptosis). In cultured cells, a rare focus of immortal cells occasionally arises from cells in M2/crisis (FIG. 3). In most cases, these cells have reactivated the expression of telomerase that is able to repair and maintain the telomeres (FIGS. 3 and 4). Under normal physiological conditions, germline and embryonal stem cells fully maintain telomere length, and this is due to the expression and activity of telomerase (FIG. 4). Pluripotent stem cells in specific tissues normally have extended

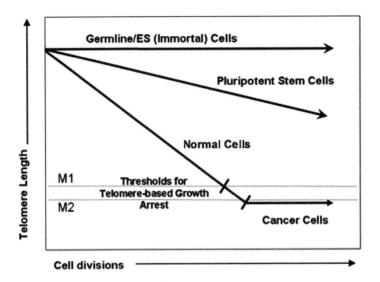

FIGURE 4. Only certain male germline cells and totipotent embryonal stem cells fully maintain telomere length. Pluripotent stem cells are telomerase competent but do not fully maintain telomere length; thus with increased age telomeres become progressively shorter. Most normal cells in the human body that divide do not express telomerase activity and thus may divide fewer times than highly proliferative pluripotent stem cells (such as those of the intestine, bone marrow, and skin).

proliferative capacity due the regulated activity of telomerase (FIG. 4). For example, in the most primitive stem cells of renewal tissues (e.g., crypts of the intestine, bone marrow cells, resting lymphocytes, basal layer of the epidermis) telomerase activity is low, while it is increased in the proliferative decendents of these cells. Thus, there are telomerase-competent cells that have low activity when quiescent (not dividing) and increased activity when proliferating (dividing). However, these telomerase-competent stem cells do not fully maintain telomere length (FIG. 4) since it has been observed that such cells obtained from older individuals have shorter telomeres than those derived from younger individuals. Thus, in germline (reproductive) cells and tumor cells, telomerase fully maintains telomere length in contrast to stem cells (with regulated telomerase activty) and most somatic cells (with no detectable telomerase activity) in which telomeres progressively shorten with increased age.

Aging is a process that turns a healthy young animal into an old one that is more vulnerable to disease and tissue or organ failure. A cell's survival and function clearly play into the well-recognized phenotypes associated with aging. For example, it is believed that when a threshold number of cells in the dermis age, our skin wrinkles; reduced numbers of bone-forming osteoblasts leads to osteoporosis; and diminished T cell performance contributes to a compromised immune system.[33] Thus, it is entirely possible that a small number of stem-like cells in a given tissue could exhaust the maximum number of permissible doublings, leading to telomere-based replicative senescence. Many factors besides replication could affect telomere length, and other factors (stress, oncogene-induced growth arrest) can also lead to inhibition of growth sometimes called premature senescence. In all of these instances, replicative and stress- or oncogene-induced senescence may function as potent tumor-suppressor mechanisms, but as a consequence also contribute to some aspects of aging Clearly further work is necessary to determine whether short telomeres are causative in some aspects of organismal aging.

It is reasonable to ask whether stopping the shortening of telomeres would prevent cellular aging? The evidence that shortening of telomeres is causal in cell aging has now been demonstrated.[53] Introduction of the telomerase catalytic protein component (hTERT) into normal human cells without detectable telomerase results in restoration of telomerase activity.[53] Normal human cells stably expressing transfected telomerase demonstrate a many-fold extension of life span, providing direct evidence that telomere shortening controls cellular aging.[53] The cells with introduced telomerase maintain a normal chromosome complement and continue to grow in a normal manner.[54] Initial concerns that the introduction of telomerase into normal cells could directly increase the risk of cancer have not proven true.[54] A theory abolishing replicative aging in all the $\sim 10^{14}$ cells in the human body would be expected to increase the risk of cancer over a lifetime, but removing the counting mechanism does not by itself confer abnormal growth properties. One way to think about this is that specific germline and embryonal stem cells maintain high levels of telomerase throughout life, and there is no increased incidence of cancers in these special cells when compared with other types of cancer. Thus, the major role of telomerase is to maintain telomere stability and keep the cells dividing. These observations provide much more direct evidence for the hypothesis that telomere length determines the proliferative capacity of human cells.

This raises the important question: can telomerase used as a product to extend cell life span? The ability to immortalize human cells and retain normal behavior holds promise in several areas of biopharmaceutical research, including drug development, screening and toxicology testing.[55] The development of better cellular models of human disease and the production of human products are among the immediate applications of this new advance. This technology has the potential to produce unlimited quantities of normal human cells of virtually any tissue type and may have most immediate translational applications in the area of transplantation medicine. In the future it may be possible to take a person's own cells, manipulate and safely rejuvenate them without using up their life span, and then give them back to the patient (with almost no potential for a risk of increases in cancer). In addition, genetic engineering of telomerase-immortalized cells could lead to the development of cell-based therapies for certain genetic disorders such as muscular dystrophy. Our group has successfully immortalized corneal epithelial cells and keratocytes with the goal of producing engineered corneas for transplantation.[56] In addition, we have used immortalized skin keratinocytes, melanocytes, and fibroblasts to make skin organotypic cultures that can be xenotransplanted.[57] We have demonstrated that these skin equivalents can be wounded using a laser and that the epithelial layer will heal the wound.[57] As is illustrated in FIGURE 5, we have recently immortalized human bronchial epithelial cells (HBECs) with hTERT and placed these in organotypic culture with normal or immortalized lung fibroblasts.[58] When placed in an air/medium interface these cells (though immortalized) still differentiate into mucous and ciliated HBECs (FIG. 5, bottom). Experiments such as this and others[59–61] clearly support the concept that engineering cells for tissue enhancements is feasible.

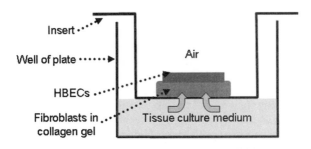

Insert

Well of plate

Air

HBECs

Fibroblasts in
collagen gel

Tissue culture medium

FIGURE 5. Organotypic (3D) cultures of human bronchial epithelial cells and lung fibroblasts. When hTERT-expressing lung fibroblasts and bronchial epithelial cells are placed in conditions that induce terminal differentiation (**top**), ciliated and mucous-secreting cells are apparent on the cell surface (**bottom**, scanning electron micrograph of the surface of hTERT expressing human bronchial epithelial cells once induced to differentiate). This suggests that ectopic expression of hTERT does not prevent normal cell differentiation.

DISCUSSION/PERSPECTIVE

There have been many recent reviews on the utility of expressing the catalytic subunit of telomerase into normal cells for cell and tissue engineering.[53–61] This leads to the possibility that in the near future we may be able to reset or rewind the telomere clock so that old cells that are running out of divisions can continue to divide. This would be applicable in some normal tissues of older individuals with chronic diseases or in younger individuals who have genetic disorders that have resulted in cells prematurely using up their telomeres (e.g., T cells in patients infected with HIV or muscle cells in children with muscular dystrophy). To be safe, telomerase therapy would be initially conducted *ex vivo* (e.g., in the laboratory environment). We would remove a few cells from a person by fine-needle biopsy, reset the telomere clock of these cells in the laboratory (thus providing the cells with greatly extended life span), and then reintroduce the cells back into the person. This would have the enormous advantage of not having to deal with immune rejection, since these are the person's own cells. In the case of a genetic disorder, we would reset the telomeres, reinsert the corrected gene, and then reintroduce the cells. It is important to recognize that in the initial experiments we could insert telomerase to elongate the telomeres and then remove telomerase. This could be easily accomplished using excisable vectors or even adenoviral genes containing hTERT that do not integrate into the genome. This would avoid potential problems with having too much growth potential (and thus increased risk for cancer). However, our results indicate that introduction of telomerase in perfectly normal cells does not increase the cancer risk very much and the cells with intact telomerase can still undergo differentiation and participate in normal tissue functions.[55–58]

One of the major difficulties in the development of any tissue/bioengineering program is the practical problem of repeatedly obtaining human material for primary cultures and the enormous variability encountered between cultures and between laboratories. The ability to immortalize a variety of specialized cell types avoids this by providing reproducible reagents that can be shared and cultured indefinitely.

However, one of the most profound opportunities that results from having an unlimited proliferative capacity is that one can now envisage genetically engineering normal human cells so that they acquire properties that make them more useful than their normal counterparts.

Concerns about insertional mutagenesis and the activation of oncogenes (or the inactivation of tumor-suppressor genes) has restricted the enthusiasm for genetic engineering. The basis for these concerns change dramatically if one is engineering an immortal normal diploid cell *in vitro*. For example, the risks are not very favorable if the probability of an insertional mutagenic event is one in a million and one is infecting ten million cells *in vivo*. However, immortal normal diploid cells can be cloned and characterized after performing genetic engineering. Even after introducing ten different vectors, the number of insertions in a cloned cell is only ten. The probability of an adverse event is thus reduced by six to seven orders of magnitude by the simple process of picking the progeny of a single cell. It is also a simple matter to use inverse PCR of circularized DNA fragments to obtain a small amount of DNA sequence flanking each insertion site, and thus map all the insertions within the human genome and know whether or not any insertions are in worrisome locations. Most importantly, the cloned cell can be fully characterized to verify that it behaves in the desired fashion and exhibits the same resistance to transformation as "normal" cells. Finally, every vector used could contain conditional suicide control genes (such as the Herpes virus thymidine kinase, or cytidine deaminase gene) that are only toxic to dividing cells after administering a pro-drug. This would provide additional redundant safety factors of built-in magic bullets to control proliferation if any unanticipated problems arising from the transplanted cells.

The implications of being able to genetically alter normal cells are limited only by our imagination. One should be able to engineer cells to avoid immune rejection, and thus make universal donor cells that would be much more widely applicable than any approach that requires isolating cells from each new patient. One could engineer cells to be much more tolerant of hypoxia, facilitating tissue engineering of larger and more complex structures *in vitro* that lack a fully developed blood supply. One could engineer cells to express factors such as factor IX (missing in hemophilia), using regulatable constructs so that the amount of expression in the body following transplantation could be controlled. In short, the ability to use telomerase to immortalize normal diploid cells opens up exciting new vistas for the ways in which tissue engineering can contribute to the amelioration of human disease and decline.

[*Competing interests*: The authors declare that they have no competing financial interests.]

REFERENCES

1. MULLER, H.J. 1962. The remaking of chromosomes. *In* Studies of Genetics: The Selected Papers of H.J. Muller. pp 384–408. Indiana University Press. Bloomington.
2. McCLINTOCK, B. 1941. The stability of broken ends of chromosomes in *Zea mays*. Genetics **26:** 234–282.
3. CAMPISI, J. *et al.* 2001. Cellular senescence, cancer, and aging: the telomere connection. Exp. Gerontol. **36:** 1619–1637.
4. CAWTHON, R.M. *et al.* 2003. Association between telomere length in blood and mortality in people aged 60 years or older. Lancet **361:** 393–395.
5. WU, X. *et al.* 2003. Telomere dysfunction: a potential cancer predisposition factor. J. Natl. Cancer Inst. **95:** 1211–1218.
6. SLAGBOOM, P. E., S. DROOG & D.I. BOOMSMA. 1994. Genetic determination of telomere size in humans: a twin study of three age groups. Am. J. Hum. Genet. **55:** 876–882.
7. SHAY, J.W. & W.E. WRIGHT. 2002. Historical claims and current interpretations of replicative aging. Nat. Biotech. **20:** 682–688.
8. SHAY, J.W. & W.E. WRIGHT. 2000. Hayflick, his limit, and cellular aging. Nat. Rev. Mol. Cell Biol. **1:** 72–76.
9. HARLEY, C.B., A.B. FUTCHER & C.W. GREIDER. 1990. Telomeres shorten during aging. Nature **345:** 458–460.
10. WRIGHT, W.E. & J.W. SHAY. 2001. Cellular senescence as a tumor-protection mechanism: the essential role of counting. Curr. Opin. Genet. Dev. **11:** 98–103.
11. FELDSER, D.M., J.A. HACKETT & C.W. GREIDER. 2003. Telomere dysfunction and the initiation of genome instability. Nat. Rev. Cancer **3:** 623–627.
12. MASER, R.S. & R.A. DEPINHO. 2002. Connecting chromosomes, crisis, and cancer. Science **297:** 565–569.
13. SHAY, J.W. 2003. Telomerase therapeutics: telomeres recognized as a DNA damage signal. Clin. Can. Res. **9:** 3521–3525.
14. DE LANGE, T. & T. JACKS. 1999. For better or worse? Telomerase inhibition and cancer. Cell **98:** 273–275.
15. HAHN W.C. *et al.* 1999. Inhibition of telomerase limits the growth of human cancer cells. Nat. Med. **5:** 1164–1170.
16. HERBERT, B.-S. *et al.* 1999. Inhibition of human telomerase in immortal human cells leads to progressive telomere shortening and cell death. Proc. Natl. Acad. Sci. USA **96:** 14276–14281.
17. KELLAND, L.R. 2001. Telomerase: biology and phase I trials. Lancet Oncol. **2:** 95–102.
18. KEITH, W.N. *et al.* 2004. Drug insight: cancer cell immortality—telomerase as a target for novel cancer gene therapies. Nat. Clin. Pract. Oncol. **1:** 1–9.
19. NEIDLE S. & G. PARKINSON. 2002. Telomere maintenance as a target for anticancer drug discovery. Nat. Rev. Drug Disc. **1:** 383–393.
20. SHAY J.W. & W.E. WRIGHT. 1996. Telomerase activity in human cancer. Curr. Opin. Oncol. **8:** 66–71.

21. VONDERHEIDE, R.H. 2002.Telomerase as a universal tumor-associated antigen for cancer immunotherapy. Oncogene **21:** 674–679.
22. WHITE, L., W.E. WRIGHT & J.W. SHAY. 2001. Telomerase inhibitors. Trends Biotechnol. **3:** 146–149.
23. SHAY, J.W. & W.E. WRIGHT. 2002. Telomerase: a target for cancer therapy. Cancer Cell **2:** 257–265.
24. MEEKER, A.K. *et al.* 2004. Telomere length abnormalities occur early in the initiation of epithelial carcinogenesis. Clin. Cancer Res. **10:** 317–326.
25. RUDOLPH, K.L. *et al.* 2001. Telomere dysfunction and evolution of intestinal carcinoma in mice and humans. Nat. Genet. **28:** 155–159.
26. O'SULLIVAN, J.N. *et al.* 2002. Chromosomal instability in ulcerative colitis is related to telomere shortening. Nat. Genet. **32:** 280–284.
27. FARAZI, P.A. *et al.* 2003. Differential impact of telomere dysfunction on initiation and progression of hepatocellular carcinoma. Cancer Res. **63:** 5021–5027.
28. OKUDA, K. *et al.* 2000. Telomere attrition of the human abdominal aorta: relationships with age and atherosclerosis. Atherosclerosis **152:** 391–398.
29. BENETOS, A. *et al.* 2001. Telomere length as an indicator of biologic aging: the gender effect and relation with pulse pressure and pulse wave velocity. Hypertension **37:** 381–385.
30. VON ZGLINICKI, T. *et al.* 2000. Short telomeres in patients with vascular dementia: an indicator of low antioxidative capacity and a possible risk factor? Lab. Invest. **80:** 1739–1747.
31. EPEL, E.S. *et al.* 2004. Accelerated telomere shortening in response to life stress. Proc. Natl. Acad. Sci. USA **101:** 17312–17315.
32. YANG, L. *et al.* 2001. Telomere shortening and decline in replicative potential as a function of donor age in human adrenocortical cells. Mech Ageing Dev. **122:** 1685–1694.
33. AVIV, A. 2004. Telomeres and human aging: facts and fibs. Sci. Aging Knowl. Environ. **51:** 43.
34. GREIDER, C.W. & E.H. BLACKBURN. 1985. Identification of a specific telomere terminal transferase activity in Tetrahymena extracts. Cell **43:** 405–413.
35. WRIGHT, W.E. *et al.* 1996. Telomerase activity in human germline and embryonic tissues and cells. Dev. Genet. **18:** 173–179.
36. KIM, N.W. *et al.* 1994. Specific association of human telomerase activity with immortal cells and cancer. Science **266:** 2011–2015.
37. SHAY, J.W. & S. BACCHETTI. 1997. A survey of telomerase activity in human cancer. Eur. J. Cancer **33:** 787–791.
38. MORIN, G.B. 1989. The human telomerase terminal transferase enzyme is a ribonucleoprotein that synthesizes TTAGGG repeats. Cell **59:** 521–529.
39. NUGENT, C.I. & V. LUNDBLAD. 1998. The telomerase reverse transcriptase: components and regulation. Genes Dev. **12:** 1073–1085.
40. ZOU, Y. *et al.* 2004. Does a sentinel or groups of short telomeres determine replicative senescence? Mol. Biol. Cell **15:** 3709–3718.
41. SHAY, J.W. & W.E. WRIGHT. 2004. Telomeres are double-strand DNA breaks that are hidden from DNA damage responses. Mol. Cell **14:** 420–421.
42. YAMAGUCHI, H. *et al.* 2005. Mutations in TERT, the gene for telomerase reverse transcriptase, in aplastic anemia. N. Engl. J. Med. **352:** 1413–1424.
43. MITCHELL, J.R., E. WOOD & K.A. COLLINS. 1999. A telomerase component is defective in the human disease dyskeratosis congenita Nature **402:** 551–555.
44. SHAY, J.W. & W.E. WRIGHT. 1999. Telomeres in dyskeratosis congenita. Nat. Genet. **36:** 437–438.
45. SHAY, J.W. & W.E. WRIGHT. 2004. Mutant dyskerin ends relationship with telomerase. Science **286:** 2284–2285.
46. VULLIAMY, T. *et al.* 2001. The RNA component of telomerase is mutated in autosomal dominant dyskeratosis congenita. Nature **413:** 432–435.
47. HEISS, N.S. *et al.* 1998. X-linked dyskeratosis congenita is caused by mutations in a highly conserved gene with putative nucleolar functions. Nat. Genet. **19:** 32–38.
48. FOGARTY, P.F. *et al.* 2003. Late presentation of dyskeratosis congenital as apparently acquired aplastic anemia due to mutations in telomerase RNA. Lancet **362:** 1628–1630.

49. STEINERT, S., SHAY, J.W. & W.E. WRIGHT. 2000. Transient expression of human telomerase (hTERT) extends the life span of normal human fibroblasts. Biochem. Biophys. Res. Com. **273:** 1095–1098.
50. OUELLETTE, M.M. *et al.* 2000. Subsenescent telomere lengths in fibroblasts immortalized by limiting amounts of telomerase. J. Biol. Chem. **275:** 10072–10076.
51. HEMANN M.T. *et al.* 2001. The shortest telomere, not average telomere length, is critical for cell viability and chromosome stability. Cell **107:** 67–77.
52. SHAY, J.W. & W.E. WRIGHT. 2004. Senescence and immortalization: role of telomeres and telomerase. Carcinogenesis **25:**1–8.
53. BODNAR, A.G. *et al.* 1998. Extension of lifespan by introduction of telomerase in normal human cells. Science **279:** 349–352.
54. MORALES, C.P. *et al.* 1999. Lack of cancer-associated changes in human fibroblasts after immortalization with telomerase. Nat. Gen. **21:** 115–118.
55. SHAY, J.W. & W.E. WRIGHT. 2000. The use of "telomerized" cells for tissue engineering. Nat. Biotech. **18:** 22–23.
56. ROBERTSON, D.M. *et al.* 2005. Characterization of growth and differentiation in a telomerase-immortalized human corneal epithelial cell line. Invest. Ophthalmol. Visual Sci. **46:** 470–478.
57. VAUGHAN, M.B. *et al.* 2004. A reproducible laser-wounded skin equivalent model to study the effect of aging *in vitro*. Regeneration Med. **2:** 99–110.
58. VAUGHAN, M.B. *et al.* 2005. A three-dimensional model of differentiation of immortalized human bronchial epithelial cells. Differentiation In press.
59. THOMAS, M., L. YANG & P.J. HORNSBY. 2000. Formation of normal functional tissue from transplanted adrenocortical cells expressing telomerase reverse transcriptase (TERT). Nat. Biotechnol. **18:** 39–42.
60. THOMAS, M., S.R. NORTHRUP & P.J. HORNSBY. 1997. Adrenocortical tissue formed by transplantation of normal clones of bovine adrenocortical cells in scid mice replaces the essential functions of the animals' adrenal glands. Nat. Med. **3:** 978–983.
61. POH, M. *et al.* 2005. Blood vessels engineered from human cell. Lancet **365:** 2122–2124.

Mind–Body Medicine

Stress and Its Impact on Overall Health and Longevity

L. VITETTA,[a] B. ANTON,[b] F. CORTIZO,[c] AND A. SALI[c]

[a]Centre for Molecular Biology and Medicine, Epworth Hospital Medical Centre, Melbourne, Australia

[b]PathLab, Melbourne, Australia

[c]National Institute of Integrative Medicine, Swinburne University, Melbourne, Australia

ABSTRACT: The belief that adverse life stressors and the emotional states that can lead to major negative impacts on an individual's body functions and hence health has been held since antiquity. Adverse health outcomes such as coronary heart disease, gastrointestinal distress, and cancer have been linked to unresolved lifestyle stresses that can be expressed as a negative impact on human survival and ultimately a decrease of the human life span. Psychological modulation of immune function is now a well-established phenomenon, with much of the relevant literature published within the last 50 years. Psychoneuroimmunology and psychoneuroendocrinology embrace the scientific evidence of research of the mind with that of endocrinology, neurology and immunology, whereby the brain and body communicate with each other in a multidirectional flow of information that consists of hormones, neurotransmitters/neuropeptides, and cytokines. Advances in mind–body medicine research together with healthy nutrition and lifestyle choices can have a significant impact on health maintenance and disease prevention and hence the prolongation of the human life span.

KEYWORDS: mind–body medicine; psychoneuroimmunology; psychoneuro-endocrinology; psychosocial stress; longevity

MIND–BODY MEDICINE

The concept of the mind and human thought is not a novel one. Many tentative explanations have been advanced, even in the form of art—notably in Michelangelo's *Creation of Adam*, in which man's mind and thought are depicted by a sagittal section of the brain that is strikingly similar to current anatomical drawings of the brain (FIG. 1).

The most important factor in why a person becomes ill lies in the brain. Stress and pleasure play a critical role in wellness and disease, with stress contributing

Address for correspondence: Dr. Luis Vitetta, Associate Professor and Senior Research Fellow, Centre for Molecular Biology and Medicine, Epworth Medical Centre, 185-187 Hoddle St. Richmond, Melbourne, Victoria, Australia 3121. Voice: +61 3 9426 4200; fax: +61 3 9426 4201. lvitetta@cmbm.com.au

Ann. N.Y. Acad. Sci. 1057: 492–505 (2005). © 2005 New York Academy of Sciences.
doi: 10.1196/annals.1322.038

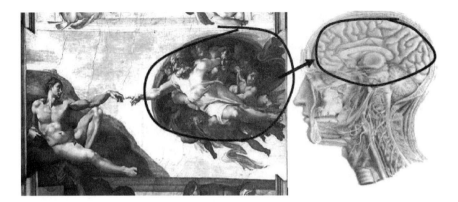

FIGURE 1. Comparison of a portion of Michelangelo's *Creation of Adam* with a modern anatomical drawing of the brain.

significantly to the risk of disease.[1–4] It is becoming increasingly clear that pleasure is also important. Even if a person is stressed, if they are also feeling good, stress will have fewer ill effects. Other factors also play an important role such as diet, smoking, and genetics. An additional important factor seems to be the ability and opportunity that persons have to express their feelings about how stressed they are or how they are feeling about themselves. Stress is less likely to cause problems if a person has some form of emotional outlet for it. In many cases, people appear calm on the outside but they are in turmoil on the inside. This may then manifest as a disorder such as migraine, irritable bowel syndrome, rheumatoid arthritis, or multiple sclerosis.[5–7]

Evidence that has emerged within the past several decades and that continues to accumulate strongly indicates that the state of the human mind—which associates psychosocial factors with emotional states such as depression and with behavioral dispositions that include hostility and psychosocial lifestyle stresses—can directly and significantly influence human physiologic function and, in turn, health outcomes.[8–11]

Moreover, the placebo effect is further evidence that there is a significant connection between the human psyche and positive thinking. In a recent review, the placebo effect was investigated to determine wether the power of belief and conscious expectation, to which the term placebo response has been applied, can change the neurochemical environment in key areas of the brain responsible for movement (corpus striatum), pleasure (nucleus accumbens), physical pain, and the psychological pain of human sadness stemming from separation (anterior cingulate).[12] The authors concluded that the placebo response will tend to be relatively robust with pain disorders, depressive disorders, and Parkinson's disease. It was further hypothesized that the mesolimbic–mesocortical dopamine system also has control of the hypothalamic–pituitary–adrenal and amygdalalateral hypothalamus–locus coeruleus sympathetic nervous system stress response axes. This then suggests that belief and positive expectation can modify the stress response and thus may lead to placebo responsiveness of many psychophysiological disorders such as hypertension, angina, inflammatory bowel disease, and asthma.[12]

There is now clear scientific evidence that indicates that the brain regulates immune function through efferent autonomic and neuroendocrine pathways.[1] Moreover, there are numerous afferent pathways by which the immune system modulates brain function. Psychoneuroimmunology, a term first coined by Robert Ader in the 1980s, refers to these connections between the psyche, the nervous system, and the immune system.[1,2] The development of the science of psychoneuroimmunology provides an enormous opportunity to understand the role of the mind in the cause of disease. Psychoneuroendocrinology, which refers to the connections between the psyche and the endocrine system, is equally important. Thus, behavior, the nervous system, and the endocrine system all influence the immune system. A feedback system also exists by which it is able to connect with all aspects of the brain. The details of the mechanisms involved in psychoneuroimmunology are the subject of much of the current research.

STRESS AND HEALTH

The history of the concept of stress in relation to disease processes reaches back to the nineteenth century and through to the twentieth century and beyond. It has been and continues to be the subject of intense debate and research with reports that stress may have ameliorating or detrimental capacities. The pioneering work of Hans Selye is often credited with establishing the scientific fact that there is a significant relationship between pathophysiological processes and the onset of chronic diseases.[13,14] Selye successfully advanced the concept that stress was critically important in physiology and medicine, and that today we recognize that many contemporary theories relating to the etiology of chronic diseases have a stress component as a significant precipitating variable. Moreover, recently it has been shown that stressful life events can have a significant negative impact on longevity, by playing an important role in the onset of cardiovascular disease, immunological disorders, and pathophysiological consequences of normal aging.

Stress describes the effects of psychosocial and environmental factors on physical or mental well being. Stressors and stress-related reactions have been documented and recognized and exposure to chronic social stress has been associated with many systemic and mental disorders. Hypotheses from different research groups support the notion that health consequences are more likely to occur when unpredictable stressors of a social nature chronically induce physiological and behavioral adjustments that may create wear and tear on the underlying physiological functions.[15] When stressors challenge an organism's integrity, a set of physiological reactions is elicited to counteract the possible threat and adjust the physiological setting of the organism to the new situation. This has become known as the stress response.[16]

The stress response that initiates the neuroendocrine system has been extensively studied in the sympathetic–adreno–medullary system, which is under the control of the central nervous system.[16] A further component of the stress response is the hypothalamic–pituitary–adrenal axis, which is diagrammatically illustrated in FIGURE 2, and is an important modulator of the brain–immune–endocrine–neurotransmitter interconnection cycle.

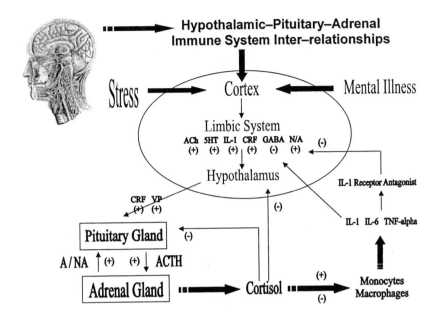

FIGURE 2. Diagrammatic representation of the inter-relationships between the hypothalamic–pituitary–adrenal axis and the immune–endrocrine responses. Ach = acetylcholine; 5HT = 5 hydroxytryptamine; CRF = corticotrophin-releasing factor; GABA = gammaaminobutyric acid; A = adrenaline; NA = noradrenaline; IL = interleukin; TNF = tumor necrosis factor. (Modified from Song and Leonard.[1])

These two systems are normally operating within a fine-tuned state of balance known as homeostasis, which is established in order to maintain the organism's integrity even under extremely challenging conditions. Further, these two systems have a dependence on chemical transmitters for effective communication. This is observed when electrical signals along nerve pathways are converted to chemical signals at the site of synaptic junctions between neurons. The chemical message that is produced by the activity of immune-system cells can communicate not only with other parts of the immune system but also with the brain and nerves. Alternatively, chemicals released by nerve cells can act as signals to immune cells.

There is also an important endocrine component to these systems, where the brain has a central and prolific role as an endocrine organ in the body and produces many hormones that act both on the brain and on tissues throughout the body. In particular, a key hormone shared by the central nervous and immune systems is corticotropin-releasing hormone (CRH). CRH is produced in the hypothalamus and other brain areas and is essential in uniting the stress and immune response. Specifically, the hypothalamus releases CRH into a specialized bloodstream circuit that conveys the hormone to the pituitary gland. CRH causes the pituitary gland to release adrenocorticotropin hormone (ACTH) into the bloodstream, which stimulates the adrenal glands to produce cortisol, an extensively studied stress hormone.[16,17]

Cortisol is a steroid hormone that has many actions that include: increasing the rate and strength of heart contractions and sensitizing blood vessels to the actions of norepinephrine; it has effects on many metabolic functions as well. These actions assist the body in coping with stressful situations. Moreover, cortisol is a potent immunoregulator and anti-inflammatory agent. Cortisol also has a crucial role in preventing the immune system from overreacting to injuries and damaging tissues. Cortisol inhibits the release of CRH by the hypothalamus, thus keeping this component of the stress response under a fine control. Hence the body's brain-regulated stress response and its associated immune response directly link the activity of CRH and cortisol.There are CRH-secreting neurons of the hypothalamus that send fibers to regions in the brain stem that help regulate the sympathetic nervous system, as well as to other brain stem areas such as the locus coeruleus. The sympathetic nervous system, which mobilizes the body during stress, also innervates immune-active tissue, such as the thymus, lymph nodes, and the spleen, and further assists to control inflammatory responses throughout the body. Stimulation of the locus coeruleus leads to behavioral arousal, fear, and enhanced vigilance. Other brain CRH-secreting neurons are present in the central nucleus of the amygdala and send fibers to the hypothalamus, the locus coeruleus, and other parts of the brain stem. These CRH-secreting neurons are targets of messengers released by immune cells during an immune response. The additional recruitment of these CRH-secreting neurons allows immune signals not only to activate cortisol-mediated restraint of the immune response, but also to induce behaviors that assist in recovery from illness or traumatic injury. Further, chemical transmitters are also involved in tensing muscles. Thus the tenser a muscle becomes, the more alert and stressed the brain becomes and the tenser the muscles become, which creates further stress. An additional and important connection is that CRH-secreting neurons also have connections with hypothalamic regions that regulate food intake and reproductive behavior.[16,17]

The immune response is an elegant surveillance mechanism that is finely tuned in a series of cascade cellular events whose ultimate goal is to protect the organism from foreign substances, such as bacteria and viruses. The discovery of small protein molecules known as cytokines (e.g., interleukin-1 and interleukin-2), which are elaborated by different types of white blood cells, allowed for further understanding of how the immune response is structured and coordinated. Cytokines from the body's immune system can send signals to the brain via several mechanisms, including crossing the brain–blood barrier via the bloodstream. This permeability is essential for communication with brain areas. Cytokines can attach to their receptors in the lining of blood vessels in the brain and stimulate the release of secondary chemical signals in the brain tissue around the blood vessels. Cytokines can also signal the brain via direct nerve routes, (e.g., the vagus nerve) and a multitude of connections with abdominal organs result. The activation of the brain by cytokines from the peripheral parts of the body induces the behaviors of anxiety and cautious avoidance associated with the stress response's principle activity—maintaining the organism's integrity during recovery from stressful activities or from traumatic injury.[18]

The disruption of communication between the brain and the immune system leads to greater susceptibility to inflammatory disease and, frequently, to increased immune complications.[18,19] Laboratory animals whose brain-immune communications have been disrupted (through surgery or pharmaceuticals) are highly vulnerable to lethal complications of inflammatory diseases and infectious diseases.

Further, in studies with laboratory animals and in human pharmacological and surgical trials there has been evidence of a causal link between an impaired stress response and susceptibility to inflammatory disease.

The stress response hence represents a complex effector system that is susceptible to pathophysiological factors and conditions and has an input into many biological processes. Therefore the stress response can exert ameliorating or detrimental effects, where mediators of stress physiology can have both protective and harmful effects in organs such as the heart, the immune system, or the brain. Adverse effects on such organs may well have significant negative shifts in overall survival.[16–19]

Humans living in the 21st century are subject to an ever-increasing degree of homeostatic control between the interrelationships of external/environmental and internal (behavioral, cognitive, emotional) stressors, the physiological responses to these challenges, and the prevalence of disease/illness. This dynamic balance of continual adjustment to stressors maintains stability. Also, the cascade of adaptive responses imprinted within the human genetic makeup may themselves turn into stressors capable of triggering and maintaining disease processes.

MIND–BODY MEDICINE: THE CONNECTION TO DISEASE

Social stressors and lifestyle choices can be a significant trigger for disease, which can then affect longevity. Psychosocial stress is a personal response perceived through the complexities of social interactions. These interactions can either be negative and increase or be positive and lessen psychological stress. These interactions can affect the hormonal responses that are elicited as a result, which can then significantly alter immune responses. Moreover, stress can significantly reduce physical and mental tolerances (immune potential) of humans and it can induce the progression of existing illnesses or it can cause latent disorders to become active. Therefore, the control and suppression of stress is very important in the improvement of quality of life and the prevention of diseases. The physiological changes that ensue with unmanageable stress may then have a negative impact on health.[20] The social-psychological network of stressors that we experience can affect our susceptibility to inflammatory and infectious diseases as well as the course of these and other diseases and in so doing affect the human life span.[20–23]

In humans, loneliness is associated with a threat, or an epinephrine-like pattern of activation of the stress response that results in an elevation of blood pressure, whereas exercising is associated with a challenge pattern of high blood flow and cardiac output. Studies have shown that people exposed to chronic social stresses for more than two months have increased susceptibility to the common cold.[24]

It is of significant interest that in Australia there is a disparity in the rate of suicide deaths between rural and city areas.[25–27] Australia has one of the highest per capita young male youth suicide rate as well as one of the highest rates of antidepressant use.[25,26] Young males are experiencing grave difficulties with life changes that occur, whether in the work or home environments. In rural areas unemployment and isolation play major roles in the rate of suicide. Further a recent study evaluated the association of glycemic control and major depression in type 1 and 2 diabetes mellitus.[28] Type 1 insulin–dependent diabetes mellitus patients with a lifetime history of depression showed a significantly worse glycemic control, whereas patients with type 2 non-insulin–dependent diabetes mellitus with a similar history did not have a significantly worse glycemic control.[28] Hence depression can be a significant marker for increased risk of ill health and a decrease in the human life span.[29]

This link between depression and ill health is further strengthened in studies between depression and idiopathic urinary incontinence. It has been shown that this association may be due to altered serotonin functioning and may help explain the efficacy of serotogenic-based antidepressants in the treatment of urge incontinence.[30] Also, depression has been associated with functional disability in patients with coronary artery disease (CAD). Depressive symptoms are highly prevalent in patients with CAD and have been shown to increase their rate of cardiac-associated death and rate of mortality due to all causes. Depressed CAD patients have been shown to have an 84% greater risk of cardiac death 5–10 years later when compared to nondepressed CAD patients.[31]

Evidence continues to support the fact the physical disease results from emotional distress. School examination–induced stresses increase viral infection susceptibility in adolescents.[32,33] Stress from lack of control in the workplace or from life events increases susceptibility to cardiovascular disease.[34] A recent meta-analysis of biofeedback and pelvic floor exercises for female stress incontinence has been reported.[35] The results showed that biofeedback may be an important adjunct to pelvic floor muscle exercises. Further, interstitial cystitis, which causes bladder symptoms, has been shown to be stress related and these patients significantly have an elevated level of norepinephrine in urine.[36]

Other studies have shown that the immune responses of long-term caregivers, such as spouses of Alzheimer's patients, become blunted.[37] Immune responses during marital discord are also blunted in the spouse (usually the wife) who experiences the greatest amount of stress and feelings of helplessness.[38] In such studies the levels of stress hormones are elevated in the affected spouse.[37,38] Conversely, a positive supportive environment of extensive social networks or group psychotherapy can enhance immune response and resistance to disease—even cancer.[39,40] Studies have shown that women with breast cancer, for instance, who receive strong, positive social support during their illness have significantly longer life spans than women without such support.[41]

Moreover, recently our group has shown in preliminary data that psychosocial treatment programs for cancer patients annexed to traditional approaches offered by mainstream medical institutions in less orthodox approaches, such as that delivered by The Gawler Foundation in Melbourne, Australia, may constitute the holistic care provision that active cancer support programs should provide. With a design focus to improve quality of life and, if possible, to affect length of life, this holistic health approach, which is complementary to mainstream medical treatment, adjusts the patient's focus on mind–body medicine modalities, such as relaxation and meditation, that then are combined with low-fat vegetarian diets, positive thinking, and drawing on effective support. This approach provides a positive environment that significantly contributes to enhancing the quality of life of chronically ill patients and could significantly affect longevity.

Certainly, our preliminary data showed that a significant reduction in scores determined by the Profile of Mood States (POMS) for total mood disturbance, tension-anxiety, depression-dejection, anger-hostility, fatigue, and confusion-bewilderment scores and showed increase scores for vigor (all $P < 0.01$). Scores on the helplessness-hopelessness subscale of the mini-Mental Adjustment to Cancer (mini-MAC) were significantly reduced and scores on the fighting spirit subscale were significantly increased ($P < 0.01$). Total scores on the Functional Assessment of Chronic Illness Therapy-Spirituality (FACIT-Sp) scale were significantly increased as were scores on the physical, emotional, functional, and spiritual subscales (all $P < 0.01$). Salivary cortisol levels were significantly decreased ($P < 0.05$).[42]

Several epidemiological studies have shown that social and emotional support can protect against premature mortality, prevent illness, and assist recovery from illness.[43,44] A recent report found that smokers treated with nortriptyline and cognitive behavioral therapy were significantly aided in smoking cessation. It is not surprising that this treatment was successful, as other research has demonstrated that smokers are more likely to be depressed.[45] Vigorous exercise facilitates smoking cessation when combined with a cognitive behavioral program. Moreover, exercise has been shown to work as an antidepressant.[46]

Nutrition is an important factor that significantly contributes to how long a person will live. Decisions about poor dietary choices, in addition to smoking, lack of exercise, and excessive drinking are known to be important causes of disease.[47] The size and frequency of meals are fundamental aspects of nutrition that can have profound effects on the health and longevity of laboratory animals. Moreover, caloric restriction is the most potent and reproducible environmental variable capable of extending the life span in a variety of animals from worms to rats. In humans, excessive caloric intake is associated with increased incidence of cardiovascular disease, diabetes, and certain types of cancer. It is a major cause of disability and death in industrialized countries.[48] The type of nutrients consumed also have significant correlations to numerous kinds of cancer (TABLE 1).[49–51]

The worsening global epidemic of obesity has increased the urgency of research aimed at understanding the mechanisms of appetite regulation. An important aspect of the complex pathways involved in modulating energy intake is the interaction between hormonal signals of energy status released from the gut in response to a meal, and appetite centers in the brain and brainstem. The challenges of the obesity epidemic are not limited to concerns about bulk and weight. The disabilities caused by obesity are physiologic and psychosocial. The increased waist-to-hip girth is associated with increased risk of cardiovascular disease, hyperlipidemia, hypertension, and diabetes.[52] Obesity also has been related directly to increased risk of sleep apnea, cancer, gallbladder disease, musculoskeletal disorders, severe pancreatitis, bacterial panniculitis, diverticulitis, infertility, urinary incontinence, and idiopathic intracranial hypertension. The psychosocial factors and quality of life in the obese population also have been documented. Although there is some debate, the obese have been found to be twice as likely to suffer from anxiety, impaired social interaction, and depression when compared with the nonobese population.[52]

Several studies indicate that adherence to a Mediterranean diet is associated with a reduction in total and cardiovascular mortality.[53–56] High intake of olive oil is considered a hallmark of the traditional Mediterranean diet, resulting in high intake of monounsaturated fatty acids and lower intake of saturated fatty acids. The replacement of saturated with monounsaturated lipids is strongly associated with a significant reduction in coronary heart disease risk, through a mechanism involving reduction of LDL cholesterol, without a reduction of HDL cholesterol or an increase in triacylglycerols.[57]

TABLE 1. Specific cancer relationships based on epidemiological studies[a]

Cancer site	Incidence association
Pancreatic	smoking
	possibly meat
	cholesterol
Esophageal	alcohol
	tobacco
	combined use
Stomach	salt-preserved foods
	possibly BBQ
	grilling
Colorectal	saturated fat
	possibly eggs
	grilling
	sugar
Liver	hepatitis B
	aflatoxins
	alcohol
Lung	smoking
	possibly alcohol
	saturated fats
	cholesterol
Breast	obesity
	early-puberty alcohol consumption
	possibly meat
	saturated fats
Endometrial	obesity
	estrogen therapy
	saturated fats
Cervical	folate deficiency
	smoking
Bladder	smoking
	possibly artificial sweeteners
	coffee
	alcohol
Prostate	high saturated fat intake

[a]Adapted from Cummings[49] and Bingham and Willett.[50,51]

THE PSYCHE AND ITS LINK TO HEALTH/DISEASE
AND PREMATURE DEATH/LONGEVITY

Characteristics of an unhealthy lifestyle, such as too little physical activity, smoking, irregular and unbalanced diet, and excessive consumption of alcohol and coffee are assumed to have relationships with major health problems such as cardiovascular disease, cancer, and osteoporosis[58] and thus significantly reduce the human life span. Moreover, hostility, anger, and aggressive behavior play a central role in the centuries-old hypothesis that emotions and aspects of personality influence physical health.[59]

The idea of personality traits may be as old as human communication and language itself. In Aristotle's (384–322 B.C.) *Ethics*, character traits such as vanity, modesty, and cowardice are described as key determinants of human moral and immoral behavior. He also described individual variations in these traits, often referring to excessive, defective, and intermediate levels of each. The role of personalities for the development of chronic diseases has received much attention and the role of behavioral medicine in the area of cardiovascular disease has successfully introduced the type A or coronary-prone behavior pattern evident from large epidemiologic studies.[59] This pattern of chronic disease development was shown to be marked by both psychologic and physiologic hyper responsiveness. Type A individuals appear to be hostile, easily angered, competitive, and hard-driving. More recently, oncologists who study human behavior have similarly attempted at hypothesizing a type C or biopsychosocial cancer risk pattern: it was noted these individuals have a profile consistent with denial and suppression of emotions, in particular anger. Further, additional features of this pattern are a sense of too much niceness that borders on a pathological problem, avoidance of conflicts, exaggerated social desirability, harmonizing behavior, over-compliance, over-patience, as well as high rationality and a rigid control of emotional expression. It is further conceptualized that this pattern could be usually concealed behind a facade of pleasantness that appears to be effective as long as environmental and psychological homeostasis is maintained. However the facade can collapse in the course of time under the impact of accumulated strains and stressors, especially those evoking feelings of depression and reactions of helplessness and hopelessness as encountered during everyday life. As a prominent feature of this particular coping style, excessive denial, avoidance, suppression, and repression of emotions and basic needs appears to weaken the human organism's natural resistance to carcinogenic influences.[59,60] Yet conflicting data fail to conclusively elucidate the causal role of the Type C personality on cancer development.[61]

As a general view of mechanisms that link an individual's personality to health, the constitutional vulnerability model suggests that basic, possibly genetically determined, biological, individual differences are responsible both for the personality phenotype and the increased risk of disease.

Recent clinical reviews present cumulative clinical evidence that lends strong support to the notion that medicine should indeed adopt a biopsychosocial rather than an exclusively biologic–genetic model of health.[62,63] This is the domain that includes mind–body medicine, which is at the center of the provision of holistic health. However, this may be only a modern version of an already-practiced art that seemed lost in history.

Throughout history, spa and sanatorium mountain resorts have played an important role in the history of medicine; in almost every culture there have been archaeological discoveries of the existence of hot springs and spa treatment.[64] In Homeric times (1000–700 B.C.), bathing was used primarily for cleansing and hygienic purposes. By the time of Hippocrates (460–370 B.C.), bathing was considered more than a simple hygienic mesasure; it was a healthy practice considered beneficial for most diseases. The hypothesis then elaborated by Hippocrates proposed that the cause of all diseases lay in an imbalance of bodily fluids. Hence, to regain the balance, a change of habits and environment was advised, which included bathing, perspiration, exercise/walking, and bodily massage. The baths were often combined with sports and education, the precursors of modern gymnasiums and meditation centers.

The scientific evidence emerging within the past several decades strongly suggests that psychosocial factors from emotional states that include depression and behavioral dispositions that range from hositlity to psychosocial stress can directly influence both physiologic function and health outcomes. It then follows that inflammatory responses are modulated by a bidirectional communication flow between the neuroendocrine and immune systems and the brain. Also, many lines of research have established multiple pathways by which the immune system and the central nervous system communicate. Hormonal and neuronal mechanisms by which the brain regulates the function of the immune system and, conversely, cytokines, which allow the immune system to regulate the brain provide the basis for mind–body medicine modalities such as relaxation and meditation that impart a positive influence on homeostatic balance. In a healthy individual this bidirectional regulatory syustem forms a negative feedback loop that keeps the immune system and contral nervous system in homeostatic balance. Changes to these regulatory systems have been postulated to potentially lead to either overactive immune response, inducing inflammatory disease, or oversuppression of the immune system and increased susceptibility to infectious disease.[62–65] Relaxation and meditation in daily recreatrional activities may consititute the requisite trigger for a regularizatrion of homeostatic balance.

Practices that link the mind through brain-guided activites to physiological functioning extend our understanding of the biopsychological model and the role emotional well-being has in the brain, immune system, and endocrine organs. Hence, a link is established between psychological and lifestyle-induced stressors and a broad array of adverse health outcomes that can be traced to hypothalamic–pituitary–immune system–endocrine mediators. It may well be feasible to view attendance at spa and sanatorium mountain resorts as reversal programs of deleterious health outcomes in which the ultimate aim is to revitalize the body. The physiological underpinnings of this relationship comprises an exciting area of medicine that aims at restoring the disrupted communication system, whether through pharmacotherapy alone or, more importantly, in conjunction with the relaxing effects of a spa and meditation in an integrative approach, the framework of mind–body medicine.

As evidence continues to accumulate within the field of psychoneuroendocrinology and psychoneuroimmunology, the brain has truly an overarching role in health and disease[1–3] and may hence be viewed as the *conductor of the body orchestra*. Clinical research in health and in disease prevention as well as diseases associated with chronic inflammations that significantly affect an individual's mood or level of anxiety and depression will further crystallize the role of the brain and the mind in human health. When the body is working well, nutrition and mental health are also good. If one then feels good about oneself, quality of life is enhanced, and a long life is made possible.

Finally, the plethora of scientific evidence that is currently available overrides the boundaries that have restricted the elucidation of the interrelationships between the mind, brain, and body and that will be looked back upon as artifacts.

[*Competing interests*: The authors declare that they have no competing financial interests.]

REFERENCES

1. Song, C. & B.E. Leonard. 2001. Fundamentals of Psychoneuroimmunology. Wiley. New York.
2. Kiecolt-Glaser, J.K., L. McGuire, T.F. Robles & R. Glaser. 2002. Psychoneuroimmunology and psychosomatic medicine: back to the future. Psychosom. Med. **64:** 15–28.
3. Young, E. *et al.* 1998. Psychoneuroendocrinology of depression: hypothalamic–pituitary–gonadal axis. *In* Psychoneuroendocrinology. C.B. Nemeroff, Ed.: 309–324. The Psychiatric Clinic of North America. Saunders. Philadelphia.
4. Plotsky, P.M. *et al.* 1998. Psychoneuroendocrinology of depression: hypothalamic–pituitary–gonadal axis. *In* Psychoneuroendocrinology. C.B. Nemeroff, Ed.: 293–308. The Psychiatric Clinic of North America. Saunders. Philadelphia.
5. Ernst, E. *et al.* 1998. Complementary therapies for depression: an overview. Arch. Gen. Psychiatry **55:** 1026–1032.
6. Mufson, L. *et al.* 1999. Efficacy of interpersonal psychotherapy for depressed adolescents. Arch. Gen. Psychiatry **56:** 573–579.
7. Faba, G.A. *et al.* 1998. Prevention of recurrent depression with cognitive behavioral therapy. Arch. Gen. Psychiatry **55:** 816–820.
8. Rozanski, A., J.A. Blumenthal & J. Kaplan. 1999. Impact of psychological factors on the pathogenesis of cardiovascular disease and implications for therapy. Circulation **99:** 2192–2217.
9. Kiecolt-Glaser, J.K. 1999. Norman Cousins Memorial Lecture. Stress, personal relationships, and immune function: health implications. Brain Behav. Immun. **13:** 61–72.
10. Salovey, P., A.J. Rothman, J.B. Detweiler & W.T. Steward. 2000. Emotional states and physical health. Am. Psychol. **55:** 110–121.
11. Baum, A. & D.M. Posluszny. 1999. Health psychology: mapping biobehavioral contributions to health and illness. Annu. Rev. Psychol. **50:** 137–163.
12. Fricchione, G. & G.B. Stefano. 2005. Placebo neural systems: nitric oxide, morphine, and the dopamine brain reward and motivation circuitries. Med. Sci. Monit. **11:** 54–65.
13. Seyle, H. 1973. The evolution of the stress concept. Am. Sci. **61:** 692–699.
14. Esch, T. 2002. Health in stress: change in the stress concept and its significance for prevention health and lifestyle. Gesundheitswesen **64:** 73–81.
15. Tamashiro, KL, M.M. Nguyen & R.R. Sakai. 2005. Social stress: from rodents to primates. Front. Neuroendocrinol. **26:** 27–40.
16. Leonard, B. 2000. Stress, depression, and the activation of the immune system. World J. Biol. Psychiatry **1:** 17–25.
17. Harbuz, M.S. & S.L. Lightman. 1992. Stress and the hypothalamo–pituitary–adrenal axis: acute, chronic, and immunological activation. J. Endocrinol. **134:** 327–339.
18. Chrousos, G.P. 1995. The hypothalamic–pituitary–adrenal axis and immune-mediated inflammation. N. Engl. J. Med. **332:** 1351–1362.
19. Bailey, M., H. Engler, J. Hunzeker & J.F. Sheridan. 2003. The hypothalamic–pituitary–adrenal axis and viral infection. Viral Immunol. **16:** 141–157.
20. Broadhead, W.E. *et al.* 1983. The epidemiologic evidence for a relationship between social support and health. Am. J. Epidemiol. **117:** 521–537.
21. Chrousos, G.P. & P.W. Gold. 1992. The concepts of stress and stress system disorders. Overview of physical and behavioral homeostasis. J. Am. Med. Assoc. **267:** 1244–1252.
22. Rosengren, A. *et al.* 1993. Stressful life events, social support, and mortality in men born in 1933. Br. Med. J. **307:** 1102–1105.
23. Maunder, R.G. & J.J. Hunter. 2001. Attachment and psychosomatic medicine: developmental contributions to stress and disease. Psychosom. Med. **63:** 556–567.
24. Takkouche, B., C. Regueira & J.J. Gestal-Otero. 2001. A cohort study of stress and the common cold. Epidemiology **12:** 345–349.
25. Aguglia, E. *et al.* 2004. Stress in the caregivers of Alzheimer's patients: an experimental investigation in Italy. Am. J. Alzheimers Dis. Other Demen. **19:** 248–252.

26. KIECOLT-GLASER, J.K. *et al.* 1993. Negative behavior during marital conflict is associated with immunological downregulation. Psychosom. Med. **55:** 395–409.
27. CANTOR, C.H. *et al.* 1999. Australian suicide trends 1964–1997: youth and beyond? Med. J. Aust. **171:** 137–141.
28. CALDWELL, T.M., A.F. JORM & K.B. DEAR. 2004. Suicide and mental health in rural, remote, and metropolitan areas in Australia. Med. J. Aust. **181:** 10–14.
29. LESTER, D., P. WOOD, C. WILLIAMS & J. HAINES. 2004. Motives for suicide: a study of Australian suicide notes. Crisis **25:** 33–34.
30. DE GROOT, M. *et al.* 1999. Glycaemic control and major depression in diabetes with type 1 and type 2 diabetes mellitus. J. Psychosom. Res. **46:** 425–435.
31. TSUTSUMI, A. 2005. Psychosocial factors and health: community and workplace study. J. Epidemiol. **15:** 65–69.
32. ZORN, B.H. 1999. Urinary incontinence and depression. J. Urol. **162:** 82–84.
33. STEFFESN, B.C. *et al.* 1999. The effect of major depression on functional status in patients with coronary artery disease. J. Am. Ger. Soc. **47:** 319–322.
34. SIVONOVA, M. *et al.* 2004. Oxidative stress in university students during examinations. Stress **7:** 183–188.
35. CHIU, A., S.Y. CHON & A.B. KIMBALL. 2003. The response of skin disease to stress: changes in the severity of acne vulgaris as affected by examination stress. Arch. Dermatol. **139:** 897–900.
36. MÖLLER, J. *et al.* 2005. Work related stressful life events and the risk of myocardial infarction. Case-control and case-crossover analyses within the Stockholm heart epidemiology programme (SHEEP). J. Epidemiol. Comm. Health **59:** 23–30.
37. WEATHERALL, M. 1999. Biofeedback or pelvic floor muscle exercises for female genuine stress incontinence: a meta analysis of trials identified in a systematic review. Br. J. Urol. Int. **83:** 1015–1016.
38. STEIN, P.C. *et al.* 1999. Elevated urinary norepinephrine in interstitial cystitis. Urol. J. **53:** 1140–1143
39. DAVIS, L.L. *et al.* 2004. Biopsychological markers of distress in informal caregivers. Biol. Res. Nurs. **6:** 90–99.
40. DE KLOET, E.R. 2004. Hormones and the stressed brain. Ann. N.Y. Acad. Sci. **1018:** 1–15.
41. KISSANE, D.W. *et al.* 1998. Psychological morbidity and quality of life in Australian women with early-stage breast cancer: a cross-sectional survey. Med. J. Aust. **169:** 192–196.
42. REAVLEY, N., L. VITETTA, F. CORTIZO & A. SALI. 2004. A preliminary report of the effect of psychosocial support on the psychological and physical wellbeing of a heterogeneous group of cancer patients. Psycho-Oncology **13:** 51.
43. YATES, B.C. 1995. The relationships among social support and short- and long-term recovery outcomes in men with coronary heart disease. Res. Nurs. Health **18:** 193–203.
44. FROST, M.H. *et al.* 2000. Long-term satisfaction and psychological and social function following bilateral prophylactic mastectomy. J. Am. Med. Assoc. **284:** 319–324.
45. HALL, S.M. *et al.* 1998. Nortriptyline and cognitive behavioural therapy in the treatment of cigarette smoking. Arch. Gen. Psychiatry **55:** 683–690.
46. MARCUS, B.H. *et al.* 1999. The efficacy of exercise as an aid for smoking cessation in women. Arch. Int. Med. **159:** 1229–1234.
47. MYINT, P.K. *et al.* 2005. Relation between self-reported physical functional health and chronic disease mortality in men and women in the european prospective investigation into cancer (EPIC-Norfolk): a prospective population study. Ann. Epidemiol. Epub ahead of print Jul. 6, 2005.
48. MATTSON, M.P. 2005. Energy intake, meal frequency, and health: a neurobiological perspective. Annu. Rev. Nutr. **25:** 237–260.
49. CUMMINGS, J.A. & S.A. BINGHAM. 1998. Diet and the prevention of cancer. Br. Med. J. **317:** 1636–1640. Review.
50. WILLETT, W.C. 1999. Goals for nutrition in the year 2000. CA Cancer J. Clin. **49:** 331–352.
51. WILLETT, W.C. 2000. Diet and cancer. Oncologist **5:** 393–404.

52. PENDER, J.R. & W.J. PORIES. 2005. Epidemiology of obesity in the United States. Gastroenterol. Clin. North Am. **34:** 1–7.
53. TRICHOPOULOU, A. *et al.* 2005. Modified Mediterranean diet and survival: EPIC-elderly prospective cohort study. Br. Med. J. **330:** 991. Epub Apr 8, 2005.
54. PSALTOPOULOU, T. *et al.* 2004. Olive oil, the Mediterranean diet, and arterial blood pressure: the Greek European prospective investigation into cancer and nutrition (EPIC) study. Am. J. Clin. Nutr. **80:** 1012–1018.
55. TRICHOPOULOU, A., T. COSTACOU, C. BAMIA & D. TRICHOPOULOS. 2003. Adherence to a Mediterranean diet and survival in a Greek population. N. Engl. J. Med. **348:** 2599–2608.
56. LASHERRAS, C., S. FERNANDEZ & A.M. PATTERSON. 2000. Mediterranean diet and age with respect to overall survival in institutionalized, nonsmoking elderly people. Am. J. Clin. Nutr. **71:** 987–992.
57. ASCHERIO, A. 2002. Epidemiologic studies on dietary fats and coronary heart disease. Am. J. Med. **113:** 9–12.
58. HAVEMAN-NIES, A. *et al.* 2002. Dietary quality and lifestyle factors in relation to 10-year mortality in older Europeans: the SENECA study. Am. J. Epidemiol. **156:** 962–968.
59. SMITH, T.W. & L.C. GALLO. 2001. Personality traits as risk factors for physical illness. *In* Handbook of Health Psychology. A. Baum, T. Revenson & J. Singer, Eds.: 139–172. Lawrence Erlbaum. Hillsdale, NJ.
60. SMITH, T.W., K. GLAZER, J.M. RUIZ & L.C. GALLO. 2004. Hostility, anger, aggressiveness, and coronary heart disease: an interpersonal perspective on personality, emotion, and health. J. Pers. **72:** 1217–1270.
61. NAKAYA, N. *et al.* 2003. Personality and the risk of cancer. J. Natl. Cancer Inst. **95:** 799–795.
62. SMITH, A. & K. NICHOLSON. 2001. Psychosocial factors, respiratory viruses, and exacerbation of asthma. Psychoneuroendocrinology **26:** 411–420.
63. ASTIN, J.A. *et al.* 2003. Mind-body medicine: state of the science, implications for practice. J. Am. Board Fam. Pract. **16:** 131–147.
64. VAN TUBERGEN, A. & S. VAN DER LINDEN. 2002. A brief history of spa therapy. Ann. Rheum. Dis. **61:** 273–275.
65. ESKANDARI, F., J.I. WEBSTER & E.M. STERNBERG. 2003. Neural immune pathways and their connection to inflammatory diseases. Arthritis Res. Ther. **5:** 251–265.

Bio-Identical Steroid Hormone Replacement

Selected Observations from 23 Years of Clinical and Laboratory Practice

JONATHAN V. WRIGHT

Tahoma Clinic, Renton, Washington 98055, USA

ABSTRACT: To maximize the safety and efficacy of human hormone replacement therapy, it is suggested that exact molecular copies of human hormones ("bio-identical" hormones) be administered in physiologic quantities and proportions, following physiologic timing and routes of administration. It is also suggested that physicians return to the practice of monitoring hormone therapy by precise laboratory measurement levels of the hormones administered. This paper also presents clinical and laboratory data concerning appropriate proportions of bio-identical estrogens, the physiologic and supraphysiologic nature of commonly employed doses, estrogen levels achieved by varying routes of administration, and the significant effects of iodine on estrogen metabolism and cobalt on estrogen excretion.

KEYWORDS: bio-identical hormones; estradiol; estrone; estriol; estrogens; menopause; hormone replacement therapy

> *The physician is only the servant of Nature, not her master. Therefore it behooves medicine to follow the will of nature.*
>
> —PARACELSUS

In contemporary medical therapeutics, "copy nature" is a dictum less often honored than ignored. The relatively recent outcomes of the Women's Health Initiative[1] in the United States and the Million Women Study[2] in the UK have demonstrated once again that replacing human molecular substance with inexact molecular copies has undesirable consequences.

If human hormones are to be replaced during peri-menopause and postmenopausally, it is entirely logical to copy nature, replacing human hormones with exact molecular copies. It is also entirely logical to copy nature in every other parameter possible, including quantities of bio-identical hormones given, routes of administration, and timing of administration of bio-identical hormones.

Use of inexact molecular copies of human hormones has also led to inexcusable sloppiness of medical practice: For decades, women were given dosages of powerful steroid-hormone-mimetic substances with no follow-up testing at all to determine appropriate dosage. The use of nonhuman pseudohormones automatically precluded

Address for correspondence: Jonathan V. Wright, M.D., Medical Director, Tahoma Clinic, 801 S.W. 16th Street, Renton WA 98055. Voice: 425-264-0059; fax: 425-264-0071.
admin@tahoma-clinic.com

Ann. N.Y. Acad. Sci. 1057: 506–524 (2005). © 2005 New York Academy of Sciences.
doi: 10.1196/annals.1356.039

the normal practice of post-treatment testing, since (for example) there are no "normal" levels of equilin (the major molecular species in Premarin) or medroxyprogesterone (Provera) to be found in women of any age for use as guidelines to correct dosage of these substances. By contrast, dosing with thyroxine and/or tri-iodothyronine is almost always followed at intervals with serum testing of thyroid hormones to insure correct dosage and patient safety.

This paper will briefly review practice and research concerning:

    (1) types of estrogens used in bio-identical hormone replacement;

    (2) quantities (dose) of estrogens used; and

    (3) route of administration of estrogens.

This paper will also discuss two "problems" in bio-identical hormone replacement therapy and their correction with elements from nature:

    (4) incomplete metabolization of estrone and estradiol towards estriol; and

    (5) hyperexcretion of estrogens.

Although the focus of this paper is on practice and clinical research concerning bio-identical estrogens, in daily practice bio-identical estrogen replacement is routinely combined with bio-identical progesterone, testosterone, dehydroepiandrosterone, thyroid, and melatonin, mimicking physiologic quantities, timing, and route of administration as closely as possible.

SERUM ESTRIOL, ESTRADIOL, AND ESTRONE IN NONPREGNANT, PREMENOPAUSAL WOMEN

Exact molecular copies of human steroid hormones are increasingly referred to as "bio-identical" hormones, to distinguish them from inexact (and often patentable) molecular copies. While a small minority of physicians in the United States and a somewhat larger number in Europe and Japan have used one or more bio-identical steroids (usually estradiol, estriol, progesterone, testosterone) for the treatment of the symptoms of menopause and/or andropause for one or two decades or more, completely comprehensive bio-identical hormone therapy (including all female and male steroid metabolites, major and minor) has not been in use since the medieval period in China, where the earliest record of such therapeutic preparations (*chhui shih*) and their uses dates to 1025 AD.

Contemporary bio-identical steroid replacement uses only one or just a few of the many steroid hormones and their metabolites. For women, the bio-identical estrogen most often used in theUnited States and Europe is estradiol; in Japan, it is most often estriol. Bio-identical progesterone and testosterone are also often used for women.

In an effort to be a bit more comprehensive, and to better copy nature, a small but growing minority of American physicians and a handful throughout the world use a combination of bio-identical estrogens (along with progesterone, testosterone, and dehydroepiandroserone). Bio-identical estrogen combinations most often used are either "triple estrogen" (estrone 10%, estradiol 10%, estriol, 80%) or "bi-estrogen" (estradiol 10–20% and estriol 80–90%). The use of these combinations in practice is supported by the following research data.

In an effort to confirm that triple estrogen as used since the early 1980s is reasonably close to "naturally occurring" estrogens in both proportions and quantities, our

TABLE 1. Serum estrogen levels and estrogen quotients[a]

Subject no.	Day of cycle	Estriol (pg/mL)	Estradiol (pg/mL)	Estrone (pg/mL)	Estrogen quotient
1	10	233	52.5	21.2	3.2
2	10	329	43.4	14.4	3.2
3	10	1254	188.8	55.0	5.1
4	10	320	40.4	23.4	5.2
5	10	369	43.9	20.4	5.7
6	10	700	30.9	68.0	7.1
7	10	374	35.9	15.1	7.3
8	10	429	25.3	33.1	7.3
9	10	941	86.7	32.5	7.9
10	10	736	21.5	70.6	8.0
11	10	1616	61.2	127.4	8.6
12	10	726	61.7	52.5	6.4
13	10	826	22.6	39.1	13.4
14	10	1614	57.9	144.4	8.0
15	10	771	11.3	35.4	16.5
16	10	1700	30.7	58.6	19.0
17	11	59.5	15.1	35.9	11.7
18	11	2408	156.8	121.9	8.6
19	12	961	32.8	71.6	9.2
20	12	1161	71.2	27.3	11.8
21	12	1271	22.2	72.0	13.5
22	14	685	24.6	43.4	10.1
23	14	988	31.1	05.8	10.2
24	14	212	23.7	12.5	5.9
25	14	1412	183.4	54.7	5.9
26	14	1146	17.4	68.6	13.3
			Average:		8.9
			Standard deviation		3.9

[a]From Robinson et al.[3] (Estriol)[(Estradiol) + (Estrone)].

laboratory (Meridian Valley Laboratory, Renton, WA) tested 26 women for estrone, estradiol, and estriol levels during the 10th to 14th days of their menstrual cycles. Participant criteria included good health, ages between 18 and 40, and not taking birth control, steroid, or steroid-mimetic medications. Results of these tests are shown in TABLE 1.

We also tested five other women meeting the same criteria stated above at varying points throughout their menstrual cycles. At all times tested, circulating estriol significantly exceeded the sum of circulating estrone and estradiol (FIGS. 1–5).

From these data, it is apparent that even in nonpregnant women, estriol is normally present in healthy, nonpregnant women in considerably larger quantities than either estrone or estradiol. Even though estriol has long been considered a "weak estrogen," important only during pregnancy (much as DHEA was considered an "un-

Day of Cycle	Estriol pg/mL	Estradiol pg/mL	Estrone pg/mL	Eq E3/(E2+E1)
7	700	68	31	7.1
10	1616	127	61	8.6
14	685	43	25	10.1
17	944	52	31	11.4
24	897	40	22	14.5

EQ Average: 10.3

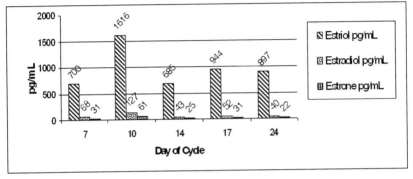

FIGURE 1. Serum estrogen concentrations on various days of the cycle: woman 1.

Day of Cycle	Estriol pg/mL	Estradiol pg/mL	Estrone pg/mL	Eq E3/(E2+E1)
7	689	36	17	13
12	961	72	33	9.2
21	1018	51	18	14.8

EQ Average: 12.3

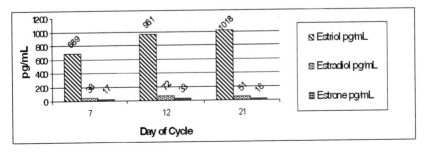

FIGURE 2. Serum estrogen concentrations on various days of the cycle: woman 2.

Day of Cycle	Estriol pg/mL	Estradiol pg/mL	Estrone pg/mL	Eq E3(E2+E1)
4	343	24	17	8.4
10	429	33	25	7.4
19	512	12	17	17.7

EQ Average: 11.1

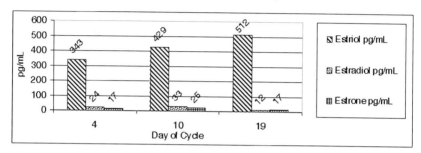

FIGURE 3. Serum estrogen concentrations on various days of the cycle: woman 3.

Day of Cycle	Estriol pg/mL	Estradiol pg/mL	Estrone pg/mL	Eq E3(E2+E1)
2	303	7	1	37.9
15	988	66	31	10.2
19	653	85	23	6

EQ Average: 18.0

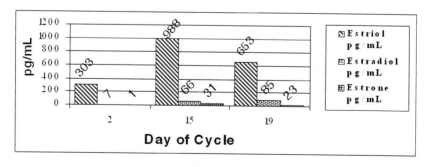

FIGURE 4. Serum estrogen concentrations on various days of the cycle: woman 4.

Day of Cycle	Estriol pg/mL	Estradiol pg/mL	Estrone pg/mL	Eq E3(E2+E1)
8	1104	68	11	14
10	736	71	22	7.9
26	786	17	10	29.1

EQ Average: 17.0

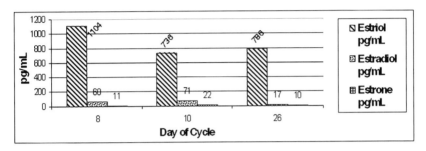

FIGURE 5. Serum estrogen concentrations on various days of the cycle: woman 5.

important steroid" for several decades), we conclude that nature must have a purpose for the presence of these relatively large proportions, even in healthy nonpregnant women.

Unless future controlled studies demonstrate unacceptable risk, we believe that estriol, estradiol, and estrone used in physiologic quantities and proportions, along with progesterone and testosterone, copying both nature's timing and routes of administration as closely as possible are the safest form of (ovarian) steroid "hormone replacement therapy."

ESTRADIOL: IMPORTANCE OF APPROPRIATE DOSE AND METABOLISM TO ESTRONE

Careful attention is also given to dose, with a deliberate effort made to copy nature by not exceeding physiologic quantities of any of the bio-identical hormones administered. Follow-up testing is always done to ascertain whether this goal has been achieved. Many practitioners deliberately target hormone levels slightly above the "normal" menopausal range, overlapping into the "low physiologic range" found in young, healthy, cycling women.

In the United States, when a bio-identical estrogen is recommended instead of an inexact molecular copy, the large majority of academic and affiliated practitioners have given estradiol alone. (Practitioners recommending "triple estrogen" or "bi-estrogen" are usually community-based.) Doses given by academic and affiliated practitioners have routinely been in the range of 1–2 milligrams per day, with no follow-up testing done to examine whether these doses compare to physiologic quantities. Unfortunately, a very widely read book by a Hollywood actress undergoing treatment for breast cancer also takes this approach. The actress writes of taking 2

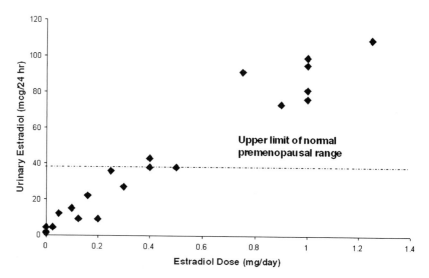

FIGURE 6. Urinary estradiol excretion after oral estradiol. (Reproduced from Friel *et al.*[4] with permission.)

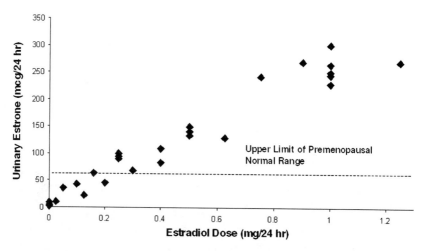

FIGURE 7. Urinary estrone output in postmenopausal women taking oral estradiol. (Reproduced from Friel *et al.*[4] with permission.)

milligrams of estradiol daily, and reports no follow-up testing. The following data illustrates the hazards of this approach.

Twenty-four-hour urinary steroid hormone profiles, including the measurement of estrone, estradiol, and estriol, were conducted for 35 postmenopausal women receiving oral estradiol at doses from 0.025–2.0 mg/day. Urinary excretion of estradiol exceeded normal premenopausal values in women taking estradiol at doses greater than 0.5 mg/day (FIG. 6). Urinary estrone excretion exceeded normal premenopausal values in women taking estradiol doses of 0.25 mg/day or higher (FIG. 7).

These results show that frequently recommended oral doses of estradiol (1–2 mg/day) result in urinary excretion of estrone at values 5–10 times the upper limit of normal in premenopausal women. Based on these measurements, a prudent dose ceiling for oral estradiol replacement therapy appears to be 0.25 mg/day. Retrospective studies associating oral estradiol with increased risk of breast cancer may reflect overdose conditions.

Under most circumstances serum and urinary estrogens correlate well. However, our laboratory measures serum levels of estrogens less frequently than urinary levels. TABLE 2 shows results for serum estrone measured in women at various phases of the menstrual cycle and pregnancy by another group, compared with serum estrone levels in women taking estradiol 1.0 milligrams daily orally. As FIGURE 7 demonstrates with urinary measurement, TABLE 2 shows that estradiol at 1.0 milligrams orally results in supraphysiologic serum levels, which are very likely unnecessarily hazardous.

Since the early 1980s, average daily "triple estrogen" doses have uniformly contained 0.25 milligrams estradiol, 0.25 milligrams estrone, and 2.0 milligrams estriol. Bi-estrogen doses have usually contained 0.25 milligrams of estradiol, and either 2.0 or 2.25 milligrams of estriol. Follow-up measurement of hormone levels in women given these doses routinely finds both blood and urine doses within the physiologic range noted above: slightly above menopausal, overlapping into the "low range" of these hormones as found in young, healthy, cycling women. (Exceptions are found among women who appear to be "estrogen hyperexcreters" as described in Cases 1, 2, and 3 below.)

TABLE 2. Estrone concentrations in pre- and postmenopausal women, in pregnancy, and with estradiol[a]

Patient status	Serum estrone (pg/mL)
Postmenopausal	14–103
Follicular phase	37–138
Luteal phase	50–114
Peri-ovulatory	60–229
Estradiol 1.0 mg/day	159–997
First trimester of pregnancy	62–715
Second trimester of pregnancy	167–1862
Third trimester of pregnancy	1039–3210

[a]From Stehman-Breen et al.[5]

INFLUENCE OF THE ROUTE OF ADMINISTRATION
ON ESTROGEN BIOAVAILABILITY

Nature introduces estrogens and other ovarian steroids into the circulation via the pelvic plexus of veins, whence they are carried to the heart and circulated through the lungs, then back to the heart again, and thence to be carried to all parts of the body, including the liver, which is thought to be the major organ of steroid hormone metabolism and detoxification. It is fairly obvious that nature does not introduce ovarian steroids into the intestinal tract, where they might have been absorbed into the plexus of veins leading to the portal vein, and thence taken directly to the liver.

To more closely copy nature, the large majority of physicians prescribing bio-identical estrogens prescribe "transdermal" preparations, usually rubbed in to areas of "thin skin" (inner thighs, wrists) on a monthly cyclic basis. While this practice is closer to the "natural route" described above, it is not as close as possible. Intravaginal administration of the transdermal preparation is the closest possible approximation of the "natural route" of ovarian hormones, introducing them first into the pelvic venous plexus. Experience has shown that this is not just a theoretical consideration.

At Tahoma Clinic we ask women using bio-identical replacement therapy to monitor their urinary estrogen levels (representing "free" and "conjugated" but not "sex-hormone globulin bound" estrogens) at regular intervals to make certain that doses used are within the physiological target range. Over several years time, many women using transdermal estrogen preparations have progressively lower urinary estrogens. Sometimes these lower levels are reflected in symptoms, but sometimes (particularly in women who have used bio-identical hormone replacement for longer periods of time) there are no symptoms of these lower estrogen levels. When the route of administration is then switched to intravaginal, with no change in dosage, the urinary levels rise once again to the target range, and any symptoms of low estrogens disappear.

We have termed this phenomenon "dermal absorption fatigue." Because it occurs often, we have switched our routine recommendation for site of bio-identical hormone administration to the intra-vaginal route.

An example of "dermal absorption fatigue" is shown in FIGURE 8.

On the basis of this and other clinical observations, it appears that intravaginally or transdermally administered doses of estrogens might be absorbed, metabolized, or excreted somewhat differently than orally administered estrogens, even if doses are the same. The following data appear to demonstrate some of these differences.

Eighteen postmenopausal women were given the same 2.5-milligram quantity of bio-identical triple estrogen as part of hormone replacement therapy. Seven took it orally or sublingually, four rubbed it into the skin (routine transdermal administration), and seven used the intravaginal route of administration. Women using triple estrogen by the vaginal or labial route applied the cream on the night before the 24-hour urine collection, to avoid potential contamination of the urine specimen.

For the three different routes of administration the median values for urinary estrone and estradiol concentrations are shown in FIGURE 9, while FIGURE 10 shows the median values for urinary estriol concentrations.

One might wonder whether route of administration of estrogens would produce similar results in both serum and urine. As a very preliminary attempt to answer this question, two postmenopausal women receiving estriol (one woman taking 1 milli-

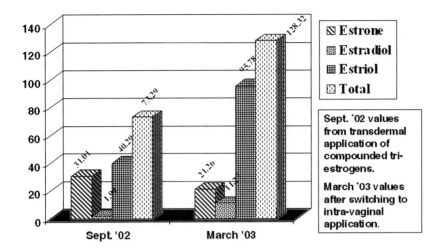

FIGURE 8. 24-hour urine comprehensive hormone levels.

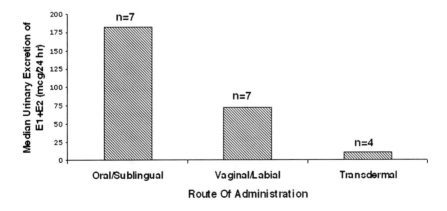

FIGURE 9. Estrone plus estradiol excretion in patients receiving 2.5-mg triple estrogen therapy daily.

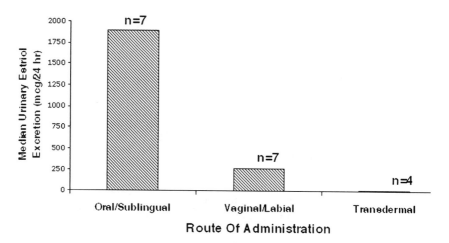

FIGURE 10. Estriol excretion in patients receiving 2.5-mg triple estrogen therapy daily.

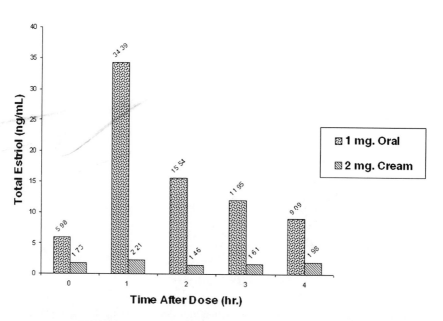

FIGURE 11. Serum estriol: 1-mg oral dose vs. 2 mg creme.

gram orally, the other 2 milligrams transdermally) were tested for total serum estriol immediately before administration of a dose, and at 1, 2, 3, and 4 hours post dose. The results are shown in FIGURE 11, which demonstrates that serum total estriol concentrations are in fact much higher in the woman who took 1 mg orally than in the woman who took 2 mg transdermally, in agreement with the urinary excretion data presented here.

Admittedly, the data presented in this section are very small. However, clinical experience over 23 years is in accordance with this laboratory data. It appears reasonable to hypothesize preliminarily (pending large-scale studies) that the route of bio-identical hormone administration closest to nature's route is preferable over the long term.

BIO-IDENTICAL ESTROGEN TREATMENT FAILURE AND ESTROGEN HYPEREXCRETION: CORRECTION WITH PHYSIOLOGIC-DOSE COBALT

Practitioners who have prescribed bio-identical hormones for menopausal or postmenopausal women for any period of time have experienced occasional treatment failures. In this small percentage of women, despite carefully escalating doses of bio-identical estrogens (and other accompanying steroid hormones), hot flashesand other symptoms of menopause are not relieved, or are relieved only minimally.

In most cases these treatment failures occur in menopausal or postmenopausal women who have previously taken Premarin or other non-bio-identical "hormone" replacement. In frustration, most of these women may return to taking Premarin or other more risky pseudohormones, since "at least they take care of my symptoms."

Over the years, it has become obvious that almost all treatment failures with bio-identical hormone replacement have a common biochemical "signature." When given an "average effective treatment dose" of bio-identical estrogens, women who ex-

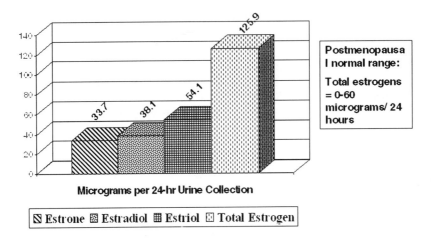

FIGURE 12. Typical normal response with complete symptom relief using triple estrogen: 2.5 mg on days 1–25 (progesterone 25 mg).

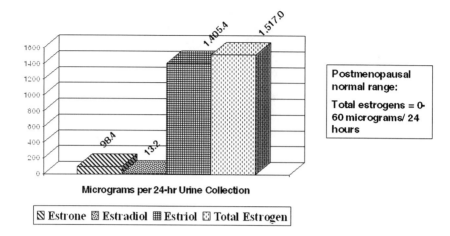

FIGURE 13. Example of hyperexcretion prior to cobalt supplementation while using triple estrogen: 2.5 mg on days 1–25 (progesterone 25 mg).

perience treatment failure hyperexcrete estrogens, as compared with women who respond to treatment. FIGURES 12 and 13 show typical estrogen excretion patterns in two women (a "responder" and a "treatment failure") using the exact same dosage and route of administration of bio-identical estrogens.

Along with many other substances, estrogens are metabolized and "detoxified" by cytochrome P450 (and other) cytochrome enzymes. It seems logical to hypothesize that non-responders to bio-identical estrogen replacement may have upregulated cytochrome enzymes that may be gradually downregulated to normal function. A variety of minerals, both essential and toxic elements, are known to affect cytochrome function.

Because of its safety in quantities ordinarily found in dietary surveys, cobalt was chosen as a possibility for "downregulating" possibly overactive cytochrome P450 enzymes. The results in three typical cases follow.

Case 1

The patient, R.W., was a 49-year-old white woman with a prior history of chronic fatigue syndrome.[a] In December 1997, she saw Dr. Lamson concerning menopausal symptoms, including moderately severe hot flashes, insomnia, and anxiety. She collected a 24-hour urine specimen, upon which a comprehensive steroid determination was performed, with results reported in January 1998. Estrogens were reported as: estrone: 12.30 mg/24 h; estradiol: 22.60 mg/24 h; estriol: 1.85 mg/24 h; and total estrogens: 36.75 mg/24 h.

[a]The data in this first case were collected by my clinic at the Tahoma Clinic, Davis Lamson, N.D., and his patient R.W.

The patient was started on triple estrogen [E1:E2:E3/10:10:80], 1.25 milligrams daily, with no ensuing symptom relief. The same estrogens were increased to 2.5 milligrams daily. Symptoms continued unabated.

While still using the 2.5-milligram daily dosage of estrogens, the patient collected another 24-hour urine specimen, upon which the same steroid determinations were performed. This time results were: **estrone:** 55.7 μg/24 h; **estradiol:** 53.6 μg/24 h; **estriol:** 808.8 μg/24 h; and **total estrogens:** 918.0 μg/24 h.

Along with her continuing triple estrogen therapy, R.W. then started cobalt supplementation, 400–500 micrograms daily. Two weeks later, she reported that her hot flashes and other menopausal symptoms ceased for the first time. However, she also reported that if she skipped a daily dose of cobalt,the hot flashes promptly returned, only to disappear with resumption of the cobalt.

Slightly over a month after starting the cobalt, she reported that on another "skipped dose" occasion, she had not only return of hot flashes, but also "spotting," which continued at a low level for two weeks, even though the hot flashes ceased as before with resumption of cobalt.

As she continued cobalt (along with the triple estrogen at 2.5 milligrams daily, taken on a 25/28 day cyclic basis) she had no further hot flashes. She omitted occasional cobalt dosages without recurrence of hot flashes.

After 10 weeks of supplemental cobalt along with the continuing triple estrogen, her 24-hour steroid analysis yielded the following estrogen values: **estrone:** 11.3 μg/24 h; **estradiol:** 50.4 μg/24 h; **estriol:** 41.7 μg/24 h; and **total estrogens:** 103.4 μg/24 h.

As noted above, this estrogen total is within the anticipated range for postmenopausal women supplemented with 2.5 milligrams of triple estrogen daily.

Case 2

The patient, J.S., was a 63-year-old woman[b] who had been on bio-identical hormone replacement therapy for approximately three years. She was still having night sweats and difficulty sleeping, as well as difficulty with memory and concentration. Twenty-four hour urinary estrogens were: **estrone:** 98.4 μg/24 h; **estradiol:** 13.2 μg/24 h; **estriol:** 1405.4 μg/24 h; and **total estrogens:** 1517.0 μg/24 h.

The patient continued to take triple estrogen 2.5 milligrams daily (cyclic basis), and was advised to start cobalt, 500 micrograms daily. Eleven weeks later, her estrogens were: **estrone:** 8.8 μg/24 h; **estradiol:** 95.6 μg/24 h; **estriol:** 32.2 μg/24 h; and **total estrogens:** 136.5 μg/24 h.

Her symptoms of menopause were relieved.

Case 3

The patient, J.K., was a 52-year-old woman with symptoms of hypothyroidism and menopause, the latter including night sweats, vaginal dryness, anxiety, irritability, and mood swings. Her last menstrual period was four months previously.

---

[b]The data for the patients in Cases 2 and 3 were collected by my former Tahoma Clinic colleague, Joni Olehausen, N.D., and her patients J.S. and J.K.

She was already taking bi-estrogen (estriol 1 milligram, estradiol 250 micrograms) twice daily on a cyclic basis with a progestin. Her sex hormones were evaluated with the 24-hour urine testing for sex steroids with the result below: **estrone:** 396.9 µg/24 h; **estradiol:** 101.3 µg/24 h; **estriol:** 1405.4 µg/24 h; and **total estrogens:** 1903.5 µg/24 h.

Menopausal symptoms continued. On a return visit, six months later, she was advised to start on cobalt, 500 micrograms daily, while continuing her bio-identical estrogens. Ten weeks later, her estrogen values were as follows: **estrone:** 212.8 µg/24 h; **estradiol:** 73.3 µg/24 h; **estriol:** 25.9 µg/24 h; and **total estrogens:** 311.9 µg/24 h.

Although the patient's total estrogens had not yet diminished to a more expected range, except for continuing difficulty in sleeping, her menopausal symptoms were improved. Unfortunately, she has not come for follow-up evaluation.

Although many individuals need only 500 micrograms of cobalt daily to help normalize symptoms of menopause and reduce apparent estrogen hyperexcretion, some have needed up to 1120 micrograms daily. In nearly all cases, cobalt supplementation has been discontinued without need for resumption once estrogen excretion has declined to normal.

In both Cases 2 and 3, although total estrogen excretion declined dramatically after the ingestion of cobalt, the relative proportion of estriol as a fraction of total excreted estrogens also declined dramatically. This is addressed in the next section.

APPARENT STIMULATION OF THE ESTRADIOL → ESTRONE → ESTRIOL → PATHWAY BY IODINE AND IODIDE

While estrone and estradiol are both physiologic estrogens, they are generally conceded to be slightly pro-carcinogenic. Estriol has been long considered as either neutral or anti-carcinogenic. Accumulating research shows that in the presence of more pro-carcinogenic estrogens, estriol becomes anti-carcinogenic.

The most prominent proponent of the "estriol hypothesis" of anti-carcinogenesis, gynecologic oncologist, Henry Lemon, M.D., has argued that the relative proportions of estrone, estradiol, and estriol were at least as important as the absolute quantities of each hormone. He developed and named the Estrogen Quotient (EQ), the quantity of estriol divided by the sum of estrone and estradiol [EQ = E3 / (E1 + E2)]. He argued that a higher estrogen quotient means lower cancer risk and a better chance of recovery form cancer should it occur.

The following is one of the best examples available of the iodine-estrogen interaction, specifically the effect of iodine and iodide in stimulating the estradiol → estrone → estriol pathway towards estriol, and thus a higher estrogen quotient (EQ). The following case history is taken directly from chart notes and laboratory tests compiled between 1980 and 1992.

Case 4

23 April 80: The 34-year-old patient's initial visit was for fibrocystic breast disease. Prior to this visit, she had detected three breast lumps; she stopped caffeine intake and the lumps disappeared. However, her breasts were still uncomfortable, so she sought medical attention. She reported that her menstrual periods are regular, but lighter than before. Her last menstrual period was on 13 April. She took thyroid 5 years ago, and was now taking supplemental kelp. Her examination disclosed minimal but tender fibrocystic breast disease.

2 May 80: 24-hour urinary estrogens showed the following values: **estrone:** 19.3 µg/24 h; **estradiol:** 17.8 µg/24 h; **estriol:** 27.4 µg/24 hours; and **estrogen quotient (EQ):** 0.74.

She was asked to start on iodo-niacin tablet (135 milligrams iodide, 15 milligrams niacin; no longer available) once daily, along with other supplementation.

30 May 80: Her breast tenderness was much improved;patient instructed to continue iodo-niacin.

22 December 80: No further fibrocystic problems unless patient eats beef, and then her breasts become a little tender.

6 August 81: 24-hour urinary estrogens (same supplement program) were as follows: **estrone:** 44 µg/24 h; **estradiol:** 42 µg/24 h; **estriol:** 84 µg/24 h; and **EQ:** 0.97.

20 August 81: Fibrocystic disease remained abated. T3, T4, and TSH were checked as iodine was to be continued.

22 September 82: Fibrocystic disease remained improved. Patient can only use iodo-niacin 3 times weekly or her nose drips. Thyroid function tests to be checked to make sure there is no suppression; estrogens also to be re-checked.

25 September 82: 24-hour urinary estrogens were: **estrone:** 11 µg/24 h; **estradiol:** 9 µg/24 h; **estriol:** 29 µg/24 h; and **EQ:** 1.48

9 October 82: Thyroid function test results were normal; patient to continue program, including iodo-niacin 3 times weekly.

[*Summer 1983:* Patient subsequently reported stopping iodo-niacin at this time as her breasts were better.]

7 February 84: 24-hour urinary estrogens were: **estrone:** 32 µg/24 h; **estradiol:** 11 µg/24 h; **estriol:** 21µg/24 h; and **EQ:** 0.5.

7 March 84: Patient has a small nodule in her left breast, probably fibrocystic, for which a mammogram is scheduled. Results of thyroid function test remain normal. Since iodo-niacin has become unavailable, Lugol's iodine solution is recommended at dose of 6 drops daily (~ 37.5 milligrams iodide and iodine).

15 October 84: 24-hour urinary estrogens were: **estrone:** 40 µg/24 h; **estradiol:** 4 µg/24 h; **estriol:** 44 µg/24 h; **EQ:** 1.0.

17 December 84: Mammogram "benign." Patient's breasts were much softer again since resumption of iodine per Lugol's solution.

25 March 87: 24-hour urinary estrogens were: **estrone:** 15 µg/24 h; **estradiol:** 2 µg/24 h; **estriol:** 76 µg/24 h; and **EQ:** 4.47.

5 May 87: Doing very well. No interim problem with fibrocystic breast symptoms. Therapy continued as is.

[*"Sometime in 1988":* As patient's fibrocystic breast disease symptoms had been gone since 1984, she cut back the Lugol's iodine to 1 drop daily.]

11 September 89: 24-hour urinary estrogens: **estrone:** 19.2 µg/24 h; **estradiol:** 9.6 µg/24 h; **estriol:** 11.5 µg/24 h; and **EQ:** 0.4.

12 October 89: On the basis of these follow-up lab reslts with the lowered EQ the importance of resuming Lugol's iodine, 6 drops daily, was emphasized.5

9 January 91: Thyroid function tests normal. In 1990 patient switched from Lugol's iodine [6 drops daily] to SSKI, 6 drops daily; these contain iodide total of 108 milligrams). First appearance of "hot flashes" and other menopausal symptoms.

22 February 92: 24-hour urinary estrogens: **estrone:** 2 μg/24 h; **estradiol:** 1 μg/24 h; **estriol:** 9 μg/24 h; **EQ:** 3.0

22 April 92: No changes. Patient not interested in bio-identical hormones for symptoms of menopause.

Case 5

26 August 85: The patient, an asymptomatic postmenopausal woman, made an initial visit for pelvic examination and Pap smear. Because she had a family history of estrogen-related cancer, 24-hour urinary estrogen determination as recommended.

28 August 85: 24-hour urinary estrogens:**estrone:** 3 μg/24 hours; **estradiol:** 16 μg/24 hours; **estriol:** 1 μg/24 h; **EQ:** 0.05.

29 October 85: Results of Pap smear unremarkable, but because of very low estrogen quotient, Lugol's iodine, 8 drops daily (equivalent to 50 milligrams iodine and iodide daily), was recommended and her thyroid function was to be checked before next visit.

31 December 85: Results of 24-hour urinary estrogens: **estrone:** 8 μg/24 h; **estradiol:** 1 μg/24 hours; **estriol:** 14 μg/24 h; **estrogen quotient:** 1.55

Case 6

[day obscured] February 1993: The patient, a postmenopausal woman without symptoms of menopause, presented for a general health check, pelvic exam, and Pap smear.Because she has a family history of estrogen-related cancer, testing of 24-hour urinary estrogens was recommended.

2 March 93: Results of 24-hour urinary estrogens: **estrone:** 10 μg/24 h; **estradiol:** 5 μg/24 h; **estriol:** 15 μg/24 h; **EQ:** 1.0.

19 April 93: Follow-up visit concerned results of lab tests and Lugol's iodine, 8 drops daily, was recmmended/

28 September 93: 24-hour urinary estrogens: **estrone:** 20 μg/24 h; **estradiol:** 13 μg/24 h; **estriol:** 46 μg/24 h; **EQ:** 1.39.

Case 7

The following case illustrates that large quantities of iodine are sometimes needed to change the estrogen quotient. Of course, thyroid function should be monitored.

31 January 84: The patient's initial visit was for fibrocystic breast disease. Her history showed that she had had one breast cyst drained and was taking 800 IU vitamin E daily, believing it helped a little. Mammogram in November 1983 was "negative." Breast exam disclosed 2+ fibrocystic changes bilaterally. Myers' iodine treatment for fibrocystic breast disease and a 24-hour urinary estrogen test was recommended.

22 March 84: 24-hour urinary estrogen levels were as follows: **estrone:** 13 μg/24 h; **estradiol:** 11 μg/24 h; **estriol:** 45 μg/24 h; **EQ:** 0.67.

1 May 84: The patient's breasts were better with Myers' treatment, but are improving more slowly than usual; there is sufficient improvement to switch to iodoniacin, 1 tablet daily.

9 October 84: Breasts are generally better, but recent small tender lump was felt in right breast. Examination disclosed a small tender lump in the right breast, although her overall fibrocystic problem is less than it as in January. Mammogram taken as a precaution and the patient was to continue iodo-niacin.

7 April 85: 24-hour urinary estrogens: **estrone:** 13 µg/24 h; **estradiol:** 2 µg/24 h; **estriol:** 10 µg/24 h; **EQ:** 0.67.

As the patient's EQ is no better, she was to increase iodo-niacin to 2 tablets, 3 times daily and to have her thyroid function monitored.

16 November 86: 24-hour urinary estrogens: **estrone:** 14 µg/24 h; **estradiol:** 3 µg/24 h; **estriol:** 33 µg/24 h; **EQ:** 1.94.

Results of thyroid function test were within normal limits.

DISCUSSION

Since 1982, I've reviewed several thousand 24-hour urinary steroid analysis reports (which include estrone, estradiol, and estriol), both from individuals I have worked with in my practice and from other practitioners who have had specimens sent to the Meridian Valley Laboratory. While a majority of women have a "favorable" estrogen quotient, a significant minority have relatively low levels of estriol, with the majority of these having higher estrone than estradiol levels. When estradiol alone is used as a supplement for symptoms of menopause, most women whose preliminary tests show low estriol fractions continue to have lower estriol fractions, with relatively greater increases in estrone and estradiol.

Japan has the highest average daily dietary iodine and iodide intake world-wide. It is likely not a coincidence that Japanese women following traditional Japanese diets have both the highest average serum estriol levels world-wide, and also the world's lowest rate of breast cancer. It would likely be prudent to copy nature in her 3000+-year Japanese dietary iodine experiment, and include at least "Japanese quantities" of iodine in our recommendations to women who want to reduce their risk of cancer, especially women with low estrogen quotients.

CONCLUSION

The above data have been gathered from 23 years of bio-identical hormone replacement therapy and clinical laboratory testing of this therapy, contained in a background of 35 years of medical practice using natural substances and homeopathy magnets. Certainly such data are not a substitute for large-scale, controlled studies; but they can possibly be used to provide some direction and clues for future research. Until such research studies are done, these few data appear to confirm the observations of Paracelsus and the many, many other less famous physicians who have noted that copying nature as closely as possible is the best and safest practice of medicine.

ACKNOWLEDGMENTS

I am grateful to Patrick Friel, M.S., for data collection, correlation, and his constant research efforts, both library and laboratory. I am grateful to former staff of Meridian Valley Laboratory including Brian Schliesmann, M.T., and Lynn Robinson,

M.T., for data collection and correlation. I thank Carrie Scharf for invaluable assistance with data presentation.

[*Competing interests*: The author declares that he has no competing financial interests.]

REFERENCES

1. WRITING GROUP FOR THE WOMEN'S HEALTH INITIATIVE INVESTIGATORS. 2002. Risks and benefit of estrogen plus progestin in healthy postmenopausal women: principal results from the Women's Health Initiative Randomized Controlled Trial. JAMA **288:** 321–333.
2. BERAL, V. & MILLION WOMEN STUDY COLLABORATORS. 2003. Breast cancer and hormone-replacement therapy in the Million Women Study. Lancet **362**(9382): 419–427.
3. ROBINSON, L., B. SCHLIESMAN & J.V. WRIGHT. 1999. Comparative measures of serum estriol, estradiol and estrone in non-pregnant menopausal women: a preliminary investigation. **4:** 266.
4. FRIEL, P., C. HINCHCLIFFE & J.V WRIGHT. 2005. Hormone replacement with estradiol: conventional oral doses result in excessive exposure to estrone. Alt. Med. Rev. **10:** 36–41.
5. STEHMAN-BREEN, C., G. ANDERSON, D. GIBSON, *et al.* 2003. Pharmacokinetics of oral micronized beta-estradiol in postmenopausal women receiving maintenance hemodialysis. Kidney Int. 2003. 64: 290–294.

Can We Delay Aging?

The Biology and Science of Aging

B. ANTON,[a,c] L. VITETTA,[b,c] F. CORTIZO, AND A. SALI[c]

[a]*Path Lab, Melbourne, Australia*

[b]*Center for Molecular Biology and Medicine, Epworth Hospital Medical Centre, Melbourne, Australia*

[c]*National Institute of Integrative Medicine, Melbourne, Australia*

ABSTRACT: Long before the fountain of youth, mankind has had an interest in staying young. As we move into the 21st century, that interest has not only continued, but it has become an obsession. While no one can really prevent normal, chronological aging, there are things we can do to slow down "pathological aging." After all, aging is about accelerated inflammation, depletion, and wear and tear. With the marked increase in life expectancy and life span, clinicians need to be aware of the effects of aging on the provision of treatment modalities. Appropriate interventions individualized to the patient can help to "compress morbidity" by shortening the period of functional decline common in old age. Therefore, the "health span" will come closer to matching the life span. Disease and disuse are far more likely explanations for functional decline and the onset of common chronic conditions in older persons than is "true" natural or normal aging. Regardless of your genetic inheritance, you can accelerate aging by lifestyle choices and environmental conditions to which you expose your genes. There are even ways to reverse the problems associated with aging. Getting older does not have to mean growing older. Welcome to the world of preventative gerontology, better known as *anti-aging medicine*!

KEYWORDS: anti-aging medicine; longevity; pathology of aging; gene therapy; cell therapy; inflammation

Until quite recently, the notion of reversing human aging was mere fantasy, absent of any scientific support. Throughout history, going as far back as the *Epic of Gilgamesh* 4700 years ago, we have dreamed of being able to cure aging and the diseases that accompany it; but every claim of a "fountain of youth" has proven to rely on nothing more than false hopes and—more often than not—an urge to profit at the expense of the gullible. The fact that we never really understood aging made it extremely unlikely that we could learn to slow, prevent, or reverse the process.

Address for correspondence: B. Anton, Path Lab, 68 Burwood Highway, Burwood, Victoria 3125, Australia. Voice: +61-3-8831-3020.
billa@pathlab.com.au

Ann. N.Y. Acad. Sci. 1057: 525–535 (2005). © 2005 New York Academy of Sciences.
doi: 10.1196/annals.1356.040

HISTORY AND AGING

Today, we stand at a unique point in history, much like where we were in 1870 with regard to infectious disease. At that time, few had heard of Pasteur or Koch, and well-known scientists ridiculed the idea of microbes being dangerous or causing disease. Time passed, however, and, once ridiculed or not, we now take the concept of infectious disease for granted. In fact, much of what is good about modern medical care—sterile technique, antibiotics, immunizations, for example—derives from this single, powerful conceptual revolution that began 135 years ago. Before we came to grips with the fact that microscopic creatures could harm and even kill us, effective intervention in most common diseases was also fantasy. In those days, treatment for tetanus infection—"lockjaw"—was a matter of early cauterization to remove "devitalised tissue" (using a red-hot iron rod or boiling oil), amputation (without anaesthesia) if things got worse, followed finally by hope and prayer, though nothing really improved the deadly outcome. We think of malaria, cellulitis, tetanus, pneumonia, and yellow fever as a short list of infectious diseases; to the physicians of those times, each of these diseases was independent and unique, without shared mechanism and without hope of effective treatment.

Today, we have much the same conception (and misconceptions) of aging and age-related diseases. We think of cancer, atherosclerosis, osteoporosis, osteoarthritis, skin aging, and immune senescence as all unrelated, except chronologically. You get these diseases as you get older, not because they have anything in common, but "just because you get older." Even pathologists rarely consider common mechanisms, cellular events that link each of these diseases at the genetic level. After all, what could osteoarthritis and atherosclerosis, aging skin and Alzheimer's possibly have to do with one another except that they happen to old people? Yet, not only do they have a great deal in common, but also it is precisely this common thread that will allow effective intervention both in age-related diseases and in aging itself.

THEORIES OF AGING

Many theories of aging have been proposed, yet science has not produced a universal theory of aging. Why do we age? Current theories of aging at the cellular and molecular level generally revolve around two themes: aging is programmed, and aging is accidental.

To better understand the process of aging we need to focus on the differences between "normal" aging and "pathological" aging. Pathological aging is the aging process that is brought on by the presence of disease, such as adult-onset diabetes or arthritis, and that may later bring on cardiovascular disease or osteoporosis. This is not considered normal aging. These conditions are due to heredity or lifestyle. However, developing cataracts is considered normal aging, because if you live long enough, you will develop them.

Every time the earth circles the sun, we are one year older chronologically, yet we know that the rate at which people age biologically varies widely. The loss of muscle tone, circulation, immune capacity, skin elasticity and joint flexibility, for example, occurs much more quickly in some people than in others. This is due to gene expression. Genes do not change, but their expression does:

[genotype] + [(diet, lifestyle, and environment)] = [phenotype]

Therefore, we are not prisoners of our genetic destiny. There is a level of plasticity in gene expression.

There is a large paradigm shift that needs to occur in our thinking: aging is a disease that can be prevented or reversed. Changing variables that affect genes such as diet, lifestyle, and stress can have an impact on genetic expression. We age because our hormones decline; our hormones do not decline because we age. Here we need to remember that hormones are required to trigger genes in cells to manufacture the necessary proteins, peptides, and other hormones. Lower volumes will yield lower manufacture.

The goal of anti-aging medicine is to increase the health span, not just the life span. After all, who would want to live longer with chronic debility or cognitive impairment? Anti-aging medicine consists of modalities and therapies involving diet, nutrition, exercise, destress, functional balance, system reserves, and a balance in anabolic to catabolic metabolism. Some of the familiar modalities of biochemistry and pathology associated with anti-aging medicine include glycation, inflammation, oxidation, methylation, endocrine deficiencies and imbalances, and decline of immunity. Aging is fundamentally a metabolic process. While clinicians deal constantly with the myriad symptoms of aging, the best results are not obtained until the underlying decrease in anabolic metabolism has been corrected.

PATHOPHYSIOLOGY

As humans age, all functions and characteristics are modified. However, there is no clear consensus about what aging actually is—what "naturally" occurs with the passage of time—versus the effects of disuse and disease. Optimal health management requires an understanding of aging processes that are assumed to be natural and inevitable and knowledge of the most accepted theories of how and why we age prematurely, usually due to many reversible risk factors. Although at present there is a significant emphasis on development of new technologies, especially in the area of human genomics and stem cell research, to provide greater clarity on genetic and cellular aging mechanisms, adaptation of traditional, allopathic, complementary, and alternative integrative modalities to reflect current understanding of the effects of aging and of how to postpone or prevent premature aging is now possible.

Earlier studies attempted to identify the effects of natural or normal aging free of disease as distinct from the development of age-related diseases such as cancer, cardiovascular disease, diabetes, osteoporosis, and neurodegenerative diseases such as dementia and Alzheimer's disease. Results from such studies suggest that the effects of aging are extremely "plastic" and variable from person to person. McEwen's concept of the "allostatic load" suggests that each person's signature of aging is a result of interactions among genetic makeup, lifestyle, diet, and environmental challenges. Thus, a combination of holistic parameters can be used to assess cumulative physiologic and psychological challenges over the life span and to predict how today's health care practitioners will respond not only to new challenges such as management of age-related conditions and application of treatment modalities in old age, but also to longevity itself. Accordingly, a broader pattern of aging processes or

TABLE 1. Biomarkers of aging[a]

Loss of strength
Reduced flexibility
Decreased cardiovascular endurance
Increased body fat (and resultant loss of lean muscle mass, or sarcopenia
Reduced resting energy expenditure
Lower kidney clearance
Reduced cell-mediated immunity
Increased hearing threshold
Reduced vibratory sensation
Compromised near visiion and dark accommodation
Reduced taste and smell acuity
Increased autoantibodies
Altered hormone levels

[a]Adapted from Evans and Rosenberg.[17]

biomarkers should be considered in determining how "well" a particular person is aging and how to customize interventions to reduce the effect of a chronic condition, thereby preventing certain processes from leading to the development of one or more ailments that are more likely to appear in old age.

TABLE 1 lists various physiologic biomarkers found more often with suboptimal aging. Another finding, from longitudinal studies, is that normal aging appears to be a phenomenon of gradual rather than precipitous change. Rapid decline is more likely to occur with the onset of a specific age-related pathologic process.

OLD CELLS AND FREE RADICALS

To understand the common mechanism, we need first to understand how aging occurs. To many, aging is simply a matter of wear and tear, often expressed in the scientific terminology of free radical damage to proteins and DNA or of reactive oxygen molecules and mitochondria. Some scientists view getting old as the same thing that happens to a car, as it gathers rust, loses power, and falls apart. The problem with the car analogy is that organisms are not cars. What car can continually repair itself for decades? You live in a body that actively resists wear and tear by continually repairing itself, replacing lost cells and damaged proteins, making new mitochondria and new molecules, fixing DNA, and remaking itself from top to bottom.

And yet this body that continually repairs itself grows old. The problem lies in the fact that it stops repairing itself. There is always free radical damage, but older cells stop doing much about it. Every single one of your cells divided and ultimately came from two joined cells, one from each of your parents (with the mitochondria from your mother), whose cells in turn came from their parents, and so on back as far as life has been around. Following your cells (and their mitochondria) back through your maternal line, we quickly realize that you are part of a line of cells that is three and a half billion years old. You look pretty good, considering that free radical damage has been after your cells for several billion years. Why have *those* cells not aged and died? Perhaps the problem is not just free radical damage, but something about fertilization and having so many cells. But there are multicellular organisms that never age and single-celled organisms that do. In fact, the reason that your cells age is that they *allow* themselves to do so.

Some cells, cancer cells or the germ cell lines that created you, never age. Other cells, such as most (though not all) of the cells of your body do age, although at varying rates. All of these cells—aging or not, at different rates or not—are exposed to free radical and other damage, yet only certain cells age. The difference is that aging cells slow down their repair (and other) processes, whereas cells that do not age continue to deal with the damage, quite literally forever.

Let us look at what kinds of damage we are talking about, even just narrowing it down to free radical damage. Almost all (about 92%) of free radicals are made in your mitochondria. The first problem, then, is trying to avoid making free radicals. Unfortunately, since we need oxygen to survive, we cannott avoid making at least a few free radicals as we make ATP, the molecule that fuels almost everything in your cells. Worse yet, as your cells age, they make more and more free radicals for the same amount of ATP. In other words, your cells get sloppier as they get older.

The second problem is keeping the free radicals away from things that you need. It is bad enough to make free radicals within the mitochondria, but the last thing you want is to expose your DNA and critical cell proteins to attack from these dangerous free radicals. Luckily, your cells (like all eukaryotic cells) hide the DNA in a safe place—the nucleus—and tries to keep the free radicals in another—the mitochondria. But as your cells get older, the lipid membranes begin to leak: the free radicals begin to escape from the mitochondria.

The third problem is catching and breaking down those escaping free radicals. Your cells use vitamin E, superoxide dismutase, and a number of other mechanisms to deal with free radicals. Unfortunately, as you get older, all of these mechanisms become a bit less available. As a result, free radicals roam about more freely and do more damage in older cells than they did in younger cells.

Finally, no matter how good your cells are otherwise, there is always *some* damage that your cells have to deal with. In the case of DNA, you repair it. In the case of everything else, you replace it. Unfortunately, as your cells age, all of this slows down, too. The result is a gradual increase in the likelihood of damaged DNA, proteins that do not work, and membranes that leak (as above).

Together, these four problems are a guarantee that your cells will slowly fall apart and fail to work, resulting in tissues that do not work, a body that does not work—in other words, problems for you. The obvious question is what we might be able to do about any of this? You could try to fix any one of these problems. For example, you might use caloric restriction to limit the production of free radicals. Or you could increase your dietary vitamin E to help scavenge the ones that escape. Both of these and most other approaches deal with only a single part of the problem and, worse yet, with problems only *after* they have occurred. The best approach would be to deal with all of the problems, not just by "cleaning up after them," but by stopping the entire problem at the cause. But is there really a single place to intervene?

GENES AND CELL REPAIR

Curiously enough, all of the problems come together in one single place: gene expression. All of the changes listed above, and a lot of others, occur because the pattern of gene expression changes as we age. Your genes are just the same, but what they do certainly is not. Just as the difference between a muscle cell and a skin cell is in the pattern of gene expression, so, too, is the difference between a young cell and an old one. But what controls that pattern and, more importantly, can we do anything about it?

The list of things that affect gene expression is enormous. Every cell affects its neighbors; and depleted hormones, diet, activity, infections, methylation, glycation, inflammation, and a host of other things affect gene expression. In fact, the list is practically infinite: almost everything affects gene expression to some degree in a cell somewhere in your body. Even the much smaller list of things that control the change in pattern of gene expression between young cells and old ones is remarkably long. Luckily, however, we know of one thing that appears to be the *major* controlling of that change—namely, the telomere.

The telomere is a long piece of DNA at the end of each one of your chromosomes. Because of the way DNA is replicated, every time one of your cells divides, it loses a small part of its telomere. This gradual loss causes a change in the proteins around the telomere, which in turn causes an indirect change in gene expression throughout the rest of the chromosome. The overall result is simple: every time your cells divide, they get a little bit older. Although some of your cells—nerve and muscle cells, for example—do not divide very often, this does not protect them. In each case the cells that do not divide (and so do not age much) are dependent on cells that divide quite a bit. In the case of heart muscle cells, for example, it is not the heart that ages, but the arteries supplying the heart. In the coronary arteries that supply the heart muscle, the cells lining the vessels—the vascular endothelial cells—not only divide, but do so all the more in the face of smoking, high blood pressure, diabetes, and other things known to cause atherosclerosis. In short, the *reason* that most cardiac risk factors cause heart attacks is that they make the cells that line your arteries divide and age.

In each organ, we can trace aging diseases to aging cells. In Alzheimer's disease, it is the microglia that appear to be the culprit. In arthritis, it is the chondrocytes that make up the cartilage in your joints. In your bones, the osteoblasts age, resulting in osteoporosis. In your immune system, the lymphocytes age and result in poor immune function. In your skin, the fibroblasts and keratinocytes age and result in thin and wrinkled skin. In every organ, in every tissue, in every disease, we find dividing cells, aging, changing, and failing.

None of this would be of much importance if we could not prevent the failure; but, as it turns out, we can. The first study that showed we could prevent aging in cells came out only a few years ago. Since then, the same result has been repeated in a host of other laboratories and other cell types. At the cellular level, reversing aging is well within our current ability.

None of us, however, are mere cells, but tissues, organs, and bodies—vast collections of cells, each cell with a specific function and each dependent upon all other cells. While we can reverse aging in cells, can we go further and reverse aging in tissues or entire organs? In a sense, we already have. We can now reset aging in "reconstituted" human skin. If we take a mouse and transplant human skin cells (keratinocytes and fibroblasts) onto it, the cells layer out and grow human skin. If we use young human cells, we get young human skin, with thick and deeply interdigitated layers, strongly bound together between the dermis and epidermis. If we use old human cells, we get old human skin, with thin and barely adherent layers, weakly bound between dermis and epidermis and prone to sloughing off at the least pull. But if we take old human cells and reset the pattern of gene expression, the result is, once again, young skin; the skin is thick, the layers have deep interdigitations, and the cells are typical of young skin both in terms of their gene expression and their histology. The age of your skin is not a matter of how old the cells are, but of how old the gene expression is.

REVERSING AGING

Just as the telomere is the key to the altered pattern of gene expression in aging cells, so, too, is it the key to resetting gene expression in cells and in reconstituted human skin. Here, as always, the question is not, what causes aging? but rather, what is the single most effective point to intervene in aging? The issue is not academic, but concrete. How can we most effectively and efficiently prevent or treat the diseases of aging? In treating arthritis, we could (and do) replace the affected joints, but this is painful, expensive, and not entirely effective. In treating heart disease, we could replace the heart itself, but this is not only painful and expensive, but remarkably risky as well. In treating the genes that underlie these and other age-related diseases, we could—just as with hips and hearts— replace the affected part. But just as in hips and hearts, so, too, with genes: why not simply make the normal part work the way it was intended to work? The difference between a young cell and an old cell is not the superoxide dismutase gene, nor should we replace this or other genes. The difference between a young cell and an old cell is that this and other genes are not being expressed in the right amounts and at the right times. All of this can, and has been, reset by using telomerase both in the laboratory and in reconstituted skin.

The current question is, what is the best way to reset gene expression to that of normal young cells? We could replace the telomerase gene, which would then express normal telomerase, reset the genes, and rejuvenate normal cell function. Even better, however, would be to control the existing telomerase gene in each of your cells, turning it on and off as needed. This is the role of a telomerase inducer currently under development. Either of these techniques—inserting another copy of the normal telomerase gene or using a telomerase inducer—should do the trick.

Gene insertion has already been used in other contexts, and human trials using telomerase are not far off. Using this technique, a gene gun can be used to fire millions of copies of the human telomerase gene (hTERT) into human skin. While the "take" for this technique is normally fairly low, it would be sufficient. Dermal and epidermal cells would take up the hTERT gene and begin expressing it, resetting gene expression and returning to normal young adult cell function. Current plans call for attempting this in four different types of patient: those with Fanconi's anemia, those with dyskeratosis congenita, normal older patients in a wound care center, and children with Hutchinson–Gilford progeria. In the first two diseases, patients are known to have difficulty maintaining normal telomere function. In Hutchinson–Gilford progeria, the cells lose telomere length early in life, at least in the blood vessels, skin, hair follicles, and joints. The result is that these children have atherosclerosis, thin skin, little hair, and arthritis, usually dying by age 13 of a heart attack or stroke. Small wonder we might want to try fixing the problem!

In the case of normal older patients, we may try inserting a normal hTERT gene into the skin, particularly around the pressure sores these patients typically have. These are the result of poor innervation (so the patient is often unaware of sitting on them for hours), poor circulation (so they easily get infected and have a poor oxygen supply), and poor skin function (so the cells are slow to divide and heal the lesion). If we can repopulate the skin with healthy cells, the sores may heal more quickly and fully than is normally the case in the elderly.

The real question, however, is, what happens if we try these approaches in normal, older patients even without skin sores? Moreover, we could try a similar approach in coronary arteries (the cause of heart disease), glial cells in the brain (which may underlie Alzheimer's dementia), chondrocytes in the joints (which cause osteoarthritis), osteoblasts in the bones (which fail in osteoporosis), lymphocytes in the blood (which cause immune aging), and other sites. Both these trials and trials using telomerase inducers are likely to begin within the next few years. Only when we are finally able to intervene in the fundamental causes of aging—the altered pattern of gene expression that permits your cells to finally succumb to free radicals and a host of other problems—will we finally be able to reverse human aging and prevent the suffering that accompanies the diseases of aging.

ENDOCRINOLOGY

Aging is also regarded by many physicians and scientists as a manifestation in hormone levels over the course of adulthood. In longevity medicine, we are trained that with aging all our organs and glands are shrinking. Even our skin is shrinking. In addition, we are losing bone and about one-half pound of muscle a year.

Not merely limited to menopause, andropause (male menopause), somatopause (adult growth hormone deficiency), and other sex hormone related diseases, hormones are now implicated in conditions such as obesity, osteoporosis, fibromyalgia, chronic fatigue syndrome, cancer, attention deficit, and more.

Conventional medicine treats only severe hormone depletion, ignoring the mild and moderate deficiencies; whereas the anti-aging medicine model focuses on mild, moderate, and severe deficiencies. It stands to reason that symptoms and severity of symptoms will be proportional to the level of deficiency for each hormone. Only recently has the science of endocrinology started focusing on advanced testing methods, taking into account reference ranges based on age as well as gender, free or bioavailable hormone levels versus bound hormone levels, and ratios between major antagonistic hormones to titrate patients and achieve an optimum balance. Titration and optimum balance have been used with thyroid and insulin hormones, but strangely enough this concept has not been universally applied to all other hormones in conventional medicine until now.

In order to adequately address hormone decline in our adult patients, anti-aging physicians must first understand the hormonal cascade, an intricate interplay between the signals, pathways, and production and delivery systems that are responsible for our youthful hormonal state. The hormonal cascade involves estrogen, progesterone, testosterone, thyroid hormone, dehydroepiandrosterone (DHEA), cortisol, human growth hormone (hgh), and melatonin.

Second, anti-aging physicians must be able to readily identify symptoms of hormone-related decline in our adult patients. On a related note, we must also be familiar with the laboratory tests that are able to objectively confirm, or rule out, hormone imbalances.

TESTING PROGRAMS

A series of all-encompassing testing and analysis programs are designed specifically to determine the body's rate of aging, alleviate symptoms of aging, and return your body to a healthier state of functioning. As you age, your "cellular soup"—the molecules, nutrients, and chemicals that circulate in and among the cells of the body and determine your genetic potential—becomes deficient in various nutrients and minerals. This can cause a breakdown in functions throughout the body and, over time, promote accelerated aging. Each individual body responds differently to the aging process and therefore has different requirements.

Treatment programs and protocols are based on assessing the patient's current status and determining if the patient falls in one of the following categories: accelerated aging, premature aging, normal aging, reduced aging. Programs are designed to help patients progress from accelerated aging to reduced aging.

Based on one's individual needs, the anti-aging practitioner can then design a customized treatment regimen to address the specific symptoms. Remarkable breakthroughs in modern science indicate that aging can be effectively managed with the application of new options including hormonal replacement therapy, nootropics, and brain entrainment techniques. These therapy programs use these approaches, combined with comprehensive lab tests (blood, skin, saliva, and urine), diet, exercise, nutrition, and high-tech meditation, to alleviate aging symptoms.

Augmenting the body's ability to repair DNA damage, stimulating the immune system, and keeping the neurohormonal centers of the nervous system functioning are key components in retarding age. Some of the documented outward effects of longevity programs include measurable improvements in mental speed, clarity of thought, skin tightness, sexual function, plus a general improvement of energy levels.

Ongoing research in the field of longevity includes the use of nutrient precursors and hormonal precursors to augment the body's ability to make more of its own hormonal compounds as well as to limit DNA damage. Some doctors feel that the essence of all anti-aging treatment centers on protecting DNA from damage and augmenting its ability to repair itself and produce the key peptides and proteins that are responsible for all cell growth and regeneration. Others consider accelerated aging a nutritional deficiency syndrome.

TREATMENT PROGRAMS

With the onset of the 21st century, it is apparent that we stand on the horizon of a revolution in anti-aging therapies and technologies. One of the most promising techniques that within the next 5–10 years will allow us to markedly expand the quality of health and human longevity is the concept of stem cell treatments. Recent advances by major corporations, both private and public, have documented that we are already capable of taking what are termed *stem cells* from an individual and selectively copying these cells along with their DNA components to restore the rejuvenative properties that our bodies lose as we age. This revolutionary technology of therapeutically cloning our own cells and giving them back to our cells will allow us to selectively focus on the components that deteriorate with the aging process, uniquely for each individual. We all inherited certain genetic weaknesses—some a poor immune system, some a poor nervous system, some a poor cardiovascular system. With these new technologies we will be able to selectively restore these faulty genetic mechanisms.

REFERENCES

1. FOSSEL, M. 1996. Reversing Human Aging. William Morrow. New York.
2. SANDARS, N.K., ED. 1960. The Epic of Gilgamesh. Penguin Books. London.
3. KING, L.S. 1991. Transformations in American Medicine. Johns Hopkins University Press. Baltimore.
4. HAYFLICK, L. 1998. How and why we age. Exp. Gerontol. **33:** 639–653.
5. OLOVNIKOV, A.M. 1971. Principle of marginotomy in template synthesis of polynucleotides [in Russian]. Doklady Akademii Nauk SSSR **201:** 1496–1499.
6. WATSON, J.D. 1972. Origin of concatameric T7 DNA. Nature: New Biol. **239:** 197–201.
7. SHELTON, D.N., E. CHANG, P.S. WHITTIER, *et al.* 1999. Microarray analysis of replicative senescence. Curr. Biol. **9:** 939–945.
8. FUNK, W.D., C.K. WANG, D.N. SHELTON, *et al.* 2000. Telomerase expression restores dermal integrity to in vitro-aged fibroblasts in a reconstituted skin model. Exp. Cell Res. **258:** 270–278.
9. FOSSEL, M. 2003. Cells, Aging, and Human Disease. Oxford University Press. New York.
10. GIAMPAPA, V., R. PERO, M. ZIMMERMAN. 2004. The Anti-Aging Solution: 5 Simple Steps to Looking and Feeling Young. John Wiley, Hoboken, NJ.
11. CASSEL, C.K. 2001. Successful aging: how increased life expectancy and medical advances are changing geriatric care. Geriatrics **56:** 35–39.
12. KIRKWOOD, T.B. & S.N. AUSTAD. 2000. Why do we age? Nature **408:** 233–238.
13. INTERNATIONAL LONGEVITY CENTER–USA. 1999. The aging factor in health and disease: the promise of basic research on aging. Workshop Report. International Longevity Center–USA, Ltd., February 1999.
14. BRODY, J.A. & E.L. SCHNEIDER. 1986. Diseases and disorders of aging: an hypothesis. J. Chron. Dis. **39:** 871–876.
15. MCEWEN. B. 1999. <www.macses.ucsf.edu/Research/wgal.htm>.
16. INTERNATIONAL LONGEVITY CENTER–USA. 2000. Biomarkers of aging: from primitive organisms to man. Workshop Report. International Longevity Center–USA, Ltd., October 2000.
17. EVANS, W. & I.H. ROSENBERG. 1991. Biomarkers: The 10 Keys to Prolonged Vitality. Fireside. New York.
18. Merck Manual. 2001. The aging body, section 1: fundamentals. <www.merck.com>.
19. KANE, R. *et al.* 1994. Essentials of Clinical Geriatrics, 3rd edit. McGraw-Hill. New York.
20. Bland, J.S. 1998. The use of complementary medicine for healthy aging. Altern. Ther. Health Med. **4:** 42–48.
21. SCHWARZBEIN, D. The Schwarzbein Principle: the Program: Losing Weight the Healthy Way, an Easy 5-Step No Nonsense Approach. Health Communications. Deerfield Beach, FL.
22. NATIONAL CENTER FOR HEALTH STATISTICS.

23. HAZZARD, W. *et al.* 1990. Principles of Geriatric Medicine and Gerontology, 2nd edit. McGraw-Hill. New York.
24. KLIGMAN, E.W. 1992. Preventive geriatrics. Geriatrics **47:** 39–50.
25. GOODWIN, J.S. 1999. Geriatrics and the limits of modern medicine. N. Engl. J. Med. **340:** 1283–1285.
26. BALLENTINE, R. 1999. Radical Healing. Harmony Books. New York.
27. RAKEL, D. 2003. Integrative Medicine, Ch. 80. Saunders. Philadelphia.
28. KLATZ, R. & R. GOLDMAN. 2003. The New Anti-Aging Revolution Stopping The Clock. Basic Health Publications. North Bergen, NJ.
29. CHANSON, P. *et al.* 2004. Endocrine Aspects of Successful Aging: Genes, Hormones and Lifestyles. Springer-Verlag. New York.
30. HERTOGHE, T. 2002. The Hormone Solution: Stay Younger Longer. Harmony Books. New York.
31. COMFORT, A. 1979. The Biology of Senescense, 3rd edit. Churchill-Livingstone. Edinburgh.
32. STREHLAR, B.L. 1962. Time, Cells and Aging. Academic Press. New York.
33. DHABI, J.M. & S.R. SPINDLER. 2003. Biology of Aging and Its Modulation, Vol. 3: Aging of Organs and Systems. R. Aspinwall, Ed. Kluwer Academic Publishers. Dordrecht, the Netherlands.
34. WARTIAN SMITH, P. 2003. HRT: The Answers. Healthy Living Books. Ann Arbor, MI.
35. ROTHENBERG, R. & K. BECKER. Forever Ageless, Wilcox Literary Agency. Healthspan Institute. Encinitas, CA.
36. MOUNTZ, J.D. & H-C. HSU. 2003. Impact of Immune Senescense on Human Aging, Vol. 23, No. 1. W.B. Saunders. Philadelphia.
37. KYRIAZIS, M. 2000. The Anti-Aging Plan, Stay Younger Longer. Element Books. UK.
38. WINTERS, R. 1997. The Anti-Aging Hormones, Three Rivers Press. New York.
39. KLATZ, R. 1999. Hormones of Youth. American Academy of Anti-Aging Medicine. Chicago.
40. AHLGRIMM, M. & J. KELLS. 1999. The HRT Solution, Optimizing Your Hormone Potential. Avery Publishing Group. Garden City Park.
41. YANICK, P. & P. YANICK, JR.. 1998. ProHormone Nutrition. Longevity Institute International. New Century Press. San Diego, CA.
42. LAVIN, N. 2002. Manual of Endocrinology and Metabolism. Lippincott Williams & Wilkins. Philadelphia.
43. GRIFFIN, J. & S. OJEDA. 2000. Textbook of Endocrine Physiology, 4th Edit. Oxford University Press. Oxford and New York.

New Horizons for the Clinical Specialty of Anti-aging Medicine

The Future with Biomedical Technologies

RONALD KLATZ

American Academy of Anti-Aging Medicine (A4M), Chicago, Illinois 60614, USA

Department of Internal Medicine at the Central American Health Sciences University of Medicine, Belize City, Belize

Graduate School of Medicine at Swineburne University, Victoria, Australia

ABSTRACT: Anti-aging medicine is a medical specialty founded on the application of advanced scientific and medical technologies for the early detection, prevention, treatment, and reversal of age-related dysfunction, disorders, and diseases. It is a health care model promoting innovative science and research to prolong the healthy life span in humans. As such, anti-aging medicine is based on principles of sound and responsible medical care consistent with those applied in other preventive health specialties. Because it embraces the use of biomedical technology, anti-aging medicine offers a hopeful model of health care in which healthy human life spans of 120 years and longer may be achieved—if we employ anti-aging therapeutics today, and encourage the continued expansion of biomedical technologies to prevent, treat, and cure diseases.

KEYWORDS: anti-aging medicine; biomedical technologies; life enhancement; life extension; human life span

INTRODUCTION TO ANTI-AGING MEDICINE

Historian Alan Kan remarked that "The best way to predict the future is to invent it." Anti-aging medicine—its 13-year history, as well as its future horizons—is a case study on future-forward innovative thinking that benefits humankind and future generations.

In 1992, a group of 12 pioneering physicians and scientists convened to discuss the wide-ranging ramifications of rapidly emerging important discoveries aimed at identifying the mechanisms of deterioration and vulnerability to age-related diseas-

Address for correspondence: Dr. Ronald Klatz, American Academy of Anti-Aging Medicine (A4M), 1510 West Montana Street, Chicago, IL 60614. Voice: 773-528-4333; fax: 773-528-5390.

rklatz@worldhealth.net

Ann. N.Y. Acad. Sci. 1057: 536–544 (2005). © 2005 New York Academy of Sciences.
doi: 10.1196/annals.1356.041

es. As such, this group introduced a new definition of aging. In this new perspective, the frailties and physical and mental failures associated with normal aging are caused by physiological dysfunctions that, in many cases, can be altered by appropriate medical intervention. As an extension of this redefinition, this group proposed an innovative model for health care that focused on the application of advanced scientific and medical technologies for the early detection, prevention, treatment, and reversal of age-related dysfunctions, disorders, and diseases. "Anti-aging medicine" and the American Academy of Anti-Aging Medicine (A4M) were born.

In the years that have followed, anti-aging medicine has achieved international recognition. Anti-aging medicine is now practiced by thousands of physicians in private medical offices, as well as at some of the most prestigious teaching hospitals around the world.

Universally, those involved in health care, or those whose fields of expertise intersect with health care issues, support anti-aging medicine as a health care model promoting innovative science and research to prolong the healthy human lifespan. Public policy organizations and government agencies are now embracing anti-aging medicine as a viable solution to alleviate the mounting social, economic, and medical woes otherwise anticipated with the aging of the population of nearly every nation on the planet. In 2001, the World Future Society—a nonprofit educational and scientific organization founded in 1966 as a neutral clearinghouse exploring the impact of social and technological developments on the future—noted that anti-aging medicine was an effective solution to the growing aging population worldwide. It states that "geriatrics may ... be suffering from competition arising in a new health-care subspecialty: antiaging."[1] Citing an "aging baby-boom generation [that] is bringing a potential medical crisis to the fore: a critical lack of doctors who specialize in treating elderly patients," the World Future Society refers to antiaging medicine as embracing "a realignment of priorities from the problems of the elderly to the opportunities of longer lives."[1] The publication also notes the steady rise in the number of members of the A4M and certified anti-aging physicians and health practitioners, while the number of certified geriatricians is on the decline.

Similarly, the highly respected Global Aging Initiative of the Center for Strategic and International Studies issued its support of anti-aging medicine in its Summary Report of the Co-chairmen and Findings and Recommendations of the CSIS Commission on Global Aging. In its Economic Restructuring and Labor Policy section it stated that governments should "pursue an integrated strategy designed to raise productivity by ... providing financial support, and creating a favorable tax and regulatory environment for research and development in the industrial, new services and health sectors, including disease prevention, anti-aging medicine, and other innovative technology."[2]

The A4M is a nonprofit educational medical organization dedicated to the scientific premise that diseases and disabilities of human aging are largely preventable, treatable, and perhaps even reversible. The A4M is an international body of physicians, scientists, academicians, and government- and university-affiliated officials. A4M's membership numbers 14,500 located in more than 78 nations. The A4M serves as an advocate for the new clinical specialty of anti-aging medical science and acts as a conduit to physicians, scientists, and the educated public who wish to benefit from the almost daily breakthroughs in biotechnology that promise both a greater quality as well as quantity of life.

DEMOGRAPHICS OF GLOBAL AGING

One of humankind's greatest achievements is the extension of the human life span. Since 1950, average life expectancy worldwide has increased by 20 years, to stand at 66 years.[3] Over the next 50 years, the United Nations projects a steady increase in life expectancies for all countries, regardless of the extent of their economic development. While more-developed regions experience the lowest mortality and have higher levels of life expectancy at birth than less-developed regions (and the least developed countries), the gains in life expectancy will result in a worldwide life expectancy of 76 by the year 2050.[3]

According to the Population Reference Bureau, "the world's older population has been growing more numerous for centuries, but the pace of growth has accelerated." Population Reference Bureau statisticians observe: "The global population age 65 or older was estimated at 461 million in 2004, an increase of 10.3 million just since 2003. Projections suggest that the annual net gain will continue to exceed 10 million over the next decade—more than 850,000 each month. In 1990, 26 nations had older populations of at least 2 million, and by 2000, older populations in 31 countries had reached the 2 million mark."[4] By 2030, the Population Reference Bureau predicts that "more than 60 countries will have at least 2 million people age 65 or older."[4]

Global greying is intrinsically tied to human longevity, for the demographic shift described in the preceding paragraph has a direct impact on the health status across the population. In short, people are living longer, and the health needs of this booming older population are now drawing much attention. All diseases fall into four categories. The first three—inherited genetic disease, infectious disease, and trauma—account for only 10% of the cost for treating all disease in most nations. Ninety percent of all health care dollars are spent on extraordinary care in the last two to three years of life, specifically to treat aging-related diseases. As observed by Wilmoth and colleagues, whereas in the early 1900s, deaths were primarily attributable to infectious disease and sanitation-related diseases, by the late 1900s, a much different list of the leading causes of death emerged. Wilmoth and colleagues conclude from their case study of Swedish men and women dating from 1861 to 1999, that the continued upward trend in longevity in the twenty-first century will rely on the ability of science to "prevent and cure ailments such as coronary heart disease, stroke, and cancer."[5]

Concurrent with the longevity revolution, we are experiencing the technology revolution. New medical technologies are continually being developed or refined such that we can improve health care, increase its accessibility, encourage independent living, and promote extended productivity and vitality. Emerging biotechnologies are leading the way to innovations, including plentiful sources for replacement organs and tissues, interventions that help us to maintain our physical strength and mental acuity, and the end of today's incurable diseases with nanochips or genetic therapies. By embracing the adoption and utilization of technology, we foster both quantity and quality of life.

As such, anti-aging medicine has become the focal point of the new biotechnological revolution: anti-aging medical technologies have the potential to halt the degenerative physiology we now call "normal human aging." In the following section, we present a timeline of key anti-aging biotechnologies making recent headlines.

ANTI-AGING BIOTECH IN THE HEADLINES

In the realm of clinical anti-aging medicine, there are four key biotechnologies with near-term implications for the extension of the healthy human lifespan: (1) regenerative medicine innovations to regrow damaged or diseased tissues and organs; (2) stem cell technology to permit development of a supply source for human cells, tissues, and organs for use in acute emergency care as well as treatment of chronic, debilitating disease, (3) genetic engineering advancements that allow scientists to alter genetic make-up to eradicate disease; and (4) nanotechnology to enable scientists to use tiny tools to manipulate human biology at its most basic levels

Recent advancements in these biotechnologies have been in the news:

- Immature neural stem cells were induced to mature into adult brain cells. This new technique, conducted by the McKnight Brain Institute, successfully replicates the actual process of brain cell maturation with a precision never before accomplished. Lead researcher Dr. Bjorn Scheffler commented that "We can basically take these cells and freeze them until we need them. Then we thaw them, begin a cell-generating process, and produce a ton of new neurons." This advancement has tremendous potential for neurodegenerative diseases, such as Parkinson's disease and Huntington's disease, in that regeneration of parts of the brain will soon be possible.[6]

- A Japanese woman was cured of diabetes via a donor transplant of insulin-producing islet cells from her mother; no rejection occurred.[7]

- In the latest follow-up of eight Alzheimer's disease patients who had received injections in 2001 and 2002 of genetically modified tissue designed to boost nerve growth factor (NGF), it was reported that in 6 of the 8 patients, the implants "have successfully slowed their disease." Memory tests suggest that gene therapy slowed cognitive decline by as much has 50%, and brain scans showed that the patients' brains were more active than before. In addition, and most interestingly, in the one patient who died (of hemorrhage during the operation), a post-mortem found that "some of the brain tissue that had been dying off as result of AD had started to rejuvenate" with tissue regeneration located precisely surrounding the site of the injected cells.[8]

- Clemson University scientists have modified ink-jet printers to deposit a "bio ink" of cells, growth factors, and degradable gel (acting as scaffold for growth) to form three-dimensional tubes of living tissue. Current research is now focused on incorporating functional blood vessels into organs as they are "printed," and future prospects include the large-scale mass production of organs such as human hearts, livers, kidneys, and other organs, along with the essential nutrients they need to survive.[9]

- Researchers from the University of South Florida found that stem cells from human umbilical cord blood reduced heart attack damage in rats. Injected directly into rats' hearts after suffering induced heart attacks, the stem cells greatly reduced the amount of heart damage and restored heart pumping function to near normal. Drugs were not needed to prevent rejection. Dr. Robert Henning, lead researcher, predicts that "umbilical cord blood stem cells could offer a new way to limit or repair heart attack damage in people."[10]

- At the Hospital de Egas Moniz (Lisbon, Portugal), doctors have performed a total of 41 procedures in which tissue from a patient's nasal cavity, which is rich in stem cells, was implanted in the spinal cord at the site of an injury that caused paralysis. Dr. Carlos Lima and colleagues report that 10% of patients are now able to walk and all patients have reported some degree of increased feeling in paralyzed limbs since surgery. No infections or deaths were reported.[11]

- Dr. Seung Kim and colleagues at Stanford University coaxed immature brain cells to develop into insulin-producing islet cells that are lacking in diabetes. They added a "cocktail" of brain chemicals to brain stem cells, thereby coaxing them to become cells able to produce insulin in response to blood sugar levels. Transplants of modified stem cells into kidneys in mice were found to release insulin in response to rising blood sugar levels. After 4 weeks, the cells remained alive and continued to produce insulin, and none had turned cancerous.[12]

- Dr. Woo Suk Hwang and colleagues from Seoul National University, who cloned the first human embryo to use for research, report they have used the same technology to create batches of embryonic stem cells from 9 patients (6 adults and 3 children, with spinal cord injuries, juvenile diabetes, and a rare immune disorder). Their work demonstrates that embryonic stem cells "can be derived using nuclear transfer from patients with illness, regardless of sex or age." It also shows that it is possible to provide a source for tailored tissue to cure various diseases, in that stem cells can now be matched to patients and their medical conditions.[13]

- Dr. Alison Murdoch and colleagues from Newcastle University (UK) cloned the country's first human embryo. In their study, 3 clones survived and grew in the lab for 3 days; 1 clone survived for 5 days. In Great Britain, therapeutic cloning is allowed, where there is also a nonbinding ban on human cloning; supporters of therapeutic cloning point to its potential future applications in fighting disease.[14]

- RNA interference (RNAi) is a technology for silencing genes gone awry. Scientists correct defective RNA splices that, when expressed, cause disease. It is estimated that about 30% of all human genetic diseases involve RNA splicing errors, and other diseases, such as cancer, involve defective genes that may be targeted and silenced through RNAi. Dr. Norbert Perrimon and colleagues at Harvard University attached modified RNA molecules to an altered cholesterol molecule readily absorbed by cells. The modified RNA-carrying cells cut total cholesterol levels by 37%.[15]

THE ARRIVAL AT PRACTICAL IMMORTALITY

To summarize the importance of the profiled biotechnological advancements in medicine, we cite the speculations produced in a study of the next fifty years of cardiology, released by the American College of Cardiology at its annual meeting of 2000. Their prediction:

It is the year 2024. You are 75 years old, and you discover that a man next to you on an airplane has a pig heart, and his arteries are swarming with "smart dust" that sends continuous reports on his condition to his doctor's computer. That's not so strange, because you have a pig heart, too. And by 2049, when you are 100, many of your organs will be replaced. Plus you'll feel better than you did at 50 because "nanolabs" in your blood can manufacture and supply drugs whenever they are needed.[16]

The value of biotechnology and its application to human health and longevity has been well stated by the board of editors of *Scientific American*. In the magazine's March 2001 issue, the editors remark that "Thanks to modern technology and medicine, people have taken much more control over their differential survival. ...[I]lls are not the barriers they once were. Our technology may exert the greatest influence."[17]

Dr. Jim Oeppen of Cambridge University (United Kingdom) and Dr. James Vaupel of the Max Planck Institute for Demographic Research (Germany) have observed that maximum life expectancy has risen by a quarter of a year, each year, for the past 160 years. Commenting on their demographic analysis, Drs. Oeppen and Vaupel remarked, "If life expectancy were close to a maximum, then the increase in the record expectation of life should be slowing. *It is not.*"[18]

Indeed, practical immortality—healthy human life spans of 120 years and longer —may be achieved if we employ anti-aging therapeutics today and encourage the continued expansion of biomedical technologies to prevent, treat, and cure diseases. Medical knowledge and technology doubles every 3.5 years and gains in human longevity are directly proportional to the cumulative sum of advancements in the biotech fields.

INNOVATIVE VISION OF THE FUTURE OF MEDICINE

We are amidst an exciting period in the accelerating biotech revolution. In order to effectively deliver the biomedical and biotechnological advancements that constitute the anti-aging medical clinical specialty to patients, it is necessary to establish a premier, world-class facility at which pioneering discoveries in clinical and research objectives in life enhancement and life extension may be freely pursued by leading anti-aging scientists and physicians in the field.

The World Center for Anti-Aging Medicine in an innovative vision of the future of medicine. The World Research Center functions as a world-class, university-affiliated research and treatment facility unique in its focus on the investigation and application of diagnostic and treatment protocols that extend the length and enhance the quality of the human life span. Researchers yielding discoveries that will revolutionize the very early detection, treatment, and rejuvenation of aging-related disorders will work alongside clinicians delivering multitherapeutic interventions designed to slow, stop, and/or reverse the process of human aging. As a testing ground for the most promising and innovative biotechnological advancements pertaining to aging intervention, the World Research Center will be the undisputed pinnacle of cutting-edge medicine.

The World Center for Anti-Aging Medicine features state-of-the-art clinical applications for assessing biomarkers of aging, various technologies to ascertain "biological age" that measure the performance of the body's cells, tissues, and organs and the early detection of cancer and infectious diseases (some of the most costly and burdensome illnesses today).

Additionally the center features such technologically advanced therapeutics as:

• bio-identical hormone replacement therapy (BHRT), aimed at arresting age-related declines in hormone levels such that the natural peaks achieved in youth are maintained through life;

• dietary supplementation, to improve or maintain peak physical performance and mental acuity as we age;

• natural detoxification, to cleanse toxins from the body;

• intravenous chelation, to remove heavy metals from the body;

• aesthetic procedures, including Botox® injections, cosmetic fillers, dermabrasion, and mesotherapy, to reverse the physical signs of aging;

• hyperbaric oxygen therapy, to promote oxygen levels available to tissues and retard premature cellular death;

• immune restoration, including reversal therapies for stimulation of the thymus gland, which becomes involuted as we age and thereby contributes to a weakening of the immune system;

• stem cell therapeutics to replace cells in the skin, organs, sex glands, immune system, blood-forming system, muscles, and other systems that decline due to age-related changes and thereby progressively weaken such cells and cause cell death;

• gene therapies, along with DNA and RNA repair techniques, to counteract the otherwise immutable onset of aging-related diseases;

• sports medicine rehabilitation, spinal and musculoskeletal rehabilitation, and physical therapy, to counteract the cumulative wear-and-tear on muscles and bone as we age;

• high-tech surgical robotics, employed for invasive therapies to reduce human errors and enhance post-surgical recovery.

Combined, this multidisciplinary approach towards multimodal medical care will yield invaluable insights and permit the deployment of the latest interventions to slow, stop, and quite possibly, reverse aging.

CONCLUDING REMARKS

Aging is a global dilemma. While the world's total population grows at an annual rate of 1.7%, the segment over age 65 increases by 2.5% per year. Developed nations, thanks in large part to their adoption of diagnostic techniques affording screening and early detection of disease, have experienced a profound transformation of their demographics: nearly 20% of the developed world is more than 60 years old. In the next 20 to 30 years, the World Health Organization projects that elderly populations in developed countries will increase by 30 to 140%, and in developing countries this bracket will grow by 200 to 400%.[19]

The World Health Organization's Ageing and Health section states that "in the absence of appropriate policies to deal with population ageing, resources are often ill spent." Worldwide, in developed and developing nations, public policy as well as resource allocation fail to provide for the medical, social, and economic needs of a rapidly expanding group of older citizens.[19]

With a worldwide life expectancy now standing at 78.59 years (weighted average),[20] and a projected world population exceeding 9 billion by the year 2050,[21] the field of anti-aging medicine is witnessing unprecedented growth and acceptance. Anti-aging medicine is a medical specialty founded on the application of advanced scientific and medical technologies for the early detection, prevention, treatment, and reversal of age-related dysfunction, disorders, and diseases. It is a health care model promoting innovative science and research to prolong the healthy human life span. As such, anti-aging medicine is based on principles of sound and responsible medical care that are consistent with those applied in other preventive health specialties. The phrase "anti-aging" is, as such, a euphemism for the application of advanced biomedical technologies focused on the early detection, prevention, and treatment of aging-related disease. Anti-aging medicine is:

- *Scientific:* Anti-aging diagnostic and treatment practices are supported by scientific evidence and therefore cannot be branded as anecdotal.

- *Evidence-based:* Anti-aging medicine is based on an orderly process for acquiring data to formulate a scientific and objective assessment upon which effective treatment is assigned.

- *Holistic:* Anti-aging medicine utilizes an organized framework for the head-to-toe diagnostic assessment, and subsequent design of a treatment regimen.

- *Synergistic:* Anti-aging medicine recognizes that often, a multimodal, multi-therapeutic approach (including nutritional supplements) may deliver greater rejuvenative effects than could be achieved by single therapies alone.

- *Well-documented:* Research is published in peer-reviewed journals including *Aging, American Journal of Cardiology, Journal of the American Geriatrics Society, Journal of the American Medical Association, The Lancet,* and many others.

Anti-aging medicine unites physicians and scientists from around the world to forge a model for clinical health care that advances life-enhancing and life-extending technologies. As the founding organization in the world's fastest growing medical specialty, the American Academy of Anti-Aging Medicine (A4M) is committed to the pursuit of excellence in all aspects of leadership in the anti-aging medical field that it provides to 14,500 members in 78 nations. The flagship World Center for Anti-Aging Medicine serves as a microcosm for a hopeful, helpful vision of life-enhancing and life-extending clinical health care and medical research that creates a better world for every man, woman, and child. We welcome your participation in this project, which will directly and indirectly improve the lives of hundreds of thousands of people around the globe.

REFERENCES

1. WAGNER, C. Aging versus antiaging: geriatrics is in trouble while antiaging medicine takes off. Futurist **35:** 8–9.

2. Summary Report of the Co-Chairmen and Findings and Recommendations of the CSIS Commission on Global Aging. August 29, 2001. Washington, DC: Global Aging Initiative. Center for Strategic and International Studies.
3. World Population Prospects: The 2000 Revision—Highlights (ESA/P/WP.165), Population Division, Department of Economic and Social Affairs. United Nations. February 28, 2001.
4. KINSELLA, K. & D.R. PHILLIPS. 2005. Global Aging: The Challenge of Success. Popul. Bull. **60:** 1–40
5. WILMOTH, J.R., L.J. DEEGAN, H. LUNDSTROM & S. HORIUCHI. 2000. Increase of maximum lifespan in Sweden 1861–1999. Science **289:** 2366–2368
6. "Brain cells are matured in lab." BBCNews.com. June 14, 2005.
7. MATSUMOTO, S., T. OKITSU, Y. IWANAGA, et al. 2005. Insulin independence after living-donor distal pancreatectomy and islet allotransplantation. Lancet **365:** 1642–1644.
8. TUSZYNSKI, M.H., L. THAL, M. PAY, et al. 2005. A phase 1 clinical trial of nerve growth factor gene therapy for Alzheimer disease. Nat. Med. **11:** 551–555. Epub Apr 24, 2005.
9. XU, T., J. JIN, C. GREGORY, et al. 2005. Inkjet printing of viable mammalian cells. Biomaterials **26:** 93–99.
10. HENNING, R.J., H. ABU-ALI, J.U. BALIS, et al. 2004. Human umbilical cord blood mononuclear cells for the treatment of acute myocardial infarction. Cell Transplant. **13:** 729–739.
11. "Experimental Portuguese stem cell procedure offers hope to paralyzed." Agence France-Presse. Feb. 7, 2005.
12. HORI, Y., X. GU, X. XIE & S.K. KIM. 2005. Differentiation of insulin producing cells from human neural progenitor cells. PLoS Med. **2:** 103. Epub Apr 26, 2005.
13. "Scientists clone human stem cells from patients." Reuters, May 20, 2005.
14. "UK scientists clone human embryo." BBC News, May 20, 2005.
15. WOHLBOLD, L., H. VAN DER KUIP, A. MOEHRING, et al. 2005. All common p210 and p190 Bcr-abl variants can be targeted by RNA interference. Leukemia **19:** 290–292.
16. RAEBURN, R. 2000. "Oh, so you have a pig's heart too." Business Week March 27, 2000.
17. BOARD OF EDITORS. 2001. The future of human evolution. Sci. Am. March 2001.
18. OEPPEN, J. & J. VAUPEL. 2002. Broken limits to life expectancy. Science **296:** 1029–1103.
19. World Health Organization's Ageing & Health. www.who.int/ageing/scope.html; accessed Oct 18, 2000.
20. CIA World Factbook. March 2005.
21. World Population 1950 to 2050. U.S. Census Bureau, International Data Base, April 2005.

The Need for Anti-aging Medicine

The Challenges Faced to Incorporate Preventative Medicine into the Clinic and into Society

PHILIP MICANS

IAS Group, Les Autelets, Sark GY9 0SF, United Kingdom

ABSTRACT: The world's population is getting older and this fact is made clear by the aging demographics of both developed and developing countries. Projections estimate that there will be very sharp rises in the numbers of people over the age of 50, and very small increases expected in those under 15. From these facts, we judge that the costs of traditional health care based on the current "nationalistic" model will struggle to be met unless they undergo a radical alteration. A number of possible solutions are debated, ending in a conclusion that a liberal society is likely to choose preventative medicine as the next health care model for the twenty-first century. To this end, the concept of anti-aging preventative medicine is tackled, including some of the difficulties involved with bringing this new arm of medicine into the clinical environment. Also discussed is the challenge of biological age measurement, to assist in the determination of the actual level of health in a "healthy" person, with a view to an optimal health, rather than just normal health. Anti-aging medicine is a new concept, one that enables individuals to work with health care professionals in a different way and to take better care of themselves. It is in everyone's interests to understand that there is a need for preventative medicine in our societies and that this information is conveyed effectively to the public.

KEYWORDS: aging; preventative medicine; anti-aging; biomarkers; biological aging measurement; health care cost; bio-identical; demographics; orthomolecular; inner age; visualization; holistic; age management; optimal health

It is generally known that the average life expectancy is getter longer, increasing year on year. FIGURE 1 illustrates the life expectancy for the USA from 1900 to 2050[1] and similar trends can be presented for all the developed and indeed developing countries. Presently, the average man of 65 years of age in the developed world can expect to live to be 79 and the average woman can expect to live to be 82. The changes taking place can be exacerbated when you realize that a child born in 1997 can expect to live 29 years longer than one born in 1900, representing a 60% increase.[2]

Yet this could just be a "drop in the bucket" compared to what some scientists believe is already possible. Many believe that with today's existing knowledge about the human biological system and the technology already available many more people

Address for correspondence: Philip Micans, IAS Group, Les Autelets, Sark GY9 0SF, UK. Voice: +44 207 117 0107; fax: +44 208 181 6106.

phil@antiaging-systems.com

Ann. N.Y. Acad. Sci. 1057: 545–562 (2005). © 2005 New York Academy of Sciences.
doi: 10.1196/annals.1356.042

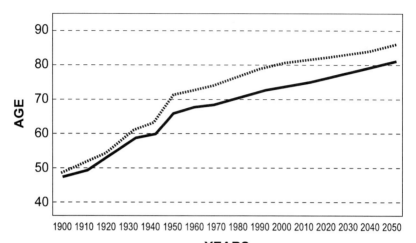

FIGURE 1. Life expectancy for the USA from 1900 to 2050. (Source: US Department of Health and Human Resources. January 2000.)

will be able to live beyond 100, perhaps even beyond 120. The theory is that the longer you live, the more chance you have to expose yourself to new technologies that can further extend life span.[3,4]

Naturally, none of this is worthwhile unless at the same time we are also enhancing or maintaining a good standard of health, thus improving the optimal health span as well as the life span. Furthermore, as we will see, simply extending life span based upon the costs and impracticalities of the current standard health care system is unlikely to be feasible.

THE LIFE SPAN OF THE BABY BOOMERS

Not only is the average life span becoming much longer (thus increasing the numbers of elderly individuals in the population), but we are also about to see the impact on society of the "baby boomers," those individuals born between the Second World War and the Korean War. They represent the largest single section of the population and are now beginning to move into their geriatric period. Because of their numbers, they have had major impacts on every single area of commerce and society —now expect this same impact for all their geriatric requirements.

Taking Italy as an example, we can study their current and forecasted populace.[5] FIGURE 2 is for the year 2000, here we can see a typical apple-shaped graph with still seemingly reasonable numbers of the working age-group, (i.e., income generating). FIGURE 3 is the graph for 2025 and it changes quite dramatically—it looks more like one of those old-fashioned spinning tops for children—and shows many more people in the over-50s age groups. In the projections for 2050 (FIG. 4), we see that the demographics look like an inverted pyramid, with for the first time, massive numbers in their 60s, 70s, and 80s. But we don't have to wait until 2050, as it is estimated that the 65-plus age group will start to experience a surge of growth from 2010.

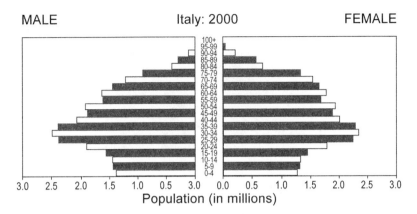

FIGURE 2. Aging demographics for Italy in the year 2000. (Source: US Census Bureau.)

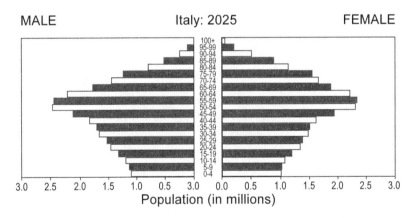

FIGURE 3. Aging demographics for Italy in the year 2025. (Source: US Census Bureau.)

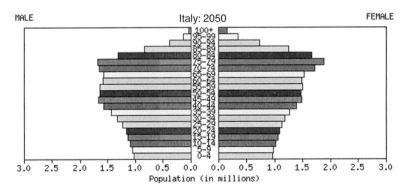

FIGURE 4. Aging demographics for Italy in the year 2050. (Source: US Census Bureau.)

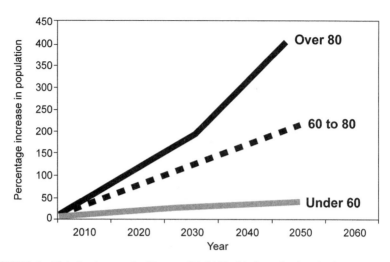

FIGURE 5. Global aging trends. (Source: World Health Organisation Ageing and Health.)

The elderly population is growing at a far greater rate than any other part of society (FIG. 5). The size of the group of individuals over the age of 65 is expected to grow at a rate of 200% between 2000 and 2050. While the growth rate of the group of those between the ages of 15 and 64 will expand at 16% and the group of those under 15 years of age at only 5%. But the single biggest increase is attributable to those over 80, with an expected increase of more than 400%! Indeed, this 80+ age bracket is by far the fastest-growing population group. To put this in perspective, in in the year 2000 there were 600 million people in the world aged 60 and over and by 2025 there will be 1.2 billion, rising to 2 billion in 2050. This is not just a developed-world issue. Today, about two thirds of all older people are living in the developing world; by 2025, that will increase to three quarters.[6]

AGING FACTORS

While it is common knowledge that the older one is, the more likely we are to be treated for a disease or ailment, FIGURE 6 helps highlight this. It shows by groups of varying ages, their percentage of the total drug consumption. This particular study looks at 1990, 2000, and 2001 and while there are variations between the years, it is clear that the overall trend is up, and that apart from birth and childhood diseases, the older you are, the more likely you are to be prescribed medication. For example, someone who is 75 years old is likely to use four times as many drugs as a 45 year old.[7] The question we have to ask ourselves is, what does this mean to society if we let these facts meet our previous graph showing the over-65s increasing by 200%?

Another factor that advances with age is disability (FIG. 7). While the graph shows similar results for France, Germany, the Netherlands, and the United Kingdom, the consensus is that the older we are, the more likely we are to suffer from some form of disability—many of which may need some type of care with its con-

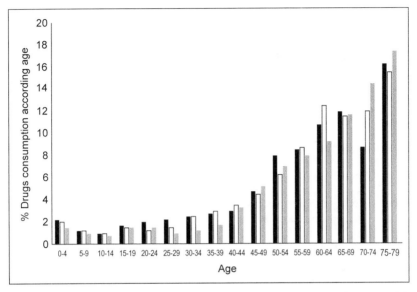

FIGURE 6. Drugs in use by age group: (Source: Lehman Brothers Equity Research.)

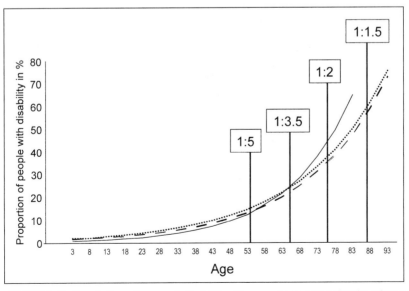

FIGURE 7. Correlation of age and disability. (Source: Association for the advancement of assistive technology in Europe.)

current cost. Indeed, at the age of 55 there is a 1 in 5 chance of having a disability, increasing to 1 in 2 at the age of 75 and so on (FIG. 7).[8] Again, let us pose the question, what happens if we let these figures meet those in the aging population demographics already mentioned?

Consider that old-age dependency rates will rise in every major world region over the next 20 years, and that the burden in 2025 is expected to be at least 50% larger than it was in 1998.[9] In just 14 years society will consider 1 in 5 of its population as elderly. An aging populace could have benefits. For example, an increase of just 6 years in life expectancy in the United States between the years of 1970 and 1990 translated into an additional gross national product of $57 trillion dollars, and those figures were based on a 1992 report.[10]

For those who say we can't afford to introduce preventative medicine, consider that in June 2000 the University of Chicago estimated that curing heart disease would be worth 48 trillion dollars and that curing cancer would be worth 47 trillion dollars.[11] Perhaps the question should be, can we afford not to introduce preventative medicine!

THE FINANCIAL BURDEN OF AGING

Looking more closely at one example of an disease associated with aging, we can begin to see more clearly the financial costs involved. Alzheimer's disease is currently the twelfth leading cause of death and it is perhaps an accurate example of an "aging disease," because it very rarely affects anyone under the age of 60. At present, there is a 1 in 7 chance of being diagnosed with Alzheimer's disease if you are over the age of 65. That figure rises to a massive, and rather alarming 1 in 2 chance if you are over the age of 85. In fact, the risk of Alzheimer's disease doubles every five years after the age of 65. In the United States alone, 80 to 100 billion dollars is spent every year on Alzheimer's health care or lost in earnings. It has been estimated that curing Alzheimer's disease could be worth as much as 1.5 trillion dollars.[12]

If we take into account all the age-related diseases and disorders, is it possible to get an overall view of the financial costs involved? Essentially all diseases fall into one of four categories (1) inherited or genetic disease; (2) infectious disease; (3) trauma; and (4) degenerative diseases. We suggest that degenerative diseases attributable to aging—for by literally being older the chance of acquiring these diseases dramatically increases.

According to the statistics, only 10% of all health-care costs are spent on the first three disease categories. In other words, 90% of the total health-care budget is spent on degenerative disease. In the United States alone, that is 700 billion dollars every year![13] Thus, if we really want to make a big impact on health-care in the world, we must focus on the degenerative diseases of aging. If we can slow or prevent the signs of aging from occurring, we could eliminate the majority of diseases and therefore the majority of its cost, both in financial and human terms.

But why are health costs so high? Clearly there are a number of factors here, including the need for highly trained staff, specialist equipment and premises, and liability insurance. All these factors are clearly present in the treatment of a disease, and obviously it could all be significantly reduced if far fewer people were actually sick in the first place!

Looking more closely at drug development, we see that the pharmaceutical companies are mainstays of the Fortune 500. Indeed, 10 drug companies registered more profits than the 490 remaining Fortune 500 companies combined![14] We are reminded that high margins/profits are required for research and development and the approval process. Naturally, not all drugs become approved and thus the average cost of approving a new drug—just in the United States—was estimated in 2002 to be $800 million.[15]

When going for approval, the most important question is whether the molecule be patented. After all, who is going to risk millions of dollars to approve a molecule that anyone can sell? But as it is so difficult to patent a natural molecule, it means the accent is upon the artificial. Once approved, a flood of press releases and marketing follows with the blessing of the authorities, which leads to greater public awareness.

This is the way things work and it greatly impedes the "small guys" and the "natural molecule brigade." The commercial reality is claims can't be made about substances that aren't approved and approval is expensive and requires strong patents.

PROBLEMS FOR THE WIDESPREAD INTRODUCTION
OF PREVENTATIVE MEDICINE

There are a number of issues that impede the introduction of preventative/anti-aging medicine. These include patents (already discussed), approval, research, time, insurance, scope, and measurement.

Approval

Apart from the prohibitive cost of the approval process, there is also the fact that products have to be tested to correspond to a given disease or disorder. There is no current acceptance of disease prevention, or of aging itself, therefore if the category fails to exist it is clearly difficult to start an approval process! There will have to be a change to the categories to accept perhaps just biochemical changes. For example, rather than treating an end-point disease, perhaps the hormonal change itself or the reduction of inflammatory markers could be good enough to prove a substance's worth, thus leaving the physician free to use these tools as they see fit. Another issue is the fact that most often substances for approval are tested singularly. While more often than not a combination of substances work synergistically together. This is especially true of natural molecules and is the foundation of a holistic approach to medicine. Unfortunately, the tradition is that single-item substances are tested individually to prove their lone merit, even though the results could be so much greater when combined with its methylation agents, enzymes, or other substances.

Research

In preventative medicine, we really need negative studies. By this I mean, we want to keep healthy people healthy. The types of clinical trials needed must study healthy people using the substances in question and showing that over time the people who were in the trial were less prone to disease. Again, all to often the trials run are the exact opposite, starting with a sick person and trying to make them healthier.

That is not the ultimate goal of preventative medicine and therefore it becomes another hurdle in the battle of the current approval mechanism.

Time

As usual it is not on our side. Once again the ultimate goal of anti-aging medicine is to extend both optimal health and the life span. Trials in humans need to be run over decades to prove their worth, although many are willing to accept that when we see positive results in a short-to-medium period we can't afford to wait for the endpoint—otherwise the data is of little relevance to today's generation.

Insurance

Very little tends to be reimbursable through health insurance at present. Therefore as preventative/anti-aging medicine is available to cash-only patients, the numbers of people are limited.

Scope

Asking the question, what do you prevent in your healthy patient? raises numerous issues of the vastness of the possible scope. In its widest sense the goals and aims are enormous, obviously because aging is affecting every part and every system of our bodies. Of course, one can concentrate on specific areas, perhaps those that are most likely to be lethal or are of the greatest concern to the patient themselves. In its broadest sense, using all the skills and technologies available to us—including examination, observation, blood work, scanners and good old-fashioned questions and family background—the testing could potentially become enormous and is clearly limited by the issues of cost and convenience.

Measurement

We need to develop techniques to measure healthy people, to try to determine just how healthy the healthy patient is! Furthermore, we need to move away from the idea of "normal" and think more along the lines of "optimal." A major criticism of anti-aging medicine is that it is suggested that advice and substances are given ad hoc, and that, by definition, healthy patients don't feel much difference in some cases, so the advice and recommendations are taken on "faith." So is there any way that we can show that the use of various protocols or products, such as the bio-identical hormones, are not only affecting the level of that particular hormone, but are also affecting the aging status of the patient as a whole? These types of problems may be solvable with the use of biomarkers, (also known as biological aging measurement), which would bring the time-honored tradition of measure-treat-measure to the preventative medicine field.

THE BIOMARKERS OF AGING

A biomarker or biological aging measurement is simply a measure that changes over time. It is assumed that they can act as a measure of aging, at least in a biological sense, because certain parameters change with age. If one has obtained data from a large cross-section of the community, then the patient's results can be compared to the average in their chronological age group to see if they are biologically younger or older.

A major premise of anti-aging medicine is to return particular biomarkers toward a youthful level, perhaps to those of someone in their mid 20s. So for example, if growth hormone injections are given to older patients, we can state that the patient is biologically younger—at least for the biomarker growth hormone. It is in this way that we refer to a biological age, rather than to unchangeable chronological age.

Biomarkers are obtained through physiological and biochemical means and include, for example, weight, height, blood pressure, waist-to-hip ratio, lung capacity tests, eyesight exams, hand-grip strength, skin elasticity and thickness, body flexibility, static balance, hair baldness and grayness evaluations, body mass index, as well as cholesterol, fibrinogen, creatinine, albumin, DHEA, growth hormone, T3, estrogens, and androgens.

Most biomarker data can be captured relatively easily. The methods usually include a clinical assessment to gather the required physiological data; blood analysis; 24-h urine analysis; saliva analysis; and, in some rare cases, other methods, such as hair analysis. As more data becomes available it is clear that gene testing will play a big role too.

What is important to remember, is that whatever or however the measurement is tested, the procedure must conform to the published clinical trials to which the data is being compared, otherwise it is of little value (unless of course it is being gathered with the purpose of becoming published in its own right).

BIOMARKER HISTORY

When looking at the literature, it is clear that biomarkers are not necessarily a new invention. In fact, these "age averages" for the population have been gathered for many more years than most people imagine. Indeed, one can argue that biomarkers have been captured since the 1880s, maybe even before then. However, as the first gerontological and geriatric societies, institutes, clinics and journals were formed from the 1950s onward, it was probably from that period that biomarker data began to be gathered and evaluated in a modern sense. Some of the useful reference publications the interested reader may be interested in obtaining are shown below as references 16–10.

CLINICAL APPLICATION ISSUES

Looking at the status of the actual measurement of aging and health today, what challenges do clinics face in adopting preventative medicine? First, we can recognize that there is a lack of scientific measurement within the industry, with no clear path to follow. Second, any solution must be quantitative and standardization is lacking within the field. Finally, consider that any measurement that takes place has to be seen to be changeable within a relatively short period of time. From the patient's perspective, results have to be achievable within a reasonable time frame.

Ward Dean noted in his book "It is assumed that with increased automation and computerization, that data will be more easily collected and analyzed, to rapidly improve the accuracy and cost effectiveness of aging measurement. I hope this book will be useful in assisting to accomplish this goal."[19]

With Dean's foresight, we can now consider that an internet-based software system would have the advantages of being flexible and therefore easier to change and update as developments take place. It would also be able to incorporate data from international research of all kinds, therefore offering a powerful portal to the physician. This would assist with measurements, but also make any published data that may be of assistance in the treatment of the patient available. Data could be acquired globally through the internet and would allow for biological age measurement relationships and patterns to be spotted much more quickly, leading to research being published more quickly and offering new information much more quickly than at present. One such system has been attempted under the name of Inner Age,[20] developed under the medical guidance of Dean[21] along with other physicians including Marios Kyriazis.[22]

Over a three-year period, more than 200 biomarkers were assessed, all of which had been published in numerous international journals. In total, it is estimated that the review included data from more than half a million subjects from 1000 studies.[23] The end result is that the Inner Age medical panel identified three specific categories of biomarkers: full biomarkers, candidate biomarkers, and general health indicators.

FULL BIOMARKERS

The following criteria has been used to classify an Inner Age full biomarker. In other words, while trawling through the published studies, only those that could meet these criteria were accepted for inclusion in this category: the results had to be based on human studies; the numbers of subjects in the study had to be significant and representative of a cross-section of society and races; the data had to have a relatively narrow standard deviation; the biomarker graph must result in a clear association with aging (see FIG. 8 for an example)[24]; the published data had to cover the age range of 30 to 70 years to allow for biological age determination of the vast majority of preventative medicine patients who might visit an anti-aging clinic; and finally, the measurement to obtain the data should be convenient and cost effective to obtain and usable under normal clinical conditions.

From the original 200 biomarkers extracted from the published clinical studies, and first deemed to be suitable, only 30 met the above criteria, and it is these that have been classified as full biomarkers.

CANDIDATE BIOMARKERS

Candidate biomarkers are the result of human studies. They also result in an association with aging. However their sample sizes may be unrepresentative. For example, the study may have been conducted on only Japanese women or just athletes. Also there may be no standard deviation available at present and the range of data may not be available for ages 30 to 70 years. There may be other statistical challenges. For example, FIGURE 9 highlights prolactin changes in men with age.[25] This type of graph is difficult to incorporate for statistical reasons, after all the same measure-

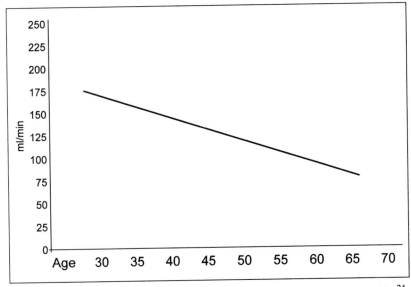

FIGURE 8. Creatine clearance changes in men with age. Source: Andres & Tobin.[24])

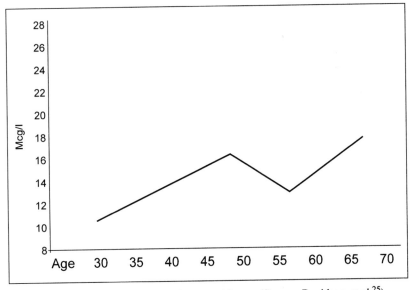

FIGURE 9. Prolactin changes in men with age. (Source: Davidson *et al.*[25])

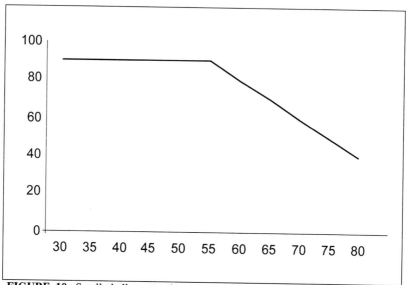

FIGURE 10. Smell challenge testing, changes in men and women with age. (Source: Doty *et al.*[26])

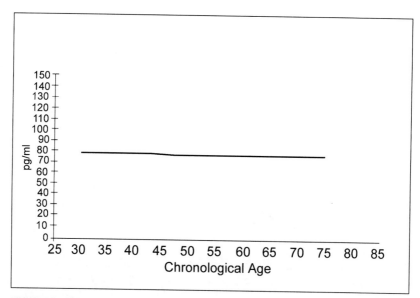

FIGURE 11. Melanocyte stimulating hormone, no variability or changes with aging. (Source: Dean.[27])

ment could result in different biological ages! FIGURE 10 represents the results of a smell test.[26] Until the age of 55 it is not possible to use smell as a biological age measurement, given that the results for the ages of 25 to 55 are the same. However, past that age it is very effective.

But, it is useful to continue to capture candidate biomarkers, to see if the individual patient's results are completely out of kilter with the norm expected or to continue to capture additional data, so that in the future they may be suitable to become a full biomarker. The Inner Age panel has currently identified 40 candidate biomarkers, some of which will undoubtedly become full biomarkers once further data are accumulated.

GENERAL HEALTH INDICATORS

General health indicators do not result with an association to aging, as indicated by FIGURE 11.[27] They might have large standard deviations and their sample size may be unrepresentative. FIGURE 11 has a flat line, this therefore makes it impossible to work out a biological age: Is the patient biologically 35, 55, or 75 years old? Unless further evidence can be provided, these types of measurements cannot be used to assess biological age. However, it can be accepted that such measurements are general health indicators and may highlight the trigger for a specific disease.

MEASURING BIOLOGICAL AGE

A typical current procedure to measure a biological age first involves physiological testing. At this time, more specialized tests may also be incorporated, such as dental examinations, cognitive functions, and auditory and smell tests.

It is true that some biomarkers are more important biological age markers than others, but the more data captured, the better the outcome: In particular, additional data about the patient can be used for future reference. After all, what is thought of as unimportant today, may be tomorrow's cancer risk marker. Additional data may also have a greater influence on the overall biological age assessment, in other words it leads to further accuracy. Additional data can be used to spot trends faster and accelerate our learning through shared research. Finally, additional tests allow the patient to feel and experience a comprehensive examination, thereby providing "value for money."

The second stage, which in theory could be optional, but adds great value as already stated, are the biochemical examinations. This is the collection of blood and urine and so on. Once evaluated at the laboratory, the patient's data can be entered into the software to create biomarker visualizations.

BIOLOGICAL AGE VISUALIZATIONS

FIGURE 12 shows a patient's biomarker visualization. This specially designed body map from the Inner Age system is designed to show a lot of detail quickly. Clearly, this is a male patient, and when the patient's details were entered, including

his date of birth, the system calculated his current chronological age, which is 48.3
With this in mind, the system cross-referenced his test results with the statistical data
for his age group and produced a legend. This legend groups by color those biomarkers that are in excess, up to 12 years more than his chronological age, and other
shades indicate approximately 5 years more than his chronological age, as well as
areas close to chronological age. Furthermore, additional colors indicate biological
age of down to 12 years less than chronological age.

Hence immediately, both the patient and the physician can determine those areas
that require attention first. This gives reasonable goals for the patient to aim for. Furthermore, we find that these body maps are welcomed by the patients as they understand more easily why they are undertaking certain programs, and indeed stick to
them more fervently. Whereas before much was being undertaken on faith, now patients can see results. As we are visual creatures, much more data can be evaluated
much more quickly in such a form, although naturally the tabulated and graph data
still remains available.

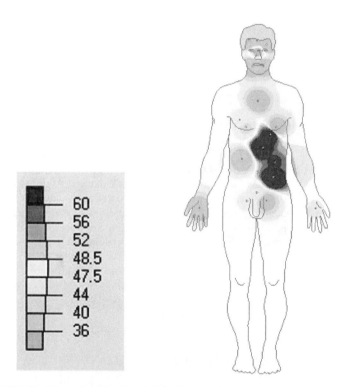

FIGURE 12. Biomarker visualization (body map). Varying colors (red = high/ green =
low, all shown here in grayscale) are overlaid onto a body map to indicate the measurements
where the patient is above or below his biological age. (Source: Data on file IAS Group.[20])

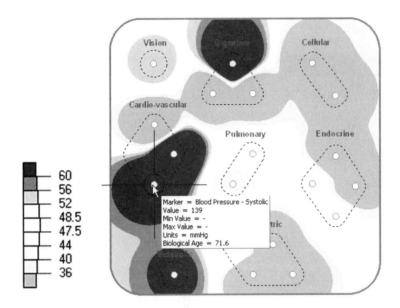

FIGURE 13. Biomarker visualization (topological map). Varying colors (red = high/ green = low, shown here in grayscale) are overlaid into measurement groupings to indicate the systems where the patient is above or below his biological age. (Source: Data on file IAS Group.[20])

FIGURE 13 shows another view of the same data in FIGURE 12. while the body map has more impact for the patient, the topological map is actually a better tool for the physician. Here the same data are grouped into biological systems, such as vision, cardiovascular, and pulmonary. The same legend represents the same biological age groups and uses the same color codes. Only now, for this patient it is clearer that the primary aging issues lie within cardiovascular and pulmonary regions. (Note: Each white dot represents a measurement and its full details can be revealed on the live system). We've always known that the human body doesn't age uniformly, now this kind of technology is making this fact obvious.

Remember what we are doing here is comparing the measurement to the average in their age group and showing them whether they are biologically the same, younger, or older. Clearly the goal is to be as biologically young as possible, and while the overall biological age is an interesting feature that everyone wants to know, it is the very high (i.e., dark red) measurements that are the most important to look at, as they indicate that a measurement is well out of the norm.

Accurate biological age measurement can represent the earliest signs of change, seeing changes even before the appearance of so-called disease markers. Furthermore, the possibility of the accumulation of hundreds of varying measurements, displayed in this visual manner makes it easier to achieve the goal of health evaluation and holistic medicine.

WHAT IS THE ANSWER TO WORLD AGING?

In the very near future, the world has to ask itself some basic questions. Can we expect the future working generations to pay 70% or even more in taxes to support the same health-care program for a "top-heavy" society? It is estimated that if the policy of early retirement continues, (i.e., retirement between the age of 55 and 60 years) in just 20 years there will only be two people working for each retired person! Thus, it appears that older people will be a necessary component of the workplace, continuing to contribute to society, if only in order to support themselves. That's another reason why we need a fit, lucid, and agile older society.

Another basic question is whether we should continue as we are. Can we allow an ever-increasing elderly population to meet an already stretched health-care budget and simply increase the numbers of people who are ending their lives in suffering? There are already discussions in the United Kingdom about whether or not certain drugs, particularly ones that are expensive and perhaps have a small benefit, should be given to the elderly. These questions have currently been raised in regard to certain senile dementia drugs and cancer drugs. They likely represent the "thin end of the wedge" and we are all going to hear a lot more debate on such matters.

Another question arises from those who state "there are too many people already." If one of the answers is a low birth rate, are we ready for societies that dictate who, when, and indeed if you can have children? The only other answer to population control is a high death rate. Do you wish to live in a society that considers euthanasia necessary—perhaps dependent merely upon chronological age or on a ratio of your health-care cost and age?

The alternative is to introduce anti-aging medicine and revolutionize health-care by concentrating on the true prevention of disease. We could reduce the overall cost of health, adopt the holistic use of natural substances and allow our elderly to live in dignity and self reliance, by remaining lucid and useful to the community at large.

ANTI-AGING MEDICINE

Whether or not one likes the terminology, anti-aging medicine itself represents the ultimate preventative medicine. The model is based upon the very early detection, prevention, and reversal of aging-related disorders. The science of anti-aging medicine is truly multi-disciplinary, for it is represented by advances in the fields of biochemistry, biology, and physiology and enhanced by contributions from mind-body medicine, molecular genetics, and the new emerging medical technologies. Anti-aging medicine has its foundation in what Nobel laureate, Linus Pauling described as orthomolecular.

Furthermore, anti-aging medicine accepts that aging diseases and disorders can and should be prevented, rather than simply treated. Today there are literally thousands of physicians, scientists, and researchers around the world involved in the research and treatment of aging. Huge strides are being made in the understanding and control of the aging process. Our challenge is to bring together the international research and to utilize whatever molecules and techniques may be necessary for the long-term health of the patient according to that science.

Ultimately, that means changing the way the patent system and the approval system operates—big steps—but if we remember to educate and prove to the public that this is the common sense they have been looking for, then usually no one stands in the way of a ground-up revolution for very long.

CONCLUSION

We stand at the threshold of a new paradigm: for the first time in history we understand some principles of why we age, how we can measure biological aging, and how we can slow aging and treat its effects. Through the use of lifestyle choices, chelation, nutrition, hormones, drugs, and emerging technologies, we now have the ability to delay, reduce and even prevent the appearance of numerous disorders and diseases.

I envisage in the decade to come that thousands of anti-aging clinics will be established. People will attend these clinics for regular checkups and through the use of biological aging markers, an individual's rate of aging and risk from particular degenerative disease will be measured. Steps will then be taken to slow and eradicate these biological aging signs before they become diseases, and therefore difficult and expensive to treat.

In other words, the traditional recognition and diagnosis of disease will change forever. One could consider these aging but otherwise "healthy" individuals attending anti-aging clinics, in much the same way as individuals visit the dentist today, for their preventative checks and measures.

However, as is usual with all great advances, mankind will probably experience the vested interests of the establishment and dogma that together will attempt to slow down or perhaps even prevent the wide-scale use of anti-aging medicine. After his discovery, Christopher Columbus said: "Human progress has never been achieved with unanimous consent. Those who are enlightened first are compelled to pursue the light in spite of others."

If we too can grasp the fundamental fact that we need to be pro-active about aging instead of just tolerating it, then mankind will be able take into its hands the possibility to radically alter the way we think about and approach "health." If we don't allow dogma, vested interests, and other imposed restrictions to stand in our way, then the door is already open to us.

REFERENCES

1. U.S. Department of Health and Human Services. January, 2000.
2. United States Census Bureau. 1995. US Department of Commerce. May 1995.
3. American Academy of Antiaging Medicine. <www.worldhealth.net >.
4. DE GREY, A. 2004. Journal of Rejuvenation. Mary Ann Liebert Publications.
5. US Census Bureau <www.census.gov/ipc/www/idbpyr.html>.
6. World Health Organisation Ageing & Health <www.who.int/hpr/ageing>.
7. Lehman Brothers Equity Research.
8. Association for the advancement of assistive technology in Europe.
9. United States Census Bureau. May 1995.
10. United States Department of Commerce. 1992.
11. *The Economist*. June 3, 2000.

12. National Vital Statistics Reports, Vol. 47, No. 20. June 30, 1999.
13. United States Health-Care and Finance Administration, 1996.
14. Physicians for a national health program, June 25, 2003.
15. Life Extension Magazine, May 6, 2002. Life Extension Foundation,
16. BALIN, A. 1994. Human Biologic Age Determination. CRC Press.
17. The biological markers of aging. US National Institute on Health.
18. SPIRIDUSO, W. 1995. The Physical Dimensions of Aging. Human Kinetics Publishers.
19. DEAN, W. 1988. Biological Aging Measurement. Center for Bio-Gerontology.
20. Inner-Age website <www.inner-age.com >.
21. DEAN, W. Center for Biogerontology, Florida, www.wardeanmd.com
22. KYRIAZIS, M. British Longevity Society. www.anti-ageing.freeserve.co.uk
23. Inner-Age system, data on file. IAS Research Group. www.iasgroup.com
24. ANDRES, R. & J.D. TOBIN. 1976. The effect of age on creatinine clearance in man. A cross section and longitudinal study. J. Gerontol. 31: 155–163.
25. DAVIDSON et al. 1983. Hormonal changes and sexual function in aging men. J. Clin. Endocr. Metab. 57: 71–77.
26. DOTY et al. 1984. Smell identification ability changes with age. Science 226: 1141–1143.
27. DEAN, W. 1988. Melanocyte stimulating hormone, no variability or changes with aging. Biol. Aging Measurement page 393.

Science Is a Never-Ending Quest

A Brief Critique and Appreciation of Stromboli IV

NOVERA "HERB" SPECTOR

INTRODUCTION

Twenty years ago, an intrepid band of many of the world's leading scientists and clinicians, all experts in cancer and aging, gathered under the watchful eye of "Mother Stromboli," the nearby volcano who constantly and audibly grumbled her skepticism about the august proceedings being held below. Each expert proved (scientifically) that one or another (or several) peptides, hormones, cytokines, trace elements, vitamins, curious chemical compounds, or behavioral procedures could definitely reverse aging and/or cancer. In summarizing that meeting, I drew up the entire list as ingredients for an infallible prescription, called the "Stromboli cocktail,"[1] which can be seen in *Neuroimmunomodulation: State of the Art* (Vol. 521 of the *Annals*). The ingredients included volcanic ash, thymus—whole, desiccated calf, and all fractions—haloperidol, SOD, selected catalases, activated charcoal, epsom salts, butterfly wings, milk of magnesia, beta-carotene, vitamin A, melatonin, arginine, thioproline, PCPA, vitamin E, enkephalins, zinc, selenium, DHEA, estrogens, prolactin, growth hormone, 3,4-dimethyl pyridine, and levamisole. It was noted that this cocktail should be taken with social support, exercise, sex, and electrical stimulation of the hypothalamus. Several years later, at the second Stromboli congress, more experts added at least twelve more (proven and guaranteed) substances to make the second Stromboli cocktail," consisting now of 42 ingredients, for which I proposed the structural formula seen in FIGURE 1.

To our astonishment, despite these amazing and well-documented discoveries, many humans (and numerous other species) are still aging and dying of cancer. To better approach these problems, I proposed a number of question (problems) for further research[2] that are slightly modified here:

1. Is there an evolutionary survival value to individual death?
2. Why do different species have different "maximum" life span?
3. How immutable are these maxima?
4. Why do some cells in multicellular animals seem to be immortal, while others age?
5. How many viable stem cells remain in the aging brain?
6. Under what conditions can neurons proliferate?
7. Is there a master aging clock, and if so, does it reside in the hypothalamus, the epiphysis (the pineal), the hypophysis (the pituitary), the heart, the bone marrow, or the genes?

Address for correspondence: Novera Spector, 4014 G, Layang Circle, Carlsbad, CA 92008. Voice: 760-720-0906.

Ann. N.Y. Acad. Sci. 1057: 563–571 (2005). © 2005 New York Academy of Sciences.
doi: 10.1196/annals.1356.043

8. By tinkering with this clock, can we get the gears to move more smoothly, and can we reset the clock to make it last longer?

9. Can we improve immune function in the aging individual without increasing the risk of autoimmune diseases?

10. Is there a link between cancer and other diseases of old age?

11. What common mechanisms do different types of cancer share?

12. What common features are shared by lymphocytes, macrophages, glial cells, neurons, hepatocytes, and cardiac cells?

13. In the feedback loops between the nervous and immune systems, how many are open and which ones are closed? Are there any positive feedback loops that are not fatal?

14. (a) What afferent messages are transmitted to the central nervous system by means of hard-wired (neuronal) fibers? (b) Are these of equal importance with chemical messengers in the circulation? Such messages, by means of nerve fibers, would have the advantage of much greater speed of transmission, and would not be subject to impedance by the blood–brain barrier. (c) How are various forms of energy transduced into nerve membrane depolarization, and how are these messages decoded in the CNS?

15. What do the afferent terminals look like?

16. How do ultradian, circadian, menstrual, lunar, and other chronobiologic factors coalesce into the lifetime rhythm? How do their alterations affect life span?

17. What are the links among the pineal, superchiasmatic nucleus, and immune functions?

18. What common features are shared in the transduction of signals across all cellular and intercellular membranes?

19. Are we shortening our lives by caloric overload?

20. Would it be reasonable to simplify our nomenclature to make our thinking about biological problems easier? For example, could we stop using the horrible word "stress"? This word means all things to all people, and is used to mean "distress," "eustress," "stimulus," and "strain," in which case *the stimulus is totally confused with the response*. Also, could we clarify such terms as lymphokine, monokine, neuromodulator, immnomodulator, and so on? Does the word cytokine encompass all of these? What about aging versus senescence?

21. Does senescence begin with conception? When does aging begin?

22. To what degree and by what mechanisms does the macroenvironment influence the microenvironmens and *vice versa*?

23. Will we ever have a single equation that quantitatively describes all of the factors and their interactions in the twin cascades of anabolism and catabolism, from conception to death (a sort of biological "unified field theory")?

24. If, by pharmacologic, genetic, and other mechanisms, we can get a lab mouse to live 4 years, have we raised the pre-set life-span maximum? How does this affect the potential life span of a mouse in the wild. How does this affect the practical or potential maimum life span of a human in today's quasi-civilized society?

25. Is there a death gene or a set of death genes? If we were to eliminate or to alter this gene (or genes), how would this alter the biological balance of the individual or of society?

26. Can the biological problems of aging and cancer be solved while we live under the increasing threat of extinction of our species? Is there a lonk between molecular biology and sociobiology?

At this meeting, I will attempt: (*a*) to formulate a new, improved Stromboli cocktail; to (*b*) examine how many of the above problems have been solved; and (*c*) to examine what old and new problems we now face in the continuing battle against cancer and aging.

FIRST VS. THE FOURTH STROMBOLI CONFERENCE

At the end of Stromboli I, I was able to summarize and even critique most of the papers presented. It would be impossible to do the same for this meeting because (1) the problems are now more complex and (2) I lack the expertise to make an intelligent critique of all the different fields of research covered.

However, I will make a few comments, based on my (subjective, to be sure) impressions of Stromboli IV. But, of course, you must read the papers themselves to make your own judgments. Walter Pierpaoli has again done a splendid job of assembling highly qualified scientists and clinicians from the four corners of the world. About one-third of the papers were superb, and most of the papers were excellent, adding to the scientific basis for future clinical improvements in our treatment of aging and cancer. There were also a few reports that presented only nice anecdotes, but not science, and finally there were a very few that shamelessly promoted their own commercial ventures and/or products.

"MacMedicine"

I was delighted to hear Suzie Schuder talk about MacMedicine, one of the major horrors of modern clinical practice. With the encouragement of Wall Street and the "big pharmas," doctors give the standard dose to each patient, regardless of the condition of the "substrate." No time is taken to know the biomedical individuality of each patient, which is so essential to good medical practice! Does it make any sense to give every patient "two green pills and one white pill" regardless of the patient's age, body weight, endocrine status, etc.? Does it make sense to give the same "standard dose" to a 90-pound old woman as to a young 280-pound football player? Too many physicians do not realize that the old joke "take two aspirins and call me in the morning" is really a bad joke, not least because they fail to realize that even one aspirin might cause some aspirin-sensitive person to bleed to death!

Similarly, failing to heed the caution that Franz Halberg, the "father of chronobiology," has been shouting for half a century: the same dose of a medicine might cure in the evening, but kill in the morning.

Thank you, Dr. Schuder, for pointing out the hazards of modern MacMedicine. Most of us at this meeting would like to see some of the selfless passion of laboratory science grafted onto the hearts of clinical practitioners.

Neuroendocrine Immunomodulation

At this conference, we have seen some new and fascinating aspects of the neuro-*endocrine* parts of NIM (neuroimmunomodulation). Very valuable data have been presented, but, as with all the new data on genetics (particularly various RNAs), we must be careful not to take a narrow, limiting view of the physiology of life, which is in reality extraordinarily complex. Universities and medical schools are divided into departments: biochemistry, immunology, endocrinology, genetics, etc., but the whole living organism is *not* thus divided: all the divisions are interconnected and *interdependent*. Taking too narrow a view runs the all-too-common risk of failing to understand the whole.[3,4]

Science vs. Politics and Science vs. High Finance

We have heard some talk at this meeting about "science vs. politics" and "science vs. high finance." We must be aware that, for many in the worlds of politics and big business, profit is their most important, and in many cases, sole motivation. We, scientists, must ever be on guard not to lose our virginity at the hands of rapacious capitalists, who all too often have no regard for human welfare, for the preservation of the environment, or for the safety and health of future generations. Drug companies are loath to finance research that does not promise more and higher profits.

In my own experience, I have shown, many years ago, with my colleagues Hiramoto and Ghanta in the U.S., and Fabris *et al.* in Italy, that we can repeatedly reverse "incurable" cancer and aging in mice, using only classical (Pavlovian) conditioning techniques (no drugs, bacteria, virus, or other antigens)[5-8] but l am still waiting, despite approved protocols, after many years, for adequate funding to start even pilot studies in humans.

The Clear and Present Danger

Dear auditors and readers, forgive me for injecting this personal note here, but I believe that it is relevant.

As funding for bombs and destruction increases astronomically and funding for health research is still woefully inadequate, there is great danger for the hopes of civilization. While I do not have answers, I urge all scientists and health practitioners to be aware of these problems.

Behavioral Physiology

We have heard at this meeting references to the importance of psychological and psychiatric treatments of the problems of aging and cancer. This is a refreshing antidote to the purely reductionist approach to all biology! Perhaps in the distant future we will be able to understand all problems of health and disease in terms of the simplest "molecular" functions, but because we are not there yet, bridges that span our chasms of knowledge still need to be built, and are indeed useful.

We look back in horror today at the prevailing medical practices of the eighteenth century, including the concept that most diseases could be cured by phlebotomy. Indeed, well-intentioned physicians thus killed George Washington and nany other famous and less-famous people—by bleeding them to death.

If our species survives, we will, in the not-too-distant future, look back with equal horror at the common practice of treating cancer by destroying the immune system. Poisonous drugs and whole-body irradiation start a race to see whether the cancer or the patient is killed first!

Luckily, there is a growing movement to use the so-called alternate system of boosting the patient's immunity. This was predicted by the growth of the interest in NIM, whose pioneers have been closely associated with these Stromboli conferences, including Walter himself, Nicola Fabris, and "Brana" Jankovic, co-organizers of Stromboli conferences, as well as myself, often called "father of neuroimmunomodulation," a pleasant and flattering, albeit not entirely accurate, designation.

OLD AND NEW PROBLEMS IN AGING AND CANCER RESEARCH

At the second Stromboli conference, based on the existing state of the art, I posed about two dozen problems for future research—these have been listed on the first three pages of the present article.

Most of these problems remain to be solved, but progress has been made on some of them:

For example, question 6 asks: Under what conditions can neurons proliferate?

Until recently, despite mounting evidence that new neurons can form in the adult brain, this remained very controversial. In the 1960s, in my own lab, I watched neurons in culture which, when resuspended after being plated out, withdrew their dendrites, balled up and then divided. There was evidence from Altman's lab in the United States, and from some Russians, that new neurons were observed in the adult mammalian cortex, but, for the most part, these results were generally not believed. Eventually, irrefutable evidence from birds, and later mammels showed that, indeed, new neurons are formed in the cerebral cortex. In recent studies Katica Jovanova Nesic of Serbia as well as Fred "Rusty" Gage of the Salk Institute, and others, have demonstrated that embryonic neurons planted in the adult mammalian brain, not only can proliferate, but also can repair and restore thermoregulatory and immune functions. Still, much more work remains to be done.

Question 10 asks whether there is a link between cancer and other diseases of old age. At this conference, excellent evidence, at the molecular level, has answered this question in the affirmative.

Question 13 asks how many feedback loops between the nervous and immune systems are open, and which ones are closed. It also asks whether there are any positive feedbacks that are not fatal.

We know that in a healthy organism most feedbacks are negative, part of the normal checks and balances, in feeding, drinking, thermoregulation, hormone secretions, etc. We need to ask ourselves, with all these "natural" hormonal replacements in the elderly, are we closing down the brain's normal capacity to regulate production by graded negative feedbacks? Knowing that in the aging mammal many hormones, as well as various enzymes, trace minerals, etc., are in short supply, how can we achieve that delicate balance during "replacement" therapy that is necessary for optimal function?

The answer to question 19 remains as it was at Stromboli II—Yes, we are still shortening our life spans with caloric overload.

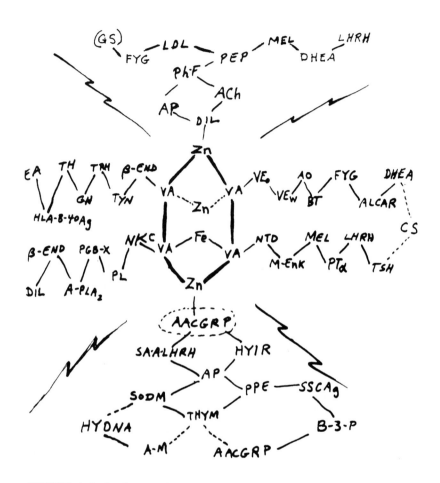

FIGURE 1. R_x for the second Stromboli cocktail: 10% solution in fine Sicilian wine (or Stromboli rain water) p.o.—10 cc/kg wet body wt, as needed for relief of signs or symptoms of aging and/or cancer. Key to the molecular structure: AACGRP = anti-calcitonin gene-related-peptide (hypothetical); ACH = acetylcholine; ALCAR = acetyl-l-camitine: AO = all other antioxidants; AP = assorted other peptidcs; A-PLA2 = anti-PLA2; b-End = beta endorphin; BT = bretylium tosylate; CS = chronic starvation; DHEA = dihydroepiandosterone; DIL = dilantin; EA = ethanolamine; Fe = iron; FYG = fresh young genes; GH = growth hormone; (GS) = gene stabilizers (hypothetical?); HLA-B40-Ag = HLA-B-40-Ag; HYDNA = healthy young DNA; HYIR = healthy young insulin receptors; LDL = low-density lipids and lipoproteins; LHRH = luteinizing hormone-releasing factor; Mel = melatonin; M-Enk = metenkephalin; NKC = natural killer cells; NTD = nootropic drugs; PEP = polypeptide extract of pineal; PGB-X = PGB-X; PhF = phenformin; PL = assorted phospholipids; PPE = polypeptide pineal extract; PT = pro-thymosin alpha; SA-A-LHRH = superactive analogues of LHRH; TH= T3, T4; TRH = thyrotropin-releasing hormone; TSH = thyroid-stimulating hormone; TYN = thymulin; VA = volcanic ash; VA = more volcanic ash; VE = vitamin E, lipid-soluble; VEw = vitamin E, aqueous; Zn = zinc; Zn = more zinc. Note: this has 42 ingredients: at least a dozen more than in the original Stromboli cocktail. (From Spector.[2] Reproduced by permission.)

Question 24 must still be pondered by gerontologists, sociobiologists, and philosophers.

Finally, the problems posed in question 26—whether the biological problems of aging and cancer can be solved under threat of imminent species extinction—are, without doubt, the most pressing and most difficult of all, and yet they are the ones to which the least attention is paid.

I challenge all, or any of, my colleagues to solve them!

STROMBOLI COCKTAILS

At the conclusion of Stromboli I, in 1986, on the basis of irrefutable scientific evidence presented at the meeting by many of the world's leading experts in aging and cancer, I drew up a list of the infallible cures and preventions proposed at that meeting. I put together the 27 items listed on the first page of this article and called it the Stromboli cocktail. Presumably, if we took this cocktail we would neither age nor get cancer. Strangely, perhaps because no one took the whole R_x, we still age and we still get cancer!

Then, three years later, at Stromboli II, based on further irrefutable scientific evidence, I added a dozen or so more ingredients to form the second Stromboli cocktail, even formulating a possible structural picture of a single molecule, incorporating all 42 ingredients (FIG. 1). Inexplicably, no pharmaceutical company has yet picked this up and put it on the market as a (gigantic) super-pill that contains the fountain of eternal youth and health!

At this meeting, based on *still further* irrefutable evidence, I found it necessary to add at least 15 new ingredients to our infallible cocktail. So the third Stromboli cocktail is the most powerful medicine (so far) ever developed in the history of science. Its additional ingredients include: two dozen RNAs; tailored germ cells; cultured pineal-cytes; caloric restriction mimetics; assorted chelating agents; good clock genes; good kinases; *more* melatonin; telomere-lengthening devices; *really good* mitochondria; hibernation factor(s); friendly chaperones; and genetically screened saliva.

The take-home lesson from all this is the old German proverb: "Ve get too soon oldt und too late schmart!"

EPILOGUE

We are gathered here from all around the world
With many a peer
Science banner unfurled!
Barring a sudden cardiac arrest,
Our science is a never-ending quest.
But if we should falter,
We'll be poked by Walter
And not only that, I swear, forsooth
There's many a slip twixt data and truth!
But even with good statistics and control,
We must remember that for each our role
Is to forge ahead and hardly ever rest
For science is a never-ending quest!

ACKNOWLEDGMENTS

First, I'd like to offer thanks to the following persons on behalf of all the participants in this endeavor:

(1) my old friend and "younger brother," Walter Pierpaoli, for once again putting together this remarkable gathering of leading scientists, old and young, from all corners of the world, and for affording the ambience in which we could enjoy a week of intensive and fruitful discussion;

(2) the wonderful Russo family, for their excellent and friendly hospitality in providing the great "Sirenetta";

(3) my old (but younger) professor, S.M. "Don" McCann, for coming half-way around the world despite many severe illnesses, to give us the benefit of his scientific acumen;

(4) the brilliant and ever-friendly Linda Mehta of the New York Academy of Sciences for her helpful encouragement to get the final volume printed and online in good time;

(5) to the extended Pierpaoli clan, who aided in the planning, organization and execution of this conference;

(6) to the Parretti family, who did so much financially and organizationally to ensure that this meeting would be a great success; and

(7) to the many very efficient and good-natured (but unfortunately, anonymous) volunteers who aided in every aspect and in every day of the meeting.

I would like to pay tribute to the great, now departed, pioneers, my friends and colleagues and the leading scientists who did so much to put together the first Stromboli conferences: "Brana" Janković and "Bill" Regelson. I would also like to acknowledge the late Vladimir Dilman, who was a pioneer in aging research.

Finally, I would like to honor Nicola Fabris, the co-organizer of the first two Stromboli conferences and my colleague in research at the Gerontology Research Institute in Ancona; he has been ill for many years, but is present here in spirit.

REFERENCES

1. SPECTOR, N.H. 1988. Rmbunctious remarks and a look towards the future. *In* Neuroimmunomodulation: Interventions in Aging and Cancer. W. Pierpaoli and N.H. Spector, Edds. Ann. N.Y. Acad. Sci. **521:** 323–335.
2. SPECTOR, N.H. 1991. Two dozen problems in aging and cancer: the second Stromboli cocktail: further rambunctious remarks. In Physiological Senescence and Its Postponement. W. Pierpaoli and N. Fabris, Eds. Ann. N.Y. Acad. Sci. **621:** 441–446.
3. SPECTOR, N.H., M. PROVINCIALI, G. DI STEFANO, *et al.* 1994. Immune enhancement by conditioning of senescent mice. Comparison of old and young mice in learning ability and in ability to increase natural killer cell activity and other host defense reactions in response to a conditioned stimulus. *In* Neuroimmunomodulation: The State of the Art. N. Fabris *et al.*, Eds. Ann. N.Y. Acad. Sci. 741: 283–291.
4. SPECTOR, N.H. 1999. The NIM revolution. Rom. J. Physiol. 36: 127–143.
5. GHANTA, V. R. HIRAMOTO, B. SOLAVSON & N.H. SPECTOR. 1985. Neural and environmental influence on neoplasia in conditioning of NK activity. J. Immunol. 135: 848–852.
6. FABRIS, N., B. MARCOVIĆ, N.H. SPECTOR & B. JANKOVIĆ, EDS. 1995. Neuroimmunomodulation: The State of the Art. Ann. N.Y. Acad. Sci. Vol. 741.

7. SPECTOR, N.H. 1998. Neuroimmunomodulation. *In* Encyclopedia of Neuroscience, 2nd ed. G. Adelman & B. Smith, Eds. The Health Foundation. New York.
8. SPECTOR, N.H. 2004. Neuroimmunomodulation. *In* Encyclopedia of Neuroscience, 3rd edit. G. Adelman & B. Smith, Eds. The Health Foundation. New York.

Index of Contributors